APPLIED ANIMAL NUTRITION

A Series of Books in Agricultural Science

Animal Science

Editors: G. W. Salisbury, E. W. Crampton

2ND EDITION

APPLIED ANIMAL NUTRITION

The Use of Feedstuffs in the Formulation of Livestock Rations

E. W. CRAMPTON
Macdonald College of McGill University, Quebec

L. E. HARRIS
Utah State University, Logan, Utah

W. H. FREEMAN AND COMPANY
San Francisco

Printed in the United States of America

Library of Congress Catalog Card Number: 68-10996

Standard Book Number: 7167-0814-0

9876

FOREWORD TO SECOND EDITION

The developments that have taken place in the field of animal nutrition in the twelve years since the publication of the first edition of *Applied Animal Nutrition* have not changed our basic premises so much as they have extended their application to livestock ration formulation.

This second edition includes the first textbook use of a new system of nomenclature for products used in animal feed mixtures. It presents information on the use of the caloric system of describing biological energy, preferred by many to the total digestible nutrient (TDN) system. The concept of expressing dietary requirements as nutrient-to-calorie ratios is proposed as more rational in the light of modern nutritional knowledge than the consideration of nutrients as independent entities. The quantitative data are given in both the metric and avoirdupois systems in anticipation of the early universal adoption of the former for scientific use.

The system for naming feed products, proposed by Harris in 1963, was standardized by the National Research Council Committee on Feed Composition in 1966, and adopted for use in all NRC Tables of Feed Composition. The nomenclature features a unique name and reference number for each of the more than 5,500 animal feeds of North America: the name of a given product gives precise information about its (1) genus or parent material, and (2) its species, variety or kind; (3) the part of the parent material that constitutes the feedstuff; (4) the processes to which it has been subjected in preparation; (5) for forages, the stage of physiological maturity and (6) the cutting or crop it represents within the year of harvest; (7) its quality, grade, and/or guarantees, and (8) its feed class code number. In tabulations, feeds are listed alphabetically, with the common area names for any given feed integrated into the listing and cross-referenced to the "official" NRC name.

An improved and systematic nomenclature for feeds has been long overdue, not only to describe the origin and nature of a product and make more meaningful the analytical data, but also to accommodate the rapidly expanding number of by-products from the processing of human foods that are or can be made edible and acceptable for livestock: for example, citrus pulp, seeds, and rind; poultry feathers; laboratory-synthesized vitamins; purified amino acids; defluorinated rock phosphate; fermentation products.

A development closely related to and facilitated by the adoption of a systematic nomenclature has been the machine retrieval and manipulation of feed compositional data, to produce least-cost feed combinations containing the amounts and proportions of the nutrients that the feeding standard indicates are needed for a particular ration. Thus we now can "balance" the rations, not only for three or four nutrients, but for all of them for which we have requirement data. This advance has brought into prominence the current paucity of data on the nutrient composition of all but a few feeds.

The so-called linear programming of rations is but an extension, albeit an important one, of the principle of flexible formulae for balanced rations dealt with in detail in the text. Each formula is based on the premise that there is no one best formulation—that the adequacy and balance of nutrients rather than those of the nutrient sources are the nutritionally critical features of a fully acceptable ration.

As the number of nutritionally important components of livestock feeds on which data are available has increased from the six of the Proximate Analysis to the eighty or more now recognized, the problem of the format of a functional working table in which to record these items has taken on new significance. The NRC's Committee on Animal Nutrition sub-committee on Feed Composition* devised the tabulation format that is used in the NRC *Encyclopedia of Feed Composition* and this format has been followed in this edition of *Applied Animal Nutrition*. It assembles systematically and consecutively all of the data currently available on a feeding stuff. There are no blank spaces in the table, yet the format lends itself to expansion and, at the same time, preserves the unity of each feed and its position in the alphabetical arrangement of the tabulation as a whole. Feedstuffs are classified according to their principal contribution to the feeding characteristics of the ration: (1) dry, (2) fresh, or (3) ensiled forages; and (4) energy, (5) protein, (6) mineral, (7) vitamin carriers, (8) additives. To identify (as in linear programing

---

* E. W. Crampton and Lorin E. Harris.

of rations), the class of a particular feed, its class code number has been made the first digit of its six-digit reference number.

In accordance with the trend in scientific publications, the metric system of weights and measures has been incorporated into the quantitative parts of the book. In many tables both the original units and the corresponding metric value are shown. Where the weight units are only generally descriptive, we have rounded values, using, for example, 450 kilograms (instead of 454) to describe a 1000-pound animal. This sort of approximation will, of course, disappear as the avoirdupois system falls into disuse.

The broad general objectives of the second edition do not differ from those of the first edition, since we believe them to be as relevant now as when outlined initially in 1956.

Our sincere thanks are due Dr. J. Malcolm Asplund, Utah State University, for his help with the Feed Composition Table.

E. W. Crampton
L. E. Harris

September, 1968

FOREWORD TO FIRST EDITION

The subject of animal nutrition is concerned with the application of scientific knowledge to the day-to-day feeding of livestock. Much remains to be discovered before we can construct a completely satisfactory overall pattern of nutrition. Nevertheless, we already have sufficient information to justify the belief that apparent gaps between animal husbandry, in its broadest sense, and such sciences as chemistry, physics, and physiology are not truly discontinuities, but merely voids in our information. As discoveries in the sciences have gradually filled these voids, there has been a tendency to change the grouping of subject matter for the teaching of animal nutrition, which has become, in fact, an integration of the relevant phases of all those sciences that underlie the principles of animal feeding. One of its most urgent tasks, therefore, has become the assessment of the relative significance and importance, from its peculiar viewpoint, of advances in all sciences that bear on this phase of animal husbandry.

In their concern with integrating basic scientific advances into their subject, animal nutritionists have all too often stopped short of practical application. Nutrients are considered in regard to their physiological roles; but problems of getting these nutrients to the animal day by day in accordance with the exigencies of practical livestock management tend, it would seem, to be regarded as proper to some other field.

The author feels that such a tendency is regrettable. He believes that the teaching of animal nutrition should bring closer together the theory of nutrition and the practice of animal feeding. This gap is not adequately covered by recapitulation of the specifications and behavior in rations of the hundreds of edible products used in feeding animals. Nor should we shirk our responsibility by discussing theories of animal nutrition, and leaving it for some phase of animal care and management to apply the theories to feeding practice. For example, the problems of ration formulation, together with those of ingredient procurement, processing, and mixing have become so broad in scope and so intricate in application that the commercial preparation of "balanced rations" and of specialized ration supplements is no longer merely a matter of convenience to the feeder.

The present book has been designed expressly to help bridge this gap between animal nutrition and livestock feeding practice; it is an attempt to extend fundamental animal nutrition into what we may call "applied animal nutrition."

Students taking this course should have as prerequisites, as far as practicable, those subjects of a college undergraduate curriculum necessary for an understanding of fundamental animal nutrition, as well as many of those that deal with animal care and management.

The subject matter of the text can partly be deduced from the paragraphs above. The author, in attempting a critical consideration of feedstuffs and their use, accepts on the one hand the facts presented and discussed in animal nutrition; and on the other, presumes that livestock feeding practice is a part of the subject of animal management. He also assumes that a catalogue of feedstuffs is an important part of the subject matter of reference books on feedstuffs rather than a desirable feature of a text dealing with problems of the assembly of nutrients into rations. But he does believe firmly, that to present a coherent and reasonably complete treatment of applied animal nutrition, he cannot be bound by traditional subject limitations.

The reader will find the subject matter treated under four main sections, plus an appendix. The first section is devoted to definition and critical appraisal of the terms and expressions used in describing feedstuffs. Section II deals with the nutritional requirements of animals, with special attention to the biological basis for feeding standard data. The nature of feeding standards and their limitations as guides in ration formulation are also considered in some detail. Section III features a classification of feeds. The discussion of the properties and functions of key feeds is intended to establish a sound basis for feed substitution in ration formulation. A classification of roughages according to available energy is a feature of this section.

The last section (IV) has to do with the problems of ration formulation. The translation of feeding standards into terms of meal mixtures, and the development of the concept of flexible formulae for meal mixtures, mineral mixtures; and mixed supplements intended as all, or as a part of the rations of farm livestock (cattle and swine) receives careful consideration in this section. It is through such formulae that the facts of nutrition and the characteristics of feeds are eventually brought together in terms that are immediately useful in feeding practice.

Finally, pertinent comments on feed legislation and a selected table of feed composition appear as an appendix.

E. W. Crampton

April, 1956

CONTENTS

II NUTRITIONAL REQUIREMENTS OF ANIMALS

Chapter 7. Requirements for Protein 165

Chapter 8. Requirements for Minerals, Vitamins, and Miscellaneous Additives 181

Chapter 9. Feeding Standards 191

III THE NUTRITIONAL CHARACTERISTICS OF SOME COMMON FEEDS

Chapter 10. A Classification of Feeds 223

Chapter 11. Basal Feeds 233

IV RATION FORMULATION

V NUTRIENT NEEDS OF ANIMALS

TABLE

TABLE

GLOSSARIES

INDEX

SECTION **I** # WHAT ARE FEEDSTUFFS?

The dilemma facing the authors of a text about feedstuffs and their use in practice is where to begin. Seemingly, no matter where we start, there is something that should already have been explained or considered. A part of this difficulty arises because some of the same terms that are used to describe animal needs are employed to describe the properties of the products used to meet these needs. And so it may be logical to define first a variety of terms, all of which eventually you will need to understand. We shall interpret "definition of terms" broadly, to include, where it seems desirable, much more than names or expressions. To make the definitions useful, we may have to expand them into somewhat detailed discussions of the significance and use of the term under consideration.

Not all of the nutritional terms used in this book, however, will be thus defined, for we shall assume that the reader already has at least an elementary knowledge of fundamental nutrition and hence of the biochemistry, physiology, bacteriology, and mathematics that underly this subject. Perhaps we may sum up the whole matter by saying that this section will deal with what feedstuffs are, how they are described, and what significance is to be attached to their descriptions in applied livestock feeding.

In this section we do not intend to catalog by name the hundreds of feeding stuffs known and used in the rations of animals. Our objective is rather to consider a number of the terms, particularly those other than the name, that we employ in describing feeds, and to examine what each of these really tells about the nutritional properties of the product. Some of the terms are applied to all feedstuffs as, for example, the proximate analysis, *either complete or in part; while others apply only to specific products or perhaps to a limited group of products.*

CHAPTER 1

Terms and Definitions for Feeds

History and Development of the
NRC Nomenclature for Feeds

In 1963 the Committee on Animal Nutrition of the United States National Academy of Sciences–National Research Council, realizing the need for a systematic feed nomenclature that would (1) describe feeds accurately, (2) be adaptable for coding to permit machine retrieval of data for specific tabulations, and (3) be useful internationally, appointed a subcommittee (consisting of the present authors) to study the problem of feed naming and to arrange a program to assemble on a continuing basis and record on IBM cards the currently available feed-composition data for North America. The Canadian National Committee on Animal Nutrition was invited to collaborate in the project in the person of the chairman of its subcommittee on feed composition.

This 1963 project began as the culmination of a series of related efforts to provide an encyclopedia of the composition of feedstuffs for the use of both the feed industry and those in professional agriculture. In 1946 the NRC Agricultural Board set up a Committee on Feed Composition under the chairmanship of Dr. R. V. Boucher, then head of the Biochemistry Department at Pennsylvania State Agricultural College. W. R. White, then Chief of the Feed Section of the Canadian Department of Agriculture, Plant Products Division, and E. W. Crampton, Professor of Nutrition at Macdonald College (McGill University) and Secretary of the Quebec Provincial Feed Board, were invited to sit with this committee.

The 1946 committee undertook to collect analytical chemical data on the feeds of the United States and Canada. It reported annually to the NRC Committee on Animal Nutrition, although the latter group did not define the terms under which the Feed Committee functioned. In 1956 a tabulation of the analytical data on concentrate by-products was published as NRC publication no. 449; and in 1959 publication no. 585, covering grains and roughages, appeared. These two bulletins contained the analytical chemical data, to the extent they were available, for about fifty "nutrients" of over six thousand differently named feeds. It was the most comprehensive collection of data on feed composition ever made. Following publication of these two bulletins, the committee was discharged, with no provision made for updating the data as new values became available.

The format used for tabulation in these two publications, although satisfactory for a reference work, was not ideal as a working table for ration formulation. To remedy this inconvenience, the NRC Committee on Animal Nutrition in 1958 appointed an ad hoc committee, consisting of the chairmen of the Dairy Cattle, Beef Cattle, Sheep, Swine, and Horse Feeding Standard Committees, with E. W. Crampton of Canada as chairman, to select data for an abridgement of publications 449 and 585 that would include some 1,500 of the most common feeds of North America, and to arrange for its publication in a more usable form. This abridgement appeared in 1959 as NRC publication no. 659. It was financed jointly by the NRC Committee on Animal Nutrition and by the Canadian Department of Agriculture through its National Committee on Animal Nutrition.

Supplies of this bulletin were exhausted in 1963, at which time there was some question as to whether or not it should be reprinted without revision. The NRC Committee on Animal Nutrition had already decided that, after January 1965, the NRC Animal Requirement Bulletins were to show all data in the metric system, which meant that the feed-composition tables should also be in the metric system. In addition it was desirable to update the data where possible, and to identify Canadian data as such. Ultimately it was decided to begin preparing at once a systematic feed nomenclature, which would be used to rename all feeds recorded in NRC publications 449 and 585, starting with the abridged tables in publication 659.

The feed-naming system (hereinafter referred to as the NRC nomenclature) adopted by the Crampton-Harris committee was based on a scheme proposed by L. E. Harris* in which the full name is built up from a possible

---

* Lorin E. Harris, "Symposium on Feeds and Meats Terminology. III. A System for Naming and Describing Feedstuffs, Energy Terminology, and the Use of Such Information in Calculating Diets," *J. Animal Sci.,* XXII (1963), 535.

eight components, which together describe the morphological and/or physical nature of the product. The revised edition of the *Joint United States–Canadian Tables of Feed Composition* (NRC publication no. 659) became available in February 1965 as NRC publication no. 1232, in which the feeds are renamed according to the NRC nomenclature, and the quantitative data are in the metric system.

Following the printing of the revised edition, the Crampton-Harris committee, after screening the 6,500 feeds of NRC publications 494 and 585 to eliminate duplication, proceeded: (1) to rename some 5,700 different products according to the NRC nomenclature; (2) to assemble available analytical data on grains, by-product feeds, and roughages of Canadian origin, only a few of which were included in publications 494 and 585; (3) to convert all data to metric units and compute the standard deviation for each nutrient; (4) to code and record all data on IBM cards; and (5) to establish a final format for tables that would show in one place the NRC name of the feed, its American Association of Feed Control Officials (AAFCO), Canada Feeds Act (CFA), and/or local names, its permanent reference number, its country and area of origin, all analytical data on both an as-fed and a dry basis, and the coefficient of variation of each "nutrient" recorded. This job was completed in the fall of 1965.

At this point, it may be well to point out the basic reason for efforts to assemble an encyclopedia of the "nutrient" makeup of feedstuffs. We recognize, of course, that animals have no requirements for specific feeds. Feeds are merely the carriers of the nutrients and the potential energy we must provide in a satisfactory diet. No feed has been found that is nutritionally complete for, or balanced to the needs of, a given animal; and many feeds contain undesirable substances that may interfere with the availability of desirable components, or may even be toxic to an animal. Feeds differ widely in chemical composition, availability, and cost. Hence to assemble least-cost, nutritionally adequate, and physiologically acceptable rations requires specific knowledge of both the assets and the possible liabilities of the individual feedstuffs that could go into the mixture that is to be compounded.

In the 1965 project we made provisions to record for each feed the compositional data on: (1) 53 individual minerals, vitamins, and amino acids; (2) the components of the proximate analysis; (3) the fatty acid component of the ether extract; (4) the important celluloses and hemicelluloses of the carbohydrates; and (5) the digestible, metabolizable, and net usable fractions of the gross energy, of the crude protein, and of the ether extract.

If the premise is valid that, in general, all these data are of use in a fully adequate description of the potential nutritional properties of a feed, then it is surprising that there are not complete data for a single feed in our compilation of over 5,500. But when we realize that for 70 per cent of the 1,500 most commonly used feeds we have data on less than ten of these "nutrients," we begin to appreciate how inadequate is our knowledge of the composition of feeds. Ten nutrients actually means the proximate analysis plus four other "nutrients" (as vitamins, minerals, amino acids, etc.)

The NRC Nomenclature for Feeds

The dictionary says, in effect, *"To name is to identify."* There are on record names of about 6,500 feedstuffs. Their names were usually given to them by the persons using them, and identified them more or less locally. These "common names" were given with no thought of any purpose other than physical identification. It was not surprising, then, to discover that when 6,500 names for feeds were examined, about 20 per cent of them were other names for the same products called by one name in one area but by another in a different part of the country. For example, rolled oat groats is often known as rolled oats or as table oatmeal. The milling by-product of rolled oat groats is named oat middlings by the Canada Feeds Act, but feeding oatmeal by the American Association of Feed Control Officials in the United States.

The great proliferation of by-products suitable for animal feeding that has resulted from advances in food technology, coupled with the lack of any systematic scheme for naming such products, has led to confusion in feed identification and has emphasized the need for a modernized nomenclature that would provide a *unique identification* for each feed and define precisely its morphological and/or physical nature. In most cases common feed names give little information of nutritional significance, not because of any deliberate withholding of facts, but because the potential usefulness of such information as a part of the name, both to the feed trade and to the ultimate user of the product, has not been recognized. Furthermore, insofar as two people discussing feeds should be sure they are talking about the same feedstuff, not only should the official name of each feed be unique, but the words or terms used in the names should have consistent meanings.

For example the term "meal," unless otherwise qualified, normally denotes a coarsely ground grain, as ground barley or ground oats. But rice meal is the mill-run by-product of the polishing of brown rice, and is a mixture of rice bran, germ, and broken kernels. Oatmeal is rolled oat groats; feeding oatmeal is the mill-run by-product of rolling oat groats. Citrus meal is dehydrated citrus-pulp screenings. Linseed meal and peanut meal come from flax seeds and peanut kernels, respectively; both are ground and the fat extracted. Whale meal is ground whale meat with the fat extracted. Fish meal is ground fish, either with or without the fat extracted.

The term "feed" is another whose common meaning varies. We find that "feed barley" and "barley feed" are not synonymous. "Feed barley" names a grade or quality; it refers to whole barley that is rejected for use as seed or for malting. "Barley feed" refers to a combination of barley hulls and middlings, which are the by-product obtained when barley grain is dehulled and the groat scoured down to a rounded (pearl-shaped) particle called pearl barley. Or again, we find corn feed meal to be the siftings from making cracked corn, but corn gluten feed to be the product obtained when the bran from the wet milling of corn has been returned to the gluten part of the original grain.

There are also problems with names for plant products that are derived from a common or area name for a plant rather than from its botanical name. For example, although the term "corn" is confusing only in international communications, the fact remains that "corn" is strictly an area name. "Corn" denotes a small, hard seed of any cereal grass used for food, especially that of the most important cereal crop of a particular region. In England, "corn" refers to wheat, in Ireland and Scotland, to oats. In Canada, Australia, and the U.S.A., it refers to Indian corn; "Indian corn" is itself a regional name for the seed of the maize plant.

Perhaps the best example of the area-name problem arises with wheat milling by-products (see Fig. 1-1). From the outer coats of the wheat berry

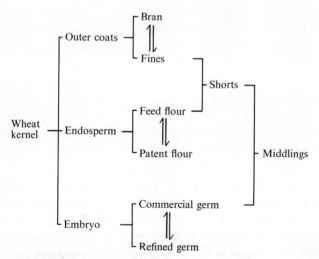

Figure 1-1. *Wheat milling products and by-products.*

and the endosperm (exclusive of wheat bran and patent flour) are made fine bran siftings or fines, and "impure" or feed-grade flour. From the embryo comes wheat embryo or commercial wheat germ (as well as refined germ). The fines and feed flour produced are often combined into the product originally called shorts; if the germ is also included the product is called middlings. In fact, middlings consists of all the by-products of milling flour, and is often sold as such in normal or "mill-run" proportions, which, by proximate analysis, means it contains about 8 per cent crude fiber, 3 per cent fat, and 17 per cent protein (see Table 1-1).

TABLE 1-1 *Makeup of Wheat Flour By-Products*

By-product	Parts per 100 of wheat	Per cent protein	Per cent fat	Per cent fiber
Fines	18	16	2	10
Feed flour	5	13	2	2
Commercial germ	3	28	12	2
Mill-run	26	17	3	8

TABLE 1-2 *Expected and Reported Analyses of Wheat Flour Milling By-Products*

Product	Type of analysis	Per cent protein	Per cent fat	Per cent fiber
Mill-run	expected	16.8	3.2	7.5
	reported	17.0	4.4	8.9
Middlings	reported	19.1	5.1	8.9
	reported	20.4	4.7	5.6
	reported	20.2	4.0	2.2
Shorts	expected	15.3	2.0	2.3
	reported	20.4	4.7	5.6

Table 1-2 makes it quite evident that these three "middle products" of wheat flour milling are not likely to be combined in mill-run proportions, and that the terms shorts and middlings are significant only in that they designate wheat milling products. The specific samples in the table consist of undisclosed proportions of the three intermediate fractions made by screening and bolting the three successive crackings of wheat berry in reducing it to table flour. The "yield" of any particular fraction apparently depends on the economics at the time. (For the proposed NRC solution to this problem of naming wheat milling by-products, see examples 6–9, p. 12.)

The Ideal Feed Name

An ideal feed name should precisely (1) describe that feed, genotypically and morphologically, (2) define its quality or grade if applicable, and (3) indicate its place in the classification of feeds; that is, it should give information that makes its chemical composition more intelligible to the nutritionist. The NRC names have eight potential parts, each of which contributes specific information that is significant for understanding the probable place the product might take in a ration. These eight parts are as follows.

Origin. The first term of the NRC name does not usually refer to the part actually eaten. It refers instead to the parent substance from which the material that is eaten originates, and is the name of the plant, animal, mineral, or other original feed product. The term "plant" is not used as an origin term, but "animal" is often used to designate feeds produced from animals, especially mixtures of several kinds of animal products, or products from an unspecified "animal" (as opposed to fish or fowl, which are also used as origin names). If the feed name always begins with the origin term, all names can be listed alphabetically for official definitions and in tabulations of analytical data.

Variety or Kind. If the variety or kind of the original source of the feed is nutritionally significant, this information is included as the second term of the name. For example, the animal feed product "milk" can come from the cow or the goat; indeed, we may wish to distinguish here between Holstein and Jersey. With maize we should know whether we are dealing with a white or a yellow variety. Similarly, we may need to indicate soft versus hard wheat, or western versus eastern barley, or new versus old oats. These differences are shown in the variety or kind term of the name. If there is no information for which this second term is needed, it is either omitted or replaced by the expression "kind unspecified."

Part Eaten. The third component of a feed name is the actual part of the parent material that is consumed. This may seem to be a rather obvious item, referring, for example, to such parts of plants as leaves, stems, or seeds, or to the meat trimmings, milk, or bone from animals. However, food technologists are today disassembling many of our natural foods and are subsequently reconstituting many of the parts thus formed into new products for human consumption. This extensive fractionation of plant seeds and animal by-products provides innumerable products that are useful as feeds for animals.

These products include not only the orthodox components of grain (such as bran, germ, or flour, and the aerial parts making up hay), but also such products as:

blood	distillers grains
blood albumen	distillers grains
blood with stick	without solubles
bone	distillers grains
bone and blood	with solubles
bone ash	distillers residue
bone black	distillers residue
bone phosphate	grain
distillation solubles	grain and cobs
distillation stillage	grain and cobs with
distillation residue	husks
without solubles	grain screenings
distillation residue	grain without starch
with solubles	

See Appendix Table A-2-2 for definitions of all parts eaten which are used in Appendix Table A-3-1 on Composition of Feeds.

Processes and Treatments. Some parts of parent feed materials have had no processing (for example, pasture grass), but most of them have had something done to them, either to preserve them, or to make them more palatable, or to change their form or state to improve them as feedstuffs (by heating, grinding, sifting, bolting, extracting, digesting, pelleting, etc.). The particular processes involved are often exceedingly important: nutrients in some products are damaged by too much heating; others become available only after heating. Some products are made more digestible for some species of animals by heating or denaturing, but other products are made less digestible by such treatment. Grinding may make a product more useful or less useful; it certainly has a pronounced effect on the proportions of fatty acids produced by rumen microflora, which in turn has implications for some products' fattening properties as compared to their growth- or product-promoting properties.

See Appendix Table A-2-2 for definitions of all processes described in Appendix Table A-3-1 on Composition of Feeds.

Stage of Maturity. The stage of maturity applies only to roughage products, generally speaking, but with them it is perhaps the most important factor in determining their nutritive values. We know that grass in the vegetative state has a high nutritive value for ruminant animals, but with advancing maturity, plant structures become lignified; the presence of lignin greatly interferes with the attack of rumen microflora on the cellulose and the parts which the cellulose encloses, thus reducing the digestibility of the product appreciably. Probably the greatest problem in preparing hay for winter feeding is arranging to get the material cut at a stage of maturity when it is most nutritious. Glossary A-2-3 in Appendix 2 defines the stages of maturity, and gives the term for each stage that is preferred by the NRC nomenclature.

Cutting or Crop Number. This part of the name refers to whether the roughage material was first cut (or first crop), second cut, third cut, etc. This has some significance in that plants may differ somewhat in composition if they have regrown after a first cutting, after a second cutting, or after subsequent cuttings.

Grade or Quality Designation. A great many products, such as hays and grains, have for many years been graded by official government standards. These different gradings are significant in interpreting the nutritive value of the product. In the NRC nomenclature the grade or quality category has been extended to include such other things as minimum weights per bushel, or guarantees about the product's content of fat, fiber, or protein. Where official grades of any of the feedstuffs listed have been established by government regulations these are included.

Classification. All feeds have been segregated into groups classified according to the scheme in Table 1-3.

According to the NRC nomenclature, products which when dry normally contain more than 18 per cent crude fiber are classified as roughages. Products which contain 20 per cent or more protein are classified as protein supplements. Products which contain less than 20 per cent protein and less than 18 per cent crude fiber are classified as energy feeds.

The components of the NRC feed nomenclature, with examples of how they apply to certain feeds, are shown in Table 1-4. In such tabular form, the components stand out quite clearly; we can particularly see where we omit the components that have no significance or are unknown.*

Normally the name would be written in linear form, with the components

---

* For abbreviations, see the Glossary of NRC Feed Nomenclature Abbreviations, pp. 464–465.

following in serial order and separated by commas, but with no other punctuation, as follows.

1. Alfalfa, Ranger, leaves, dehy pltd, early blm, cut 1, gr US 1, (1).
2. Corn, yellow, grain wo germ, grnd, (4).
3. Soybean, seed, solv-extd grnd, mn 44 prot mx 7 fbr, (5).
4. Animal, carcass res w bone mx 35 blood, dry- or wet-rend dehy grnd, mn 40 prot, (5).
5. Fish, sol, dehy, mn 60 prot, (5).
6. Wheat, flour by-prod, mil-run, mx 9.5 fbr, (4); common names are mill-run, wheat mill-run, and wheat middlings.
7. Wheat, flour by-prod, c-sift, mx 7 fbr, (4); common names are wheat shorts, shorts, and wheat middlings.
8. Wheat, flour by-prod, f-sift, mx 4 fbr, (4); common names are wheat middlings, red dog, and shorts.
9. Wheat, flour by-prod, c-bolt, mx 2 fbr, (4); common name is feed flour.

TABLE 1-3 *NRC Classification of Feeds*

Code and class	Typical products
1. Dry forage or roughage	hay straw seed hulls fodder (aerial pt w ears, husks, or heads) stover (aerial pt wo ears, husks, or heads)
2. Succulent forage or roughage	pasturage range plants soiling crops
3. Silages	grain crop silage grass silage haylage
4. Energy feeds	grains and seeds: low cellulose; high cellulose mill by-products: low cellulose; high cellulose fruits nuts roots
5. Protein supplements	animal by-products marine by-products avian by-products plant by-products
6. Mineral supplements	natural or pure elements
7. Vitamin supplements	natural or pure substance
8. Additives	antibiotics coloring materials flavors hormones medicants

TABLE 1-4 *The Components of the NRC Feed Nomenclature*

Component	Examples		
1. Origin	alfalfa	corn	soybean
2. Variety	ranger	yellow	—
3. Part	leaves	grain wo germ	seed
4. Process	dehy, pltd	grnd	solv-extd, grnd
5. Maturity	early blm	—	—
6. Cutting	cut 1	—	—
7. Grade	US 1	—	mn 44 prot, mx 7 fbr
8. Class	(1)	(4)	(5)

Long Names

NRC names are long, but they are not intended for use in ordinary conversation. The NRC name is the final identification and physical description of a unique feed. As such it is the single designation under which the analytical data can be given, and thus solves the problem of tabulating data for the many feeds that have two or even four different common or area names. Since all known common names are integrated alphabetically with the NRC names, and each is cross-referenced to its NRC name, it becomes immaterial which name of a feed one uses in everyday practice, so long as everyone understands what product is referred to by the name used at the time.

Consider the analogous case of an animal. In a herd of 100 head, one cow has the official pedigree name, "Macdonald Supreme Beauty Ann Sue 2nd 1106596." In the barn, she is called "Sue 2nd" by the herdsman, "old 32" by the milker, and a rather select name by the stableman who has to clean her. These common names suffice for the local dealings with and about her. But her specific identity and performance are officially recorded only under her pedigree name.

Looked at in this way, the idea of recording the composition and biological data for corn gluten feed under the more informative NRC name of: "Corn, gluten w bran, wet-mil dehy grnd, (5)" does not seem so unreasonable or impractical. This in no way need restrict or discourage the continued use locally of the common names of these products.

Format for Tables of Feed Composition

As already mentioned, the incorporation of descriptive material in the name of a feed inevitably leads to longer names, and hence increases the

TABLE 1-5 *Format for Tables of Feed Composition*

Feed name or nutrient	Mean analysis		Coeff. of var. (per cent of mean)
	Per cent as fed	Per cent dry	

OATS, *Avena sativa*
Oats, cereal by-prod, *mx 4 fbr, (4)*
    Feeding oatmeal (AAFCO)
    Oat middlings (CFA)
    Oatmeal (local)
Ref. no. 4–03–303: US region 06			

Dry matter	91.0	100.0	1
Ash	2.3	2.5	18
Crude fiber	4.0	4.4	24
Ether extract	5.8	6.4	22
Protein, crude (N × 6.25)	15.8	17.4	12

space needed to print tables. To reduce this space requirement to a minimum, we prepared a glossary of abbreviations (pp. 464–465), which were used extensively in the NRC tabulations. The format to be used in presenting the analytical data on the feeds is illustrated in Table 1-5. This sample table, which is intended particularly to show the format for beginning the tabulation for each feed, shows only the first few entries of analytical data.

The first, wide column gives the feed name or the "nutrient." The next two columns give the mean analytical values for the material "as fed" and as "dry matter." The third column gives the coefficient of variation expressed as a per cent of the mean. The standard deviations have been calculated for all means based on more than five samples.

Following the column headings is a space for the details of the name, first the name of the parent material, then, in the case of plant products, the scientific name. Next comes the full NRC name (it may take more than one line if necessary), the AAFCO name, the CFA name, and the local names, if any. Below the names we find the reference number—in this case 4–03–303. The first digit indicates the feed classification, i.e., this is an energy feed. This reference number will be permanently associated with this particular NRC name, whatever the feed may have been called elsewhere. It is particularly useful where linear programming of rations is involved, because the reference number can be used rather than the name. On the same line with the reference number is entered the code for where the samples of the product originated; these came from United States feed region 06.

Below the space for the information on the name, the first column lists the components for which analytical data have been recorded. The analytical data are continued to include the protein components, their digestibility coefficients, and their gross energy, digestible, metabolizable, net energy, and TDN values, for cattle, sheep, and swine separately. Then follows data on minerals, amino acids, and any special analyses that we may happen to have. Altogether the tables provide for eighty components. The table column for any one feed is as long as is necessary to include all of the analytical data that are available for it. When the data have been completely recorded, a space is made for the name of the next feed in alphabetical order, which is followed by the analytical data for it.

Notice that all of the official names are in alphabetical order by origin, or by origin and variety within class, and that all other names are also entered in their appropriate alphabetical position, with a cross reference to the NRC name for that product. Thus one can find the name of any product, whether it be the common, regional, CFA, or AAFCO name, but the analytical data for that product will be found only with the official NRC name. As one becomes more familiar with the nomenclature and the scheme of tabulation, the information wanted can usually be found directly, merely by remembering that the feeds are listed alphabetically by origin and variety within class. Thus the feed called "feeding oatmeal" in the AAFCO publication can be found by turning directly to "oats."

Feeding Stuffs

This term is, in general, synonymous with feed, food, or fodder, although it is broader, covering all materials included in the diet because of nutritional properties. It embraces not only the naturally occurring plant or animal products and the by-products prepared from them, but also chemically synthesized or otherwise manufactured pure nutrients or prepared mixtures of them used as supplements to natural foods. Thus, while wheat germ meal is a livestock feed or a human food, thiamine hydrochloride is a pure nutrient, which may be chemically synthesized and used as a supplement to feeds. It is not a feed, but it is foodstuff. A feeding stuff therefore is any product, whether of natural origin or artificially prepared, that when properly used has nutritional value in the diet.

Ration and Diet. A ration is a 24-hour allowance of a feed or of the mixture of feeding stuffs making up the diet. The term carries no implications that the allowance is adequate in quantity or kind to meet the nutritional needs of the animal for which it is intended. It merely refers to a daily allocation of provisions and is the usual basis for food accounting in the armed forces and in institutions.

Some confusion occasionally arises about the meanings of the two terms *ration* and *diet*. Is there any distinction between them? According to Webster a ration is "a fixed daily allowance of food for one person (or one animal) in an army or navy." A diet is "what a person or animal usually eats and drinks; daily fare." The distinction made by some in using *diet* only for human food and *ration* only for animal feeding is not logical, especially since we refer to the rations of laboratory animals as diets. In this sense of their meanings, the terms are synonymous and are so used in this book.

The introduction of the term *ration* into livestock feeding language was accompanied by a less precise definition of its meaning. The feeder was usually as much interested in a nutritionally adequate ration as in a statement of a quantity of some food. To meet this need the term *balanced ration* was coined to refer to a feed mixture just sufficient to meet the 24-hour requirements of a specified animal; the *balance* referred to the proportion of carbohydrate, fat, and protein in the ration.

We recognize, of course, that some hundred or more known nutrients may be involved in an adequate diet, and that a ration balanced in terms of the primary or energy-yielding nutrients may still be sadly deficient for the nourishment of the animal. Furthermore, in practical husbandry, rations are not prepared for individual animals. Rather some mixture of feeds is prepared, and animals of the same feeding group are either given portions of it or are allowed to help themselves to it.

The original implication of the ration as a 24-hour allowance of food for one animal was never compatible with the practical use of feedstuffs. Ration appears to be firmly established in the feeding lingo, but its definition has undergone an important change. As most often used since about 1940 in the feed trade and by the feeder, ration or even balanced ration refers to a mixture of feedstuffs prepared for the feeding of some specified class or group of animals and intended to constitute either the entire dietary allowance or some definite and specified portion of it. Referring to the mixture as a balanced ration implies that the mixture is nutritionally adequate for the feeding of the animals specified when used according to recommendations. These recommendations are important since the ration as it is may not be a complete ready-to-feed mixture. It may require the addition of basal feeds, as in the case of a supplement; or it may be intended for feeding over a restricted period only, as in the case of medicated mixtures; or it may be suitable for feeding in conjunction with forage of a certain kind, as in the case of cattle feeds.

Basal (Energy) Feeds

This term appears to be of Canadian origin and was first used to designate the whole group of grains and grain by-products which contain not more than 16 per cent protein and 18 per cent fiber. The feedstuffs of this category form the basis of normal livestock meal rations.*

Nutritionally, basal feeds are mainly concentrated sources of energy, being especially rich in starches and sugars. They have been described as carbonaceous concentrates. In everyday feeder's language they are the low-protein concentrates, such as corn, barley, oats, wheat, and the by-products milled from them, that do not contain enough of the embryo or the gluten layers to appreciably increase the protein of the feed over that of the parent grain.

Basal feeds average between 10 and 14 per cent crude protein and something less than 5 per cent ether extract. The chief gross difference between basal feeds, which is of significance in their practical use, lies in their digestible energy content, which, in turn, is likely to be inversely proportional to their crude fiber content.

Inasmuch as feeds of this category normally constitute from 60 to 90 per cent of practical livestock rations (exclusive of roughage), it is evident that an important consequence of substitutions among these feeds is a change in the useful-energy value of the ration, and hence in the *quantity of the ration that must be fed to meet the animal's requirements.* Undoubtedly more feeding problems are traceable to failure to meet energy requirements than to any other single cause.

Supplements

A supplement is a feed or a feed mixture used with another to improve the nutritive balance or performance of the total and intended to be (1) fed undiluted as a supplement to other feeds, or (2) offered as free choice with other parts of the ration separately available, or (3) further diluted and mixed to produce a complete feed.

Feeds of this type contain large amounts of protein, of some mineral element, or of some particular vitamin. A mixed protein supplement is by

---

* The Canada Feeds Act (1967) defines a basal feed as a mixture that has too many energy ingredients to be classified as a chop feed, but that does not qualify as a "complete" feed. A "complete" feed must provide all the nutritional requirements except water needed to maintain normal health or to promote production. The word concentrate, used nutritionally to describe energy feeds, will no longer be accepted as a descriptive name synonymous with supplement.

convention a mixture of feeds that carries 30 per cent or more protein. However, single feeds that contain 20 per cent or more protein are included in the supplement category. Any mineral or vitamin carriers added to the ration are normally referred to as supplements.

The term supplement describes its definition. For example, in order to properly balance the basal feeds, additional protein must often be incorporated in the final mixture. This is normally done by adding feeds that are richer in protein than the basal feeds. Consequently, such feeds are commonly referred to as protein supplements.

Some protein supplements are high in protein because by nature they contain very little carbohydrate. This would be true of many animal by-products, such as meat meal, liver meal, or dried milk, as well as of marine products such as fish meal, since the original fat has been largely removed. Nevertheless, some fish meals may have appreciable amounts of oil remaining in the meal, which not only dilutes the protein but may limit the use of the product, as we shall point out later.

On the other hand, most of the high-protein feeds are of plant origin and contain a high concentration of protein because some parts of the original product have been removed by milling. Thus, we find the protein supplements from oil-bearing seeds are nothing more than the seed from which the oil has been extracted, with the result that the residue carries relatively more protein than the whole seed. Another group of protein supplements are high in protein because the starch has been removed from the original seed.

Concentrates

A concentrate is officially defined as a feed used with another to improve the nutritive balance of the total and intended to be further diluted and mixed to produce a supplement or a complete feed. However, in feeding practice a concentrate is usually described as a feed or a feed mixture which supplies primary nutrients (protein, carbohydrate, and fat) and contains less than 18 per cent crude fiber (see *Forage and Roughage below*). In the feed trade, however, the term concentrate has been almost universally used for commercially prepared supplements. In this sense the term concentrate refers to a concentration of proteins, of minerals, or of vitamins in excess of those found in the basal feed. Such concentrates are usually mixtures, which frequently supply several of the individual nutrients with which the basal feeds must be fortified in order to make adequate rations. There is seldom any difficulty in interpreting the terms *concentrate* and *supplement,* because the context in which they are used indicates what they mean.

Forage and Roughage

In farm usage a forage or a roughage is normally considered to be material making up fodder, such as hay, silage, pasturage, etc. The distinguishing characteristic of forage is usually a high fiber content, which for hays frequently runs between 25 and 30 per cent of the dry weight. There is, however, another group of feeds that in physical appearance might be classed as concentrates, but which nevertheless are nutritionally more like forages, since they are high in fiber and relatively low in useful energy. For example, it seems inconsistent to class alfalfa hay as forage, and the meal made from it by grinding, as a concentrate. It is also difficult to justify the practice of labeling oat hulls or oat feed as a concentrate in spite of the fact that they are a part of ground oats. When any of these products are sold unmixed with other feeds there is no particular difficulty about their feeding value. When, however, they are used as ingredients in "balanced" meal rations, merely naming them in the list of ingredients may fail to warn the feeder that a low-energy product is involved.

One solution of this problem has been put into effect in Canada, where, by definition in the *Feeds Act,* roughage is any material suitable for feeding livestock which contains more than 18 per cent crude fiber. Thus, in Canada, such products as alfalfa meal, ground oat hulls, oat feed, some samples of corn bran, oat mill feed, some grain screenings, etc., are actually classified officially as roughages, even though their common form might lead one to think of them as concentrate feeds.

Nutrients

A nutrient is defined by Morrison as any food constituent, or group of food constituents of the same general chemical composition, that aids in the support of animal life. We must interpret this definition somewhat more broadly than was originally intended, because we now have to include substances that are not of food origin. Although foods are parcels of nutrients usually mixed with nonnutrient material, a complete ration may be more than a combination of foods. It may include synthetically produced vitamins, chemically prepared inorganic salts, or perhaps amino acids recovered from hair. What animals eat in terms of the products actually consumed is fundamentally of much less importance than the quantity and assortment of the nutrients furnished by the rations made available to them.

One cannot be sure today that a list of the presently recognized nutrients found in feeds, or in animal tissues, represents a complete list of the operating needs of the body. For practical purposes, however, we do know what the majority of these nutrients are, and enough about most of them so that we understand what their general functions are, and consequently why they must be supplied to the animal.

In the practical feeding of animals we recognize that some of the nutrients are nonspecific in function and thus may be essential only as members of a group of nutrients having similar functions. However, even where some nutrients can be classed together on the basis of certain common properties, individual members of these groups may have specific functions not duplicated by other members of the same group. In order to systematize a consideration of the nutrients you may find it useful to look at them grouped into categories which have nutritional or at least unique descriptive characteristics. Such a listing is shown in Table 1-6.

In this list only the twelve essential amino acids are detailed, and no attempt is made to indicate the wide assortment of fatty acids, or the different sugars and other carbohydrates, that the chemist is able to isolate from feedstuffs. Nor is the list of vitamins as complete as present information would permit. The list is intended rather to include those nutrients with which the feeder may have to concern himself in the preparation of satisfactory rations.

We have attempted to classify the inorganic elements into those that often have to be added in ration formulation, and those that are usually abundant in any diet. The feeder is cautioned that this part of the classification cannot be made fixed, because geographic areas differ in both the qualitative and the quantitative occurrence in feeds of some of the elements.

It is difficult to place these few mineral elements in any order of priority or importance. In point of quantity, calcium and phosphorus will stand first, if we assume that common salt is to be supplied as a matter of routine and quite independently of any quantities of sodium and chlorine that may be in the feedstuff. Information concerning how much iron, cobalt, and copper feeds contain will be desirable insofar as these elements are required, but one may not need to worry about the iodine content of any of the feedstuffs. Evidence of the need for supplemental iodine is quite clear for young farm animals born with the well-known symptoms of iodine deficiency. Since few animals subsist entirely on feeds grown in the particular district where the animals themselves are kept, knowing the iodine content of the feeds of such districts is of relatively little help in predicting the need for iodine supplementation. Furthermore, supplementing the rations of pregnant females with iodine is largely a routine matter wherever goiter is found, and is usually accomplished by purchasing iodized salt.

We shall deal with the question of the quantities of the several nutrients needed by various animals in a later section.

Table 1-6 *Nutrients to Be Considered in Ration Formulation*

Main groups	Subgroups	Nutrient group	Specific nutrients
Nitrogenous	proteins	amino acids	Lysine, trytophane, histidine, leucine, phenylalanine, iseolucine, threonine, methionine, valine, arginine, glycine, glutamic acid
	nonproteins	amino acids	
Nonnitrogenous	lipides	neutral fats fatty acids sterols	nonspecific sources of energy; linoleic, linolenic, arachidonic, acetic, propionic, butyric; mother substance of vitamin D
	carbohydrates	starches and sugars cellulose hemicellulose	nonspecific sources of energy; essential food for some microflora as cellulose
	other	lignin	not a nutrient—a hindrance to bacterial breakdown of cellulose
Vitamins	fat-soluble	carotenes	vitamin A; α, β, γ, and hydroxycarotenes
		antiricketic sterols	vitamin D_2 or calciferol; vitamin D_3 or irradiated 7-dehydrocholesterol
		tocopherols	vitamin E; α, β, and γ isomers of tocopherol
		antihemorrhagic naphthaquinones	vitamin K; menadione
	water-soluble	vitamin B-complex	thiamine, riboflavin, niacin, pyridoxine, pantothenic acid, vitamin B_{12}
Inorganic elements	essential	often required as supplements	calcium, phosphorus, sodium, chlorine, iron
		needed as supplements in specific geographic areas	iodine cobalt selenium
		for normal bone development	fluorine
	toxic	when consumed above requirement	fluorine, selenium
		in excess amounts	arsenic, molybdenum

Enzymes

An enzyme is defined as one of a number of complex organic substances capable by catalytic action of transforming some other compound. In the sense that concerns us here, enzymes are the biological units that make possible the assembly and/or dissociation of the chemical units that are involved in the changes that foods, nutrients, and tissues undergo in their metabolization. All chemical changes within the body are enzyme-catalyzed, from which it follows that partial or complete enzyme failure is at once reflected in deranged metabolism. There is a direct relation between enzymes and nutrition. Enzymes are primarily combinations of amino acids, some of which, of necessity, are of dietary origin. Since they must function continually, dietary protein is obviously essential.

Perhaps because digestive ferments are so often cited as examples of enzyme activity, enzymes have been thought of as liquid. We should note, however, that almost all cell proteins are constituents of active enzyme systems. In fact, it is doubtful that any large part of the protein of the cells is primarily structural material. For example, the musculature of the body is essentially an enzyme system.

Muscle is about 75 per cent water. Of its dry substance some 80 per cent is protein, most of which consists of the enzyme system actomyosin. Thus, for practical purposes the dry weight of muscle, and consequently a large part of the weight of the body, is an enzyme system, whose function is to facilitate the removal of phosphorus from its position in the energy-rich adenosine triphosphate, thus releasing energy for metabolic purposes. In this action the physical shape of the actin part of the actomyosin changes from globular to fibrous form by undergoing successive hydration and dehydration. When these muscles are attached to bones, this change from the short globular form to the long fibrous form causes body movement. In the meantime, the adenosine triphosphate is restored by the addition of more phosphorus to the depleted adenylic acid molecule. The energy needed to effect this phosphorylation comes from dietary carbohydrates, fats, and proteins. Thus is the potential energy of foods put to work.

The relation of this enzyme system to protein nutrition is clearly evident on examination of the makeup of the enzyme myosin. Amino acids account for more than 70 per cent of rabbit myosin, as follows:

Cystine	1.39%
Methionine	3.40
Serine	3.57
Threonine	3.81
Tyrosine	3.40
Aspartic acid	8.90
Glutamic acid	22.10
Arginine	7.00
Lysine	10.30
Histidine	1.70
Trytophane	0.82
Glycine	1.90
Alanine	5.10
	73.39%

Although enzymes themselves are amino acid complexes, the coenzymes and the enzyme activators making up the other part of the active enzyme systems may involve mineral elements and/or vitamins.

Enzyme systems vary considerably in their complexity. The component parts of the system acting on a substrate, such as pyruvic acid, include:

1. Substances necessary to produce an environment in which the reaction can proceed. Here we have water, materials to establish the correct pH, a suitable redox potential, and an appropriate ion concentration. This requirement is not very specific and the parts of it are not a part of the coenzyme.
2. A protein, which is the enzyme or apoenzyme. This is inactive without its specific cofactors.
3. Cofactors, including specific inorganic divalent ions (magnesium, calcium, manganese, cobalt, and zinc), and specific organic compounds—the coenzymes. These latter are highly specific and many of them are complex molecules in which one of the vitamin B complex members is involved (such as thiamine pyrophosphate or the pantothenic acid that contains coenzyme A).

A good example of such a system is the "tricarboxylic cycle" of energy metabolism. Normal pyruvic acid metabolism, in which this product is oxidized to CO_2 and H_2O, can proceed only in the presence of thiamine, pantothenic acid, riboflavin, and niacin; catalytic amounts of some C_4 dicarboxylic acid (as oxalacetic acid); some divalent ion (such as Mn^{++} or Mg^{++}); inorganic phosphate to form the phosphate ester linkage between apoenzyme and coenzyme; some hydrogen transport system (as dipyridal nucleotide or a flavoprotein); and oxygen.

The pantothenic acid is a part of coenzyme A necessary to condense acetyl phosphate with oxalacetic acid; thiamine is necessary for the oxidative decarboxylation of pyruvate; nicotinic acid is a part of the coenzymes di- and triphosphopyridine nucleotide needed for the dehydrogenation of iso-citric and malic acids; and riboflavin is a component of the flavoprotein enzymes, which appear to be the hydrogen acceptors in the conversion of succinic to fumaric acid (see Fig. 1-2).

We should point out that while certain metabolic disorders may yield to the administration of one or more of the B vitamins, these vitamins are not active by themselves. To relieve the syndrome caused by a lack of them, the vitamins must be incorporated into larger molecules, which may then function as coenzymes in some of the metabolic machinery. It has been suggested that the specific function of the vitamin is to form the protein-coenzyme bond.

There are a large number of separate reactions involved in the metabolism of carbohydrates, fats, and proteins, but the types of reactions are relatively few. Nevertheless, each vitamin appears to be specific for a given type of reaction.

This brief consideration of the makeup of some of the enzymes involved in the metabolism of food should clarify several points regarding the critical need for amino acids, mineral elements, and vitamins, as well as explain why a single deficiency might be reflected in a wide variety of clinical symptoms; and, conversely, why any one of several possible deficiencies might result in the same clinical picture. It is the difficulty in identifying the exact deficiency that is so often misleading in the diagnosis of the cause of some condition.

For example, you will note from the chart of energy metabolism that the reaction between the "active acetate" (acetyl Co A) and acetoacetic acid is reversible. If there is any disturbance in the smooth operation of the "Krebs Cycle," the fate of acetyl Co A is likely to be toward the formation of aceto-acetic acid instead of a condensation with oxalacetic acid to form citric acid. The increased production of acetoacetic acid results in the formation of acetone, which is then excreted both in the urine and by respiration to relieve the acidosis that the body cannot tolerate. Here we find, then, that the deficiency of any one of several of the B vitamins, or of phosphorus, or of certain minerals, or of enough carbohydrate to maintain the necessary oxalacetic acid (especially in the case of high fat intake), or the presence of an enzyme poison such as fluorine, might lead to secondary acetonemia. The diagnosis of acetonemia is relatively easy—but to spot its specific cause is another matter. Obviously a single criterion of faulty metabolism is seldom a reliable index of its possible nutritional cause.

Toxicity

At first thought this may seem to be an anomalous term to use in connection with a consideration of nutrients or even with feedstuffs. One might argue that if a nutrient is any edible material that aids in the nourishment of an animal, such material can hardly have toxicity as one of its attributes. Nevertheless, there are substances that when used at certain levels are essential to complete nutrition, but above those levels are harmful enough to be classed as toxic.

The difficulty lies in the definition of toxicity. The word is of Greek origin and denotes a poisonous substance. And the common implication is that the

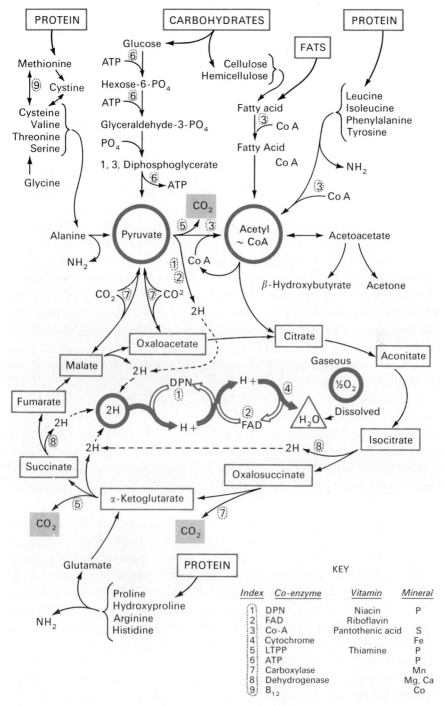

Figure 1-2. *Some important pathways of energy metabolism.*

consequence of poisoning is death. Veterinarians often qualify the term with such adjectives as mild or acute to denote the possible degrees of damage to the body's functioning.

In nutrition there are situations where *toxic* seems the best descriptive term available and yet where *poisonous* would not be a synonym. For example, it is generally accepted that linolenic acid is an essential nutrient. Its complete absence from the diet for a sufficiently long period of time leads to the death of the animal. This fatty acid, however, is one (perhaps the only one) that causes "flavor reversion," which is characteristic of some edible lipides. Fats are sometimes heated to destroy much of this acid through *polymerization;* and we now know that one or more of the resulting products, probably cyclic monomers, when ingested even in small amounts (i.e., on the order of 2 per cent of the diet) will cause the death of young rats. In smaller concentrations or with older animals the only clinical result may be a reduction of feed intake plus a decline in the gains per unit of food eaten. These products of the heat polymerization of trienoic fats are conveniently termed toxic, meaning in usual cases that they interfere with normal food metabolism. They are not toxic in the sense of being corrosive poisons. They are toxic in the sense that they inhibit an enzyme system which otherwise would effect degradation of some molecule to useful metabolic products, or would permit the molecules to be harmlessly discarded (i.e., a detoxifying reaction). Such materials are often called enzyme poisons.

If we pursue this line of argument far enough, we arrive at the point where we must conclude that an essential nutrient such as a vitamin may cause toxic effects by its absence! And here, of course, we come to the justification for the term *deficiency disease.*

However, it is in another sense that the term toxic must sometimes be used with feedstuffs. Some forage plants normally contain at certain periods of development *cyanogenic substances* that render the forage nutritionally harmful. The Sorghums, Johnson's (quack) grass, and Sudan grass are among those that may contain high concentrations of glucosides, which, in the digestive tract of the animal, may be broken down by enzymes of the forage itself to yield free prussic acid in toxic amounts. Young forage of these species and also forage following wilting or after a severe check of growth, as from frost, is more dangerous than other samples. Cattle and sheep are especially susceptible to damage from such forages and may die very quickly after eating them.

Most of the cases of toxic nutrients or feedstuffs seem to be associated with one or another of four mineral elements. *Molybdenum* is entirely a forage problem. Forage raised on some soils carries enough molybdenum to interfere with the animal's use of dietary copper. This effect appears to be its only association with toxicity. It is not required by the animal, and in the presence of adequate copper is readily eliminated without any harmful effects. With inadequate copper, ingested molybdenum replaces some phosphorus in the bone complex, leading to thin and fragile bones; but most of the toxic symptoms are those of copper deficiency. The antidote is an increased copper intake.

Selenium is another mineral element that in certain geographic regions becomes a nutritional problem. Both the forage and the seeds from the plants grown on selenium-containing rocks are dangerous as livestock feeds. In South Dakota, Montana, and Wyoming there are rather extensive areas of shale soils that contain dangerous levels of this mineral element. Excessive ingestion of selenium causes alkali disease, sometimes also called blind staggers and bobtailed disease. One general characteristic of early clinical symptoms is a loss of hair from the mane and tail of horses, the switch of cattle, and the body hair of swine. Many severely affected animals die or have to be destroyed, although if removed soon enough from the hazard areas they may recover.

Toxically, the mineral element most likely to be a problem in concentrate mixtures is *Fluorine*. Here is a substance that some believe to be a desirable "nutrient" in small amounts (on the order of 1 ppm in the drinking water), but which when ingested in larger amounts (on the order of 30 ppm or more in the dry diet) may show definite clinical evidence of toxicity. Species differ somewhat in their tolerance to fluorine intake. We do not have to be much concerned with the fluorine content of concentrate feeds that are by-products of plants, since plants that pick up fluorine from the soil concentrate it more in their leaves than in their seeds. Thus the feeder should give some attention to the fluorine content of forage in areas that may be contaminated; but far more important he should know the fluorine content of mineral supplements, particularly those that are used to supply phosphorus. Fluorine is a likely component of all rock phosphates that are used for livestock feeding, and fluorine is also found in feeding bone meals. In the case of bone meal the fluorine content appears to be increasing with time, probably because more and more livestock from which the bones are derived have been fed fluorine-containing phosphates in the course of their lives. Thus meat animals that can tolerate for fairly long periods of time a high fluorine content often receive rations in which the phosphorus is provided in fluorine-containing rock phosphates. There may be no particular economic damage to these animals, but their bones eventually find their way into feeding bone meal, and thus animals that are fed bone meal as a source of calcium and phosphorus receive, occasionally, undesirable amounts of this toxic element. No defense has yet been found against the harmful effects of the ingestion of excess fluorine and, consequently, the only safe course is to try to keep it out of the ration.

The *Arsenic* problem is usually one of accidental ingestion, except for the use (principally with poultry) of medicated feeds or supplements where an arsenic compound is sometimes one of the materials employed. Some arsenicals have been found to have antibiotic-like effects in growth stimulation of young animals.

No definite discussion of food toxicity is in order in this book and the comments above, which arose out of an attempt to define *toxicity* in applied nutrition, are more by way of illustration than of a guide to the animal husbandman. The feeder should be aware of the possibility that some feedstuffs under some conditions may be correctly classed as *toxic* in the sense here used.

SUGGESTED READINGS

Cook, C. W., L. A. Stoddard, and L. E. Harris. *The Nutritive Value of Winter Range Plants in Great Basin as Determined with Digestion Trials with Sheep.* (Utah Agr. Expt. Sta. Bul. 372, 1954.)

Crampton, E. W., and Lorin E. Harris. *Joint United States–Canadian Tables of Feed Composition.* (Natl. Acad. Sci., Natl. Res. Council, Pub. 1232, 1964.)

Harris, Lorin E. "Symposium on Feeds and Meats Terminology. III: A System for Naming and Describing Feedstuffs, Energy Terminology, and the Use of Such Information in Calculating Diets," *J. Animal Sci.,* XXII (1963), 535.

Harris, Lorin E. 1966. *Biological Energy Interrelationships and Glossary of Energy Terms.* First Revised Ed. (Natl. Acad. Sci., Natl. Res. Council, Pub. 1411, 1966.)

Harris, L. E., J. Malcolm Asplund, and Earle W. Crampton. *Feed Nomenclature and Methods for Summarizing Feed Data.* (Utah Agr. Expt. Sta. Bul. 479, 1968.)

Harris, Lorin E., Earle W. Crampton, Arlin D. Knight, and Alice Denney. "Composition Data. II: A Proposed Source Form for Collection of Feed Composition Data," *J. Animal Sci.,* XXVI (1967), 97.

Knight, Arlin D., Lorin E. Harris, Earle W. Crampton, and Alice Denney. "Collection and Summarization of Feed Composition Data. III: Coding of a Source Form for Compiling Feed Composition Data," *J. Dairy Sci.,* XLIX (1966), 1548.

Miller, D. F., H. E. Bechtel, K. C. Beeson, R. V. Boucher, L. E. Harris, D. F. Huffman, H. L. Lucas, F. B. Morrison, E. M. Nelson, B. H. Schneider. *Composition of Concentrate By-product Feeding Stuffs.* (Natl. Acad. Sci., Natl. Res. Council, Committee on Feed Composition, Pub. 449, 1956.)

Miller, D. F., H. E. Bechtel, K. C. Beeson, R. V. Boucher, L. E. Harris, D. F. Huffman, H. L. Lucas, F. B. Morrison, E. M. Nelson, B. H. Schneider. *Composition of Cereal Grains and Forages. (Ibid.,* Pub. 585, 1958.)

Morrison, Frank B. *Feeds and Feeding.* 23d ed. (Clinton, Iowa: Morrison Pub., 1967.)

Schneider, Burch H. *Feeds of the World, Their Digestibility and Composition.* (West Virginia Agr. Exp. Sta., 1947.)

Schneider, Burch H., Henry L. Lucas, Mary Ann Cipolloni, and Helen M. Pavlich. "The Prediction of Digestibility for Feeds for Which There Are Only Proximate Composition Data," *J. Animal Sci.,* XI (1952), 77.

CHAPTER 2

The Proximate Analysis of Feeds

For many nutrients that are required by animals there are direct chemical procedures by which we can establish how potent feedstuffs are in these nutrients. We have no particular problem in using such data. But there are other feed fractions which we can isolate chemically, but which are combinations of nutrients that have some common property permitting a chemical analysis of the group. The nutritional significance of such nutrient groups depends on factors that are not indicated by the proportion of the feed comprising the group. For example, the several feed fractions separated by the proximate analysis are all combinations of nutrients, and are consequently of varying nutritional significance.

The proximate analysis is probably the most generally used chemical scheme for describing feedstuffs, in spite of the fact that the information it gives may often be of uncertain nutritional significance, or may even be misleading. We should, therefore, consider in some detail the nature, the peculiarities, and the limitations of the proximate analysis as a description of the nutritional properties of feedstuffs.

This scheme of analysis was devised by workers at the Weende Experiment Station in Germany. According to it a feedstuff is partitioned into six fractions:

water	nitrogen-free extract
ether extract	crude protein
crude fiber	ash

When a chemist is asked to run a standard feedstuffs analysis on a given food, he proceeds to determine, chemically, five of these proximate principles. The last one, nitrogen-free extract, he then determines by difference. The sum of the crude fiber and the nitrogen-free extract represents the total carbohydrate of the feed.

This plan of description groups together a variety of substances in terms of some of their common chemical characteristics. It is not, as is sometimes erroneously supposed, an analysis of the nutrients of the food. Each of the components, except water, represents a combination of substances, some of which are nutrients or combinations of nutrients, and some of which are not of any nutritional value to the animal at all.

Water

Water, the simplest of all substances in foods, is not the simplest to determine. Usually it is recorded as the loss in weight of a sample as a result of oven-drying it to a constant weight at atmospheric pressure and at a temperature just above the boiling point of water. For many biological products such as feeds and the excreted feed residues, such drying results in a loss of volatile fatty acids, and of some sugars that decompose at temperatures above 70°C. Such substances will obviously be counted as water. Drying at lower temperatures *in vacuo* sometimes helps correct this inaccuracy, but necessitates vacuum ovens. Such ovens are usually of relatively small capacity, which may be a problem if much routine work is involved.

The direct determination of water is often done by distillation, using toluene as the product immiscible with water. The distillate is received in a calibrated sedimentation tube with a side arm that returns the medium to the distilling flask while the water is trapped in the calibrated tube. An apparatus for a continuous extraction of water and fat from biological materials was designed in the Department of Nutrition, Macdonald College (see Fig. 2-1).

The significance of the water content of feeds depends on the kind of feed and the amount of water. The greatest difference in nutritive value between many feeds, as fed, is traceable to differing moisture content. For example, using digestible energy (DE) and total digestible nutrients* (TDN) as a measure of overall energy value, the cereal grains and some of the tubers show almost the same feed values per unit of dry matter (see Table 2-1). Dry matter, therefore, becomes a common denominator for the comparison of foods, particularly in terms of energy value. This applies, however, to all other "nutrients" (protein, fat, calcium, etc.).

* See Chapter 3.

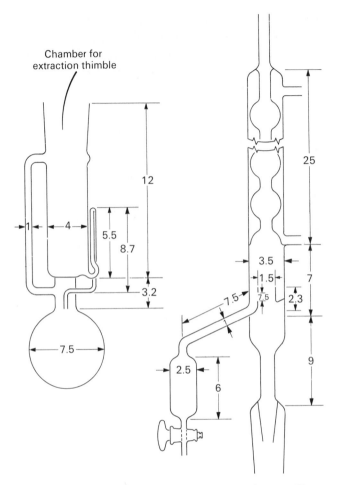

Figure 2-1. *Apparatus for extraction of fat and water. The part on the right fits into that on the left. (All measurements in centimeters.)*

It is customary to report the proximate principles on a moisture-free basis in tables of food composition. Dry matter is then given separately for any further information that it alone conveys. This plan is not always followed, and you must be sure in making comparisons between feeds in nutrient content that the figures are expressed on comparable bases so far as dry matter is concerned.

The problem of food storage is complicated by high moisture content. Foods containing more than 14 per cent moisture cannot be stored in bulk. They are likely to mold, and spontaneous combustion may also take place. The problems of moisture content in relation to storage are serious for some

TABLE 2-1 *Feed Values per Unit of Dry Matter*

Food	Per cent water	Energy			
		DE (kcal/kg)		TDN	
		As fed	Dry	As fed	Dry
Corn, dent, grain, gr 2 US mn 54 wt	12	3841	4365	87	99
Barley, grain	11	3375	3792	76	86
Oats, cereal by-prod, grnd, mx 2 fbr (Oat meal)	9	4052	4453	92	101
Melons, whole	94	212	3527	5	80
Potato, tubers, fresh	75	911	3704	21	84
Sweet potato, tubers, fresh	68	1157	3616	26	82
Beet, sugar, roots	87	441	3395	10	77
Apples, fruit, raw	82	587	3263	13	74

grains. Corn grain, for example, is officially graded on the basis of moisture content. The grade of shelled corn, its water content, and its feeding value as indicated by digestible calories and TDN are shown in Table 2-2.

TABLE 2-2 *Grades and Water Content of Corn*

Grade of shelled corn	Per cent water	Energy	
		Dig. kcal/kg	TDN
1	12.8	3630	82.5
2	14.8	3545	80.6
3	16.5	3475	79.0
4	18.8	3380	76.8
5	21.5	3265	74.2
Soft	30.5	2920	66.7

Moisture content in newly harvested grains is also likely to be higher than that in the same feed after several months of storage in barns or elevators. Sometimes the change in moisture content complicates the estimate of the quantity of farm grain on hand. If one takes the initial value as the basis of calculation one is likely to find the grain loses at least 5 per cent of its initial weight during the course of a winter storage, and one may consequently over-estimate the feed available.

Calculating the relative cost of food per unit of nutritional value frequently involves the consideration of water, since water is not a nutrient in the usual sense. Thus, sometimes grains or grain by-products are offered for sale at what appear to be bargain prices, but on investigation one finds that these samples are high in moisture content and that the price per unit of dry matter is no particular bargain after all. This situation arises with such feeds as brewers' grains and distillers' grains, or other products that, in the process of manufacture, have been wet and must be dried by artificial heat. The cost of drying is considerable, and when the material can be sold in a partially dried state it can be reduced in price.

The moisture content of forages that are to be preserved by ensiling is also of some importance, since it affects the ease with which ensiling can be effected. For grass silages the newly cut forage must frequently be wilted to reduce its moisture content to about 70 per cent. On the other hand, more mature forages often require the addition of water during the ensiling process in order to facilitate the packing needed to exclude air from the mass.

One further problem with moisture content is that of dustiness, which is often encountered when excessively dry feeds are ground in preparation for feeding. Animals universally dislike dusty feeds, and consequently eat rations prepared from powdery ingredients less readily than rations with a more granular composition. Very dry grain and such products as alfalfa tend to produce an excessively dusty ration, which may be partially relieved by adding small quantities of water, or by steaming, or by feeding the product as a wet or moistened mash. Another corrective for dusty feeds is to add something like molasses, which tends to stick the dusty particles together.

Quite aside from all we have said above concerning the significance of water in feed, we should not forget that, though it is not counted as a nutrient, water is nevertheless a dietary essential, and that its amount in relation to caloric intake is perhaps more critical than this relationship is for most nutrients. We discuss quantitative requirements for water in another section, but we want to point out here that the first response of the animal to a restriction of water intake is a restriction of voluntary food intake. This may amount to as much as a 30 per cent restriction in food consumption and in the efficiency of the feed as measured by the gain made per unit of feed eaten. Consequently, the water content of feeds, particularly those that are normally high in water, such as silage, fluid milk, pasture forage, and roots, is an important consideration in their feeding value and in the overall requirement of the animal for water. Water that is not supplied with food must be supplied from some other source, and some foods owe their chief nutritional properties to their water content.

On the other hand, forcing animals to consume too much water may limit intake of other nutrients because of the limited capacity of the digestive system. Thus, we find that milk as the entire diet of animals past weaning age is so bulky because of its water content that they are unable to consume as much of the other nutrients from the dry matter of the milk as they would otherwise take; consequently, they do not grow as rapidly as other animals not forced to take such a bulky diet. Much the same situation may occur in the slop-feeding of pigs if the meal-water mixture is made too thin and of food so fine that it stays in suspension.

Feeding standards do not include the water requirement for animals, an omission that perhaps should be corrected. One reason for the omission is obviously the difficulty of stating the water requirement for specific animals, inasmuch as it differs with the composition of the animal's diet. In the absence of any better information the approximate water that should be provided as beverage, where it is impossible to furnish water *ad libitum,* may be calculated from the caloric requirement. The digestible calorie intake for any animal is not far from one hundred times the body weight (in kg) raised to the three-quarter power, i.e., $100 \ (W_{kg}{}^{.75})$. The quantity of water to be provided as beverage is, for average conditions, about one milliliter for each three kcal calculated from the above formula, i.e., $33 \ (W_{kg}{}^{.75})$. The balance of the day's water is provided by metabolic water and that which is contained as free water in the foods eaten.

Ether Extract

In most cereals and animal products used for food, the fat or oil obtained by mechanical extraction is almost the same chemically as that extracted by ether. It consists of:

glycerides of fatty acids	chlorophyl
free fatty acids	alkali substances
cholesterol	volatile oils
lecithin	resins

The chlorophyls, alkalies, volatile oils, and resins are not classed as nutrients but are, nevertheless, found in the ether extract of feeds.

The ether extract will differ in composition among different foods, and it may be quite unlike fat or fatty oil. In particular, ether extracts from different feeds will likely differ in sterol content, and since this has no energy value it will affect the energy value of the product. For all practical purposes the food value of ether extract will be more nearly correlated with its content of neutral fat than with any other component, barring possibly a few products such as alfalfa in which the vitamin A content is of particular importance. Thus we must remember that the food value of ether extract is not constant, in spite of the fact that in practice it has been taken as yielding 9.35 kcal of gross energy or approximately nine kcal per gram of metabolizable energy as consumed in the diet.

The useful energy of dietary ether extract is its gross energy minus that found in the subsequent fecal excretion (i.e., the loss in digestion). However, when feces are extracted with ether, soaps that may have been formed in the intestinal tract from free fatty acids and calcium will not be removed. This incomplete recovery of the fecal fat gives erroneously high values for the digestibility of the ration fat, particularly in practical diets containing relatively high calcium content. The fat content will be the major factor causing differences in the gross energy of various foods and feeds, because carbohydrates and proteins yield about five kcal per gram, while the ether extract yields a little over nine kcal.

Estimates of ether extract in feeds may be made either by weighing the extracted material after evaporating the solvent, or by recording the loss in weight of the moisture-free sample following its extraction by anhydrous ether. The second method makes it possible to employ a multiple-sample extraction procedure. The samples are weighed into Alundum crucibles, ovendried, reweighed, placed in groups in the extraction chamber, extracted, and again weighed following extraction of the sample. The samples, without transfer from the extraction crucibles, can then be ignited in a muffle furnace for the determination of crude ash. This procedure, unless very exact values are necessary, is probably preferable to the extraction of single samples by standard Soxhlet procedures.

Nutritional Significance of Ether Extract. The nutritional significance of ether extract in the diet is a matter on which there is still some disagreement. To the extent that it is normally the source of the essential fatty acid (linolenic) it is admittedly indispensable. However, an animal requires only a few grams of ether extract from most feeds to supply all of the necessary linolenic acid. Many authorities believe that the only other function of this fraction of feeds is as a nonspecific source of energy. Before accepting this view, one should remember that ruminant animals obtain much of their energy needs from the short-chain acids (acetic, propionic, and butyric) arising from the breakdown of ration cellulose by rumen microflora; and factors that alter the *proportions* in which these acids are produced are reflected in milk-fat production. It may be that some fatty acids have unique roles in metabolism other than as nonspecific sources of energy.

Deuel,* has reviewed the recent evidence on this question and has arrived at the conclusion that dietary fat *per se* has specific nutritional functions quite unrelated to essential fatty acids and in addition to its function of providing calories in concentrated form. There can be little doubt that the addition of fats to many diets or rations lowers the heat increment of such diets and hence results in a greater energy efficiency for a given calorie intake. In other words, the higher efficiency of fat calories appears to be the result of a reduced heat loss on the higher-fat diets.

Deuel summarizes his conclusions as follows: "It is as yet a moot question whether the beneficial effects of fats, nutritionwise, are to be ascribed solely to the EFA (essential fatty acids) which they contain, or whether triglyceride fats have a specific nutritive value *per se*. The beneficial effects of high-fat diets on pregnancy, lactation, and possibly on growth can probably be largely, if not entirely, ascribed to their EFA content. On the other hand there is no positive evidence that the improved caloric efficiency resulting from the associative dynamic effects or the sparing action of fats on certain phases of protein metabolism are necessarily functions of EFA. Some of the beneficial effects of fat are undoubtedly to be traced to the establishment of definite enzyme patterns when high-fat diets are consumed over an extended period. . . . It may be that increased capacity for work, noted in rats receiving high-fat diets, is a reflection of the establishment of such new enzyme patterns."

The lipide content is of importance in the selection and use of feeds. For example, the bulkiness of whole milk is compensated for by the high energy of its fat content. Without this component, milk would not be useful as the early diet of the mammal, since its bulk relative to the capacity of the digestive tract would make it impossible for the young animals to consume enough energy for their early growth needs.

As we shall see later, oil meals prepared by heat and pressure may contain up to 10 per cent fat, and as a consequence their energy is sometimes higher than that of basal feeds. The effect of more complete removal of the oil, as by solvent extraction, on their available energy yield is indicated in Table 2-3.

It is the fat portion of feeds that is the most unstable. This feature makes the storage of high-fat feeds a problem. Rancid feeds are objectionable, because they have usually lost appreciable quantities of such nutrients as vitamin A or its mother substance, carotene, and may also have suffered oxidative destruction of some of the essential fatty acids. In addition, the chemical changes may have caused the formation of undesirable substances such as amines. Furthermore, the rancidity may proceed to the point of heating

---

* Harry J. Deuel, Jr., "Fat as a Required Nutrient of the Diet," *Proc. Fed. Am. Soc. for Exp. Biol.,* XIV (1955), 639.

and the actual combustion of the feed. The greater the unsaturation of the fats present, the greater the danger from this damage. All plant oils are subject to easy rancidity. These fatty materials as they occur in the seeds of the grains are quite stable, but if the grain is ground the lipases are likely

TABLE 2-3 *Effect of Fat Removal on Energy of Oil Meal Feeds*

Feed and preparation	Per cent fat in meal	Energy DE (kcal/kg)	TDN
Soybean, seed			
none	18.0	3690	84
mech-extd grnd	4.7	3294	75
solv-extd grnd	0.9	3139	71
Flax, seed			
none	36.0	4320	98
mech-extd grnd	5.2	3210	73
solv-extd grnd	1.7	3129	71
Peanut, kernels			
mech-extd grnd	5.9	3570	81
solv-extd grnd	1.2	3367	76

to be activated by the heat and moisture of the process and the fatty fractions then quickly become rancid.

Fat-free or even very low-fat rations are frequently found to be less acceptable to animals than those of greater fat content. Just where this fat level should be set in this connection is not by any means clear, and probably depends, in part, on species of animal as well as on age, and perhaps on conditioned preference. In studies at Macdonald College Nutrition Department one of the consequences of the fortification of low-fat rations (less than 2 per cent ether extract) with up to 20 per cent of some nine different fats and oils was that young pigs weaned at ten days and puppies weaned at 14 to 21 days learned to eat from self-feeders diets that carried 7 per cent or more fat more quickly than they did the 2 per cent fat control diets. Young guinea pigs weaned at two days of age, however, did not show this preference. As we might expect, once these young animals had learned to eat the diets, the effects of the fat additions were largely traceable to the greater energy value of the mixtures.

The observation that animals accustomed to a corn ration dislike its replacement by barley (but not by oats), while the reverse is not true, has led to the general belief that the low fat of barley is the factor responsible. This is no doubt a conditioned preference, for animals that have been raised on barley rations consume voluntarily as much of these rations as do comparable animals raised on corn mixtures.

It may also be worth noting that some feed manufacturers prefer to fortify their ready-to-feed mixtures with vitamin A by using low-potency fish oils. The oil thus added is helpful with mixtures that, when ground, are otherwise inclined to be dusty.

The nutritional usefulness of ether extract obviously depends in part on the extent to which it is *digested*. Although neutral fats are normally over 90 per cent digested, this may not be true with the ether extract from animal feeds. The assumption that, of the 9.35 kcal per gram of potential energy found in fats of foods typical of the human diet, approximately nine kcal is available for body metabolism cannot be safely applied in the case of ether extract obtained from feeds typical of animal rationing. The problem is clearly illustrated in a tabulation of the *coefficients of digestibility* of ether extract for a few animal feeds (see Table 2-4).

TABLE 2-4 *Coefficients of Digestibility of the Ether Extract of Certain Feedstuffs*

Feedstuff	Per cent ether extract	Coefficient of digestibility	
		By cattle (per cent)	By swine (per cent)
Alfalfa, hay, s-c	1.8	36	14
Barley, grain	1.5	60	44
Grains, brewers grains, dehy	7.3	89	60
Corn, dent, grains, gr 2 US	4.0	87	46
Cotton, seed w some hulls, mech-extd grnd (Cottonseed meal)	10.2	92	90
Fish, whole or cuttings, cooked mech-extd dehy grnd (Fish meal)	7.4	97	81
Corn, grits by-prod (Hominy feed)	7.7	96	—
Flax, seed, mech-extd grnd (Linseed meal)	6.3	89	62
Cattle, milk, fresh	4.6	100	97
Oats, grain	3.5	82	82
Soybean, seed, mech-extd grnd (Soybean meal)	6.0	84	—
Soybean, seed, solv-extd grnd (Soybean meal, solvent extd)	1.2	38	58
Animal, carcass res w blood, dry- or wet-rend dehy grnd (Tankage)			
12 per cent fat	11.9	100	96
2 per cent fat	2.0	—	73
Wheat, grain	1.8	—	80
Wheat, bran, dry-mil	5.0	62	58

There are several factors that affect the digestibility figures of ether extract, the first of which is the fact that we are dealing with apparent digestibility rather than the true digestibility. By definition the apparent digestibility of a nutrient is the percentage of the intake that is not recoverable in the feces. With ether extract the problem is complicated by the fact that some ether-extractable material is synthesized in the digestive tract, presumably by microorganisms, and, consequently, appears in the feces. It is designated metabolic fecal fat. Obviously anything that increases the fecal fat is reflected in a lowering of the apparent digestibility coefficient for that nutrient. This proportion of fecal fat may be an appreciable value, particularly in the case of low-fat intakes. To the extent that metabolic fecal fat is of bacterial origin and perhaps synthesized initially from dietary carbohydrate rather than dietary fat, there is an error in the usual assumptions concerning the significance of digestible ether extract. (Low-fat diets often give negative digestibility because of metabolic fat.)

A second factor influencing the apparent digestibility of ether extract is the proportion of neutral fat in the ether extract consumed. This is highly variable among different feeds, particularly if the feeds are forages, where the neutral-fat component of the ether extract may be relatively small. Most of the other ether extractives are nondigestible, and consequently the apparent digestibility of ether extract becomes a low value.

Schneider has examined the digestibility coefficients of over twenty thousand digestion trials and from his data has determined partial regression coefficients that can be used to predict the coefficient of digestibility for the ether extract of feeds where the proximate analysis of that feed is known. He has also prepared partial regression coefficients that can be used in the absence of other data for predicting probable digestibility of ether extract.

The equation for calculating the probable coefficient of digestibility of the ether extract of a specific sample of a concentrate feed where both the average digestion coefficient and the proximate composition of the feed are known is:

$$Y = \bar{Y} - 1.399(x_2 - \bar{x}_2) + 1.706(x_3 - \bar{x}_3) + 17.317(x_4 - \bar{x}_4).$$

In this formula Y is the apparent digestibility of the ether extract and $\bar{Y}$ is the average digestion coefficient for fat of the same feed as given in some table of feed composition and digestibility. The values $\bar{x}_2$, $\bar{x}_3$, and $\bar{x}_4$ are the percentages of crude fiber, nitrogen-free extract, and fat, respectively, also taken from a table of food composition for the feed in question. The values x_2, x_3, and x_4 are the corresponding percentages of crude fiber, nitrogen-free extract, and ether extract from the analysis of the particular sample that is being investigated. The numerical values in this equation are the partial regression coefficients applicable to the three proximate principles with which we are dealing.

As an example of how this formula works, let us calculate the probable percentage digestibility of the ether extract of a sample of barley showing the following composition: crude fiber 15 per cent; nitrogen-free extract 65 per cent; ether extract 3 per cent. The average digestibility of the ether extract of barley grain fed as the sole ration to swine, taken from Schneider's *Feeds of the World,** is 44 per cent, and the average crude fiber, nitrogen-free extract, and ether extract figures are 10.7, 66.6, and 1.9 per cent, respectively. Solution of the equation is:

$$Y = 44 - 1.399(15.0 - 10.7) + 1.706(65.0 - 66.6) + 17.317(3.0 - 1.9);$$
$$Y = 54\%.$$

This example illustrates how the proximate analysis affects the apparent digestibility of ether extract. In this particular example, the differences in the ether extract between the sample and the average for all feeds of that class are particularly important in affecting the apparent digestibility of the ether extract.

A large portion of this change in apparent digestibility of ether extract is traceable to the metabolic fecal fat, which presumably is constant. Consequently, as the percentage of fat inherent in the feed increases, the relative effect of the metabolic fecal fat becomes smaller, and the apparent digestibility therefore increases. Students of digestibility have failed to stress this effect of metabolic fecal fat on the apparent digestibility and hence on the presumed available portion of the dietary fat, and only recently has it become recognized as an important consideration in the calculation of the available energy of a feedstuff.

We now realize that using a figure such as nine kcal per gram of useful energy from dietary fat in animal feedstuffs is likely to be highly erroneous. Even with feeds in which the ether extract is largely neutral fat, the apparent digestibility is oftentimes much below the assumed 90 per cent or more on which the human dietary fat energy has been calculated. Incidentally, this error creeps into the TDN values when these are converted by any constant figure to approximate caloric values. Certainly the value of four kcal per gram as the energy equivalent of TDN is open to much question in this connection. The significance of the ether extract for feedstuffs is therefore highly variable, and were it not for the fact that most feedstuffs are relatively low in ether extract, the problem of evaluating feeding value would be much more uncertain than it is.

* B. H. Schneider, *Feeds of the World* (Charleston, W.Va.: Jarrett, 1947).

Schneider has also developed regression equations that make it possible to predict the probable digestibility of the ether extract of feedstuffs for which no basic digestibility data are available, provided we can determine the proper category in the feed classification for the product under question. Let us suppose, for example, that we have a sample of very light-weight immature frosted barley. Let us suppose that no feeding or digestion trials have been done with this product, and all we have in the way of chemical description is the proximate analysis. The equation proposed by Schneider for dealing with this situation in order to estimate the probable digestibility of the ether extract is:

$$Y = C + b_2X_2 + b_3X_3 + b_4X_4.$$

We may assume that our sample analyzes: crude fiber 15 per cent, nitrogen-free extract 40 per cent, ether extract 2 per cent. The equation with the values for the coefficients and constants inserted is:

$$Y = -162.9 + 3.240(15.0) + 1.946(40.0) + 16.204(2.0);$$
$$Y = -1.9.$$

Here we have a negative value for the probable apparent digestibility of the ether extract. We find this situation occasionally, and again it is a reflection of the effect of metabolic fecal fat. One would assume that the high crude fiber in this feed had probably depressed the true digestibility of the dietary ether extract and that consequently most of the intake was recovered in the feces, to which has been added the metabolic fecal fat, so that the sum total recovered is greater than the intake, and we therefore obtain an apparent digestibility that is negative.

Crude Fiber

Unless otherwise specified, crude fiber refers to the residue of a feed that is insoluble after successive boiling with dilute alkali and dilute acid in accordance with the procedures originally proposed by the Weende Experiment Station and officially recorded in the procedures of the Association of Agricultural Chemists. The definition of crude fiber is of some importance because of the many procedures that have been proposed as improvements in the original Weende crude-fiber method. These modifications give values for crude fiber that differ from those of the Weende procedure. The chief difference is usually that the residues contain more of the lignin, and hence the total values are larger than in the original method.

The biological and hence nutritional significance of the crude fiber figure is far from clear and certainly is not precise. Crude fiber is the portion of

the total carbohydrate of a food that is resistant to the acid and alkali treatment mentioned above, and the original supposition was that it therefore
represented an indigestible portion of the feed. We have since learned that
the Weende crude fiber may be a misleading index of the overall digestibility of
a feed, for the simple reason that in an appreciable number of cases the
crude fiber itself is at least as highly digested as the soluble carbohydrate
usually referred to as nitrogen-free extract.

The data in Table 2-5 illustrate this situation. The reason for the relatively

TABLE 2-5   *Digestibility of Crude Fiber by Ruminants*

Class of feed	Total cases recorded	Per cent of cases where the fiber is digested by ruminants as completely as is nitrogen-free extract
Dry roughages	115	42
Green roughages	65	23
Silages	27	29
Concentrates	95	11
Total and average	302	28

high digestibility of crude fiber by ruminant animals lies in the fact that
the largest component (perhaps 95 per cent) of crude fiber is cellulose, and
we know that the microorganisms of the rumen are able to break down
cellulose for their own needed energy, and that in the process they produce
acetic(and some butyric and propionic) acid, which is absorbed from the
rumen and supplies energy to the host. What many do not realize so fully is
that much the same situation holds true for species other than the ruminants.

It may surprise some to note (see Table 2-6) how much crude fiber is
digested by various species. The rations are those typical for the species cited,
and thus the levels of crude fiber in them will not be the same. The data do
not mean that unlimited quantities of fiber can be digested to the extent shown,
nor that there is no optimum level of ration fiber for any given species, but
they do show that the digestion of cellulose, and hence of crude fiber, is often
of sufficient magnitude to necessitate its consideration in estimating the energy
value of foods in the diet.

Another reason for considering crude fiber is that it is correlated with the
bulkiness of a feed, especially when the feed is ground. As we shall see later,
bulkiness of ration is an index of net energy (NE) value, and hence of feeding
value. Bulkiness is one index that the feeder himself can use for predicting the
relative feeding values of certain feed mixtures.

TABLE 2-6   *Digestibility of Crude Fiber by Various Species*

Species	Where digested	Per cent of contained crude fiber digested
Ruminants	rumen	50–90
Horse	caecum	13–40
Pig	caecum	3–25
Rabbit	caecum	65–78
Rat	caecum	34–46
Dog	caecum	10–30
Man	small and large intestine	25–62
Poultry	caecum	20–30

The physical role of crude fiber in a ration cannot be overlooked. The indigestible residue of feeds of plant origin is largely crude fiber. Thus it is material of this nature that gives to rations their physiologically effective bulk. The normal peristaltic movements of the intestinal tract are dependent, in part, on internal distention, which is furnished by food residues that have not been attacked by the digestive agents or that, though attacked, have not yielded absorbable fractions. Some such residues are hydrophylic and by their water-holding capacity help maintain a moist, soft condition of the fecal mass, and thus facilitate its easy passage through the large bowel. In particular, hemicellulose residues appear to be hydrophylic and probably owe some of their laxative properties in monogastric species to this characteristic. Sometimes the crude-fiber residue is apparently attacked by microflora with a resulting formation of gas, some of which is not absorbed. The result is an increase in fecal mass due to entrapped gas. This extra bulk may stimulate peristalsis. The bacterial action also results in the formation of free fatty acids, and it has been suggested that in the monogastrics these fatty acids in the large bowel are irritants that stimulate peristalsis.

Probably the fiber of some foods is of such a nature that it is physically irritating to the point of being objectionable or even actively harmful. Thus the diarrhea that they may cause initially may be followed by constipation from excessive bowel constriction, as in spastic colonic conditions. Herbivora are largely unaffected by the physical nature of the ration crude fiber, perhaps because the organisms that attack it normally produce absorbable fatty acids, and because herbivora can, by belching, void gases which in other species must pass through the large bowel with the fecal mass. Bulk of ration with the herbivora is of more importance as an index of net energy value than because of possible laxative properties.

Nitrogen-Free Extract

In order to understand clearly the nature and makeup of the fraction of a feed designated as nitrogen-free extract, we should first consider total carbohydrate. Carbohydrates are a group of substances formed by photosynthesis in the plant and containing carbon, hydrogen, and oxygen, with the last two in the proportions of water. Carbohydrates are the chief sources of potential energy in livestock rations, but the different members of the group differ in their yield of net energy because of differences in their digestibility and, to a lesser degree, in the end product they yield on breakdown in the digestive tract.

The carbohydrates of foods can be listed as follows:

Monosaccharides	Disaccharides
Dioses	Sucrose
Trioses	Maltose
Tetroses	Trehalose
Arabinose	Lactose
Xylose	Trisaccharides
Ribose	Raffinose
Rhamnose	Polysaccharides
Fucose	Starch
Hexoses	Dextrin
Glucose	Glycogen
Mannose	Mannan
Galactose	Araban
Fructose	Xylan
Sorbose	Cellulose

Within this family of carbohydrates there are a few which may have some specific nutritional function other than as sources of energy. For the most part, however, the sugars (i.e., mono-, di-, and trisaccharides) and the starches (starch, dextrin, glycogen), on digestion, eventually yield blood sugar (glucose), and accordingly are often classed together as one "functional" group. The polysaccharides also include xylan, araban, and mannan, which are sometimes classed as hemicelluloses. Hemicelluloses, together with cellulose, are found as principal parts of the plant cell-wall structure. Lignin, though not itself a carbohydrate, forms a physical combination with cellulose often referred to as lignocellulose, and thus is considered in nutrition with the true carbohydrates. Foods of plant origin usually contain a combination of most of these four kinds of carbohydrates.

The above classification of carbohydrates, which is based on chemical structure, is not as useful in applied nutrition as is one that groups them on a "functional" basis. Such a grouping is indicated in Table 2-7.

TABLE 2-7   *A Functional Grouping of the Carbohydrates of Feeds*

Group	Principal useful endproduct of digestion	Important nutritional functions
Sugars and starches	sugar	1) nonspecific and highly digestible source of energy via pyruvic acid
Hemicellulose	acetic acid	1) nonspecific energy via acetyl phosphate 2) stimulant to intestinal peristalsis (hence laxative to monogastrics)
Cellulose	acetic acid	1) nonspecific energy via acetyl phosphate 2) chief component of ration bulk 3) source of energy for cellulose-feeding microflora of digestive tract
Lignin	none	1) inhibits microflora "digestion" of cellulose, hence a liability

The Weende scheme of proximate analysis separates these four functional groups of carbohydrates into two categories: crude fiber and nitrogen-free extract. The makeup of these two fractions will be apparent from Table 2-8,

TABLE 2-8   *Effects of the Weende Crude-Fiber Procedure on Fat-Free Food*

Constituent	Boiling with 1.25 per cent H_2SO_4	Subsequent boiling with 1.25 per cent NaOH
Protein	partial extraction	complete extraction
Starches and sugars	hydrolysis and extraction	—
Cellulose	slight effects	slight effects
Hemicellulose	variable extraction	extensive but variable extraction
Lignin	slight effects	extensive but highly variable extraction

which gives the effects of the crude-fiber procedure on a dry, fat-free feed sample. In this table note that the sample contains protein, starches plus sugars, cellulose, hemicellulose, and lignin. (The ash may be disregarded here.) We shall consider that the last four items constitute the total carbohydrate. Also, since all of the protein is "dissolved" by the acid-alkali treatments, it too may be disregarded for the present. It is evident that any insoluble material remaining after the boiling will be carbohydrate, and will be made up of: (1) all the original cellulose, (2) variable proportions of the hemicellulose, and (3) a small, though again highly variable, proportion of the lignin. These three together constitute *crude fiber*.

Indirectly (that is, by difference) we see that nitrogen-free extract is a mixture of all the starches and sugars of the sample, plus some hemicellulose, and much of the lignin. In actual fact nitrogen-free extract is the difference between the original weight of the sample and the sum of the weights of its water, ether extract, crude protein, crude fiber, and ash, as determined by their appropriate analyses. Thus its numerical value will be affected by the chemical errors in the analyses of all five of the separate fractions, as well as by the lack of precision of the crude-fiber procedure in separating the functional categories of the carbohydrates.

Nevertheless, nitrogen-free extract is a practically useful index of the noncellulose portion of feed carbohydrates, and is primarily a nonspecific source of energy to the animal. Its digestibility is variable, though is ordinarily a little higher than that of the protein, fat, or crude fiber of the same feed. Some idea of the differences in digestibility of these proximate fractions, as well as of their variability among feeds, is indicated by data in Table 2-9. On the basis of these figures one might estimate that nitrogen-free extract from dry roughages will yield to an animal about three kcal per gram, and from dry

TABLE 2-9 *Digestibility of Proximate Fractions of Typical Forages or Roughages and Grains*

Feedstuff	Crude protein (per cent)	Ether extract (per cent)	Crude fiber (per cent)	Nitrogen-free extract (per cent)
Forages or roughages				
Alfalfa, hay, s-c	72	31	45	69
Clover, hay, s-c	62	43	56	64
Timothy, hay, s-c	47	47	55	61
Mixed hays	50	47	61	62
Average	58	42	54	64
Grains				
Barley, grain	72	66	42	84
Corn, grain	78	87	30	99
Oats, grain	78	82	37	83
Wheat, grain	78	72	33	92
Average	76	77	36	90

concentrate feeds about four kcal per gram, of digestible (and also of metabolizable) energy.

The importance of nitrogen-free extract as a source of energy to the animal lies in the fact that this fraction makes up about 40 per cent of the dry weight of forage feeds, and 70 per cent of the basal feeds. In general, the proportion of nitrogen-free extract in concentrate feeds will be inversely related to the protein content, so that protein supplements may have as little nitrogen-free extract as the forages.

Crude Protein

Crude protein is the figure usually obtained by multiplying the nitrogen of the feed by the factor 6.25. In some cases crude protein is now determined by multiplying the nitrogen by some other factor, depending on the percentage of nitrogen that is known to be in the protein of that particular feed. Such special figures are commonly employed for wheat and for milk products, but unless they are specified we should assume that the protein content given in a table of feed analysis represents nitrogen times 6.25.

From the amount of publicity that has been given to protein content in feeds and in rations, it is not surprising that in the feeder's mind protein frequently looms up as the most important component of a feedstuff or of a ration. Nothing could be further from the truth. In fact, altogether too much relative importance has been given to protein in practical feeding operations.

We should not assume, however, that protein is not an important consideration in feeds; it is important for several reasons. The usually accepted classification of feedstuffs is essentially one based on protein content. Basal feeds are low-protein feeds. Protein supplements are high-protein feeds. Legume roughages are higher in protein than other forages, which in turn are higher in protein than straw. Therefore, by knowing the protein content of a feed we can get some idea of the class of feed to which it must belong, even though we do not know its other characteristics.

Probably the protein content of a feed is also an indirect measure of its digestible energy, because the protein component of feeds is usually highly digested as compared, for example, with the coarser carbohydrates. Consequently, forages that are high in protein are almost sure to be correspondingly lower in crude fiber. Therefore such a product is more digestible than one that has a high fiber content and a lower protein content. To a limited extent this holds true with the concentrate feeds.

The overemphasis on protein has largely concerned the minimum protein necessary in satisfactory rations for livestock. Before we knew as much about the great variety of nutrients required by animals as we do now, the feeder's practical experience frequently led him to believe that additions of protein supplements to his grains, or to some mixture that he had previously been feeding, gave improved results because of the increased protein in the diet. Now we know that when proteins are added to feeds in the form of high-protein feedstuffs, many other factors besides proteins are also added to the ration. We can understand why when we realize that most high-protein feeds are high in this nutrient merely because either starch or oil has been removed from the original product. Consequently, not only protein but all of the other nutrients in the feed except starch (or oil) are more concentrated than in the original product. The fact that animals use surplus protein in the ration quite acceptably as a source of energy has often been forgotten. Tests, however, have indicated that, when the addition of a protein supplement has given beneficial effects in feeding practice, very often addition of sugar to the same ration has given equally good results; the supplementary feeding has been beneficial because of added energy rather than because of added protein. The tendency today is to reduce the supposed protein requirements in view of our greater knowledge of the nutrients that are actually needed to make a ration more acceptable.

Protein serves both as a source of energy and as an index of the total amino acids of the feed. We must recognize the limitations of this figure in both respects. As a source of energy protein is subject to a loss of about 20 per cent, because the body fails to "burn" the urea formed from deaminized amino acids. When this loss is added to that from incomplete digestion, we find that only about 60 per cent of the potential energy of a feed protein becomes available to the animal to meet caloric needs. Thus it is evident that when the ration is to be enhanced in terms of energy, protein is not the preferred source. As for protein's being a measure of the total amino-acid complex, we must recognize that not all proteins contain 16 per cent nitrogen, and that estimating the protein as 6.25 times the nitrogen in the sample leads to inaccuracies of varying degrees.

The nitrogen content of the protein of different feeds ranges from 16 to 19 per cent (see Table 2-10). In general, the proteins of the oil seeds show the most marked discrepancies from the figure of 16 per cent. On the average, the nitrogen figures and the corresponding conversion factors to protein for feeds might be taken as in Table 2-11.

The solution that would give greater accuracy, however, is not entirely straightforward. In many cases it is an estimate of the apparent digestible protein that is needed to assess feeding value, and by definition this is the portion of the protein intake not recovered in the feces. But a part, and often

TABLE 2-10 *Protein Content of Selected Foods (Showing Errors from Use of a Constant Factor 6.25)*

Food	Per cent N in protein	Per cent N in feed	Calculated		
			N × 6.25	Specific factor	N × factor
Wheat endosperm	17.5	1.39	8.7	5.70	7.9
Wheat, bran	15.8	2.45	15.3	6.31	15.5
Wheat, grain	17.2	2.06	12.9	5.83	12.0
Barley, grain	17.2	1.68	10.5	5.83	9.8
Corn, grain	16.0	1.54	9.6	6.25	9.6
Cottonseed meal	18.9	5.82	36.4	5.30	30.8
Flaxseed meal	18.9	5.82	36.4	5.30	30.8
Peanut kernels	18.3	4.13	25.8	5.46	22.5
Milk	15.8	0.53	3.3	6.38	3.4
Eggs; meat	16.0			6.25	
Gelatin	18.0	1.46	91.4	5.55	81.2

Source: Jones, D. B., U.S.D.A. circular 183 (1931).

a large part, of fecal nitrogen is of metabolic origin, to which the conversion factor for the feed protein may not apply. Consequently, the factor more appropriate for the feed nitrogen may, if applied to the fecal nitrogen, compound the final error in the estimate of the digestible protein. The practical aspects of the problem may be of less importance than we might think at first. Feeding standards and feeding experience have established with usable accuracy the protein levels needed in rations for various feeding situations; and the feeding standards that have been developed for individual amino acids now make it possible, where necessary, for the feeder to adjust protein intake in terms of its quality more accurately than he could by any estimate based on nitrogen figures alone.

Thus, although N × 6.25 may be a somewhat rough estimate of total

TABLE 2-11 *Selected Conversion Factors for Proteins*

Origin	Per cent N in protein	Conversion factor
Oil seed proteins	18.5	5.4
Cereal proteins	17.0	5.9
Plant leaf	15.0	6.6
Animal or fish	16.0	6.25

protein, and little or no estimate of its quality, it is still reasonably satisfactory as an index of the need for protein and of the protein included in practical livestock rations. Quality of protein is of relatively minor importance in the rations of herbivorous animals, and for other farm stock (except possibly for very young animals) the presence of adequate essential amino acids can be reasonably well assured by including small quantities of feeds of animal or marine origin in their rations.

Ash

Ash is the inorganic residue from the firing of a sample at about 600°C. The nutritional significance of the ash figure will depend, in part, on the food under consideration.

Crude ash may enable one to calculate the calcium and/or phosphorus in certain products such as bone meal, or, indeed, in feeds of animal or marine origin, with sufficient accuracy for practical use, since the makeup of such ash is relatively constant. For plant materials, however, the figure for crude ash has little direct nutritional use except where it is necessary for the calculation of carbohydrate or of nitrogen-free extract by difference, or occasionally where we want the figure for total organic matter. The reason ash from plant materials is a poor index of any of the inorganic nutrients is that the ash component of plant materials is highly variable, not only in total amount but in its component parts. Many foods are high in silica, an element that is of no nutritional value (although it may be a liability in such materials as the hull of coarse grains), but which, nevertheless, may be a factor in the total crude ash reported in the sample. We should also note that with some types of food the ash figure may be much magnified because of adhering sand or mineral material, as, for example, in pasture forage where soil has been splashed onto the forage.

Another Look at the Proximate Analysis

The student of feeds, after careful and detailed scrutiny of the faults and limitations of the proximate principles of the so-called "standard feed analysis," is sometimes left with a feeling that the whole scheme is a delusion as a useful description of the nutrient properties of feeds, and that it might better be abandoned. In order to reestablish perspective, we should look again at this method of feed examination.

In brooding over the fact that values for crude protein, crude fiber, ether extract, etc., lack the precision which a chemist would demand in the analysis of a single technical entity, we tend to forget the equally important fact that these feed fractions are not chemical entities but mixtures of substances that cannot be uniquely characterized chemically. The surprising thing is that any scheme for chemically describing the overall nature of a feed can succeed in giving as practically useful a picture as does this Weende method. For example, this analysis establishes unerringly the category in which a feed belongs. It is an adequate guide to its fatty and/or watery nature and hence to its stability in storage. Proper interpretation of the figures for the carbohydrate fractions gives one adequate information about the class of animal the feed will be most suitable for. The proximate analysis has been the key to determining the useful digestible or metabolizable energy of a feed. Indeed, the feed industry has been based on it. Our modern knowledge of nutrition has been supplementary to rather than a replacement for it.

Our difficulty lies, first, in demanding something from the analysis that with our expanding knowledge of nutrition we think is desirable, but which this scheme was never intended to give, and second, in our failure, because of uncritical or perhaps erroneous thinking and deduction, to properly interpret the figures this analysis yields. The Weende analysis does not define the nutrient content of feeds. It is an index of nutritive value only because the fractions that it isolates are correlated with some of the properties of feeds that have nutritional significance. Consequently, it is a useful descriptive device in establishing the characteristics of feeds. As with any other specialized tool, to use it correctly and to its full potential requires much other nutritional knowledge and judgment. An appreciation of its design, weaknesses, and limitations, though often stressed in destructive criticism, is more correctly an aid in making full legitimate use of a scheme of feed description which has broad and basic value.

These comments about the proximate analysis are not intended to minimize its shortcomings as an indicator of feeding value. It is encouraging that, since about 1960, efforts to find alternative schemes of greater precision and specificity by which to evaluate the nutritionally available energy of feeds have shown promise, especially for forages. One of these efforts,* which appears to have many advantages over the Weende method, makes use of the concept that the dry matter of feeds of plant origin can be considered as consisting of two principal parts: cell walls and cell contents.

Plant cell contents consist of sugars, starch, soluble carbohydrates, pectin, nonprotein nitrogen, protein, lipids, and miscellaneous other water-soluble

---

* P. J. Van Soest and L. A. Moore, "New Chemical Methods for Analysis of Forages for the Purpose of Predicting Nutritive Value," *Proceedings of the Ninth International Grassland Congress,* São Paulo, Brazil, I (1965), 739.

materials, including minerals and several vitamins. For these, individually and collectively, the true digestibility (or nutritional availability) is almost complete, averaging 98 per cent. Thus the cell content of a given feed can be considered as a unit nutritionally.

The fecal dry matter recovered from animals consuming feeds that consist of plant cell components contains, in addition to the some 2 per cent of the ingested cell contents, a further quantity of material amounting, on the average, to about 13 per cent of the weight of the dry feed eaten. The origin of this latter component of the feces is *endogenous*. It is usually referred to as metabolic fecal matter. It is principally of bacterial and intestinal mucosa origin.

The cell *walls* of feeds of plant origin are not nutritionally uniform, in the sense that their principal components (cellulose, hemicellulose, silica, lignin, etc., singly or in such combinations as nitrogen-hemicellulose or lignocellulose) differ widely in nutritional availability depending on the kind and maturity of the plant as well as on the age and species of the animal fed. Cell-wall nitrogen as a hemicellulose complex appears not to be digestible. Some of the interrelationships between the nutritional availability and amounts of the several cell-wall components have been elucidated by employing, in lieu of the Weende crude-fiber procedure, a buffered neutral detergent reagent (sodium lauryl sulfate) to dissolve the cell contents and thus isolate definitively all the fibrous portions of the feed, i.e., the cell wall. This latter, in turn, is digested with an acid detergent (cetyltrimethyl ammonium bromide) in H_2SO_4 to dissolve the nitrogen-hemicellulose complex, leaving "pure" lignocellulose. The lignin is recovered by the 72 per cent H_2SO_4 method. It has been shown that the degree to which cellulose is used depends on its degree of lignification. The biologically available (or true) digestibility of feeds of plant origin is the sum of the digestibilities of the cell contents and of the cell wall. In other words it is the neutral-detergent solubles times their percentage of digestibility plus the neutral fiber times its percentage of digestibility.

Van Soest and Moore at the U.S.D.A. have found high correlations of the *in vivo* digestibility of the cell contents (neutral-detergent solubles or NDS), of the cell wall (neutral-detergent-insoluble fiber or NDF), and of the lignin with the *in vitro* data. Their formula for predicting the true digestibility of forage feed (as dry matter) is:

(1) $0.98 \, \text{NDS} + 147.3 \, \text{NDF} - 78.9 \, (\log \text{lignin})$,

where NDS, NDF, and lignin are expressed as percentages of a unit weight of feed. If the constants of this equation are, in fact, constants for feeds in general, it means that we will have a valid, entirely *in vitro* method of describing

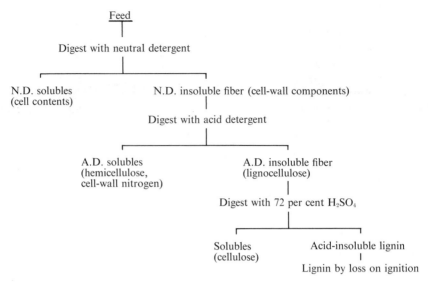

Figure 2-2. *Partition of forage feeds.*

the nutritionally available energy of feedstuffs. A general formula for predicting the *in vivo* apparent digestibility of cattle feed (as dry matter), recently proposed by Van Soest and Moore, is:

$$(2) \quad 0.98 \, NDS + W(147.3 - 78.9 \log \text{lignin}) + 12.9\%,$$

in which 12.9 per cent is the constant percentage of the weight of the feed due to metabolic fecal dry matter, and the other factors are as in formula (1). Given the apparent digestibility of the dry feed, one computes digestible kcal per gram of feed as gross kcal per gram times per cent of digestibility, and can estimate metabolizable kcal as 0.82 times digestible kcal.

The *in vitro* steps are in principle:

1. Digest dry sample with neutral detergent reagents to dissolve cell contents.
2. Recover on filter insoluble residue as neutral-detergent fiber (NDF).
3. Calculate neutral-detergent solubles (cell or seed contents) by difference (weight of dry sample minus NDF).
4. Digest NDF with acid detergent reagent to dissolve resistant nitrogen.
5. Recover on filter insoluble residue as nitrogen-free lignocellulose or acid-detergent fiber (ADF).
6. Digest ADF with 72 per cent H_2SO_4 to dissolve pure cellulose.
7. Recover lignin on filter and record by direct weight or by loss on ignition.

PROBLEMS ————————————————————————

1. Using a well-known feed such as corn, soybean meal, or wheat bran, prepare a table listing, for each of the nutrients indicated in Table 1-6, the quantity present in the feed. In a parallel column, list the quantity of each of the proximate principles that is contained in the same feed, in such a way that nutrients included in each proximate fraction are shown. How much useful information on the nutrient makeup of the feed is: (1) masked by the Weende analysis scheme, (2) omitted entirely by this analysis?

2. Using the combination of feed shown as "recommended" in the flexible formula for a 15 per cent protein swine ration (Chapter 19, page 377), calculate from the average DP and TDN of these feeds the DP and TDN of the mixture. Then, using for these feeds their proximate analysis, and Schneider's regression equations, compute the digestibility of the protein, fat, fiber, and nitrogen-free extract of these feeds. You now have data from which to compute the DP and TDN of the ration mixture. How closely do your figures obtained by the two methods agree?

3. From Schneider's *Feeds of the World*, Table 3, prepare a list of the different coefficients of digestibility of crude fiber for corn, barley, and oats which have have been reported from feeding tests with swine. Compare these in each case with the corresponding TDN figure. What explanation can you offer for these figures? Do you think an average coefficient of digestibility should be regarded as a "constant" for a given feedstuff?

SUGGESTED READINGS ————————————————

Deuel, H. J., "Fat as a Required Nutrient of the Diet," *Proc. Fed. Am. Soc. for Exp. Biol.,* XIV, no. 2 (1955), 639.

Jones, D. Breese, *Factors for Converting Percentages of Nitrogen in Foods and Feeds into Percentages of Protein,* U.S.D.A. Circular No. 183 (1931).

Morrison, F. B., *Feeds and Feeding* (Clinton, Iowa: Morrison, 23d ed., 1967).

Schneider, B. H., H. L. Lucas, M. A. Cipolloni, and H. M. Pavlech, "The Prediction of Digestibility for Feeds for Which There are only Proximate Composition Data," *J. Animal Sci.,* XI, no. 1 (1952), 77.

Schneider, B. H., H. L. Lucas, H. M. Pavlech, and M. A. Cipolloni, "Estimation of the Digestibility of Feeds from Their Proximate Composition," *J. Animal Sci.,* X (1951), 706.

Winton, A. L., and K. B. Winton, *Structure and Composition of Foods.* (N.Y.: John Wiley, 1932), vol. I.

CHAPTER 3

Schemes for Describing the Energy Value of Feeds

The Caloric System

In November 1958 the NRC Committee on Animal Nutrition passed a resolution to start using the caloric system, along with the total digestible nutrient (TDN) system, to describe the energy values of feeds, rations, and nutrient requirements of animals. All current reports on the nutrient requirements of domestic animals contain DE, ME, and TDN values. The *United States–Canadian Table of Feed Composition* (NRC publication 1684) contains data on the DE and ME contents of feeds. As soon as sufficient data can be obtained on the GE, DE, ME, and NE values of feeds, rations, and nutrient requirements of animals, the TDN values will be dropped in favor of the caloric system. It is appropriate therefore to introduce this chapter with an attempt to establish in the reader's mind a working concept of the biological energy interrelationships recognized in applied nutrition.

At the outset we must remember that *conventionally* we determine the fractions of biological energy by measuring only the potential energy of:

1. The food ingested by an animal.
2. The fecal recovery.
3. The urine recovery.
4. The methane-gas recovery (using a calorimeter or by formula for ruminants).
5. The increased heat loss in the fed animal over that of the animal while fasting, i.e., the heat increment (HI).

What this means is that we assume:

> digestible energy (DE) = food energy minus total energy in feces;
> metabolizable energy (ME) = DE minus total energy in urine (and
>   the methane energy for ruminants); and
> net energy (NE) = ME minus the total heat increment.

In other words the DE, ME, and NE values of a food are conventionally obtained *by difference*.

This method assumes that all the energy of feces and urine is of food origin and is wastage. Neither assumption is valid. A part of the organic material of feces is of endogenous origin, and consists of intestinal mucosa, "spent" enzymes, and microflora; and urine contains, besides products from incomplete metabolism of ingested food, an irreducible minimum of urea and creatinine that are obligatory end-products of fasting metabolism. Although these two urine components are sometimes shown in charts as separate entities, they are in fact portions of the same metabolic pool, and are separable in metabolism trials only by comparing the excretion levels of animals observed successively fasting and on feed.

When animals are in an environment below their critical temperature, the heat increment may be used to keep the body warm. Thus both the ME and the NE of foods as ordinarily computed somewhat underestimate the true values. Whatever error this represents, it is small compared with errors in the figures obtained for the DE of feeds, especially those consumed by herbivora, because the fecal energy neglects entirely the loss to the animal, via the intestine or by belching, of food energy in the gaseous products of digestion (of cellulose), chiefly methane. This credits the "gas loss" energy to the DE fraction. To emphasize that commonly used DE, ME, and NE figures are not true values, they are, in precise writing, called *apparent* energy; however, in common usage the term apparent is omitted. When the biological energy fractions have been partitioned as they should be and corrected for the energy losses in urine and feces arising from endogenous sources, the resulting values are always designated as *true* energy fractions, i.e., as TDE, TME, or TNE (see Figs. 3-1 and 3-2). Notice that in the true energy distribution system (Fig. 3-2) the heat of fermentation becomes a digestion loss instead of being a part of the heat increment. The gaseous products of digestion also become a digestion loss. Metabolic fecal energy and endogenous urinary energy become part of the maintenance requirement. Thus true digestible energy (TDE), true metabolizable energy (TME), and true net energy for maintenance (TNE$_m$) may be calculated.

Figure 3-1. *Conventional biological partition of food energy.*

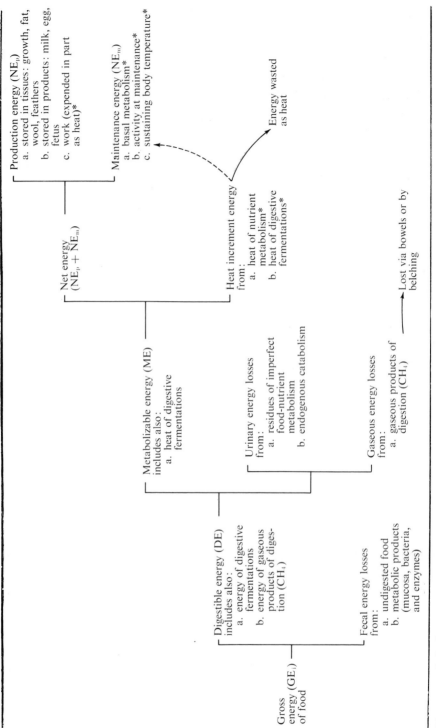

*Processes producing heat.

Figure 3-2. *True biological partition of food energy.*

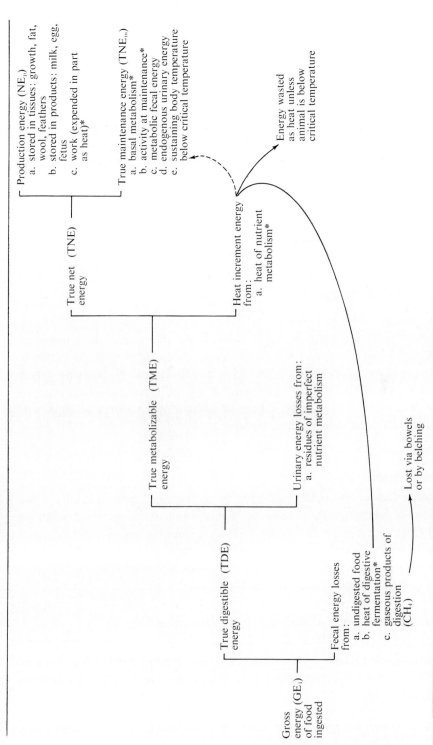

*Processes producing heat.

In Appendix Table A-3-1 the energy contents of the feeds appear according to the conventional system as follows:

feed energy (GE)
$\downarrow$ — fecal energy
digestible energy (DE)
$\downarrow$ — urinary and methane (for ruminants) energy
metabolizable energy (ME) or for poultry (ME_n)
$\downarrow$ — heat increment
net energy (NE_{m+p})

     net energy for maintenance (NE_m)
     net energy for production (NE_p) or (NE_{gain})

Glossary of Energy Terms

A glossary of terms* used in the caloric system of describing the energy values of feeds will perhaps clarify some of the interrelationships indicated in Figs. 3-1 and 3-2, and provide a condensed statement of the caloric system as it is used in considering both the needs of animals and the nutritive values of their diets. The letter abbreviations should be carefully noted, since in most discussions of the energy content of foods or of the energy needs of animals only the letter abbreviations are used.

BASIC TERMS

calorie (cal). As usually used in nutrition literature, a small calorie is the amount of heat required to raise the temperature of one gram of water to 15.5 from 14.5 degrees centigrade. However, since the specific heat of water changes with temperature, a calorie may be defined more precisely as 4.1840 international joules. We will get around the old distinction between "calorie" and "Calorie" by using "kcal" for the latter.

Kilocalorie (kcal). A kilocalorie is 1,000 small calories.

Megacalorie (Mcal). A megacalorie is equivalent to 1,000 kilocalories or 1,000,000 calories. A megacalorie is equivalent to a therm, but megacalorie is the preferred term.

* L. E. Harris, *Biological Energy Interrelationships and Glossary of Energy Terms* (Natl. Acad. Sci., Natl. Res. Council, Pub. 1411, 1966).

Gross Energy (GE). GE is the amount of heat, measured in calories, that is released when a substance is completely oxidized in a bomb calorimeter containing 25 to 30 atmospheres of oxygen. A similar term is heat of combustion.

Metabolic Body Size ($W_{kg}.^{75}$). Metabolic size is defined as the weight of the animal (in kg) raised to the three-fourths power. (See Appendix 2.)

ENERGY TERMS USED IN THE CONVENTIONAL PARTITION
OF BIOLOGICAL ENERGY

Gross Energy Food Intake (GE_i). GE_i is the gross energy of the food consumed. GE_i = dry weight of food consumed times GE of food per unit of dry weight.

Fecal Energy (FE). FE is the gross energy of the feces. It consists of the energy content of the undigested food and of the metabolic (body) fraction of the feces. FE = dry weight of feces times GE of feces per unit of dry weight.

Apparent Digestible Energy (DE). $DE = GE_i - FE$. Similar terms are apparent absorbed energy or apparent energy of digested food. Notice (in Fig. 3-1) that in this scheme the *fecal losses* do not include the energy of the digestive fermentations or the energy of the gases which have been formed during digestion. Consequently the DE "by default" is greater than the true digested energy of the food alone. For ruminants, where large quantities of methane are involved, the discrepancy between apparent and true digestible energy may be considerable.

Coefficient of Apparent Digestibility of GE. This is computed as

$$\frac{100(GE_i - FE)}{GE_i}$$

Gaseous Products of Digestion (GPD). GPD includes the combustible gases produced in the digestive tract by the fermentation of the ration. The energy of these gases (methane) can be estimated from the gross energy of the diet.*

Methane makes up the largest portion of the combustible gases produced. Ruminants produce the most methane; however, nonruminants also produce

* Kenneth Lyon Blaxter, *The Energy Metabolism of Ruminants* (London: Hutchinson, 1962), p. 200.

methane. Hydrogen, carbon monoxide, acetone, ethane, and hydrogen sulfide are produced in trace amounts. The loss of energy as methane gas reaches significant proportions only in ruminants.

Urinary Energy (UE). UE is the gross energy of the urine. It includes the energy content of the nonoxidized portion of the absorbed nutrients and the energy contained in the endogenous (body) fraction of the urine.

Metabolizable Energy (ME). $ME = GE_i - FE - GPD - UE$. In the figure obtained for ME by the conventional scheme, there still remains (incognito, so to speak) the energy of the digestive fermentations, since only the gaseous products of digestion have been subtracted from the DE (see Fig. 3-1).

Nitrogen Balance (NB). NB is the nitrogen in the food intake (NI), minus the nitrogen in the feces (FN), minus nitrogen in the urine (UN); i.e., $NB = NI - FN - UN$. The nitrogen balance must be calculated to account for the nitrogen retained in or lost from the body tissues and to adjust the ME accordingly. For precise work, the nitrogen lost by perspiration and other epidermal excretions should be taken into account. For some types of research, the nitrogen in the products synthesized, such as milk, eggs, or wool, should also be considered.

Nitrogen-corrected Metabolizable Energy (ME_n). ME_n is the total ME as corrected for nitrogen retained or lost from the body. For birds and mammals with simple stomachs, the gaseous products of digestion do not need to be considered. For mammals the correction is made as follows. For each gram of urinary nitrogen derived from the catabolism of body proteins (equal to negative nitrogen balance), 7.45 kcal are added to the ME, and for each gram of nitrogen retained in the body (equal to positive nitrogen balance), 7.45 kcal are subtracted from the ME; i.e., $ME_n = GE_i - FE - GPD - UE \pm (NB \times 7.45\ kcal)$. Since this value was obtained with dogs, it may not be entirely correct for other animals. For animals synthesizing products such as milk or eggs, no correction is made for the nitrogen in these products. A similar term for nitrogen-corrected metabolizable energy is catabolizable energy. For birds, the factor most often used is 8.22 kcal, because it represents the energy equivalent of uric acid per gram of nitrogen. Sometimes the factor 8.7 kcal is used, because it gives approximately the average energy content of urine per gram of nitrogen. There is relatively little difference between these factors, but the basis for choosing between them should be made clear and the differences among species should be recognized.

Heat of Fermentation Corrected Metabolizable Energy (ME$_{hf}$). ME$_{hf}$ is the gross energy food intake minus fecal energy, gaseous products of digestion, heat of fermentation, and urinary energy.

$$ME_{hf} = GE_i - FE - GPD - HF - UE.$$

Heat Increment (HI). HI is the increase in heat production following consumption of food when the animal is in a thermally neutral environment. It consists of increased heats of fermentation and of nutrient metabolism. The energy of the heat increment is wasted except when the temperature of the environment is below the critical temperature. This heat may then be used to help keep the body warm, and so becomes part of the net energy required for maintenance (see Fig. 3-1). A method that gives consistent results for measuring the HI is to subtract the heat production of the animal when fasting from the heat production of the animal when fed. If it is not feasible to fast the animal, the heat production may be determined by feeding it at two or more levels of nutrient intake and calculating the difference in heat production. The levels fed should be somewhere near that of the physiological function to which the data are to apply. Similar terms for heat increment are calorigenic effect, thermogenic action, and sometimes specific dynamic effect.

Heat of Fermentation (HF). HF is the heat produced in the digestive tract as a result of microbial action.

Heat of Nutrient Metabolism (HNM). HNM is the heat produced in intermediary metabolism as a result of using absorbed nutrients.

Net Energy (NE). NE = ME − HI. It includes the amount of energy used either for maintenance only or for maintenance plus production. NE can also be expressed as the GE of the gain in tissue and/or of the products synthesized, plus the energy required for maintenance. Below the critical temperature some of the HI is also part of net energy (see Fig. 3-1). When reporting NE, one must state clearly which functions are included. For example, there may be NE values for maintenance plus production (NE$_{m+p}$), NE for maintenance only (NE$_m$), or NE for production only (NE$_p$). The subscripts are suggested because there is often confusion in the literature about which energy fractions are contained in net energy.

Net Energy for Maintenance (NE$_m$). NE$_m$ is the fraction of total NE expended to keep the animal in energy equilibrium. In this state, there is no net gain or loss of energy in the body tissues. The NE$_m$ for a producing animal may be different from that for a nonproducing animal of the same weight, because of changes in amounts of hormones produced and differences in voluntary activity. This difference may be charged to maintenance, but in practice it is usually charged to the production requirement.

Net Energy for Production (NE_p). NE_p is the fraction of NE required (in addition to that needed for body maintenance) for involuntary work, for tissue gain (growth or fat production), or for the synthesis of a fetus, milk, eggs, wool, fur, feathers, etc. It should always be clearly stated which production fractions are included, such as

net energy for eggs	NE_{egg}
net energy for fur	NE_{fur}
net energy for gain	NE_{gain}
net energy for milk	NE_{milk}
net energy for pregnancy	NE_{preg}
net energy for wool	NE_{wool}
net energy for work	NE_{work}

Basal Metabolism (BM). BM is the chemical change which takes place in the cells of an animal in the fasting and resting state when it uses just enough energy to maintain vital cellular activity, respiration, and circulation as measured by the basal metabolic rate. It is a constant for adult homiotherms and can be computed as kcal/24hr. $= 70(W_{kg}^{.75})$. For the *measurement of basal metabolism,* the animal must be under basal conditions, i.e., in a thermally neutral environment at post-absorptive state, conscious, and quiescent. In the case of ruminants, since it is difficult to determine just when they reach the post-absorptive state, terms such as fasting heat production (FHP) and fasting heat catabolism (FHP + UE lost during fast) may be preferred. The length of the fasting period should be specified. Experimentally, it has taken from 48 to 72 hours postprandium to obtain valid fasting metabolic values.

Energy of Voluntary Activity (VAE). VAE is the amount of energy needed in getting up, standing, moving about to obtain food, grazing, drinking, lying down, etc. (See *Net energy for maintenance* for differences between nonproducing and producing animals.)

Heat to Keep Body Warm (HBW). HBW is the additional heat needed to keep the animal's body warm when the temperature of the environment is below its critical temperature. The critical temperature for an animal is defined as that environmental air temperature below which its heat production increases. The heat increment and the heat of digestive fermentation, in total or in part, can be used for keeping the animal warm.

Heat to Keep Body Cool (HBC). HBC is the extra energy expended by the animal when the temperature of the environment is above its zone of thermal neutrality. Above the critical air temperature for an animal, the rate of metabolism remains rather constant with a rise in air temperature, until the air becomes so hot that the body temperature increases. This then causes greater heat production by speeding up the body functions (panting, respiration rate, heart rate, etc.) in spite of the animal's already being too hot. If the animal suffers so much from heat that appetite fails, then less total heat may be produced because of the decrease in heat increment.

Total Heat Production (HP). The total heat production of an animal consuming food in a thermally neutral environment = HI + BM* + VAE = HF + HNM + BM + VAE (see Fig. 3-3).

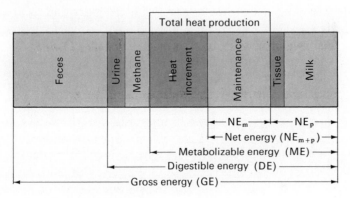

Figure 3-3. *Total heat production and energy use by a lactating cow. [Adapted from data of W. P. Flatt.]*

HP may be measured by direct or indirect calorimetry. In direct calorimetry, HP is measured by use of an animal calorimeter. In indirect calorimetry, heat production may be calculated by the formula:

$$HP_{kcal} = 3.866(\text{liters } O_2) + 1.2(\text{liters } CO_2) - 0.229(UN_g \times 6.25) - 0.518(\text{liters methane}).$$

The factor 6.25 in the formula converts the nitrogen to protein. This formula may be applied to ruminants, nonruminants, and birds; however, nonrumi-

---

* In ruminants this may be fasting heat production.

nants and birds produce little methane, so the methane component may be left out.

Heat production may also be determined by the comparative-slaughter technique, which uses the equation $HP = ME - NE_p$. By using the comparative-slaughter method to measure NE_p and metabolism trials to determine ME, total heat production can be determined. The portion of the total heat production used for maintenance (NE_m) may be estimated by feeding at two or more levels and extrapolating the data to zero energy intake. Heat production may also be estimated from carbon balance and nitrogen balance.

Relation of Environmental Temperature to Heat Production. Figure 3-4 is a simplified schematic diagram that illustrates the hypothetical relationships between environmental temperature and rates of heat production of animals. The actual values vary with the species of animal, type of ration

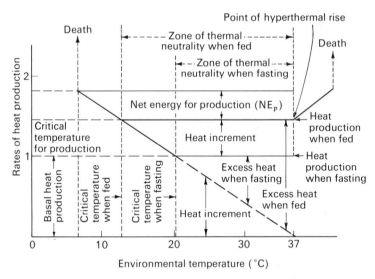

Figure 3-4. *Relationships between environmental temperature and heat production.*

being consumed, level of feed intake, and other factors. The zone of thermal neutrality is a region between the critical temperature and the point of hyperthermal rise. It varies with the amount and balance of food consumed, which is reflected in the amount of heat increment. It also represents the environmental temperature when the animal is not using energy to keep its body warm (HBW = 0) or cool (HBC = 0). When the environmental temperature goes below the critical temperature of the fasting animal, the HI of the food decreases in proportion to the decrease in temperature, because part of this heat is used as part of the NE requirement of the animal. The HI is zero when the minimal heat of the fed animal (BM + VAE + HI) just meets the thermostatic heat requirement, at the critical temperature of the fed animal. Below the critical temperature of the fed animal, the rate of heat production increases in proportion to the decrease in environmental temperature, following the thermostatic heat requirement.

Figure 3-4 also contains an area indicating NE_p, which is that part of GE_i used for growth, milk, eggs, etc. Within the zone of thermal neutrality, the rate of production (in terms of energy) is independent of changes in environmental temperature. As the environmental temperature of the fed animal decreases, the rate of production (with constant food intake) decreases. When the temperature decreases to the third critical temperature (the minimum temperature for production), production is reduced to zero. Below this temperature, the animal loses body substance and starves to death, even though it may eat to capacity. Its food capacity is insufficient to meet its thermostatic heat requirement. When the environmental temperature rises above the point of hyperthermal rise, if the animal cannot dissipate heat, it dies. In practice, voluntary feed intake by an animal is usually reduced when temperature increases to the point of hyperthermal rise.

Energy Balance (EB). EB is the relation between the GE_i and the energy output. It is calculated as, $EB = GE_i - FE - UE - GPD - HP$. For precise work, the energy content of the perspiration and of the epidermal excretions should also be subtracted from the GE_i. If the EB is positive, the animal is in positive energy balance; if it is negative, the animal is in negative balance. For birds and animals with simple stomachs, the gaseous products of digestion need not be considered.

Carbon Balance (CB). CB is the relation between the carbon from food intake and the carbon output. It is calculated as the carbon content of the food consumed, minus the carbon in feces, urine, gaseous products of digestion, and carbon dioxide. For precise work, the carbon content of the perspiration and of the epidermal excretions should also be subtracted from the carbon content of the consumed food. If the carbon content of the food intake is greater than that of the products (tissue, fetus, eggs, milk, wool, etc.), the animal is in positive carbon balance. If this fraction is negative, the animal is losing carbon from its body. For birds and simple-stomached mammals, the gaseous products of digestion need not be considered.

Nutrient to Calorie Ratio. There is acceptable evidence that the energy needs of animals and their requirements of the several nutrients are quantitatively correlated,* not necessarily directly, by cause and effect, but there is an optimum balance between them. For those nutrients that are needed to metabolize energy, it is logical to consider that the amount of energy metabolized "determines" their requirements. Hence, it is logical to express nutrients in weight per unit of energy needed. For example, the protein to calorie ratio should perhaps be expressed as grams of protein per 1,000 kcal metabolizable energy (*g protein/1,000 kcal ME*). If the ME is corrected for nitrogen balance, then the abbreviation should be *g protein/1,000 kcal ME$_n$*. This may easily be extended to other nutrients, as *g calcium/1,000 kcal ME* or *mg riboflavin/1,000 kcal ME*, and so on. If one wants to express the concentration (percentage) of a nutrient in the weight of a feed, feed mixture, or diet, one must first compute the weight of feed in weight units (g, kg, lb, etc.) that equal 1,000 kcal ME or the total kcal required. The amount of nutrient required for 1,000 kcal ME or for the kcal required is then divided by the weight of feed which contains 1,000 kcal ME, or by the weight of feed which contains the total kcal ME, respectively. These calculations are illustrated in the following example.

A cow weighing 500 kg and yielding 20 kg of milk containing 4 per cent butterfat requires:

	Total	Per 1,000 kcal ME
Energy (ME)	29,180 kcal	1,000 kcal
Digestible protein	1,220 g	42.0 g
Phosphorus	46 g	1.6 g
Weight of dry feed: 1,800 kcal/kg	16.2 kg	0.56 kg
Digestible protein concentration	7.5%	7.5%
Phosphorus concentration	0.28%	0.28%

To convert the digestible protein to per cent of crude protein, divide by the protein digestion coefficient times 100. For birds and swine, one may want to quote requirements on a crude-protein basis instead of a digestible basis.

TRUE ENERGY DISTRIBUTION

Under the true energy distribution system, metabolic fecal energy and endogenous urinary energy are part of the maintenance requirement and the HF and GPD become part of the digestion losses (Fig. 3-2).

---

* E. W. Crampton, "Nutrient to Calorie Ratios in Applied Nutrition," *J. Nutrition,* LXXXII (1964), 353.

Fecal Energy, Metabolic (FE$_m$). FE$_m$ is the amount of energy contained in the metabolic (body) fraction of feces (i.e., abraded intestinal mucosa, digestive fluids) that is not derived from unabsorbed ration residues. This fraction measures part of the maintenance requirement and is continually replaced. Because producing animals consume more food than comparable nonproducing animals, their food requirements are larger, and hence their FE$_m$ fraction is larger. In practice, this difference may be considered part of the production requirement.

True Digestible Energy (TDE). TDE is the food intake gross energy minus fecal energy of food origin (FE minus FE$_m$) minus energy in gaseous products of digestion minus heat of fermentation.

$$TDE = GE_i - (FE - FE_m) - GPD - HF$$

or

$$TDE = GE_i - FE + FE_m - GPD - HF.$$

In the second formula FE$_m$ is a plus item because it is part of the maintenance requirement. A similar term is true absorbed energy (TAE).

Urinary Energy, Endogenous (UE$_e$). UE$_e$ is the amount of energy contained in the endogenous (body) fraction of the total urine, and consists of urinary energy not directly of food origin. This fraction measures part of the maintenance requirement and is continually replaced. If hormonal control increases the basal metabolism in producing animals, this fraction may be larger for them (see *Net energy for maintenance,* page 63).

True Metabolizable Energy (TME). TME is the food intake gross energy minus fecal energy of food origin (FE minus FE$_m$), minus energy in gaseous products of digestion, minus heat of fermentation energy, minus urinary energy of food origin (UE minus UE$_e$).

$$TME = GE_i - (FE - FE_m) - GPD - HF - (UE - UE_e)$$

or

$$TME = GE_i - FE + FE_m - GPD - HF - UE + UE_e.$$

In the second formula FE$_m$ and UE$_e$ are plus items because these fractions are part of the maintenance requirement.

Nitrogen-Corrected True Metabolizable Energy (TME$_n$). TME$_n$ is the food intake gross energy minus fecal energy of food origin (FE minus FE$_m$), minus energy in gaseous products of digestion, minus heat of fermentation energy, minus urinary energy of food origin (UE minus UE_e); the total is then corrected for nitrogen retained or lost from the body.

$$TME_n = GE_i - (FE - FE_m) - GPD - HF - (UE - UE_e)$$
$$\pm (NB \times 7.45 \text{ kcal})$$

or

$$TME_n = GE_i - FE + FE_m - GPD - HF - UE + UE_e$$
$$\pm (NB \times 7.45 \text{ kcal}).$$

See ME_n above for explanation of factors to use for birds in place of 7.45.

True Net Energy (TNE). TNE is the intake gross energy minus the fecal energy of food origin $(FE - FE_m)$ minus energy in gaseous products, minus heat of fermentation energy, minus urinary energy of direct food origin $(UE - UE_e)$, minus heat of nutrient metabolism.

$$TNE = GE_i - (FE - FE_m) - GPD - HF - (UE - UE_e) - HNM$$

or

$$TNE = GE_i - FE + FE_m - GPD - HF - UE + UE_e - HNM.$$

True Net Energy for Maintenance (TNE_m). TNE_m is the sum of the energy required for basal metabolism, voluntary activity, metabolic fecal energy (body origin) and endogenous urinary energy (body origin). The net energy for a producing animal may be different than that for a non-producing animal of the same weight (see *Net energy for maintenance*).

$$TNE_m = BM + VAE + FE_m + UE_e.$$

Below the critical temperature and above the point of hyperthermal rise, the heat to keep the body warm or the energy to keep the body cool must also be considered.

Physiological Fuel Values

Physiological fuel values (PFV) are expressed in calories, and are units used in the United States to measure food energy in human nutrition. The system for determining PFV was developed by the classic investigations of W. O. Atwater and his associates at the Connecticut (Storrs) Agricultural Experiment Station. Working with human subjects, he determined the PFV of a wide range of foods.

The most common method for calculating the caloric value of food involves the determination of the average amount of protein, fat, and carbohydrate in a gram of food. The contents of protein and fat are determined by chemical analyses, and the percentage of carbohydrate is taken to be the remainder after the fat, protein, ash, and moisture have been deducted from 100. This so-called total carbohydrate, therefore, includes fiber as well as any noncarbohydrate residue present. The PFV in 100 g of a sample of food could then be calculated, using the established values of 4-9-4 (kcal of PFV per gram of carbohydrate, fat, and protein, respectively). This procedure assumes a constant digestibility for all proteins, fats, and carbohydrates of all foods. The basis for calculating these 4-9-4 values is shown in Table 3-1. The gross energy given in this table is an average value for

TABLE 3-1 *The Basis for Calculating PFV per Gram of Human Food*

	Average gross energy per g (kcal)	Assumed digestion coefficient (per cent)	Digestible energy per g (kcal)	Assumed loss in urine (kcal)	PFV per g (kcal)
Protein	5.65	91	5.14	1.25 (.91) = 1.14	4
Fats	9.40	96	9.02	none	9
Carbohydrates	4.15	96	3.98	none	4

all sources of each of the three nutrients, proteins, fats, and carbohydrates. The bomb calorimeter is not used in this procedure to determine the heat of combustion of each food.

The origin of the 1.25 kcal subtracted from each gram of digestible protein to estimate the PFV in kcal/g is as follows. For every gram of nitrogen in the urine, there was unoxidized organic matter sufficient to yield 7.9 kcal. Assuming that all of the energy in the urine is derived from protein, "digestible protein" was substituted for the urinary nitrogen in the calorie to nitrogen ratio by multiplying it by 6.25. Since one gram of nitrogen in the urine is assumed to represent the breaking down of 6.25 grams of digestible protein consumed, there would be 7.9 kcal of energy lost in the urine, or 7.9/6.25 = 1.264 kcal/g of digestible protein. The 1.264 was rounded off to 1.25, since most of the values were expressed to the nearest .05. Since this 1.25 kcal is estimated for 1 gram of absorbed nitrogen, it must also be multiplied by the digestion coefficient (assumed here to be 91 per cent) for it to be applicable to the digested and absorbed protein. Thus, 5.14 (the kcal of digestible protein) minus 1.14 = 4.00 kcal, the specific factor to apply to ingested protein.

In 1947, the Atwater system for arriving at the PFV of foods was carefully reviewed by a committee of experts of the Food and Agriculture Organization of the United Nations, and the factors were found to be satisfactory when correctly used. This system is the one used as the basis of the caloric values in the tables of food composition published by that organization. The heats of combustion (gross calories) of the fat, protein, and carbohydrate in a food are adjusted to allow for the losses in digestion and metabolism found in humans, and the adjusted calorie factors are applied to the amounts of protein, fat, and carbohydrate in the food. Using this procedure, the energy factors that Atwater derived from his digestibility experiments have been expanded and modified to take into account additional experiments conducted with human subjects since his time. Examples of the current factors are shown in Table 3-2. Most of these factors were developed before 1950 and were used in calculating the data listed in the 1950 publication of Agriculture Handbook no. 8. (See Harris.*)

Total Digestible Nutrients

With our increased knowledge of the needs of the body, the inadequacy of the relatively simple scheme of proximate analysis has become more and more apparent. Not only are some of the substances within the several groups of widely different biological significance, but the scheme omits much essential information concerning the nutritional properties of foods. However, it is from the proximate analysis that the figures for total digestible nutrients are calculated.

From the standpoint of the nutritionist the proximate analysis is often most useful as a basis for estimating the overall available energy of a feed. The information we can get from the protein figure as such is, of course, important, but not in as exact a sense as was at one time believed. All of the fractions of the dry matter of a feed separated by the Weende analysis, excepting ash, are potential sources of energy, the carbohydrate, protein, and ether extract yielding approximately 4.3, 5.6, and 9.3 kcal gross energy per gram. If only approximate values are needed the gross energy of a feed may thus be calculated from its proximate analysis.

The animal, however, is not able to obtain all of the potential energy from the food it eats. The most important, and also most variable, difference between feeds is in how completely they are digested. The different proximate principles may release anywhere from 10 to 95 per cent of their gross energy, as measured by the quantities of protein, fat, and carbohydrate recoverable

* L. E. Harris, *Biological Energy Interrelationships and Glossary of Energy Terms* (Natl. Acad. Sei., Natl. Res. Council, Pub. 1411, 1966).

from the appropriate fecal excretion of the ration consumed. We shall deal with the limitations of this method of measuring the apparent digestibility of these nutrients at a later point. Here we are concerned with it only in order to consider part of the presently used method of arriving at the value termed total digestible nutrients (TDN).

The thinking behind the use of TDN is entirely straightforward. If we add together the digestible portions of the crude fiber, nitrogen-free extract, protein, and ether extract of a feed, each weighted in accordance with its appropriate caloric value, we can get a figure representing the total digested energy expressed in terms of calories for that feed. Assuming a feed analyzing 10 per cent protein, 3 per cent ether extract, 10 per cent fiber, 65 per cent nitrogen-free extract, 2 per cent ash, and 10 per cent water, we might calculate the caloric yield as in Table 3-2.

TABLE 3-2 *Calculation of Caloric Yield of a Feed from Its Proximate Analysis*

Proximate principle	Content (per cent)	Average caloric values per gram	Gross kcal	Average digestibility of fraction (per cent)	Digestible kcal
Protein	10	5.6	56.0	75	42.0
Ether extract	3	9.3	27.9	90	25.1
Fiber	10	4.3	43.0	50	21.5
N-free extract	65	4.3	279.5	90	251.6
Ash	2				
Water	10				
Total	100		406.4		340.2
Per gram		4.1			3.4
Per cent digestible energy (as calories)					83%

In this example we find that the hypothetical feed contains a potential or gross energy content of 4.1 kcal per gram. When the apparent digestibility of each energy-yielding fraction is considered, the amount of energy apparently absorbed into the body following its digestion becomes 3.4 kcal or 83 per cent of the gross energy. Essentially, we would get this result by direct measurement, using a bomb calorimeter and measuring directly the caloric value of the feed and of the subsequent appropriate feces voided. Neither the value of 3.4 kcal per gram nor the figure of 83 per cent digested energy is the TDN value that we would obtain by the actual procedure used in arriving at the TDN.

By definition the TDN in 100 units of feed is calculated as:

TDN = dig protein + dig N-free extract

+ dig fiber + 2.25 (dig ether extract).

Using our example we would find:

TDN = $10(0.75) + 65(0.90) + 10(0.50) + 3 \times 0.90(2.25) = 78\%$.

The discrepancy between the two calculations arises from several sources, appreciation of which should make clear some of the limitations of TDN as a term descriptive of the useful energy of a feed.

The method for calculating TDN appears to come from a combination of the digestible nutrient figures for animal feeds and the Atwater* scheme for arriving at the metabolizable calories yielded from foods in human nutrition. The metabolizable calories from dietary protein are less than the digestible calories because of the potential energy of the urea excreted via the urine. Atwater found this loss amounted to 1.25 kcal per gram of digested protein. This results in a metabolizable energy value for food protein of

$5.65(0.92) - 1.25 = 4$ kcal per gram.

Again, in the case of the human, if we compare the digested calories from one gram of fat of the diet with those from the digested carbohydrate, or with those from the digested protein less its appropriate urinary loss, the ratios are about 9 to 4, or 2.25 to one. Since the apparent digestibility of fat and of carbohydrate are about the same (in the human diet), the ratios of their digestible energy is also about 2.25 to one $(9 \div 4 = 2.25)$. Hence, metabolizable fat can be expressed as its "carbohydrate equivalent" by multiplying by 2.25. Having thus expressed the fat, protein, and carbohydrate in equicaloric values, we can add them to give a "total metabolizable energy" in terms of weight rather than calories.

The use of such a method with farm animal feeding appears to date back to 1900, when Hills of the Vermont State Agricultural Station suggested that the fat (ether extract) and the carbohydrate be combined in cattle feeding standards into one quantity by multiplying the fat by 2.25, a figure obtained from the ratio of the digestible caloric values of 9 for fat and 4 for carbohydrate. Some ten years later the Vermont workers and Woll and Humphrey of the Wisconsin Station added digestible protein to the digestible

---

* L. A. Maynard, "The Atwater System of Calculating the Calorie Value of Diets," *J. Nutrition,* XXVIII (1944), 443.

fat-carbohydrate "carbohydrate equivalent" and coined the term *total digestible nutrients.*

The literature gives no further explanation of this term, but we may assume that it was intended to apply only to digested nutrients. If this is true, then the caloric value of the protein should be its heat of combustion minus only its fecal loss. In the Atwater scheme the average apparent digestibility of protein is taken as 92 per cent and the digested energy per gram from protein would correspondingly be calculated as 5.65 (0.92) = 5.2 kcal. However, the digestibility of the protein of animal rations is considerably lower than 92 per cent, approaching more nearly 75 per cent for the grains constituting the entire ration and 85 per cent for protein supplements as measured indirectly. If some constant value for digestibility must be assumed for the protein of rations as fed, 80 per cent might be used for those not including roughage, or 60 per cent for rations including the normal forage component for cattle. For rations without roughage then we could estimate the digested energy from protein to be 5.65 (0.80) = 4.5 kcal, or for rations with roughage 5.65 (0.60) = 3.4 kcal per gram. In neither case is the value equal to 4 kcal per gram.

Furthermore, one might question the 4 kcal per gram for carbohydrate, since its mean digestion coefficient for feeds with appreciable quantities of crude fiber will be lower than the 98 per cent found by Atwater for the carbohydrate of human diets. A digestibility value of 80 to 90 per cent for nitrogen-free extract and of 50 to 75 per cent for crude fiber would be more in line with the experimental data, though the values for crude fiber will be highly variable between feeds and between species of animals. In spite of these facts, the custom has long been followed of considering 4 kcal as the digestible *caloric value of a gram of TDN.* It obviously assumes that both the digestible protein and fat have, by the method of calculation, been acceptably expressed on a digestible carbohydrate-equivalent basis, an assumption that may be questioned. If we assume the above digestibilities for the carbohydrate fraction the correct values would be: $85 \times 4.1 = 3.5$ or $62 \times 4.1 = 2.5$, which is less than 4 kcal per gram.

The caloric value of TDN of individual feeds will differ according to their proximate composition. An average for all feeds appears to be in the neighborhood of 4.4 kcal per gram or about 2,000 kcal per pound TDN. Such a factor would need to be used with discretion for individual feeds.

For example, the apparent digestibility of each of the energy-yielding fractions is based on the assumption that all of each fraction that is not recovered in the feces has been absorbed and hence is potentially available to the animal. Aside from the problems of metabolic fecal products, there is another that is completely disregarded in the TDN calculations, namely, the energy lost as belched or otherwise voided gas, resulting from carbohydrate fermentation in the digestive system. In the case of the ruminant this can amount to 20 per cent of the gross energy of the crude-fiber fraction of the ration. Thus TDN values for roughages consistently and appreciably overestimate the usable energy of such feeds by ruminant animals.

It may be noted here that in North America there is a growing trend away from using the total digestible nutrient system for expressing the useful energy of feeds and the energy needs of the animal. Modern feeding standards state digestible energy (DE) or metabolizable energy (ME) needs in terms of kilocalories; and tables of feed composition include DE and/or ME values. Where analytical values are not yet available, such tables frequently give DE and/or ME figures computed from the commonly accepted average equivalent:

$$DE \text{ (kcal/kg)} = \frac{TDN \text{ \%}}{100} \times 4409$$

ME (kcal/kg) for ruminants = DE (kcal) $\times$ 0.82

$$ME \text{ (kcal/kg) for swine} = DE \text{ (kcal/kg)} \left[\frac{(96 - 0.202 \times \text{protein \%})}{100} \right]$$

1 lb TDN = 2,000 kcal DE = 1,640 kcal ME

Net energy values for some cattle feeds including net energy for maintenance (NE_m) and net energy for gain (NE_{gain}) may be estimated from formulas developed by Lofgreen.[*]

Log F = 2.2577 − 0.2213 ME
NE_m = 77/F
NE_{gain} = 2.54 − 0.0314 F

The terms used in the formulas are those which have been defined above and are on a dry matter basis (moisture free).

1. ME is the metabolizable energy in kcal/g of dry matter (DM) (or Mcal/kg DM).
2. F is the grams of dry matter per unit of $W^{.75}$ required to maintain energy equilibrium.
3. NE_m is the net energy for maintenance in kcal/g DM (Mcal/kg DM).
4. NE_{gain} is the net energy for gain in weight in kcal/g DM (Mcal/kg DM).

To convert NE_m and NE_{gain} to kcal/kg the values are multiplied by 1,000.

[*] G. P. Lofgreen and W. N. Garrett, "A System for Expressing Net Energy Requirements and Feed Values for Growing and Finishing Beef Cattle," *J. Animal Sci.*, XXVII (1968), 793.

Starch-Equivalent Values

While the total digestible nutrient scheme of describing quantitatively the useful energy values of feeds was being developed in North America, Kellner in Germany was devising a somewhat different method of describing the energy value of feeds, also based on the Weende analysis. His feeding standard* appeared in 1905 in German and in English in 1909. Kellner objected to the total digestible nutrient plan because it did not measure the energy actually available to the animal, since it neglected metabolic processes in which energy escaped in the urine and also as wasted heat.

In order to arrive at the net energy which a feed might yield, Kellner fed adult oxen a maintenance ration and then determined that one pound of starch fed in excess of maintenance produced 0.248 lb of body fat. Other pure carbohydrates gave similar values. But digestible true protein and digestible fat gave different values: 0.235 for true protein, and between 0.474 and 0.598 for ether extract. Kellner took the figure for body fat produced by starch as 1, and expressed the protein as $0.235/0.248 = 0.9476$ units starch equivalent, and the ether extract as $0.598/0.248 = 2.41$ units starch equivalent. When expressed as percentages, protein had 94.76 per cent and ether extract 241 per cent of the value of starch in producing body gains in adult steers.

Experiments with actual feedstuffs revealed that ether extract did not produce the full predicted value, but instead varied from 191 per cent for that of roughage to 212 per cent for the fats of cereal grains. Experiments also showed that the net nutritive value of various concentrate feeds was about the same as that of the pure nutrients when they were added separately to the maintenance rations.

On this basis, Kellner calculated the starch-equivalent values for concentrate feeds from their proximate analysis. He found, however, that for feeds containing much crude fiber his starch equivalents were too high. As an adjustment he proposed to reduce the computed value by an "evaluating factor" that depended on the fiber content.

The starch-equivalent values for the proximate principles eventually arrived at were:

Dig true protein	0.94
Dig fat in fodders	1.91
Dig fat in grains	2.12
Dig fat in oily seeds	2.41
Dig carbohydrate + fiber	1.00

* See O. Kelner, *The Scientific Feeding of Animals* (N.Y.: Macmillan, 1913).

The evaluating factors ranged from 30 to 95 per cent. As an example of the computation we may use rapeseed cake.

Dig true protein	$24.8 \times 0.94 = 23.3\%$ starch equivalent
Dig fat	$6.3 \times 2.41 = 15.2$ starch equivalent
Carbohydrate	$20.5 \times 1.00 = \underline{20.5}$ starch equivalent
	59.0% starch equivalent
Evaluating factor for fiber	$\underline{95\%}$
	56% starch value

This result we interpret to mean that it requires 100 lb of rapeseed cake, fed above maintenance, to yield as much energy to the body as 56 lb of starch.

TDN vs. Starch Equivalent. The essential differences between the TDN and the starch equivalents as measures of useful feed energy will be apparent if the equations are set down together.

$$TDN = CHO + protein + 2.25 \text{ (fat)}.$$
$$Starch\ equivalent = [CHO + 0.94\ \text{(true protein)} + 2.41\ \text{(fat)}]$$
$$\times \text{ fiber correction.}$$

Using these formulae, we calculate that for rapeseed cake TDN = 59.4 per cent, but the starch equivalent = 56.0 per cent. We can see that as a description of the useful energy of a feed these two schemes are not markedly different.

Direct Determination of Caloric Value

To avoid some of the difficulties and inaccuracies inherent in the proximate analysis for carbohydrate fractions, most laboratories are now determining the gross caloric value of a feed by combustion in a bomb calorimeter. Subsequent similar determinations on the fecal residues from such diets make it possible to correct the gross caloric value for the apparent loss in digestion and obtain a figure that has been called digestible calories, which is a better measure than total digestible nutrients for expressing the useful energy of feedstuffs. With this procedure we need to make no assumptions concerning the source of the energy, and, of course, none of the errors inherent in the determination of the proximate principles are involved in this procedure.

Among the objections to the use of a bomb calorimeter for routine feed analysis have been that it is time-consuming and that there is no permanent record of the temperature changes. The second of these objections has been largely overcome in the nutrition laboratory at Macdonald College by the use of a recording potentiometer equipped with a maganin-nickel thermocouple. The advantages afforded by the recording potentiometer are that the temperature change is easily determined mechanically, and is permanently recorded.

The bomb is ignited with an electric ignition unit and is rotated in its water bath by an electric motor driving through an elastic pulley, as shown in Fig. 3-5. The recording apparatus in use at the laboratory is a Brown

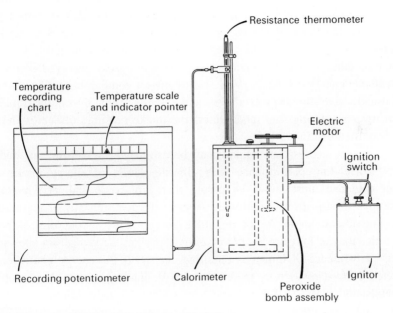

Figure 3-5. *An improved recording bomb calorimeter.*

Electronic Strip Chart Recorder and Indicating Potentiometer (Minneapolis-Honeywell Regulator Co. Model no. 153X11W-X-28P4, using Chart no. 5947-1). The scale is graduated from 18°C. to 28°C., and readings can be estimated with good accuracy to the nearest hundredth of a degree. The six-inch, steel resistance thermometer is connected to the potentiometer by a 36-inch conductor cable, both of which are supplied by the instrument manufacturer.

An analysis of variance was carried out for about a thousand determinations of feed and feces samples.* The values for 95 samples showed a standard deviation of 3.0 per cent of the mean. A comparable set of values determined for some 900 samples, with the temperature rise read from the usual type of thermometer, showed a standard deviation of 3.4 per cent of the mean.

Significance of the Energy Determination

Before we turn to considering methods for describing feeds in terms of protein, we should indicate the reason why we have paid so much attention to ways of describing the energy values of feeds. Practical experience has established that a below-average performance by farm livestock is most usually caused by faulty feed intake. In practical terms this is almost the same as saying "inadequate energy intake," for unless there is a marked difference in fat content and a lesser difference in protein level, the dry matter of different feeds contains about the same gross energy. Differences in *available* energy are likely to be small between comparable rations, which will usually have similar protein levels and much the same overall digestibility. Furthermore, altering feed intake alters the intake not only of energy but of all the nutrients of the ration.

The energy need of an animal may be estimated by calorimeter studies or by correlating performance with differing levels of energy intake. In either case one must know the usable energy of feeds and feed mixtures to be able to adjust feeding practice to meet the requirements of the animal fed. Too liberal an intake of digestible energy leads to deposits of body fat, which may or may not be important or desirable; too meager an allowance limits performance of functions requiring energy. If the energy restriction is effected by feed restriction, there may also be inadequate intake of protein, minerals, and vitamins, if these have been adjusted in the ration to correspond to the energy requirement.

In a more fundamental sense, the quantity of energy metabolized determines the necessary quantities of the many nutrients that function as enzymes or enzyme systems in the intermediary metabolism of energy. *Indeed, it is becoming evident that the nutrient requirements of animals can be better expressed per 1,000 kcal of digestible or metabolizable energy than per unit weight of the ration.* Thus we might truthfully say that a satisfactory working knowledge of the usable energy of feeds is the starting point in the successful use of feeds. It is at the same time one of the difficult things to come by.

* Macdonald College Nutrition Dept., unpublished data.

Feed Efficiency

Another term we use to describe the usefulness of feeds is *feed efficiency*. Inasmuch as this term has certain limitations, it may be well to discuss it in some detail.

Trials intended to evaluate feeds or combinations of feeds ordinarily use live weight gains and feed consumption records as the principal criteria. The magnitude of the gains made by animals on such trials is undoubtedly influenced partly by the quantities of the ration they have consumed and partly by the nature of that ration. If we divide the gain figure by the feed consumption figure, we get "gain produced per unit of feed," which has been taken as a method of adjusting the gain for equal feed intake, on the assumption that any difference in gain which remains after such adjustment is legitimately credited to differences in kind of feed. Such a ratio is frequently used as a quantitative measure of feed efficiency. There are two fundamental weaknesses of this use of such figures. One of these is statistical and the other is biological.

In the usual calculation of the gain–feed ratio, the biological importance of the amount of feed voluntarily eaten is sometimes minimized, if not altogether overlooked, perhaps partly because of careless thinking about the causes of variations in feed intake. The willingness of animals to eat and the extent to which they will consume food are in part regulated by the nutritional character of the diet or perhaps more fundamentally by the state of their intermediary metabolism. An outright nutrient deficiency, or a nutritionally significant imbalance in nutrients, may be expected to disturb intermediary metabolism. It is a recognized fact, established by experimental evidence, that imbalances caused by excessive quantities of protein, or a deficiency of some vitamin B-complex member, or by restriction of water, etc., actually result in an increased total heat loss. And, since the basal metabolism is a constant, this must be a reflection of increased heat increment loss. Similarly, as the diet becomes more perfectly adjusted to the requirements of the moment, the heat increment loss declines. In any case, any increase in heat loss means a reduction in the energy available to the body for its maintenance or for productive purposes. This may be diagrammed as in Fig. 3-6.

It appears that the body's defense mechanism against ingesting materials that it can dispose of only at an extra energy cost is usually reduced appetite and the consequent reduction of food intake. Therefore those quantitative differences that animals exhibit in feed intake become an important index of the nutritional properties or adequacies of the feed or of the ration. It is

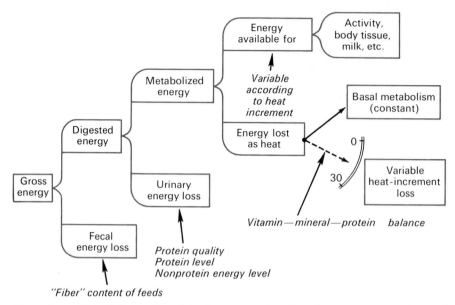

Figure 3-6. *Biological partition of food energy and heat loss. Important causes of several losses are indicated by italic type.*

almost universally true that nutritionally poor diets are consumed in smaller quantities than the same diets made adequate by suitable amendments. Consequently, when we adjust gains to equal feed intake and judge feed efficiency from such calculations, we should not forget the importance of the actual feed intake as a fundamental index of nutritional value. Where differences in feed intake between comparative lots are relatively small, establishing feed efficiency by comparing gains per unit of feed eaten or feed required to produce a unit of gain may be a highly useful procedure. If, however, there is a wide discrepancy in feed intake, then simple comparison of the gain–feed ratios may lead to implications that are not justified biologically.

These implications raise the statistical side of the question. Calculation of the gain–feed ratio implies that gain is directly proportional to feed intake over the whole range of intake from zero to whatever is maximum. Biologically, such an assumption is obviously untrue, because some of the feed consumed does not produce gain, but must be used to maintain weight already attained. If we want to eliminate the effects of differing feed intake on gains, it is statistically more acceptable to adjust gains to an equal feed intake by regression than by simple ratio, and it would seem logical in any given test to adjust the gains of different comparable lots to the average feed intake of the whole test. This statistical device is relatively simple to calculate and permits a more accurate measure of the differences in gain from causes other than variability in feed intake than is possible by the simple gain–feed ratio.

While we cannot at this point consider the mathematics involved, it may be useful to present a graph of the results of an analysis of a factorial test intended to measure the relative usefulness of feeds of animal origin and of plant origin as sources of protein for growing rats. The gain figures and the feed–consumption figures have been used in the one case to calculate feed efficiency as a gain–feed ratio (see Fig. 3-7). This is shown in curve A. The

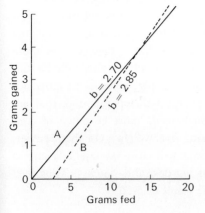

Figure 3-7. *Gain-feed relationships: simple proportion and regression.*

same data are also used to calculate the regression of gain on feed and the regression line is shown as curve B. Note that these two curves cross at a point that represents the mean feed and mean gain for the test. You can also see that either method of adjusting gains to equal feed will give about the same answer for those groups that are close to the mean; but for groups that deviate significantly from the mean, there will be an increasing error if the simple gain–feed ratio is used as the measure of feed efficiency. Curve B indicates that in this test approximately 2½ grams of feed produced no gain; it is presumably the quantity of feed that was sufficient to maintain the animals. The efficiency with which feed did produce gain is, then, indicated by the slope of curve B, which is greater than that of curve A.

Another comment may be worthwhile in connection with the interpretation of any measure of feed efficiency that is calculated entirely from the gain and the feed consumed. Inasmuch as more rapid gain relative to feed intake gives a figure indicating increased efficiency, it is very easy to fall into the error of assuming that in all cases those feeds that induce the most rapid gains are necessarily the feeds of highest feeding value. Obviously such feeds are of highest feed efficiency if efficiency means only gain per unit of feed eaten. This meaning is probably the objective with dairy cattle feeding and perhaps with the feeding of horses, where maximum production per unit of feed used or maintenance of weight per unit of feed used is the economically important factor.

In the case of bacon hogs, however, the feed that produces the greatest gain per unit of feed is often not the desirable one. If such feeds are fed *ad libitum* in the finishing ration of bacon pigs they result in overfat carcasses, which are penalized both by consumer demand and by dollar return to the producer. It is, of course, possible to restrict the allowance of such feeds and, consequently, capitalize on their high available energy content. Restricting feed, however, requires individual penning, which is not practical for the pig feeder. If group-fed pigs have their total feed limited in amount, the usual result is uneven performance of the group. Some pigs will be more aggressive and still be full-fed, while the more timid animals will be undesirably restricted in intake. One solution for the pig feeder is to introduce less efficient feeds—feeds that have lower energy value per pound, that are bulky per unit of their available energy. Such feeds must be eaten in considerably larger amounts to produce the same live weight gain that heavier, higher-energy feeds will produce. Such feeds will also be less efficient in producing gains but more efficient in producing desirable carcass. Thus feed efficiency must be interpreted in terms of the result desired.

PROBLEMS

1. In order to provide each 100 kcal for basal metabolism, how much difference in the metabolizable calorie requirement would it make to so balance the ration that the heat increment was reduced from 25 per cent to 5 per cent of the total heat loss? Assume that the caloric demand for the nonproductive activity of an idle adult is about 25 kcal for each 100 kcal of basal metabolism. Can you suggest economical conditions where it would not pay to feed a "balanced ration"?

2. Given a ration of the following specifications:

	Per cent composition by weight	kcal per gram	Per cent digestibility
Protein	15	5.6	75
Fat	3	9.3	70
Carbohydrate	70	4.0	85

Assume that when fed to an animal the ration efficiency was 30 units gain per 100 units air-dry feed consumed. What would you expect, other factors constant, the efficiency to be if the fat content were increased to 20 per cent at the expense of carbohydrate? If equal quantities of digestible calories were consumed, how much difference (in percentage) would there be in daily feed intake?

SUGGESTED READINGS

Armsby, H. P., *The Nutrition of Farm Animals* (N.Y.: Macmillan, 1930).

Armstrong, D. G., "Calorimetric Determination of the Net Energy Value of Dried S. 23 Ryegrass at Four Stages of Growth," in *Proceedings of the Eighth International Grasslands Congress, Reading, England* (*1960*), pp. 485–89.

Blaxter, K. L., *The Energy Metabolism of Ruminants* (London: Hutchinson, 1962).

Brody, Samuel, *Bioenergetics and Growth* (N.Y.: Reinhold, 1945).

Brody, S., M. Kleiber, H. H. Mitchell, E. G. Ritzman, and E. B. Forbes, "Recommendations of the Conference on Energy Metabolism of the Committee on Animal Nutrition," *Report of the Conference on Energy Metabolism* (Washington, D.C.: NRC Committee on Animal Nutrition, 1935).

Colovos, N. F., H. A. Keener, H. A. Davis, B. A. Reddy, and P. P. Reddy, "Nutritive Value of the Dairy Cattle Ration as Affected by Different Levels of Urea and Quality of Ingredients," *J. Dairy Sci.,* XLIV (1963), 696.

Cook, C. Wayne, L. A. Stoddart, and Lorin E. Harris, "Determining the Digestibility and Metabolizable Energy of Winter Range Plants by Sheep," *J. Animal Sci.,* XI (1952), 578.

Crampton, E. W., L. E. Lloyd, and V. G. MacKay, "The Calorie Value of TDN," *J. Animal Sci.,* XVI (1957), 541.

Crampton, E. W., "Nutrient to Calorie Ratios in Applied Nutrition," *J. Nutrition,* LXXXII (1964), 353.

Flatt, W. P., and K. A. Tabler, "Formulae for Computation of Open-Circuit Indirect Calorimeter Data with Electronic Data-Processing Equipment," in *Proceedings of the Second Symposium on Energy Metabolism.* (Publ. 10 of the European Assoc. of Animal Producers, 1961) pp. 39–48.

Forbes, E. B., Max Kriss, W. W. Braman, C. D. Jeffries, R. W. Swift, R. B. French, R. C. Miller, and C. V. Smythe, "Influence of Position of Cattle, as to Standing and Lying, on the Rate of Metabolism," *J. Agr. Research,* XXXV (1927), 947.

Harris, L. E., *Pasture and Range Research Techniques* (Ithaca, N.Y.: Comstock, 1962), pp. 75–89.

Harris, L. E., "Symposium on Feeds and Meats Terminology: III. A System for Naming and Describing Feeds, Energy Terminology, and the Use of Such Information in Calculating Diets," *J. Animal Sci.,* XXII (1963), 535.

Harris, L. E., *Biological Energy Interrelationships and Glossary of Energy Terms* (Natl. Acad. Sci., Natl. Res. Council, Pub. 1411, 1966).

Kleiber, Max, *The Fire of Life, An Introduction to Animal Energetics* (N.Y.: John Wiley, 1961).

Kriss, Max, "Evaluation of Feeds on the Basis of Net Available Nutrients," *J. Animal Sci.,* II (1943), 63.

Lofgreen, G. P., D. L. Bath, and H. T. Strong, "Net Energy of Successive Increments of Feed above Maintenance for Beef Cattle," *J. Animal Sci.,* XXII (1963), 598.

Lofgreen, G. P., "A Comparative-Slaughter Technique for Determining Net Energy Value with Beef Cattle," in *Proceedings of the Third Symposium on Energy Metabolism* (London: Academic Press, 1965), pp. 309–17.

Mitchell, H. H., "The Evaluation of Feeds on the Basis of Digestible and Metabolizable Nutrients," *J. Animal Sci.,* I (1942), 159.

Moustgaard, J., and Grete Thorbek, I. *Undersogelser over Jodkaseinets Virkning Pa Maelkesekretionen og Stofskiftet Hos Koer.* 204 beretning fra Forsogslaboratoriet (Kobenhavn, 1945).

Proceedings of the First Symposium on Energy Metabolism (Publ. 8 of the European Assoc. of Animal Producers) 1958.

Proceedings of the Second Symposium on Energy Metabolism (Publ. 10 of the European Assoc. of Animal Producers) 1961.

Proceedings of the Third Symposium on Energy Metabolism (Publ. 11 of the European Assoc. of Animal Producers), London: Academic Press, 1965.

Rubner, Max, "Der Energiewert der Kost des Menschens," *Zeitschrift für Biologie,* XLII (1901), 261.

Swift, R. W., "The Caloric Value of TDN," *J. Animal Sci.,* XVI (1957), 753.

Van Es, A. J. H., *Between-Animal Variation in the Amount of Energy Required for Maintenance of Cows* (Wageningen, Holland: Centrum voor Landbouwpublikaties en Landbouwdocumenatie, 1961).

CHAPTER 4

Schemes for Describing the Protein Value of Feeds

Basic Terms Used in Describing Protein

It will be helpful first to define the principal fractions recognized in protein metabolism, together with their abbreviations (see also the terms defined on pp. 89–91 and 94–99).

CP	Crude protein is usually given as N $\times$ 6.25. In special cases factors other than 6.25 may be used.
NI	Nitrogen intake.
FN	Total feces nitrogen; unabsorbed feed N plus metabolic fecal nitrogen.
FN_m	Metabolic fecal nitrogen; consisting principally of "spent" digestive enzymes, abraded mucosa, and bacterial nitrogen.
DP	Apparent digestible nitrogen or crude protein.

$$DP = (NI - FN).$$

TDP	True digestible nitrogen or protein.

$$TDP = NI - (FN - FN_m).$$

UN	Total urinary nitrogen; wasted food nitrogen plus UN_e.
UN_e	Endogenous urinary nitrogen; urinary nitrogen of body origin. It is the urinary excretion of nitrogen that occurs in the absence of any dietary nitrogen intake. It is analogous to the energy equivalent of Basal Metabolism, and quantitatively $= 146_{mg}(W_{kg}^{.75})$.

BV   Apparent biological value of dietary protein or nitrogen is the portion of digestible nitrogen that is not accounted for by FN + UN.

$$BV = \frac{NI - (FN + UN) \times 100}{NI}.$$

CNL   Cutaneous nitrogen loss; composed of the nitrogen of shed hair, cutaneous debris, and sweat.

TBV_{m+p} True biological value for maintenance and production; the portion of the digestible nitrogen not accounted for by $(FN - FN_m) + (UN - UN_e) + CNL$.

$$TBV_{m+p} = \frac{NI - (FN - FN_m) - (UN - UN_e) + CNL \times 100}{NI - (FN - FN_m)},$$

or

$$TBV_{m+p} = \frac{(NI - FN - UN) + (FN_m + UN_e + CNL) \times 100}{NI - FN + FN_m}.$$

In the latter equation, FN_m and UN_e are shown as plus items, since they must be replaced. These together with CNL represent the net minimum requirement of nitrogen. Hence, in this equation, $(NI - FN - UN)$ is nitrogen balance or production nitrogen, $(FN_m + UN_e + CLN)$ is nitrogen needed for maintainance, and $(NI - FN + FN_m)$ is true absorbed nitrogen, whence,

$$TBV_{m+p} = \frac{\text{maintenance N} + \text{production N} \times 100}{\text{true absorbed N}}.$$

Measures of Protein Quality

Somewhat analogous to the problem of describing the overall available energy value of feeds is that of describing the nutritive worth of the protein complex. It is generally accepted today that the usefulness of feeds as sources of protein depends primarily on two factors, the total concentration of the proteins (i.e., nitrogen $\times$ 6.25) and the distribution of the amino acids making up the proteins. There is ample evidence that imbalance among amino acids of the ration results in inadequate protein nutrition. This deficiency may appear only as a minor decrease in feed efficiency for some such function as growth or milk production, or it may be severe enough to cause the death of the animal.

The relative usefulness of the protein of a particular feed in meeting the animal's protein (nitrogen) needs is often referred to as its quality, and terms that describe protein quality quantitatively include *biological value, protein value, chemical score,* and *replacement value.*

Biological Value (BV) of Proteins. In the basic terms used for describing proteins, we have described two biological values. The first is a measure of the amount of nitrogen retained (nitrogen balance) that is used for production. We have called this apparent biological value (BV). The formula for it is given above.

The classical biological value may be defined as the percentage of true absorbed nitrogen that is utilized for maintenance and production, or for maintenance only. The amounts of nitrogen utilized reflect accurately a "perfect" assortment of amino acids. Surpluses of absorbed individual amino acids over the maximum that could make up this "perfect" assortment are presumed to be deaminized and the nitrogen component excreted via the urine. We have called this true biological value (TBV). Thus TBV is precisely defined by its (Thomas-Mitchell)* formula:

$$TBV = \frac{NI - (FN - FN_m) - (UN - UN_e) \times 100}{NI - (FN - FN_m)}.$$

Note that the total feces nitrogen is corrected for the metabolic fecal nitrogen (FN_m) and the urinary excretion for the endogenous nitrogen (UN_e). (In working out the formula Mitchell did not account for cutaneous losses.) We presume that the metabolic fecal nitrogen is not a direct dietary residue, and hence we must disregard it in measuring the true digestibility of the dietary intake. We also assume that Folin's endogenous urinary nitrogen measures normal wear and tear of nitrogenous body tissue and does not, therefore, represent surpluses of amino acids discarded because of imbalance. Although Folin's concept of the significance of endogenous and exogenous urinary nitrogen is not altogether accepted today, the endogenous excretion must be used in obtaining a true biological value by this method. Perhaps we should not concern ourselves too seriously over the theory's shortcomings, since other methods of measuring the quality of food protein are now more often employed. A more precise formula is:

$$TBV_{m+p} = \frac{(NI - FN - UN) + (FN_m + UN_e + CNL) \times 100}{NI - FN + FN_m}.$$

The so-called Thomas-Mitchell biological value as a practical measure of the quality of a protein has two serious limitations. The first is that it is hard to measure the UN_e and, to a lesser extent, the FN_m. In practice the former is sometimes calculated on the basis of body size according to the equation:

$$UN_e \text{ (in mg)} = 146 \, (W_{kg}^{.75}).$$

* H. H. Mitchell, "A Method of Determining the Biological Value of Protein," *J. Biol. Chem.*, LVIII (1924), 873.

If the values for UN$_e$ are to be determined experimentally, it is necessary to prepare and feed a nitrogen-free diet. To prepare such a diet that will be voluntarily eaten for long enough to obtain the necessary data is almost impossible with any but laboratory animals.

The second difficulty arises from the fact that the level of nitrogen (protein) fed modifies the calculated TBV independently of the amino-acid balance, because deamination appears to proceed somewhat according to the law of mass action. Thus to obtain maximum TBV there must be a minimum of dietary protein furnished. This automatically means that production rations, including those for growth, where liberal protein feeding is necessary for maximum performance, show low biological values as compared with maintenance allowances. Hence TBV data for individual feeds will change according to the rations in which they are used. In order to compare them for different feeds, they must have been determined at the same protein levels of intake, and to standardize this, such rations often are adjusted to 10 per cent protein.

The utilization of high BV ration proteins in actual feeding according to the level of protein ingested is shown in Fig. 4-1. Here we should note that

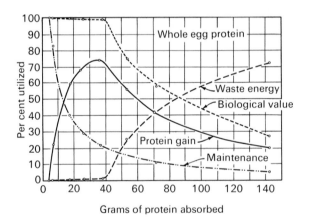

Figure 4-1. *Effect of amount of protein ingested on its use.* [*After Barnes,* J. Nutrition, *XXXII* (*1946*), *535.*]

when up to 40 grams of whole egg protein is ingested per day, essentially all of it is used as protein. At five grams intake all is used for maintenance. But as intake rises to ten grams per day only 65 per cent of it will be used for maintenance, and 35 per cent will be used for tissue synthesis (i.e., as protein gain to the body). Thus at this level the TBV is 100 per cent. With increasing intake the curve for the proportion used for maintenance steadily declines while the protein gain rises. These two uses account for all the intake until the daily consumption becomes 38 to 40 grams, which appears to be the maximum that can be utilized as protein. (Note that the protein is whole egg protein, which is considered to have an almost perfect distribution of amino acids.) Above this level there is a surplus that will be deaminized with a consequent rise in exogenous urinary nitrogen and an increase in heat increment loss. Since less than all the protein ingested remains in the body (i.e., for maintenance plus protein gain), the TBV declines. Thus, whereas at 38 grams intake the TBV is about 100 per cent, at 70 grams it is only 60 per cent, and the other 40 per cent is metabolized for other purposes.

We should remember that the category marked on the graph as protein gain refers to all utilization of protein for tissue synthesis other than the maintenance component. Hence, it includes body tissue gained by growth and production of milk, eggs, or wool. This graph makes it clear that age, class of animal, and production all influence the effective TBV of a protein as much as any "abnormalities" in amino-acid balance.

TBV figures are not now used in applied feeding, though they have served a useful purpose in establishing differences between protein sources in terms of nutritional usefulness, and undoubtedly their limitations stimulated the development of more satisfactory methods of description. The term biological value, however, is still used to denote protein quality.

Protein Replacement Value (RV). Because of the limitations of the Thomas-Mitchell biological values, Murlin[*] at Rochester University devised a slightly different measurement, which indicates the extent to which a given protein can replace some other protein in terms of nitrogen balance. His formula for obtaining this measurement is:

$$\text{Per cent of RV} = 100 - \frac{100\,(NB_1 - NB_2)}{NI_1}.$$

Two nitrogen balances (see Table 4-1) are conducted with nitrogen intakes essentially equal. The numerical difference between the two balances is expressed as a percentage of the nitrogen intake of the test having the larger nitrogen consumption. (If the intakes are nearly equal, the mean intake of the two tests may be used.) As measured by nitrogen retention, this percentage figure represents the extent to which the one protein has failed biologically to match the other. The complement of this percentage figure is the RV for the protein being examined.

An example may clarify the calculations. Assume a comparison between

[*] J. R. Murlin, and H. A. Mattill, "The Digestibility and Nutritional Value of Cereal Proteins in the Human Subject," *J. Nutrition*, XVI (1938), 15.

TABLE 4-1 *Typical Work Sheet for Nitrogen Balance*

	Food eaten (24 hr dry-matter basis)	as recorded during the test period ÷ days	1767 g
Intake data	Per cent N in dry matter	from chemical analysis of a sample of the feed	3.01%
	Grams of NI (24 hr basis)	1767 × 0.0301 =	53.2 g
	Grams excreted (24 hr, dry-matter basis)	moist feces × per cent dry matter ÷ days	556 g
Output in feces	Per cent N in dry matter	from chemical analysis of composited aliquots of daily collection	2.55%
	Grams N voided in feces	556 × 0.0255 =	14.2 g
	Volume urine (24 hr basis)	Total ml urine voided for test period ÷ days	3712 ml
Output in urine	Per cent N in urine	by chemical analysis on composited daily aliquots	0.52%
	Grams N voided in urine	3712 × 0.0052	19.1 g
Balance (grams/24 hr)		53.2 − (14.2 + 19.1) =	+19.9 g

two rations, the protein of which in one is from whole egg, and in the other from soybean meal. The daily nitrogen intake in each ration is 50 grams, and the nitrogen balances are +20 and +16 grams, respectively. We then calculate:

$$\text{Egg RV of soybean protein} = 100 - \left(\frac{20 - 16}{50}\right) \times 100$$

$$= 100 - \left(\frac{400}{50}\right)$$

$$= 92\%.$$

This calculation shows that soybean protein failed to equal egg protein by $[(20 - 16)/50] \times 100$, or 8 per cent, and so we say it has an egg replacement value of 92 per cent.

Replacement values are free from the implications about protein metabolism that biological values carry; may be employed in a relative sense, to describe the relative usefulness of any two proteins used under comparable conditions; and if tested against egg or milk proteins (whose biological values are nearly 100) have about the same numerical value as biological values determined under optimum conditions.

To interpret RV figures used in describing protein quality, one must also consider the apparent digestibility of the protein, since a nitrogen balance is dependent on both digestibility and intermediary metabolism. It is theoretically possible for two proteins, each measured against egg protein, to have similar replacement values, in the one case because of high utilization of a relatively poorly digested fraction, in the other because of high digestibility but poor metabolic utilization. In practice, however, we find more often that poor digestibility is correlated with low metabolic use, or else the digestibility is essentially the same in comparative cases.

A series of tests in which egg replacement values of cereal protein were determined with rats will illustrate the type of data to be expected with this technique (see Table 4-2). These data indicate that cream-of-wheat protein is

TABLE 4-2   *Egg Replacement Values of Cereal Foods as Affected by Heating*

Food	Nitrogen output (g/day)			Apparent digest- ibility	NI (g/day)	NB (g/day)	Egg NI (g/day)	RV
	Urine	Fecal	Total					
Egg	4.49	.97	5.46	83	5.95	+0.49	4.64	
Precooked oats	4.60	1.34	5.94	77	5.84	−0.10		87%[a]
						diff. = −0.59		
Egg	4.49	.78	5.27	86	5.85	+0.58	4.56	
Cream of wheat	5.66	.89	6.55	85	5.87	−0.68		72%
						diff. = −1.26		
Egg	4.40	.89	5.29	84	5.83	+0.54	4.55	
Shredded wheat	4.74	2.00	6.74	66	5.84	−0.90		68%
						diff. = −1.44		
Egg	4.38	.81	5.17	86	5.88	+0.69	4.59	
Puffed wheat	5.37	1.85	7.22	69	5.94	−1.28		57%
						diff. = −1.97		

[a] $100 - \dfrac{0.59 \times 100}{4.64} = 87\% \text{ RV}$

as fully digested as is egg protein but has only 72 per cent the overall biological value. Shredded wheat and puffed wheat are about equally less digested than egg, but the lower RV of the more intensely heated puffed wheat indicates a further damage to the amino acids from the processing.

Another example of the use of replacement values and digestibility data in estimating the protein quality of a feedstuff is from studies at Macdonald College of malt sprouts as a hog feed. In one of the balance trials the linseed meal RV of malt-sprout protein was determined at three levels of protein in the ration, as shown in Table 4-3. In these data we find a situation where the protein of a corn and malt-sprouts ration is not equal (in terms of RV)

TABLE 4-3 *NB and Linseed Meal RV of Malt Sprouts as a Supplement to Corn*

	10.3 per cent protein in ration		12.3 per cent protein in ration		14.2 per cent protein in ration	
	Malt-sprout ration	Linseed meal ration	Malt-sprout ration	Linseed meal ration	Malt-sprout ration	Linseed meal ration
NB[a]	5.23 g	5.02 g	5.55 g	7.20 g	5.02 g	7.34 g
Per cent total N excretion as:						
UN	70	78	68	77	65	77
FN	30	22	32	23	35	23
RV of malt sprouts protein	104%		84%		84%	
Per cent of total N supplied by supplements	27		45		58	
Per cent of total digested N excreted as UN	66	70	69	65	74	68

[a] Average of four pigs per group.

to a corn and linseed meal ration, except when only about a quarter of the total protein is derived from the supplements. The data reveal that the proportion of the digested protein that was excreted in the urine was essentially the same in all tests (the average difference being only 2 percentage points). The conclusion is, therefore, that the poorer digestibility of the malt-sprout protein, not its amino-acid balance, is responsible for the overall lower RV of malt sprouts in the pig ration when enough is used to provide 50 per cent or more of the total protein in the ration.

Replacement values are useful in comparing protein values under nutritional states other than those in which protein is the limiting nutritional factor. This method is nevertheless a biological method and hence is time-consuming; it is hard to interpret because of animal variability; and finally it is an overall measurement that does not indicate the specific cause of the figure obtained and hence gives no direct clue as to how to improve a protein having an undesirably low value.

Chemical Scores for Protein. To overcome the economic problems with biological methods for describing protein quality, two schemes have been proposed, both of which arrive at a numerical value for quality by considering the relative amounts of the amino acids present in the protein as determined by chemical (or in some cases biological) analysis.

Block and Mitchell * conceived of poor protein quality as caused primarily by a relative shortage of some one essential amino acid. They took the makeup of whole egg protein as the standard or ideal, and determined the percentage by which the amino acids of a comparative protein were individually deficient or in excess of those of the standard (see Table 4-4). Excesses were considered innocuous. The amino acid in greatest deficit was considered the limiting amino acid and the complement of its percentage deficit was their *chemical score* for that protein. Thus a protein whose lysine is only 40 per cent of that in egg protein would have a chemical score of 60, providing this was the amino acid in greatest deficit.

TABLE 4-4 *Calculation of Mitchell's Chemical Score for Wheat as an Index of Protein Quality*

Amino acid	Per cent in egg protein	Per cent in wheat protein	Per cent deficiency in wheat
Arginine	6.4	4.2	−34
Histidine	2.1	2.1	0
Lysine	7.2	2.7	−63[a]
Tyrosine	4.5	4.4	−2
Tryptophane	1.5	1.2	−20
Phenylalanine	6.3	5.7	−10
Cystine	2.4	1.8	−25
Methionine	4.1	2.5	−39
Cystine + methionine	6.5	4.3	−34
Threonine	4.9	3.3	−33
Leucine	9.2	6.8	−26
Isoleucine	8.0	3.6	−55
Valine	7.3	4.5	−38

[a] Chemical score for wheat, based on amino acid in greatest deficit, is $100 - 63 = 37$.

An extensive series of feeding trials with rats established that the chemical scores for a considerable range of proteins correlated well with the relative values of these proteins for growth, and also with biological values. The correlations are by no means perfect, though they are statistically significant.

On theoretical grounds we might argue that nutritional relationships among food proteins should be better revealed by a determination of amino-acid

* R. J. Block, and H. H. Mitchell, "The Correlation of the Amino-Acid Composition of Proteins with Their Nutritive Value," *Nut. Abst. and Revs.*, XVI (1946), 249.

composition than by biological estimation. Nevertheless, the results of such determination were found to vary relatively greatly, so that while chemical score appears to be a useful measurement for separating proteins into categories of usefulness, there are likely to be discrepancies in individual cases that would be of practical importance. Another problem involves the protein that is deficient in several essential amino acids. Correction of one deficiency still leaves a combination that is biologically imperfect.

Oser's EAA Index Method of Chemical Evaluation. Oser,[*] although approving the general principle of chemical score as outlined by Mitchell and Block, takes the stand that all essential amino acids should be considered, rather than the single one that is most deficit with respect to some standard. His argument is based on the view that each essential amino acid is essential in its own right—that each is equally essential. (This view neglects the possibility of different rates of synthesis for those amino acids that can be thus formed, though at a suboptimum rate.) One may picture the protein quality, in terms of the ten amino acids that are commonly classed as essential, of

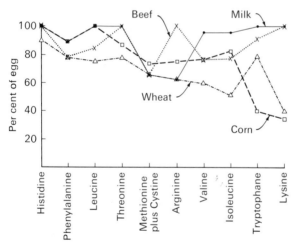

Figure 4-2. *Chemical scores of four protein sources.*
[*After Oser,* J. Amer. Dietetic Assn., *XXVII (1951), 396.*]

milk, beef, wheat, and corn by means of a graph of their respective egg ratios (see Fig. 4-2). The egg ratio is merely the percentage by which the amino acid in a feed protein departs from that found in whole egg protein, with excesses taken as 100 per cent. The order of arranging the essential amino

---

[*] B. L. Oser, "Method for Integrating Essential Amino Acid Content in the Nutritional Evaluation of Protein," *J. Amer. Dietetic Assn.,* XXVII (1951), 396.

acids is immaterial, but in this example it follows the general increasing deficiency in wheat flour. In this graph the lowest point for each of the four proteins would represent the Mitchell–Block chemical score.

When we examine the common feeds by this method the data suggest what is well-established analytically, that the protein complex of the cereal grains as a class is of lower biological value than egg protein and other animal proteins because of shortages (in descending order) of lysine, tryptophane, isoleucine, valine, arginine, and methionine plus cystine. Supplementation with lysine, while correcting the first deficiency, still leaves several other amino acids to become, in turn, the limiting factors in biological quality.

In order to take into account all the essential amino acids, Oser has proposed an essential amino acid (EAA) index, which is the geometric mean of the ten egg ratios found by comparing the content of the ten essential amino acids in a feed protein with that found in whole egg protein. Algebraically the index is expressed as:

$$\text{EAA index} = \sqrt[10]{\frac{100a}{a_e} \times \frac{100b}{b_e} \cdots \frac{100j}{j_e}},$$

in which $a, b, \ldots j$ are the per cent of essential amino acids in the food protein and $a_e, b_e, \ldots j_e$ are the per cent of the respective amino acids in whole egg protein. For computation it is convenient to express the equation in logarithmic form as:

$$\log \text{EAA index} = \frac{1}{10}\left(\log\frac{100a}{a_e} + \log\frac{100b}{b_e} \cdots + \log\frac{100j}{j_e}\right).$$

An example of Oser's computation of white flour protein is shown in Table 4-5.

Oser's indices are closely correlated with the biological values, much more closely than the Mitchell–Block chemical scores are. Thus, biological values can be predicted from the EAA index with reasonable accuracy, the mean deviation of the index from BV being on the order of $+2.8 \pm 4$. With this scheme for describing protein quality we can estimate the effect of fortifying proteins with one or more pure amino acids, or the effect of combining proteins of dissimilar amino acid distribution. In this respect it would appear to be more useful than the Mitchell–Block chemical score. However, we should point out that proteins of quite different amino acid makeup may have identical EAA index rating. For example, a protein with a very marked relative shortage of one or two amino acids might show an EAA index no lower than another with a more balanced assortment. With the first protein, strong fortification with the one or two deficient acids would be called for

TABLE 4-5 *Calculation of Oser's EAA Index for White Flour Protein*

Amino acid	Egg ratio	Log egg ratio
Lysine	27.1	1.4330
Tryptophane	53.5	1.7267
Isoleucine	54.5	1.7364
Valine	56.9	1.7551
Arginine	59.1	1.7716
Methionine plus cystine	60.9	1.7788
Threonine	62.8	1.7980
Leucine	76.1	1.8814
Phenylalanine	87.3	1.9410
Histidine	91.7	1.9624
1/10 × sum of logs		1.7784
EAA Index (i.e., antilog)		60.3

to raise the nutritional value, whereas with the second such amendments would be only partially effective. Thus such methods require more than the index value alone if they are adequately to describe the protein quality of a feed.

Application of Protein Quality in Feeding Practice

We sometimes find it useful, after being absorbed in some specific feature of a problem, to sit back and have an overall look at the situation in order to establish perspective and balance. This procedure applies particularly well to the question of protein quality. That certain amino acids are needed for the synthesis of the protein of one tissue or another in the body is unquestioned. Some, perhaps half, of these the body can readily put together from stored amino radicals plus nonnitrogenous fragments of carbohydrate or fatty acid metabolism. The others must be supplied to the bloodstream preformed, and in quantities and assortments dependent upon the kind and extent of the physiological functions requiring them.

From this point on we must recognize species differences. Herbivorous animals, beyond the early juvenile ages, are enabled by their symbiotic microflora to acquire enough of these essential amino acids almost, if indeed not entirely, independently of the nature of the nitrogen of the ration. Inorganic sources such as ammonium sulfate or urea nitrogen are apparently as acceptable as protein nitrogen to these organisms for their synthesis of the whole assortment of amino acids. From the feeder's standpoint, description, either qualitative or quantitative, of the protein quality of feeds is thus of no particular importance in the preparation of rations for or in the feeding of herbivorous livestock. A simple statement of protein content indicated by "N × 6.25" is all that he needs. This means that we do not need to concern ourselves with the protein quality of any roughages or concentrates that are of use only in the feeding of herbivores.

But for other species, the omnivores and carnivores, the assortment of amino acids needed is not constant, but depends on the physiological activity at the time. Digestive enzymes are not composed of the same amino acids as the enzyme myosin, or milk protein, or the protein of wool or hair. Not only do the rates and extent of the demands for amino acids fluctuate hour by hour, but the special functions of some of them, such as detoxication and the production of immune bodies, are not continuous. Thus there cannot be a fixed biological quality for any protein.

The question thus arises as to whether, for most practical feeding, one can ever hope to provide animals with a perfect protein, or if it were possible, whether it would be feasible. For a herd of swine, for example, how many different feed mixtures will it be economical or even feasible to employ? Couldn't we just group feedstuffs into categories based on the nature of their protein, so that those short of lysine are separated from those lacking sulfur-containing acids, and so on? Such a practice would help one avoid the unwitting choice of several feeds with a common deficiency, as well as help one obtain the benefits of supplementary sources rich in any acids deficient in certain feeds. After all, except for proteins with gross deficiencies of essential amino acids, the chief advantage of using rations whose proteins have high biological values is in the possibility of using less total protein—an economic problem. This possible saving must therefore be balanced against the cost of restricting feed selection in ration formulation, as well as against the practicability of adjusting feeding practice carefully enough to realize any difference in the potential efficiency of the ration with the higher protein quality.

Supplementary Values of Protein

It is generally agreed that the biological value of a digestible protein, fed in amounts not exceeding requirements and in a diet that provides enough energy, depends primarily on its assortment of essential amino acids. As proteins of increasingly poorer biological value are fed, there are correspondingly larger and larger quantities of surplus amino acids that will not be utilized to meet protein needs of the body. Rather, they may be deaminized and eventually oxidized for their energy. If, however, two or more proteins, each of imperfect biological value because of different specific amino-acid shortages, are combined in the ration, each may supplement the amino acids of the other, so that the biological value of the mixture is greater than that of the separate components.

We can illustrate this situation by a ridiculously oversimplified, purely hypothetical example. Assume that some essential protein tissue, such as an enzyme, must be made up of five amino acids, *A, B, C, D, E,* in the proportions 48, 10, 4, 32, 6. We might describe such a tissue as

$$A_{48}B_{10}C_4D_{32}E_6,$$

and if such an assortment were supplied to the animal as the ration protein, we might expect all of it to be used. That is, it would have a biological value of 100 per cent. Now assume that we offer this animal a protein of the following description:

Protein I $\quad A_{26}B_{28}C_2D_{34}E_{10}.$

Its usefulness is reduced because of a relative shortage of *A* and *C;* it will permit only half as much of the enzyme to be synthesized as the ideal protein would. The remaining amino acids have no biological value as far as the synthesis of our enzyme is concerned, because the amino acids must be in the proportions needed in the enzyme; the enzyme can be synthesized only to the extent that the amino acid in shortest supply (in this case *C*) will allow. Since protein I has only 2 per cent of *C,* its amino acids can be synthesized into the enzyme only in the proportions:

$$A_{24}B_5C_2D_{16}E_3.$$

This synthesis uses up only 50 per cent (notice that the subscripts add up to 50) of the amino acids; this 50 per cent thus has a BV of 100 per cent. The other 50 per cent of the amino acids, which are in the proportions

$$A_2B_{23}C_0D_{18}E_7,$$

are burned for energy and have a BV of 0. This residue of amino acids is often referred to as the supplementary fraction of a protein. It is all the amino acids that could not be used in protein tissue synthesis because of the absence of one or more essential ones.

One of two things can be done to salvage some or all of this residue. Pure amino acids may be fed with the protein to correct its deficiency of acids *A* and *C.* Such amendments to correct methionine shortage are now economical, and are commonly employed where this deficiency is the cause of the low biological value. The practical usefulness of this procedure generally is limited by presently available knowledge of specific requirements and of the amino-acid assortments of many feeds. Experimentally, the consequences of such amendments were demonstrated by Pearce *et al.*,* who fed mice on a purified diet, except for protein that was derived in different tests from casein, fibrin, or oxidized casein. The results are summarized in Table 4-6.

TABLE 4-6 *Effect of Supplementing a Deficient Protein with the Missing Amino Acids*

Protein fed	Supplement	Per cent of ingested amino acids excreted in urine
Casein	none	3–9
Fibrin	none	1–5
Oxidized casein	{ methionine cystine tryptophane	10 or less
Oxidized casein	{ methionine cystine	13–48[a]
Oxidized casein	{ tryptophane cystine	10–38

[a] These averaged 25 per cent of the amino acids (other than the deficient one).

Evidently when a protein low in some amino acid is fed, only that quantity of the other amino acids needed to match the one in short supply can be used, and the surplus is excreted in the urine. Thus, given a protein deficient in both tryptophane and methionine, high excretions of other amino acids occur. This ceases when tryptophane and methionine are added to the diet. In these tests the excretion fell from an average of 25 per cent to about 10 per cent when the amino-acid deficiency was made good. The test also indicated that the fate of the surplus amino acids from deficient protein is not entirely one of deamination and subsequent oxidation for energy.

The second and more often-used procedure for improving the biological value of a feed mixture is to employ protein mixtures. If two or more proteins of low biological value but of differing amino-acid assortment are combined, they may supplement each other's deficiencies. To return to our example, assume we have available a second protein as follows:

Protein II $\qquad A_{46}B_{18}C_6D_{20}E_{10}.$

* E. L. Pearce, H. E. Sauberlich, and C. A. Baumann, "Amino-Acids Excreted by Mice Fed Incomplete Proteins," *J. Biol. Chem.*, CLXVIII (1947), 271.

This protein has a surplus of C but still is of relatively low biological value, as we can see if we compare it with the ideal. The proportions needed to synthesize the enzyme are:

$$A_{48}B_{10}C_4D_{32}E_6.$$

Taking C as the unit, these proportions reduce to:

$$A_{12}B_{2.5}C_1D_8E_{1.5}.$$

In protein II, D is the amino acid in shortest supply compared to the ideal, so it will be the limiting factor in synthesizing the enzyme. Multiplying through the ideal proportions by 2.5 (since 2.5 times 8 gives us the 20 units of D that are available in protein II), we find that the enzyme can be synthesized in the proportions

$$A_{30}B_{6.25}C_{2.5}D_{20}E_{3.75},$$

or, raising to the next whole number,

$$A_{30}B_7C_3D_{20}E_4.$$

Here the subscripts add up to 64; 64 per cent of the amino acids will be synthesized into the enzyme, so protein II has a BV of 64 per cent. The residue will be in the proportions

$$A_{16}B_{11}C_3D_0E_6;$$

thus only 36 per cent of protein II, as compared with 50 per cent of protein I, is wasted. Furthermore, this wastage differs from that of protein I. The residue from I contains no C, while that from II contains no D. Besides that, protein II residue has a considerable quantity of A, while that of protein I has almost none. Mixing these two proteins should, therefore, result in a biological value of the combination greater than the mean biological value of proteins I + II, which is

$$\frac{50 + 64}{2} = 57.$$

In terms of our example we may illustrate by mixing equal quantities of proteins I and II and comparing the average amino acid assortment with our ideal:

$$
\begin{aligned}
\text{Ideal:} &\quad A_{48}B_{10}C_4D_{32}E_6 . \\
\text{Protein I:} &\quad A_{26}B_{28}C_2D_{34}E_{10}. \\
\text{Protein II:} &\quad A_{46}B_{18}C_6D_{20}E_{10}. \\
\text{Mixture of I + II:} &\quad A_{36}B_{23}C_4D_{27}E_{10}. \\
\text{Used for synthesis:} &\quad A_{36}B_7C_3D_{24}E_5 = 75\%. \\
\text{Used for energy:} &\quad A_0B_{16}C_1D_3\,E_5 = 25\%.
\end{aligned}
$$

Thus by mixing proteins I and II we obtain a protein complex, 75 per cent of which may be used for our enzyme synthesis as compared with 57 per cent expected as the mean of the two proteins we have employed.

Mitchell gives a numerical rating for the supplementary value of the protein of feeds. It is the complement of the biological value, i.e., it is the percentage of the digested protein that is not usable for the protein needs of the body. In our example, protein I would have a supplementary value of 50 per cent, protein II one of 36 per cent, and the mixture one of 25 per cent.

It does not always follow that proteins having the lowest biological values will in actual fact have the greatest supplementary value, for obviously the enhancement of biological values by mixtures depends on mutual supplementation of amino-acid deficiencies. Proteins with quantitatively similar deficiencies will not likely show effective supplementary values if combined. Thus, in practice, mixtures of proteins of plant origin are not so effective in providing high protein quality in the ration as combinations of plant and animal or marine proteins.

PROBLEMS

1. The 1955 NRC swine standards give the quantities that young pigs require of essential amino acids. Tabulate this list and compare it with the corresponding figures for the amino-acid makeup of egg protein. Are we justified in assuming the "egg distribution" of essential amino acids is biologically ideal?
2. Compare with the swine requirement in the same way the amino-acid makeup of fish meal, milk powder, soybean meal, corn, and oats. Does this comparison give any qualitative suggestion as to desirable feed combinations for young pigs?
3. Compute the essential amino acid index (see page 96) for the no. 1 swine ration shown in Table 8, p. 16, of the 1955 NRC swine standard. Plot the curve listing the amino acids in the same order as in the graph on page 96. What suggestions do you have for improving the biological value of the protein mixture of this ration?

SUGGESTED READINGS

Block, R. J., and H. H. Mitchell, "The Correlation of the Amino-Acid Composition of Proteins with Their Nutritive Value," *Nut. Abst. and Revs.,* XVI (1947), 249.
Bosshardt, D. K., W. Paul, K. O'Doherty, and R. H. Barnes, "Influence of Caloric

Intake on the Growth Utilization of Dietary Protein," *J. Nutrition,* XXXII (1946), 641.

Carbery, M., I. Chatterbee, and M. A. Hye, "Studies on the Determination of Digestibility Coefficients," *Indian Journal of Veterinary Science and Animal Husbandry,* IV (1934), 295.

Cipolloni, Mary Ann, B. H. Schneider, H. L. Lucas, and H. M. Pavlech, "Significance of the Differences in Digestibility of Feeds by Cattle and Sheep," *J. Animal Sci.,* X (1951), 337.

Murlin, J. R., and H. A. Mattill, "The Digestibility and Nutritional Value of Cereal Proteins in the Human Subject," *J. Nutrition,* XVI (1938), 15.

Oser, B. L., "Method for Integrating Essential Amino Acid Content in the Nutritional Evaluation of Protein," *J. Amer. Dietetic Assn.,* XXVI (1951), 396.

Schneider, B. H., *The Total Digestible Nutrient System of Measuring Nutritive Energy* (Pullman: Washington Agricultural Experimental Station, Scientific Paper no. 1250, 1954).

Schneider, B. H., H. L. Lucas, H. M. Pavlech, and M. A. Cipolloni, "Estimation of the Digestibility of Feeds from Their Proximate Composition," *J. Animal Sci.,* X (1951), 706.

Schneider, B. H., H. L. Lucas, M. A. Cipolloni, and H. M. Pavlech, "The Prediction of Digestibility for Feeds for Which There Are Only Proximate Composition Data," *J. Animal Sci.,* XI (1952), 77.

CHAPTER 5

Measuring the Intake and Utilization of Energy and Nutrients of Feeds

We have discussed several terms and expressions which are commonly used in describing feedstuffs. Some of these describe the physical nature of the feed; some describe feeds in chemical terms. All these descriptions have one feature in common: they are based on the assumption that if a feed has this or that nutritional characteristic the animal can make use of it. We assume, for example, that because there is 15 per cent protein or 60 per cent carbohydrate in a ration that the animal has all of it to use. Unfortunately, this assumption is not true. In fact we must consider that ingested food is still outside the body until it has been absorbed through the wall of some part of the digestive system and has gotten into the circulating fluids of the body.

The term *digestibility* is usually taken to imply that nutrients, or parent substances of nutrients, which are attacked by some digestive enzyme in the digestive system or are broken down by microflora, are absorbed. Consequently, the term digestion as ordinarily employed implies both digestion and absorption, and we shall follow this concept in our discussions in this book.

We could, of course, go much further than mere digestibility, to include some of the reactions in intermediary metabolism in describing nutritive properties of feeds, but usually information about how much of a nutrient in a feed is digested is of the greater practical value. One reason why this information is so important is that the digestibility of feeds is the most variable of the factors that affect the processes of utilization. Another reason is that incomplete digestion often represents the greatest loss between the quantity of the nutrient present and the amount finally utilized by the animal.

By definition *apparent digestibility* of the dry matter or of some constituent nutrient of a feed is merely that fraction of the intake that is not recovered in the feces. When this unrecovered fraction is expressed as a percentage of intake, it is called the coefficient of digestibility (the word *apparent* is usually omitted). Obviously this presumes that the feces does not contain any portion of a nutrient that has once been digested and absorbed, but we can make this assumption for only a few nutrients.

For example, sugar when ingested is normally all absorbed, and in the course of its metabolism its by-products are completely eliminated either as carbon dioxide (through respiration) or as water formed from the hydrogen, which combines with molecular oxygen brought to it by respiration. Thus, sugar would be 100 per cent digestible, since no sugar would be found in the feces. Crude fiber is a more resistant carbohydrate. Any portion of it that escaped breakdown and absorption would be found in the feces, but any portion of it that was broken down and absorbed would be eliminated in the same manner as sugar.

On the other hand, a nutrient such as phosphorus behaves somewhat differently. After ingestion a portion of it may remain unabsorbed in the digestive system, be passed along with the feces, and hence be recovered in the stool. Some of it will have been absorbed and will have entered the circulating fluids. A part of it may be incorporated into soft tissues and some into the skeleton. Concurrently, there is some mobilization of phosphorus from the bone and some is released from various soft tissues. A part of this is re-excreted into the intestine and so also finds its way into the feces. Another portion of it is excreted as phosphates in the urine. Consequently, with nutrients of this type, digestibility is not a very satisfactory measure of availability. The feces might contain quantities of the nutrient that had been absorbed and metabolized, and were now being eliminated.

With nutrients such as protein, there is another problem with digestibility. Dietary protein is attacked in the alimentary canal by digestive fluids and also by microorganisms. Consequently, in addition to any undigested protein from the diet itself, the feces may contain protein of bacterial origin. Furthermore, some previously digested and absorbed protein will have been metabolized into compounds that later are reexcreted into the digestive tract as digestive enzymes, some of which will be passed out of the body in the feces. Thus, feces will contain nitrogenous material from three sources, i.e., bacterial nitrogen, spent digestive-fluid nitrogen, and undigested diet-residue nitrogen. There may also be small amounts of nitrogen coming from abraded intestinal mucosa, the amount of which is influenced by the quantity of dry matter consumed. Except for that represented by the digestive fluids, the nitrogen that was absorbed will eventually be excreted via the urine. Thus we see that apparent digestibility of protein (or of nitrogen) is not as accurately indicative of the extent to which the dietary protein was really digested as it is with carbohydrate, but in practical feeding it is nevertheless an index of the extent to which the dietary protein is potentially useful.

Dietary fat is perhaps in the same category as dietary protein, in that some of the fat of the feces may actually have been synthesized by bacteria, and thus may erroneously be considered a diet residue. Some of the bacterial fat may have been synthesized from carbohydrate or protein rather than from diet fat itself. However, since most animal diets contain only 3 or 4 per cent fat, the error introduced is never large in terms of the total energy involved.

We can understand, therefore, that digestibility is not an equally useful descriptive term for all nutrients. In fact, digestibility is not ordinarily determined at all for the mineral elements or the vitamins. Instead, mineral balances are carried out in which all of the output is balanced against the total intake in an attempt to measure the fraction of the intake that is retained in the body. Thus, we find that coefficients of digestibility are ordinarily considered only for total dry matter, total energy, protein, fat, carbohydrate, and the fractions of carbohydrate that comprise nitrogen-free extract, cellulose, hemicellulose, and crude fiber.

The coefficients of digestibility of the various organic fractions of the proximate analysis have been determined for most of the commonly used feeding stuffs. It is from such coefficients, together with the data from the proximate analysis, that we calculate the digestible nutrients of feeds. In practice, the coefficients of digestibility of these proximate principles and the calculated total digestible nutrients for feeds have been taken as constants, and have therefore been used as a means of describing the feeding values in terms of available energy and digestible protein. Feeding standards have been set up in which the requirements of animals are also described in these terms. The facts are, however, that digestion coefficients are not biological constants, and consequently values for total digestible nutrients are not constants. We must use such descriptive terms with some understanding of their limitations if we are not to arrive at erroneous conclusions. It will be desirable at this point, therefore, to consider some of the problems involved in determining the coefficients of digestibility in order that we may more fully realize their limitations.

Technical Problems with Determining Coefficients of Apparent Digestibility with Penned Animals

As we have already indicated, a measurement of the digestibility of a nutrient is essentially a bookkeeping job. The *conventional* method for calculating digestibility requires an accurate record of feed intake and of feces output. From this information, together with a chemical analysis of the nutrient, the digestibility is actually calculated. An example for protein is given below.

Per cent digestibility of protein =

$$\frac{\left[\begin{array}{c}\text{dry wt of} \\ \text{diet eaten}\end{array} \times \begin{array}{c}\text{per cent of pro-} \\ \text{tein* in diet}\end{array}\right] - \left[\begin{array}{c}\text{dry wt of} \\ \text{feces voided}\end{array} \times \begin{array}{c}\text{per cent of pro-} \\ \text{tein in feces}\end{array}\right]}{\text{dry wt of diet eaten} \times \text{per cent of protein in diet}} \times 100.$$

Suppose we have the following data from a digestion trial:

Amount feed eaten	100 g,
N in feed	3%,
Amount feces voided	25 g,
N in feces	2%.

Then the per cent dig of N $= \dfrac{[(100 \times 3\%) - (25 \times 2\%)] \times 100}{100 \times 3\%}$

$$= \frac{(3 - 0.5) \times 100}{3} = \frac{250}{3}$$

$$= 83.3\%.$$

The first real technical problem involved in the determination is to get a satisfactory measure of the feces belonging to the measured feed intake. For the omnivora and carnivora the feces belonging to a given food intake are often identified by the use of markers. Markers are usually colored substances that can be consumed as a part of the first meal of a digestion test and again as a part of the first meal after the conclusion of the test. Their function is to color the feces subsequently produced from those feedings. The animal is housed in a suitable stall or pen during the term of the trial and the feces are collected beginning with the first excretion that is colored by the first marker and continuing until the appearance of the second marker. This quantity of feces is taken to represent the residue from the diet consumed from the first

---

* Since protein is determined from a N analysis, the computations are ordinarily carried out with %N figures rather than the "N $\times$ 6.25" protein values.

marked meal to the last unmarked meal, inclusive. Commonly used markers are iron oxide, bone black, and chrome green.

The marker system for such animals is reasonably satisfactory, as measured by the reproducibility of the digestibility coefficients in repeated tests. However, we get some variation in results, part of which is due to the fact that markers sometimes diffuse into adjacent unmarked meals, and consequently the separation of feces is not completely accurate. The errors, however, are usually random and are minimized by using several animals and averaging the results.

Marker methods are not satisfactory for use with herbivorous animals, because the material from adjacent feedings is mixed, either in the rumen or in the caecum, and markers can give no sharp division between feedings. Markers are sometimes eliminated by cattle over a period of four days after some particular feeding. For herbivorous animals, therefore, experimenters have resorted to *time collections* for the determination of most of the digestion coefficients that are now published.

The assumption we make in all digestion studies where we use time collections is that if a constant daily intake of a diet can be arranged over a sufficiently long period, the daily output of feces will also remain relatively constant, and that within a fixed time interval the feces collected represent quantitatively the output from the ration consumed over an equal period of time. This assumption is valid only when time-collection periods are of several days duration, in order that fluctuations from day to day may be balanced out. Thus, whereas with marker methods we may have to continue the feeding of the test meal no longer than a week, we may have to continue the feeding for as much as three weeks for cattle in order to minimize errors in the estimation of the appropriate quantity of feces to be used in the digestibility calculations.

Index Methods. We can sometimes avoid the need for quantitative collection of feces and quantitative records of feed intake by using index substances. Index substances are materials that may be consumed by, or administered to, an animal, but that are entirely inert in the digestive system, and are completely and regularly excreted, uniformly mixed with the fecal material. Where this method is used the digestibility is determined from differences in the concentration of the index substance in the feed and in its concentration in the corresponding fecal output.

Chromic oxide (Cr_2O_3) or chrome green is the most commonly used index substance at the present time for carnivora, omnivora, and avian species. When prepared in the form of slow-release pellets, it is also used for ruminants, though with less precise results because of its irregular fecal excretion. In 1964, some studies reported the successful use of polyethylene as an index for digestion trials with ruminants. Where the same combination of feed is fed at all feedings, the index may sometimes be mixed in a fixed proportion with the ration as a means of getting it into the animal. With rations made up of combinations of roughage and grain in differing proportions, or with diets in which the supplements and the basal feeds are fed separately, it is sometimes better to administer the index substance in a capsule. The quantity to be administered will depend on the amount of feed eaten, for the concentration of the index substance in the feed must remain constant over the test period.

The arithmetic for calculating the apparent digestibility of a ration component (such as nitrogen) by the index method depends on one premise, that the amount of index substance in the feed and the amount voided in the feces is the same over equal periods of time. To demonstrate:

Let I = dry matter intake,

O = dry matter output,

c_i = concentration of index in the feed dry matter,

c_o = concentration of index in the feces dry matter,

n_i = concentration of nitrogen in feed dry matter, and

n_o = concentration of nitrogen in feces dry matter.

Then, by definition, $I \cdot c_i = O \cdot c_o$ and $\dfrac{O}{I} = \dfrac{c_i}{c_o}$.

The digestibility of the dry matter ingested $\quad = \dfrac{I - O}{I} \times 100$

$$= I - \frac{O}{I} \times 100,$$

and, by substitution, digestibility of dry matter $= I - \dfrac{c_i}{c_o} \times 100$.

Digestibility of nitrogen $\quad = I - \dfrac{O \cdot n_o}{I \cdot n_i} \times 100,$

and, by substitution, $\quad = I - \dfrac{c_i \cdot n_o}{c_o \cdot n_i} \times 100.$

Now let us try an example.

Index substance in feed, c_i, = 1%,

in feces, c_o, = 5%.

Nitrogen in feed, n_i, = 10%,

in feces, n_o, = 5%.

$$\text{Digestibility of nitrogen} = I - \frac{1}{5} \times \frac{5}{10} \times 100$$

$$= I - \frac{5}{50} \times 100$$

$$= I - 0.1 \times 100$$

$$= 90\%.$$

Quantitative Feed Records. There are also some problems in obtaining a quantitative record of feed intake. Let us suppose, for example, that 100 units of some ration are offered to an animal that is to serve as a subject for a digestion trial. For some reasons that may not be known, this animal may consume 95 units of this material and refuse the balance. If it is dry matter for which we are determining the digestibility, there is no particular problem, for we can simply recover the uneaten feed, weigh it, subtract it from the total offered, and have a record of the dry substance that was actually consumed. If, however, we are attempting to determine the digestibility of some nutrient such as protein, then the problem may be complicated by the fact that the concentration of protein in the refused portion of the feed may not be the same as in the part that was eaten. Perhaps this problem also can be solved by recovering the uneaten feed, analyzing for nitrogen and calculating its probable protein, and deducting that from the protein of the total feed offered.

Such correction of feed offered to take account of portions refused may be satisfactory if we are merely concerned with theoretical consideration of what a given intake of a nutrient actually does in terms of the response of an animal. But if we are attempting to evaluate a food in terms of the usable nutrients it will furnish to an animal, we cannot disregard portions that are refused. If these portions are consistently refused, because they are to all intent and purposes inedible portions, then knowing the digestibility of a selected portion of the protein of the feed is of little practical use.

Let us suppose that an allowance of corn stover or straw is offered to an animal, and varying quantities of it are refused. To determine the digestibility of the portion eaten without regard to the uneaten portion gives a much higher valuation to the product than is warranted. This is in effect what happens when the nutrients of refused feed are subtracted from the total offered in determining the digestibility. Some investigators feel that the refused feed and the nutrients which it contains should be *added to the corresponding nutrients in the feces*. This, in effect, charges against the value of the feed not only that which appears in the feces but that which is not edible.

Indirect Digestibility. Thus far in our consideration of digestibility we have been concerned with problems where the feed in question constitutes the sole ration. However, for most animals there are relatively few feedstuffs that can constitute the entire ration for long enough to determine their digestibility. When mixtures are fed there is no particular problem in determining how much of a nutrient came from one feed and how much from another. It is not possible, however, to make any such separation of the feces. Consequently, it is not possible to measure directly the digestibility of a feed that is fed in combination with some other feed. We must resort to *indirect methods,* which are sometimes referred to as partial digestibility.

To determine the digestibility of some feed or feed component by indirect procedures, two or more digestion trials are necessary. In one of these the diet without the foodstuff in question is fed and the digestibility of its nutrients determined. In a subsequent test the same diet (often referred to as the basal diet) is fed mixed with the food that is to be tested. The total fecal output is measured as usual. Then, on the assumption that the nutrient in the basal diet shows the same percentage of digestibility as it did in the first test (when it constituted the whole of the diet), we estimate the amount of nutrient in the feces of the second digestion trial that presumably belong to the basal portion of the ration. The remaining fecal nutrient is considered to have come from the food being tested (see Fig. 5-1).

Supppose, for example, we wish to test the dry-matter digestibility of a dairy cow meal mixture. A meal mixture is not usually fed as a sole diet to dairy cows, since without roughage they soon refuse food. On the other hand, a roughage such as alfalfa meal can constitute the entire ration. Consequently, we conduct our digestion trial with alfalfa as the basal feed. Assume that in this particular case 40 per cent of the dry matter consumed was recovered in the feces.

Now we must conduct a second digestion trial, and in it we might feed ten units (by weight) of alfalfa plus ten units of a meal mixture, from which a total of five units of dry feces is produced. We now assume that from the ten units of alfalfa eaten in this second test, 40 per cent of it has reappeared in the feces, i.e., that of the five units of feces produced in the second digestion trial, four units of it came from the alfalfa. The other one pound of fecal dry matter we consider to be a residue from the meal mixture. The digestibility of the meal mixture would then be calculated as:

$$\frac{10 - 1}{10} \times 100 = 90\%.$$

The assumption in this procedure, that mixing two foods together does not alter the digestibility of either over what it would have been if fed alone, is often unwarranted. For example, with some poor-quality roughage we may

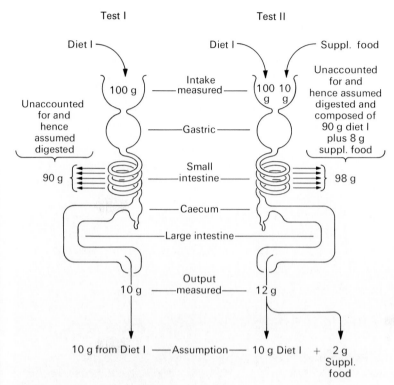

Figure 5-1. *A schematic diagram of the facts and assumptions involved in calculating the digestibility of a food.*

find that the digestibility of its dry matter is relatively poor. Inasmuch as the digestibility of such material is largely dependent on the activity of microflora which break down the cellulose and make the resulting fatty acids available to the animal, it is conceivable that a meal supplement, especially one that contains suitable food for bacteria, if fed with the roughage might result in a more complete digestibility of the cellulose portion of the poor roughage. *By the method of indirect digestibility any improvement in the digestibility of roughage would actually be credited to the meal supplement,* whereas in reality it may have been an improvement in the digestibility of the roughage itself which was involved.

Fig. 5-2 attempts to illustrate this situation. In this figure we assume that molasses as food B added to some material called food A may depress the digestibility by diverting bacteria from the job of breaking down cellulose, whereas the adding of food C to food A may actually increase the bacterial activity and hence increase the digestibility of food A. This interaction between foods is sometimes referred to as associative digestibility.

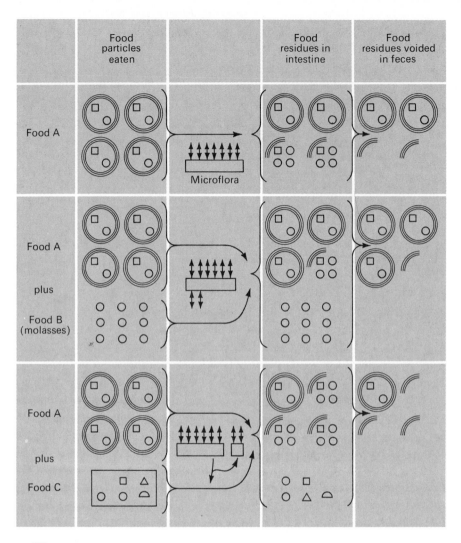

| Food
particles
eaten | | Food
residues in
intestine | Food
residues voided
in feces |

Microflora

= Food particle in which protein (□) and soluble carbohydrate (○)
are "encased" by cellulose

Figure 5-2. *A schematic representation of the hypothetical process that causes the associative effect in the digestion of foods. (See text for full explanation.)*

The Arithmetic. In actually working out the coefficient of digestibility of a feed that is fed in combination with some basal diet, we may understand the problem more easily if we remember that the coefficient determined for the combined diet is actually the mean of the digestion coefficient of the basal diet, *B,* and that of the supplemental food, *S,* being tested, each weighted by the percentages, *b* and *s,* in which the basal and supplemental foods are present in the combined diet. That is:

Digestibility of feeds	Parts in 100 used	Weighted amounts of feeds
B	b or $(100 - s)$	$B \cdot b$ or $B(100 - s)$
$\dfrac{S}{}$	$\dfrac{s}{}$	$\dfrac{S \cdot s}{}$
Totals $\quad T$	100	$T \cdot 100$
Weighted mean of T		$T \cdot 100/100$

If we can determine values for B and for T in digestion trials, we can compute S:

$$S \cdot s = T \cdot 100 - B(100 - s)$$
$$= 100T - 100B + B \cdot s$$
$$= 100(T - B) + B \cdot s.$$
$$S = \frac{100(T - B)}{s} + \frac{B \cdot s}{s}$$
$$= \frac{100(T - B)}{s} + B.$$

Let us suppose that by experiment the following values have been determined:

$$T = 91\%,$$
$$B = 90\%,$$

for a mixed diet containing 20 per cent supplement. The digestibility of the supplement will then be:

$$S = \frac{100(91 - 90)}{20} + 90$$
$$= 95\%.$$

In cattle rations one must usually employ hay as the basal ration, in which case the coefficients for the hay (B) in the first trial, and for the hay plus meal (T) in the second trial might be:

$$T = 60,$$
$$B = 50.$$

Let us assume that the proportion of meal was 40 per cent of the total dry ration. Then the digestibility of the meal mixture would be:

$$S = \frac{100(60 - 50)}{40} + 50$$
$$= 75\%.$$

Effect of Variability

We have made no mention so far of the fact that comparable animals vary in their digestibility of identical rations. Some investigators believe that if coefficients of digestibility can be reproduced within five percentage units they are acceptably accurate. In experiments at Macdonald College variability in the coefficients of digestibility of dry matter for five different species as measured by the standard deviation has been found as follows:

Species	Standard deviations of coefficients of digestion of dietary dry matter
Human	±0.93
Rat	±0.85
Guinea pig	±2.80
Swine	±0.52
Sheep	±1.33

The question to be answered is, how much difference must be observed between the digestibility of a basal diet (B) and that of the combination of basal plus supplement (T) before we can legitimately claim that the supplement (S) really differed at all from the basal diet in digestibility? This question can be answered from the standard deviation of the coefficients of digestibility. The standard deviation of a difference between two coefficients is equal to the standard deviation of a single coefficient times the square root of 2.

If we demand a probability of being right in 95 per cent of the cases (odds of 19 to 1), we must approximately double this value. We can increase the accuracy of our figures by using several animals in each group to determine the digestibility figures. The final equation for the necessary difference between the digestibility of the total diet (T) and that of the basal diet (B) can be written as:

$$T - B = \pm \left[\frac{\text{S.D.}}{\sqrt{n}} \times \sqrt{2} \times 2 \right],$$

where n is the number of animals in the trial group. Assuming that we have four sheep in a group the numerical value for the necessary difference between T and B works out as:

$$T - B = \pm \frac{1.33}{\sqrt{4}} \times 1.414 \times 2$$

$$= \pm 1.86.$$

Thus we see that for identical diets we shall have to expect digestion co-efficients based on an average of four sheep per group to differ by almost two percentage units, because of the operation of all of those factors that eventually lead to variation in digestibility. If we now go back to our example of the 60-40 combination of hay and grain, in which the digestibility of the basal diet (the hay) was 50 per cent and the digestibility of the mixture (T) was 60 per cent, using four sheep per group for the test, we can calculate how much the digestibility coefficient for the total ration fed must differ from that of the basal diet before we can claim there is any real difference between the supplement and the basal diet. The calculation is:

$$T - B = \frac{100(\pm 1.86)}{40} = \pm 4.7\%.$$

This calculation tells us that we might find in two replicate tests as much as ± 4.7 percentage points difference in digestibility between *identical diets*. Therefore, in our example, we can interpret the digestibility of the supplement as follows:

$$\text{Probable digestibility of } S = 100 \frac{[(60 - 50) \pm 1.86]}{40} + 50$$

$$= 100 \frac{(10 \pm 1.86)}{40} + 50$$

$$= 70.3\% \text{ to } 79.7\%.$$

This, of course, is the same thing as saying that the probable ($P = .05$) digestibility of the supplement is $75\% \pm 4.7\%$.

It will be evident from the formulae presented above that the magnitude of the necessary difference between the digestibilities of the supplement and basal diets needed to cover random variability is much influenced both by the number of animals for which the mean digestibility coefficients are determined, and by the proportions of supplement used in the final diet fed. Using the variations in the digestion coefficients of dry matter for sheep we have calculated a series of values to show these effects quantitatively, in Table 5-1.

The example above, in which the digestibility of a meal mixture is estimated by feeding first a basal diet of hay alone and subsequently a mixture of hay and meal, is reasonably clear-cut. There are many cases, however, in which rather unexpected consequences of the inherent variability of digestion coefficients come to light. Let us assume, for example, a pig ration in which the digestibility of the basal diet is 75 per cent. If, to this we add 5 per cent molasses, and feed this new mixture in order eventually to calculate the probable digestibility of the dry matter of molasses, assuming that the S.D.

TABLE 5-1 *Nonsignificant Variation (P = .05)*
Applicable to Indirectly Determined Mean
Coefficients of Digestibility for Sheep

Proportion of supplement in ration fed	Number of sheep per group		
	1	5	10
10%	±37.6	±16.9	±11.9
20%	±18.8	±8.4	±6.0
30%	±12.5	±5.6	±4.0
40%	±9.4	±4.2	±3.0
50%	±7.5	±3.4	±2.4

of the dry-matter digestibility coefficients for swine is ±0.52, we discover that we cannot prove that the molasses differs in digestibility from the basal meal mixture. The allowance which we must make above and below the digestibility of the dry matter of the meal mixture without the molasses turns out to be approximately 29 percentage points, and 29 percentage points plus 75 per cent is more than 100 per cent. In other words, if we have only one pig on which to calculate the digestibility we may expect to get differences in successive tests as great as ±29 percentage units between the molasses-containing diet and the plain diet due to errors of the test alone. If we can add 10 per cent molasses to the basal diet, then the allowance we must make is approximately 15 percentage points, from which we would estimate that the digestibility of the molasses was not less than 90 per cent. If four pigs were available in each group, and we used 5 per cent molasses, we would have the same result as if we had used only one pig in a group and had 10 per cent molasses.

These figures should make it clear that digestibility coefficients, particularly those that we must determine indirectly, are not constants, and consequently that total digestible nutrient values also are not constants. There is inherent in all such figures a certain minimum of variability, which we must recognize in any sound interpretation of values that involve their use.

This, of course, does not mean that digestion coefficients and calculated digestible nutrient values are not useful in measuring feed values. It does explain, however, why under many conditions the calculated digestible energy value for a feed or particularly for a ration may lead us to expect a certain performance of the animals that we fail to get. A part of this failure is because the values (such as digestible kcal, or TDN) that appear in feeding standards are subject to the same uncertainties as are those for the rations that we prepare to meet these standards; and it is not surprising, therefore, that all too often a ration that by calculation contains certain digestible energy values and certain digestible protein values still does not permit or encourage an animal to grow or to produce in accordance with expectations.

Measuring Intake and Digestibility of Rations of Animals on Range or Pasture

To determine the nutritional value of an animal's diet it is necessary to know the amount of each kind of feed consumed and its digestibility. These data are impossible to obtain completely for animals grazing range or pasture. As a result, feed intake and digestibility must be estimated by indirect means.*

In making the indirect estimates, indicators are used. These include *internal indicators,* which occur in the plant and are used to estimate apparent digestibility (those that have been used include lignin, plant chromogens, nitrogen, "normal acid fiber," the methoxyl group, and silica); and *external indicators,* which are fed to the animals to estimate fecal output without a feces bag, and include chromic oxide (chromium sesquioxide), iron oxide, and monastral blue.

An important foundation point is that if the digestibility of a component of a diet is known and if its fecal output is known, then the intake can be calculated. The fecal output of a grazing animal can be measured by fitting an animal with a feces bag to collect the total output, or by feeding an external indicator to the animal. The only purpose of the external indicator is to permit estimation of the fecal output without using a feces bag.

Calculating Forage Intake and Digestibility

Two techniques highly useful to the range nutritionist for calculating forage intake and digestibility by grazing animals are the *ratio technique* and the *fecal index technique.* There are five formulae that may be involved in computing one or the other of the items needed.

The following data illustrate the use of the formulae for sheep grazing on shadscale:

1. Fed	10.0 g	chromic oxide per day
2. Shadscale	13.0 %	lignin [dry basis]
	7.7 %	protein [dry basis]

---

* Lorin E. Harris, "Range Nutrition in an Arid Region," Faculty Honor Lecture Series, Utah State University (Logan, Utah: 1968).

3. Feces   1.15 %   chromic oxide [dry basis]
                         22.6 %   lignin [dry basis]
                         6.0 %   protein [dry basis]

4. Feces output   870.0 g   [dry basis—collected with bag or calculated by formula (5)]

Dry matter consumed

$$= \left[\frac{\text{wt internal indicator in fecal output}}{\%\ \text{indicator in forage}} \right]$$

$$= \left[\frac{870\ \text{g} \times 22.6\%\ \text{lignin in feces}}{13\%\ \text{lignin in shadscale}} \right]$$

$$= 1512\ \text{grams dry matter consumed} \tag{1}$$

Apparent digestion coefficient (%)

$$= 100 - \left[100 \times \frac{\%\ \text{internal indicator in forage}}{\%\ \text{internal indicator in feces}} \times \frac{\%\ \text{nutrient in feces}}{\%\ \text{nutrient in forage}} \right]$$

for protein

$$= 100 - \left[100 \times \frac{13.0}{22.6} \times \frac{6.0}{7.7} \right]$$

$$= 55.2\%\ \text{apparent digestibility of protein} \tag{2a}$$

for dry matter

$$= 100 - \left[100 \times \frac{13.0}{22.6} \times \frac{100}{100} \right]$$

$$= 42.5\%\ \text{apparent digestibility of dry matter} \tag{2b}$$

Indigestible DM (%)
$$= 100 - \%\ \text{digestibility of DM}$$
$$= 100 - 42.5$$
$$= 57.5\%\ \text{apparent indigestible DM} \tag{3}$$

DM consumption

$$= \frac{\text{DM in feces (g)} \times 100}{\%\ \text{indigestibility of DM}}$$

$$= 1512\ \text{grams DM consumed} \tag{4}$$

Fecal DM output (g)

$$= \frac{\text{external indicator fed (e.g., g chromic oxide)} \times 100}{\%\ \text{external indicator in feces grab sample DM}}$$

$$= \frac{10 \times 100}{1.15}$$

$$= 870\ \text{grams fecal DM} \tag{5}$$

The Ratio Technique.   If the herbage ingested is properly sampled, and the internal indicator is completely indigestible, dry matter intake can be calculated as in equation (1).

The apparent digestion coefficient for each nutrient of the diet is calculated according to the general equation (2).

Of the internal indicators mentioned above only lignin, silica, and the methoxyl group are completely indigestible. The ratio technique works well only if an accurate sample of the forage consumed by the animal is obtained and the indicator is completely indigestible.

The most widely used internal indicators for the ratio technique are chromogen and lignin. Chromogen* appears to be a good indicator for succulent green forage during the summer; lignin† is a good indicator for winter range plants.

Fecal Index Technique.   After basic data is developed, digestibility is estimated by feces analyses only. To obtain data for this method, it is necessary to clip forage and feed it in a conventional digestion trial in which the forage intake and the fecal output of a few animals are quantitatively measured using a digestion stall. The feed and feces are then analyzed for an internal indicator, and for gross energy, organic matter, nitrogen, silica-free dry matter, or dry matter. The internal indicator does not need to be indigestible, although that would be ideal. Correction is made for digestibility in the regression equation, as indicated below.

At the same time the conventional trial is being conducted animals equipped with fecal bags or animals fed an external indicator to calculate total fecal output are grazed on the pasture, and the concentration of the internal indicator in their feces is determined. Regression equations for the data on the animals fed in the conventional digestion trial are then calculated as indicated in Figure 5-3, and are used to calculate the digestibility of the forage.

Once a regression equation has been established for a particular set of pasture conditions, it is not necessary to have animals fed clipped forage for subsequent trials. Digestibility can be calculated from the concentration of the internal indicator in the feces. However: *A regression equation determined for one set of conditions does not work under all sets of conditions.*

* J. T. Reid, P. G. Woolfolk, W. A. Hardison, C. M. Martin, A. L. Brundage, and R. W. Kaufman, "A Procedure for Measuring the Digestibility of Pasture Forage Under Grazing Conditions," *J. Nutrition,* XLVI (1952), 255.

† C. W. Cook, and L. E. Harris, "A Comparison of the Lignin Ratio Technique and the Chromogen Method of Determining Digestibility and Forage Consumption of Desert Range Plants by Sheep," *J. Animal Sci.,* X (1951), 565.

Each research worker should therefore determine his own regression equation or should make sure that someone else's equation applies to his own experimental conditions. The regression formula for the line will not always be linear, but the steeper the slope the more reliable the estimate will be.

Using the data previously noted, the dry matter intake can be calculated as the complement of the per cent indigestibility of the dry matter [equation (3)], and the dry matter consumption then computed as in equation (4).

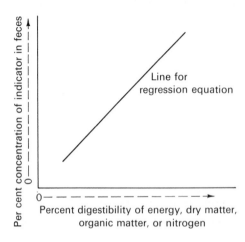

Figure 5-3. *Schematic diagram of regression equation to predict digestibility from concentration of internal indicator in feces.*

Estimation of Total Feces by Grab Samples. If it is not feasible to determine the total feces output of a grazing animal by using a collection bag, it can be done by feeding the animal a measured amount of an external indicator and subsequently collecting grab samples of feces. These are small samples of feces removed from the rectum manually or collected at random from the pasture at various times during the day. Calculating the total fecal dry matter output is illustrated by equation (5). At least 50 grab samples should be taken from each animal to keep the error below 5%, and the chromic oxide should be fed for at least 10 days before the start of sampling.

Obtaining Samples of Actual Forage Ingested

Under winter range conditions in the Great Basin, a sheep's diet is composed of a variety of browse species, and of range grasses. In the winter these species are mature and relatively high in lignin and low in chromogen. With mixed species of this type, it is impossible to collect forage samples that represent the animal's diet in the field. To obtain the necessary data for the forage nutrient analyses, the "before and after" method has been used. The lignin ratio technique is then used to find the dry matter intake and its digestibility.

Figure 5-4. *A sample (A) containing many units is collected for browse species such as big sagebrush (Artemisia tridentata). Before-grazing units (B), and after-grazing units (C) consisting of the current year's growth, are picked off the samples.*

The Before and After Method. This consists of measuring the dry weight and chemical composition of the available forage of each species on a pasture before and after grazing.* The difference represents the animal's diet. The

* C. W. Cook, L. E. Harris, and L. A. Stoddard, "Measuring the Nutritive Content of a Foraging Sheep's Diet under Range Conditions," *J. Animal Sci.,* VII (1948), 170.

amount of available forage is measured by collecting, along transect lines, plant units from each species before and after grazing.

The unit for most browse plants consists of twigs of the current year's growth; for bunch grasses, the entire clump; for semibunch grasses, only the individual stem; for sod grasses, either measurements on a $\frac{1}{16}$ square foot of sod or measurements of individual tiller stems; for annuals and for most forbs, either the entire plant or an individual stem; and for some coarse broad-leaved forbs, only the leaf and leaf stem.

Table 5-2 illustrates the method of calculating the botanical and chemical composition of the sheep's diet by this method.

The total intake of forage and the digestibility of its protein and crude fiber in a diet such as that shown in Table 5-2, can be obtained by the lignin ratio method [see equation (1)]. This "before and after" method is particularly applicable on the winter range of the Intermountain region. During the winter, the plants are relatively dormant and corrections do not need to be made for growth between collections. The plant species (predominantly browse) are far enough apart so that shattering and trampling are not problems.

Esophageal Fistula Method.   A representative sample of a grazing animal's diet is most satisfactorily gathered by the foraging animal itself by means of an esophageal fistula.* The fistula, as used today, is fitted with a plastic cannula having a cap which can be closed when collections are not wanted.† During collection, a plastic bag replaces the cap of the cannula and the sample is obtained as the animal swallows the food it has grazed.

The method has many advantages but also some disadvantages. Correction usually has to be made for saliva which also collects in the bag, and there may be considerable contamination of the minerals.** Intake and digestibility of the sample are calculated by the ratio technique using equations (1) and (2).

Rumen Evacuation Method.   In this method, the sample of the diet is obtained by introducing a cannula in to the rumen and removing its contents.‡

* D. T. Torell, "An Esophageal Fistula for Animal Nutrition Studies," *J. Animal Sci.,* XIII (1954), 878.

† C. W. Cook, J. L. Thorne, J. T. Blake, and James Edlefsen, "Use of an Esophageal-Fistula Cannula for Collecting Forage Samples by Grazing Sheep," *J. Animal Sci.,* XVII (1958), 189.

** D. L. Bath, W. C. Weir, and D. T. Torell, "The Use of the Esophageal Fistula for the Determination of Consumption and Digestibility of Pasture Forage by Sheep," *J. Animal Sci.,* XV (1956), 1166.

‡ A. L. Lesperance, V. R. Bohman, and D. W. Marble, "Development of Techniques for Evaluating Grazed Forage," *J. Dairy Sci.,* XLIII (1960), 682.

The rumen is then quickly rinsed with water which is immediately removed by a small pump. The animals are then allowed to graze for 0.5 to 2 hours, the forage sample is then removed from the rumen and reticulum and the original contents, removed earlier, are then returned to the rumen.

A sample collected by this method is contaminated with saliva for which corrections must be made if the sample is to be used for chemical analysis. This method gives a total collection figure and, for coarse dry grass sampling, may be preferred to the esophageal method unless the esophageal plug is removed at sampling times.

Coefficients of Apparent Digestibility and the Proximate Analysis

Schneider* has examined critically the sources of variability in digestibility data and has shown that the apparent digestibility of a feed is influenced by the proximate analysis. His finding means that feeds of the same name, but which differ in chemical makeup, will show differing digestibilities. This fact is particularly important for forages, where chemical composition changes radically with the stage of maturity although the feed is still called by the one name. It is also important in feeds whose chemical composition may be altered in the milling process. Thus, there are different protein levels in the cottonseed meals. Furthermore, there are marked differences in chemical composition of cereal grains. The protein content of barley, for example, runs all the way from 9 per cent to about 19 per cent, and other grains show a similar magnitude of variability in this nutrient. Any change in the content of protein must be reflected in a change in the composition of one or more of the other proximate principles.

Schneider† has shown that somewhere between 30 and 50 per cent of the total variability in the digestibility of differing samples of the same feed can be traced to one or another, or a combination, of the proximate principles. For animal feedstuffs he has worked out values to be used in regression equations for predicting the digestibility from their proximate analysis. The equation in simple form is:

$$Y = C + b_1x_1 + b_2x_2 + b_3x_3 + b_4x_4.$$

In this equation C is a constant that is specific for the nutrient and the class of feed under consideration. The b's are the partial regression coefficients, and

---

* B. H. Schneider, H. L. Lucas, H. M. Pavlech, and M. A. Cipolloni, "The Value of Average Digestibility Data," *J. Animal Sci.*, IX (1950), 373.

† B. H. Schneider, and H. L. Lucas, "The Magnitude of Certain Sources of Variability in Digestibility Data," *J. Animal Sci.*, IX (1950), 504.

TABLE 5-2 *An Illustration of the Method of Calculating the Botanical and Chemical Composition of Sheeps' Diet on a Winter Range*

	1	2	3	4	5	6	7
Species and sample	*Average density[a] per 100 sq m*	*Units per sq m*	*Average wt per unit*	*Wt per sq m density*	*Total wt per 100 sq m*	*Amount consumed*	*Utilization*
	sq m	no.	g	$\frac{g}{(col\ 2 \times 3)}$	$\frac{g}{(col\ 4 \times 1)}$	g/100 m²	%
Atriplex confertifolia							
Before grazing	1.3	4370	.1231	538	699		
After grazing	1.3	4370	.0921	402	523		
Diff. (ingested)						176	25.2
Eurotia lanata							
Before grazing	3.2	2282	.5264	1201	3843		
After grazing	3.2	2282	.3321	758	2426		
Diff. (ingested)						1417	36.9
Gutierrezia sarothrae							
Before grazing	2.4	6480	.0946	613	1471		
After grazing	2.4	6480	.0854	553	1327		
Diff. (ingested)						144	9.8
Sporobolus cryptandrus							
Before grazing	1.7	180	3.443	6197	10535		
After grazing	1.7	180	2.843	5117	8699		
Diff. (ingested)						1837	17.4
Totals (ingested)						3574	
Per cent in diet							

Species and sample	Diet	Protein content		Crude fiber content		Calcium content		Phosphorus content	
	8	9	10	11	12	13	14	15	16
	%	%	g/100 m² (col 9 × 5)	%	g/100 m² (col 11 × 5)	%	g/100 m² (col 13 × 5)	%	g/100 m² (col 15 × 5)
Atriplex confertifolia									
Before grazing		6.88	48	33.3	233	2.59	18	.069	4.8
After grazing		6.50	34	35.9	188	2.19	11	.062	3.2
Diff. (ingested)	4.93		14		45		7		1.6
Eurotia lanata									
Before grazing		10.12	389	36.7	1410	1.50	43	.099	3.8
After grazing		9.00	218	39.0	946	1.37	33	.095	2.3
Diff. (ingested)	39.66		171		464		10		1.5
Gutierrezia sarothrae									
Before grazing		6.62	97	24.0	353	1.13	17	.079	1.2
After grazing		6.53	87	24.8	329	1.13	15	.076	1.0
Diff. (ingested)	4.03		10		24		2		0.2
Sporobolus cryptandrus									
Before grazing		4.13	535	39.9	4203	0.48	51	.069	7.3
After grazing		4.02	350	43.5	3784	0.40	34	.065	5.7
Diff. (ingested)	51.38		185		419		17		1.6
Totals (ingested)	100.00		380		952		36		4.9
Per cent in diet		7.85		26.80		1.36		.096	

[a] This represents normal ground cover of each species without artificial rearrangement of the foliage to arrive at the per cent density.

the x's the moisture-free percentages of protein, crude fiber, nitrogen-free extract and ether extract, respectively, for the sample of feed for which one wants to estimate the digestibility from the proximate analysis. The b values for a considerable number of feeds have been worked out by Schneider.*

An example of the application of this equation is as follows. For a feed analyzing

Protein	6.5%,
Crude fiber	34.2%,
Nitrogen-free extract	47.4%,
Ether extract	2.1%,

we substitute in the equation above the values for b and for x as given in Schneider's table. We find the digestibility estimated as:

$$Y = 215.8 - (2.489 \times 6.5) - (2.820 \times 34.2) - (.891 \times 47.4) = 61.9\%.$$

The extent to which the differences in chemical composition may affect the apparent digestibility are indicated by two samples of linseed meal, the one carrying 35 per cent protein and the other 32.6 per cent. Using the appropriate constants from Schneider's tables again, we find that the one with 35 per cent protein would be estimated at 78.3 per cent digestibility, while the digestibility of the other would be 86.3 per cent. This calculation is just another way of saying that with equal dry-matter intake, increases in the concentration of protein (i.e., increases in the per cent of protein in the ration) are not matched by equivalent fecal protein output. This effect is exaggerated by the fact that the amount of metabolic fecal protein is related to the amount of dry matter eaten, but not to the amount of protein. The situation may be illustrated by data from a rat test,† in which a highly digested protein was fed in varying levels. The FN was almost entirely metabolic and bacterial. The intake-output relations are shown in Fig. 5-5 and Table 5-3. Here the fecal output becomes a smaller and smaller fraction of the intake as intake increases, and hence the per cent apparently digested increases with the per cent of protein in the ration. In comparison, studies with pig rations showed that natural fiber depresses digestibility of protein more than protein level increases it, and thus rations of increasing protein and decreasing fiber will

---

* B. H. Schneider, H. L. Lucas, M. A. Cipolloni, and H. M. Pavlech, "The Prediction of Digestibility for Feeds for Which There Are Only Proximate Composition Data," *J. Animal Sci.,* XI (1952), 77.

† E. W. Crampton and B. E. Rutherford, "Apparent Digestibility of Dietary Protein as a Function of Protein Level," *J. Nutrition,* LIV (1954), 445.

TABLE 5-3 *Apparent Digestibility of Protein and Per Cent of Protein in Diet Combinations (All figures are percentages)*

Basal diets	Supplement 0 Basal diet 100		Supplement 20 Basal diet 80		Supplement 40 Basal diet 60		Supplement 60 Basal diet 40		Supplement 80 Basal diet 20		Supplement 100 Basal diet 0	
	Cheese	Egg	Cheese	Egg	Cheese	Egg	Cheese	Egg	Cheese	Egg	Cheese	Egg
100% Shredded wheat												
Protein	13	11	17	18	22	26	26	33	32	40	36	—
Apparent digestibility	80	79	82	83	86	86	88	89	88	91	90	—
20% Shredded wheat and 20% methocel												
Protein	10	9	15	15	20	24	25	31	31	39	36	42
Apparent digestibility	73	68	80	79	85	86	87	87	89	90	90	94
60% Shredded wheat and 40% methocel												
Protein	8	7	13	14	18	22	24	30	30	38	36	47
Apparent digestibility	71	64	80	79	84	85	86	87	88	91	89	94

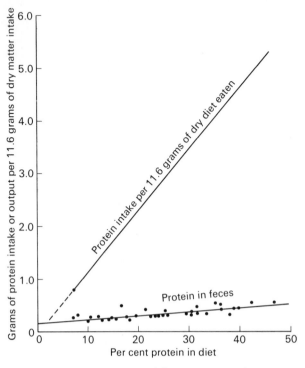

Figure 5-5.   *Feces protein at differing amounts of daily protein intake.*

show a different effect on protein digestion than combinations in which fiber increases as protein decreases.

It should be clear that descriptions of feeds in terms of the digestibility of their nutrients can be given only in general terms. Average values will be useful but cannot be taken as constant characteristics applicable to all samples.

PROBLEMS

1. Tabulate from the data in *Feeds of the World* (Schneider) the TDN values for corn, barley, oats, and wheat in such a way as to compare the differences in the figures, for both cattle and swine, according to whether the feed constituted the entire ration (direct determination) or was fed along with another feed and the digestibility therefore determined indirectly. How well do these figures compare within species?

2. What proportion of the feeds given in the appendix tables of this book could be fed as the entire ration to cattle; to growing pigs?
3. Do you think typical or average TDN values should be relied on to predict the useful energy values of feed mixtures intended for livestock feeding? Be careful—the answer here will require thought.

SUGGESTED READINGS

Bath, D. L., W. C. Weir, and D. T. Torell, "The Use of the Esophageal Fistula for the Determination of Consumption and Digestibility of Pasture Forage by Sheep," *J. Animal Sci.,* XV (1956), 1166.

Bohman, V. R., L. E. Harris, G. P. Lofgreen, C. J. Kercher, and R. J. Raleigh, *Techniques for Range Livestock Nutrition Research,* Ut. Agr. Expt. Sta. Bul. no. 471 (1967).

Carbery, M., I. Chatterbee, and M. A. Hye, "Studies on the Determination of Digestibility Coefficients," *Indian Journal of Veterinary Science and Animal Husbandry,* IV (1934), 295.

Cipolloni, Mary Ann, B. H. Schneider, H. L. Lucas, and H. M. Pavlech, "The Significance of the Differences in Digestibility of Feeds by Cattle and by Sheep," *J. Animal Science* X (1951), 337.

Cook, C. W., and L. E. Harris, "A Comparison of the Lignin Ratio Technique and the Chromogen Method of Determining Digestibility and Forage Consumption of Desert Range Plants by Sheep," *J. Animal Sci.,* X (1951), 565.

Cook, C. W., L. E. Harris, and L. A. Stoddard, "Measuring the Nutritive Content of a Foraging Sheep's Diet Under Range Conditions," *J. Animal Sci.,* VII (1948), 170.

Cook, C. W., J. L. Thorne, J. T. Blake, and James Edlefsen, "Use of an Esophageal-Fistula Cannula for Collecting Forage Samples by Grazing Sheep," *J. Animal Sci.,* XVII (1958), 189.

Corbett, J. L., J. F. D. Greenhalgh, I. McDonald, and E. Florence, "Excretion of Chromium Sesquioxide Administered as a Component of Paper to Sheep," *Brit. J. Nutrition,* XIV (1960), 289.

Crampton, E. W., and B. E. Rutherford, "Apparent Digestibility of Dietary Protein as a Function of Protein Level," *J. Nutrition,* LIV (1954), 445.

Harris, Lorin E., "Range Nutrition in an Arid Region," Faculty Honor Lecture Series, Utah State University (Logan, Utah: 1968).

Lesperance, A. L., V. R. Bohman, and D. W. Marble, "Development of Techniques for Evaluating Grazed Forage," *J. Dairy Sci.,* XLIII (1960), 682.

Lucas, H. L., "Algebraic Relationships Between Digestion Coefficients Determined by the Conventional Method and by Indicator Methods," *Sci.,* CXVI (1952), 301.

Reid, J. T., P. G. Woolfolk, W. A. Hardison, C. M. Martin, A. L. Brundage, and R. W. Kaufmann, "A Procedure for Measuring the Digestibility of Pasture Forage Under Grazing Conditions," *J. Nutrition,* XLVI (1952), 255.

Schneider, B. H., *The Total Digestible Nutrient System of Measuring Nutritive Energy*. Washington Agricultural Experimental Station, Scientific Paper no. 1250 (Pullman, Wash.: 1954).

Schneider, B. H., and H. L. Lucas, "The Magnitude of Certain Sources of Variability in Digestibility Data," *J. Animal Sci.,* IX (1950), 504.

Schneider, B. H., H. L. Lucas, H. M. Pavlech, and M. A. Cipolloni, "The Value of Average Digestibility Data," *J. Animal Sci.,* IX (1950), 373.

Torell, D. T., "An Esophageal Fistula for Animal Nutrition Studies," *J. Animal Sci.,* XIII (1954), 878.

SUMMARY OF SECTION **I**

Before we leave this section, let us recapitulate some of the terms that we have been discussing. The important ones are:

Name. Classification. Moisture. Protein: total; digestible; quality—biological value, replacement value, chemical score, essential amino acid index. Ether extract: total; digestible. Ash: total; calcium; phosphorus; toxic elements. Carbohydrate: crude fiber; cellulose; hemicellulose; lignin; digestibility. Nitrogen-free extract: total; digestible. Available energy: total digestible nutrients; starch equivalents; digestible calories; metabolizable energy; net energy.

Are you not surprised at the number of terms which are used in trying to describe the nutritive properties of feedstuffs? Feedstuffs are complex combinations of nutrients. In the compilation of tables of feeds, the NRC Committee on Feed Composition is recording about eighty items, including amino acids, mineral elements, vitamins, the fractions of the proximate analysis, and a selected group of carbohydrates.

More important than a mere knowledge of the different descriptive terms used is familiarity with the precise meanings and the limitations of these terms. Through carelessness, and sometimes from slipshod thinking, we can be led to misuse or misinterpret some of these terms, and so may fail to employ the feed in such a way that its full value is obtained or its possible adverse effects avoided. For example, the fallacy of the common assumption that TDN is a constant for a given feed is clearly evident when one looks critically at the nature of this term. Many nutritionists feel that rations are often inefficient because of inadequate descriptions of the properties of the individual feeds quite as much as because of ignorance of the requirements of the animal.

Nowhere is a faulty premise likely to lead to an unexpected result more surely than in the problem of ration formulation. A faulty assumption as to the nutritive properties of a feed or ration may be the whole explanation of an unpredicted and usually undesirable performance of the animals fed.

The significance of full and accurate descriptions of feedstuffs will be increasingly evident as we proceed first to consider the nutrient needs of animals, then to a study of feedstuffs themselves, and finally to take up the problems of compounding feed combinations of specified nutritive properties. Actually at this point we are closing only our formal examination of the nature of the terms used in describing feedstuffs, we begin now to use them to understand the subject matter that lies ahead.

III THE NUTRITIONAL CHARACTERISTICS OF SOME COMMON FEEDS

Many different feedstuffs are used at one place or another in the feeding of livestock. Morrison lists analyses for 173 dry roughages (not including grades of the same feed), 143 green roughages and roots, 46 silages, and 432 concentrates. Schneider records the analyses and digestibility of some 2,300 feedstuffs. In 1968 the Committee on Feed Composition of the NRC Committee on Animal Nutrition of the U.S. National Research Council published an encyclopedia of some 7,500 feeds with the available data on their composition. In this tabulation varieties and grades of feeds of common plant or "animal" origin are listed as separate feedstuffs, whereas in most tabulations all analyses of feeds of one common name are averaged under one heading, such as "Barley, western, all analyses" (see also Chapter 1).

With such reference data readily available there is little reason to attempt here any extensive discussion of only a few key feeds. The plan of this section is rather to discuss first the general nutritional properties of feeds by groups (i.e., basal feeds, protein supplements, and so on). We shall augment this discussion by considering in detail the characteristics of one or more key feeds in that group. Finally, we shall consider the unique properties of some other common feeds belonging to the group—properties that determine the extent

And so it is not really surprising that the result of our search for more accurate and complete data about the nutrient requirements of livestock has been not so much to change the general specifications of animal rations as to increase the scope of feeding standards to include facts about previously unknown nutrients. Thus the performance of modern rations in the feed lot is more predictable, because the unknowns are fewer; more efficient, because the balance of nutrients is more satisfactory; and more economical, because previously unrecognized surpluses are more completely eliminated. Quite apart from the greatly expanded scope of the modern feeding standard over its earlier prototype, a fundamental fact of nutrition has emerged from our more complete knowledge: it is that in applied feeding the basic need of animals is for energy, and that this need is the basis for most, and perhaps all, of the other nutrient requirements.

In considering animal requirements, and before examining the details of actual feeding standards, we should therefore discuss thoroughly the basis for, and the present evidence concerning, the energy needs of livestock. Then, because protein has so often (though erroneously) been taken to be the nutrient of greatest practical significance in successful livestock feeding, we should examine critically the present role of this nutrient in the formation of rations. We shall then mention briefly the minerals and vitamins necessary to augment the data of feeding standards. Finally, we shall discuss the characteristics of livestock feeding standards, and examine those for dairy and beef cattle and for swine.

CHAPTER 6

The Energy Requirements of Animals

In this and subsequent chapters (except for quotations, where the original terms will be used), the weights of animals, of feeds, and of biological products will be given in both avoirdupois and metric units. We will assume the following approximations as practical working equivalents:

> *1 kg = 2.2 lb;*
>
> *450 kg = 1,000 lb.*

Energy values may be shown as TDN, as digestible energy (DE), or as metabolizable energy (ME), assuming:

> *1 lb TDN = 2,000 kcal DE or 1,640 kcal ME;*
>
> *1 kg TDN = 4,400 kcal DE or 3,600 kcal ME.*

For more accurate conversion values see Appendix Table A-1-2 and page 407.

No one can feed experimental animals or be associated with the practical feeding of commercial herds without soon realizing that their performance is critically linked with the quantities of feed individual animals consumed daily. The buyer of "feeder" cattle scans the animals for characteristics believed to indicate good feeding ability. Judging scorecards have emphasized broad muzzles, deep hearts, and blocky conformation, with the implication that these indicate good feeding ability. Horsemen, sheepmen, and hogmen have also stressed feeding ability as an important characteristic in desirable animals.

There is justification for this emphasis on feed intake in relation to performance. Production by livestock, whether it is of milk, wool, or body tissues, requires energy in excess of maintenance, and feed intake is almost synonymous with energy intake. The most common nutritional cause of poor performance of livestock is too meager intake of feed. Many nutrients are required because of their function in energy metabolism; others that appear to be needed in proportion to body size are therefore indirectly required in proportion to feed consumption. Since natural feeds not only yield energy to the animal but also carry assortments of specific nutrients, increasing feed intake often automatically corrects other deficiencies that may have appeared with too meager feed consumption.

The factors that regulate, or at least affect, the willingness of animals to eat and that influence the extent of their consumption are by no means fully understood. Perhaps the one most certainly known fact is that the taste of the feed is seldom the major important factor, though there are some exceptions. Bitter weed seeds, for instance, or musty or dusty materials are often refused by animals. Aroma may sometimes be a factor because of the stimulating effect it has on early gastric secretions in simple-stomached animals. Some tastes may also be acquired, and ration changes may therefore cause temporary food refusal.

The more important problem, however, is related not so much to complete feed refusal as to the unwillingness of animals to "eat their fill." Why does voluntary feed intake increase when an unsatisfactory quality of the protein component is corrected either by feed substitution or by amino-acid amendment to the ration? Why does feed intake decline when animals are on restricted water intake? These are not problems of taste or smell, and they are not solved by the inclusion of condiments or flavoring substances.

It is an attractive hypothesis that the body's defense mechanism against further ingestion of substances that disturb its normal metabolism is to decrease feed intake. In any case, deficiency of essential nutrients, perhaps excesses of certain nutrients, or the accidental ingestion of enzyme poisons, such as fluorine, often show their first adverse effects in precarious appetite and lowered feed intake.

That adequate feed or energy intake is necessary for normal performance is an accepted fact. With self-feeding, the problem scarcely exists (assuming a satisfactory ration), but for those animals to which regulated quantities of food are offered at each feeding it becomes important to know "how much is enough." The answer is not always indicated by appetite. In a milking herd, consumption must be adjusted to production for reasons of economy, and the energy need of the bacon pig differs from that of the fat hog at certain stages of growth.

In view of its special importance, we should deal at this point in some detail with the fundamental basis of the quantitative energy requirement of animals. This may be the more pertinent since feeding standards usually present little if any of the biological background underlying the requirements we have to consider in a discussion of energy requirements.

Partition of Food Energy in Digestion and Metabolism

Figure 6-1 attempts to show the interrelationships among the fractions of food energy during their utilization by the animal. The gross energy is merely the total potential energy of the food consumed. It is normally determined for a food by combustion in a bomb calorimeter, or it can be determined by summing the average heats of combustion of the three energy-yielding components of the food, carbohydrate, lipide and protein.

During digestion and absorption the gross energy is broken down, and a part of it escapes the body via the large bowel in the forms of gas (chiefly methane), undigested food residues, and the energy-yielding metabolic products that have originated in the digestive system. These latter include abraded and sloughed mucosa cells, principally the worn-out absorptive cells of the villi of the intestine, the microflora residue which has escaped digestion by the host, and digestive enzymes which have escaped reabsorption following their excretion into the lumen of the gut, where they have helped prepare the food residues for absorption.

The solid fecal residues are fairly simply recovered and their energy determined chemically. The gas, however, and the heat of fermentation frequently escapes without record, and consequently the digestible energy, computed as $GE_i - FE$, is frequently too high, since not all of the unused energy (i.e., the methane) has been recorded. There is recent evidence that a part of the carbon of the rumen methane is salvaged by absorption and is eventually incorporated into useful metabolites. It is also possible (as discussed in Chapter 3) to compute from the gross energy the probable amount of methane which has been produced from a given ration. At this point we should just note that the true digestible energy of a ration is frequently lower than everyday calculations might indicate. Also, when we mention digested nutrients we usually assume or imply that these nutrients have also been absorbed. Usually this assumption is warranted, but differences in completeness of absorption cause some of the variations in digestion coefficients for the same food by different but comparable animals. (This was also discussed in Chapter 5.)

The digested and absorbed nutrients then enter intermediary metabolism. Intermediary metabolism can be roughly defined as the series of changes

Figure 6-1. *Partition of food energy in digestion and metabolism.*

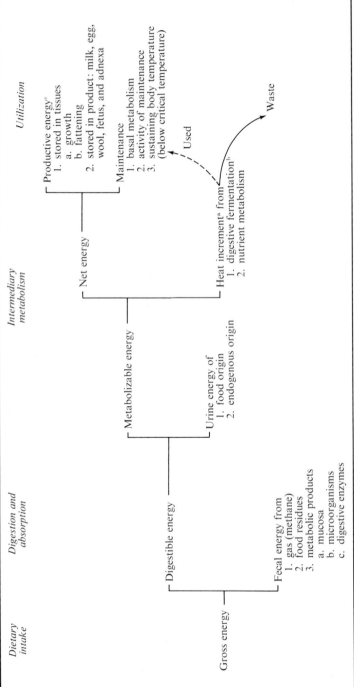

Dietary intake

Digestion and absorption

Intermediary metabolism

Utilization

Gross energy

Digestible energy

Metabolizable energy

Net energy

Fecal energy from
1. gas (methane)
2. food residues
3. metabolic products
   a. mucosa
   b. microorganisms
   c. digestive enzymes

Urine energy of
1. food origin
2. endogenous origin

Heat increment[a] from
1. digestive fermentation[b]
2. nutrient metabolism

Productive energy[c]
1. stored in tissues
   a. growth
   b. fattening
2. stored in product: milk, egg, wool, fetus, and adnexa

Maintenance
1. basal metabolism
2. activity of maintenance
3. sustaining body temperature (below critical temperature)

Used

Waste

[a]Heat increment is an inevitable fraction of the energy cost of production. It varies in amount with the balance of the diet and with the level of productive energy expenditure. Heat increment is used (and may be sufficient) to sustain normal body temperature when the animal is in an environment below the critical air temperature.

[b]Digestive fermentation is actually a part of the energy lost in the digestion tract, but it is usually listed here in the conventional partition of food energy.

[c]Available only after all maintenance needs are satisfied.

which the absorbed materials undergo in the course of the synthesis of tissues needed by the body, including enzymes, and the molecular changes necessary to release the potential energy of the molecules and capture it in forms that can be used to "drive" the metabolic machinery.

Those portions of the absorbed nutrients which have served their useful purpose must of course be discarded, and the channel for the removal of such materials is the kidneys. Some of this material is directly of food origin, such as the nitrogenous components of amino acids; some of it is of endogenous origin, arising from the catabolism of some of the tissues in the body. In practice, therefore, metabolizable energy is normally computed as $DE - UE - GPD$. It is in fact the usable portion of the ingested energy.

Metabolizable energy is used for a variety of purposes, which are outlined in Figure 6-1 under the heading "Utilization." It is difficult to arrange the possible uses of the ME in any order of importance, but it is obvious that one constant requirement for metabolized energy is for the maintenance of the animal body itself. Figure 6-1 shows the energy used for maintenance as consisting of two parts; the first is basal metabolism, and represents the minimum energy required to operate the metabolic machine under basal conditions (which have already been defined). Basal metabolism is a direct and constant function of the metabolic size of the animal itself:

$$BM_{24\ hr} = 70(W_{kg}^{.75})\ kcal.$$

The second part of the energy used for maintenance is needed to sustain a certain unavoidable amount of activity by animals that are at "maintenance" living (i.e., animals that are not using energy for "work" or for the production of any tissues or products). With most animals, including the human, basal metabolism represents 75 per cent of the metabolizable energy used for maintenance. We can then compute the maintenance energy requirement as basal metabolism increased by a third:

$$Maintenance_{24\ hr} = 1.33 \times 70(W_{kg}^{.75})\ kcal.$$

After basal metabolism and maintenance activity requirements have been satisfied, most of the rest of the energy of the diet is available for production. Figure 6-1 shows that productive energy can conveniently be divided into three portions. One of these portions represents the energy stored in tissues, such as in growth or fattening. A second portion of it may be found in products that leave the body, such as milk, eggs, wool, fetus and its adnexa. The energy of these products is permanently lost to the body, whereas the energy stored in tissues can be reused by catabolizing the tissues, as happens when intake is below the requirements for maintenance. The third portion of productive energy is that necessary for the increased activity of the body as in work or in pleasurable activities that exceed the minimum activity of maintenance. This energy expenditure, of course, is irreversible, and if the animal persists in activity requiring energy in excess of maintenance without ingesting adequate energy components in the diet, it will lose body weight.

The Maintenance Energy Requirement of Adult Animals

Although the metabolism of energy-yielding nutrients in the body is not a direct oxidation as in a furnace, the body nevertheless uses oxygen in proportion to the rate of its metabolism. By simultaneously measuring the heat escaping from the body and the oxygen consumption of that animal over the same period of time, we can establish a caloric equivalent of each liter of oxygen. This calculation has been made for many species and under many conditions, and within a relatively small error (less than 3 per cent) we can assume that for each liter of oxygen consumed, the body will lose 4.825 kcal of heat as a result of metabolism. In order to maintain an animal in energy equilibrium this quantity of energy must be replaced in the form of food supplied to the body. Using suitable equipment to measure oxygen consumption, we can then compute the amount of energy an animal requires to maintain energy equilibrium under conditions that may be defined.

Basal metabolism represents the irreducible energy cost of maintenance of an animal during rest in a thermally neutral environment and in a postabsorptive state. To maintain an animal in energy equilibrium under practical everyday conditions, the ration must contain, in addition to this minimum for basal metabolism, increments sufficient to cover any additional caloric needs occasioned by expenditures for activity, production, or both, to allow for fecal and urinary losses, and to balance the energy wastage incidental to food utilization. To be useful in applied feeding, energy requirements must be stated in terms of rations. Consequently, the caloric equivalent of the above categories must be evaluated, as well as the basal metabolism, and their total expressed in terms of rations.

Basal-Energy Metabolism. We refer to the minimum energy cost of the automatic body processes representing the excess of endothermic over exothermic reactions as basal metabolism. Energy used in circulation, excretion, secretion, and respiration accounts for perhaps 25 per cent of this cost, the balance being required for maintaining muscle tone and body temperature. Basal metabolism has been measured for animals of many different sizes, and from such data have come two facts: the basal heat production is affected by the weight of the animal; and the metabolism of small animals is greater than that of large animals per unit of body weight.

Theoretical considerations suggest that basal metabolism might be related to the surface area of the body. If the heat losses were affected by radiation, surface might be a factor, and the ⅔ power of body weight (i.e., $W^{2/3}$ or $W^{.66}$) is a better index of surface than is weight to the first power (i.e., $W^{1.0}$). However, we must note that exterior body surface is not a constant in living animals; nor can it be measured satisfactorily. Furthermore, the casual factors of the heat production, and hence of the basal heat loss, are not dependent on external body surface. Consequently, we may conclude that the relation between surface area and basal metabolism is not directly one of cause and effect. Rather, we should consider that W^n is a measure of physiological body size, or metabolic size, and that the value of the exponent, n, should be determined from the data in question. The relationships may be expressed mathematically as:

$$C = bW^n$$

or

$$\log C = \log b + \log W^n.$$

If C is kcal of basal metabolism and W^n is metabolic size, then the ratio C/W^n should be a statistical constant, b. Basal metabolism data for adult animals of species ranging from mice to elephants were plotted by Brody[*] on log-log paper and the regression fitted by the method of least squares. The slope of the curve proved to be 0.73, and the value of b, the ratio $C/W^{.73}$, was 70.5. Thus the data indicated that, on the average, kcal of basal metabolism = $70.5(W^{.73})$. The numerical value of b depends on the units of measurement used. When metabolism is in kcal for 24 hours and weight is in kilograms, $b = 70.5$; when weight is in pounds, $b = 39.5$.

Brody states "The direct control of the metabolic curve resides not in the external surface but in the neuro-endocrine system, which [for geometric and mechanical reasons discussed in his text] tends to vary in size with surface area rather than with simple body weight. So it comes about that the size of the neuro-endocrine components, the surfaces, the heat dissipation, and the heat production all tend to vary in parallel. They may all be said to vary with W^n, and the value of n tends to be near 0.7. It will be shown presently that the quantity of milk-energy production and of egg-energy production likewise tends to vary with $W^{.7}$, as does basal-energy metabolism and endogenous protein metabolism. This brings out the broad significance of the

[*] Samuel Brody, *Bioenergetics and Growth* (New York: Reinhold, 1945).

proposed reference base $W^{.7}$, which may be termed 'physiological' weight in contrast to $W^{1.0}$, which is the 'physical' or gravitational weight. In the meantime, it is suggested that $W^{.7}$ be tentatively adopted as reference base for basal-energy metabolism, endogenous nitrogen excretion, milk-energy production, egg-energy production, and related processes." Note that both Brody and Kleiber now recommend that the equation be written

$$\text{kcal of basal metabolism} = 70(W_{kg}^{.75}),$$

and consider it to be a biologic constant applicable to all homiotherms. This has been accepted by nutritionists generally.

It may help at this point to review the energy categories with which we deal in animal feeding. Fig. 6-2 is an attempt to show these categories in their proper relations, and from it we can see that we have in basal metabolism the starting point for calculating either the metabolizable, the digestible, or the total energy that must be returned to the animal as food. The basal metabolism value depends on the biological size of the animal. The increments to be added to this value depend on (1) the nutritional balance of the ration, (2) the activity and production of the animal, (3) the protein level of the ration, (4) the nature and extent of the cellulose or crude fiber of the ration, and (5) the energy of the fecal excretion.

Heat Increment of Feeding. The biological origin of this energy category has been variously assigned. That it represents the work of digestion has been disproved by feeding a variety of foods in the digestion of which no heat increment (HI) was found, as well as by the fact that amino acids made available to the body for metabolism by injection give rise to the same quantity of HI as when they are consumed normally. There is evidence to suggest that it represents a decrease in the efficiency of the energy metabolism of food eaten as compared with that occurring under basal conditions. This conclusion is supported by the observation that the proportions of energy-yielding nutrients in the ration influence the HI. For example, absence of an essential amino acid results in an increase in HI. This may be explained on the hypothesis that under these conditions the normal synthesis of protein is hampered, and consequently a larger quota of amino acids must be deaminized. The disposal of such amino acids may be less efficient in terms of energy expenditure than is protein synthesis, and since the body must remain at constant temperature there is an increase in the heat loss.

This effect is not limited to energy-yielding nutrients. The lack of common salt in the ration reduces the utilization of the metabolizable energy of corn by increasing the heat loss from the body. Limitation of phosphorus causes similar effects. Thus possibly HI is a reflection of the degree to which the ration eaten fails to meet exactly the nutrient needs of the body at that time. Whatever the specific cause, the fact remains that the heating effect of feeding is large and, to a considerable degree, unpredictable. For cattle Fig. 6-2 illustrates the average utilization of energy of a typical ration fed at maintenance level.

The values in Fig. 6-2 are Brody's analysis of data available in 1934,

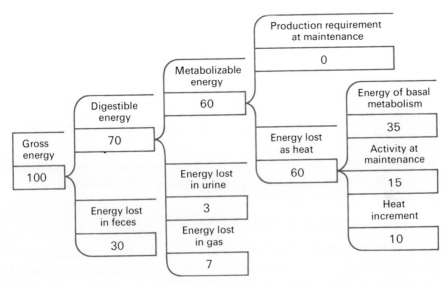

Figure 6-2. *Approximate partition of ration energy by cattle at maintenance intake.*

except for the partition of the heat loss category. The figures for HI represent the average for the HI of cattle rations at maintenance levels, and that for activity is calculated from the energy of basal metabolism. According to Mitchell, the activity of animals at maintenance increases basal requirements by 25 to 50 per cent. Of all of these figures, that for HI is the most variable, ranging from 3 to about 20 per cent of the gross energy of the feed as the level of nutrition increases from half to three times maintenance.

Energy Lost as Gas. The extent of the gas losses will depend largely on the quality and quantity of cellulose fed. As the proportion of roughage in the total ration decreases (as from a maintenance all-roughage ration to a milking ration containing only two-thirds or less of its dry weight as roughage), the gas loss also decreases. In the usual digestion trial the gas loss is counted as a digestible nutrient since it is not in the feces. Consequently, the apparent digestible energy of a high roughage ration is overestimated by usual figures for either TDN or for digestible energy. This error will be discussed in more detail later.

Fecal Energy Losses. With omnivora, because of the nature of their usual foods, the utilization of ration energy is appreciably higher than with herbivora. Digestibility approaches 85 per cent, and gas losses are lower; hence we may find that the sum of the basal metabolism plus the "activity" energy requirement at maintenance is about 75 per cent of the gross ration calorie requirement. The activity at maintenance for the smaller species is very likely relatively higher than for the larger animals. Mitchell places the "activity of idleness" of small animals at double that of the larger species. We may reasonably assume that the total heat loss by adult swine at maintenance living is 70 per cent of the total energy, and may be partitioned as basal metabolism 45 per cent, activity plus HI 25 per cent, of the gross energy needed; for cattle the corresponding values may be more nearly 60, 35, and 25 per cent.

Calculation of Adult Maintenance Energy Requirement. Practical adult maintenance energy requirements of animals generally may be calculated from the formula:

$$\text{kcal} = a \times b(W_{kg}{}^{.75}),$$

where $W^{.75}$ is "metabolic size of animals"; $b = 70$, the kcal required per unit of metabolic size for resting metabolism; and a, the factor to convert the "resting" caloric requirement to that for maintenance. We may think of it as an "activity factor" covering the energy cost of the incidental activity of nonproducing adult animals being maintained at constant body weight.

The numerical value of factor a will vary, depending on whether the requirement is to represent metabolizable energy or digestible energy. If the first, it will include the energy cost of incidental activity and the HI loss at maintenance levels of food intake. If the second, it will be larger, since it must include the average urinary losses of energy from protein metabolism, and the digestive tract (energy) losses of fermentation gases from carbohydrate (cellulose). The latter loss is important only for herbivora, and consequently the factor a varies also with species of animal. In other words, the terms of our equation are constants only in the sense that they hold for average conditions. It has been estimated that the caloric requirement calculated by this formula will have a coefficient of variation of about ± 7 per cent of the mean. That is, the true maintenance energy needs of an individual animal will in two cases out of three be within ± 7 per cent of this calculated average caloric figure. The suggested value of *a* in our equation to calculate digestible calories is 2 for herbivores and 1.4 for omnivores. If we desire finally to obtain the requirement in terms of TDN, we must divide the calorie figure by the caloric value of TDN (i.e., 4.40 kcal per gram) as has already been discussed.

Summary of Maintenance Energy Requirements of Dairy Cattle Given in Feeding Standards. Since 1810, when Thaer's Table of Hay Equivalents* was published, feeding standards of one sort or another have appeared periodically. Most of them were for dairy cattle, where feeding according to need was of greater economic importance than it was with stock whose production was the gain in body size or where maintenance of weight was the only criterion needed. Some of these tables received wide acceptance; others were short-lived. Many are of historical interest, and the evolution of feeding standards, as indicated by the features of the better-known ones of each era, parallels the advancement of scientific feeding of livestock. The early standards dealt largely with energy requirements, though this was not specifically expressed and perhaps not at the time recognized. These tables, after about 1850, included a figure for protein as well, but in the light of present knowledge, essential information given about protein was meager.

The record of the total energy needed by a given animal, regardless of the particular terms in which it was actually stated, has been surprisingly constant. The calculations from a few such standards are summarized in Table 6-1. The calculated TDN values equivalent to the original terms of Table 6-1 are approximations only. However, they illustrate something of interest and perhaps of significance: that, generally speaking, the maintenance energy requirement given in standards based on *energy balance* is about 20 per cent lower than in standards based on *feeding trials* with the TDN needed to maintain body weight, as calculated from the amount of rations eaten and their TDN content.

It is worth noting that Morrison in his 22d edition gives a minimum "recommended" allowance that is some 10 per cent below his previous figures. This reduced value, together with the statement from the committee responsible for the NRC standard that its allowance is liberal because underfeeding is the most common cause of unsatisfactory dairy cow performance, leads to the conclusion that the most probable daily requirement for the energy maintenance of an average 1,000 lb adult dairy cow is not far from 6.5 lb

* A. D. Thaer, *Principles of Agricultural Crops, Animal Husbandry*, IV (1812).

TABLE 6-1 *Daily Energy Requirements for the Maintenance of Adult Dairy Cattle (Calculated for a live weight of 1,000 lb)*

Feeding standard		Original terms	Approximate minimum	
			TDN (in kg)	DE (in Mcal)
1909	Kellner	2.631 kg starch equivalent	2.82	12.4
1917	Armsby	6 therms net energy	2.64	11.6
1931	Forbes and Kriss	8.487 therms metabolizable energy	2.36	10.4
1935	Brody	$2 \times 70\,(W^{.73})$ kcal	2.77	12.2
	Average		2.65	11.7
1903	Haecker	7 lb dig carbohydrate + 0.1 lb dig fat + 0.7 lb dig protein	3.23	14.2
1912	Savage	7.925 lb TDN	3.23	14.2
1915	Morrison	7.925 lb TDN	3.23	14.2
1936	Morrison	7.925 lb TDN	3.23	14.2
1950	Morrison	7.00–7.93 lb TDN	3.18	14.0
1950	NRC	8.00 lb TDN	3.28	14.4
1966	NRC	7.00 lb TDN	3.18	14.0
	Average		3.23	14.2
	Overall average		2.94	12.9

(2.94 kg) TDN or very nearly 13 therms of digestible energy. In equation form we may write it as

$$\text{lb TDN} = \frac{1.72 \times 70(W^{.75})}{1,816}.$$

We should recall that the energy requirement forms the basis for the requirement of many other operating needs of the body. It is therefore essential that this figure be established with as little error as possible. *Using as a standard an intake that is in excess of the true requirement may lead to difficulties from unexpected sources.* In practical feeding a single meal mixture is prepared for all animals of the same feeding category, and allotments are made to individuals. All nutrients are thus in fixed proportion to the total feed, which is essentially the same as saying they are in fixed proportion to the energy of the feed. If, for example, in preparing a mixture, one were to use a standard that was actually in excess of needed energy by 20 per cent, but was correct for protein, then a cow being fed enough to keep her at constant weight would receive too meager an allowance of protein. Intentionally liberal estimates of *nutrient* requirements are thus less serious than is the use of too large a figure for *energy.*

How Much Feed? It may be necessary to consider the *amounts* of food needed to supply specified quantities of digestible energy, because of the way in which tables of nutrient requirements are actually used in applied feeding. The data making up a feeding standard are primarily useful as guides for the formulation of meal mixtures from which animals of the same feeding category will be offered appropriate quantities. The formulation is seldom an inflexible recipe. It changes as frequently and as rapidly as the availability and price of the ingredients fluctuate.

Examination of formulae recommended by various agencies reveals a surprising uniformity (sometimes by geographic area) in the calculated DE content of meal mixtures intended for any one feeding category (such as milking cows, nursing sows, or laying hens). The reasons for this are more deep-seated than the simple fact that the between-mixture differences might be expected to be smaller than that between single feeds. At the moment it will suffice to state that pattern meal mixtures intended for milking cows, as well as typical feed combinations designed for most feeding classes of swine, contain very close to 75 per cent TDN (3,300 kcal DE/kg) when they are based on corn, and about 70 per cent (3,080 kcal DE/kg) when the basal feeds are largely barley or oats. Thus, it is feasible to express the daily energy requirements of animals (whether for maintenance or for producing stock) in terms of equivalent pounds of feed, and specifically in terms of a quantity of meal mixture, after suitable adjustments are made for feeds other than meal that are used (as in the case of herbivores). The general equation may be written:

$$\text{lb meal required daily} = \frac{1.87 \times 70(W^{.75})}{\text{per cent of TDN in feed to be used}},$$

or

$$\text{kg meal required daily} = \frac{1.87 \times 70(W^{.75})}{\text{kcal/kg of feed to be used}}.$$

The significance of a separate statement of the maintenance energy requirement varies with the class of stock involved. Idle adult farm animals are normally fed maintenance rations, but often the ration consists of roughage alone, and is fed essentially *ad libitum*. If the animals do not maintain their weight, supplementary concentrate feeds may be given. For swine the meal allowance is restricted to a quantity that maintains live weight. Feeding standards for energy are seldom used under these conditions.

Where producing cattle are to be fed, it is possible, and with dairy cattle customary, to arrive at the final feed requirement by adding together the maintenance and production needs. In practice the producing stock (other than dairy cattle) is sometimes grouped by weight, age, and/or production category, and the total requirement stated as one figure. Thus pregnant heifers and pregnant mature beef cows are separately grouped, each by live weight, but lactation allowances are given separately. Similarly, young and adult pregnant swine and young and adult nursing sows are listed separately. The fundamental basis of the maintenance energy needs for all adult stock is the same, and the feeder who has a working knowledge of the principles involved will be able to make sound adjustments in feed allocations to individual animals.

Energy Cost of Production and Work

The energy cost of muscular activity as in work, and that for the production of milk or wool or the synthesis of body fat, must also be taken into account in the feeding of livestock. We should call attention to the fact that, except for adult breeding cattle, the maintenance needs of animals are not in practice dealt with as separate values, but are combined with those for production into one figure. Nevertheless, we should look into the energy requirements of milk and fat production in order to appreciate how differing rates of production eventually affect the total feed allowances.

The partition of the digestible calorie intake between maintenance, milk production, and body gain in weight cannot be made directly. However, Brody and Proctor have attempted to estimate this partition by using the statistical device of partial regression. They have assumed for this purpose that TDN is used for three purposes in the body, maintenance, body gain, and milk production. The basic equation to indicate the relationships is

$$\text{TDN consumed} = b_1(x_1 - \bar{x}_1) + b_2(x_2 - \bar{x}_2) + b_3(x_3 - \bar{x}_3),$$

where

$x_1 = $ lb 4 per cent fat milk produced per lactation,
$x_2 = $ average metabolic body weight (i.e., $W^{.75}$) during lactation,
$x_3 = $ change in body weight during the lactation, and
b_1, b_2, b_3 are the units of TDN required for each unit of $x_1, x_2, x_3,$ respectively.

Brody and Proctor applied this method to 243 yearly lactation records of Holstein and Jersey cows in the Missouri Agricultural Experiment Station, and derived the following figures:

$$\text{TDN} = 0.305(x_1) + 0.053(x_2) + 2.1(x_3).$$

Assuming 1,816 kcal per lb TDN, we may calculate that each pound of milk of 4 per cent fat content produced requires $(.305 \times 1,816) = 553$

kcal. Since a pound of 4 per cent fat milk contains 340 kcal, the net efficiency of production is $(340 \div 553) = 61$ per cent; to produce milk requires, in addition to maintenance of the cow, 1.61 times the energy contained in her milk. (Experiments have shown that efficiency of production is independent of body size.) Forbes and Kriss of the Pennsylvania State Agricultural Experiment Station have estimated from their data that the energy cost of milk production is 1.67 times the metabolizable energy consumption, which agrees satisfactorily with Brody's value based on TDN.

Kriss has reported that the efficiency of metabolizable energy for bodyweight increase in adult cattle is 59 per cent, which, if expressed in terms of digestible energy, might be taken as approximately 50 per cent ($\frac{60}{70} \times 59$), if we assume that metabolizable energy is 82 per cent of digestible energy for cattle. On this basis, double the calories deposited as body fat would be required in the ration in excess of the maintenance energy.

Energy Cost of Work. The energy cost of work (muscular activity) is considerable. As measured by rate of oxygen consumption, it has been found that the energy cost of standing above lying is about 9 per cent in man, cattle, and sheep. The horse does not use more energy for standing because of the special anatomical arrangement of his suspensory ligaments. However, walking causes an energy expenditure of about 100 per cent over standing among several species. Sustained work uses energy in relation to the energy requirement at rest about as follows:

Activity	Ratio of O_2 consumption of activity to that of standing
Walking	2
Sustained heavy work (6–10 hrs per day)	3–8
Maximum activity per day	20
Maximum energy during maximum brief effort	100

In horses, an average expenditure can be worked out on the assumption that the normal load is 10 per cent of the body weight and that the speed is about 2.2 miles per hour. The figures in terms of TDN are computed by Brody according to the formula:

TDN per day $= 0.053 \, (W_{kg}{}^{.75}) + 1.27$ (horsepower per hour).

The daily requirement thus calculated according to the size of the horse and the hours worked per day, and expressed in terms of DE, are shown in Table 6-2.

TABLE 6-2 *Megacalories of DE Required for Maintenance and Work According to Total Hours Worked per 24 Hours*

	Mcal of DE needed for horse weighing:		
Hours working	*1,000 lb (454 kg)*	*1,400 lb (636 kg)*	*1,800 lb (818 kg)*
0 (maintenance)	16	20	26
2–4	4	6	8
5–8	10	14	18
8–12	16	22	30

The exact values for a given animal are subject to wide variation, as might be expected since it is obvious that many factors other than body weight, weight of load, and rate of movement are involved. Thus far no simple classification has been devised to describe severity of work which we can use for estimating energy requirements. The problem may be more academic than practical, because the criterion in actual feeding is maintenance of body weight. The ordinary plan of allowing quantities necessary for maintenance of weight is satisfactory if we take the precaution of adjusting the allowances day by day to correspond to marked changes in activity (as on idle days). Table 6-2 can at least give some idea of the quantities that would be involved.

Energy Requirements of Growth

Sometimes we can express the energy needs of adult animals in two separately measurable quantities, one for maintenance and another for specific activity in addition to maintenance. With growing animals overall increase in weight is continuously variable, and in practice one must not only feed to maintain the weight and size attained, but provide enough more to permit further gain in weight. The increase in weight of animal normally follows a characteristic pattern related to age, which, in turn, may reflect changes in nutrient needs. Hence, so-called normal growth curves are often used as indices of requirements of growing animals. We should therefore consider briefly the normal growth of animals.

The Nature of Growth. Although the overall change in body weight with advancing age can be represented as a smooth curve, the different tissues do not individually grow with equal rapidity. For example, muscles increase 48 times in size from birth to maturity, while the skeleton only makes a little more than half this change. Curves of skeletal growth, therefore, when plotted against age, are flatter than curves of total weight gain. Hammond,* in his studies of the effect of age and nutrition on change in body proportions, found that the body proportions of fat-type pigs of 100 lb (45 kg) weight were essentially the same as those of bacon-type pigs of 200 lb (90 kg) liveweight.

Of particular interest is the effect of level of nutrition on development at different stages of maturity. Pigs, following weaning, held on meager rations for 16 weeks, and then full-fed to a slaughter weight of 200 lb (90 kg), showed relatively poor muscle development and heavy layers of external fat. Others fed liberally following weaning and then, beginning at 16 weeks, given restricted intake showed the greatest development of lean tissues and the least fat when killed at 200 lb (90 kg). The growth curves (age-weight graphs) of the pigs from these two regimens were strikingly different. That for the "low-high" nutritional program showed a continuously increasing upward trend; that for the "high-low" regimen was sigmoid, showing a steadily accelerating slope to 16 weeks followed by a segment of less rapid weight gain.

Measures of Growth. At the outset we may state that body weight alone cannot be an accurate measure of tissue growth. An animal may remain at constant body weight while simultaneously losing water or fat and growing in skeleton. In practical livestock husbandry a combination of body weight plus a measurement such as height is often of greater use as a guide to feed allowances than either alone. Thus, standards of normal weight-for-height-for-age make it possible to distinguish between well-grown individuals and those that are heavy because of excessive fat deposits. Such a standard for two breeds of dairy cattle is already in use. Some of the figures are shown in Table 6-3 and plotted in Fig. 6-3. These data can be plotted for use with intermediate values. The importance of the problem of accurately measuring true growth varies with species and cases. For animals being raised for breeding herds, the desirability of a more accurate assessment of nutritional needs than is necessary for short-lived "meat" animals will perhaps justify the more elaborate schemes.

The General Nature of Growth Curves. If we record the live weight of an animal periodically from birth to maturity, and plot these weights against

* J. Hammond and G. N. Murray, "The Body Proportions of Different Breeds of Bacon Pigs," *J. Agr. Sci.*, V (1937), 394.

TABLE 6-3 *Weight-Height-Age Relationships in Normally Grown Dairy Heifers*

Weight		Height at withers		Age in months	
(lb)	*(kg)*	*(inches)*	*(cm)*	*Jersey*	*Holstein*
100	45	28.5	71	1.0	0.5
200	90	34.0	86	6.0	3.0
300	135	38.0	96	7.5	5.0
400	180	41.0	104	9.0	7.0
500	225	43.5	110	12.5	8.0
600	275	46.0	117	18.0	12.0
700	320	47.5	120	26.0	16.0
800	365	49.0	125	40.0	20.0

time at relatively short intervals, we get a curve depicting the trend of "growth." If we thus measure a sufficient number of comparable animals, the line representing the average weights for age will be a smooth curve of sigmoid form. The curve will have two major segments. The first is of increasing slope and extends from birth to puberty. The second is of decreasing slope and runs from puberty to maturity. The point of inflection, which coincides with puberty in most species, is of biological significance in that it marks

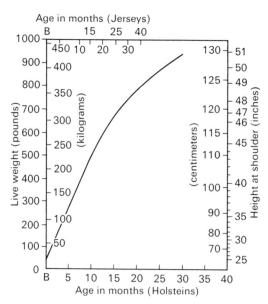

Figure 6-3. *Weight-height-age relationships in dairy heifer calves.*

a point of physiological age equivalence between species. At this age, rate of potential growth is at a maximum, as is feed efficiency as measured by feed required per unit of body increase. Puberty occurs with most animal species when they have attained about one-third of their mature size. The normal curves of animals beyond puberty coincide if plotted on a physiologically equivalent age basis.

Problems in the Use of Growth Curves. Much research has been undertaken to establish normal growth curves, and relevant data have been published for most species of animals. In nutrition such curves are used primarily as standards against which to gauge the adequacy of nutrient allowances. In fact, so-called normal growth curves are often the entire basis for the allowances set down in dietary and feeding standards.

Obviously if rates of growth are in part hereditary, discrepancies in attained body size between growing animals of identical ages will increase with advancing age. Consequently, any classification for growing animals will include a relatively wide range of weights at any specified age, and a growth curve based on averages at advancing ages will not necessarily depict satisfactorily the expected normal growth of individuals.

One must bear in mind also that, although daily gain in body weight is the most often-used criterion for adequate nutrition of growing animals, it does not follow that maximum gain is the desired objective. Indeed, rapid gain may be an index of a faulty nutritional regimen, as is so well illustrated by Hammond's studies on the relation of level of nutrition and age to body proportions. A rapid gain on market pigs after a weight of about 100 lb (45 kg) indicates rapid fattening, which may be damaging to bacon carcass quality.

Estimating Net Energy Requirements for maintenance (NE) and Net Energy for Production (NE_p). If net energy (NE) is to be used as a measure of the energy value of feeds and of the energy requirements of animals, it is necessary to consider the terms in which NE values can be expressed.[*] There are three expressions of NE.

1. NE for maintenance alone (NE_m).
2. NE for production alone (NE_p).
3. NE for maintenance and production (NE_{m+p}).

It is important that one be aware of which measure is being used, since their numerical values differ considerably.

---

[*] G. P. Lofgreen, "The Net Energy System to Evaluate Feeds," (Livestock Nutrition Short Course, Utah State University, Dec. 1964).

Net Energy for Maintenance Alone (NE_m). The NE required for maintenance is normally considered to be equal to the basal heat production of an animal, or the heat produced when the animal is consuming no feed. Lofgreen* determined heat production at 'no feed intake' by an indirect method involving a comparative slaughter technique. At the beginning of a feeding trial a sample group of animals is slaughtered to determine initial body composition. At the conclusion of the feeding period the remaining animals are slaughtered and the body composition determined. From a comparison of the "initial" and "final" body composition, the gain in weight of protein and of fat can be deduced. Since we know the energy value of protein and fat, it is a simple matter to calculate the amount of total energy gained (NE_{gain}). If NE_{gain} is measured at various levels of feed intake, total heat production (HP) can be determined from the following equations.

$$NE = ME - HI \tag{1}$$
$$NE = NE_m + NE_{gain} \tag{2}$$
$$NE_m = NE_{BM} + NE_{activity} \tag{3}$$
$$HP_{total} = NE_{BM} + NE_{activity} + HI \tag{4}$$
$$NE_m + HP_{total} = ME - HI \text{ (at a given feed intake)} \tag{5}$$
$$NE_{BM} + NE_{activity} + HP_{total} = ME - HI \tag{6}$$
$$NE_{BM} + NE_{activity} + HI \quad = ME - NE_{gain} \tag{7}$$
$$HP_{total} = ME - NE_{gain} \tag{8}$$

If ME is determined by an energy balance study and NE_{gain} is determined by comparative slaughter tests, heat production (HP) can be computed by the difference [equation (8)].

For example, in Table 6-4 we present values of metabolizable energy intake (ME), energy retained (NE_{gain}), and total heat production (HP_{total}) for groups of beef animals fed two widely differing rations at three levels of intake. For each ration, the low level of feeding was intended to approximate maintenance, and the high level was free choice feeding. Heat production was determined from equation (8) with NE_{gain} measured by the comparative slaughter method and ME from balance studies. When the log of heat production is plotted against metabolizable energy intake, with each being expressed per unit of metabolic body size ($W^{.75}$), the regression obtained can be extrapolated to the point of "zero energy intake," giving an estimate of heat produced at that point. Extrapolation of the data in Table 6-4 to the

* *Ibid.*

TABLE 6-4 *Daily Metabolizable Energy Intake and Heat Production by 700-pound Steers Fed Either an Alfalfa Hay Ration or a Concentrate Ration, Each at Three Levels*

| Item | Level of feeding | | |
	Maintenance	Intermediate	Free choice
Alfalfa hay ration			
Intake of Dry Matter, lb	7.35	11.92	19.45
Metabolizable Energy (ME), kcal	8,019.	12,695.	20,714.
$(ME)/W^{.75}$	59.	94.	152.
Energy Gain (NE_{gain}), kcal	−47.	1,054.	2,810.
Heat Produced (ME + NE_{gain}), kcal	8,066.	11,641.	17,904.
$(HP)/W^{.75}$	59.3	85.6	131.6
$\log{[(HP)/W^{.75}]}$	1.77	1.93	2.12
Concentrate ration			
Intake of Dry Matter, lb	6.27	10.79	15.68
Metabolizable Energy (ME), kcal	8,088.	13,919.	20,227.
$(ME)/W^{.75}$	59.5	102.3	148.5
Energy Gain (NE_{gain}), kcal	624.	2,827.	5,430.
Heat Produced (ME − NE_{gain}), kcal	7,464.	11,092.	14,797.
$(HP)/W^{.75}$	54.9	79.7	108.7
$\log{[(HP)/W^{.75}]}$	1.74	1.91	2.03

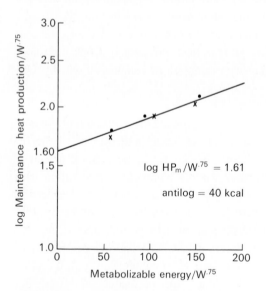

Figure 6-4. *Determining the net energy for maintenance of steers.*

point of zero energy intake yields a value for HP_m of approximately 40 ($W^{.75}$). (See Fig. 6-4). By way of illustration, the metabolic body size of a 700-pound feeder is $700^{.75} = 136$ lb. The NE_m requirement of a 700-pound steer would therefore be $40 \times (700^{.75}) = 5,440$ kcal (5.44 Mcal) per day. From the equations describing the relationship of heat production to ME intake for the two kinds of rations it can be determined at what point the ME intake will equal heat production. For example, it is seen that on the alfalfa hay ration an intake of 70 kcal per pound of metabolic body size produced 70 kcal of heat. For the 700-pound steer used as our example, the ME intake at energy equilibrium would therefore be $70 \times 136 = 9,520$ kcal (9.52 Mcal). The ME content of the alfalfa hay was 1.065 Mcal per pound of dry matter, thus requiring 9.52/1.065 or 8.94 pounds of hay to maintain energy equilibrium. This quantity of hay therefore has an NE_m equal to the heat production at no feed intake, or 5.44 Mcal per day. Expressed per 100 pounds of dry matter, the NE_m of this alfalfa hay is thus $5.44/0.0894 = 60.9$ Mcal. The NE_m of the high-concentrate ration, determined in the same manner, was 86.0 Mcal per 100 pounds of dry matter.

Another interesting fact can be illustrated by these data. From the equations describing the relationship of ME intake to heat production the effect of an increase of 100 Mcal of ME intake from each of the two rations can be determined. Increasing the ME intake 100 kcal above energy equilibrium (from 70 to 170 kcal/$W^{.75}$) by feeding more alfalfa hay results in an increase in heat production of 87 kcal. On the other hand, increasing the ME intake 100 kcal by feeding more high concentrate brings about an increase in heat production of only 56 kcal. This greater heat production from high-fiber feeds is the primary reason for the overevaluation of roughage feeds by DE, ME, and TDN.

Net Energy for Production Alone (NE_p or NE_{gain}). The NE_{gain} requirement of growing-finishing cattle can be determined by a "difference trial." This method involves feeding at two levels of intake and measuring the increase in energy gain that results from the increase in feed. Such a trial is illustrated in Table 6-5. The feed used was the high-concentrate ration with an NE_m of 86.0 Mcal per 100 pounds of dry matter that was used in the heat-production test. It will be noted that increasing the feed intake by 7.94 pounds resulted in an increase in weight gain of 2.11 pounds and in increase in energy gain of 4.5 Mcal, which represents the NE required per pound of weight gain in addition to the NE required for maintenance and therefore represents the NE_{gain} requirement. From a series of feeding and slaughter trials the NE_m and NE_{gain} requirements for growing-finishing beef cattle can be estimated.

TABLE 6-5 *Determination of NE_{gain} Requirements per Pound of $W^{.75}$*

Item	Restricted	Free choice	Differences
		Level of feeding	
Mean body weight, lb	546	693	
Mean metabolic body size, lb ($W^{.75}$)	113	135	
Dry matter intake, lb/$W^{.75}$	7.27	15.21	7.94
Empty weight gain, lb/$W^{.75}$	0.29	2.40	2.11
Total energy gained (NE_{gain}), Mcal/$W^{.75}$	0.72	5.27	4.55

Energy Requirements for Growing Animals. Gains in weight by growing animals in accordance with some normal growth standard is the most commonly used criterion for adequate feed intake. Feed intake is almost synonymous with energy intake, since unless feeds differ markedly in fat content, they are of quite similar gross energy, and for any one species of animal the types of feeds used in the ration are enough alike so that differences in digestibility of the dry matter of typical rations are not wide.

To determine energy requirements on the basis of the growth of animals, we conduct a bio-assay type of feeding trial. Comparable groups of animals are fed different amounts of a ration, the digestible energy content of which is known. The intake of energy by the animals that grow at rates comparable to those of the normal growth curve is taken as the energy requirement.

To arrive at a feeding standard by this method is time-consuming and costly. It is necessary to observe many different groups in order to obtain values not biased by the hereditary characteristics of any one group or strain of individuals. Mitchell has suggested a somewhat different approach to the problem of establishing the energy needs of young animals. He considers that the total energy requirement may be partitioned into a maintenance fraction, an activity fraction, and a new-tissue fraction. He cites the data for a chicken given in Table 6-6.

TABLE 6-6 *Calculations of Caloric Requirement of a Growing Cockerel (Mitchell)*

Weight of bird (lb)	(g)	Basal metabolism (kcal)	Activity (50 per cent of BM) (kcal)	Tissue formed (kcal)	Total (kcal)
0.5	227	37	18	15	70
1.0	454	55	27	19	101
2.0	908	72	36	21	129
3.0	1,363	94	47	19	160

In his scheme the energy requirement for maintenance is actually the basal metabolism plus 25 or 50 per cent, the smaller value being for larger animals and the larger for smaller species. The "growth" requirement is the caloric value of the new tissue formed as determined by slaughter tests and analysis of tissues. Of critical importance is the magnitude of the activity factor, data for which are limited at present. Mitchell's plan in reality treats true growth as a "production" (in excess of maintenance). His treatment is entirely logical from the standpoint of nutrient needs, and has the theoretical advantage of making it possible to use already well-established methods for calculating the maintenance requirements on the basis of the attained size of the animal and its relation to basal metabolic rate.

Efficiency of Energy Utilization by Growing Animals. In general, the energy stored in the successive equal-weight increments becomes larger with advancing age because of the change toward lower water and higher fat content of body tissues with increasing age. Thus with advancing age more energy intake is needed per unit of body gain. This increased feed cost of gain is strikingly seen in the case of the market pig that for the whole of his life is "growing."

Age of pig in months	*Gain in body weight per 100 lb (or 100 kg) of feed*	
	(lb)	*(kg)*
2–3	40–50	17–22
3–4	25–30	11–13
4–6	18–20	8–9

Such figures are considerably modified by the rate of gain of the animal. In slower-gaining animals, where a tissue of relatively low fat content is being formed, the gain per pound of feed will be appreciably higher than with animals that are rapidly fattening. From feeding standards that indicate both the expected feed intake and the expected weight increase, we can calculate the expected feed efficiency as feed required per unit of gain. This figure may be useful in judging the adequacy of the nutrient combinations represented by the feed mixtures fed. If the animals are eating less than the standard calls for, as in intentional feed restriction, we might expect that their gains would also be restricted. If the gain–feed ration remains "normal" one could assume that the nutrient assortment met requirements. If low feed intake is a voluntary reaction by the animal, then faulty nutrient assortment may be the cause of the restriction, but feed efficiency is usually also adversely affected.

Normal Growth Curves, Daily Feed Intake, and Expected Daily Gains of Young Animals. From the previous discussion we can see that any average statement of the energy requirements of young animals will be but a rough guide to the needs of specific individuals or of small groups of comparable animals. It may be useful, however, to record typical data that are in use by the livestock feeder, data against which he frequently checks his own results.

In order to set a workable figure for the daily feed that growing animals may be expected to consume corresponding to their live weights, we have to indicate as well the energy value of the rations referred to, particularly for herbivorous animals, where with advancing maturity the ration normally changes from one of relatively high TDN to one including more roughage material. Growth data and suggested corresponding feed requirements for young cattle and swine are shown in Table 6-7.

Conclusions: Where Do We Stand Now?

In the preparation of a modern feeding standard, the figure we finally set down for any particular category of animal as its probable daily need for energy, whether it be expressed as calories or TDN or starch equivalent, we must base on such considerations as are presented in this chapter, together with a review of the data from comparative feeding trials designed to elucidate energy needs. The discussions here, of necessity, have been abbreviated and the factors often simplified. Nevertheless, we should now understand that any decision on the probable true energy needs is not a simple matter of averaging figures. Perhaps some will conclude that the whole matter is a hopeless jumble —that we can find no solution at present. Such a conclusion is a defeatist attitude and is unjustified in the light of the record of steady progress in ration efficiency, which parallels the development of feeding standards.

The fact that the problem is complicated illustrates forcefully one of the so-often disregarded characteristics of all feeding standards, that the figures are guides to the probable needs of a hypothetical average animal of the particular feeding category in question, but are not to be taken as defining in any precise way the quantity of a diet component that a specific animal will need for its optimum performance. Indeed, there may be no such figure; the needs of an individual may vary almost continuously for causes either not detectable clinically or not recognized as related to the problem.

This possibility in no way defeats the purpose of the feeding standards nor minimizes the need for critical evaluation of the data on which they are based. In practical feeding, we can make adjustments in total feed allowance when desirable to meet the peculiar needs of individual animals within feeding-

TABLE 6-7 *Daily Air-Dry Feed Required by Growing Cattle and Swine According to Weight Attained*

Weight attained (kg)	Weight attained (lb)	Growing heifers — Daily gain (g)	Growing heifers — Daily feed (kg)	Growing heifers — Roughage as per cent of daily feed	Growing market pigs — Daily gain (g)	Growing market pigs — Daily feed (kg)	Dig kcal per kg feed — Meat-type	Dig kcal per kg feed — Bacon-type
10	22				275	.550	3,500	3,500
20	44	275	.350	0	440	1.000	3,500	3,500
30	66	375	.550	0	525	1.500	3,300	3,300
40	88	475	.850	0	630	1.900	3,300	3,300
50	110	500	1.000	0	680	2.200	3,300	3,300
75	165	550	2.000	0	770	2.360	3,300	3,100
100	220	650	2.800	30	860	3.540	3,300	3,100
200	440	700	5.200	60				
300	660	600	7.200	75				
400	880	600	8.800	100				
500	1,100	600	9.600	100				
600	1,320	600	10.000	100				

standard categories. These adjustments are left to the pig or the beef critter or the sheep wherever self-feeding is practiced. With the dairy cow and the horse regulation of the concentrate feeds allows the feeder to use his skill in feeding, but even here different but comparable animals consume differing quantities of forage and thus are not likely to consume identical quantities of nutrients. But to establish the *proportions* of nutrients generally best suited for a feeding group, we have to set down average or at least typical values for each nutrient. When viewed in this light the significance of feeding standards is at once apparent.

PROBLEM

1. Using the formulae on pages 146 and 148, and the figure for net efficiency of milk production (p. 150), prepare a table giving the TDN and the pounds of meal needed by milking cows of weights of 900, 1100, and 1300 lb that are fed roughage of 50 per cent TDN at a rate of 2 lb per 100 lb body weight, and are producing 4 per cent milk in daily amounts of 25, 40, and 55 lb. How does your feeding standard compare with that of NRC?

SUGGESTED READINGS

Brody, Samuel, *Bioenergetics and Growth* (N.Y.: Rheinhold, 1945).

A series of individual pamphlets published by the NRC Committee on Animal Nutrition, 2101 Constitution Ave., Washington, D.C., on the nutrient requirements of domestic animals: *Poultry* (1966); *Swine* (1964); *Dairy Cattle* (1966); *Beef Cattle* (1963); *Sheep* (1964); *Horses* (1966).

Crampton, E. W., "Growth and Feed Consumption of Bacon Hogs," *Sci. Agr.,* XIX (1939), 736.

Forbes, E. B., R. W. Swift, L. F. Macy, and M. L. Davenport, "Protein Intake and Heat Production," *J. Nutrition,* XXVIII (1944), 189.

Feeders' Guide and Formulae for Meal Mixtures, Quebec Provincial Feed Board (1955).

Hammond, J., and G. N. Murray, "The Body Proportions of Different Breeds of Bacon Pigs," *J. Agr. Sci.,* V (1937), 394.

Harris, L. E., *Biological Energy Interrelationships and Glossary of Energy Terms* (Washington, D.C.: NRC publ. no. 1411, 1966).

Kleiber, Max, "Body Size and Metabolism." *Hilgardia, California, Agr. Exp. Sta. Pub.,* VI (1932), 315.

Kleiber, Max, "Body Size and Metabolic Rate," *Physiol. Revs.,* XXVII (1947), 511.

Kriss, Max, "A Comparison of Feeding Standards for Dairy Cows, with Special Reference to Energy Requirements," *J. Nutrition,* IV (1931), 141.

Morrison, F. B., *Feeds and Feeding* (Ithaca, N.Y.: Morrison, 21st. ed., 1948).

Matthews, C. A., and M. H. Fohrman, "Beltsville Growth Standards for Jersey Cattle," U.S.D.A. tech. bull. no. 1098 (1954).

Maynard, L. A., *Animal Nutrition* (N.Y.: McGraw-Hill, 1947).

CHAPTER 7

Requirements for Protein

Fundamentally, it does not seem logical to attempt to rank the nutrients animals require in terms of importance. If they are requirements, they must all be present, and the consequences of deficiencies are undesirable. From the standpoint of practical feeding, however, and especially in the sense of daily requirements, we can rank nutrients in the order of their importance. Deficiencies of energy cause prompt and usually economically important modifications in the performance of the animal. Reactions to unfavorable protein levels are almost as prompt. Appreciable deficiencies in total protein or in the quality of protein are reflected in decreased voluntary food intake and in less efficient use of the food that is eaten. Marked surpluses of protein over optimum requirements also reduce the efficiency of the ration by increasing the heat increment loss from the body.

Most of the mineral elements, and most of the vitamins, are accumulated in the body, and are available for varying lengths of time to carry on their nutritional functions. Consequently, deficiencies of them frequently do not result immediately in altered response by the animal. But storage of protein appears to be limited, and consequently the losses of protein from the body must be made good continuously through dietary intake.

Because the protein level of the ration is normally adjusted by selection of appropriate feedstuffs, and because the quantities required bear a close relationship to the energy requirements, we need to discuss the fundamental basis of protein requirements in some detail. The quantities of protein presumed to be needed daily by animals of various feeding categories are specifically set down in current feeding standards. Such standards, however, do not discuss the problems with determining the requirements or the factors that may modify requirements, an understanding of which may be useful in getting the most out of our feedstuffs.

We should first point out that in the minds of many livestock feeders lack of protein is the cause of most of the unsatisfactory results of livestock feeding. This attitude has undoubtedly arisen, in part, from the emphasis placed on "balanced rations" at a time when balance was considered to be primarily a matter of the proportion of protein to carbohydrate equivalent. In practical livestock feeding undoubtedly many cases of unsatisfactory performance of livestock have been improved by more liberal allowances of foods of high protein content. It was quite natural, then, to ascribe such beneficial effects to protein; but recent evidence shows clearly that the improvement was frequently due to the other nutrients in the high-protein feeds. It is also worth pointing out that in many cases the use of high-protein feeds actually increased the available energy of the mixture, by replacing low-protein feeds of high fiber or other low available-energy content. For example, the introduction of linseed meal into rations as a partial replacement for oats not only raises the protein level of the mixture but also reduces the fiber content and increases the DE of such a mixture.

Fate of Dietary Protein

The basis for the requirement of protein in the ration eventually traces back to the losses the body suffers in nitrogenous end-products. In order to maintain the animal in nitrogen (or protein) equilibrium the intake must be sufficient to balance the output loss. It may be helpful, therefore, to consider the fate of the protein ingested by an animal in its ration. Fig. 7-1 is a flowsheet diagram of the different channels through which this nutrient finally disappears from the body.

Referring to Fig. 7-1, we should note that the losses of protein from the body of an adult, nonproducing animal may be grouped into four categories:

1. Undigested and hence unabsorbed diet-protein residues.
2. Nitrogenous residues from the intestinal microflora.
3. Spent digestive-fluid nitrogen.
4. Urinary excretion of the nitrogen wastage from intermediary protein catabolism.

Animals that are storing protein, such as in growth, will also have requirements for tissue anabolism, and animals producing protein-containing prod-

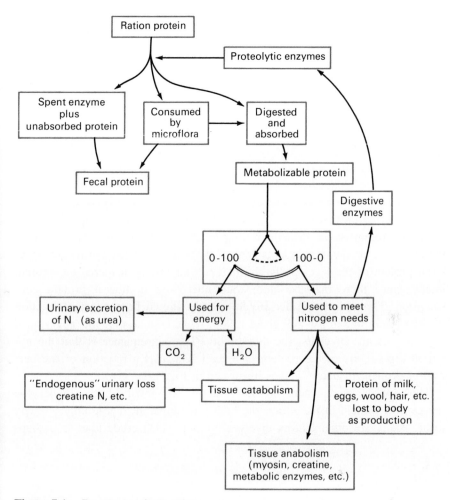

Figure 7-1. *Protein metabolism flow.*

ucts, such as eggs, milk, wool, or hair, will have a further requirement to meet these needs. To maintain an animal in protein equilibrium, these losses must be replaced. Their sum represents the protein requirement of that animal. To arrive at a working figure for the sum, we have to consider at least the more important factors that influence the losses.

The nitrogen (or its protein equivalent) of the first three categories above appears in the feces undifferentiated in terms of its origin. Two of these categories do not represent undigested protein of the diet. They are, rather, the chief components of metabolic fecal nitrogen. In applied livestock feeding the exact origin of these two fractions, or even of the total protein they represent, is of relatively minor importance. Of much greater significance is the relation they bear to the protein intake requirement.

By definition the apparent digestibility of protein is the percentage of the intake that is not recovered in the resulting fecal excretion. To the extent that any portion of the fecal protein is not an unabsorbed diet residue, the calculated apparent digestibility of protein will be smaller than the true digestibility. Because of the technical difficulties in measuring true digestibility (that is, in measuring the metabolic fecal nitrogen), and because of the fact that the metabolic losses must ultimately be made good by protein intake in any case, it is universally customary to use the apparent digestibility coefficient in dealing with applied feeding.

This practice, however, introduces a problem frequently overlooked in considering the protein value of feeds and consequently in dealing with statements of protein needs. Protein from spent digestive fluids is proportional to the total feed intake, but independent of the amount of protein intake. Bacterial protein of the feces probably increases slightly with increasing protein intake, since there may be more total microflora in liberal protein consumption. The undigested diet-protein residue should be a constant fraction of the protein intake for any given ration, if we assume that the quantity of protein does not affect its true digestibility. The consequence is that the apparent digestibility of the protein of a food is in part a function of the per cent of protein it carries. This is illustrated in Fig. 7-2.

Thus we see that, where no other factors are involved, the higher the per cent of protein in the ration the higher its apparent digestibility is likely to be. There are other factors involved, among which crude fiber is perhaps

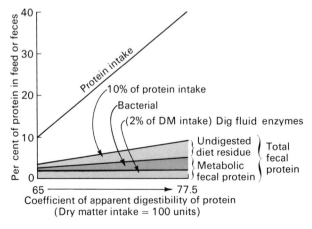

Figure 7-2. *Relationship between a constant dry-matter protein intake and apparent digestibility.*

the most important. Increases in crude fiber in the ration tend to depress the digestibility of the protein. Consequently, if increases in protein are accompanied by increases in crude fiber as well, we may find little change in apparent digestibility, since the increasing protein is balanced by increasing fiber insofar as the effect on apparent digestibility of the protein is concerned. But if increasing the protein decreases the crude fiber of the ration, the apparent digestibility of the protein will increase. Schneider made a quantitative study of the relationships between the proximate principles of a feed and the apparent digestibility of the several components, using the statistical device of multiple correlation and regression. He has published the following general equation for predicting the apparent digestibility of protein of a feed from its protein (X_1), fiber (X_2) and nitrogen-free extract (X_3) content:

Coefficient of digestibility $(\%) = C + b_1 X_1 + b_2 X_2 + b_3 X_3.$

Values for C and for the partial regressions (b) are specific for the class of feed, and the appropriate figures are given in his paper.* Only three of the four proximate principles are used in any one case.

The effect of the amount of protein on its digestibility is illustrated by the calculations for two feeds of differing (35 vs. 12 per cent) protein content:

Feed A   dig = 4.7 + 1.233(35) + .755(8) + .626(52)

    = 86.4%.

Feed B   dig = 62 + .821(12) + 1.20(8) + [−.204(77)]

    = 65.7%.

From one point of view this question of the apparent digestibility of protein is academic, since the range of protein levels ordinarily supplied in the final mixed rations of livestock is not great enough to distort the apparent digestibility sufficiently to affect adversely the preparation of rations.

Referring again to Fig. 7-1, we can see that the protein which is eventually digested and absorbed may be used entirely to meet the protein needs of the body, or it may under some circumstances be very largely used for energy, and the amino groups excreted in some form in the urine. The proportion of the digested protein that is actually used to meet protein needs in the body is the biological value of the protein. Two major factors appear to regulate the biological value of the protein of the diet. The first one is the level of energy nutrition. If the energy requirements of the animal are not fully met by nonprotein sources, then enough of the digested protein will be diverted to

* B. H. Schneider, H. L. Lucas, M. A. Cipolloni, and H. M. Pavlech, "The Prediction of Digestibility for Feeds for Which There Are Only Proximate Composition Data," *J. Animal Sci.*, XI (1952), 77.

meet the energy deficiency. When the energy requirement is adequately supplied from nonprotein sources, the chief factor that will control the proportion of the digested protein used for protein needs will be the particular amino-acid assemblage of that protein. As the proportions of amino acids approach the ideal, the true biological value will increase.

Biological values of proteins, however, are not constants; they will be higher under minimum intake than when liberal allowances are fed. This variation is partly because there is an upper limit to the needs at a given time, and surpluses of ideal mixtures of amino acids will be no more useful than leftovers from poor assortments. Furthermore, the assortment needed is not constant but depends on the tissue to be synthesized. We find a clear example of this in the amino-acid makeup of wool as compared with that of skeletal muscle (see Table 7-1). Marston suggests that the makeup of the amino-

TABLE 7-1   *Amino-Acid Composition of the Protein of Diet, Wool, and Muscle*

Amino acid	Per cent of amino-acid N in total N in:		
	Diet protein	Wool protein	Muscle protein
Cystine	1.2	9.0	0.8
Methionine[a]	1.4	0.4	1.9
Threonine[a]	3.4	4.7	3.9
Arginine	12.5	20.0	15.3
Histidine[a]	2.8	1.1	4.1
Lysine[a]	5.2	3.2	10.4
Tryptophane[a]	1.5	0.6	1.0
Phenylalanine[a]	2.6	1.9	2.6
Tyrosine	2.4	2.0	1.5
Leucine[a]	6.2	7.2	5.3
Isoleucine	3.1	—	4.2
Valine[a]	4.1	3.4	4.3

[a] Essential amino acids.

acid pool available to the tissues in the circulating fluids determines, in part, not only the total but the kind of tissue that will be formed in the body. Thus wool growth can be favored by arranging for relatively higher intakes of the sulfur-containing amino acids.

We can see that the proportion of the digested and absorbed protein from the ration, which ultimately is used to meet the nitrogen losses from the body's metabolism, is a variable and is determined largely in an otherwise adequately nourished animal by the quality of the ration protein in terms of its amino-acid makeup. For the moment it makes little difference whether there is a "pool" of amino acids to which catabolic residues and dietary intake contribute and from which quotas may be drawn for anabolism as needed, or whether some other scheme actually operates to permit the continual synthesis of enzymes and other nitrogen-containing molecules indispensable for body function. The fact is that given an energy-adequate, nitrogen-free dietary regime for adult, nonproducing animals of all species, we find an irreducible minimum daily urinary nitrogen excretion, which represents a steady loss of protein to the body from its metabolism. To attain nitrogen equilibrium this loss must be replaced from dietary sources.

In this connection we must interpret "dietary sources" somewhat liberally in order to include the herbivorous group of animals, whose bodies obtain the necessary assortment of amino acids indirectly from the diet through the synthesis of proteins by the intestinal microflora, so that the feeder of herbivorous animals does not need to arrange his ration to contain all of the essential amino acids. He must, however, supply about the equivalent of these amino acids in total protein in order to provide the necessary nourishment for the microflora in the digestive tract.

Fate of Dietary Protein vs. Dietary Energy. The terms and expressions used in describing the fate of dietary protein obscures the similarity between the steps in the utilization of protein and energy by the body. It may facilitate an understanding of both if they are presented to show comparable fractions in the schemes. We can chart in the same format the fates during biological use of the energy and of the protein (as nitrogen) of the ration (see Figs. 7-3 and 7-4). We easily recognize the almost parallel partition of these diet components, though the actual metabolic changes are unique for each. We note that for both the metabolic fecal losses include bacteria residues, abraded mucosa, and spent enzymes; that the urinary losses contain a component of endogenous origin; and that the portion of the true metabolizable energy (or of the protein) which is available for production is the amount remaining after the maintenance needs are satisfied. These maintenance needs for energy and for protein (or nitrogen) are proportional to (or dependent on) the metabolic size of the animal, $W_{kg}^{.75}$. We also see that true biological value is, in fact, another term for true metabolizable protein. By definition both terms refer to the amount of the digested and absorbed dietary protein that is used to meet the needs of the body for maintenance, production, or both. Finally, we see that true net nitrogen utilization corresponds to true net energy.

Figure 7-3. *Biological partition of dietary energy.*

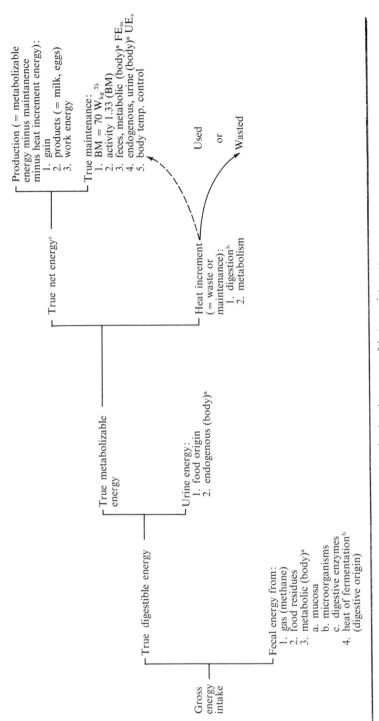

[a]Feces, metabolic (body), and urine, endogenous (body) energy must be replaced, so they are part of the true maintenance requirement for energy and are listed under maintenance also.

[b]The heat of fermentation occurs in the digestive tract, but it is conventionally listed as part of the heat increment (digestion), so it is shown twice.

[c]True net energy corresponds to net protein utilization in Fig. 7-4.

Figure 7-4. *Biological partition of dietary nitrogen (protein × 6.25).*

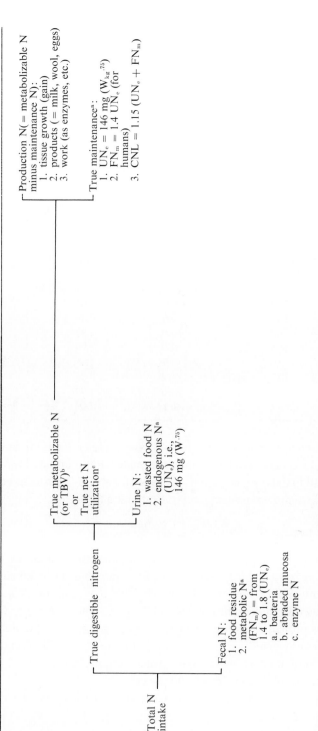

[a]FN_m and UN_e are part of the maintenance needs for nitrogen and are also shown under maintenance.
[b]TBV is merely metabolizable nitrogen as a per cent of true digestible nitrogen (or protein) and shows how much of the true absorbed nitrogen is used for maintenance or maintenance plus production.
[c]True metabolizable nitrogen or true net nitrogen utilization is the utilization of N for maintenance or maintenance plus production and conforms to true net energy in Figure 7-3.

Quantitative Considerations. Quantitatively this minimum protein metabolism of adult animals, as represented by the minimum endogenous urinary nitrogen output, is related to their metabolic size. This relation is evident from the fact that there is, at all body weights, an almost constant ratio between basal metabolism and so-called endogenous urinary nitrogen. These represent, respectively, minimum energy and minimum protein expenditure. Brody has assembled data showing this ratio (see Table 7-2).

TABLE 7-2 *Basal Metabolism and Endogenous Urinary Nitrogen at Different Body Weights*

Body weight kg	Basal metabolism kcal/kg	Endogenous urinary nitrogen mg/kg	Ratio $N_{mg}/kcal$
1	70.5	146.0	2.07
10	38.2	76.6	2.00
20	31.8	63.0	1.98
40	26.5	52.0	1.96
80	22.0	42.7	1.94
100	20.7	40.2	1.94
500	13.5	25.6	1.90
All weights	223.2	446.1	2.0

The resting energy metabolism, as we have already seen (Chapter 6), equals $70(W_{kg}^{.75})$ kcal. Realizing that the endogenous urinary nitrogen is expressed in milligrams, we can from the data of Table 7-2 write an equation for the probable minimum protein requirement in grams as

$$\text{Minimum protein} = \frac{2 \times 70(W_{kg}^{.75}) \times 6.25}{1,000}.$$

Examination of the appropriate data from the literature led Brody to conclude that the regression of endogenous urinary nitrogen on metabolic body size was 146 mg, which agrees well enough with the value of $(70 \times 2) = 140$ mg calculated from the ratios of Table 7-2. This value, 146 mg or 140 mg, converted to its protein equivalent and expressed as grams, would represent the maintenance requirement of animals for digestible protein if all of that which was digested were used to meet body protein requirements, that is, if the protein had a true biological value of 100 per cent.

The effective biological value of ration proteins, or even of the digested portion of them, is much less. There does not appear to be any way of predicting what the actual true biological value of a dietary protein will be; and even if there were, it seems unlikely that feeding practice could be adjusted with enough precision to take advantage of such information. For economic reasons, substitutions of feeds are frequently made at the expense of some quality in the protein mixture. What we really need is a working biological value that will be applicable to a protein mixture found in a typical balanced ration and fed according to acceptable practice.

One way of estimating this wastage due to imperfect biological value might be by comparing the digestible protein allowance that feeding standards give for the maintenance requirements with the basic minimum protein figure as indicated by our calculations above. When we do this for cattle we find that the "minimum" requirement is about one quarter of the digestible crude protein figure arrived at by averaging that in the more important feeding standards in use today. Specifically, the calculations indicate that if the minimum protein requirement calculated from the formula given above is multiplied by 3.4, the result agrees with the average digestible protein required to maintain 1,000 lb (450 kg) dairy cows according to current feeding standards. The calculations are:

$$\left.\begin{array}{l}\text{DP equivalent (in grams) of} \\ \text{24 hr } UN_e \text{ for a } 1,000 \text{ lb} \\ \text{(450 kg) cow}\end{array}\right\} = \frac{146(456^{.75}) \times 6.25}{1,000} = 90 \text{ g}$$

$$\left.\begin{array}{l}\text{Average DP for maintenance} \\ \text{for a } 1,000 \text{ lb (450 kg) cow} \\ \text{(from feeding standards)}\end{array}\right\} = 305 \text{ g}$$

$$\text{Ratio (total/endogenous)} \qquad = 3.4$$

Let us speculate on how we make up this 3.4 figure. Obviously we must include the wastage due to imperfect biological value, and some workers in protein nutrition have suggested that probably on the average the effective biological value of animal ration protein is not far from 50 per cent. If we should take this figure we would have to double the calculated minimum endogenous protein equivalent calculated from our original formula. The other losses will, for the most part, be associated with metabolic fecal protein losses; and it has been estimated that under minimum conditions these total about the same as the endogenous urinary nitrogen excretion. However, there is no very satisfactory way of arriving at a reliable figure here, and perhaps the best we can do is to determine our result by difference. For what suggestive information they may give, some of these relationships have been worked out in Table 7-3. These relationships can be expressed in an equation as:

$$\text{Grams of DP per day} = \frac{146(W_{kg}^{.75}) \times 6.25}{1,000} \times (1 + 1 + 1.4).$$

TABLE 7-3 *The Maintenance Requirement of Protein by a 1,000-lb (450-kg) Adult Cow*

Category	Determined as	DP (g)	Multiple of endogenous
Endogenous urinary loss	$\dfrac{146\,(W^{.75}) \times 6.25}{1,000}$	90	1
Allowance for TBV at 50%	equal to endogenous urinary loss	90	1
Other unspecified N losses (FN$_m$)	by difference from feeding trial maintenance data	115	1.4
Probable DP required per day	average of feeding trial standards	305	3.4

This equation is a general one, applicable to all species to which the corresponding energy requirement equation applies.

As with energy, it is often desirable in practice to express protein need in terms of the total instead of the digestible value. One practical reason for this practice is the fact that all commercially prepared rations or mixed-protein supplements must guarantee a minimum total crude protein. If animal requirements are shown only in terms of digestible crude protein, we must make a conversion if the standard is to be useful as a direct guide for the preparation of rations.

Some may argue that conversion is not feasible, because the digestibility of protein of different feeds is variable, and no satisfactory working figure for converting digestible to equivalent total protein is therefore possible. This view may be questioned on the basis of published data, with the possible exception of roughages. Table 7-4 gives the coefficients of apparent digestibility of the protein in the more commonly used concentrate feeds; from it we can see that individual feeds do not deviate widely from the averages of 79 per cent.

The standard deviation of individual feeds from this mean is ±6.44 percentage points. If we can assume that an average meal mixture contains not less than four feeds, then it follows that the standard deviation of the average digestible protein calculated for the mixture will be ±2.13 percentage points, and only once in twenty cases would we find that this average coefficient of digestibility of protein would differ by more than $\left(\dfrac{\pm 6.44}{\sqrt{9}} \times 2\right) = \pm 4.26$ percentage points from our original average of 79 per cent. This variability is no greater than we may expect between the digestibility coefficients of different samples of the same feed. We can thus often

TABLE 7-4 *Coefficients of Apparent Digestibility of Protein of Concentrate Feeds*

Feed[a]	Coefficient of digestibility of protein	Feed[a]	Coefficient of digestibility of protein
Barley	79	Malt sprouts	91
Barley feed (low grade)	81	Milo	78
Brewers' grains	74	Oats	78
Buckwheat	72	Rice (rough)	75
Corn	78	Rye	79
Corn gluten feed	86	Soybean meal	90
Cottonseed meal	81	Wheat	79
Distillers' grains	69	Wheat bran	75
Emmer	81	Shorts	85
Fishmeal	81	Wheat screenings	72
Hominy feed	64	Average	
Linseed meal	84	(and standard deviation)	79 ± 6.44

[a] Common names. For NRC names, see Chap. 1.

give valid working figures for the total crude protein that corresponds to digestible crude protein. On this basis we may also state the probable requirements of animals in terms of total protein by assuming that the digestible value is equivalent to *about* 80 per cent of the total crude protein.

The digestibility of the protein of roughage feeds is lower than that of the concentrate feeds, and is about twice as variable as measured by the standard deviation of eighteen common roughages. The figures are shown in Table 7-5. Not only is the protein digestibility of forages rather variable, but in practical feeding a mixture of roughages will seldom be fed at any given time. Consequently it is less feasible to attempt to express protein requirements in terms of total protein where the ration is to be made up essentially of roughage.

Protein Requirements for Growth

The general characteristics of the growth of young animals have been discussed in a previous chapter dealing with the energy requirements of growth. We must realize, however, that the energy requirements for growing animals do not cover the requirement for the deposit of tissue. Tissue growth in young animals is largely protein in nature—if for the moment we can neglect the water and the skeleton growth. Hence an important consideration in the rations of growing animals is the protein.

TABLE 7-5 *Coefficients of Apparent Digestibility of Protein of Some Roughage Feeds*

Feed	Coefficient of digestibility of protein	Feed	Coefficient of digestibility of protein
Alfalfa	71	Millet	29
Bermuda grass	51	Native western hay	30
Bluegrass hay	63	Oat hay	56
Brome grass	60	Orchard grass	50
Buffalo grass	55	Sorghum	38
Alsike	67	Soybean	69
Red clover	61	Sudan grass	51
Cowpea hay	69	Timothy	47
Mixed grasses	50	Average	
Kafir fodder	50	(and standard deviation)	54 ± 12.3

The amounts of protein that should be in the rations for growing animals will be affected by the size of the animal and by the rate at which new protein tissue is being formed. And if we look at feeding standards individually, there does not seem to be any rhyme or reason to the figures shown as requirements. Most of the figures for protein requirements for growing animals have been determined in actual feeding trials, and consequently are not to be taken as mere estimates. On the other hand, if the information that is available concerning protein requirements is to be useful in the practical feeding of animals, we must find some common basis for expressing these requirements in terms of the per cent of protein in rations.

We have already shown that the energy and protein requirements bear a definite relation to each other for adult animals, and it seems reasonable to believe that the same thing might be true for growing animals at any particular stage of growth. Guilbert and Loosli attempted to bring some order out of the masses of data in the various feeding standards, and in 1951 published their report under the title "Comparative Nutrition of Farm Animals." * These authors found that at physiologically equivalent ages the nutrient requirements for protein, calcium, and phosphorus are similar for the various species if expressed as a percentage of the TDN requirement. The ratio of these nutrients to the energy intake changes with advancing age because of the change in the composition of the growth increments. By expressing the digestible protein as a percentage of the TDN requirement, they could pre-

* H. R. Guilbert and J. K. Loosli, "Comparative Nutrition of Farm Animals," *J. Animal Sci.*, X (1951), 22.

pare a generalized recommendation for protein applicable to most species of farm animals. The figures they present are in the first four lines in each of the two halves of Table 7-6. These form the basis, in part, of the data of subsequent lines.

TABLE 7-6 *Calculation of Per Cent of Protein Needed in Meal Mixtures for Growing Swine and Dairy Cattle*

	Per cent of adult weight attained					
	10	20	30	40	50	60
Swine						
Average weight in lb[a]	50	100	150	200	250	300
Total daily feed in lb[a]	3	5.3	6.8	7.5	8.3	8.8
Daily TDN in lb[a]	2.3	4.0	5.1	5.6	6.2	6.6
DP as % of TDN[a]	18	16	14	13.5	12.1	11.4
Pounds feed	3	5.3	6.8	7.5	8.3	8.8
grain[b] roughage[c]	0	0	0	0	0	0
Pounds protein[d]	0.52	0.8	0.89	0.94	0.94	0.94
Per cent total protein needed in meal mixture[e]	17	15	13	12.5	11.3	10.7
Dairy cattle[f]						
Average weight in lb[a]	140	280	420	560	700	840
Total daily feed in lb[a]	3.5	8.0	11.4	14.0	17.0	19.5
Daily TDN in lb[a]	2.8	5.0	6.6	8.0	9.2	10.1
DP as % of TDN[a]	20.7	14.8	12.3	10.6	9.6	9.0
Pounds feed	3.5	4.0	4.0	4.0	2	1.5
grain[b] roughage[c]	0	4.0	7.4	10.0	15	17
Pounds protein[d]	0.73	0.92	1.02	1.06	1.11	1.13
Per cent total protein needed in meal mixture[e]	21	15	14	12	11	9

[a] From Guilbert and Loosli, *J. Animal Sci.*, X(1951), 22.
[b] Grain = 75% TDN with protein 80% digestible.
[c] Roughage = 50% TDN and 6% DP.
[d] Pounds protein $= \dfrac{\text{TDN} \times (\% \text{ protein in TDN})}{80\%}$, i.e., $\dfrac{2.3 \times 0.18}{0.80}$ 0.52.
[e] Per cent of protein $= \dfrac{\text{lb protein} - (\text{lb roughage} \times 6\%)}{\text{lb meal}}$, i.e., $\dfrac{0.52 - 0}{3} = 17\%$.
[f] Applicable to growing beef cattle not being fattened.

The important part of Table 7-6 at the moment is the data in the last line of each half, which indicate the approximate percentages of total crude protein that should be in the meal mixtures intended for the two classes of stock according to the stage of development of the animals.

Protein Requirements for Lactation

We can state the protein requirements for lactation somewhat more easily than we could those for growing animals, because the problems involved are less complicated. Lactation represents a direct loss of protein to the body, which obviously must be replaced, and we can calculate the extent of this loss in a relatively straightforward way. Experimental evidence indicates that animals can adjust themselves to lactation over a relatively wide range of protein intake, which seems to range all the way from one to six times the quantity of protein that has been excreted in milk. Many dairy cattle research workers feel that if we add to the maintenance requirement enough protein to replace that lost to the body in the milk produced plus about 25 per cent, we meet the minimum requirements for lactation. No distinction is made between milks of differing fat content insofar as protein requirement is concerned. Nor is there any evidence that feeding more protein than is called for by 1.25 times that of the milk is harmful. The only reason for feeding minimum allowances appears to be one of economy.

SUGGESTED READINGS —————————————

Brody, Samuel, *Bioenergetics and Growth* (N.Y.: Rheinholdt, 1945).

The series of pamphlets published by the NRC Committee on Animal Nutrition on the nutrient requirements of domestic animals (see *Suggested Readings* for Chap. 6).

Crampton, E. W., and B. E. Rutherford, "Apparent Digestibility of Dietary Protein as a Function of Protein Level," *J. Nutrition,* LIV (1954), 445.

Morrison, F. B., *Feeds and Feeding* (Ithaca, N.Y.: Morrison, 23d ed., 1967).

Mitchell, H. H., "The Validity of Folin's Concept of Dichotomy in Protein Metabolism," *J. Nutrition,* LV (1955), 193.

Schneider, B. H., H. L. Lucas, H. M. Pavlech, and M. A. Cipolloni, "Estimation of the Digestibility of Feeds from Their Proximate Composition," *J. Animal Sci.,* X (1951), 706.

Schneider, B. H., H. L. Lucas, H. A. Cipolloni, and H. M. Pavlech, "The Prediction of Digestibility for Feeds for Which There Are Only Proximate Composition Data," *J. Animal Sci.,* XI (1952), 77.

CHAPTER 8

Requirements for Minerals, Vitamins, and Miscellaneous Additives

Minerals

Feeding standards have only recently included data on the needs of animals for mineral elements, and were largely limited to calcium and phosphorus. Nevertheless, information has been accumulating about all the dozen or more elements which we now consider essential. These are conveniently grouped into macro and micro elements, the former being the ones ordinarily discussed in terms of pounds or kilograms, the latter those described in terms of grams, milligrams, parts per million, or other very small units or quantities.

All the NRC standards give the quantities of calcium and phosphorus believed needed by animals, and sometimes also the common salt requirements. The quantities are much less certainly established than are those for energy and protein. In general they are on the liberal side, especially for phosphorus. Nutritionally these liberal values may not be particularly serious, because animals appear to be tolerant of a considerable excess of both calcium and phosphorus, and probably of several other elements as well. We have only to examine the figures for the calcium and phosphorus content of feeds used in livestock feeding to realize that many rations which have been successfully fed contain these two elements in amounts and proportions that differ widely from the presumed requirements. For feeds commonly fed to cattle and sheep, this variation is graphically shown in charts adapted from some prepared by Dr. G. Bohstedt* of the University of Wisconsin (see Figs. 8-1,

---

* G. Bohstedt, "Mineral Requirements for Beef Cattle and Sheep," *Proceedings Semi-Annual Meeting Nutrition of the American Feed Manufacturers Assn.*, (Nov. 1951), 8.

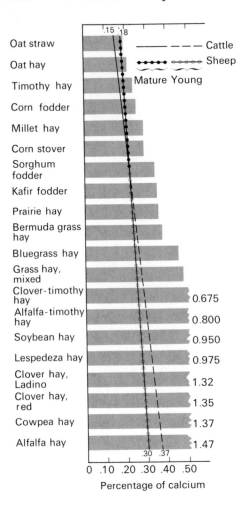

Figure 8-1. *Calcium content of common forages, and recommended percentages of calcium in rations for cattle and sheep. Diagonal lines represent the minimum to maximum feeding-standard requirements.*

TABLE 8-1 *Calcium and Phosphorus Requirements for Cattle and Sheep*

Stock	Calcium		Phosphorus	
	Min.	Max.	Min.	Max.
Fattening yearlings and 2-year old steers and cows	0.15	0.26	0.15	0.21
Calves	0.26	0.37	0.21	0.28
Well-grown lambs and yearlings, pregnant ewes	0.18	0.24	0.16	0.19
Growing lambs and nursing ewes	0.24	0.30	0.19	0.22

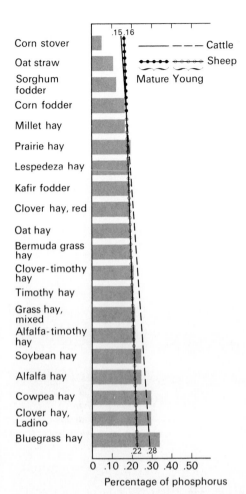

Figure 8-2. *Phosphorus content of common forages, and recommended percentages of phosphorus in rations for cattle and sheep. Diagonal lines represent the minimum to maximum feeding-standard requirements.*

8-2, and 8-3, and Table 8-1). The wide tolerance, where adequate drinking water is available, of all farm stock for salt is well-known to feeders generally.

Excesses of Ca, P, and NaCl each lead to different problems. Salt is commonly supplied as an ingredient of a mixed mineral supplement. Here it acts not only as a nutrient but as a condiment to encourage free consumption of the mineral mix, which otherwise might not be readily eaten. Obviously as the proportion of salt in the mix increases, the quantities of the other ingredients must decrease. This excess of salt could easily defeat the purpose of the mineral supplement, especially for calcium or phosphorus, which are needed in amounts approaching those of salt. Economically there may be a temptation for the manufacturer of salt-containing mineral mixtures to include such a large percentage of salt that the mix is no adequate protection

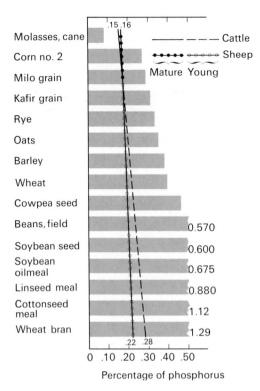

Figure 8-3. *Phosphorus content of common concentrates, and recommended percentages of phosphorus in rations for cattle and sheep. Diagonal lines represent the minimum to maximum feeding-standard requirements.*

against calcium and phosphorus shortage. The latest revision of the Canada Feed Act includes a regulation dealing with this situation. It specifies that if salt is an ingredient of a mixed mineral supplement that also supplies calcium and phosphorus, it cannot exceed 25 per cent in a swine, or 33 per cent in a cattle or sheep, supplement.

The excessive use of phosphorus becomes a problem where the carrier contains fluorine, as many carriers do, especially those in the lower price groups. We discuss flourine problems in Chapter 13.

The problem of excessive calcium in a mineral mixture usually concerns the effects on the quantities of phosphorus (and perhaps salt) that can be present, and arises mostly because ground limestone is cheaper than phosphorus carriers. In general no supplementary calcium (without phosphorus) is needed with herbivorous rations, but swine and poultry rations that do not include roughage may need more calcium than phosphorus. This species difference is also recognized in the regulation in the Canada Feed Act covering mineral mixtures, by restrictions on the maximum and/or minimum ratios between calcium and phosphorus permitted in mineral supplements.

The Trace Minerals. The quantitative data here are limited, and those responsible for the NRC standards are not yet prepared to specify "requirements." But as a guide to the use of the trace minerals in rations, we have assembled the tentative figures given by various research workers for the several elements (see Table 8-2).

Vitamins

Another expansion of feeding standards has been the consideration of vitamin needs. Here we find a sharp species difference, since we need give no attention to the B-complex with herbivora. With cattle, sheep, and horses, more vitamin A (and sometimes D) is now often thought to be needed than is supplied by their forage and grain. The quantities are given in the NRC standards and need no further comment here (see Chapter 9). The needs of these species for vitamin D is recognized, but no figures except for young animals have been suggested.

For swine the requirement figures cover, in addition to vitamins A and D, five B fractions for all classes, as well as choline and B_{12} for the very young animals. Of the required vitamins the feed will probably have to provide sources of carotene, riboflavin, and pantothenic acid in rations for different classes of this species.

Antibiotics

Although we cannot discuss in any detail the physiological roles of the nutrients needed by livestock, we should relax our position somewhat in order to consider the problem of whether to fortify rations with antibiotics. Since any decision we may make on this question depends in part on what function these products have, and since the usual scientific or animal-husbandry literature gives no broad summary of the present knowledge, we shall attempt here to present a condensed statement of what now seems to be known about these substances. Probably the most comprehensive review of the nutritional effects of antibiotics is given in the *Fine Chemicals Technical Bulletin no. 3* (1955) of the American Cyanamid Company under the

TABLE 8-2 *Summary of Some Recommendations on Livestock Needs for Trace Minerals*

Element	Cattle: Beef (1963)	Cattle: Dairy (1966)	Sheep (1964)	Swine (1967)
Cobalt	0.15–0.22 mg/100 kg of wt; 0.06–0.11 mg/kg feed.	75 g/100 kg salt; 0.1 to 1.0 mg/cow/day.	0.18 mg/100 kg of wt; 125 g/100 kg salt. Tolerance: 180 mg/100 kg of wt/day.	1.94 mg/kg feed as growth stimulator where B_{12} is inadequate.
Copper	5 mg/kg feed; 25 g $CuSO_4$/100 kg salt.	1 kg $CuSO_4$/100 kg salt. Mineral mix: Bone meal 50.0 kg; Salt 46.7 kg; $CuSO_4$ 2.5 kg; CuO 0.8 kg; $CoCO_3$ 60 g	5 mg/day/ewe; 50 g $CuSO_4$/100 kg salt. Tolerance: 20 mg/100 kg of wt/day.	10 mg/kg feed.
Iodine	1 g KI/100 kg salt.	1 g KI/100 kg salt.	1 g KI/100 kg salt.	1 g KI/100 kg salt; 4.5 mg KI/100 kg of wt; 0.2 mg KI/kg feed.
Iron		50 mg/kg feed.		80 mg/kg feed; 8 mg/pig of 5 kg of wt.
Magnesium		20 mg/kg of wt (calves).		400 mg/kg feed.
Manganese		20 mg/kg feed.		40 mg/kg feed.
Zinc				50 mg/kg feed.

authorship of T. H. Jukes and W. L. Williams of the Lederle Laboratories Division. We have taken from this publication most of the material summarized here.

The most concise statement of a reason for feeding antibiotics is by Dukes and Williams: "It is logical to expect that the rate of growth of an animal will be increased when a pathogenic or debilitating infection has been eliminated." That this is, in fact, the function of antibiotics as feed supplements is now reasonably well-established. The lines of evidence that their growth effect is due to their antibacterial action in the intestinal tract has been summarized by these authors as follows:

1. The antibiotic growth effect is shared by a group of substances of diverse chemical characteristics. The only known property which these substances have in common is their antibacterial potency.
2. Aureomycin, penicillin, and streptomycin did not increase the growth of chick embryos when they were injected under sterile conditions. "Germfree" chicks did not show a growth response to antibiotics, and neither did chicks which were kept under conditions which tended to reduce or eliminate contact with bacteria. The growth response thus seems to depend on an antibacterial effect.
3. Injected aureomycin and penicillin at best increased growth much less than when they were included in the diet. Any growth-increasing effects produced by injection could be explained by the excretion of a part of the injected dose into the intestine.
4. Bacitracin and streptomycin, which are not readily absorbed from the gut, have a growth-increasing effect when added to the diet of chicks. The effect thus seems due to their activity in the intestine.
5. Despite its wide antibacterial potency, chloramphenicol has only a weak antibiotic growth effect, which may be due to the fact that it is absorbed through the gastric wall and hence does not tend to reach the intestinal tract when added in small quantities to the diet.
6. Hydrolysis with penicillinase destroyed the growth-increasing effect of penicillin on chicks. The antibacterial potency was also destroyed by this treatment.

We do not know how antibiotics affect the intestinal flora. There is evidence to support at least three possibilities: (1) they may inhibit or destroy organisms that produce subclinical infections; (2) they may increase the number or activity of organisms that synthesize growth factors which are ultimately available to the host; (3) they may inhibit organisms that compete with the host for nutrients. Of these the first is already well-supported by experimental evidence from studies with poultry and pigs. There is evidence also that including one or another of the now common antibiotics in the ration can be expected under most practical conditions to increase feed intake, and often, though not always, increase feed efficiency in producing gains. The more-often-noted effect of the increased feed intake is a corresponding increase in weight gain. Growth of pigs is commonly stimulated by 10 to 15 per cent with feed efficiency improved about 5 to 7 per cent. Such difficulties as diarrhea are usually easily controlled, and unthrifty pigs often show marked improvement in performance.

Another consequence of antibiotic supplements has been a decline in the amount of protein needed in the pig ration. This effect must be an indirect one, and it may be related to the nutrient needs of the different types of microorganisms in the intestinal tract of pigs. In this connection, however, we should again point out that protein has been erroneously credited or blamed for many effects in livestock feeding. The amounts of protein recommended and used by some feeders, and indeed by some investigators, have been appreciably higher than those others have found adequate. The reduction made possible by antibiotic feeding has been more apparent where the higher protein recommendations were being followed previously.

The increased feed consumption and consequent increased gains have led to some misgivings concerning pigs in the "finishing" stages. Increased feed intake at this stage normally results in fatter carcasses—a condition incompatible with high-quality bacon. Reports from several experiments suggest that this effect is negligible, or even that no such effect exists. However, all biological reasoning leads to the conclusion that extra energy intake above maintenance needs must result in fattening in proportion to the ratio of energy intake to energy of tissue formed. Experimental evidence from stations in areas where more stringent specifications of fat and lean in an acceptable bacon rasher are used, has led investigators in such areas to recommend the removal of antibiotics from the rations of bacon hogs after they have reached the "finishing stage" (i.e., weights of about 125 pounds or 58 kg).

Experiments show that the beneficial effects of antibiotics in poultry and pig feeding are also likely in calf and lamb feeding. Calf scours in particular are usually well-controlled by antibiotics.

It is not appropriate here to discuss in further detail the nature or extent of the responses of animals to antibiotic feeding. We have said enough to indicate the desirability of including one or another of these substances in the rations of young animals kept under usual farm conditions. The beneficial effects are likely to be inversely proportional to the degree of sanitation of the housing quarters. Antibiotics show no effects in environments in which bacteria are absent. Better than average sanitation therefore tends to nullify in part the growth effect of antibiotics. This conclusion should not be interpreted to mean that sanitation can justifiably be neglected merely because antibiotics are used. There are other reasons for sanitation.

The problem we must now consider has to do with the quantities to be used. The evidence is not complete, nor is there agreement among recommendations. So far as swine feeding is concerned the committee that is responsible for the NRC swine standards has stated its position:

A summary of all the antibiotic work conducted with swine to date shows the following effects: *Growing pigs:* (1) It increases growth rate an average of 10–20 per cent. The higher over-all nutritive value of the ration, the less the improvement of growth due to antibiotic additions. (2) It may increase efficiency of feed utilization up to 5 per cent. With certain poor-quality rations, the efficiency of feed utilization may be increased considerably more. (3) Antibiotic feeding tends to reduce the amount of protein supplement needed in the ration for swine. It is not yet clear how much of the protein-sparing effect is due to the antibiotic and how much to a ration more adequate in B_{12}, riboflavin, niacin, pantothenic acid, and other nutrients. Evidently, protein requirements have been high in the past because protein supplements were supplying factors other than amino acids. (4) The greatest beneficial effect from antibiotics is observed during the early growth period. Older animals are benefited by antibiotic supplementation, but the improvement in gain is not as great as for younger animals. (5) When antibiotics are fed to a weight of 45–50 kg and then dropped from the diet, the pigs slow down in rate of gain. Thus, for continued maximum growth it is necessary to feed antibiotics from weaning to market weight. (6) When antibiotics are fed to fattening pigs (i.e., market pigs over 45 kg), a fatter carcass is likely to be produced. (7) Runty pigs and young pigs suffering from scours respond markedly to antibiotic feeding. (8) Antibiotic feeding results in a more uniform group of pigs by market time. (9) Antibiotics increase bloom and in some cases are effective in preventing a skin dermatitis (unidentified type) with certain rations. (10) Antibiotics give a favorable response with pigs both in drylot and on pasture. (11) At present there are insufficient data to conclude that combinations of antibiotics are more effective than any single one used at the appropriate level. (12) A level of 5 milligrams of antibiotic per pound of total ration (10 g per ton), or 55 milligrams per kg of a protein supplement (50 g per ton) which is intended to be fed free-choice with grain, is probably quite near the desirable level of antibiotic feeding. This level may vary somewhat, depending on the antibiotic, the ration used, and environmental conditions.

Breeding stock and suckling pigs: (1) There is some evidence that the feeding of antibiotics to pregnant sows, at 20 to 35 milligrams per kg of ration, may increase the birth weight and livability of pigs. (2) Antibiotics are not transferred through the milk of sows in sufficient amounts to show a marked stimulation in the growth of nursing pigs. (3) Feeding antibiotics to pigs in creep rations during the suckling period will increase the weaning weight 2 to 4 kg at 56 days of age.*

* NRC, *Nutrient Requirements of Swine* (1967).

SUGGESTED READINGS

Bohstedt, G., "Mineral Requirements for Beef Cattle and Sheep," *Proceedings Semi-Annual Meeting Nutrition of the American Feed Manufacturers Assn.* (Nov. 1951), 8.

Guilbert, H. R., and J. K. Loosli, "Comparative Nutrition of Farm Animals," *J. Animal Sci.,* X, no. 1 (1951), 22.

Jukes, T. H., and W. L. Williams, "Nutritional Effects of Antibiotics," *Pharm. Revs.,* vol. V, no. 4 (1953).

CHAPTER 9

Feeding Standards

According to orthodox definition a *feeding standard* is a table which records what is believed to be the daily need of a specific animal for one or more of the recognized nutrients. Standards have been proposed for most of the kinds of animals that are kept in captivity, and some of them date back at least to 1810.

In 1943 the Committee on Animal Nutrition of the United States National Research Council embarked on a long-term project of preparing and keeping up-to-date by periodic revision a set of modern feeding standards for the several species of animals which are kept in captivity, including not only the so-called farm animals and poultry, but also fur-bearing animals, dogs, and some laboratory animals. To deal with each species, a small committee of specialists is appointed to review the literature and from it prepare a table of nutrient needs. The subcommittee proposals are then reviewed by the parent NRC Committee on Animal Nutrition. When finally approved, the standards, called *Nutrient Requirements for Swine* (or *Poultry,* or *Beef Cattle,* etc.) are published in pamphlet form. Besides up-to-date tables on the nutrient needs of animals, the pamphlets include pertinent information not suitable for tabulation, as well as extensive references to the research literature which has helped establish the requirements for specific nutrients.

Since these standards are under constant review and revision, to attempt to duplicate them would be largely a waste of effort. The basic tables from these NRC standards for the nutrient requirements of the several feeding categories of swine, dairy cattle, and beef cattle can be found on pp. 418–439.

Before referring further to them, however, we should consider feeding standards in general in order that they may be more correctly used in the preparation of animal rations.

The Evolution of Feeding Standards

Tables of requirements have, over the years, become more specific and more inclusive as our knowledge of nutrient needs and of the factors that determine and modify them has expanded. Whereas early standards dealt chiefly with energy and protein for average animals, modern tables include not only caloric needs, but also those for amino acids, vitamins, and minerals, and in addition may be highly specific in defining the animal to which the requirement figures apply. The evolution of the feeding standard, however, has been more a matter of additions than of quantitative revision. It is worth noting that in the important standards for dairy cows that have been proposed since about 1900, there have been no important changes in the energy or protein figures, either for maintenance or for milk production. Significantly, the recommended quantities of these two nutrients are essentially the same in all recent standards.

The Nature of Early Standards.   The philosophy underlying feeding standards is of interest and also of importance in using the figures. Many of the standards in use (and most of the earlier ones) reflect the research, observations, and ideas of one man. We might take as an example the tables that were presented by Kellner. In his description of how he arrived at the maintenance energy and protein figures, he also states a typical philosophy regarding standards.

He did his work on the maintenance of cattle with oxen at rest. He chose them because "the feeding of working oxen at rest is the simplest task which the owner of cattle has to undertake; they are practically sexless, are little subject to nervous influence, and except for slight growth of hair, hoof, and horn, do not increase in size, nor do they perform any utilizable work." His aim was to feed them "so that the body does not need to supply any of its substance for the production of energy and that, on the other hand, there shall be no excess of food to be made into fat." *

Kellner then proceeded to feed two groups of resting oxen, using an "insufficient" ration for one and a "sufficient" ration for the other. The results of this typical experiment are shown in Table 9-1. The exact rations he used are not particularly important if we assume for the moment that the

* Kellner, *The Scientific Feeding of Animals* (N.Y.: Macmillan, 1913).

TABLE 9-1 *Example of Kellner's Energy Balance Studies to Establish Maintenance Needs of Cattle*

Animals	Digestible nutrients consumed				Gain or loss of	
	Protein (kg)	*Fat (kg)*	*CHO (kg)*	*Starch equivalent (kg)*	*Flesh (g)*	*Body fat (g)*
III	0.35	0.10	6.17	4.00	−51	+139
IV	0.34	0.10	6.08	3.96	−57	+45
B	0.28	0.12	6.63	4.52	−144	−172
C_1	0.42	0.21	6.58	4.60	−6	+1
Average	0.35	0.13	6.37	4.28	−65	+3
V	0.60	0.07	6.45	5.04	+48	+235
VI	0.57	0.07	6.31	4.64	+26	+263
20	0.65	0.20	6.78	5.76	−7	+158
A	0.56	0.18	6.62	5.44	+66	+227
C_2	0.54	0.28	7.26	5.04	+161	+37
Average	0.59	0.20	6.68	5.20	+59	+184

digestible protein, fat, and carbohydrate have the same usefulness regardless of their origin. Kellner points out that such an assumption is not always justified—a fact we appreciate more fully today. He corrected in part for this assumption by expressing the total energy intake in terms of starch equivalents (see Chapter 3).

The conclusions Kellner arrives at from these data are of interest. He states "It suffices for maintenance to give 0.5 kg digestible protein and 5.2 kg starch equivalent. As, however, in practice it is not advisable, owing to the individuality of the animals, to restrict the feeding to the absolute minimum, it is preferable to reckon per day 0.6–0.8 kg digestible protein and 6.0 kg starch equivalent. . . . This standard assumes that . . . coarse fodder will be used and any deficit in protein will be made good by . . . oilcakes, etc. Such rations contain sufficient mineral substances, for investigations . . . have shown that daily consumption of 50 g phosphoric acid and 100 g lime per 1000 kg body weight will . . . meet all requirements."

These data and comments by Kellner illustrate several characteristics inherent in most feeding standards. For example, we can see from the data of Table 9-1 that the variability between animals in nutrient intake and in their response in body weight is considerable. A large number of feeding trials conducted by Crampton in 1933 showed that for normal feeding trials the variability between animals within lots as measured by the standard deviation was relatively constant for different trials. The average coefficients as a percentage of the means are shown in Table 9-2. If we apply the appropriate

TABLE 9-2 *Probable Average Per Cent of Variability Between Animals Within Lots in Gain, Feed Consumption, and Gain per 100 Units of Feed Eaten*

Measurement	Hogs and cattle (S.D. in %)	Sheep (S.D. in %)
Live weight gains	17	21
Feed consumption (90% dry matter)	13	16
Feed required per 100 units gain	11	13

coefficient to the means of Table 9-1 (assuming that variation in feed required per 100 units of gain is a direct reflection of variation in nutrient need), the probable necessary intake per 1,000 lb weight, in protein and in starch equivalent, to meet the maintenance needs of 95 per cent of cattle under practical feeding might be taken as 0.72 and 6.34 lb, respectively (see Table 9-3).

TABLE 9-3 *Probable Maximum Protein and Starch Equivalent Requirement of Kellner's Oxen*

Measurement	Mean ± S.D. × 2	Probable maximum need	Kellner's maximum
Protein (g)	268 ± 59[a]	325	365
Starch equivalent (kg)	2.35 ± .52[b]	2.88	2.74

[a] $268 \times 11\% \times 2 = 59$. [b] $2.35 \times 11\% \times 2 = .52$.

In his standard, Kellner has increased his observed average protein to a maximum of 365 g and the starch equivalent to 2.74 kg per 450 kg of body weight. In the former his margin is roughly 35 per cent over his observed average, or 10 per cent over the probable maximum need to nourish a 450-kg ox adequately. For energy, however, he raises his standard only 15 per cent above his mean, which is at least 5 per cent below what one such animal in twenty might be expected to need for its maintenance.

These figures are cited to call attention to three points. First, the figure in a feeding standard, given as the requirement for a nutrient, is usually considerably above the actual average intake of the group of animals whose performance was judged satisfactory. It is set at an amount the *investigator believes is safe*—one he believes guarantees that all comparable animals will be adequately provided for.

Second, the magnitudes of the margins over the intakes observed to be satisfactory for different nutrients are not consistent, either with the normal variability of the animals described, or with the variability of the nutrient in different samples of the kinds of feed used in the feeding tests. There is nothing particularly wrong with feeding rations that are more liberal than is actually needed, but such excesses should not be made part of *requirement figures*. Such figures would become a mixture of true requirement plus some extra allowance, the magnitude of which represents an opinion that has no experimental basis. When a feeder has to adjust feeding to produce some particular result, such standards are of uncertain value.

The third point concerns *other nutrients,* consideration of which was not a primary objective of the particular comparative feeding test. Kellner says, in effect, that neither phosphorus nor calcium need be considered, because the feeds used already supplied more than the quantities *other studies* had shown to be enough. This statement raises an important question in connection with some modern standards, namely, whether the results of different and unrelated studies can be simply piled up to give a recipe for a nutritionally complete diet.

The Significance of Balance Between Nutrients and Energy

The answer is not straightforward or simple, but the basic factors involved are: (1) whether the nutrients in question are interrelated or interdependent in function, and (2) whether seemingly comparable animals will react the same to identical rations. We should consider both questions.

With respect to energy and protein, Brody's classic studies in this field * have brought together data from which it is evident that irrespective of species there is a close relationship between the fasting protein (nitrogen) and fasting energy catabolism in the adult (see Chapter 7). When we base our calculations on adult maintenance, we find that about 10 per cent of the total calories should be derived from protein. Research by Forbes† *et al.* clearly demonstrates that increasing the protein between equicaloric rations for maintenance causes a *decrease* in the efficiency of the ration by increasing the heat increment production and decreasing the metabolizable energy. On the other hand, a greater concentration of protein is required for the most efficient rations

---

* S. Brody, *Bioenergetics and Growth* (N.Y.: Reinhold, 1945).
 † E. B. Forbes, R. W. Swift, L. F. Macy, and M. T. Davenport, "Protein Intake and Heat Production," *J. Nutrition,* XXXVIII (1944), 189.

where growth or production is involved. The requirement is related to rate of growth or level of production, as is the increased energy demand. Consequently, the protein demand is related to energy need, though the proportions differ according to the extent and nature of the production or growth. Thus, it is clear that there is an optimum protein–energy ratio for each particular condition. Distorting this ratio by an excess of protein, contrary to being beneficial, actually reduces the efficiency of the diet.

Closely related to the energy requirement is that of several of the B-complex vitamins, whose functions are primarily as components of enzyme systems involved in energy metabolism. Thiamine, for example, appears to be required at about 0.23 mg per 1,000 metabolizable calories. This figure is arrived at from analysis of many independent feeding trials. Niacin is also required in proportion to calories, but the situation is complicated by the fact that tryptophane is a possible precursor of niacin. Therefore, before we can make a satisfactory statement of niacin requirement, we must know the amino-acid intake. Since the tryptophane complication was unknown when early niacin studies were undertaken, values for niacin needs may change in feeding standard revisions. Riboflavin requirement, on the other hand, appears not to be directly related to caloric requirement or to muscular activity, but rather to protein intake. Calcium need is largely determined by skeletal size in the adult, by rate of skeletal growth, or by the production of calcium-containing products. In practical feeding, phosphorus is usually supplied in some ratio to calcium.

That the nutrients are interrelated in function is thus clearly evident. Whether or not imbalance, especially in the direction of excesses of individual nutrients, is seriously harmful to the overall usefulness of the ration, is not so clear. For some it is obvious that relative excess is undesirable. For example, if the caloric value is greatly in excess of those nutrients that regulate and facilitate energy metabolism, the ration efficiency will suffer and we may expect malnutrition. The problem less often considered is whether excesses should be avoided in vitamins and minerals to keep them, individually or collectively, in proper relation to calories or protein.

Feeding Standards as Hypothetical Rations

A modern feeding standard is based on the results of many feeding trials, and is a statement of what is believed to be the requirement for some selection of nutrients, each treated as though it were an independently acting entity. The novice might visualize the standard as a formula, with each nutrient a separate ingredient to be mixed into a batch for feeding. However, most feeds contain a wide assortment of nutrients. The proportions in which they occur are seldom identical even between feeds that practical experience as well as experimental tests have proven to be interchangeable in the ration. The obvious conclusion must be that, although for maximum efficiency each nutrient may have an optimum concentration in the ration in relation to other nutrients, there is no way in which an "ideal" ration as described by a feeding standard can be prepared.

We do not mean to say that the selection of feedstuffs for a "balanced" ration should be made without reference to specific nutrient contents. By appropriate choices, we can make a nutrient assortment more nearly like the "ideal" than we could by blind substitution. Protein levels, for example, are normally adjusted quite closely to standard requirements, and by careful choice of feeds we can easily avoid gross shortages of amino acids. Rather, our point is that in substitution one must usually choose between a variety of evils. Replacing corn with oats raises protein and increases tryptophane, but reduces available energy because it increases fiber. In practical ration formulation we make a rough balance of nutrients according to a feeding standard by selecting available feedstuffs within the limitations of relative costs, and then make further needed adjustments by adding pure nutrients or concentrates of such nutrients to eliminate any apparent remaining deficiencies.

Biological Tests of Feeding Standards. The fact is that to a large extent feeding standards describe purely hypothetical rations. None of the standards has ever been critically tested as a complete working unit. Indeed, research men have shown an unexplainable lack of curiosity about whether a ration compounded in accordance with a given standard will prove to be ideal from the standpoint of the animal fed, even though the authors of such standards proclaim them as guides to feeding, and feed manufacturers use them as a basis for the fortification of feed mixtures intended to be properly balanced for feeding.

Still, this omission is not altogether mere negligence on the part of those setting up feeding standards. To test adequately a feeding standard for farm livestock is no mean undertaking. The classes of dairy animals for which nutrient needs are described in the NRC table of nutrient requirements comprise 25 different liveweight groups, 7 subcategories for level of milk fat, plus a supplementary section for late pregnancy, which effectively adds at least 3 additional groups. To test the swine standard would involve feeding trials with 9 weight classifications for market pigs and 6 for breeding stock. If, for each group of animals, a critical test of the optimum level of each of the nutrients recorded were to be made, the project, in terms of number of animals needed and the facilities for their use in controlled feeding trials, becomes fantastic. We do not mean, however, that feeding standards should be accepted blindly or that no testing should be done, though there may be some misconceptions about the extent of the applicability of such tests.

The 1949 Canadian Pig Feeding Test

There is a unique situation in Canada today in the system of swine advanced registry test feeding stations. When it was decided to set up a scheme of advanced registration for Canadian swine based on the breeding performance of purebred sows and on the feed lot and slaughter records of their progeny, feeding stations were established in each province of the Dominion (except New Brunswick and Newfoundland). Two male and two female pigs from each litter to be tested are sent to one of the stations, to be fed from weaning time (approximately) to a market weight of 200 lb. Pigs to be tested are nominated when the litters are two weeks old and are shipped to arrive at the feeding stations when 60–70 days old. The rations fed at the different stations are mixed to one formula. The formulae are reviewed and established periodically by a Feed Committee of the National Swine Board, and are based on comparative feeding trials conducted under its supervision at several Canadian universities.

In 1949 the question whether or not self-feeding should be introduced in these feeding stations was raised. Before making a decision the Feed Committee arranged for a Dominion-wide comparative test. In this study five of the advanced registry feeding stations, one each in British Columbia, Alberta, Saskatchewan, Ontario, and Nova Scotia, fed groups of purebred Yorkshire pigs from their respective areas on a meal mixture prepared according to committee specifications by one milling company in Ontario and shipped to each of the cooperating stations. There was one mixture for pigs from 75 to 125 lb weight. After this weight half the pigs at each station were fed to market weight on a normal finishing mixture; the others received this finishing ration diluted with 25 per cent bran as a possible means of restricting fattening. The study involved two complete replicates at each station and was one of a series to establish official rations for bacon carcass production. When the pigs reached a market weight of 200 lb, they were slaughtered at local abattoirs, and their carcasses were graded, cut, measured, and scored by government graders in accordance with the requirements of the official swine advanced registry carcass-scoring scheme.

Although this study involved only purebred Yorkshire pigs, it included animals comparable in age and sex but coming from geographic areas as far apart as 3,000 miles. Thus, types and strains within one breed, such as develop within local areas, were fed on identical rations under comparable conditions.

An examination of the nutrient makeup of these rations, which were intended to be adequate, and of the performance of the pigs is enlightening in a general consideration of feeding standards. In Tables 9-4 and 9-5 we

TABLE 9-4 *Rations Used in Canadian Advanced Registry Carcass Test (1949–50)*

Ingredient	Grower (lb)	Normal finisher (lb)	Diluted finisher (lb)
Barley	430	460	360
Wheat	170	180	140
Oats	250	280	220
Wheat bran	—	—	200
Tankage	75	40	40
Fish meal	22.5	12	12
Linseed meal	37.5	20	20
Salt	7.5	4	4
Limestone	7.5	4	4
Total	1,000	1,000	1,000

give the ration formulae; their calculated contents of those nutrients listed in the U.S. swine standard; and the feed intake, gain, and feed efficiency data for (1) the growing ration, (2) the normal finisher, and (3) the bran-diluted finisher. Included for comparison are the figures for a hypothetical nutrient mixture suitable for market pigs as calculated from the U.S. swine standard. The discrepancies are calculated as the per cent by which the Canadian test rations or the performance of the pigs exceeded (+) or failed to meet (−) the standards.

In the first third of Table 9–5 the nutrient content of 1,000 lb of the mixture fed to the pigs from weights of about 75 lb to 125 lb is shown in comparison with that set out in the U.S. standard for 100-lb market pigs. The "poor fit" will no doubt surprise many. First, the protein level is some 20 per cent higher than "needed." Here is a fundamental problem. The standard calls for 14 per cent crude protein. The Canadian basal feed mixture for reasons of economy was based on barley, wheat, and oats, and contained 12.9 per cent crude protein. In order to include in the mixture enough proteins of animal and marine origin to insure the protein quality believed to be needed, it was impossible to keep the protein of the final mixture down to 14 per cent. Had the basal mixture consisted of corn, it would have

TABLE 9-5 Comparisons of Test Rations with Feeding Standard

Nutrients per 1,000 lb of mixture		Grower ration (100 lb pig)			Normal finisher (150 lb pig)			Diluted finisher (150 lb pig)		
		As fed	U.S. Standard "requirement"	Difference (%)	As fed	U.S. Standard "requirements"	Difference (%)	As fed	U.S. Standard "requirements"	Difference (%)
Protein	lb	170	140	+21	147	130	+13	155	130	+19
TDN	lb	692	750	−8	698	750	−7	670	750	−11
Calcium	g	4261	2948	+45	2403	2500	−4	2439	2500	−2
Phosphorus	g	3080	2041	+51	2392	1510	+58	3242	1510	+115
Salt	g	3405	2268	+50	1816	2280	−20	1816	2280	−20
Thiamine	mg	2588	491	+427	2668	500	+433	2819	500	+460
Riboflavin	mg	1363	1000	+36	1362	1000	+36	1322	1000	+32
Niacin	mg	20476	5000	+309	20096	5000	+302	38104	5000	+662
Ca-Pantothenate	mg	4073	4498	−9	4017	4500	−11	5487	4500	+22
B$_{12}$ (added)	µg	0	5		0	—		0	—	
Carotene	mg	140	756	−81	2	442	−99	2	442	−99
Vitamin D	IU	36000	90000	−60	0	90000	−99	0	90000	−99
Daily feed	lb	4.7	5.3	−11	7.4	6.8	+9	7.1	6.8	+5
Daily TDN	lb	3.4	4.0	−15	5.1	5.1	0	4.8	5.1	−6
Daily gain	lb	1.3	1.6	−19	1.6	1.8	−11	1.5	1.8	−16
Feed per 100 gain	lb	383	332	+15	455	378	+20	485	378	+28
TDN per 100 gain	lb	274	250	+10	318	283	+12	324	283	+14

been possible to make 15 per cent of it a 40 per cent protein supplement and still not exceed 14 per cent protein in the mixture, but with the basal feeds economically available in Canada, the use of 15 parts of mixed protein supplement in 100 resulted in 17 per cent protein in the completed mixture.

The use of coarse grains, barley, and oats has diluted the TDN to 69 per cent as compared with about 75 per cent considered normal for U.S. hog rations. The consequences are twofold. Either the daily feed intake must be increased above "normal" or the gains must be expected to fall below standard. And in either case, the feed required per pound of gain will increase, i.e., the feed efficiency will decrease. The pigs in this test, however, ate not more, but less, feed per day than called for in the standard. Thus, TDN intake was doubly depressed from that expected, and gains were 19 per cent lower than "normal." There was no report of goosestepping among any of the pigs, and so one questions whether the 9 per cent deficiency of pantothenic acid was detrimental. Nor was there clinical evidence of vitamin D shortage. Concerning vitamin A there is room for argument, except for the the fact that pigs of this age could hardly have become depleted to the point of growth restriction unless they had been nursing very vitamin-deficient mothers, a situation that seems unlikely in view of the widely different sources of the pigs and the fact that they were from high-class breeding herds. All of the other nutrients considered were consumed in excess of standard requirements. It would seem, therefore, that the feeding standard was not only liberal in its requirement figures, but too high in the values for normal feed intake and expected live-weight gains for pigs of this category.

In the second third of Table 9-5 similar data are presented for the ration fed these pigs from the time they weighed 125 lb until they reached 200 lb market weight. Here the ration nutrients are clearly much the same, but these heavier pigs have voluntarily eaten 9 per cent more feed per day than is indicated as "normal," and by so doing have consumed the quantity of TDN that is called for in the standard. There can be no argument, therefore, that below-standard concentrations of pantothenic acid, carotene, and vitamin D have checked feed intake with these pigs. Gains are 10 per cent slower, however, and hence the TDN efficiency is below "expected" to about the same extent. In the last third of Table 9-5 we see what happens when wheat bran is introduced. Protein is now almost 20 per cent over standard, and TDN is reduced still further. Even with feed intake 5 per cent above normal, TDN consumption is less than standard figures. This was the effect desired. The bran was employed specifically to reduce fattening rate by reducing TDN intake. Gains have actually been depressed 16 per cent from standard and about 7 per cent from that with the undiluted finisher.

Importance of Energy Intake. In a critical consideration of these data, we can see at once that the response of the pigs has been more closely related to the energy intake than to any other ration factor. The greater concentrations of protein, of calcium and phosphorus, and of several of the vitamins have not prevented daily gains from being smaller than expected from the standard.

In swine feeding we should keep in mind that market pigs are fed to produce a carcass, of which the consumer recognizes two kinds—fat ones and lean ones. The hogs on this Canadian trial were all subjected to official slaughter test. The details of the carcass measurements are of no importance, but the overall result is, and of all the items considered and recorded, that of carcass grade is of the most direct significance. Carcasses that are classified as Grade A bring a cash bonus in addition to the regular grade price. Excessive fat deposits have consistently been the principal fault with Canadian hog carcasses that has kept them out of Grade A. The dilution of the energy value of the finishing ration with bran was done specifically to counteract the tendency for self-fed hogs to become too fat.

We may summarize the results briefly as follows. The altered ration had no effect on any carcass measurements except those directly related to depth of fat. The lower-energy ration effected a 7 per cent reduction in depth of back fat, which was reflected by an overall increase in carcass score of 4 percentage points, and an increase of 27 per cent in the proportion of Grade A carcasses. The increase in grade was relatively uniform between stations, and again indicates the particular sensitivity of animals to changes in energy intake.

Variation in Response to Feed. Before we finally draw conclusions about feeding standards from the example just discussed, we should consider one further feature of these tests. In this Canadian study it was possible to measure the variability of the response of pigs between replicates (between groups of pigs within a local geographic area) and between stations (between widely separated geographic areas). The first variations would be typical of differences between batches of pigs on the same farm or even between different farms in a small area keeping the same breed of pigs and following similar feeding practice. The variability between pigs of different stations will measure overall differences between different types or strains of pigs within one breed, differences due largely to hereditary background. The results of such analysis are shown in Table 9-6.

In this table the maximum and minimum values above and below the means (i.e., the fiducial limits) are twice the standard deviations for the

TABLE 9-6 *Variability in Response of Pigs from Different Geographic Areas to Identical Rations*

Ration group and data measured	Means from Canadian test pigs (lb)	Fiducial limits at ±2 × S.D. for total variability		Per cent of variability traceable to*		Expected mean value (lb) from U.S. feeding standards
		Mini-mum (lb)	Maxi-mum (lb)	Repli-cate	Sta-tion	
Growing pigs						
Feed	4.7	3.1	6.3	5	85	5.3
Gain	1.3	0.8	1.7	10	88	1.6
Feed per 100 lb gain	383	307	459	6	67	332
TDN per 100 lb gain	274	216	332	30	57	250
Pigs on normal finisher						
Feed	7.4	5.5	9.3	5	73	6.8
Gain	1.6	1.3	2.0	4	75	1.8
Feed per 100 lb gain	455	397	513	1	58	378
TDN per 100 lb gain	318	283	353	2	59	284
Pigs on diluted finisher						
Feed	7.1	6.1	8.2	9	60	6.8
Gain	1.5	1.2	1.8	4	80	1.8
Feed per 100 lb gain	485	413	557	1	44	378
TDN per 100 lb gain	324	279	371	1	43	284

* Balance of variability due to "interaction" between replicate and station.

variance for all causes and represent the limits for 95 per cent of the population. In other words, while only one lot in 20 would normally exceed these limits either way, one could expect individual lots that statistically belonged to this population to range in response anywhere between these values.

We should note two things here. First, in every case the figure given in the U.S. standard lies within the fiducial limits of this study. Second, for voluntary feed intake and live-weight gains, from 57 to 88 per cent of the differences between the separate feeding lots were traceable to stations, while within a geographic area the performance of the pigs was relatively consistent. Here we find much food for thought about the structure of feeding standards and their application as guides to ration formulation.

Margins of Safety in Feeding Standards

The problem about the structure of feeding standards themselves is: can data on the requirement of separate nutrients taken from independent tests justifiably be averaged without taking precautions against a bias of the data by different types, strains, and breeds of animals that are otherwise comparable?

It is well-known that different experimental stations develop facilities for different types of studies. Thus, one group of workers will be intensively studying some vitamin B problem, while another devotes its time to amino-acid requirements. The probability is that the inherent differences in animals between such stations will be such that the requirements for those at the one station would not closely fit those of another station. It also follows statistically that if the values stated as requirements are not unbiased averages, then margins defined by standard deviations above and below the means will not much improve the situation.

This undesirable feature of feeding standards which are prepared by assembling scattered available data for the several nutrients individually is, to a large degree, unavoidable. In this respect standards prepared entirely from observations of a single herd may perhaps show more consistency for the *relative needs* of the separate nutrients, especially in relation to the energy intake. An attractive hypothesis is that more consistent standards would result if the pertinent nutrient data from appropriate tests were expressed in terms of some measure of energy rather than in absolute values. The fact that energy intake is so closely correlated with size and performance of animals often justifies its use as a base on which other nutrient needs may be expressed, even if no proven biological cause and effect is known.

Description of Feeding Groups

A problem with the use of feeding standards, clearly shown by the results of the Canadian studies, has to do with the description of the animals or the feeding groups. In many standards, at least by implication, we get the impression that not only is gain the most important criterion of optimum nutritional intake but that maximum gain is the objective. This is obviously not justifiable. Some of this misconception could be eliminated by more complete description of the animals and of the criteria for optimum conditions. If gains are to be indicated they should perhaps show upper as well as lower acceptable limits. Perhaps we need to include other criteria not now used to describe adequately the kind of animals for which the nutrient list is intended. Thus, a multiple standard of height-weight-age might be more fully descriptive of the size of growing cattle than weight alone. With pigs, subclassification by breed or type might help, in that long-bodied pigs show differences from short-bodied on slaughter, and should not necessarily be considered as of equivalent physiological age and development at equal live weights. Certainly there should be subgroups for market pigs beyond weights of 125 lb (55 kg) to describe separately hogs intended for fresh pork and for bacon.

Concerning the discrepancies between the concentration of the several nutrients in the rations fed and that calculated from a feeding standard, we must draw conclusions cautiously. Certainly, there is no evidence that any detrimental effect is caused by surpluses or imbalances of any of the B-complex vitamins. In practical feeding, however, little can be done about excesses of many nutrients because of the wide differences between basal feeds in concentrations of these nutrients and the economic need to frequently substitute feeds in formulating feeding mixtures.

The conclusion we can draw from these considerations of feeding standards may be rather generally stated as follows.

1. As distinct from tables of recommended allowances, which usually are admittedly liberal in their quantities, feeding standards that claim to represent the actual needs of animals give values for the various nutrients which are high enough that there is seldom any justification for intentionally adding margins of safety to avoid nutritional deficiency.

2. The variation between animals in their response to identical assortments of nutrients makes it questionable whether standards should be used to measure the adequacy of rations fed to individuals or even to small groups; rather, the data of feeding standards should be used as a guide for the relative nutrient makeup of meal mixtures for animals of specified feeding categories. Daily allowances of such mixtures will be adjusted to the capacity of the individual animal or group or to the desired objective.

3. Surpluses of nontoxic minerals and of vitamins are probably of little practical importance in a nutritional sense.

4. Appropriate amounts of (and perhaps balance between) protein and energy components must be maintained in feeding practice since these may influence promptly and markedly the animal's performance.

Feeding Standards as Indispensable Guides

A word of caution may be in order at this point. The discussions above are not to be taken as condemning feeding standards. Advances in feeding practice over the past years have been greatly facilitated by the development of feeding standards and their wider use. It should be obvious that they are still imperfect, however, and hence if used blindly may lead to faulty nutrition. Progress in scientific feeding of livestock parallels the refinement and expansion of feeding standards, and although many practical livestock feeders never use such standards directly, the extension of the knowledge of the nutrient needs of stock by schools, clubs, and meetings has resulted in the use of more efficient rations generally.

We should not think of feeding standards as final statements of animal requirements. Modern ones, such as those issued by the various subcommittees of the NRC Committee on Animal Nutrition, are under continuous review and revision. Usually the revisions are downwards for specific nutrients, because where the requirement for a nutrient is first published by an investigator it is usually liberal. As additional studies give more certainty to the figure, the average value gradually finds its proper level, which is often appreciably below the first estimate. One must not therefore be surprised if the ration as made up in accordance with the currently used standard turns out to differ from a later revision of that standard. Nutritionally any excesses which are involved are fortunately not likely to be serious.

Modern Feeding Standards

One of the advances in standards is the realization that they must be revised every few years if they are to be considered up-to-date in describing nutritionally adequate rations. For example, the NRC swine standard was first published in 1944, and was revised in 1950, 1953, 1959, 1964, and 1967. The beef cattle standard of 1945 was revised in 1950, 1958, and 1963.

The life of a particular edition of any one of the ten standards now available in the NRC series depends on how rapidly new and relevant data become available from the research that is continuously proceeding in the numerous laboratories in the world. The editions are reprinted in numbers estimated to be sufficient for only a year, and the money from their sale is returned to a revolving fund, so that revisions do not present a serious financial problem. The continuous review of the published research data, and the preparation from it of the standard—or its revisions—for any particular species of animal, is done by a subcommittee of nutritionists who are appointed by the NRC Committee on Animal Nutrition and serve without pay. This service, first undertaken in 1944, makes available at a low cost to students, researchers, and industry the best available information on the nutrient and energy needs of animals, and represents one of the major continuing projects of the NRC Committee on Animal Nutrition.

One of the far-reaching modifications incorporated in the 1966 and later NRC feeding standards was the change to the metric system of weights in all editions of all standards prepared after January 1, 1965. This was in line with the action of all scientific societies in North America in requiring that, in papers submitted to their journals, weights be in the metric system and temperatures in the centigrade scale. The changes in the feeding standards will arise not so much from the use of grams (or kilograms) *per se,* but from the change for those nutrients needed in small quantities. The custom has been to use metric units for the "micro" ingredient and avoirdupois for the carrier, for example, "Thiamine: 0.5 mg/lb feed; 50 mg/100 lb feed." These quantities can be equally well expressed entirely in the metric system as:

50 mg/454,000 mg feed;

50 mg/454 g feed;

50 mg/0.454 kg feed;

5 mg/0.0454 kg feed;

5 mg/45.4 g feed;

and so on. Mixing weight systems, as mg per lb, means we must weigh some ingredients in metric units and others in avoirdupois. With research reports now using the metric system only, it will save confusion if requirements are stated exclusively in the metric system. There are decided advantages to a 100-unit scale of weight, especially where ration formulation is done by machine calculation. The only disadvantage is the temporary one of having to remember conversion factors and formulas.

The *Feeding Standard for Dairy Cattle* (1966) is the first to be published in which the metric system has been used exclusively. In this edition the NRC nomenclature for feeds has also been used for the first time. Since 1960 revisions of the standards for dairy and beef cattle, for sheep, and for swine have been published. For the basic tables of the daily nutrient requirements (in slightly abbreviated form), which illustrate the feeding categories of each class of stock and give their nutrient needs, see pp. 419–426.

A Broader Use of Feeding Standards

Entirely apart from the details of the requirements for specific species, there is another application of feeding standards in general that should be noted. To understand this, we must go back to the facts of the physiology and chemistry of animal functions. First we recognize differences in the anatomy of the digestive systems of different animal species. These are of importance *physically* largely in the accommodation of foods of differing bulk. Foods are bulky in proportion to the cellulose (or, for practical purposes, crude fiber) they contain. The dry matter of adequate rations for simple-stomached species usually contains 6 per cent or less of cellulose; that for ruminants may run as high as 35 per cent.

That the herbivora appear to utilize cellulose carbohydrates as a source of energy is an illusion: no mammalian or avian species synthesizes any cellulase, and hence cannot digest to an absorbable end-product the ingested cellulose. The explanation of the apparent utilization of cellulose by cattle, sheep, and other herbivora is the presence of a microflora they harbor in the rumen and/or caecum. These organisms synthesize cellulase enzymes that split the cellulose of their hosts' ration, producing a mixture of acetic, propionic, and butyric acids. These short-chain acids are absorbed into the blood stream of the host directly from the site of their formation. Thus these digestion products from cellulose do not pass through the areas of the intestinal tract where the other ration digestion products (sugars, nonvolatile fatty acids, amino acids, minerals, vitamins, etc.) are normally absorbed.

This special arrangement does not change the intermediary metabolism of these short-chain fatty acids, however. They merely get out of the digestive system and into the blood through a different "door." Not all of the end-products of normal digestion are subsequently absorbed at the same site in the intestine, either; rather, they are absorbed at points along the tract according to the pH at which their specific absorption is favored. Overall, digestion as the process of unpacking, sorting out, and preparing the diet components for absorption is thus the same for all homiotherms.

With a few known exceptions, the specific nutrients ultimately required by different homiotherms, and the intermediary metabolism of them, are also the same; and the metabolism itself appears to function with about the same degree of efficiency between species and between animals within species. Thus it is no coincidence that the adult basal metabolism of the some 30 species of homiotherms studied, including carnivora, herbivora, and omnivora, has been found to be a constant, $70(W_{kg}^{.75})$ kcal. Nor is it merely coincidental that the digestible protein required for each 1,000 kcal of digestible energy in the ration is constant at about 19 grams for the *maintenance* of adult cattle, sheep, horses, rats, pigs, and man; or that the normal growth curves of homiotherms beyond puberty coincide when plotted at equivalent physiological ages.

It is sound biology to consider all warm-blooded animals, including man, to be the same kind of metabolic machine. If so, it follows that *nutrient and energy needs must be the same* for individuals between and within species if they are of comparable physiological age, size, activity, and production. In other words, it should be possible to write one master feeding standard applicable to homiotherms generally.

Recently Crampton* proposed that energy might be a useful "least common denominator" for describing the nutrient needs in feeding standards. Expressing the requirements of the several nutrients as amounts per 1,000 kcal of energy would emphasize the balance of nutrients to each other and to the energy of the ration. Establishing the optimum nutrient balance in diets was the apparent objective of well over half the research in applied nutrition in North America during the early 1960's, yet relatively little of it dealt with nutrient–energy balance, in spite of the fact that the last NRC poultry standard predicated all the figures in its tables of nutrient requirements on diets yielding stated concentrations of metabolizable energy per pound of air-dry ration.

All feeding standards for the daily needs of animals are expressed as absolute quantities required per animal per day, or as percentages of a given quantity of food or food mixture to be offered to the animal. Thus we compound rations for specified animals to contain, by weight, 16 per cent protein, or 5 per cent fat, or 0.6 mg thiamin per pound of feed, or 75 per cent TDN per pound of ration. We think in terms of the weight of the ration mixture or of the individual feed ingredients, although we also specify the quantity of our final ration mixture that we expect the animal to eat, and that it must eat to get the necessary amount of each nutrient the ration contains. Rations that are not eaten in the expected quantities reduce the intake of each component nutrient. In general, animals eat to meet their energy needs. If to increase the energy concentration in the ration we add fat, we may cause a reduced feed intake, and so a reduced nutrient intake, unless the nutrients are in ratio to energy rather than to weight of ration.

Protein and energy requirements are closely interdependent. Expressing the protein needs as a per cent of the ration necessarily ties the protein level to feed intake rather than energy. The amount ingested of those vitamins and minerals that are needed to metabolize energy also depends on energy intake rather than on amounts of ration eaten. There is also evidence that differences in performance of animals fed rations of similar nutrient levels but of different energy concentrations can usually be removed by equalizing the energy concentrations of the two mixtures. Thus there is sound biological thinking behind the thesis that energy, rather than feed intake, should be the dietary component to which the nutrient needs are adjusted.

Given this general philosophy concerning the place of energy in total nutrition, it is interesting to examine some modern feeding standards and note how remarkably consistent certain species are in their protein needs relative

* E. W. Crampton, "Nutrient to Calorie Ratios in Applied Nutrition," *J. Nutrition,* LXXXII (1964), 353.

to energy, regardless of differences in size or production. (There are insufficient data as yet to compare nutrients other than protein between species.)

In all of the data, energy will be expressed as kilocalories of metabolizable energy (ME). For farm animals the ME will be computed on the assumption that one pound of TDN equals 1,640 kcal of metabolizable energy.* The nutrients will be expressed in appropriate weights per 1,000 kcal ME. We will presume that all the nutrient and energy requirements shown in the standards are acceptably accurate, but we will keep in mind that discrepancies may be evidence that the standards themselves are incorrect.

NRC Swine Standard. We may begin with the NRC 1964 swine standard, which illustrates very nicely the trends in relations of digestible protein, calcium, phosphorus, thiamine, and riboflavin to ME for market pigs and for breeding stock. The standard describes five feeding categories of market pigs and five of breeding stock. Both bred and nursing sows of 300 lb and of 500 lb were described, but data for the weight groups were combined when it was found the nutrient requirements were identical. From Table 9-7, it

TABLE 9-7 *Nutrients per 1,000 kcal ME for Swine*

Class	Nutrients per 1,000 kcal ME				
	Dig protein (g)	Calcium (g)	Phosphorus (g)	Thiamine (mg)	Riboflavin (mg)
Weight classes in lb (kg)					
10–25 (5–10)	56	2.8	2.1	0.5	1.1
25–50 (10–25)	46	2.3	1.7	0.4	1.4
50–75 (25–35)	44	2.4	1.8	0.4	1.0
75–125 (35–55)	39	1.8	1.3	0.4	0.9
125–220 (55–100)	35	1.8	1.3	0.4	0.9
Bred sows	26.6	2.2	1.5	0.4	1.2
Nursing sows	26.6	2.4	1.5	0.4	1.2
Boars	26.6	2.3	1.6	0.5	1.2

seems probable that the 25–50 and 50–75 lb groups can use rations of the same nutrient-energy balance, as can the 75–125 and 125–155 lb groups. The riboflavin requirement for 25–50 lb pigs should be reexamined. All breeding stock evidently can use one ration formula.

* TDN = 2,000 kcal DE per lb = 4,400 kcal DE per kg = 1,640 kcal ME per lb = 3,608 keal ME per kg. For more accurate conversion factors see Appendix Table A-1-2 and p. 207.

Thus, if pigs of these groups are fed amounts that meet their energy needs, rations compounded to provide the nutrient–energy balance shown in the standard will automatically supply the needed amounts of the several nutrients. Three mixtures, each having a specific nutrient–energy balance, are needed to carry growing pigs from weaning at 10 lb to market at 225 lb. One ration combination will satisfy the requirements of all breeding stock regardless of weight over 300 lb. The *daily energy requirement* can be expressed in a graph, from which needs of pigs of any weight can be read directly (see Fig. 9-1).

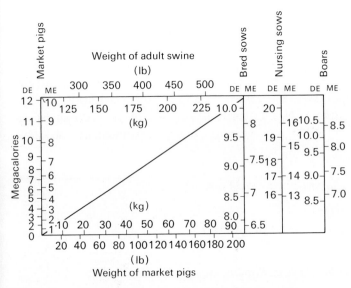

Figure 9-1. *Metabolizable energy required daily by swine.*

Adult Farm Animals. In Table 9-8 we present the maintenance protein requirements for adult ruminants and swine according to live weights of adult females. It is significant that the digestible protein per 1,000 calories ME is essentially independent of varying adult weights and species, averaging about 25 g. This suggests that the metabolic machine is the same for all farm animals. The figure for adult swine maintenance suffers from the fact that it is of necessity computed. Normally adult sows are either pregnant, lactating, or being fattened for market, and there are no satisfactory experimental data for their true maintenance protein requirements. Working figures may be computed according to the equation:

$$\text{dig protein (g)} = \frac{140(W_{\text{kg}}^{.75}) \times 6.25(a + b)}{1,000},$$

TABLE 9-8 *Digestible Protein Requirements of Adult Ruminants and Swine*

Adult weights						Grams of dig protein required per 1,000 kcal ME			
Cows		Ewes		Sows		Dairy cattle	Beef cattle	Sheep	Swine
(lb)	(kg)	(lb)	(kg)	(lb)	(kg)				
800	365	—	—	300	135	23.1	25.1	—	26.3
1,000	455	100	46	500	230	23.6	24.6	23.2	26.3
1,200	550	120	55	—	—	24.3	24.6	24.0	—
1,400	635	140	64	—	—	24.5	—	24.4	—
1,600	730	160	73	—	—	24.4	—	23.3	—

where 140 is the milligrams of endogenous nitrogen excreted, and a and b are factors to adjust for the biological values of protein and for metabolic fecal nitrogen, respectively. The former (TBV) is usually assumed to be 50 per cent, and the latter (FN_m) to equal 40 per cent of the endogenous excretion. Thus for a 500-lb sow we might compute maintenance digestible protein to be:

$$\frac{140 \times 58 \times 6.25 \times 2 \times 1.40}{1,000} = 14.2 \text{ g,}$$

and kcal ME as $93(58) = 539.4$ kcal, whence the digestible protein per 1,000 kcal $= (14.2 \times 1,000)/539.4 = 26.3$ g.

TABLE 9-9 *Digestible Protein Requirements for Growing Stock*

Per cent of adult weight attained	Grams dig protein required per 1,000 kcal ME			
	Dairy cattle	Beef cattle	Sheep	Swine
5	55	—	—	56
10	55	—	—	46
15	46	—	—	44
20	42	—	—	—
30	40	—	—	39
40	34	37	28	—
50	31	35	24	35
60	28	31	23	—
70	26	—	—	—
80	25	25	23	—
90	25	25	—	—

Growing Farm Animals. Table 9-9 suggests that when the growing animals of these mammalian species are compared in terms of the per cent attained of their adult weight, the protein–energy ratios of these rations are probably the same, and fit into a common pattern of declining protein needs relative to energy. With the dairy stock and swine, well-defined periods of development seem to be evident; during these periods the stock are normally fed (1) milk-replacer rations, (2) starter rations, (3) growing rations, and (4) adult maintenance rations (see Table 9-10). Data for growing beef stock are too meager to be of much value in such comparisons. It is interesting that the swine requirements for growth agree quite well with those for dairy cattle. The data suggest that rations should meet the needs of the weight categories of the four species for protein, as shown in Table 9-10.

Dairy Cattle. For dairy cows, the so-called adult maintenance nutrient and energy requirements are partly academic, because few adult dairy cows are fed for long at maintenance. A more realistic grouping for adult females would be: (1) pregnant, (2) lactating, (3) lactating and pregnant.

The protein–calorie ratios for dairy cows have been examined for the production categories (2) and (3) above. The 1966 NRC dairy cattle standard gives data for nine weight groups of cows, each producing milk up to 80 lb, at each of eight fat percentages. The digestible protein requirements per 1,000 kcal of metabolizable energy for 108 out of 450 possible combinations were computed. The variables used were: (1) maintenance plus lactation, or maintenance plus lactation plus pregnancy; (2) weights of 800, 1,200, and 1,600 lb; (3) milk of 3, 4, and 5 per cent fat; and (4) milk produced daily in amounts of 20, 30, 40, 50, 65, and 80 lb. A simple analysis of variance in this population of 108 combinations produced a mean and S.D. of 32.1 $\pm$ 1.45 g of protein per 1,000 kcal ME. The least significant difference for p = .05 thus becomes $\pm$2.9 g per 1,000 kcal ME.

This least significant difference can be translated into terms more often used in dairy-cattle feeding. For example, many practical dairy-cow rations consist of two parts of roughage to one part of grain. For heavier grain feeding, we might have roughage to grain ratios of 1:1, or 1:2, or 1:5, and so on, in the total ration. If we assume that average air-dry roughage will contain 50 per cent TDN and a typical meal mixture 75 per cent TDN, we can calculate TDN values for the roughage–meal rations, and from that, compute the kcal ME per pound of ration (assuming 1,640 kcal ME per pound TDN). In Table 9-11 we note that $\pm$3.0 g of digestible protein per 1,000 kcal ME is equivalent to a variation of less than one percentage point of protein in the ration.

TABLE 9-10 *Feed Groups for Farm Mammals*

Type of ration	Inclusive weights									
	Dairy cattle		Beef cattle		Sheep		Swine	
	(lb)	(kg)	(lb)	(kg)	(lb)	(kg)	(lb)	(kg)
Milk replacers	150	68	—	—	—	—	20	9
Starters	150–300	68–136	—	—	—	—	20–60	9–28
Growers	300–700	136–320	400–600	180–275	40–60	18–28	60–120	28–55
Adult	700–1,000	320–450	600–1,000	275–450	60–100	28–45	120–200	55–91

TABLE 9-11 *Computation of Least Significant Difference for DP per 1,000 kcals ME in Terms of Per Cent of DP in Ration*

Roughage–meal ratio of ration	Per cent of TDN in ration	Kcal ME per lb ration	Grams of ration per 1,000 ME kcal	Per cent of DP in ration equivalent to ±3 g DP per 1,000 kcal[a]
2:1	58	951	477	±0.63%
1:1	62	1,017	447	±0.67%
1:2	66	1,082	420	±0.71%
1:5	71	1,164	390	±0.72%

[a] In general, a change of 1 g DP 1,000 kcal will equal a change in the protein percentage of the ration of 0.23 percentage points.

Specifically it appears that deviations in the requirements from their mean of 32 g of digestible protein per 1,000 kcal ME, because of (1) size of cow, (2) per cent fat, or (3) pounds of milk produced by lactating (or lactating and pregnant) cows, do not exceed 0.7 percentage points in the protein required when it is expressed as a percentage of the air-dry weight of the ration. In other words, rations compounded to supply 32 g of digestible protein per 1,000 kcal ME will meet the needs of 97.5 per cent of cows in milk, if consumed in amounts to meet their metabolizable energy needs.

There are insufficient figures for the calcium and phosphorus requirements to make it worthwhile trying to compare them on the basis of caloric needs for the three ruminant species. It is interesting, however, that Missouri workers have recently stated that the available phosphorus could be conveniently expressed in relation to the digestible energy requirement, as approximately 0.63 mg phosphorus per kcal of digestible energy.

The Energy Requirement. Conventional feeding standards tabulate with the nutrients how much of them the animals of each feeding group need for energy. If the nutrients are to be expressed per 1,000 kcal ME, it becomes necessary to present separately the data from which the total days' energy needs can be arrived at. Since these needs vary with the weight of the animal and with production performance, they lend themselves to graphic presentation. A single multiple-scale graph can describe the energy requirements of all the categories of dairy cattle, for example, that require different rations.

Figure 9-2 is one made up to describe the megacalories of metabolizable energy required by dairy cattle per day. This six-scale chart gives the ME required per pound of milk produced according to its per cent butterfat and the total amount of milk produced daily. The ME for lactation is the pounds of

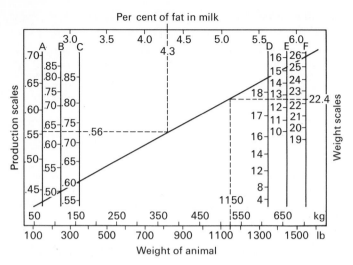

Figure 9-2. *Megacalories of metabolizable energy required daily by dairy cattle. (A) Production up to 50 pounds. (B) Production from 50 to 80 pounds. (C) Production to 80 pounds or more. (D) Growing stock. (E) Adult cow maintenance. (F) Maintenance in late pregnancy.*

milk times the ME per pound of milk. For example, 40 lb of 4.3 per cent milk requires 40 × .56 Mcal = 22.4 Mcal ME daily. To maintain an adult cow of 1,150 lb weight requires 12.7 Mcal daily; this increases to 22.4 Mcal during late pregnancy. Her total ME requirement therefore is 22.4 + 12.7 = 35.1 Mcal, or, if she is over six months pregnant, 22.4 + 22.4 = 44.8 Mcal daily.

The complete standard for a given species would tabulate each of the nutrients required per 1,000 kcal of energy for each group to be fed a ration of identical nutrient makeup (as starter, grower, or finisher rations for swine), and would also have a chart for finding the daily energy requirement for the individual of that group. The amounts of the ration to be fed depend on the energy concentration of the ration mixture, as is true for conventional standards. In practice, the actual feed allowances for individuals within a group are ordinarily judged by rules of thumb derived from the more precise data of the standard.

Use of Feeding Standards

Our purpose in examining feeding standards in terms of the ratios of nutrients to energy was to try to show how the feeding standards might be made more broadly useful. Their most obvious use is in describing exactly how much specified animals need of specific nutrients, in terms of the other nutrients and of the energy of the ration. This information is required by the research worker, the feed manufacturer, and the agricultural extension agent. But feeding standards could be of much broader and more fundamental use in biology. Evidence is rapidly accumulating supporting the thesis that homiotherms are all one kind of metabolic machine. This leads to the premise that, when described in appropriate terms of metabolic size, individuals of equivalent activity, age, breed, sex, and production require quantitatively and qualitatively similar amounts of energy and nutrients. The forms of this ration may differ widely, as may the formulation, but the nutrient and energy specifications of the formulation should be essentially the same.

We do not mean to imply that there is no genetic variability between individuals in their optimum nutrient intake needs or, more correctly, in their performance on equal intakes. The biological variation between *comparable individuals* of a population in performance on equal nutrient intakes appears to be on the order of 16 per cent of their average. The problem is in recognizing or defining comparable individuals. We recognize this variability, as well as the similarities, when we group animals that are to receive identical rations.

SUGGESTED READINGS

The NRC pamphlets on the nutrient requirements of domestic animals: *Poultry* (1965); *Swine* (1967); *Dairy Cattle* (1966); *Beef Cattle* (1963); *Sheep* (1964); *Horses* (1966).

Kellner, O., *The Scientific Feeding of Animals* (N.Y.: Macmillan, 1913).

Kriss, Max., "A Comparison of Feeding Standards for Dairy Cows, with Special Reference to Energy Requirements," *J. Nutrition*, IV (1931), 141.

Morrison, F. B., *Feeds and Feeding* (Ithaca, N.Y.: Morrison, 23d ed., 1967).

SUMMARY OF SECTION II

This section has been an examination of feeding standards—those tables in which are recorded the quantities of certain nutrients that we believe should be provided in the rations of the animals of different feeding categories. It should be evident that even the latest and most modern standards are incomplete in the nutrient requirements. The missing values are gradually being filled in as our knowledge increases, but even the data that are available must be used with judgment. We must never forget that we are dealing with living animals in applied nutrition, and that their most obvious characteristic is their variability.

This variation applies to the structure and functioning of every tissue. Not only are two animals different, but the same animal differs functionally from one time to another. Thus there is no way of describing accurately nutrient requirements in fixed figures. We must use averages, which may not depict the needs of any one animal at any particular time, but work very well as guides for the formulation of meal mixtures suitable for groups of animals, since the variations of animals of a group tend to average out.

Fortunately the differences in the nutrient needs between animals, and also for one animal at different times, are correlated with differences in energy needs, so that by adjusting the daily allowances of total food according to appetite, or to size, or to production rate, the adjustment for the nutrient need is also made. Were it not for this relation between the need for energy and that for the nutrients, the problems of feeding practice for herds and flocks would make anything like scientific rationing practically prohibitive.

The key place occupied by energy in the "operating needs" of an animal makes it especially important that our estimate of the energy requirement be as accurate as our knowledge will permit. For this reason we dealt critically with the problem of energy need. Our present beliefs regarding this requirement or that of any nutrient will have to change as more facts are learned. Feeding standards are not final, and are not likely to be for many years to come. They are nevertheless already indispensable guides to the formulation of workable rations, even though we may expect gradually to learn how better to balance the nutrients to meet more exactly the nutrient needs of our livestock, and hence to improve their performance.

Those who expect to prepare successful livestock rations must, therefore, use feeding standards, and use them properly. They are one of the basic "tools of the trade"; and as with any other occupation, the competence of the workman can be judged by his knowledge of and skill in the use of his tools.

III THE NUTRITIONAL
CHARACTERISTICS
OF SOME COMMON
FEEDS

Many different feedstuffs are used at one place or another in the feeding of livestock. Morrison lists analyses for 173 dry roughages (not including grades of the same feed), 143 green roughages and roots, 46 silages, and 432 concentrates. Schneider records the analyses and digestibility of some 2,300 feedstuffs. In 1968 the Committee on Feed Composition of the NRC Committee on Animal Nutrition of the U.S. National Research Council published an encyclopedia of some 7,500 feeds with the available data on their composition. In this tabulation varieties and grades of feeds of common plant or "animal" origin are listed as separate feedstuffs, whereas in most tabulations all analyses of feeds of one common name are averaged under one heading, such as "Barley, western, all analyses" (see also Chapter 1).

With such reference data readily available there is little reason to attempt here any extensive discussion of only a few key feeds. The plan of this section is rather to discuss first the general nutritional properties of feeds by groups (i.e., basal feeds, protein supplements, and so on). We shall augment this discussion by considering in detail the characteristics of one or more key feeds in that group. Finally, we shall consider the unique properties of some other common feeds belonging to the group—properties that determine the extent to which they individually may replace the key feeds in rations, and the consequences of such substitutions.

With a sound understanding of the general nutritional nature of the feeds making up the several groups, together with enough description of a particular product to assign it correctly to its proper category in the classification and to disclose any pronounced peculiarity it may have, we can incorporate feedstuffs that are not well-known into rations with reasonable assurance that we will make the most of their potential feeding values, and avoid major undesirable consequences of improper use.

CHAPTER 10

A Classification of Feeds

Feeds composed of naturally occurring products, and of many of the by-products of the milling or other processing of such materials, are, figuratively speaking, packages containing many, and perhaps usually most, of the nutrients recognized as needed by animals. The amounts and proportions in which these operating necessities are present differ between feeds, and so, strictly speaking, no two feeds are nutritionally alike. But in the everyday practical formulation of rations, feeds of generally similar properties are considered as potential substitutes, and exchanges are made within mixtures in accordance with market price, availability, and to some extent feeder preference. Thus it becomes expedient to establish categories of feeds within which substitutions are justified by similar nutritional properties.

Forages, Roughages, and Concentrates

The term "roughage" is usually used to designate feeds that are high in fiber and low in net energy. Technically, however, feeds such as hay, straw, silage, and pasture are "forages," a term that is perhaps more descriptive of these types of feeds than is "roughage." Feeds such as almond hulls, peanut shells, and oat hulls could be technically described as roughage. However, in this book the terms "forage" and "roughage" are used interchangeably.

The first broad grouping of feeds is logically based on their physical bulkiness—a classification that separates products that are relatively low in concentration of energy yielding nutrients from those that are high in them. Such a groupings brings together under one heading forages and roughages, which are the feeds that are essentially whole plants, such as hay, straw, silage, and pasture. The rest of the feeds are grouped under the heading of concentrates. The physical differences between these two groups of feeds are such that we do not make substitutions (in the usual sense of the term) of one for the other. Although the complete diets of herbivorous animals generally include both forages and concentrate feeds, they sometimes comprise only forages. But concentrates alone do not usually provide adequate diets for such animals regardless of the nutrient assortment they contain.

Classification of Feeds Based Upon Use

When feeds are used to make up diets or feed mixtures they usually have a specific purpose depending upon their chemical makeup. In view of this, we have adopted the NRC classification. In this classification feeds are classified into eight main groups and seven subgroups:

Class
Code

(1) Dry forages and roughages

Forages or roughages are low in net energy per unit weight, usually because of their high fiber content though sometimes because their water content is high. According to the NRC nomenclature and also according to the Canada Feed Act, products that in the dry state contain more than 18 per cent crude fiber are classified as forages and roughages. Thus, in addition to forages, such products as oat hulls, alfalfa meal, dried beet pulp, brewers grains (some samples), and rice bran, which are sometimes listed among the concentrates are included in this group. Few roughages contain as much as 1,200 kcal DE per lb (60 per cent TDN), and the average is not far from 1,000 kcal per lb (50 per cent TDN).

Hay
  Legume
  Nonlegume
Straw
Fodder (aerial pt w ears w husks, or heads)
Stover (aerial pt wo ears wo husks, or heads)
Other products with more than 18 per cent fiber
  Hulls
  Shells

(2) Pasture, range plants, and forages fed green.

Included in this group are all feeds not cut and cured. For example: all feeds cut and fed green, or feeds cured on the stem, such as dormant range plants. We have used the term "fresh" as a process term for most of these feeds even though they may be dry and weathered when consumed.

(3) Silages
> Corn
> Legume
> Grass

(4) Energy or basal feeds

Products with less than 20 per cent protein and less than 18 per cent crude fiber. Throughout the text we have used the terms "basal feeds" and "energy feeds" interchangeably.

> Cereal grains
>> Low in cellulose
>> High in cellulose
> Mill by-products
>> Low in cellulose
>> High in cellulose
> Fruits
> Nuts
> Roots

Fruits, nuts, and roots are classed as energy feeds because in making up diets and feed mixtures they are included primarily to supply energy. However, many of the fruits and some roots are excellent sources of vitamins and minerals for humans.

(5) Protein supplements

Products which contain 20 per cent more protein.

> Animal
> Marine
> Avian
> Plant

(6) Mineral supplements
(7) Vitamin supplements
(8) Additives
> Antibiotics
> Coloring material
> Flavors
> Hormones
> Medicants

The code number included in parentheses is that which is given at the end of the NRC feed names. It is also the first digit of the reference number in Appendix Table A-3-1.

Variability of Feedstuffs in Chemical Makeup

As we have already mentioned, one of the purposes of a feed classification is to group together feeds of somewhat similar nutritional characteristics, thus defining products that are partial substitutes for each other. Thus feeds legitimately included in the energy feed category (4) may usually be interchanged, completely or in part, among themselves. Normally feeds of the energy category would not be exchanged with feeds of the protein-supplement category (5). However, even though feeds may be grouped together in categories of similar general nutritional properties, it does not follow that different samples of the same feed will have the same chemical makeup. For example, different samples of shelled corn may run from 8 to 15 per cent crude protein, wheat from 9 to 19 per cent, or oats from 10 to 18 per cent. This variability between samples is obviously a factor we must consider in feed substitution, assuming, of course, that the chemical makeup of the feed bears an important relationship to its feeding value. During the past few years realization of the need for considering variation, not only in feeds but also in animals and their response to feeds, has become more general. Actually very little mathematics is needed to understand the problem and to interpret the commonly used measures of variation, such as are applicable here.

Perhaps the easiest way to make clear the problem of variation as it applies to some chemical component of a feed will be to cite an example; one that is particularly applicable comes from Bulletin no. 461 of the Texas Agricultural Experiment Station, in which is given the tabulation of the distribution of 586 samples of cottonseed meal analyzed for crude fiber shown in Table 10-1.

If we were to look in a table of feed composition for cottonseed meal [Cotton, seed w some hulls, mech extd grnd, (5)], we might find the crude fiber given as 11.3 per cent for the average of all these values, but the crude fiber in different samples has run all the way from 6 to 15 per cent. Most of the samples have come closer to the average—106 out of the total, or about 18 per cent of them, are in the 11 to 11.5 per cent class. Both below and above this point the samples become progressively fewer in number.

If we were to plot the samples according to crude-fiber class, we would get a bell-shaped curve which would approximate in shape the normal frequency-distribution curve. A great many characteristics of biological material actually follow the so-called normal frequency distribution, and consequently, even though the observed figures do not fit exactly into this curve, the statistical constants which have been calculated to describe the curve are used in

TABLE 10-1 *Distribution of 586 Samples of Cottonseed Meal in Crude-Fiber Content*

Crude-fiber class (by per cent)	Number of samples in each class	Per cent of total samples in each fiber class	
6.0	2	0.34	
6.5	0	0	
7.0	2	0.34	
7.5	5	0.85	
8.0	8	1.37	25.5% below −1 S.D.
8.5	15	2.56	
9.0	25	4.27	
9.5	55	9.37	
10.0	73	12.46	
10.5	102	17.41	
11.0	106	18.09	
11.5	80	13.65	68% within ±1 S.D.
12.0	55	9.39	
12.5	22	3.75	
13.0	22	3.75	
14.0	4	0.68	6.5% above 1 S.D.
14.5	0	0	
15.0	2	0.34	
Total	586	100%	
Av.	11.29		
S.D.	±1.26		
C.V.	11%		

Source: Texas Ag. Exp. Sta. Bull. 461.

measuring the variability actually found in distributions such as that for crude fiber.

Of these measures the standard deviation (S.D.) is the most common. Without going into any consideration of how the S.D. is calculated, we should explain that it measures an interval, above and below the average, within which we may expect approximately two-thirds of the samples to fall, with one-sixth of them being above the upper limit and one-sixth of them below the lower limit of this interval. Obviously, therefore, if we pick at random any sample in the whole population we are considering, our chances of finding that it lies within the limits of ±1 S.D. are two out of three. This result is referred to as odds of two out of three, or is noted as P = .33, meaning that the chance of *not* finding the value within ±1 S.D. is 33 per cent.

Coming back to the figures in Table 10-1, we may find it interesting to see where the S.D. of the array of crude-fiber values actually falls. The calculated S.D. of the per cent of crude fiber is ±1.26 percentage points. Consequently, we should expect two-thirds of the values to fall between the midpoint of the 10 per cent class and the midpoint of the 12.5 per cent class (i.e., $11.29 - 1.26 = 10$, $11.29 + 1.26 = 12.5$). If we make such a division (last column of the table), we find that approximately 68 per cent of all the samples fall within the limits of ±1 S.D. from the mean. As it happens only 6.5 per cent of the samples have more crude fiber than the upper limit of the S.D., and 25.5 per cent have less. This degree of "goodness of fit" in practice, as compared with theory, is to be expected where small numbers are involved. One may think that 500 samples is not a small number, but it is small as compared with the infinite population on which the normal frequency curve is based. The important thing here, however, is that about two-thirds of the samples actually fall within ±1 S.D. of the mean. Thus, the probable crude-fiber content of cottonseed meal can be described very simply by saying that it averages 11.29 per cent and that two-thirds of all the values will likely fall within 11.29 ± 1.26 per cent.

It is sometimes useful in comparing degrees of variability to express the S.D. as a percentage of the mean. This percentage is called the *coefficient of variation,* and is directly comparable between sets of figures whose means are not the same. In the example cited above, the coefficient of variation of the crude fiber in cottonseed meal turns out to be $(1.26 \times 100 \div 11.29) =$ 11.15 per cent. This figure is useful in predicting the probable variation in crude fiber of cottonseed meals whose average is not 11.29 per cent, but is perhaps, because of some process change, only 8 per cent.

If, in addition to samples of the same feed, we examine a reasonably large number of samples of feeds of the same category, we may be able to establish a typical variability for the crude fiber of feeds in general. In fact, data from this Texas bulletin show that the coefficient of variation of crude fiber between different samples of a considerable number of different concentrate feeds is actually about 12 per cent. The coefficient will be somewhat larger for feeds that are very low in crude fiber, and somewhat smaller for feeds that are high in crude fiber. This difference is caused merely by an arithmetical effect, and by the fact that errors of chemical determination of crude fiber are likely to increase with samples that contain very small quantities of crude fiber because they are hard to manipulate under routine analytical conditions. For most practical purposes, an *average coefficient of variation* calculated from the proximate analysis of feeds is as useful as the coefficient of variation calculated specifically for each feed separately, no matter whether it is protein, fat, fiber, or nitrogen-free extract with which we are concerned.

We must make one further point concerning the coefficient of variation and the S.D. By definition ±1 S.D. includes two-thirds of the population under consideration; consequently, the chance that any one member picked at random will fall within these limits is two out of three. It can be shown mathematically that if the S.D. is multiplied by two (so that the limits set are now 2 S.D. above and below the mean) this interval will include 95 per cent of the population, and for any single member picked at random, the chances will be 19 out of 20 that the value will fall within these larger limits. The limits of ±2 S.D. are sometimes referred to as the fiducial limits and they define an interval within which we can expect to find all but 5 per cent of the individual values.

Unfortunately there are few data available on the normal variability of the proximate principles of feedstuffs. From what figures we can find we may expect that the variability in analysis between samples of the same feed will probably have coefficients of variation about as follows:

Protein	± 8%;
Ether extract	±15%;
Crude fiber	±12%;
Nitrogen-free extract	± 3%.

In the table of feed composition (Appendix Table A-3-1) the coefficients of variation are included for each nutrient whose average is based on five or more samples. Taking these coefficients of variation, we can calculate the probable fiducial limits of the values for the proximate fractions of protein, fat, fiber, and nitrogen-free extract in a random sample of a feed of the same name or class, according to its per cent composition. This has been done for the proximate principles in Table 10-2.

In this table the first column gives percentage figures ranging from 2 to 90. Some of these figures are in bold face type. For example, 4 per cent will be close to the average fat content of most concentrate feeds; 12 per cent is typical for the protein content of the basal feeds, 25 per cent for the protein content of the vegetable low-protein supplements, 40 per cent for the higher protein category, and 70 per cent for the nitrogen-free extract of most of the cereal grains and low-protein concentrates. In columns 2, 3, 4, and 5 are shown the limits of reliability within which we can be reasonably sure the protein, fat, fiber, or nitrogen-free extract values of any particular samples will fall according to the average figure for that feed given in tables of proximate composition. Thus, we can probably count on most samples of basal

TABLE 10-2 *"Normal" Variability in Proximate Analysis of Concentrate Feeds*

Per cent of proximate principle in feed	Probable fiducial limits ($P_{.05}$) for concentrate feeds in			
	Per cent of protein (C.V. = 8%)	Per cent of fat (C.V. = 15%)	Per cent of fiber (C.V. = 12%)	Per cent of nitrogen-free extract (C.V. = 3%)
1		0.7–1.3		
2				
4		2.8–5.2	3–5	
6				
8			6–10	
10				
12	10–14[a]	8–16	9–15	
14				
16				
18				
20				
25	21–29			
30				
35				
40	34–46			37–43
45				
50				
60				
70	59–81			65–74
80				
90				

[a] For example, $12 \times (\pm 8\% \times 2) = \pm 1.9$, or a range between 10.1 and 13.9

feeds to contain between 10 and 14 per cent protein. However, for protein supplements such as brewers' grains and gluten feeds, where the average protein content is likely to be about 25 per cent, we shall have to expect that different samples may range between 21 and 29 per cent.

The full significance of this problem of variability in composition may not be clear until we deal with the problem of feed formulation later in this book. Nevertheless, in the discussions that follow in the next two or three chapters concerning feedstuffs, you will note that the composition figures given represent *averages;* and since no data on the variability from these averages is ordinarily available, the figures in Table 10-2 may assist you in better understanding how nearly identical feeds known by the same name are likely to be.

Before leaving this question we should draw attention to a further interpretation of the variability figures that may be applicable to feeds within any one group. As we shall see shortly, a typical basal feed is defined as one that contains about 12 per cent crude protein. On the other hand, basal feeds by classification also include all feeds containing less than 20 per cent protein. Thus, because of the second definition, molasses, which contains only about 3 per cent protein, still falls in the basal-feed category.

Although it is quite common to examine the homogeneity of a population using the criterion that observations which fall outside ±3 S.D. do not belong in the category, this practice cannot legitimately be applied to feed classification. The feed classification is not a biological classification but one for convenience in feed manipulation. Consequently, we cannot legitimately use the S.D. as a final test of whether or not a given product belongs in a certain subclassification. Perhaps what we should say is that a typical basal feed will contain between 10 and 14 per cent protein and consider that samples of basal feeds which are outside this limit but still within the category are abnormal samples or special cases. Thus, wheat is a special case of a basal feed that falls outside the expected range of basal feeds. On the other hand, molasses is a low-protein concentrate, and by that definition is still a basal feed regardless of its exceedingly low protein content.

Finally, we must realize that replacing a feed in a mixture by one that differs in composition—such as fiber or protein content—does not alter the average composition by as much as the difference between the two feeds. For example, if we have five feeds mixed in equal proportions and analyzing individually 10, 11, 12, 13, and 14 per cent crude protein, and we replace the one analyzing 10 per cent protein by one containing 14 per cent protein, we change the average protein content of the mix only from 12 per cent to 13 per cent. Similarly, if we introduce a second feed of 10 per cent protein and delete one carrying 12 per cent protein, we have dropped the protein average of the mixture by about 0.5 per cent. Thus, it is only where feeds of markedly different composition are exchanged in *major proportions* that we change the average composition of the mixture sufficiently to affect the feeding value. We shall discuss this problem in application when we consider ration formulation and feed substitution.

PROBLEMS

1. Using Appendix Table A-3-1, regroup the nonroughage feeds into the appropriate categories of the feed classification. Calculate the average protein, fat, and

fiber content by categories. How many feeds fall outside the fiducial limits suggested in this chapter?
2. What are the minimum and maximum limits found between feedstuffs that are called by the same name?
3. Does the variation in composition of roughages of the same name exceed that of different samples of the same concentrate feed? Can you suggest a reason for your findings?

SUGGESTED READINGS

Fraps, G. S., *The Composition and Utilization of Texas Feeding Stuffs* (Texas Agr. Exp. Sta. Bull. 461, 1932).

Harris, Lorin E., Earle W. Crampton, Arlin D. Knight, and Alice Denney, "Collection and Summarization of Feed Composition Data. II. A Proposed Source Form for Collection of Feed Composition Data," *J. Animal Sci.* XXVI (1967), 97.

Knight, Arlin D., Lorin E. Harris, Earle W. Crampton, and Alice Denney, "Collection and Summarization of Feed Composition Data. III. Coding of a Source Form of Compiling Feed Composition Data," *J. Dairy Sci.* XLIX (1966), 1548.

CHAPTER 11

Basal Feeds

According to the notation in the outline classification, basal or energy feeds
are low-protein concentrates. The upper limit for protein is conveniently set
at 20 per cent, because this figure then includes wheat bran, which is otherwise difficult to classify. However, it is the entire seed of the cereals that is
the typical basal feed. If we take an average of the protein, fat, fiber, TDN,
Ca, and P figures for the six common grains, barley, corn, milo, oats, rye, and
wheat, we shall have a workable chemical description of a basal feed in terms
of those nutrients and proximate principles most useful in determining its
proper place in a livestock ration. Such data are shown in Table 11-1.

Chemical Characteristics

Protein. From this table it will be seen that a basal feed is likely to contain about 12 per cent crude protein, of which between 75 and 80 per cent
is digestible. (Throughout this book *digestible* refers to *apparent digestibility*
unless otherwise stated.) It is interesting to know that the average crude
protein of the basal feeds actually found in Canadian commercially prepared,
ready-to-feed rations averages just about 12 per cent, and for the rough calculation of the protein of a Canadian meal mixture it is customary to lump
together the basal feeds at a protein of 12 per cent. But for rations based
largely on corn or wheat, this shortcut is not justifiable.

In practice, one will not go far astray by assuming basal-feed protein to
be 75 per cent digestible. The quality of the protein of basal feeds is uniformly

TABLE 11-1 *Typical Composition of Cereal Grains*

Feed name	Crude protein			Ether extract (per cent)	Carbohydrate		Energy — Cattle			Energy — Swine			Minerals	
	Total (per cent)	Dig for swine (per cent)	Chemical score		Crude fiber (per cent)	N-free extract (per cent)	DE (kcal/kg)	ME (kcal/kg)	TDN (per cent)	DE (kcal/kg)	ME (kcal/kg)	TDN (per cent)	Calcium (per cent)	Phosphorus (per cent)
Barley, grain	11.6	8.2	20	1.9	5.0	68.2	3,257	2,671	74	3,080	2,876	70	0.08	0.42
Corn, grain	9.3	7.5	28	4.3	2.0	71.2	3,569	2,927	81	3,569	3,351	81	0.02	0.29
Oats, grain	11.8	9.9	46	4.5	11.0	58.5	2,982	2,446	68	2,860	2,668	65	0.10	0.35
Rye, grain	11.9	9.6	50	1.6	2.0	71.8	3,336	2,735	76	3,300	3,079	75	0.06	0.34
Sorghum, milo, grain	11.0	7.8	—	2.8	2.0	71.6	3,139	2,574	71	3,453	3,229	78	0.04	0.29
Wheat, grain	12.7	11.7	37	1.7	3.0	70.0	3,453	2,832	78	3,520	3,277	80	0.05	0.36
Average	11.4	9.1		2.8	4.2	6.8	3,289	2,698	75	3,297	3,080	75	0.06	0.34

low as measured by any scheme that rates biological value numerically. All feeds of this group show lysine as their first limiting amino acid, which is of importance in the choice of a protein supplement to be used in a balanced ration. It also explains why substitution between basal feeds is not likely to alter appreciably the protein quality of the mixture.

Fat. The cereal grains belonging to the basal feeds normally contain from 2 to 5 per cent ether extract, but a few by-product feedstuffs contain up to 13 per cent fat, as does rice feed, the mill-run by-product of the manufacture of polished rice. Oat groats contain 7 or 8 per cent fat, as does hominy feed. The fat of non-oily seeds is concentrated in the germ, and any processing that removes an appreciable proportion of the protein or carbohydrate, but not of the germ, will leave a by-product with higher fat content than the parent seed. A knowledge of the processing involved in the production of a by-product feed is often helpful in understanding the feeding properties of such a product. The official definition of feeds may partially define the processing of by-products, as will the NRC name.

We can see an example of the effect of milling on the fat (and protein) of the by-products in corn. By a series of cracking and sifting operations, a granular cornmeal (or corn grits, or table hominy) is produced for human use. The material that is too fine for the table cornmeal or table hominy, together with all of the bran and as much of the germ as separates out, is combined into hominy feed [Corn, grits by-prod, mil-run grnd, mn 5 fat, (4)]. Thus, the germ is concentrated in the hominy feed, with a consequent increase in its fat. The increase in germ protein is counteracted by the increase in the corn bran, so that the protein of hominy is slightly lower than that of corn, while the crude-fiber content is raised.

The production of starch, on the other hand, involves a wet-milling process. The corn grain, after being softened with warm water and slightly acidified, is partly macerated and then allowed to soak in water in large tanks. The germ, because of its oil content, floats to the top, where it is removed, defatted, and dried into corn germ meal. The residue from the germ separation is reground and sifted to remove the hulls, bran, tip cap, and other fibrous material. The gluten and starch are removed from the remaining mass in suspension and later separated centrifugally. The coarse residue made up of hulls, bran, etc., is combined with the defatted germ, and to this base is added enough gluten to give two products, one with 23 per cent protein, the other with 40 per cent. The former is gluten feed [Corn, gluten w bran, wet-mil dehy grnd, (5)] the latter gluten meal [Corn, gluten, wet-mil dehy grnd, (5)]. Thus, removal of most of the corn oil results in two by-products of lower fat content than the original corn, but of higher protein, because of the separation and re-moval of the starch of the corn grain. The wet-milling steps are illustrated in Fig. 11-1.

Of course, not all plant seeds are processed in this manner, but the principle of the nature of the by-products is the same. Remove carbohydrate, or fat, or protein, and the residue will be higher in the remaining fractions than in the parent material.

Ash. Basal feeds never show concentrations of calcium high enough to meet the needs of animals, and consequently are not depended on as Ca

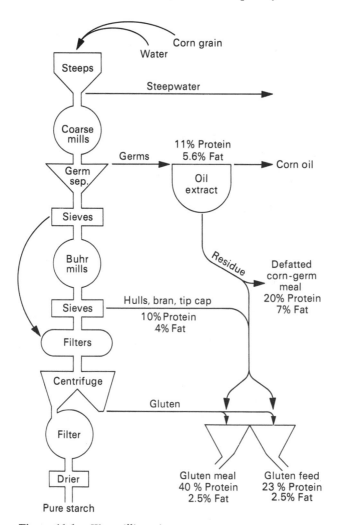

Figure 11-1. *Wet milling of corn.*

sources. In practice, they are often neglected in making calculations for calcium supplementation. The content of phosphorus, on the other hand, is enough that some classes of pigs, and sometimes cattle and sheep also, need no special supplements, but this will depend on the kind and amount of roughage also fed to the herbivorous species. The usual rule is that pig rations need calcium, while cattle rations need phosphorus fortification.

Carbohydrates. As we should expect, the fraction of basal feeds that is of primary nutritional importance is the carbohydrate. About two-thirds of the weight of the seed is likely to be starch, which will usually be about 95 per cent digested. Not only is this high concentration of easily digested carbohydrate the distinguishing feature of basal feeds, but variation in this characteristic determines the consequences of substituting among feeds of this category.

Crude Fiber. The average crude fiber of the basal feeds is about 6 per cent but individual feeds vary considerably. In this book the upper limit for concentrates is taken as 18 per cent, partly because in Canada by legal definition feeds with over 18 per cent fiber must be registered as roughages. In particular, the coarse grains (barley and oats) may show wide deviations in fiber from sample to sample, ordinarily because of either an increase in hull or a decrease in the starch filling of the groat. Differences in fiber affect markedly their available energy value and hence their relative feeding value. The most important consequence of substitution between basal feeds is usually traceable to differences in the crude fiber of the products. Fibers of different origin are often quite different nutritionally. The tabulation in Table 11-2 is instructive.

If the variation shown in Table 11-2 is all traceable to the uncertainties of the chemical procedure for this feed fraction, then the determination is of little use. However, taking the published data we find much food for thought. For example, the crude fiber of wheat, shorts, and bran is presumably the same material, though its digestibility ranges from 53 to 90 per cent. Similarly, that of oats, oat clippings, rolled oats, and oat hulls ranges from 38 to 80 per cent. It seems probable that processing which involves soaking improves the digestibility of the fiber. The digestibility of the fiber of corn grain is 57 per cent, but that of corn bran, corn gluten feed, corn oil meal, and corn distillers' grains ranges from 72 to 92 per cent, with an average of 80 per cent. Solvent extraction also appears to have improved the digestibility of the fiber of flaxseed and of soybeans.

These data are from ruminant digestion trials and may be too high for

TABLE 11-2 *Digestibility of Crude Fiber*

Crude fiber from		Coefficient of digestibility (%)
Common name	NRC name[a]	
Wheat	[Wheat, grain, (4)]	33
Wheat bran	[Wheat, bran, dry-mil, (4)]	36
Wheat shorts	[Wheat, flour by-prod, 7 fbr, (4)]	60
Oats	[Oats, grain, (4)]	32
Rolled oats	[Oats, cereal by-prod, grnd, mx 2 fbr, (4)]	80
Oat clippings	[Oats, grain, clippings, (1)]	58
Oat hulls	[Oats, hulls, (1)]	40
Barley	[Barley, grain, (4)]	45
Barley feed	[Barley, pearl by-prod, grnd, (4)]	18
Brewers' grains	[Grains, brewers' grains, dehy, (5)]	49
Malt sprouts	[Barley, malt sprouts, w hulls, dehy, mx 24 prot, (5)]	83
Flaxseed	[Flax, seed, (5)]	84
Linseed oilmeal o.p.	[Flax, seed, mech-ext grnd, (5)]	50
Linseed oilmeal solvent	[Flax, seed, solv-ext grnd, (5)]	43
Soybeans	[Soybean, seed, (5)]	37
Soybean oilmeal	[Soybean, seed, solv-ext toasted grnd, (5)]	68
Corn	[Corn, grain, (4)]	30
Corn bran	[Corn, bran, (4)]	63
Corn gluten feed	[Corn, gluten w bran, wet-mil, dehy, (5)]	78
Corn oilmeal	[Corn, germ, dry-mil solv-ext, dehy, (5)]	82
Corn distillers' grains	[Corn, distil grains, dehy, (5)]	64

[a] NRC names have been included to illustrate how much more information about a feed they give.

omnivora, but regardless of species of animal, any part of the *apparent* utilization of the fiber of these feeds that is not due to chemical error must be due to attack by digestive system microflora. One might argue that the unprocessed fiber of seeds, which in its natural state is an outer protective coating of the seed, is relatively resistant to bacterial attack. This resistance may be due to lignification, or to waxy, horny, or other weather-resistant coatings. In the milling or wet processing of such seeds, some of these coatings may be partially disintegrated or dissolved, thus exposing the cellulose to easy attack by microorganisms of the digestive system. "Digested" crude fiber, of course, yields as much energy to the animal as digested starch.

Whatever the true explanation, it is easily demonstrated in feeding trials that fibers of different origin may affect the feeding value of a ration quite differently. Add 25 per cent wheat bran to a fattening hog ration, and the rate of gain is depressed. Add malt sprouts to get an equal fiber concentration, and feed intake is increased, as is the gain of the pig. Add barley hulls, and the acceptability of the ration decreases.

Thus, although we may not be able to predict the reaction of the animals to a change in the source of crude fiber in a ration, we can usually trace the important changes in the feeding value of a ration that are caused by basal-feed substitution, directly or indirectly, to the crude fiber. It is also generally true that amount of fiber and of available energy of basal feeds or feed mixtures are negatively correlated. Thus, raising the percentage of fiber means greater bulkiness and lower available energy, which in turn demand larger amounts of feed. In other words, high-fiber feeds are relatively less efficient sources of productive energy.

Nonchemical Characteristics of Basal Feeds

Bulk. In a general consideration of characteristics of basal feeds as a group, there are some nonchemical characteristics we should mention. The first one in order of importance is probably bulkiness. It is true, of course, that all the ingredients contribute to the final weight of a feed mixture per unit of volume. But since basal feeds normally constitute more than half the total ingredients of balanced meal rations, they influence the ration in this respect more than the other components.

Bulk is variously defined, but all definitions eventually mean that per unit of volume (i.e., per quart or per bushel) a bulky feed is relatively low in its yield of biologically available energy. We can usually assume safely that among basal feeds DE or TDN is positively correlated with weight per unit of volume. The reason for this relationship is ordinarily traceable to the percentage of fiber in the feed, because of the four potential energy-yielding fractions, crude fiber is likely to be the least digestible. We get an idea of the situation from examining a few typical basal feeds, though we can interpret the figures only in general terms, for two reasons. First, figures for weight per unit volume of ground basal feeds are subject to considerable error, because of the difficulty in controlling the degree of packing of the feed when filling the measure; and second, values for the TDN of specific feeds are not constants, being modified by species of animal, by the nature of the crude fiber, and by whether the digestibility figures are determined directly or indirectly. In Table 11-3 we present typical data for TDN, weight per unit

TABLE 11-3 *Relationship of TDN, Pounds per Quart, and Per Cent of Fiber in Some Ground Basal Feeds*

Feed	TDN (swine)	Lb/quart (g/liter)	Per cent of fiber
Wheat	80	1.7 (810)	4
Corn	80	1.5 (750)	2
Rye	75	1.5 (750)	2
Barley	70	1.1 (560)	6
Oats	65	0.7 (355)	10
Standard middlings	64	0.8 (385)	7
Wheat bran	57	0.5 (255)	9
Oat mill feed	23	0.3 (150)	27

volume, and per cent of crude fiber of a few of the more common basal feeds. Fig. 11-2 shows the trends of these relationships graphically.

As we can see from these data, there is a pronounced tendency for an *increase* of one percentage point in crude fiber or a 0.12 lb (54 g) *decrease* in

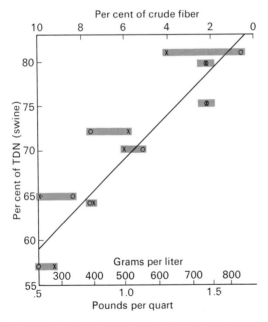

Figure 11-2. *Relationship of TDN of swine feeds to weight per unit volume and to per cent of crude fiber.* X = *intercept of TDN and per cent of crude fiber of a feed;* O = *intercept of TDN and pounds per quart of the same feed. Regression fitted by inspection.*

weight per quart (or 70 g per liter) of a basal feed to be associated with a *decrease* of about 2.5 percentage points in TDN. The degree to which these trends will hold for individual feeds is dependent to some extent on the fat content, since fat is relatively light in weight for its digestible energy value.

The significance of these relationships lies in the consequences of substitutions between basal feeds in a meal mixture formulation. Obviously the use in a meal mixture of a bulky feed in exchange for a heavier one will mean a lowering of the TDN of the mixture; consequently, more of the new mixture will be needed to meet the total energy needs of an animal. Put into other terms, bulky feeds are less efficient when we measure efficiency as feed required per unit of gain for an animal or for his production. This way of thinking about feeds, however, may lead to misconceptions, for it does not follow that feeds of relatively low efficiency (when efficiency is defined as above) are for that reason always less useful in animal feeding. There are many circumstances in which a light, bulky feed is to be preferred to a heavier, more concentrated one, as, for example, in wintering idle adult stock, or in the production of lean bacon.

Simple restriction of total feed allowance has undesirable effects on animals' behavior. They are continuously hungry and hence restless and perhaps irritable. If they are in groups, feed restriction leads to fighting for food and to the uneven distribution of the limited supply between the more and less aggressive individuals. The stockman's way of solving this management problem is often to feed a light, bulky feed in quantities sufficient to satisfy appetite, but at the same time to restrict the intake of TDN as desired. Thus, wheat bran, alfalfa meal, oat feed, etc., are sometimes deliberately incorporated in a mixture because of their low available energy. Such rations can be self-fed without the undesirable consequences of heavy intakes of more concentrated rations.

The more serious situation is where cost of feed vs. cost of TDN is involved. Ordinarily bulky feeds cost less per ton than the heavier, more fattening basal feeds. If the price is in proper relation to the TDN it may matter little which feed is used. The increased quantity of feed needed to supply the available energy will be balanced by its lower cost per pound. Unfortunately, feeders may not have the data necessary to determine the equivalent values. For many samples of feed, no data may be available.

For example, oats is a basal feed, which, because of variety, dates of seeding, and seasonal growing conditions, may vary in weight from 24 to over 40 lb per bushel. Corresponding TDN values might range from 60 to 75 per cent. Using these values a feeder would require just over two bushels of the light-weight oats to supply the TDN of one bushel of the heavy. The lighter grain would be worth only half as much per bushel or only 80 per cent as much per ton as the heavier. But, except for average or standard grades of grain, there is no information about TDN values, since each sample will have a characteristic value of its own. In such cases the graph in Fig. 11-2 might help in making some sort of estimate of relative values.

The problem of bulkiness of feeds arises again in the feeding of very young animals, which, because of limited gastric capacity, cannot consume enough of a bulky feed to meet their energy needs for the rate of growth desired. In this connection we might point out that skimmed milk is also a bulky feed, not because of relatively indigestible fiber, but because of the high content of water, which contains no energy. Where young animals that normally would depend on milk are, for whatever reason, changed to milk substitutes, it is not desirable to make the ration into a water gruel, because of the dilution of the energy value of the ration by the water. In fluid milk the energy value is maintained by its fat component. High fat in a man-made ration, however, is often a liability because of its unstable nature. Experiments with puppies weaned at two weeks, guinea pigs at two days, pigs at ten days, and calves at two weeks, all show that self-fed, dry, low-fat rations can permit as rapid gains in body weight and be nutritionally as satisfactory as liquid milk in all other ways. Fresh water, of course, must be freely available when dry rations are fed, and the nutrient makeup of the diet must be adequate. When such rations are fed as a water gruel, the progress of the young is less satisfactory, unless enough fat is incorporated to maintain, in spite of the water dilution, the energy level at that of the dry meal.

Quality in Basal Feeds

Sample-to-sample variation in quality is a special problem with basal feeds. The important feeds of this group fall into two subgroups of crude fiber. Corn, wheat, and rye or a type of plant seed that is without an enveloping hull make up one group. Barley and oat kernels, on the other hand, after threshing, remain encased in their flowering glumes, and because of this attribute, they are referred to by the trade as *coarse grains*. Because of this division of basal feeds it may be helpful in considering quality to discuss in some detail the characteristics that give various basal feeds their special nutritional properties or that require consideration in making substitutions in ration formulation.

Corn (Maize). Of the basal feeds of the low-fiber group, corn is the key feed in livestock rationing. As seen in Table 11-4, it is the lowest in crude

TABLE 11-4 *Relative Value of Basal Feeds as Carbohydrate Concentrates*

Grain	Per cent of protein (Morrison)	Per cent of net energy (Morrison)	Total feed value (Kellner)
Corn (maize)	74	100	100
Barley	91	86	98
Kafir	92	93	—
Milo	87	93	—
Oats	92	88	95
Rye	74	86	97
Wheat	100	97	95

protein and highest in available energy. Under favorable conditions of growth, an acre will produce in corn about twice as much TDN, or useful energy, as in any other cereal grain. This high production is an economic consideration and makes it clear why corn is so important a crop in areas having climatic conditions favorable for its growth.

The nutritional properties of corn cannot be dealt with so simply. Corn, like all other grains, is subject to variation in makeup because of varietal differences and the specific conditions under which it is grown and harvested. Locally produced samples may differ from published average figures for chemical composition. Perhaps the best recent information on what corn grain is, as described by the chemist, is to be had from "A Survey of Corn Grain in the United States," which was undertaken by the Committee on Feed Composition of the U.S. National Research Council and published in their Report no. 1 (1947).

In order to cover different climatic and soil conditions, the U.S. was divided into ten districts, and samples obtained from each, roughly on the basis of total production. The samples were taken by state agriculturists and were examined for 25 different nutrients, which included proximate, mineral, and vitamin analyses. The result of the statistical study has never been published, but there are, nevertheless, average data available which permit a rather comprehensive picture of corn as a source of food nutrients. In Table 11-5 the average as well as the minimum and maximum values are given for 24 nutrients.

The figures for the makeup of corn may be more meaningful if we look at them in relation to the recommended proportions of nutrients in a meal mixture for market pigs. Of course the comparisons must be general, because rations for other classes of stock will differ from those for a market pig.

We can see at once from Table 11-5 that if corn is introduced into a

TABLE 11-5 *Nutrient Composition of Corn Grain (15% moisture basis)*

Nutrient	Mean	Minimum	Maximum	Recommended in ration of 100-lb (45 kg) pig
Gross energy in kcal	3.78	3.75	4.22	
Protein (per cent)	8.6	8.2	10.9	16
Ether extract (per cent)	3.9	3.7	5.8	
Fiber (per cent)	2.0	1.9	2.1	
Nitrogen-free extract (per cent)	69.3	68.6	69.5	
Ash (per cent)	1.2			
Calcium (per cent)	0.023	0.014	0.029	0.65
Phosphorus (per cent)	0.268	0.23	0.32	0.45
Potassium (per cent)	0.27	0.15	0.37	
Iron (per cent)	0.0023	0.0017	0.0028	15.0
Magnesium (per cent)	0.10	0.09	0.12	
Sodium (per cent)	0.010	0.0008	0.028	
Chlorine (per cent)	0.058	0.048	0.065	
Iodine (per cent)	0.00006			0.0022
Fluorine (per cent)	0.00006			
Manganese (mg/lb)	2.5	1.5	3.7	18.0
Copper (mg/lb)	1.8	0.5	3.9	2.0
Cobalt (mg/lb)	0.05	0.01	0.23	
Vitamin A (IU/lb)	3289	2780	3764	800
Thiamine (mg/lb)	1.8	1.5	2.0	0.5
Niacin (mg/lb)	10.9	8.2	17.1	5.0
Riboflavin (mg/lb)	0.5	0.5	0.6	0.8
Pantothenic acid (mg/lb)	2.1	1.9	2.5	4.5
Folic acid (mg/lb)	0.04	0.03	0.04	

balanced ration it will lower the protein, calcium, phosphorus, manganese, and niacin. It is generally recognized that the quality of protein in corn will not meet nonherbivore needs. When corn is used for cattle or sheep feeding, the calcium and sometimes the phosphorus may be adequately provided by the roughage, and the quality of protein is, of course, not an important factor. *But as a source of energy,* regardless of how one chooses to measure it, *corn stands at the top among the basal feeds.* For cattle feeding, perhaps other than for adult breeding stock, the feeding problem we meet most commonly is how to provide enough energy to permit growth, production, or fattening. The common deficiency of roughage is low energy, and thus the high energy of corn is of special value, since relatively small allowances (as compared with other basal feeds, especially the coarse grains) will balance the ration.

High energy may be a liability, for there are situations where the animal or the product may be subject to damage by rations of high energy. For market-hog feeding the high energy of corn will, under full or self-feeding, produce a carcass with more fat than is desired for so-called "lean" bacon. The rashers from corn-fed carcasses are also likely to have a smaller "eye of lean," as has been shown in experiments at Macdonald College (see Fig. 11-3). This overfinish occurs merely because the more rapid gains in weight have brought the pigs to market weight at younger ages and hence with less muscle development than would be found on older pigs.

As we might expect from their nutrient makeup, wheat shows the same tendency as corn to fatten, while oats, which have five or six times as much

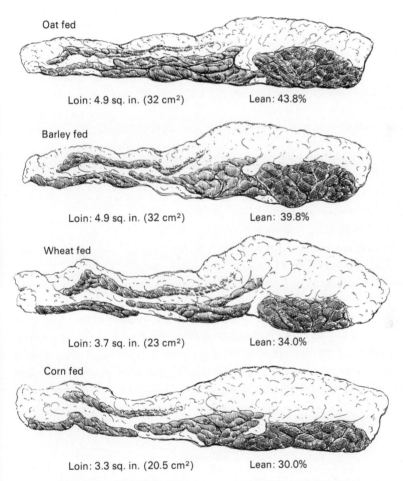

Oat fed

Loin: 4.9 sq. in. (32 cm²)   Lean: 43.8%

Barley fed

Loin: 4.9 sq. in. (32 cm²)   Lean: 39.8%

Wheat fed

Loin: 3.7 sq. in. (23 cm²)   Lean: 34.0%

Corn fed

Loin: 3.3 sq. in. (20.5 cm²)   Lean: 30.0%

Figure 11-3. *Typical bacon rasher from between third and fourth lumbar vertebrae for differing diets.*

crude fiber and about 20 per cent less TDN (for swine) produces a bacon rasher with 40 per cent more lean and a 50 per cent larger "pork chop" eye of lean.

Those who are accustomed to raising or using the coarse grains (barley and oats) frequently regard corn and wheat as uniform products. This impression is doubtless gained from the relatively low variation in the weight per measured bushel and in the crude fiber of these two grains. In the feed trade a bushel of grain is not actually weighed to establish a basis for selling or buying the product. Legal weights per bushel have been established—at values near the average for a given grain—and these are used in arriving at the price for a given weight of grain, because by custom the price quoted by grain brokers is not per pound but per bushel. A carload of 30 tons of oats would actually contain 1,500 measured bushels of 40-lb oats or 2,000 bushels of 30-lb oats. It would, however, be billed as 1,765 bushels (34 lb per bushel is the Canadian standard for oats) at, say, $1.20 per bushel. Thus weight per measured bushel, often used as an index of the relative energy value of a grain, must not be confused with the legal weight per bushel used in the grain trade.

To come back to variation in grains, a recent survey showed that individual crops of oats ranged from 15 to 50 lb per measured bushel. No such range of values is expected with corn. For fiber the U.S. survey of corn showed a range of 1.9 to 2.1 per cent. This range may be compared with data for the fiber given by Winton for the United States of 3.9 to 9.0 per cent for barley, and 8.5 to 20.0 per cent for oats. Similarly, the protein of corn is of relatively low variability, ranging essentially from 8 to 11 per cent while crops of the coarse grains will vary from 8 per cent to 20 per cent crude protein, calculated as N × 6.25.

Nevertheless, economically the variation in the protein percentage of corn may be highly important. In compounding batches of balanced rations, much more protein supplement may be needed with low-protein corn than with high-protein corn to prepare a mixture of some desired percentage of protein. Assume a ration is to be compounded with corn as the basal feed plus a mixed protein-mineral supplement containing 35 per cent protein, and that a final mix of 15 per cent protein is wanted. We calculate as follows:

Case I

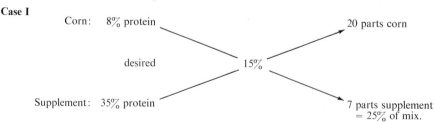

Corn: 8% protein

desired   15%

Supplement: 35% protein

20 parts corn

7 parts supplement = 25% of mix.

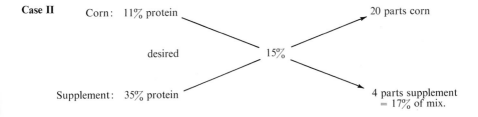

Case II

Corn: 11% protein

desired   15%

Supplement: 35% protein

20 parts corn

4 parts supplement
= 17% of mix.

Thus it is easy to understand the concern, in corn-using areas, about the protein level of this grain.

This situation may be quite different when wheat or the coarse grains are to be used. Here it is not uncommon to find samples that already contain more protein than is needed in the completed ration. The protein supplementation then is done to enhance the protein quality, and the amounts used are often the minimum needed to provide 10 per cent of the final ration protein from an animal or marine source. Thus for 1,000 lb of a 15 per cent protein mixture, it might be desired that 15 lb of the protein should come from fishmeal. This proportion would call for about 21 lb of fishmeal, and if the grain already contained 14 per cent protein, the result would be a mixture with 16 per cent protein. These considerations imply not only that corn, because of its typically low protein content, requires more protein supplementation than coarse grains or wheat, but, perhaps more important, that few livestock rations with the protein demanded by feeding standards can be made up of corn *without* protein supplementation, whereas with average barley, oats, or wheat, there is no need for protein additions to the farm grain fed to herbivora if roughage of reasonable quality is also fed.

Two other characteristics of corn should be mentioned. The one concerns its fat (or ether extract) content, which is higher than the average of basal feeds. This is both an asset and a liability. There is little doubt that a part of the acceptability of corn to animals is traceable to its fat component, not specifically because of corn oil, but rather because of the effect the fat has on the physical nature of the ground grain. Ground corn is not dusty, and unless ground to an abnormally fine modulus does not become pasty with mastication. Although there is no direct proof that the high palatability of corn to all classes of stock is traceable to the fat, it is probably significant that in feeding studies at Macdonald College the addition of about 5 per cent vegetable oil improved the acceptability of dry low-fat diets for young pigs, puppies, and guinea pigs. Without the oil the rations contained about 2 per cent ether extract. That the oil did not improve the diets otherwise is evidenced by the fact that they were no more efficient per calorie in producing weight gains than the low-fat mixtures.

The high fat level, however, can be a distinct liability, since ground corn goes rancid easily. The effect may be slight, and may represent merely a superficial loss of palatability, or it may be extensive enough to result in heating or molding with the attendant loss in nutritive value. In general, ground corn cannot be stored without risk of such damage.

The other characteristic of corn is its moisture content. Samples of corn as harvested are likely to vary more in water content than those of any other grain. They may range from 8 per cent water for fully mature corn to 35 per cent for frosted immature grain. Ear corn containing over 25 per cent water, and shelled corn containing more than about 15 per cent, will not store without damage in the usual types of cribs or bins. Aside from the effect of moisture content on storage, the nutritive value of the grain will decline as it is "diluted" with more and more water (see page 31).

The Coarse Grains. As we have already implied, it is the glume or hull that accounts for the higher fiber of the so-called coarse grains, as is clearly shown in Table 11-6, giving the pertinent data for barley and for oats.

TABLE 11-6 *Proximate Composition of Hull and Groat of Oats and Barley*

Feed	Per cent of crude protein	Per cent of fat	Per cent of fiber	Per cent of TDN (ruminant)	Approx. wt	
					Lb per quart	Grams per liter
Oats						
grain	12.6	5.2	8.9	67	1.0	500
hull	2.7	1.1	30.3	33	0.5	250
groat	15.9	5.9	1.9	92	1.9	950
Barley						
grain	11.9	2.4	4.5	76	1.5	750
hull	5.9	1.3	26.4	41	0.6	300
groat	11.6	2.0	2.4	78	1.9	950

The difficulty with these grains is that the proportions of groat to hulls are widely variable within the species, and are further modified by seasonal growing conditions. Not only do the seeds themselves vary but the crops as harvested may include, in addition to the grain intentionally planted, the seeds from an assortment of other plants of volunteer origin from a previous crop or from weed impurities in the planting grain. Corn (maize) and wheat are relatively free (or are easily freed) from such contaminants, but with barley and oats purity of sample is often a factor influencing feeding value.

Barley. Many of the problems of nutritional quality in basal feeds are particularly well-illustrated by barley as it is grown, sold, and used in Canada. This grain may be grown for malting purposes or for feeding livestock. The Canadian scheme by which the producer is paid for barley delivered to elevators involves a grading according to the purity of the crop, its variety, and its soundness. Samples, which because of admixtures of seeds from grains other than barley, or because of frost or heating damage or poor filling of kernels, are not suitable for malting, are classed as feed barley.

There are three grades for feed barley. These are partially defined in Table 11-7. As we can see from this table, No. 1 Feed barley is essentially pure

TABLE 11-7 *Partial Description of Feed Grades of Western Canadian Barley*

		Maximum tolerance of foreign material			
Grade name	Minimum lb per bushel	Per cent of weed seeds (too large to pass 4/64 screen)	Per cent of wild oats	Per cent of other grains	Per cent of total foreign material not to exceed
No. 1 Feed	46	1	4	4	4
No. 2 Feed	43	2	10	10	10
No. 3 Feed	—	3	20	20	20

barley, but because of frosting or for some other reason it is below the standard weight of 48 lb per measured bushel for malting barley. Also, in this category is found barley that, because of variety, is not suitable for malting. (Some varieties of barley peel too easily and, consequently, are not wanted in malting grades.) Barley that is still lighter in weight per bushel and that may also contain up to 10 per cent other material is classed as No. 2 Feed. The No. 3 Feed grade has no minimum weight per bushel and, furthermore, need only be 80 per cent in purity. We should note that the nonbarley material may be wild oats, or it may be "other grain," which in practice is the same as saying it may be domestic oats or wheat. All Canadian barley that goes through commercial channels to the feed industry, or to Canadian purchasers of feed barley, or is sold for export, is by law sold under one of these three categories (except that Eastern Canadian barley, though graded similarly, differs slightly in grade specifications).

Of course samples of grain as harvested seldom, if ever, contain exactly these quantities of foreign materials. The botanical makeup of the foreign material in barley as harvested (presuming pure barley was seeded) will depend largely on what crop was grown on the area the year immediately preceding and on the extent of the weed pollution. An extensive survey of the 1949 Western Canada barley crop deliveries to county elevators yielded the figures in Table 11-8 on purity and chief grain diluents.

TABLE 11-8 *Botanical Makeup of "Barley" as Harvested*

Per cent of oats	Per cent of wheat					
	0	5	10	15	20	25
0	52[a]	11	2	0.5	0.5	0.5
5	12	4	2		0.5	
10	8	2	1			
15	1	1	0.5			
20	0.5					
25	0.5					

[a] Read as "52% of barley crop contained 0% oats and 0% wheat."

Table 11-8 indicates that a little more than half the individual crops as harvested were essentially pure barley. Roughly a quarter of the crops were about 90 per cent barley, and balance of the crops on the whole would be similar in feeding characteristics to mixtures containing 80 per cent barley. Similar surveys in subsequent years revealed the same distribution of the "barleys as harvested." All commercial Canadian feed barley contains approximately the maximum tolerance of nonbarley. This is accomplished by blending at terminal elevators, sometimes with wild oats and coarse grains removed from wheat.

To describe the feeding value of "barley" as this crop actually appears in commercial channels in Canada is, consequently, not a simple matter. To be realistic we must consider under the name "barley" at least four products:

1. Pure barley (including No. 1 Feed grade).
2. Barley containing 9 per cent of an unspecified combination of oats, wild oats, wheat, or flax plus 1 per cent coarse weed seeds (No. 2 Feed grade).
3. Barley containing 17 per cent of an unspecified combination of oats, wild oats, wheat, or flax, plus about 3 per cent coarse weed seeds (No. 3 Feed grade).
4. Barley as harvested on the farm.

There is a further complication, in that the proportion of oats vs. wheat within the total tolerance of "other grains" may appreciably affect the feeding value of the barley, oats tending to reduce and wheat to increase the available energy of the final mixture.

The Canadian grading scheme is of interest here only because it brings out clearly the difficulties of describing with any simple index the feeding value of a particular sample of a coarse grain. The variability in the purity of the barley is itself an important factor, and one that neither the name nor the usual chemical analysis defines. In addition, its protein may run from 9 to about 16 per cent, its crude fiber from 2.5 to 8.5 per cent, its weight per bushel from less than 40 to over 50 lb, and its TDN from 62 to 81 per cent. With this range of variability, both botanical and chemical, it is not surprising that the performance of animals fed on rations composed chiefly of barley may not always be according to book specifications.

All barleys are, nevertheless, basal feeds and as such are used in livestock rations primarily as sources of energy. They are commonly (but not too accurately) referred to as fattening feeds. Their efficiency in this role is obviously much dependent on which barley they are. For example, for finishing hogs, given a properly supplemented barley ration, the weight of the feed required per pound of gain will increase about 2 per cent for each decrease of one pound in the weight per measured bushel; or if feeding is done by measure the corresponding increase is about 4 per cent. Thus, of 40-lb barley 20 per cent more feed by weight, or 40 per cent more by measure, will be needed to produce one unit of gain in finishing hogs than would be required of 50-lb barley. Experiments at Macdonald College indicate that for this purpose it matters little what the combination of barley and diluent is that gives a particular weight per bushel. In general, heavy barley plus oats will have the same fattening efficiency as a combination of similar bulkiness made up of light barley plus wheat. A graph showing these quantitative relations, prepared from a series of feeding tests with market pigs conducted at Macdonald College over a period of several years, illustrates these findings (see Fig. 11-4). We should expect the same type of effect

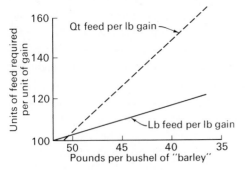

Figure 11-4. *Increase in rations needed per unit of gain for full-fed bacon pigs during last two months of feeding.*

described with cattle or with sheep, but to a lesser degree, since ruminants can better utilize crude fiber.

We should note, however, that in practical feeding it has seldom been possible to detect any effect of additions to pure barley of up to 10 per cent of other grains. On the other hand, barley samples containing 15 per cent or more of either oats or wheat are likely to show measurable differences in feeding value. The effect of wild oats is merely an exaggeration of that produced by tame oats.

As measured by the nutritional needs of animals, all barleys are deficient in salt, calcium, phosphorus, iron, iodine, and cobalt, and in vitamins A and D. Except for herbivorous animals, barley also requires supplementation with protein if it contains less than about 12 per cent protein, and in all cases to improve its quality by increasing particularly the lysine content.

Many statements to the contrary notwithstanding, there is no evidence that, once animals are accustomed to it, pure barley is less acceptable than any other entire cereal grain. Contamination with weed seeds will adversely affect its palatability, and use of such samples may explain the lower opinion some feeders have of barley than is justified by results with clean samples. Barley is frequently planted on wheat land that has become fouled with weeds, and among wheat raisers it is referred to as a cleaning crop. Thus, more often than other grains, barley as harvested may be badly contaminated with weed seeds. Barley meal made from such low-grade grain may be unpalatable, but this should not be charged to a characteristic of the barley itself.

Nutritionally the limit of its inclusion in specific livestock rations is set only by the quantities of other products that must be included in order to make good the nutritional deficiencies of the barley, except that for very young animals it may be desirable in some way to reduce the hull of the ration either by coarse grinding and sifting or by dilution with low-fiber feeds.

In practice, there are at least two uses to which "barley" is often put where the kind of "other grain" diluent may be of significance. When market pigs intended for lean bacon are finished on barley diluted with wheat, they tend to produce overfat carcasses. On the other hand, dilution of barley with oats tends to reduce the percentage of available energy and, consequently, tends to produce less fattening. Similarly, nonproducing stock being carried on maintenance rations can advantageously use the barleys of lower weight per bushel, such as oat or wild oat and light barley combinations.

Finally, it may be in order to call attention to the black sheep of the barley family—a product officially designated as *barley feed*. It consists of the mill-run residue from the production of pot and pearl barleys. The residue is barley hull plus the outer layers of the kernel that are polished off the dehulled grain to get rid of the bran and embryo portions. This product is of low feed value, having at best only two-thirds the digestible nutrients of typical barley. We mention it here because it sometimes is illegally incorporated into ground barley or into barley-containing meal mixtures. Its presence will lower the efficiency of the feed containing it, both by reducing the acceptability of the ration to the stock and by reducing available energy.

Oats. What has been said concerning the variability of barley as harvested applies, in general, to oats as a basal feed, the chief difference being that whereas barley normally contains about 6 per cent crude fiber, oats contains 10 or 11 per cent. Oats, in other words, has a lower energy value than barley. Variation between samples is fully as great as with barley, and the consequences of the differences in weight per bushel follow the same pattern as those described for barley. The botanical make-up of "as harvested" Canadian oats is shown in Table 11-9.

TABLE 11-9 *Botanical Makeup of "Oats" as Harvested*

Per cent of wild oats and chaff		Per cent of wheat and barley			
		Wheat: 0		Wheat: 5	
		Barley: 0	Barley: 5	Barley: 0	Barley: 5
Wild oats: 0	Chaff: 0	45[a]	5	7	2
	Chaff: 5	9	2	2	
Wild oats: 5	Chaff: 0	11	1	3	
	Chaff: 5	3			
Wild oats: 10	Chaff: 0	2			
	Chaff: 5				

[a] Read as "45% of crop contained 0% wild oats, 0% chaff, 0% wheat, and 0% barley."

There is no experimental evidence to support the contention put forward by some feeders that oats has any special nutritional virtue for any particular class of stock. It is true that the hull of the oat is somewhat softer and perhaps less irritating in the digestive tract than the hull of barley. Barley groats, oat groats, wheat, polished rice, and corn all are rich sources of available energy and have about equivalent feeding value in the ration. The chief differences in these grains as feeds are traceable to the proportions of the hull, more specifically, to the percentage of crude fiber.

Buckwheat. Perhaps the only other feed that requires special mention is buckwheat. First we should call attention to the problem of names of buckwheat products (see Chapter 1).

The offal of buckwheat milling consists primarily of black hulls and of middlings, the latter made up of the seed coat, the adhering endosperm, and the embryo. The hulls, which represent almost 30 per cent of the weight of the entire buckwheat, have little feeding value. The middlings are rich in protein and fat, which are derived chiefly from the aleurone layers and the embryo tissues. So-called buckwheat feed is a mixture of hulls and middlings. The proximate composition of these three products as given by Winton is in Table 11-10. We can see that entire buckwheat is a basal feed, buckwheat feed a roughage, and the middlings a protein supplement.

TABLE 11-10 *Proximate Composition of Buckwheat By-Products* (*All figures are percentages*)

	Water	Protein	Fat	Fiber	N-free extract	Ash
Entire seed	12.6	10.0	2.2	8.7	64.4	2.1
Hulls	6.5	7.8	1.4	33.6	47.1	3.6
Middlings	10.0	26.7	7.2	6.8	44.6	4.7
Flour	12.0	6.4	1.2	0.5	79.5	0.9
Feed	10.0	15.9	4.1	22.0	44.8	3.2

The one particular feature that we should mention here is that products containing the hulls are likely to contain enough of a photoporphyrine to cause light sensitization in white-skinned animals. When exposed to the sun a rash may develop of such severity as to adversely affect the performance of the animals.

Entire buckwheat is frequently incorporated into poultry scratch grain mixtures but is less often used for other classes of stock. Buckwheat middlings, however, is a common feedstuff in districts where buckwheat growing is a regular practice. The hulls, because of their woody nature, are particularly indigestible and practically useless for feeding purposes.

Wheat Bran and Other Wheat Milling By-Products. Wheat bran has had a rather checkered career as a feedstuff. Originally discarded as a worthless offal from the milling of wheat for flour, it was suggested and eventually popularized as a livestock feed by Dr. Henry of Henry and Morrison fame.

In some districts farmers began to consider this feed indispensable and shortages developed, so that bran was rationed by some feed dealers, allocating it only where other feedstuffs were also purchased. The wheels of fortune turned on occasion, however, and bran became a drug on the market. To ease their selling problems some millers made the sale of wheat flour to jobbers and bakers contingent on the purchase of bran in amounts equal to that represented by the milled flour. This practice led to the situation where the feeder bought his supplies of bran from the baker. The feeder soon realized that he was getting to be a pawn in the miller's game of disposing of one of his by-products, a game made possible by the feeder's belief that bran was a necessary ingredient of a satisfactory livestock ration. This belief is gradually being dispelled, and bran is more and more used as it should be, specifically to capitalize on its unique nutritional properties. Its light, bulky nature, its 16 per cent of high-quality protein (a chemical score equal to that of beef muscle), and its high phosphorus content give bran a unique place in livestock feeding. About 40 per cent of the wheat germ is in the bran, which accounts for its high-quality protein. Included in the herbivore ration, it provides supplementary phosphorus to correct the common shortage in the forage, and its cellulose-hemicellulose carbohydrate is, of course, an acceptable source of energy for these animals. Its bulk is often advantageous as a means of lightening a predominantly corn ration.

The bulkiness of bran is of special usefulness in the preparation of nonfattening rations, as for the bacon hog, to whom bran yields less energy than to cattle. Thus its introduction into the meal mixture of market pigs during the last two months of feeding before slaughter curtails the energy intake and the fattening of the pig, without restricting the feed. Canadian experiments and practical experience have demonstrated that hog-finishing rations diluted with 25 per cent wheat bran by weight can be self-fed to market pigs without leading to the excessively fat carcasses which otherwise result from self-feeding practices.

None of the other wheat milling by-products—shorts, middlings, feed flour—have any of the special properties of the wheat bran. They are basal feeds useful largely in proportion to their energy and nutrient values.

SUGGESTED READINGS

Ashton, G. C., "The Carcass Quality of Bacon Hogs," *Agr. Inst. Rev.,* Macdonald College Journal Series no. 273 (1950).

Committee on Feed Composition (R. V. Boucher, Chairman), *Composition of Feeds* (Report no. 1 and Supplement to Report no. 1, NRC mimeograph, 1947).

Crampton, E. W., *The Comparative Feeding Value for Livestock of Barley, Oats, Wheat, Rye, and Corn* (National Research Council of Canada, Report no. 28, 1933).

Crampton, E. W., *Barley as a Hog Feed* (Winnipeg, Man.: Canadian Cooperative Wheat Producers Limited, 1933).

Crampton, E. W., G. W. Muir, and R. W. Knox, "The Feeding Value of Canadian Western Barley for Bacon Hogs," *Sci. Agr.,* XX (1940), 365.

Crampton, E. W., "Feeding Value of Canadian Western Barley for Bacon Hogs," *Sci. Agr.,* XXII (1942), 326.

Crampton, E. W., and G. C. Ashton, "Barley vs. Wheat as a Basal Feed in Bacon Hog Rations," *Sci. Agr.,* XXV (1945), 403.

Crampton, E. W., and O. M. Ness, "Meal Mixture Suitable as the Entire Ration to be Self-fed Dry to Pigs Weaned at Ten Days of Age," *J. Animal Sci.,* XIII (1954), 357.

Crampton, E. W., G. C. Ashton, and L. E. Lloyd, "Improvement of Bacon Carcass Quality by the Introduction of Fibrous Feeds into the Hog-Finishing Ration," *J. Animal Sci.,* XIII (1954), 321.

Morrison, F. B., *Feeds and Feeding* (Ithaca, N.Y.: Morrison, 23d ed., 1967).

Winton, A. L., and K. B. Winton, *Structure and Composition of Foods* (N.Y.: John Wiley, 1932), vol. I.

CHAPTER 12

Protein Supplements

Products of Plant Origin

As we indicate in the feed classification, the protein supplements of plant origin divide quite naturally into two subgroups—one containing the feeds with 20 to 30 per cent total crude protein, the other those with 30 to 45 per cent crude protein. In order to picture certain of the characteristics of these two groups of feeds, we have entered a few of the more common products belonging to each in Table 12-1.

Insofar as we can describe them by averages, we can see that the chief difference between these two types of supplements is in protein content, the higher protein being associated with a lower carbohydrate analysis. The 20 to 30 per cent group is made up primarily of by-products of wet milling, brewing, or distilling of corn or barley. These by-products tend to be high in crude fiber. The feeds of the other group are almost entirely residues of oil-bearing seeds, which have been processed by chemical extraction or by expression to remove most of the oil. The noteworthy exception is corn gluten meal, one of the by-products of the wet milling of corn grain (see Fig. 11-1). This feed is low in fat, not because of solvent extraction, but because of a physical separation of the germ from the mash as one of the early steps in this milling process. The carbohydrate is relatively low.

The chemical scores show that the feeds in the 20 to 30 per cent group have poorer-quality protein than those in the higher-protein category. Perhaps the reason for this difference is that less of the germ proteins are re-

TABLE 12-1 *Typical Protein Supplements of Plant Origin (All figures are percentages)*

Common name and reference number	Crude protein			Crude fat	Carbohydrate		TDN		Calcium	Phosphorus
	Total	Digestible	Chemical score		N-free extract	Crude fiber	For ruminants	For swine		
5-02-903 Gluten feed	25	21	21	2	48	8	75	—	0.1	0.6
5-02-141 Brewers' grains	26	21	—	6	41	15	60	43	0.3	0.5
5-00-545 Malt sprouts	26	19	—	1	44	14	63	—	0.2	0.7
5-02-842 Distillers' grains	27	20	—	9	41	12	84	—	0.1	0.4
Average	26	20	21	5	44	12	70	43	0.1	0.5
Solvent-extracted residues of:										
5-02-048 Linseed	35	30	43	2	39	9	70	68	0.4	0.8
5-01-630 Cottonseed	42	33	50	2	30	11	63	73	0.2	1.1
5-04-604 Soybean	46	42	56	1	31	5	73	75	0.3	0.7
5-02-900 Gluten meal[a]	43	36	21	2	40	4	80	—	0.2	0.4
5-06-650 Peanut	47	42	41	1	26	13	77	82	0.2	0.7
5-04-739 Sunflower	47	43	—	3	24	11	—	—	0.4	0.7
5-03-871 Rapeseed	46	—	50	1	28	14	74	—	0.3	1.0
Average	44	34	44	2	31	10	73	75	0.3	0.8

[a] See page 236.

moved by fat extraction than by the water treatments involved in wet milling or brewing. The feeds of this lower-protein group are by-products either of corn or barley, and the chief, or at least the first, limiting factor in their quality is a deficiency of lysine. Malt sprouts, however, present an exception to this rule; its protein is a combination of the proteins found in the barley grain and those of the newly sprouted root. At the moment there is no experimental evidence of qualitative differences between these two proteins, but there is every reason to believe that the proteins of the rootlet will be similar to those of leaf. We believe also that young leaf proteins may have a more complete amino-acid makeup than those in the seed of the plant.

Solvent-Extracted Oilseed Residues. In spite of the overall better quality, the first limiting amino acid of linseed and cottonseed is lysine, but with peanut meal the sulfur-containing amino acids, methionine and cystine, are relatively the more deficient, with lysine standing second. Soybean proteins, on the other hand, are probably the most complete of any of the plant-seed proteins. Table 12-2 gives an idea of the amino-acid distribution in the protein of the important feeds of this group.

TABLE 12-2 *Partial Amino-Acid Content of Solvent-Extracted Residues of Oilseeds as a Per Cent of Dry Matter (Dried skimmilk and whole corn included for comparison)*

	Per cent of amino acid in:							
Amino acid	Dried skim milk (34% protein)	Linseed meal (33% protein)	Soybean meal (45% protein)	Cottonseed meal (42% protein)	Corn gluten meal (42% protein)	Peanut meal (45% protein)	Rapeseed meal (46% protein)	Whole corn (9% protein)
Arginine	1.46	2.04	2.61	3.11	1.30	4.46	2.45	0.36
Histidine	0.85	0.49	1.03	1.09	0.71	0.95	1.20	0.22
Lysine	2.55	0.82	2.43	1.13	0.46	1.35	2.45	0.23
Tyrosine	1.80	1.68	1.85	1.34	2.60	1.98	0.98	0.55
Tryptophane	0.54	0.63	0.63	0.55	0.25	0.45	0.55	0.05
Phenylalanine	1.93	1.85	1.85	2.77	2.77	2.43	1.75	0.41
Cystine	0.37	0.63	0.59	0.84	0.50	0.72	0.58	0.10
Methionine	1.12	0.99	0.81	0.67	2.31	0.41	0.87	—
Threonine	1.56	1.68	1.80	1.26	1.68	0.68	1.87	0.32
Leucine	5.10	—	3.60	5.88	10.50	4.50	3.00	1.94
Isoleucine	1.53	1.15	1.80	1.47	2.10	1.58	1.68	0.32
Valine	2.20	1.98	1.58	2.94	2.10	3.15	2.15	0.41
Glycine	0.13	—	0.45	2.22	1.80	2.52	2.18	—

It is evident, therefore, that supplementation of the basal feeds with any of the high-protein feeds of plant origin except soybean meal is not likely to improve biological values. Most of these feeds have a common deficiency in lysine, which sets an upper limit to their usefulness in rations of animals where protein quality must be considered.

Crude Fiber. The feeds belonging to the lower-protein category are likely to have a higher crude-fiber content than those of the oilmeal group. It is perhaps for this reason that such products as brewers' grains, malt sprouts, and distillers' grains are not as commonly thought of as hog feeds. The higher fiber content is of less direct consequence in the dairy ration.

The important factor here is the bulkiness of the feed. Bulk becomes important in practical feeding of cattle because allowances are likely to be measured by volume rather than by weight. As a ration is made bulkier by the inclusion of light feeds, the quantity (by volume) of it required to yield the amount of digestible energy called for by the feeding standard increases rather rapidly.

For example, we can assume for rough calculation that a meal mixture of standard feeds containing 75 per cent TDN will weigh about one pound per quart, and that one of 70 per cent TDN will weigh only about 0.8 lb per quart. To supply one pound of TDN we require 1.33 lb of the heavier and 1.43 lb of the lighter feed. But if we express the allowance in quarts, we find that for 1.33 quarts of the heavier mixture, we shall require 1.80 quarts of the lighter one. The increase in weight is about 7 per cent, but by volume it is 35 per cent.

Insofar as the cattle themselves are concerned, added bulkiness of ration is of little importance. The difficulty lies with the feeder, who is not always fully aware of how many more quarts of bulky ration must be fed to supply the nutrients contained in a more concentrated feed mixture.

Calcium and Phosphorus. The calcium and phosphorus content of these protein supplements should be compared with the probable concentration required in the complete cattle ration. Feeding standards indicate that approximately 0.2 per cent of the dry weight of the ration should consist of calcium, and similarly for phosphorus. Daily allowances of good-quality roughage can be expected to supply all the calcium that cattle require. The importance of the concentration of this element in the feeds of the meal mixtures is, therefore, small. In any case, these feeds will usually constitute no more than 20 per cent of the final meal mixture fed and their calcium content, therefore, will not be important in changing the calcium content of the ration.

The problem of phosphorus, however, is somewhat different. This element in feeds is quite likely to be correlated with protein content. Thus, high-protein feeds commonly provide more phosphorus than low-protein feeds. In general, the feeds of the 20 to 30 per cent protein group supply about double the concentration of phosphorus that is required in the final ration of cattle stock, and the feeds of the 30 to 45 per cent category supply somewhat more. Thus, as the protein level of the meal mixture is increased by the addition of protein supplements, the phosphorus is also augmented. As we shall explain in more detail under formulation of rations, this correlation does not necessarily mean that a phosphorus supplement can be omitted from the meal mixtures of milking cows.

Effects of Processing. We have already suggested that the by-product feeds are likely to be more constant in chemical makeup than the unprocessed basal feeds. There are nevertheless differences in the processes to which by-product feeds may have been subjected, some of which have a bearing on their effective nutritional values. Heat, for example, may be either detrimental or beneficial, depending on the feed and on the amount of heat. Soaking the product and subsequent drying may also have an effect on the availability of some of the nutrients of the resulting products.

With feeds that are by-products of brewing or distilling, the heat involved is usually that necessary to dry the product. The cost of this operation is appreciable, and in some cases suppliers offer samples that have not been dried sufficiently to ensure that the feed can be safely stored. Storage in the usual warehouse of feeds that contain appreciably more than 12 per cent moisture invites risk of spoilage. High-moisture samples should be priced so that the unit cost of dry matter is equivalent to that asked for normally dry samples.

With the oil-bearing seeds, heat is used for a somewhat different purpose. It may be applied intentionally, or it may be incidental to the process of fat extraction. In general, there are three oil-milling methods. The "old process" is more properly termed the "mechanical pressure process"; the seed is crushed into flakes and these are then subjected to steam cooking. The hot, wet mass is then spread in layers between heavy cloth and placed in a press, where as much of the oil as possible is squeezed out by pressure. The resulting cakes may then be broken into a granular form and sold as cake, or may be ground into a fine meal. In this process the residue still retains 5 per cent or more fat.

The expeller process is also a mechanical process. The seed, after cracking and drying, is heated in a steam-jacketed apparatus, and subsequently the mass is subjected to pressure in a press. A considerable amount of heat is generated from friction in this process. The residue is again ground into a meal. In the NRC nomenclature both of these processes are called mechanical extraction.

The solvent process is quite different. It employs a volatile fat solvent, in which the flakes are soaked or washed. Once the oil has been thus removed, the residue is heated to remove the last traces of the solvent. Usually only about 1 per cent fat remains in the oil meals prepared by this process. Oil meal prepared by solvent extraction may require further heating or "toasting" to improve digestibility. Whether or not this extra treatment is necessary depends on the particular protein involved.

Soybean protein is enhanced in feeding value for nonherbivorous animals by sufficient heat treatment to destroy a substance present in the soybean that otherwise inhibits proteolysis. There may also be some change in the protein molecule itself which increases the availability of the cystine and methionine. Experiments indicate that methionine in heated soybeans is more rapidly liberated by enzymic action than with an unheated product. Soybean protein is not the only one that is improved in digestibility by cooking. The proteins of the navy bean and of the velvet bean are also. Where heat does improve protein value, the temperature and time are of importance. Too severe treatment will undo the favorable effects of a milder treatment.

The proteins of most feeds, on the other hand, decrease in nutritive value when subjected to heat. Experimental evidence seems to indicate that when heating damages a protein, the damage is likely due to a destruction of lysine. Certain heated proteins are restored to their original value by additions of lysine. Lysine, in fact, is rather easily damaged, and some evidence indicates that even mild drying of some proteins of animal origin may be detrimental. To come back to the oil meals, we know that cottonseed meal and peanut meal may be damaged by heat treatment, both in digestibility and in biological value.

Block and Mitchell * come to the conclusion that food products whose unheated proteins rank lower by a biological assay than by chemical score will probably improve in biological value on heating, whereas those food proteins whose biological assays and chemical ratings show reasonable agreement are likely to be damaged in biological value by heating. Of the proteins that are ordinarily fed to livestock, only the proteins of soybean products appear to be improved by heating. The others are more likely to be damaged, primarily by destruction of lysine.

Fat. The fat content of oil-bearing seed by-products must be taken into account sometimes if they are to be used for certain classes of livestock. Most

* R. J. Block, and H. H. Mitchell, "The Correlation of the Amino-Acid Composition of Proteins with Their Nutritive Value," *Nut. Abst. and Revs.,* XVI (1946, 1947), 249.

vegetable oils, if fed to meat animals in any appreciable amounts for a month or more previous to the slaughter of the animals, tend to produce a soft, oily carcass fat. This is particularly objectionable in pigs. For hogs whose carcasses are to be made into bacon, heavy feeding of corn (of only 5 per cent fat) during the finishing period may be sufficient to cause this softening of the fat. Thus the feeding of the oilseeds as grown on the farm is not ordinarily desirable. Extraction of the oil leaves a residue that may contain from almost none to 12 per cent fat, depending on the process and the efficiency with which it is operating. Ground soybeans, ground peanuts, or other feeds of this type can be fed to cattle without undue penalty in carcass quality, but these products cannot safely be fed to finishing pigs. However, they are sometimes used for younger pigs.

Expeller oil meals will contain about 8 per cent fat. The use of solvent extraction is increasingly common, with the result that the fat content of oil meals so treated is reduced to about 1 per cent. This reduction of fat means an increase in protein and in carbohydrate concentration, but a *reduction of about 5 per cent in energy value*. The alteration of protein level is great enough to be nutritionally and economically important, but the changes in the other nutrients are not likely to have measurable effects in the final ration.

We should also call attention to the high energy values for most of the products in this category. With the exception of brewers' grains and malt sprouts, the inclusion of almost any one of the protein supplements of plant origin in the typical rations of livestock improves the TDN as well as the protein. Thus, where they are of competitive price per unit of TDN, these feeds can be included for their energy value equally as well as the basal feeds. There is no acceptable evidence that excesses of protein, such as might be caused by supplementation of this kind, are likely to be of practical significance.

Precautions. Most of the oil meals are wholesome and palatable to all classes of livestock. An exception would be *unheated soybean meal* as an ingredient in the hog ration, but the toasted product is entirely satisfactory. Still, some precautions must be used with some of these oil meals. *Cottonseed meal,* for example, must be used cautiously with any but adult cattle, because of the poisonous gossypol which may be present in grades of meal that contain appreciable amounts of the cottonseed hulls. Low grades of cottonseed meal should be especially avoided with young animals, whose susceptibility to this poisoning is greater than that of older stock, and even the high-quality products should be avoided for pig feeding.

Rapeseed meal contains glucosides, from which mustard oils may be formed in the digestive tract of animals under certain conditions. These oils are irritating and produce undesirable consequences when too much of them is included in livestock rations. In actual practice the inclusion of much over 4 or 5 per cent of rapeseed meal in livestock rations renders them unpalatable. Here, again, young animals (and possibly pregnant females) may be more susceptible than other classes to the harmful effects of rapeseed meal.

The special property that has been claimed for mechanically extracted *linseed meal* may be questioned. Raw linseed oil is sometimes used as a laxative with farm animals, and many statements have appeared to the effect that one of the beneficial effects of linseed meal can be traced to the 8 or 9 per cent of oil in the product. This was supposed to help lubricate the digestive system and to correct the constipating effects of dry hay or similar feeds. There was also the belief that cottonseed meal tended to be constipating. Experimental evidence does not support the presumed difference between linseed meal and cottonseed meal in this respect. In fact, tests indicate that the rate of passage of diet residues through the digestive system of various kinds of animals is not differentially affected by the normal use of either of these feeds.

Soybean Meal. The special role of soybean meal as a protein supplement requires comment. At least in North America soybean meal has become the key feed among the protein supplements of plant origin. The extent of its use in different parts of the U.S. and Canada at any one time is influenced by its price in relation to that of other oil meals. Because of its higher biological value, this feed has now replaced much of the meat meal, tankage, and fish meal, which were in the past the mainstay of protein quality in rations for nonherbivorous animals.

So far as the true biological value of the protein in soybean meal is concerned, it is interesting to compare its amino-acid makeup with that of the protein of milk and of linseed meal; the former is a protein of nearly perfect biological value, and the latter is a plant protein that is still the standard in many districts of North America. This comparison is given in Table 12-3.

Considering the protein quality of soybean meal this way leads to the conclusion that the chief advantage it has over linseed meal protein is its markedly greater concentration of lysine, the amino acid that is ordinarily the first deficiency in the basal feeds. The particular amino-acid distribution of this feed appears to be such that in combination with corn (and necessary mineral and vitamin supplements) it forms a ration in which little or no animal or marine protein is necessary for hog feeding. Thus where high-grade fish meals

TABLE 12-3 *Partial Amino-Acid Makeup of Soybean and Linseed Meal Dry Matter*[a]

Amino acid	Soybean meal	Linseed meal
Lysine	95 (76)[b]	32
Tryptophane	120 (96)	120
Cystine	160 (128)	170
Methionine	72 (57)	88
Isoleucine	117 (94)	75

[a] Makeup of milk taken as 100 per cent for each amino acid.
[b] Figure in parentheses is for meal corrected to 34 per cent crude protein.

or meat meals are not readily available or are not competitive in price, soybean meal offers a valuable alternate source of protein.

Protein Supplements of Animal and Marine Origin

Analogous to the high-protein feeds of plant origin is a group of edible by-products of animal or fish origin. These are usually employed to improve the total protein of basal feeds, but, in addition, they contribute a mixture of amino acids quite different from that characteristic of most proteins of plant sources. For example, plant-seed proteins are usually seriously deficient in lysine. Meat, milk, and fish proteins, however, are relatively rich in this amino acid, though they are likely to be short of the sulfur-containing cystine and methionine.

The products belonging in this high-protein group are more diverse in protein level than are feeds of any other protein category. The individual feeds frequently have unique properties affecting or limiting their use. Some of these are indicated by chemical makeup, as shown in Table 12-4.

Excluding whey powder, which really does not belong in this category but will be discussed here because of its protein characteristics, we can see that the range of protein values is from 34 to 82 per cent, that fat ranges from 0 to 15 per cent, and that calcium and phosphorus for some of the feeds are present in supplementary amounts. There are several grades of both tankage and fish meal, representing differences in processing which result in products of distinctly different characteristics as feeds. However, before dealing with individual products we should note the general feeding characteristics of the feeds of this group.

TABLE 12-4 *Composition of Typical Feeds of Animal
or Marine Origin (All figures are percentages)*

Feed	Protein		Ether extract	Ash		TDN
	Total	Digestible		Ca	P	
Meat meal	53	48	10	8.0	4.03	68
Meat and bone meal	51	45	10	11.0	5.07	65
Blood meal	80	62	2	0.3	0.22	61
Tankage				6.0	3.17	
low fat	68	60	3			65
high fat	61	45	15			77
55% protein	58	36	11			68
70% protein	73	70	12			94
Fish meal				7.9	3.60	
low ash (14)	71	66	6			78
high ash (31)	52	48	1			49
50% protein	53	49	4			55
70% protein	74	71	1			71
65% protein (oily)	68	65	10			87
Milk						
skim milk powder	34	33	1	1.2	1.00	86
whey powder	14	13	1	0.9	0.80	78

Protein Quality.   There is a remarkable similarity in the amino-acid distribution of the different feeds (see Table 12-5). All carry as much or more lysine than is found in the protein of egg (which is usually taken as the standard of excellence for amino-acid assortment). As compared with the average cereal-grain protein, animal or marine proteins have a higher lysine level by about two and a half times.

The isoleucine level of meat meal, fish meal, and milk is at least 50 per cent higher than that in the mixed proteins of cereals, but blood meal (and consequently tankage, which contains blood) is low in this amino acid. Because of their lysine and (in most cases) their isoleucine levels, the feeds of this group are valuable as supplements to the plant proteins, the combinations usually having a higher effective biological value than plant protein alone.

As a group, the feeds of this category are deficient in the sulfur-containing amino acids, cystine and methionine. Methionine can, of course, be converted *in vivo* to cystine (although the reverse is not true). Hence the combined deficiency of these two acids can be relieved by fortification of the diet with pure methionine, which is economically available as a feed supplement.

It is now believed that the biological function of methionine as a methyl-

TABLE 12-5 *Essential Amino-Acid Content of the Proteins of Certain Protein Feeds* (*As a per cent of total protein*)

Animo acid	Tank-age	Meat meal	Blood meal	Fish meal	Milk	Egg	Cereal[a]	Animal[b]
Arginine	5.9	7.0	3.7	7.4	4.3	6.4	4.8	5.7
Histidine	2.7	2.0	4.9	2.4	2.6	2.1	2.1	3.3
Lysine	7.2	7.0	8.8	7.8	7.5	7.2	3.1	7.7
Tyrosine	2.9	3.2	3.7	4.4	5.3	4.5	4.8	3.9
Tryptophane	0.7	0.7	1.3	1.3	1.6	1.5	1.2	1.1
Phenylalanine	5.1	4.5	7.3	4.5	5.7	6.3	5.7	5.4
Cystine	—	1.0	1.8	1.2	1.0	2.4	1.7	1.2
Methionine	—	2.0	1.5	3.5	3.4	4.1	2.3	2.6
Threonine	3.0	4.0	6.5	4.5	4.5	4.9	3.4	4.5
Leucine	7.7	8.0	12.2	7.1	11.3	9.2	7.1[c]	9.2
Isoleucine	2.7	6.3	1.1	6.0	8.5	8.0	4.3	4.9
Valine	5.4	5.8	7.7	5.8	8.4	7.3	5.2	6.6

[a] Average based on wheat, corn, rye, and oats; included for comparison.
[b] Average based on first five columns from the left.
[c] Corn not included in this average. Of its protein 22 per cent is leucine.

group donor can be replaced at least in part by vitamin B_{12} in its role of facilitating syntheses involving these CH_3 groups. In practice, any reduced biological value of the proteins of meat, fish, and egg (or of any other feed) caused by shortage of cystine and methionine can be so easily and effectively corrected that it can be largely disregarded (assuming, of course, the correction is made). The feeder has the option of adding methionine, or vitamin B_{12}, or both.

Ash. Another characteristic of this group of feeds is their high ash, especially their high calcium and phosphorus. Whereas the plant products contain less than 1 per cent of either of these elements, and more often only 0.25 per cent, meat and fish meals run from 5 to 11 per cent calcium and from 3 to 5 per cent phosphorus. These high levels are, of course, due to the presence of appreciable amounts of bone. In general the higher the protein in either meat or fish meals, the lower the calcium and phosphorus. In many meal mixtures the desired supplementation of the basal feeds with these two minerals is accomplished by the use of meat or fish meals in amounts needed to adjust the protein quality or quantity.

Fat. Both tankages and fish meals may have widely different fat percentages. Fat in either of these products is nutritionally a liability. It is unstable and hence complicates feed storage. The onset of rancidity not only may adversely affect palatability but may result in residues that catalyze the destruction of oxidizable nutrients in the ration, especially vitamins A and E. Also, with the feeding of oily fish meal, there is the possibility of taints in milk, egg, and flesh, as well as the production of oily (or soft) pork. Hog carcasses graded "soft" are unsuitable for bacon.

Individually some of the feeds of this grouping have peculiarities which we should note especially.

Skim milk, for example, stands out by itself in this group of feeds. Its protein is almost perfectly digested and its biological value is usually rated as next below that of egg (actually its egg replacement value is about 96 per cent). Its calcium and phosphorus are relatively low compared with feeds that contain bone. Thus it can constitute a large fraction of the ration without introducing excessive minerals. It contains no hard-to-digest components, such as the tendons and ligaments that form a part of tankage. Nor has skim milk any damaging fat content. It is often used as an important source of protein in the rations of young animals. In this role its high riboflavin is also a decided advantage. Its relative, *whey powder,* is not a protein supplement in the usual sense, but in grain mixtures where the protein level is already adequate, its exceptionally high lysine and riboflavin content can often be used to advantage, even though its total protein is about that of basal feeds. Thus, hog rations which, because of their liberal high-protein wheat, already contain 15 per cent total protein, may be fortified with needed lysine by the use of this relatively cheap but low-protein dairy by-product.

Blood meal, for another example, is unexpectedly low in digestible protein. This peculiarity is due to the fact that hemoglobin is resistant to proteolytic enzymes, perhaps because of the effect of high heat in drying the blood. Furthermore, the effective biological value of its protein is low as compared with that of the other feeds of this category.

From Table 12-4 we can see that *tankage* may be a variable product in both fat and protein. The fat level often appears to reflect the market demands for soap fats. When these are in surplus, the tankage fat levels may rise, presumably as a secondary outlet for the fat. High-fat tankages are not only less stable than low-fat ones, but much lower in protein. If the tankage is to contain only 45 per cent digestible protein, meat meal is usually a preferable choice, since it is likely to be largely muscle trimmings, whereas tankage contains appreciable quantities of gut, tendon, and connective tissue, the proteins of which are of somewhat poorer biological value than are those of skeletal muscle.

High-protein tankages are usually prepared and standardized by adding necessary amounts of blood meal to a lower-protein tankage base. Tankage may contain up to 35 per cent blood, which probably does not improve its biological value. Present-day tankage does not contain appreciable amounts of glandular organs, since these are likely to be used as sources of extracts that eventually find their way into pharmaceuticals. Loss of the glandular materials has been detrimental to the feeding value of tankage.

High-fat *fish meals* also present problems we should discuss at this point. Some fish meals are by-products of the fish-filleting business, and consist of the entire fish (sometimes including entrails) minus the fillets removed. The fat content of such materials will depend in large part on the kind of fish. Thus white-fleshed fish, including cod, haddock, hake, pollock, skate, and monk fish, can be processed into the relatively low-fat "white-fish meal." Meals from herring or pilchard, on the other hand, are not by-products of filleting, but of the fish oil industry. These meals contain considerable oil, the amount depending in part on the freshness of the fish at processing. For pilchard, operators claim that if the fish are not processed within three days of being caught, it is impossible to produce, without solvent extraction, a fish meal of less than 9 per cent fat. Furthermore, poor processing also results in a high-oil meal. Thus fish meal containing more than 9 per cent fat is less desirable as a feed not only because of its oil, but also because its high oil content is indicative of a product made from stale fish, or that is the result of bad processing. These were the factors that led to the requirement in Canada at one time that fish meals of 9 per cent or more fat be labeled as *oily*. Fish meals that are residues of oil recovery by the "sun rotting" process are invariably oily, sometimes running as high as 20 per cent ether extract. Such products should be avoided in the feeding of farm livestock.

Another matter we should comment on before leaving the subject is the various kinds of fish meal that are available. For meat meal and tankage no indication is given in the name as to the kind of animal from which the material was derived. With fish meal, however, the labeling may indicate the kind of fish involved. Thus there are herring meals, sardine meals, pilchard meals, etc., as well as whale meal. Present indications are that these products are valuable largely in proportion to their protein content and that their limitations as feeds are usually proportional to their oil content. This is, of course, indicated on the guarantee.

One final word on fish meals concerns *salt*. Since there is an upper limit to the desirable salt (NaCl) content of animal (especially poultry) rations, the salt content of fish meals is sometimes a factor limiting their usefulness. The Canadian law requires that the percentage be specified on the tag if the meal contains more than 4 per cent by weight of salt.

SUGGESTED READINGS

Block, R. J., and H. H. Mitchell, "The Correlation of the Amino-Acid Composition of Proteins with Their Nutritive Value," *Nut. Abst. and Revs.*, XVI (1946), 249.

Morrison, F. B., *Feeds and Feeding* (Ithaca, N.Y.: Morrison, 23d ed., 1967).

CHAPTER 13

Vitamin and Mineral Supplements
and Miscellaneous Additives

We have discussed the basal feeds and the protein supplements largely in terms of their major nutritional characteristics. In the preparation of modern livestock rations, it is often expedient to employ one or more products as sources of certain nutrients or desirable characteristics they may impart to the ration. These products, which may not be feeds in the usual sense of this term, include vitamin and mineral supplements as well as flavors, binders, drugs, antibiotics, and animal fats. Generally speaking they have unique uses and hence must be dealt with individually.

Vitamins

Units of Potency. The presence in edible materials of nutrient substances needed for the survival and continued health of animals was discovered long before their chemical nature was learned. They were given the general name *vitamins* and the different vitamins were identified by letters, such as vitamin A or C. The potency of a foodstuff in any one of the first few vitamins discovered was originally expressed in terms of *units*. Later in order that different research workers could correlate their findings, international reference standards for certain vitamins were agreed on. One could then measure vitamin potency of foods and express the daily need of animals in terms of these units.

Today the needs of animals and the potency of foodstuffs in a vitamin are usually expressed in terms of weight (milligrams or micrograms), though *units of potency* is still a common term in referring to vitamins A and D, and also occasionally in referring to B_1 and B_2. We have thought it desirable, therefore, to define at the outset the international standards and international units of potency (IU) for some of these vitamins.

Vitamin A. The international standard for vitamin A is pure crystalline vitamin A acetate. One international unit (IU) of vitamin A is 0.344 micrograms of pure vitamin A acetate, which is equivalent to 0.3 μg of vitamin A alcohol. The Canadian reference standard contains 10,000 IU of vitamin A in each gram. It is distributed in capsules, each capsule containing 2,500 IU of vitamin A. The U.S. Pharmacopoeia (USP) reference standard is the same as the Canadian reference standard, and the USP unit is the same as the international unit.

Provitamin A or carotene. The international standard for carotene is a sample of pure beta carotene. One international unit of vitamin A is equivalent to 0.6 μg of beta carotene; i.e., 1 mg of beta carotene = 1667 IU of vitamin A. International standards for vitamin A are based on the utilization by the rat of vitamin A and/or beta carotene. Because other species do not convert carotene to vitamin A in the same ratio as rats, it is suggested that the conversion rates listed in Table 13-1 be used.

TABLE 13-1 *Conversion of Beta Carotene to Vitamin A for Different Species*

Species	Conversion of mg beta-carotene to IU vitamin A	Per cent IU vitamin A activity (calculated from carotene)
Standard	1 = 1,667	100.
Beef cattle	1 = 400	24.0
Dairy cattle	1 = 400	24.0
Sheep	1 = 400–500	24.0–30.0
Swine	1 = 500	30.0
Horses		
Growth	1 = 555	33.3
Pregnancy	1 = 333	20.0
Poultry	1 = 1,667	100.
Dogs	1 = 833	50.0
Rat	1 = 1,667	100.
Foxes	1 = 278	16.7
Cat	Carotene not utilized	
Mink	Carotene not utilized	
Man	1 = 556	33.0

Source: W. M. Beeson, "Relative Potencies of Vitamin A and Carotene for Animals," *Federation Proc.,* XXIV (1965), 924.

Vitamin B₁. The international standard for vitamin B_1 is pure synthetic thiamin hydrochloride. The IU is the potency of 3 μg (.000003 grams) of thiamin hydrochloride.

Vitamin B₂. This is pure riboflavin, and requirements are usually expressed as micrograms per day. If expressed in Bourquin-Sherman units, 400,000 units = 1 gram riboflavin.

Vitamin D. The international standard for vitamin D is pure crystalline irradiated 7-dehydrocholesterol (vitamin D_3). One IU is 0.025 μg of the international standard. The USP reference standard is a solution of the international standard containing 400 IU in each gram of solution.

Some useful information about the vitamins as a group has been assembled in Table 13-2.

Alfalfa and Grass Meals. Because of the variability of leafy forages in carotene (pro-vitamin A) and the frequent use made of such feeds as vitamin A sources, we have thought it desirable to comment further on alfalfa and grass meals.

Sun-cured and dehydrated greenstuffs, such as alfalfa and cereal grasses, are widely used in commercial balanced rations as sources of vitamin A. Average analyses indicate that dehydrated products range in vitamin A potency from 168,750 to 75,000 IU per pound (76,000 to 34,000 IU per kg), whereas sun-cured meals are highly variable but usually inferior. In terms of replacement, eight pounds (3.6 kg) of freshly processed dehydrated alfalfa meal, containing 75,000 IU of vitamin A per pound, will provide the vitamin A equivalent of one pound (454 g) of 1386A feeding oil. Since carotene in these meals deteriorates with age, particularly in hot weather, it is important from a practical standpoint to calculate the vitamin A content by analyses at the time of mixing. This calculation will safeguard the vitamin A level of the ration where the fullest economy of the vitamin A activity of dried greenstuffs is sought.

Vitamin B₁₂. This is one of the more recently discovered vitamins and because of its peculiarities we will consider it in more detail than can be included in Table 13-2.

After several years of research by many laboratories, the vitamin-like substance that was known to be present in a number of feeds of animal and marine origin, and to be responsible for spectacular increases in the growth of young animals when these feeds were in the diet, was identified as vitamin B_{12}. It is peculiar in that it appears to be solely a product of bacterial synthesis. It is absent from plant materials. Its presence in animal tissues is a conse-

TABLE 13-2 *Data on Vitamins Frequently Added to Rations*

Name	Function; units of potency, etc.	Deficiency symptoms	Good natural sources	Other sources (synthetic concentrates)	Supplementary amounts normally added/ton meal mixture[a]
Vitamin A (animal form) Carotene (plant form)	Stimulates the formation and development of body cells—hence a growth vitamin. Also involved in dim-light vision. 0.6 μg carotene = 1 IU of vitamin A.	Retarded growth of young. Interference with reproduction. Impaired night vision.	Leafy forage, fresh or preserved against oxidative loss. Milk or fish fat (variable). Yellow corn.	Pure crystalline vitamin A acetate, 3,000,000 IU/g. Mixed carotene, 1,000,000 IU/g.	1,000,000 IU
Vitamin D (D_2 is plant form; D_3 is animal form.)	Facilitates absorption of calcium, hence is related to normal bone formation. Irradiation by direct sun (ultraviolet rays) effectively supplies animals with needed vit. D. Poultry do not utilize D_2.	Rickets in young; osteomalacia in adults. Swollen and sore joints. Poor reproduction.	Sun-cured, leafy hay for animals (but not poultry).	Fortified fish oils, irradiated sterols for both poultry and animals. Crystalline D_2 or D_3 40,000,000 IU/g.	200,000 IU

Riboflavin	Part of an enzyme necessary for the oxidation process in all living cells. It is synthesized by microorganisms in herbivora.	Slow growth, nerve degeneration, diarrhea (pigs, calves). Curled toe paralysis, low egg production (poultry).	Liquid, condensed, or dry milks. Dried leafy forage. Distillery solubles. Dried brewers' yeast.	Pure synthetic riboflavin, 400,000 BS units/g. 1 g has potency of 100 lb skim milk powder.	300 mg
Pantothenic acid	A part of Coenzyme A, necessary for the utilization of all energy-yielding nutrients. Particularly important in Canadian hog rations since typical Canadian hog rations are often too low in this vitamin.	None in herbivora. Stilted gait in pigs, especially with hind legs (often called goosestepping and confused by some with rheumatism or rickets). Dermatitis around bill and eyes in poultry.	Brewers' yeast. Dry milk or whey. Cane molasses. Alfalfa.	Synthetic calcium pantothenate (92% pantothenic acid).	2 g
Vitamin B_{12}	A cobalt-containing substance active in treating pernicious anemia. Probably active part of "the animal protein factor."	Poor growth, poor feathering, and poor hatchability. Its use usually increases gains of young by 10–20%.	Feeds of animal or marine origin. Liver meal is especially potent.	Fermentation products and by-products from making antibiotics.	5–10 mg for all young. Adult herbivora synthesize it.

* Roughly equivalent to about one-fourth of the total needs of young animals (except for B_{12}).

quence of storage by the animal before slaughter. Its presence in such feeds as tankage or fish meal may be from bacterial activity in these products following manufacture. It develops rapidly in fecal material, and the eating of feces by pigs and poultry is undoubtedly one important way they obtain it. It is synthesized by rumen and caecum microorganisms and thus is available to adult cattle, sheep, and horses. The effectiveness of B_{12} supplements in rations diminishes as the ration contains increasing amounts of such feeds as meat, fish, or milk. It is probable that in some cases the quantities of tankage and fish meal previously believed desirable can be reduced, provided some B_{12} is added from another source.

Many older recommendations, based on both practical and experimental evidence, have called for 20 per cent of the protein for pig and poultry rations to be of animal origin. This percentage is thought to be higher than necessary to meet the amino-acid demands. Using one-half the previously recommended combination of meat and fish products in a typical young pig or chick ration should supply roughly half the B_{12} believed needed in such rations, and will also result in sufficient amino-acid correction to balance the plant protein.

Enough B_{12} to supply about half the total need will probably be a useful addition to rations intended for young animals. Although the requirements of these animals are not accurately known, we have evidence that it is not far from 16 milligrams B_{12} per ton (900 kg) of ration, if the meat and fish are being used in the amounts indicated. Typical samples of meat meal or feeding tankage are tentatively reported to contain 0.09 mg per pound (0.2 mg/kg) of dry substance. Fish meals may carry double this quantity.

Minerals

With the increasing use of mineral supplements in the rations, it seems advisable to indicate the more common supplementary sources of these nutrients together with notes concerning them. This information is summarized in Table 13-3.

The Fluorine Problem. Excessive fluorine intake of animals can be caused by: (1) forages subjected to air-borne contamination in areas near certain industrial operations that heat fluorine-containing materials to high temperatures and expel fluorides; (2) drinking water high in fluoride content; (3) feed supplements and mineral mixtures high in fluoride content; and (4) vegetation growing on soils high in fluoride content. In usual feeding practice an excessive intake of fluorine is not a problem unless rock phosphates are used. Consequently, the starting point in considering this problem is obvi-

TABLE 13-3   *Sources of Minerals and Their Potency*

Nutrient	Common source	Composition or potency	Remarks
Calcium[a]	Feeding bone meal	26% Ca	Contains also 18% protein and 11% phosphorus.
	Feeding bone meal (steamed)	29% Ca	Contains also 12% protein and 14% phosphorus.
	Bone char	27% Ca	Contains also 13% phosphorus but no protein.
	Tricalcium phosphate	13% Ca	10% P.
	Dicalcium phosphate	24% Ca	20% P.
	Monocalcium phosphate	16% Ca	12% P.
	Ground limestone	24–36% Ca	Balance is likely to be carbonate and magnesium.
	Calcium carbonate	40% Ca	
	Oyster and other marine shells	38% Ca	Shells contain on the average 96% $CaCO_3$.
Phosphorus	Bone meals and Ca phosphates	(see above)	
	Rock phosphate	14% P	Rock phosphate is 75–80% tricalcium phosphate. Not advised unless guaranteed to contain less than 1% fluorine.
	Defluorinated rock phosphate	18% P	Should not contain more than 1 part fluorine to 100 parts phosphorus.
Iodine	Potassium iodide	76% I	Potassium and sodium salts may be used interchangeably.
	Sodium iodide	84% I	
	Potassium iodate	59% I	
	Iodized salt		Stabilized iodine should be used. Amounts of iodine differ but .02% and .05% are commonly sold.
Iron	Ferric oxide	35% Fe	
	Ferrous sulphate	20% Fe	Ordinary copperas, commercial grade.
	Reduced iron	80–100% Fe	May be 20% ferric oxide.
Cobalt	Cobalt sulphate	34% Co	May be administered as a drench, as cobaltized salt, or as an ingredient in the ration.

[a] As sources of calcium these products are useful in direct proportion to the calcium they contain.

ously the phosphorus requirement. This determines eventually how much supplemental phosphorus may go into the meal mixture. The supplementary phosphorus needed will obviously be the difference between the cow's total requirement and that furnished by her feeds, roughage plus meal.

Although good roughage consisting of at least half legume materials will contain about 0.20 per cent phosphorus, poor roughage comprising relatively mature nonlegume plants cannot be depended upon to contain more than 0.10 per cent phosphorus. The feed manufacturer in designing meal mixtures and the supplements of minerals to go in them must deal with the problem of poor roughage.

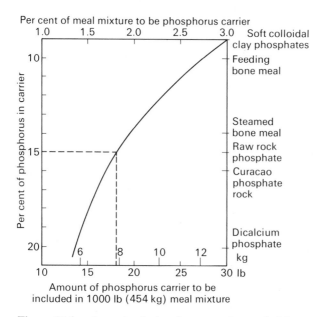

Figure 13-1. *Amount of phosphorus carrier needed in meal mixtures.*

We can calculate the probable supplementary phosphorus requirement of a 16 per cent protein dairy-cow meal mixture by taking certain typical figures for size of cow, production, and roughage fed (see Table 13-4). The quantity of phosphorus supplement that must be included will depend on the percentage of phosphorus in the carrier. These amounts can be read directly from Fig. 13-2. Of a carrier having 15 per cent P, 18 lb will be needed for 1,000 lb (or 9 kg per 500 kg) of mix.

The next problem is that of the fluorine. The Canadian Feeding Stuffs Act gives permitted tolerances in ready-to-feed meal rations for cattle of 0.009 per cent, or 90 ppm of dry matter. This is equivalent to 40 grams of fluorine in a 1,000-lb batch of feed, or 43 g in 500 kg of feed. If the per cent of fluorine in the phosphorus carrier is known, it is quite simple to calculate how much supplement can be incorporated in 1,000 lb (454 kg) of a meal

TABLE 13-4 *Supplemental Phosphorus Needed in a 16 Per Cent Protein Ration for a Milking Cow*

Daily requirement	
Maintenance of 1,000-lb (454-kg) cow	10 g
Pregnancy demands	7 g
Production of 30 lb (13.5 kg) of 4% milk	21 g
Total	38 g
Supplied daily	
In 20 lb (9.1 kg) average roughage	9 g
In 8 lb (3.6 kg) meal before supple-mentation	19 g
Supplemental phosphorus needed	
In 8 lb (3.6 kg) meal	10 g
In 1,000 lb (454 kg) meal mixture	1,250 g (2.75 lb)

mixture so that the concentration of fluorine will be 90 ppm as permitted by the Feeding Stuffs Act. Fig. 13-2 shows that if 18 lb (8 kg) of a phosphorus carrier are to be used, then it cannot contain more than 0.5 per cent fluorine. If one had a phosphorus carrier with 0.8 per cent fluorine, then only 11.5 lb of it could be used per 1,000 lb (or 5.7 kg per 500 kg) of ration.

More recent evidence indicates that these tolerances are too high. When computed on a forage or complete diet basis, the tolerances should not be higher than indicated in Table 13-5. By knowing the fluorine content of the forage and the mineral supplement, diets below these fluorine tolerances can be made up.

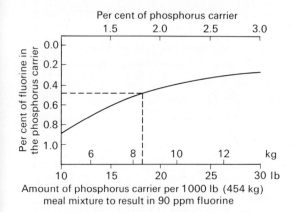

Figure 13-2. *Maximum tolerance of fluorine-containing phosphorus supplement in meal mixtures.*

TABLE 13-5 *Tolerances of Fluorine in the Forage or Complete Diet (Moisture Free) for Various Animals*

Animals	Fluorine tolerance	
	Breeding or lactating animals (F in ppm)[a]	Finishing animals with average feeding period (F in ppm)[a]
Dairy and beef heifers	30[b]	100[c]
Dairy cows	30[b]	100[c]
Beef cows	40[b]	100[c]
Sheep	50[b]	160[d]
Horses	60[c]	—
Swine[e]	70[b]	—
Turkeys[e]	—	100[f]

[a] Tolerances based on sodium fluoride or other fluorides of similar toxicity.
[b] Shupe, James L., "Fluoroses," *International Encyclopedia of Veterinary Medicine,* II (1966), 1062.
[c] Shupe, James L., Utah State University, unpublished data (1968).
[d] Madsen, Milton A., Lorin E. Harris, Ibrahim J. Alkhalisi, LeGrande Shupe, D. A. Greenwood, and Wayne Binns. *Am. Soc. Animal Prod. Western Sec. Proc.,* XXV (1954) LXXXV-1.
[e] Complete diet only.
[f] Anderson, Jay O., J. S. Hurst, D. C. Strong, H. Nielsen, D. A. Greenwood, W. Robinson, J. L. Shupe, W Binns, R A. Bagley, and C. I. Draper, "Effect of Feeding Various Levels of Sodium Fluorine to Growing Turkeys," *Poultry Sci.,* XXXIV (1955), 147.

In areas where fluorine is emitted from industrial plants and is contaminating pasture, hay, or forage used for silage, the following steps may be taken:

1. Grow grain on part of the land formerly used for these crops.
2. Increase the grain allowance in the diets.
3. Mix low fluorine hay with high fluorine hay to give a hay with less than 30 ppm of fluorine if it is to be fed to lactating or breeding cattle. If possible use high fluorine hay for finishing animals.
4. Feed phosphorus supplements with less than 100 ppm of fluorine.
5. If animals' teeth are severely damaged from fluorine, it may be desirable to chop the hay, soak small amounts of dry beet pulp before feeding, feed corn silage low in fluorine, and warm the water. (These are suggested emergency measures to be followed until the animals with damaged teeth can be sold for slaughter.)

Miscellaneous Additives

Sweeteners, Binders, and Flavors. Sweetening agents such as molasses, dextrin, and sugar are often found in meal mixtures, and fantastic claims have sometimes been made about their benefits. Most such claims can be written off as sales talk, and seldom originate from experimental stations. Sweet taste does not appear to be of any significance either in coaxing animals to learn to eat dry rations more quickly or in getting larger feed intake.

There may be some difference of opinion about whether *molasses* is a basal feed or should be classed as a special product. Its nutrient contribution to the ration is sugar. (The iron content of molasses is not usually of importance in the ordinary use of this feed.) Its protein is negligible and it contains no fiber or fat. Obviously this product is not freely interchangeable with other feeds of the basal category. Its more important contributions to the ration depend on its physical properties. Because of its sticky nature it tends to reduce the dusty, powdery nature of some finely ground feeds. In this role it often makes a feed mixture more acceptable to livestock. It is doubtful if the sweetness of molasses stimulates feed intake initially, but once accustomed to a sweetened ration animals for a time may not relish unsweetened rations. The effect of molasses in reducing dustiness can be obtained by slight moistening of a powdery feed with water, but this is only effective at the time of feeding, since the feed dries out on standing. Molasses, on the other hand, can be incorporated in the commercially prepared ration, does not adversely affect storage if not used in excess of about 10 per cent by weight, and results in a dust-free mixture acceptable to animals. Such mixtures more often than not contain products that may be powdery and heavy, and that are present in trace amounts only. The problem of maintaining a homogeneous mixture in such cases is sometimes simplified by the inclusion of 5 to 10 per cent molasses. If the feed is to be pelleted, the molasses or dextrin helps to form a durable pellet.

But molasses has another advantage. Because of its distinct flavor and aroma, it tends to mask or to dilute the flavors of other mixture ingredients. Thus the reactions of animals to the bitter taste of such feeds as rapeseed meal, ground buckwheat, or weed seeds, to the dry tastelessness of ground hay or oat hulls, or to the peculiar aroma and flavor of malt sprouts, may be modified by molasses. This use may be *all to the good* or *all to the bad,* depending on whether *we* are knowingly trying to utilize low-grade products or whether it is *someone else* trying to disguise the presence of such materials in a ration in order to pass it off as being first-quality.

Molasses in excess of 10 per cent of the mixture risks the producing of a caked and perhaps moldy condition in bagged feeds. Less than 10 per cent of molasses is enough to appreciably dilute the protein of a mixture, and this dilution must be considered when the mixture is either used as an ingredient or fed as a separate component of the ration, as when diluted and poured on poor roughage.

In connection with the problem of preparing durable pellets, it is worth noting that *sodium bentonite* may be of real assistance. Added at the rate of 2 per cent to a ground feed or mixture, it is innocuous nutritionally but facilitates the formation of a hard pellet that withstands the handling and shipping to which commercial feeds are subjected.

Feed flavors are available in wide assortments. They are usually essential oils, whose distinctive aromas will permeate the feed into which they are mixed. Their presence can be detected months after the feeds are treated. It is often claimed that use of a flavoring material will aid digestion and stimulate appetite. But animals in normal health, and for one reason or another not self-fed, will usually eat voluntarily more feed than feeders are prepared to offer them. The use of artificial flavors in "balanced rations" therefore gives one good reason to suspect that the mixture contains unpalatable ingredients; and since all high-quality feeds are palatable to the stock for which they are normally suitable, artificially flavored feeds are often suspect in the eyes of better feeders.

On the other hand, veterinarians may use flavors to cover the taste and smell of drugs in some tonics. Such use of flavor is quite a different problem, and of interest here only where a feeder has been induced to feed one or other of the many patented tonics or conditioners as a regular practice to prevent or cure ailments, real or imaginary, which he believes may adversely affect his stock.

Antibiotics are another class of "foodstuff" that must today be considered in the formulation of livestock rations. The nature of the action on the animal of the various antibiotics (aureomycin, terramycin, penicillin, etc.) as ration components is still not entirely clear. It is presumed that they affect the nature of the intestinal microflora. It is well-known that their use often results in faster gains of young animals (see also Chapter 8).

Many of the statements in the literature on antibiotics give erroneous impressions of the extent to which the use of these materials can be expected either to increase the rate of growth of the animal or to improve the efficiency of the ration consumed. A recent review of the published papers dealing with the use of antibiotics for swine showed that on the average the gains of young pigs can be expected to be increased about 15 per cent, and the efficiency of the ration improved about 5 per cent, by the inclusion of one or another recognized antibiotic. Another interesting finding in this survey was that if fish meal is a component of the ration there may be no response whatsoever to the antibiotic. Presumably fish meal already contains as much of these substances as an animal is able to utilize efficiently.

The necessary labeling of materials supplying antibiotics for feeding purposes is stipulated by the Canada Feed Act as follows: "Antibiotic feed supplement is a feeding material used for its antibiotic activity. It shall contain a single antibiotic or combination of antibiotics having growth-promoting properties. The name and amount of each antibiotic shall be declared on the label. It shall contain a minimum of one gram of antibiotic per pound."

Drugs, especially arsenicals, are sometimes added to a livestock ration because of their antibiotic-like action. Sometimes sulfonamides are employed in prophylaxis against coccidiosis (only with ruminants and swine). Although they are approved of and used in some countries, the question of their concentrations, restrictions, and precautions belongs in the field of the veterinarian. All such products are potentially harmful when improperly employed.

The effectiveness of medicated feeds for the purpose used, and the hazards to the consumer of the flesh of animals that have received such feeds, are still far from being understood, but are under active study by many research groups. It is possible that we may eventually be able to create within the animal an environment in which undesired parasites cannot thrive and at the same time be able to regulate at will some of their metabolic processes to emphasize functions desired at the time. Such developments will not, however, release the feeder from the necessity of providing rations containing the everyday operating needs of the nutrients already well-known.

Animal fats are still another type of ration additive. They are a by-product of the meat-packing industry and consist of the better grades of what are called in the meat trade "inedible fats" (tallows and greases). Tallows are fats with melting points above 40°C. Greases are fats melting below 40°C. Grades within each category are based mainly on free fatty-acid content and color, and only the top three or four grades are suitable for feeding purposes. In order to incorporate fats into feed mixtures, they are heated to about 150°–160°F. and slowly run into the mixing feed, where the fat coats the particles with a fine film.

These products are nonspecific sources of energy, and experiments indicate that they may be included in livestock rations up to about 6 per cent by weight when rations of high energy are wanted. In addition they help reduce the dustiness of finely ground feeds and facilitate pelleting by lubricating the dies through which the feed is forced in forming the pellet.

We must recognize that the protein, minerals, and vitamins of the mixture will be diluted by the addition of fat. In the preparation of such fats, an antioxidant is used to increase their stability. Since added flavorings might mask signs of rancidity, flavors should not be used in rations to which fats have been added. It would seem more sensible and economically sounder to curtail the production of excess fat on meat animals in the first place rather than try to salvage it by feeding it back to animals to produce, in turn, more surplus fat.

SUGGESTED READINGS —————————————————

Ault, Waldo C., Roy W. Rienenschneider, and Donald H. Sounders, *Utilization of Fats in Poultry and Other Livestock Feeds—Technology and Feeding Practices*. (USDA ARS Utilization Research Report No. 2, 1960).

Feeders' Guide and Formulae for Meal Mixtures (Quebec Provincial Feed Board, 1955).

Greenwood, Delbert A., James L. Shupe, George E. Stoddard, Lorin E. Harris, Harold M. Nielson, and L. Elmer Olson, *Fluorosis in Cattle*. (Ut. Agr. Expt. Sta. Special Report 17, 1964).

Hironaka, R., and J. P. Bowland, "Antibiotic Feed Supplements in Western Canadian Swine Rations," *Can. J. Agr. Sci.*, XXXIV (1954), 343.

Shupe, J. L., M. L. Miner, L. E. Harris, D. A. Greenwood, and G. E. Stoddard. 1963. "The Effect of Fluorine on Dairy Cattle. II. Clinical and Pathological Effects," *Am. J. Vet. Res.* 24(102); 964.

Wilder, O. H. M., *Storage, Handling, and Mixing of Fats into Feed* (Chicago: American Meat Institute Foundation, 1954).

CHAPTER 14

Forages and Roughages

General Characteristics of Forages and Roughages

According to the definition adopted in this book, forages and roughages are feeds with more than 18 per cent crude fiber. In practice, most roughages are forages. In applied feeding, feeds of this classification are primarily (though not exclusively) for rations of herbivorous animals, for which they play a physiological role in addition to supplying nutrients.

Like other feedstuffs, roughages can be partially described nutritionally in chemical terms. However, the concentration of many of the nutrients found in roughages is of minor importance in their use. For example, forages are good sources of several members of the vitamin B family, but since the micro-organisms of the digestive tract of herbivorous animals normally synthesize all the B-complex vitamins the animal needs, the extent of their presence in the forage is of no particular consequence. The mineral nutrients' abundance or relative deficiency in different forages is not ordinarily an important factor in the choice of forage. What is fed is what is grown, in the majority of cases, and any deficiency in mineral elements is corrected by supplementation in the form of mineralized salt licks, or by mineral fortification of the meal mixture, if such is also a part of the feeding program. Roughage that is to form the entire ration, however, may need a phosphorus supplement unless it consists of legumes.

Nor does the herbivore need to depend on the ration for any particular assortment of amino acids, though a minimum of protein equivalent is required. Here again microorganisms of the digestive tract synthesize all the amino acids needed by the host animal, apparently requiring from the forage only a source of nitrogen and carbon.

Concentration of Nutrients in Roughage. Most roughages on a dry basis yield less than 65 per cent as much metabolizable energy as grain feeds. Consequently, when the *nutrient content* of the roughage is compared directly with the requirement figures of feeding standards, the figures give an erroneous impression of the feeding value of the roughage. We are careful to adjust the composition figures of fresh milk to an equivalent dry-matter basis when we compare it with a feeding standard that is expressed in terms of a ration of 90 per cent dry matter. But we often make no adjustment for energy when we compare roughage with grain. Nutritionally all feeds should be compared on an equal energy-yield basis, for although weight of dry feed is correlated with its useful energy yield, there are large differences in the latter when grains or complete rations are compared with roughages.

When we compare feeds with each other or with an adequate ration *on a dry-matter basis,* we tacitly assume that differences in feed intake needed to yield equal amounts of useful energy is no problem, that the animals will voluntarily eat to meet their energy need if enough feed is offered. This is not necessarily true, however, where roughage is the entire ration, because the digestion of the principal energy-producing component of rough-age is not carried out by the cow or sheep directly. Rather it is done by symbiotic microorganisms in the rumen or caecum, whose activity in turn depends in part on how "well-fed" they are by the host animal's diet. As digestion rate is reduced, voluntary roughage intake is reduced. We shall return to this problem of intakes later in this chapter, but here we wish to note that the "adequate" level of some nutrient in a roughage (as in other feeds and in complete rations) depends not on its concentration in some quantity of the feed, but on its concentration relative to the useful energy of that feed. Useful energy of the dry matter of a concentrate feed is relatively constant, but that of a roughage may decline 40 per cent from fresh pasturage to the mature plant, largely because the rate of digestion is suppressed by progressive lignification of the cellulose with advancing maturity.

Table 14-1 compares four roughages, showing the concentration of energy and of nutrients as determined by the chemist, and the nutrient value per 1,000 kcal of metabolizable energy. From the latter figures it is clear that if enough of the legume roughage is eaten to meet the energy needs of the milking cow, she will have received enough of all the nutrients called for

TABLE 14-1 *Concentration of Nutrients in "Average" Roughage and Requirement of a Dairy Cow*

Energy and nutrients	Alfalfa hay	Red clover hay	Tim-othy hay	Oat straw	Daily requirement for a 400-kg cow producing 20 kg of 4% milk a day	
In terms of kcal ME/kg:	1,806	1,734	1,625	1,698	17.2 kg	34,640 kcal[a]
Dig protein (per cent)	15.6	7.9	3.1	0.7	6.7%	1,170 g
Ca (per cent)	1.48	1.41	0.32	0.30	0.33%	55 g
P (per cent)	0.23	0.19	0.17	0.09	0.25%	43 g
Co (mg/kg)	0.12	0.13	0.08		0.06%	0.1 mg
Mg (per cent)	0.29	0.39	0.15		0.13%	220 mg
Mn (mg/kg)	46.5	57.6	57.0	35.3	0.13%	220 mg
Vitamin A (IU/gm)	91.4	53.8	20.0		80	137,000 IU
In terms of 1,000 kcal ME:	1,000	1,000	1,000	1,000		1,000
Equivalent wt of feed in kg	0.550	0.575	0.615	0.590		0.580
Dig protein (g)	86.5	50.0	18.5	0.4		34
Ca (g)	8.2	8.9	1.9	1.7		1.3
P (g)	1.3	1.2	1.	0.5		1.2
Co (μg)	70	80	5			0.003
Mg (g)	1.6	2.5	0.92			0.003
Mn (mg)	25.8	36.5	35.0	21.0		0.003
Vitamin A (IU)	50,000	35,000	12,000			4,000

[a] Equivalent to 2,010 kcal ME/kg.

by the feeding standard. Nonlegume hay as the entire ration, however, will not meet the protein or phosphorus requirements, even if she voluntarily consumed the 10 per cent more of timothy or straw than of alfalfa or clover necessary to meet the caloric needs.

Significance of Energy of Roughage. In the final analysis the major factor in the use of roughages as feeds for milking cows is how much roughage the animal will consume relative to her energy needs. To the extent that less roughage is eaten than is necessary to meet energy requirements, meal feeding will be called for. The concentration of nutrients of such a meal mixture will depend both on the quantity of meal to be fed and on the specific nutrients it must furnish.

It is interesting and significant that Dr. Folke Jarl of the National Animal Experiment Station, Royal Agricultural College at Ultima, Uppsola, Sweden, uses the concentration of available energy (as starch equivalent) in 100 lb of dry matter of roughage—a figure somewhat comparable to per cent of TDN—as the index of roughage quality. He states, "The highest possible use of forage (pasture, ordinary hay, dehydrated hay, and grass silage) will depend on the concentration ratio of this forage in relation to the concentration ratio of other feeds (as concentrates). Protein content of forage need not be a limiting factor in its use, because the protein content of the concentrate mixture can be varied almost at will . . . the concentration ratio is the limiting factor in feeding large quantities of forage to high-yielding cows." *

This way of thinking about roughages—that they are primarily of use as sources of available energy—puts a somewhat unorthodox interpretation on the generally used classification of the hundreds of individual products having sufficiently different proximate analyses or botanical characteristics to be listed as different feeds. If the cow does not care which one she eats, then, in the interests of simplicity and realism, we might better classify roughages on the basis of their yield of available energy than on the basis of protein. Before going further with this argument, however, we should look briefly into the factors that influence the useful energy of roughages.

Form in Which Roughage Is Fed

Can we consider pasturage, silage, and dried forms of the same plant as nutritionally equivalent on a dry basis? Consider the data in Table 14-2

TABLE 14-2 *Energy and Protein in Fresh and Cured Alfalfa Dry Matter*

Form	ME (kcal/kg)	TDN (%)	DP (%)
Fed green	2206	61	15
Sun wilted silage	2097	58	12
Molasses silage	2133	60	12
Sun cured hay	2024	56	12

for products made from comparable alfalfa. These figures suggest that in the harvesting of hay, losses occur which reduce the digestibility of the final product. Moore and Shepherd have summarized studies at the experimental farm of the U.S.D.A. at Beltsville, Md., from which it appears that for field-cured second-cut alfalfa, one may expect, from the condition at cutting to the condition as fed, percentage losses of 35 per cent for leaves, 20 per cent for dry

* Folke Jarl, "Use of Forage in Dairy Production," *Sixth International Grassland Congress,* State College, Pa., II (1952), 1179.

matter, including leaves, and 29 per cent for proteins. These figures do not mean that such hay will be "all stem." If *as cut* the crop were 55 per cent leaf by weight, the hay *as fed* would be 34 per cent leaves and 66 per cent fine and main stem. This proportion is about the leafiness of excellent clover hay.

These losses actually reduce the TDN of the hay as finally fed *much less* than we might suppose. If we assume that the crop as cut was 55 per cent leaves, and if we use a typical TDN value of 57 per cent for alfalfa leaves and of 43 per cent for alfalfa stem meal we can calculate

$$55\%(57\%) + 45\%(43\%) = 50\%$$

as the expected TDN for the hay without any loss of leaves. Similarly, we might calculate for the crop after a 38 per cent loss of leaves,

$$38\%(57\%) + 66\%(43\%) = 47\% \text{ TDN}.$$

Leaf loss will also involve a loss in digestible crude protein, but this can be made good, if necessary, by the meal mixture. One is forced to the conclusion that, although a 20 per cent loss in dry weight in the harvesting of hay may be *economically* important, the change in energy value per pound of feed may not be excessive. It amounts to something like 6 per cent in this example. This loss is small because most of the energy of the roughage comes from cellulose or hemicellulose, which is not reduced in proportion to the harvesting losses.

The energy values of poor-quality hay are not enough below those of good hay to be the major cause of the poor results so often obtained in practice with poor hays (see Table 14-3). Recent experiments indicate that part

TABLE 14-3 *Per Cent of TDN in Good and Poor Hay*

Hay (as 90 per cent dry matter)	Good quality	Poor quality
Alfalfa	51	49
Clover	56	52
Clover and timothy	50	47
Timothy	53	49

of the cause of the poor results with low-quality hays is their failure to support a maximum microflora in the rumen, because of which digestibility of crude fiber suffers. The more serious effect, however, is a secondary one: *as rumen microflora activity declines, roughage consumption declines.* Since a cow obtains anywhere from half to almost all her TDN from roughage, any decrease in consumption is serious.

Perhaps we have said enough about the peculiarities of roughage to establish the fact that the feeding value of forages, particularly hays, depends very largely on the quality of the product. However, the term *quality* is often purely relative, and it may be well at this point to consider briefly the problem of hay grading.

Hay Grading

For those hays that are composed essentially of one kind of plant, the name (such as timothy, alfalfa, clover) is quite definite in its meaning. However, there are other names used for hays that are much less specific. *Grass hay,* for example, includes all of the grasses that are made into hay other than those that are named according to species. Grass hay may include edible sedges and rushes as well as the mixtures that are sometimes harvested from permanent pastures or from meadows that have been unplowed for some time and which have reverted to the indigenous plants of the area.

Mixed hay is a catchall term for all mixtures that are at least half plants for which standards have been separately established. Thus, for a product that was 52 per cent timothy and the balance other grasses, the term *mixed hay* would likely apply. Sometimes mixtures are given specific names such as "mixed clover and timothy" or "mixed alfalfa and timothy," particularly where such mixtures were intentionally seeded.

Within these names, grades are established to describe the quality of the product. The particular numerical rating of different grades is of no consequence to us at the moment. It will be interesting, however, to note those factors that determine the grade or quality of the different hays.

One of these is *color.* It is generally assumed that greenness indicates the feeding value of a hay. Much greenness usually indicates that the hay was cut early, was properly cured, will be palatable, and is free from dust and mold. Also, the intensity of green color is often correlated with the amount of carotene present in the product.

Livestock feeders frequently question whether as much dependence on color as an index of quality is justified as graders imply. Recent evidence indicates that the preference of cattle for different types of hay does not seem to be correlated with color, and we know that the preference of cattle for forage is a reasonably good index of its energy value.

The extent of the *foreign material* in the hay is another factor in grading for quality. Foreign material includes weeds, grain straw stubble, thistles, and other material which has little or no feeding value. Feeding tests at the U.S.D.A. indicate quite clearly that the proportion of hay refused by cattle is correlated with the amount of foreign material in it. The cattle refused almost twice as much of the offered hay in grades that contained appreciable quantities of foreign material as they did of the same sort of hay containing relatively little.

Leafiness is a grading factor in all classes of hays, because it reflects the protein content of the hay. Leaves contain from two to two and a half times as much protein as the stem of the same plant, regardless of the kind of plant. The leaves are also the richest in carotene, calcium, and phosphorus. Leafiness is much dependent on the stage of maturity of the plant when it was harvested. The more mature the plant when cut, the more easily are the fine leaves lost in handling. Consequently, hays harvested from plants cut at a relatively immature stage will generally retain a greater leafiness in the crop as fed than will plants cut when more nearly ripe.

On the other hand, *stage of maturity,* though it does have a considerable influence on the feeding value of hay, is not normally a factor included in grading hay. It has been assumed that leafiness and color, which are affected by stage of maturity, thereby become indirect indices of the maturity of the plant at cutting time. Actually an idea of the maturity at which most of the grasses were cut for hay can be rather easily determined. If the stem above the top joint can be pulled out easily from the sheath, and its lower part is found to be dark, it was tender and still growing when cut, and was cut before bloom. In the development of the grass, growth takes place at this point until just before it begins to bloom. If the top portion of the stem cannot be pulled out, then maturity can be determined by rubbing the heads in the palm of the hand and examining the maturity of the seed. Unfortunately, there is no satisfactory way of determining from the cured hay the stage at which the legumes were cut. Sometimes evidence can be found in the presence of flower petals that show color, and for very late cutting, the presence of seed pods. Maturity affects the protein and the fiber content of all kinds of hay, the protein content decreasing and the fiber content increasing with advancing age.

Though potentially useful, hay grading has never been much depended on by feeders to describe quality of roughage. In practice, feeders commonly refer to their hays as *good* or *poor*. These relative terms are often employed to compare samples without reference to the excellence of either one in feeding value. Two lots of hay may both be of low feeding value, but one is still referred to as *good* as compared with the other. We should define what constitutes *good* hay in terms of its feed value.

What Constitutes Good Hay

Maturity at Cutting. As we have already indicated, maturity at cutting is a most important factor in hay quality, and we cannot too strongly emphasize that, regardless of the species of plant, the best-quality hay can be made only from a crop cut well before maturity. The reason stage of maturity is such an important factor in hay quality is that it determines the proportion of leaf to stem in the product. The stems of forage plants are invariably of low feeding value. Every feeder recognizes that straw, which is almost entirely stem, is of low feeding value, whereas cereal grain crops cut before much stem develops can be made into excellent hay.

The rapidity with which the stem increases with maturity is such that a delay in cutting of only a few days may actually mean a large loss in feeding quality of the hay made. This is strikingly illustrated in data from experiments at Macdonald College shown in Table 14-4. It is worth noting that, in the 11

TABLE 14-4 *Proportions by Weight of Leaves and Stems in Red Clover Hay According to Stage of Maturity When Cut*

Stage of maturity when cut	Days between stages	Per cent leaves and fine stems	Per cent main stem
Prebudding	—	75	25
Prebloom	5	51	49
Early bloom	11	34	65
Full bloom	10	30	70
Late bloom	18	25	75

days from the time the heads were forming to the early bloom stage, there was a 33 per cent increase in the weight of coarse stems in the crop, and a corresponding decline in the proportion of fine leafy material.

All shallow-rooted crops used for hay are likely to be at the ideal maturity for cutting just when there is poor curing weather, since the very thing that makes for high feeding value in these hay crops is surface moisture. As soon as the moist conditions change to drier and warmer ones, the production of new leaves ceases and the plant starts to make stem and seed. It is this situation which justifies the use of grass silage or of artificial drying schemes for preserving the high feed value of forage crops raised on the farm.

In connection with the hay-grading schemes that are used, we pointed out that stage of maturity is not one of the factors directly considered. There is doubtless justification for ignoring this factor for hays to be sold commercially, but most of the hay fed to livestock is grown on the farm where it is fed. Consequently, there is no problem in knowing the stage of maturity of any particular hay. From the farmer's standpoint, stage of maturity becomes a particularly important criterion of when to cut his hay crop. It is also an index of the feeding value he can expect in the final product.

Kind of Plant. Legumes can be made into hay of higher feeding value than grain plants or nonlegume grasses. The reason is almost entirely the difference in proportions of leaf to stem. The young tender leaves of all common plants used for hay are apparently of about equal feeding value for herbivorous animals. Legumes, however, have more leaves per pound of dry plant than grasses. At the stage of maturity necessary for the best hay, the leaf in timothy is about 30 per cent, in clover 40 per cent, and in alfalfa 55 per cent of the dry weight of the plant.

Mechanical Losses in Curing. A third factor in the quality of hay is the loss in curing. We have already pointed out that this loss, in percentage, is sometimes more important economically than nutritionally. Nevertheless, the parts of the plant lost in greatest quantity are the leaves, and since this loss is one factor in determining how much of the crop will be eaten by livestock, it obviously affects the quality of the hay.

Good Hay Defined. In order that the feeder may have some practical basis for judgment in evaluating the quality of his hay, the following might be used as a standard for *good* hay. The proportion of leaf in the total weight of the dry feed should be at least 28 per cent for timothy, 35 per cent for clover hay, and 50 per cent for alfalfa hay.

In addition to being leafy, good hay should be cut in the very early bloom stage, cured rapidly, and taken into the barn without being wet. Such hay will be of high feeding value, because it will be rich in protein, high in minerals, and high in Vitamin A, but, above all, its energy yield will be at a maximum. Hay that does not meet these specifications will not be top-quality in nutritional value, regardless of any other considerations.

Low-Grade Roughages

What we have said so far applies to pasture crops, grass silage, and the ordinary hays. There is, however, another group of roughages, which, though not of as much importance as hay or silage, nevertheless represents a type of feed which is often available on the farm and, though of low-energy value, is fed in order to make some use of it. Such roughage is not ordinarily fed to producing stock, but is often made available in stacks or racks to which cattle on maintenance rations have ready access. The feeds of this group are not important in determining the nature of a suitable meal mixture (if any) to be fed also to such animals.

In this subgroup there will be some of the poor-quality hays made from miscellaneous grasses and often badly contaminated with weeds. Some hay crops that have matured to the seed stage will also be included, as will some prairie hays. Below these will come straws made from the cereals, threshed hays, and finally the sedges and swamp grasses that are harvested in order to have a roughage of some sort available under adverse conditions. The best of the products in this low-energy group yield only about three-quarters as much energy per pound of dry matter as the standard roughages, and some of them may be even lower in relative value. None of them can be depended on as sources of vitamin A, and usually the protein content will be almost negligible, perhaps 1 or 2 per cent.

Lignification. These low-energy roughages are not necessarily low in cellulose or hemicellulose; in fact, they are likely to be higher in crude fiber than the standard hays or silages. The chief reason for their low feeding value is usually lignification of the cellulose. It is believed that lignin is a protective coating laid down in the plant structure in association with cellulose to give it rigidity and durability. Lignification starts after the growth in a given part of the plant ceases. In other words, it is one of the changes that occurs in the plant structure with advancing maturity. Experiments have indicated that lignification starts at the bottom of the stem and proceeds upward, following along behind the areas that have ceased to grow.

Lignin contains a phenolic nucleus, and it is perhaps this feature that protects it from bacterial attack. Whatever the specific cause, we know that lignified cellulose is attacked very slowly, if at all, by the microorganisms of the intestinal tract. Examination of the fecal residue of roughages consumed by ruminants shows that the only particulate portions of the plant tissues that are found intact in the fecal material are those that stain with phloroglucinol, indicating that they are lignified.[*]

Not only do the stems of plants become lignified with maturity, but the leaves and husks, and even the hulls of grains such as barley and oats,

---

[*] W. J. Drapala, L. C. Raymond, and E. W. Crampton, "Pasture Studies. XXVII: The Effects of Maturity of the Plant and Its Lignification and Subsequent Digestibility by Animals as Indicated by Methods of Plant Histology," *Sci. Agr.,* XXVII (1947), 36.

are thus protected. Some plants, such as coarse rushes and sedges that appear to have no stem distinct from leaf, are also highly lignified and, consequently, their cellulose is not easily digested by the animal.

There is no evidence that the lignification of plant tissues deters their initial consumption. It seems rather that the failure to consume such low-grade roughages in amounts equal to those eaten readily of high-grade forage reflects the decreased activity of the microorganisms of the rumen (and of the caecum, where the caecum is functional). Whatever the cause, there is no dispute among feeders that livestock consume high-quality fodders better than low-grade roughages.

Some improvement in the consumption of low-grade roughages appears to be possible by supplementing such rations with additional protein and with a mineral mixture. It is believed that such supplementation provides nutrients that are deficient or unavailable in the roughages, but which are required for the full activity of the microorganisms. Under such supplementation it has been found possible to obtain about as much TDN from some of the low-grade products, such as corn cobs and straw, as is obtained normally from average-quality hay. We do not know the extent to which this represents a more active attack on the roughage by microorganisms, except by inference.

An Energy Classification of Roughages

If we attempt to group roughages according to their yield of available energy, we find that almost all the well-known ones can be fitted into six subgroups that differ by units of 5 per cent of TDN (see Table 14-5). We can give an indication of what products belong in each of these subgroups in general terms with sufficient accuracy for the practical use of the table. In general, well-grazed pasturage which is actively growing will be in the top energy class among the roughages; in energy yield such products may be the equivalent of some of the light, bulkier grains such as poorly filled oats, or of some by-products such as wheat bran. If these products are analyzed chemically, the pasturage will show a higher crude-fiber content than the grains or by-products with which they may be comparable in energy. This result merely illustrates that in about 25 per cent of the cases recorded in textbooks on feed consumption and utilization, the crude-fiber portion of the feed is higher in digestibility than the nitrogen-free extract component. Consequently, it does not always follow that feeds with the higher fiber content are also lower in available energy. Forage belonging in this top grade is readily eaten by all herbivorous animals, and grazing dairy cows will consume up to 150 lb (70 kg) per head per day—equivalent to nearly 50 lb (23 kg) of

TABLE 14-5 *Classification of Roughages by Available Energy*

Probable available energy			Forages	Daily voluntary intake of dry matter as per cent of weight of cow	Probable supplement needed for producing stock
TDN (per cent)	DE (per kg)	ME (per kg)			
60–65	2,750	2,250	Well-grazed, actively growing pasturage.	3.0 or more	P
55–60	2,550	2,075	Hay-crop silages, including cereals. Corn silage (dough stage).	3.0	ME, P
50–55	2,350	1,900	Properly cut and harvested hays from reasonably pure stands of alfalfa, clover, timothy, or mixtures of them. Average corn silage.	2.5–3.0	ME, P
45–50	2,100	1,725	Hays from nearly mature stands of otherwise suitable hay crops (see above). Dormant pasturage.	1.5	ME, prot, P, Ca
40–45	1,850	1,525	Hays from miscellaneous grasses often carrying weeds. Some hay crops in seed. Prairie hays.	1.5–1.0	ME, prot, P, Ca
35–40	1,650	1,350	Straws from cereals. Threshed hays. Sedges and swamp grasses.	1.0 or less	ME, prot, P, Ca

air-dry feed or 30 lb (13.5 kg) of TDN. Obviously, such material requires no supplementation for average production conditions.

Hay-crop silages and corn silage from corn cut in the dough stage are only a little less valuable. Any decline in energy yield will be due to a greater portion of stemmy lignified tissue. This material the animal normally avoids during grazing. Present-day management recommends that not over half the total roughage for a milking cow be supplied in the form of silage. It is evident that replacing a part of the hay with a dry-basis equivalent quantity of silage does not reduce the feeding value of the total roughage allowance. Actually, there is probably more feeding value in one pound of dry silage material than there is in one pound of most of the hay crops fed.

It is with feeds belonging in the next two categories that most of the feeding problems arise. These categories cover the majority of typical dry roughages that will be fed to producing herbivorous animals—animals that require additional energy and perhaps additional protein beyond what the roughage allowance will supply. The third category will include all the high-quality hays made from legumes and also a few of those hays that are made from early-cut nonlegume hay crops. Nonlegume hay, however, in order to get into this category, will have to be exceptionally good for its type. We should also place in this category average corn silage.

The fourth category will include in general the nonlegume hays that were too mature when harvested to be as digestible as they might have been earlier. Here also should be classed dormant pasture, pasturage available during the dry periods of summer when there is little growth; and also winter pasture, which actually is grasses cured on the root and harvested by grazing.

In the practical use of roughages the differences between these two categories will be traceable more to differences in the quantities animals will consume voluntarily than to differences in chemical composition. Whereas forages belonging to the third category will usually be consumed voluntarily by cattle at a daily rate of from 2½ to 3 per cent of their weight, those of the fourth category are likely to be consumed in quantities not much exceeding 1½ per cent of the live weight of the cow. It is this difference in the quantity consumed that is the major factor in determining the quantities of meal mixture that must also be supplied in order to meet the full energy requirements of producing animals.

With the last two categories, we are normally dealing with a feeding practice intended primarily to maintain nonproducing stock over a winter period, or at some other time when their energy requirement is relatively low. These materials are normally cheap and are fed as a means of converting them into something useful. Their efficiency is not a major consideration.

The significance of this energy classification of roughages will become more evident when the problem of meal mixture formulation is discussed later in this book.

A Nutritive Value Index for Forages

One of the frustrations for nutritionists interested in the feeding value of roughage has been that, with all their book knowledge and technical aids, no dependable scheme they could devise would consistently rank forages in feeding value, whereas the dumb ruminant animal could unerringly detect differences in nutrititve values. Indeed, it was this ability of the animal that led the nutritionists working on forage problems in 1956 at Macdonald College to think that we might be further ahead in our efforts to describe forages if, in addition to chemical and biological data, we incorporated in the description a figure for voluntary intake as representing the animal's own measure of its feeding value. As a result of such studies, Crampton *et al.*

in 1960 * formulated a numerical nutritive value index (NVI) for forages which appears successfully to rank them in effective feeding value as measured by the gains in live weight of animals subsisting on them. Their hypothesis concerning the effective feeding value of forages can be summarized as follows.

1. The feeding value of a forage depends primarily on how much it contributes toward the daily energy needs of the animal, which in turn depends almost completely on how much of the forage is voluntarily consumed.

2. Voluntary consumption of forage is limited primarily by the rate of digestion of its cellulose, rather than by its nutrients or the completeness of their utilization.

3. The rate of rumen digestion may be retarded by any one of numerous circumstances which interfere with the numbers or activity of rumen microflora, the active agents of cellulose degradation. These include lignification from advancing maturity of the forage, partial starvation of the microflora from nitrogen or mineral deficiency, and the presence of bacteriostatic agents.

4. Following alimentation, hunger recurs as a consequence of some (perhaps fixed) degree to which the rumen load has been reduced by microbial digestion, and the (largely *in situ*) absorption of end-products. The vigor of microbial attack on the ingested material governs the rate of rumen-load reduction, hence the frequency of recurring hunger, and hence in large measure of the forage consumed per unit of time.

5. The rumen load will be reduced to the point at which hunger recurs after time periods characteristic of the specific forage involved. Thus intrinsic differences between forage species or varieties, or extrinsic differences between samples or crops of forages of the same species or variety (caused by grinding, chopping, stage of maturity, anatomical part of plant fed, nutrient supplementation, etc.) may cause forage, fed *ad libitum,* to be ingested at different rates, and hence to be consumed in different amounts in 24 hours.

6. The *effective feeding values* of a forage depends jointly on how much of it an animal will voluntarily consume (daily) and the completeness of its energy digestibility. Hence both these terms should be in the description of the feeding value of the forage. One simple way of combining these two factors is: (1) to express the intake relative to some defined standard forage which has been assigned a value of 100; (2) to express the completeness of energy digestion as a percentage; and (3) to multiply the two terms together.

NVI = relative daily intake × digestibility of energy.

* E. W. Crampton, E. Donefer, and L. E. Lloyd, "A Nutritive Value Index for Forages," *J. Animal Sci.,* **XIX** (1960), 538.

The Standard Forage. The terms relative intake and standard forage require comment. Clearly one could choose any of several possible standards. An arbitrary intake figure of 3 lb hay per 100 lb live weight daily (LWD) or 3 kg per 100 kg LWD would be one choice, but the kind and quality of the standard forage would need to be defined, since these affect the voluntary intake. A more useful figure might be obtained by a direct feeding trial, using a forage that successful feeders had found would be eaten in the largest amounts by cattle or sheep, namely, a leafy legume hay, cut during early bloom, field chopped and dehydrated by aeration. Sheep-feeding trials confirmed this premise, showing that larger amounts of such legumes as bird's-foot trefoil and red clover were eaten in 24 hours than of some other similiarly prepared common roughages; and also that consumption of such forage was quite uniform, at an average of 1,232 grams per day by sheep averaging 38.1 kg in weight.

By expressing this intake per unit of metabolic size (i.e., $W_{kg}{}^{.75}$), all of the correlation between size of sheep and intake was eliminated ($r = 0.076$; $r^2 = 0.58\%$), which meant this intake per unit of $W_{kg}{}^{.75}$ was applicable to yearling sheep regardless of their size. From these findings a standard intake per unit of metabolic size was computed as: $1232/(38.1)^{.75} = 80$ grams of chopped "standard" air-dry forage. Thus the expected intake of a top-quality legume hay fed chopped was taken as eighty times the metabolic size of the animal. It is perhaps not merely a coincidence that the consumption of this legume forage turns out to be

$$\frac{(1,232 \text{ g} \times 100)}{454 \times 38.1 \text{ kg} \times 2.2} = 3.2 \text{ lb per } 100 \text{ lb LWD.}$$

Thus our figure for the intake of the standard forage is, in fact, essentially 3 lb per 100 lb LWD as normal roughage allowances to sheep. It appears from further work that the expected daily intake of the "standard forage" by cattle might be taken as 3 lb per 100 lb LWD, or 78 g per unit of $W_{kg}{}^{.75}$.

Relative Intake. In the equation for computing the NVI of a forage, the relative intake (RI) is merely the observed intake of the forage under test, divided by the expected intake of the standard forage, and the quotient multiplied by 100. It is the amount of the forage voluntarily eaten by a sheep or steer relative to each 100 units it would be expected to eat of the standard forage. For example, assume a 50-kg sheep eats 1,100 grams of a forage in 24 hours. The expected intake of the standard forage would be computed as:

$$(50_{kg}{}^{.75}) \times 80 = 18.8 \times 80 = 1,504 \text{ grams.}$$

The relative intake of this forage is then (as a whole number):

$$RI = \frac{1{,}100 \times 100}{1{,}504} = 73.$$

Energy Digestibility. The digestibility of the energy of the forage under examination is determined by a conventional *in vivo* digestion trial, using accepted methods of determining the energy in the forage and in the corresponding feces. The apparent energy digestibility (*in vivo*) of the standard forage was found to be 70 per cent.

Computation of the NVI. In the above example, assume for the forage being studied an energy digestibility of 60 per cent, from which we compute,

$$NVI = RI \times \text{per cent dig}$$
$$= 73 \times 60\% = 43.8.$$

For the standard forage, the NVI is, by definition,

$$NVI = 100 \times 70\% = 70,$$

and we estimate the effective feeding value of our example forage to be $43.8/0.70 = 62.5$ per cent as good as our standard forage (i.e., top-quality legume hay).

NVI in Application. Feeding trials have consistently shown statistically significant correlations between forage intake and body-weight gains of sheep on the order of $r = 0.5$. When the same data for the voluntary intake have been expressed in terms of the standard forage, and the relative intake multiplied by the per cent of digestibility of the forage energy, the resulting nutritive value indices for these forages have shown correlations of $r = 0.88$ to 0.94 with the corresponding body-weight gains. In our studies a change of 40 grams gain in body weight per day by sheep reflected a difference of seven or eight units of NVI between the forages. Some typical NVI data for a few common forages fed as the entire ration to sheep are shown in Table 14-6, and illustrate the differences that are found between samples depending on season and on maturity when cut.

Statistical studies have shown that the relative importance (that is, the relative statistical weight) of the RI and of the energy digestibility on the numerical value of the NVI is 70:30, respectively. This is evident in the data of Table 14-6, where the range of RI is shown to be 50 units as compared with 17 units for energy digestibility. The consistently lower feeding values of the more mature forage within species is clearly shown.

TABLE 14-6 *Nutritive Value Indices for Typical Dry, Sun-Cured Forages*

Forage	Year of harvest	Maturity at cutting	RI	In vivo dig of cellulose	NVI
Red clover	1956	unspec.	106	67	71
Alfalfa	1959	early bloom	116	61	71
Alfalfa	1959	late bloom	107	60	64
Bird's-foot trefoil	1956	unspec.	99	63	63
Alfalfa	1959	full bloom	102	60	62
Red clover	1957	early bloom	98	55	54
Brome grass	1959	early bloom	82	62	51
Alfalfa	1956	unspec.	79	63	50
Red clover	1957	late bloom	92	53	49
Brome grass	1956	unspec.	71	60	43
Brome grass	1959	full bloom	76	55	42
Brome grass	1959	late bloom	77	54	42
Timothy	1957	early bloom	66	58	38
Timothy	1957	late bloom	69	50	34

Prediction of the NVI from In Vitro Cellulose Digestion. One of the practical limitations to the usefulness of the NVI, as with any scheme for describing the feeding value of forage that involves biological measurements, is that of cost. Animal feeding trials are costly in time and equipment, and require large quantities of forage for each trial. Hence the possibility of employing an *in vitro* fermentation (digestion) of the forage cellulose by inoculum obtained from the rumen of a fistulated bovine to predict the *in vivo* NVI of the forage has been examined by the Macdonald College workers.*

These *in vitro* studies yielded two fundamental findings: (1) that the curve of cellulose disappearance with fermentation time showed an initial lag and a subsequent rate for 24 hours that was characteristic of the species of plant and its maturity, the greatest differences being evident at the end of 12 hours of fermentation; (2) that the fermentation was largely completed after 24 hours, and the marked differences between forages had largely disappeared by 36 hours.

Statistical studies on the *in vivo* and *in vitro* data revealed highly significant simple and multiple correlations, as shown in Table 14-7. The correlations make it clear that the NVI values (*Y*) determined *in vivo* can be predicted

* E. Donefer, E. W. Crampton, and L. E. Lloyd, "Prediction of the Nutritive Value Index of a Forage from *In Vitro* Fermentation Data," *J. Animal Sci.,* XIX (1960), 545.

TABLE 14-7 *Correlations of RI, DE, and NVI with the In Vitro Cellulose Digestion of Forages*

Variables correlated	Coefficients	
	r	r^2
12-hour *in vitro* cellulose digestion with:		
in vivo relative forage intake	0.83	0.69
in vivo digestible energy	0.87	0.76
in vivo NVI	0.91	0.83
12-hour and 24-hour *in vitro* cellulose		
digestion with *in vivo* NVI[a]	0.91	0.83

[a] Relative beta values for determining NVI were 12-hour fermentation = 97%, 24-hour fermentation = 3%.

from the 12-hour *in vitro* fermentation of the cellulose (X) of the forage by the simple regression equation,

$$Y = c + bX.$$

If a forage is ground for feeding, it will consistently be eaten in larger amounts than when fed chopped. This, of course, raises its RI and hence its NVI. Experiments have shown that the regression ($b = 1.23$) of the NVI on the cellulose digestibility (with rumen inoculum) was the same for ground as for chopped forage, but the constant (c) in the equation was increased by 10.9 units. Thus the general *in vitro* prediction equation of the NVI of a forage can be written as:

$$Y = c_0 + bX + c_1,$$

where c_0 is the basic equation constant (-3.5) and c_1 is the increase of 10.9 units in the NVI of a forage if it is ground instead of chopped for feeding.

A Laboratory Method for NVI. The significance of the high correlation between the *in vivo* NVI and the 12-hour *in vitro* digestion with rumen inoculum of the cellulose of a forage is that an estimation of its effective feeding value (i.e., its NVI) becomes a matter of a laboratory test involving only a fistulated steer as a source of inoculum, a sample of the forage (a kilogram is adequate), and the usual facilities of a chemical laboratory. A test requires two days, and the number of samples digested simultaneously is limited only by the number of digestion tubes in the digestion unit. A 36-tube controlled-temperature water bath requires bench space of less than a square meter. The amount of sample needed is so small that it can be supplied from experimental agronomy plot clippings, thus making the method useful in forage-production studies of the inherent nutritional differences between forage crops, and of the effects on feeding values of climatic factors, cultural practices, and harvesting methods.

Donefer *et al.** proposed in 1966 at Macdonald College an all-chemical *in vitro* technique for predicting the NVI of forages; in it a solution of 0.2 per cent pepsin dissolved in 0.075N HCl, was found to be highly correlated ($r = 0.95$) with the NVI of forages as measured *in vivo*. The regression equation for predicting the forage NVI (Y) from *in vitro* dry-matter digestibility (X) proved to be:

$$Y = -0.75 + 1.60X.$$

Correlation studies showed that the *in vitro* dry-matter disappearance values accounted for 90 per cent of the variability in the *in vivo* NVI measurements of some 35 different forage samples for which *in vivo* digestibility and the NVI had been determined. It seems probable that the above equation can be used for estimating the NVI of the forage from its *dry-matter* disappearance in the *in vitro* technique employed.

Thus it appears that there is now available an *in vitro* technique for estimating the effective feeding value of forages that constitute the sole ration of ruminants, either by employing a rumen-fermentation system or by employing an acid-pepsin digestant. The latter method is somewhat simpler, and avoids some of the problems involved in obtaining and using rumen inoculum. It also eliminates the cellulose determination.

The Caloric Equivalent of NVI.   The usefulness of the NVI as descriptive of the effective feeding value of a forage can be extended considerably if one knows (or can determine by bomb calorimeter or from its protein, fat, and carbohydrate content) its gross caloric value, which for most common forages does not deviate appreciably from 4.4 kcal per gram dry matter. Statistical examination of the published data for forages reveals that perhaps five samples in one hundred will deviate as much as one kcal from the average of 4.37 kcal per gram dry matter. Differences in plant species, stage of maturity, and methods of curing do not affect the gross energy value. These factors, however, do affect the voluntary intake and the digestibility of the feed.

The 24-hour intake of digestible calories from forage per unit of metabolic size of animal can be expressed as:

(grams intake per unit of $W_{kg}^{.75}$) $\times$ (per cent digestibility of energy)

$\times$ (gross kcal per gram).

* E. Donefer, "Collaborative *In Vivo* Studies on Alfalfa Hay," *J. Animal Sci.,* XXV (1966), 1227.

The NVI of a forage measures the relative 24-hour intake of a forage multiplied by its digestible energy. It is expressed as:

$$\text{NVI} = \frac{100 \times \text{intake}/W_{kg}^{.75}}{80} \times \text{per cent dig of energy}$$

$$= (1.25) \times (\text{intake}/W_{kg}^{.75}) \times \text{per cent dig of energy}.$$

Hence the 24-hour digestible caloric intake per unit of $W_{kg}^{.75}$ for each unit of NVI is merely the first equation divided by the second, or:

$$k = \frac{(\text{gross intake}/W^{.75}) \times (\text{per cent dig of energy}) \times (\text{gross kcal/gram})}{1.25 \times (\text{intake}/W^{.75}) \times (\text{per cent dig of energy})}$$

$$= \frac{\text{gross kcal per gram}}{1.25}$$

$$= \frac{4.4}{1.25} = 3.5,$$

whence for a given animal

24-hour dig kcal intake $= k \times W_{kg}^{.75} \times \text{NVI}.$

Since the NVI can be predicted from the equation

$$\text{NVI} = c_0 + bX + c_1,$$

we can write

$$Y = k \times W_{kg}^{.75} \times [c_0 + bX + c_1],$$

where

$Y = $ 24-hr dig kcal yield to animal from forage,
$k = 3.5,$
$c_0 = -3.5,$
$c_1 = 0$ for forage fed chopped,
$c_1 = 10.9$ for forage fed ground,
$b = 1.2$ g, and
$X = $ 12-hr *in vitro* cellulose digestion.

For example, let $X = 60$, $W = 62$ kg (so that $62^{.75} = 22$), and the forage be fed chopped. We compute:

$$Y = 3.5 \times 22 \times [-3.5 + 1.23(60) + 0]$$
$$= 77 \times 70.3$$
$$= 5,413 \text{ kcal dig energy}$$
$$= 2.7 \text{ lb TDN}.$$

Calculation of the calories of digestible energy consumed from forage per day by the animal in terms of TDN or metabolizable energy becomes highly useful in assessing the adequacy of a feeding regime employed in a given herd. After estimating the amount of useful energy from the forage which the animals would normally consume, it is merely necessary to subtract it from the total requirement for energy shown in the feeding standard to determine how many calories must come from the meal allowance; and from that to calculate the amount of meal that must be used. Until some simpler method is devised for arriving at the probable contribution of the roughage portion of the ration to the day's energy intake, this scheme gives a workable estimate for normal feeding practices.

Nutritional Liabilities in Forages

We have already noted the possibility of certain deficiencies in forages; and that in general the problem of deficiencies is not serious, because it is usually economically feasible to make up deficiencies of minerals and of such vitamins as are needed by herbivorous animals by supplementary licks or by inclusion of the nutrients in meal mixtures that are to be fed with the roughage. The problem is chiefly one of recognizing the existence of deficiency.

Ideas of what may constitute a deficiency may vary considerably, depending on experience in different localities. For example, it has been tentatively recommended that *copper* be included in a mineral supplement at a level of 0.3 per cent. On the other hand, in certain peat soil areas of southern Florida, it has been found necessary to use as much as 1.25 per cent copper in the mineral mixture to make good the deficiency in that locality. Also, we should point out that the proportions within a mineral supplement must be related to how much of it is consumed. Here we have the problem of the voluntary consumption of mineral mixtures offered free-choice, as well as the problem of mineral mixtures included in meal rations at a fixed concentration. Obviously, if very small allowances of meal are to be offered to cattle, then we shall have to include higher concentrations of minerals to meet a mineral deficiency than we would have to with more liberal grain feeding.

In the practical use of roughages in livestock feeding, it is probable that deficiencies of mineral elements can be rectified by proper supplements. However, there are a few minerals found in forages that are beneficial when ingested in small amounts, but become toxic at larger intakes.

Arsenic is such a mineral; introduced into the ration in trace amounts it is a growth stimulant. It has also been reported that as little as 25 ppm of arsenic in a ration may counteract the toxic effects of chronic selenium poisoning caused by the consumption of selenium-containing forages.

Fluorine* is another mineral element that may have both beneficial and toxic effects, depending on intake. The beneficial intakes are certainly below 10 ppm of total dry matter of food consumed or one ppm in the drinking water. The more serious problem with this mineral is related to its presence in amounts of 30–40 ppm or greater. It is ordinarily of local concern, being confined to areas in the vicinity of industrial plants where fluorine is among the effluents from the manufacturing processes. It is not unusual to find forage in such areas carrying well over 100 ppm of flourine. Experimental evidence indicates that adult breeding or lactating cattle should not be fed a forage containing more than 30 ppm if such material constitutes the entire dry-matter intake.† Cattle being fed for beef could be fed forage containing 100 ppm of fluorine if finished during an average feeding period.

Toxicity in this case must be carefully interpreted. The adverse effects of excessive fluorine intake do not appear abruptly and, indeed, there may be no clear-cut symptoms that are unmistakably the result of the fluorine ingestion. Usually when cattle ingest feed containing appreciably more than 30 ppm of fluorine there will eventually be clinical symptoms of dental fluorosis. Loss of enamel and tooth staining are regularly observed; the teeth may show uneven wear and in some cases become carious. These symptoms are themselves of less importance than the disturbance in energy metabolism. It seems probable that fluorine ingested in amounts greater than can be immobilized in the bone or excreted promptly in the urine may become incorporated into one or more of the enzymes involved in the Krebs cycle. As one result, citrate cannot be metabolized to isocitrate. Interference in this path of energy metabolism not only decreases the energy available from the feed consumed, but probably tends to lead to greater synthesis of acetoacetic acid, some of which in turn may be converted to acetone for elimination. Thus chronic fluorosis may be one of the causes of secondary acetonemia. In any case, chronic fluorosis appears to be relieved by removal of fluorine from the ration, and symptoms of semistarvation induced by failure of energy metabolism eventually subside. What may be called *acute fluorosis* is not normally induced by the consumption of fluorine-containing forages.

Molybdenum toxicity must be considered in terms of the copper level of the

---

* See also Chapter 13.

† Lorin E. Harris, Robert J. Raleigh, George E. Stoddard, Delbert A. Greenwood, J. LeGrand Shupe, and Harold M. Nielsen, "Effects of Fluorine on Dairy Cattle. III. Digestion and Metabolism Trials," *J. Animal Sci.,* XXVIII (1964), 537.

Delbert A. Greenwood, James L. Shupe, George E. Stoddard, Lorin E. Harris, Harold M. Nielsen, and L. Elmer Olson, *Fluorosis in Cattle.* (Ut. Agr. Expt. Sta. Special Report 17, 1964).

diet. The exact toxic action of molybdenum is unknown, but the experimental evidence suggests that it in some way interferes with copper utilization, and that the final result is probably a disturbance in copper rather than in molybdenum metabolism. Excessive molybdenum intake (16 ppm or more) is one of the few conditions that induces persistent scouring with ruminants.

Molybdenum toxicity in cattle has been prevented, with forages containing up to 5 ppm of molybdenum, by the administration of the equivalent of 5 ppm of copper in the ration dry matter. As the molybdenum content of the forage increases, the proportion of copper necessary to counteract its effects increases, so that with 10 ppm of molybdenum, 20 ppm of copper is needed.

Nitrates are sometimes found in some forages, particularly those made from cereal crops. Oat hay having more than 1.5 per cent potassium nitrate will be toxic to cattle. The toxicity is thought to be the result of nitrate becoming changed in the rumen to nitrite, with subsequent formation of methemoglobin.

Forages from a few areas have been found to contain *selenium*. Any soils containing 5 ppm of selenium are potentially dangerous, since crops grown on them will also contain dangerous levels of this mineral. The chronic selenium poisoning is often referred to as *alkali disease* and the acute condition as *blind staggers*. Any forage that contains 5 ppm of selenium fed over a period of time will result in chronic poisoning to livestock. The mechanism of selenium toxicity is still unknown. Experimentally it has been shown that the ingestion of small amounts of arsenic will counteract the effects of chronic selenium poisoning in cattle. The problem, however, is one that must be treated carefully, and we can give no general recommendations.

Fortunately the problem of the presence of potentially toxic materials in forages is not general, and therefore does not concern the great majority of feeders who use roughages. Those who may be in hazard zones will, of course, be well-advised to familiarize themselves with the problems and treatments.

Practical Aspects of Pasture as a Feed

On many farms pasture forage is one of the cheapest feeds available during certain times of the year. There is much misunderstanding, however, of how best to use this feed. A few comments may be useful in this connection.

Young, rapidly growing leaf growth is rich in protein (about 20 per cent).

As the plants mature the protein content drops until the dried grass may carry no more protein than straw. Heavy grazing keeps the grasses immature and hence at their highest feeding value. Pasture becomes too mature to yield its maximum value as soon as the plants begin to head out. If such growth is mowed down, it will not be wasted, for grazing stock will eat it quite readily. Mowing encourages renewed growth of leaves provided, of course, there is moisture. The feeding value of pasturage is increased by the presence of clover. In some areas wild white clover will come into many permanent pastures if they are grazed heavily enough to keep down the tall species.

Young grasses and legumes grown on lands well-supplied with phosphorus usually contain enough of this nutrient for grazing animals when pasture supplies the entire ration. The phosphorus content, however, declines with the maturity of the forage, which then may not supply enough phosphorus for grazing stock. In general, animals replenish depleted phosphorus stores while they are on pasture. It is desirable to feed a phosphorus-containing mineral mixture to all breeding stock while pasture is their principal feed. Satisfactory consumption of such a mineral mixture is assured chiefly because of its salt content. Consequently, no other sources of salt should be provided.

In order to graze pastures more efficiently, rotational grazing in some form is a sound principle. Even dividing the area into two portions and alternating from one to the other will permit alternative periods of heavy grazing followed by a short recovery period. Rotational grazing is of no value, however, unless the growth is kept down by close grazing.

Straw as a Feed

Straw from the cereal grains, particularly from oats, is readily eaten by all classes of cattle and is, in some districts, a regular part of the ration of low-producing or dry cattle. Its inclusion in *production rations* is often false economy because the net energy it will supply to the cow is small. Its maximum value will be obtained by feeding it along with high-quality legume hay. In some areas it is customary to seed clover with a grain crop. This practice results in appreciable amounts of clover in the grain at harvest time and eventually in the straw. Such straw—really a mixture of straw and clover hay—has considerable feeding value and is quite a different product from the straw of a clean grain crop. Used as one feed per day such material is probably equal to average timothy hay.

SUGGESTED READINGS

Axelsson, Joel, "Connections Between Contents of Nutrients and Digestibility of Grassland Crops," *The Annals of the Royal Agr. College of Sweden*, XVII (1950), 320.

Crampton, E. W., and L. A. Maynard, "The Relation of Cellulose and Lignin Content to the Nutritive Value of Animal Feeds," *J. Nutrition*, XV (1938), 383.

Cunningham, I. J., "Molybdenum and Animal Health in New Zealand," *New Zealand Veterinary Journal* (June 1954), p. 29.

Donefer, E., E. W. Crampton, and L. E. Lloyd, "The Prediction of Digestible Energy Intake Potential (NVI) of Forages, Using a Simple In Vitro Technique," *Proceedings of Tenth International Grasslands Congress* (1966), 442.

Donefer, E., E. W. Crampton, and L. E. Lloyd, "Prediction of the Nutritive Value Index of a Forage from In Vitro Fermentation Data," *J. Animal Sci.*, XIX (1960), 545.

Drapala, W. J., L. C. Raymond, and E. W. Crampton, "Pasture Studies XXVII. The Effects of Maturity of the Plant and Its Lignification and Subsequent Digestibility by Animals as Indicated by Methods of Plant Histology," *Sci. Agr.*, XXVII (1947), 36.

Feeders' Guide and Formulae for Meal Mixtures (Quebec Provincial Food Board, published annually).

Greenwood, Delbert A., James L. Shupe, George E. Stoddard, Lorin E. Harris, Harold M. Nielson, and L. Elmer Olson, *Fluorosis in Cattle*. (Ut. Agr. Expt. Sta. Special Report 17, 1964).

Hobbs, C. S., R. P. Moorman, Jr., J. M. Griffith, J. L. West, G. M. Merriman, S. L. Hansard, and C. C. Chamberlain, *Fluorosis in Cattle and Sheep* (Tennessee Agr. Expt. Sta. Bul. 235, 1954).

Huffman, C. F., "Roughage Quality and Quantity in the Dairy Ration. A Review," *J. of Dairy Science*, XXII (1939), 889.

Mitchell, H. H., and M. Edman, "The Fluorine Problem in Livestock Feeding," *Nut. Abst. and Revs.*, XXI (1952), 787.

Proceedings of the Sixth International Grasslands Congress (Pennsylvania State College of Agriculture, 1952), vols. I and II.

Proceedings of the Specialists Conference in Agriculture, "Plant and Animal Nutrition in Relation to Soil and Climatic Factors" (London: Her Majesty's Stationery Office, 1949).

U.S. Dept of Agriculture, *Grass, The Yearbook of Agriculture, 1948* (Washington, D.C.: U.S. Govt. Printing Office, 1948).

SUMMARY OF SECTION **III**

In Section III we have been concerned with another part of applied nutrition —the nature of feedstuffs. We have seen how diverse they are; and in order to systematize our thinking about the concentrate feeds we have suggested a functional grouping for them based largely on protein content. Plant or animal origin is also involved in a subgrouping of some high-protein products. The grouping according to protein is convenient and practical, because one of the first considerations in ration formulation is adequate protein, which is attained in practice by selecting feeds in accordance with their protein contents.

Such a classification disregards the origin of most feedstuffs, and so we lose sight of the fact that plant seeds may yield, in the course of milling, several feedstuffs, each belonging to a different functional group. For example, corn, itself a basal feed, is the parent product of two other basal feeds, (hominy and bran) and of four protein concentrates (gluten feed, gluten meal, germ meal, and distillers' grains). Similarly, wheat yields both basal feeds (middlings, shorts, and bran) and protein supplements (germ meal and distillers' grains). In neither case are the basal feeds interchangeable with the protein supplements; whereas both the basal feeds and the high-protein by-products of the one grain may be substituted for those of the other grain. The distinction as to origin may, for some feeds, be significant. In the NRC nomenclature (Chapter 1) "origin" is the first term of each name, and the basis for the alphabetical arrangement of feeds in tables of composition.

This functional grouping, by bringing together feedstuffs that serve roughly similar purposes in the ration, facilitates the preparation of formulae for meal mixtures within which a wide variety of different feeds may be used alternatively to give combinations of essentially equivalent feeding properties. This, of course, does not mean that individual feeds of a particular category have no unique properties. It means rather that, broadly speaking, basal feeds are in the mixture primarily because of their energy content, and the protein supplements because of their protein. Protein supplements fall into two subcategories, because those of plant origin in general differ in amino-acid content from those of animal source. The special properties of the individual feeds within categories may sometimes be of no nutritional importance, or sometimes their limitations are minimized by using several feeds. In the long run, however, if a feedstuff can be correctly assigned to its appropriate category in our feed-classification scheme, it can be used in ration formulation without risk of adverse consequences, even though full information about its unique properties is not on hand at the time.

Forages and roughages are distinguished from concentrate feeds by their fiber content; concentrate feeds in general contain less than 18 per cent crude fiber. Of particular importance with roughages is their yield of useful energy; and thus, DE, ME, or TDN becomes the basis for their final grouping.

Therefore, we can assign feedstuffs to their appropriate main functional categories if we know their crude-protein and crude-fiber contents. Information on whether they are of plant or animal origin will enable further subclassification of certain high-protein feeds. Information on the botanical makeup and the stage of maturity at harvest of a forage will permit a reasonable estimate of its probable ME, and hence of the roughage subcategory to which it should be assigned.

To make the best use of different feeds requires that we know as much as possible about their peculiarities and special characteristics. To this end we must ultimately examine each feedstuff critically in order to make the fine adjustment that in special cases may make one combination preferable to another. For up-to-date details on the description and performance of individual feeds, the student must consult the scientific literature. Such information will not alter the basic concepts discussed in this text, but rather will make them more useful in the practical, albeit complicated, problems of tailoring diets to given specifications.

SECTION **IV** RATION
FORMULATION

This section differs from the three preceding it in several respects. Section I dealt specifically with the definition, and sometimes with the biological significance, of a number of terms and expressions used in applied nutrition. We tried to crystallize the real meaning of such terms as ME, coefficients of digestibility, proximate analysis, biological values, and thus to clarify their usefulness as descriptive terms. The subject matter is important in its own right. Precise understanding of these terms is essential to their correct application. Failure to recognize their limitations, no less than erroneous assumptions about their meanings, or the assignment to them of attributes which they do not or only in part possess, leads only to difficulties in applied nutrition.

In Section II we gave a critical appraisal of feeding standards as quantitative statements of animals' requirements for energy and for several of the nutrients which the feeder must consider in compounding livestock rations. Obviously this is also of importance in its own right. Without an understanding of the nature of such standards and of the inherent peculiarities of the data they contain, we cannot hope to make either full or exact use of them in ration formulation.

Section III covered a third, and again somewhat autonomous, part of applied nutrition—a consideration of the nature of feedstuffs, not so much of individual feeds as of groups of nutritionally similar products, the members of which in feeding practice are to varying degrees interchangeable. It was in this section that some of the significance of the discussions in Section I should have become evident. For example, the choice between oats and barley for use in a given ration might depend on an accurate knowledge of any real difference between them in their yield of useful energy to the animal. In view of the limitations of TDN, as brought out in Section I, it should be clear that dependence exclusively on this measure as descriptive of the useful energy of these two feeds could lead to ill-advised substitutions. Again, the discussion of forages should have emphasized the uncertainties of the proximate analysis, in whole or in part, as a measure of the nutritional value of many feedstuffs.

But it is in Section IV that the facts, figures, and philosophy of the three previous parts are integrated into our ultimate objective—the formulation of nutritionally adequate rations, or of feed mixtures with defined nutritive properties. It is as components of such mixtures that most feeds are put to work. Their selection and the extent of their inclusion in mixtures will obviously depend in part on what nutrients they contain, and in part on what we need to provide the animal with.

The translation of feeding standards into ration formulae is not entirely straightforward, because, in practice, we do not prepare a day's feed for one animal. Rather we attempt to devise formulae by which "balanced" meal mixtures can be made in quantity, and from which individual animals can satisfactorily be fed by adjusting daily allowances to their energy needs. To be practicable for general use such formulae must be flexible enough to permit their being accommodated to price and availability of feedstuffs, while at the same time retaining the necessary nutritive balance and adequacy.

The idea of flexible formulae *is a departure from the traditional interpretation of feeding standards. Although in the commercial manufacture of balanced feed mixtures, limited feed substitution has been common, it has usually been on a second-choice basis. Particular consideration of feeds in the light of their interchangeability in rations has not been usual in our teaching. The idea of flexible formulae is an outgrowth of the concept of nutritionally legitimate feed substitution, and hence it is not surprising that such formulae have not previously been emphasized. Perhaps this omission is one reason why the feed industry has felt that college courses in feeds and feeding have failed to bridge the gap fully between fundamental nutrition and practical animal feeding—at least in the sense that such courses have given little guidance in problems of ration formulation.*

There is no doubt that ration formulation is an important part of livestock feeding. Alternate (with no implication of next best) choices between nutritionally similar feeds in ration formulation attained full respectability in Canada with the adoption by the Quebec Provincial Feed Board in 1942 of the flexible formula as a method of presenting the Board's official ration *recommendations, and with the sanction by the Canadian Feed Control officials of guarantees by feed manufacturers containing such statements as "corn and/or wheat," or "linseed and/or soybean oilmeal," to indicate the possible use of alternate ingredients in their commercially mixed rations.*

Inasmuch as no two natural feedstuffs are identical in nutrient content, and since the amounts and proportions of nutrients in any one feed may bear little relation to the needs of an animal, it follows that in practice it is impractical, if not impossible, to prepare a feed mixture of which a specified quantity will exactly meet the specifications for some particular animal, let alone maintain the nutrient balance when feed substitutions occur. However, there is ample evidence from practical feeding that minor excesses and fluctuations in nutrient intake, either from variation in daily intake of the same feed mixture or from legitimate substitutions in formulae, are inconsequential to the animal. On the other hand, alteration of the energy concentrations of the meal mixture by feed substitutions is promptly detected by the animal; and usually the first limiting factor in the interchangeability of feeds within a subcategory (Chapter 11) is how much the useful energy yield would be changed.

It is these facts that make tenable the whole scheme of preparing one common feed mixture for a group of animals. To make such a system work economically there must be flexible formulae, whether they are formally written or tacitly practiced in compounding the mixtures. The basic pattern of a feed mixture for some group of animals is fixed by the needs of the animal as detailed in the appropriate feeding standard. But in the choice of the individual feedstuffs within this pattern, there can be varying degrees of latitude, depending on nutritional or economic considerations. The latter may be general and locally uncontrollable, or they may arise from the policy and internal economy of a particular feed manufacturer. So long as the nutritional integrity of the final ration is guarded, the economic considerations can be given full play without prejudice to the feed, and doing so is usually to the ultimate advantage of the purchaser.

Flexible formulae offer a simple means of stating the appropriate basic formula pattern; of indicating the legitimate choices of feedstuffs; and of marking the recommended maximum and minimum limits of use of each potential selection. This section of the book deals with these matters.

Note. In this section numerical values are given in the avoirdupois system or in the metric system, or sometimes in both. Where feeding standards to which the weights refer use only pounds, the computations from them will be either in pounds or in both avoirdupois and metric, but where new calculations are involved the metric units will be used. Percentage figures will be the same in either system of weights. We hope this dual system will help the student in translating from one system of weight to the other, as he must do during the transition period through which we are now passing in applied animal nutrition. (*See Table A–1–2, p. 460, for weight unit conversion factors.*)

CHAPTER 15

Translation of Feeding Standards into Meal Mixture Specifications

Animal Feeding Categories

Feeding standards state the daily needs of specified animals for some group of the recognized nutrients. Feeding practice requires that this information be translated into terms of meal mixtures and rules of thumb for feeding. One of the first questions that must be settled, then, is how many different meal mixtures are required to feed a herd or a flock? This is the same as asking how many feeding categories there are for each class of animal. Feeding standards for swine describe requirements for six weight groupings of market pigs plus four more for breeding stock. Dairy cattle standards give separate specifications for pregnancy, and for four production categories. Beef cattle and sheep are grouped into nine and six categories, respectively.

Not all of these groups actually require meal mixtures of different makeup. Often the difference between needs of the animals is one of quantity rather than of differing concentrations of nutrients in the ration. Furthermore, economy appears to justify a reasonable latitude in the nutrient allowances to individual animals, since temporary minor surpluses of many nutrients that individual animals may get as a result of the feeding of groups on the same feed mixture do not cause obvious differences in performance.

The fundamental basis for establishing feeding categories in practice is the energy–protein requirement. Those animals of the same species that can be satisfactorily fed rations with the same percentage of protein can usually be adequately nourished with identical meal mixtures if allowances are adjusted to their individual energy needs, because most of the nutrients which need to be carefully adjusted in the ration are required in proportion to the metabolizable energy. The major nutrient that does not follow this rule is protein, and so in applied feeding the first factor determining the number of meal mixtures needed for any class of stock is how much protein is needed.

Swine Meal Mixture Specifications

Calculations from the generalized feeding standard of Guilbert and Loosli* suggest that newly weaned pigs require a ration which, on a 90 per cent dry-matter basis, should have 19 per cent total crude protein. More recent work has indicated that with the use of vitamin B_{12} the protein of swine rations may be reduced appreciably, and the 1964 recommendation of the Swine Committee of the NRC Committee on Animal Nutrition calls for 18 per cent crude protein for pigs of 25 lb weight. In 1954, Catron† at Iowa State recommended still less protein in hog rations. The data calculated or taken directly from these three sources are shown for various weights of market pigs in Fig. 15-1.

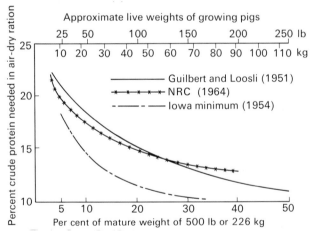

Figure 15-1. *Per cent of protein needed in meal rations for growing swine.*

It will be evident that the NRC and Iowa standards agree on 18 per cent protein for newly weaned pigs about 25 lb (11 kg) in weight. By the time they have reached a weight of 75 lb (34 kg) their protein needs have de-

* H. R. Guilbert, and J. K. Loosli, "Comparative Nutrition of Farm Animals," *J. Animal Sci.,* X (1951), 22.
† *The Midwest Farm Handbook* (Ames: Iowa State College Press, 3d ed., 1954).

clined to between 13 and 16 per cent, depending on which of the three swine standards is used. For pigs beyond 125 lb (57 kg) weight, 11 to 13 per cent protein appears to be enough.

The NRC standard does not drop below 12 per cent protein, and the Iowa table levels off at 10 per cent. However, we must remember that in practice it is desirable to include enough feeds of animal or marine origin to meet at least 10 per cent of the total protein requirement. If the basal feed used were all corn (10 per cent protein) and if 2 per cent minerals were also used, the final meal mixture would then contain about 10.5 per cent crude protein. If the basal feed contains 12 per cent protein, as coarse grains (barley or oats) do, the *minimum* protein of the mixture in which 10 per cent of the protein comes from animal sources becomes 12.5 per cent. Consequently, there is little practical use in worrying about percentages of protein in hog rations below 12 per cent.

For the pigs beyond the weaning period—that is, over 75 lb (22 kg) in weight—and until they reach weights of 125 lb (57 kg) the meal mixture evidently should be between 12 and 18 per cent protein. The NRC standard calls for 14 to 16 per cent protein at this weight range (if the ration does not carry antibiotic). Furthermore, practical experience with commercially prepared "hog growing rations" containing 15 per cent crude protein is adequate reason to consider 15 per cent satisfactory for this intermediate period of market-hog feeding. It will provide enough margin to meet the variations in need of pigs within the weight range of 75–125 lb.

There are two further categories of pigs involved under usual swine management. These include the breeding stock, the pregnant gilts and sows in one group and lactating females in the second. The NRC standard calls for 16 per cent protein for the gilts, and 14 per cent for the adults. For lactation the NRC standard calls for 15 per cent and 13 per cent for gilts and adult sows, respectively. The Quebec Feed Board recommends that pregnant and lactating sows, as well as boars, be fed a meal mixture of the same specifications as the growing pigs between weights of 75 and 125 pounds. If 15 per cent protein is accepted for the growing pigs, it will then also be adequate for the breeding stock. Thus there would seem to be no practical reason to divide the swine herd into more than three ration groups, except for a special case to be noted later.

A Method for Grouping Rations. The percentages of crude protein needed in the several different rations necessary to properly feed a herd of a given species of farm animal can be determined in another way from the NRC feeding standard for that species. Age or production groups within the species that require the same quantities of protein per 1,000 kcal of usable energy (expressed either as DE or as ME) can normally use the same ration. Since feeding standards describe the quantities of nutrients needed daily by specifically defined individuals, it is a simple matter to compute the protein–energy ratio of the rations for each group covered by the standard. This method is theoretically preferable to the conventional procedure, in that the percentages of protein is related to the intake of energy rather than of feed. If the rations of the several feeding groups have the same energy concentration, the two methods give the same answers. But if we alter the amount of energy intake by feed substitution, or by the use of edible tallow, we must maintain the protein–energy ratio, rather than the protein–dry matter ratio, to maintain the ration efficiency.

The method of computing the percentage of protein needed in swine rations to maintain the protein–DE ratio in the diet is illustrated in Table 15-1, which is based on the 1964 NRC standards. In this latest swine feeding standard, the weight data are in the metric system, and the energy is expressed in terms of digestible calories and total digestible nutrients. The latter term will gradually be dropped from all NRC feeding standards, and metabolizable energy (ME) will be used rather than DE as soon as ME values for most feeds become available.

In Table 15-1, the DE and crude-protein values, as well as the expected daily feed intakes, are used to compute the ratio of protein to each 1,000 kcal of DE. The crude protein and TDN as a per cent of the feed are also shown. If in this method we assume that protein to calorie ratios can be grouped as

60 and over	Type I ration	early weaner,
46–59	Type II ration	starter and young breeders,
40–45	Type III ration	grower, adult breeders, bacon hogs,
35–39	Type IV ration	finisher, meat-type hogs,

we find four different ration mixtures needed to properly feed all the categories of pigs in a herd containing both breeding stock and market stock. The early weaner, or preweaner used in a creep, is a rather specialized ration. The ration fed following the early weaner, until the meat-type pigs reach a weight of about 75 lb (35 kg), or the lean-bacon pigs reach about 125 lb (55 kg), is the common 16 per cent hog grower. This same ration formulation is also suitable for nursing sows and for young boars and gilts.

It is also clear from Table 15-1 that a third ration formulation can be introduced, for meat-type pigs of weights from 75 lb (35 kg) to 125 lb (55 kg), and for lean-bacon pigs from the time they attain a weight of

TABLE 15-1   *Crude Protein Levels Needed in Swine Rations*

Feeding categories by wt in lb (kg)	Daily intake			Per cent of TDN in air-dry feed	Per cent of crude protein in air-dry feed	Grams crude protein per 1,000 kcal DE	Ration types
	Feed (g)	Protein (g)	DE (kcal)				
Growing Pigs							
10–25 (5–10)	545	117	1,920	80	22	61	I
25–50 (10–25)	1,130	202	4,000	80	18	50	II
50–75 (25–35)	1,160	269	5,400	75	16	50	II
Meat Pigs							
75–125 (35–55)	2,360	332	7,800	75	14	42	III
125–175 (55–80)	3,200	392	10,000	75	13	39	IV
175–225 (80–100)	3,540	425	11,600	75	12	37	IV
Bred Sows							
300 (150)	2,500	400	8,200	75	16	49	II
500 (225)	2,950	410	9,800	75	14	42	III
Bacon Pigs							
75–125 (35–55)	2,360	375	7,200	70	16	52	II
125–175 (55–80)	3,200	425	9,400	70	14	45	III
175–225 (80–100)	3,540	495	11,000	70	14	45	III
Lactating Sows							
350 (160)	5,000	750	16,400	75	15	46	II
450 (200)	5,550	735	18,800	75	13	39	IV
Boars							
300 (135)	2,710	410	8,400	70	15	49	II
500 (225)	3,400	445	10,400	70	13	43	III

125 lb (55 kg) until they reach a marketing weight of 225 lb (100 kg). Aged boars and sows can also use this formulation. Finally we find that meat-type pigs from 125 lb (55kg) to market weight can be satisfactorily fed the fourth ration. There is no reason to make any change in the formulation for market pigs after they reach weights of 125 lb (55 kg).

Thus four rations are shown to be useful and sufficient for the five usual feeding categories of swine. However, there is one further group that should be included, though it will not be a factor on many pig farms. This extra grouping is for the pigs weaned at under three weeks, and fed man-made rations during the balance of the time that they normally would be nursed by their dams. This system of pig raising, sometimes referred to as the *pig hatchery* plan, promises to be increasingly important, because it makes possible more pigs per sow per year (about three litters in 15 months) with no greater and often a smaller feed cost per pig to an age of eight weeks (usually considered normal weaning age), plus the probability of a larger pig at eight weeks, as well.

Specific nutrient needs for these very young pigs are as yet only partially known, but empirically devised, highly satisfactory feed mixture formulae have been published by several experimental stations, and since about 1953 several such formulae have been manufactured and sold commercially. The demand for this type of pig ration is reported by feed manufacturers to be steadily increasing. The formula given in Table 15-2 was devised by Crampton

TABLE 15-2 *Dry Meal for Early Weaned Pigs*

Formula	Add to each 1,000 lb. materials to supply
370 lb Skimmilk powder	7 g antibiotic
100 lb Ground wheat	7 mg B_{12}
150 lb Ground oat groats	2 g Riboflavin (800,000 B.S. units)
100 lb Soybean oilmeal	2 g Pyridoxine
100 lb Fishmeal	4,500,000 IU Vitamin A
50 lb Brewers' yeast	100,000 IU Vitamin D
100 lb Cane molasses	
24 lb Bone flour[a]	
5 lb Fine salt (iodized)	
1 lb Ferrous sulfate	
———	
1,000 lb	

Average Analysis

Protein	30%
Fat	3%
Fiber	2%
Calcium	1.6%
Phosphorus	1.1%
Thiamine	2.0 mg/lb
Riboflavin	4.0 mg/lb
Niacin	15.0 mg/lb

[a] Finely ground feeding bone meal.

and his co-workers at Macdonald College, and was first printed in the 1953 edition of the *Feeders' Guide and Formulae for Meal Mixtures,* published by the Quebec Provincial Feed Board. It has been in commercial production since that date.

The mixture, self-fed as a dry meal or in pellet or crumble form with water available ad libitum, has been successfully used at Macdonald College for pigs weaned from their dams at 21 days after birth. Litters so fed have averaged from 30 to 35 lb per pig at 8 weeks of age.

Early weaned pigs must be given heated beds with air temperature under the hover between 75 and 80°F. Heated floors (radiant heating) are preferable since they keep the bed dry as well as warm. Pigs when weaned abruptly will usually learn to eat in 24 hours without assistance. Providing them with milk or gruel is not desirable since they learn to eat dry feed more quickly when no other feed is offered.

Dairy-Cow Meal Mixture Specifications

We cannot calculate the specifications of a meal mixture for dairy-cow feeding directly from feeding standards, for two reasons. First, requirements are for a cow of particular size, amount of lactation, and stage of pregnancy. Second, the meal allowance is only a part of the complete ration. However, as with hog rations, it is protein more than any other nutrient that determines how many different mixtures will be needed to satisfy the needs of different feeding groups.

For growing dairy animals, calculations (see Fig. 6-1) from the latest NRC dairy-cattle feeding standard indicate that up to live weights of about 25 per cent of their expected adult size, meal rations should be not less than 18 per cent total crude protein, and for the younger animals 20 per cent would not be excessive. From that stage of development on, meal mixtures of 16 per cent protein are adequate with normal feeding practice.

For the maintenance of adult nonmilking stock, meal allowances are not ordinarily needed except for heifers in calf for the first time. Here feeding standards suggest a daily allowance during the last two or three months of pregnancy of 8 lb (3.6 kg) of meal, containing 6 lb (2.7 kg) TDN and 0.6 lb (273 g) DP. Assuming a digestibility coefficient of 80 per cent for protein, and that the TDN is 75 per cent of the weight of the feed, we can see that this meal allowance needs to be only 10 per cent total crude protein. Since this is less protein than there is in a mixture of basal feed grains, there is no special meal mixture (in terms of protein) required for animals of this category.

For the milking cow, however, the story is more complicated. For the milk production itself, the proportion of TDN that should be digestible protein is about 14 per cent. Also, since the rest of her ration may not exactly meet maintenance needs, the meal mixture specifications will be influenced by the kind and amount of roughage used.

Perhaps the easiest way to describe the requirement is to calculate examples that will cover the usual situations found, and from them attempt to arrive at valid conclusions about the number of meal mixtures of differing protein levels needed to feed dairy cows generally.

The starting point is, of course, the feeding standard, but which standard? As discussed in Chapter 6, it seems probable that a 1,000-lb (450-kg) dairy cow requires about 6.5 lb TDN (13,000 kcal DE) per day for maintenance. This is approximately 20 per cent lower than the value recommended by the NRC Dairy-Cattle Standard Committee. However, they state that their figure is above the actual needs and was decided on because too low an energy intake was the most frequent cause of poor performance. Their stated reason is a valid argument for recommending liberal allowances of a ration, but a weak one for establishing a standard. Where meal mixtures are to be made for the practical feeding of groups, the margin of error should be toward the minimum energy requirement, in order that the allowances of meal sufficient to maintain the body weight of the cow shall also contain a sufficiently high concentration of protein, minerals, etc., to meet her needs. Consequently, in the calculations to be made in our examples, we shall use 6.5 lb TDN as the most probable true maintenance need per day for a 1,000-lb (450-kg) cow.

We may take the protein level as:

$$DP = \frac{3.4 \times 146(450^{.75}) \times 6.25}{1,000} = 303 \text{ g or } 0.67 \text{ lb,}$$

as discussed in Chapter 7. This figure for a 1,000-lb cow is to be compared with 0.60 given by the U.S. standard and values of 0.60 to 0.65 lb given by Morrison. For 4 per cent milk, all standards agree well enough for us to assume values of 0.32 and 0.045 lb of TDN and DP per pound of milk (or 640 kcal DE and 45 g of DP per kg of milk), respectively.

The next step is consideration of the roughage to be fed. Here we have to choose one of a number of possibilities. For high-quality (high-energy) roughage we may take a typical TDN value to be 50 per cent (2,200 kcal DE/kg) and total crude protein to be a minimum of 6 per cent. Of such roughage, average voluntary consumption can be estimated at 2.5 per cent of the weight of the cow. The steps in calculating the per cent of crude protein needed in the meal mixture are shown in Table 15-3. In the calculations of Table 15-3 we have assumed that the DP in a meal mixture is 80 per cent of the total protein, and that such a meal has 72 per cent TDN (or 72% × 2,000 × 2.2 = 3,158 kcal DE/kg). These figures are based on a typical mixture in which the basal feeds include significant amounts of barley and/or oats. Use of a heavier meal, one with perhaps 75 per cent TDN (3,300 kcal DE/kg), as might be done where the basal feeds were largely corn, would raise the per cent of protein needed to 14.2 per cent.

TABLE 15-3 *Example of Calculations of Per Cent of Protein Needed in Dairy-Cow Meal Mixtures (Requirements for a 450-kg dairy cow producing 25 kg of 4 per cent milk, fed roughage of 2,200 kcal DE/kg and 6% DP)*

Requirements	DP (g)	Equiv. protein (g)	DE (kcal)
For maintenance	303		13,000
For 25 kg of 4% milk	1,125		35,200
Total	1,428		48,200
In (450 × 2.5%) 11.25 kg hay	665		24,850
To be supplied in meal	763	954	23,350
Kg meal at 3,168 kcal/kg		7.38	
Per cent of crude protein needed in meal		13.0	

Using appropriate figures for cows of 1,000 and 1,200 lb live weight, for productions of 30 and 50 lb of 4 per cent milk, for the requirements for early and late pregnancy, and for good hay and only fair roughage, the per cent of total crude protein needed in meal mixtures has been calculated in Table 15-4 for adequate meal to meet energy needs fully, and for a meal allowance restricted to 75 per cent of the total needed to meet energy needs.

It will now be in order to examine the figures in some detail. It is possible, by averaging together all of the protein percentages that apply to the 1,000-lb cow, and then all of those applying to the 1,200-lb cow, to see what general effect size of animal has on the per cent of protein required in her meal mixture. Doing so, we find that the average per cent of protein required in the 1,000-lb cow meal mixture is about 13.5 per cent, while that for the 1,200-lb cow is 14.5 per cent. This result would lead to the conclusion that larger cows need a slightly higher percentage of protein.

For milk production, making the comparison in the same way, we find that for 30 lb (13.6 kg) of milk a meal mixture of 14.5 per cent protein is required, and for 50 lb (22.7 kg) of milk 13.5 per cent protein is enough. This result shows that for larger production, slightly less protein is required in the meal mixture.

It is assumed that, until the last two or three months of pregnancy, pregnant cows have no increased nutritional requirement that needs to be taken into account by the feeder. However, for the last two or three months of pregnancy, additional feed and nutrients must be given in order to make good the drain on the cow occasioned by the rapid growth during this period of the fetus. For per cent of protein needed in the meal mixture, we find that during early pregnancy the average figure is 15.2 per cent, but that during late pregnancy, when additional feed is given, it drops to 12.7 per cent. This drop, of course, is due to the larger meal allowance called for to meet energy demands. When we change from good hay to poor hay there is less than 1 per cent difference in the protein required in the meal mixture to satisfy the requirement.

TABLE 15-4 *Per Cent of Protein Needed in Milking-Cow Meal Mixtures*

Weight of cow	Lb of 4 per cent milk produced daily	Stage of pregnancy	Lb consumed daily of roughage (hay or equiv.)	Lb of meal (72% TDN) needed to meet energy requirement	Lb of total protein (80% dig) needed in meal allowance	Per cent of total crude protein needed in meal mixture	
						Normal	Reduced 25 per cent
1,000 lb (450 kg)	30	Early	25	3.9	0.62	15.9	20.0
			15	12.5	1.75	14.0	19.0
		Last two months	25	12.2	1.38	11.4	15.0
			15	20.8	2.50	12.0	16.0
	50	Early	25	12.7	1.75	13.7	18.5
			15	21.4	2.88	13.5	18.0
		Last two months	25	21.1	2.50	11.9	15.8
			15	29.8	3.62	13.3	16.3
1,200 lb (550 kg)	30	Early		(2.5)		(15.1)	(22.0)
			30	1.7	0.38	22.4	30.0
			18	12.0	1.72	14.4	19.0
		Last two months	30	8.6	1.12	13.0	17.5
			18	20.2	2.48	12.4	16.5
	50	Early	30	10.6	1.50	14.1	18.8
			18	20.8	2.88	13.8	18.5
		Last two months	30	18.9	2.25	11.9	16.0
			18	29.1	3.60	12.4	16.5

In Table 15-4 there is one case where calculation indicates a need for 22.4 per cent protein in the meal, because only a very small quantity of meal is needed to complete the energy requirements over that supplied in the roughage. If as much as 2.5 lb of meal were fed, the protein content necessary would be 15 per cent. This case is a special one, and need not be considered as altering the general situation.

Taking all cases into consideration, the average protein required in the meal mixture is 14 per cent. The range (deleting the case noted above) is from 15.9 per cent, where a very small allowance of meal is required, down to 11.9 per cent, where relatively heavy feeding is involved. Evidently we can conclude that the dairy-cow meal mixture need not contain over 16 per cent protein under any normal conditions, and that for most common feeding conditions 16 per cent protein will be about 2 per cent more than is actually required. This conclusion agrees very well with the fact that the commercially prepared standard milking-cow meal mixture sold in North America is likely to contain 16 per cent protein.

There is, however, another situation that we must face when feeding dairy cows. It is where the meal allowance offered is somewhat less than would be required to complete the full energy requirement of the cow. We find such situations during early lactation. Some feeders believe a cow should not be allowed to consume more than 25 lb (11 to 12 kg) of meal per day; when the cow needs more than that she has to get some of her energy by metabolizing body fat. In other words, a part of the energy requirement is provided by the cow herself in lieu of feeding the full amount of meal that would otherwise be necessary. This curtailment of the amount of meal has nothing to do with the amount of protein that is required, and consequently, where meal allowances are restricted below those that would complete the full allowance for energy, the per cent of protein in the meal mixture must be raised. Actually our table shows that if the meal allowance is cut by 25 per cent there will be an increase in the average per cent of protein from 14 per cent for normal feeding to over 18 per cent in the restricted feeding program. This is why under some conditions a dairy-cow meal mixture with 18 per cent protein has been found to give more satisfactory results than one with 16 per cent, and also why most feed manufacturers offer both a 16 per cent and an 18 per cent milking-cow ration.

The conclusion that the makeup of the meal mixture needed for cows in milk is not affected by level of production, size of cow, or stage of pregnancy is supported by data in the 1966 NRC feeding standard. From this standard, protein and energy intakes required were computed for cows weighing 400, 550, or 700 kg and producing daily 20, 27, or 35 kg of milk with 3, 4, or 5 per cent butterfat. Amounts needed for maintenance, during gestation, and during lactation were separately arrived at; and from them requirements were calculated to cover maintenance plus lactation, and maintenance plus lactation plus pregnancy. Finally the protein and energy needs were expressed as grams of digestible protein per 1,000 kcal metabolizable energy. The values arrived at for the 54 groups lend themselves to a statistical analysis

TABLE 15-5 *Analysis of Variance of Grams of Digestible Protein Required Daily per 1,000 kcal Metabolizable Energy by Cows in Milk, According to Source of Variation*

Sources of variation	D.F.	Variance	S.D.	S.D. $\times t_{(p=.05)}$
All causes	53	1.8	±1.35	±2.7
Total D.F. between groups	7	—		
Milk production levels	2	18.1		
Milk fat levels	2	2.7		
Weight groups of cows	2	9.6		
Pregnancy status	1	24.4		
Remainder and interaction	46	0.2	±.47	

TABLE 15-6 *Grams of Protein Required per 1,000 kcal ME by Cows in Milk*

Variable	Means and S.D.	Range	Limits of reliability at p = .05
All weights of cows	32.6 g	32.1–33.2 g	
All milk levels	31.9	30.8–33.4	
All fat levels	32.7	32.1–32.7	
Open or pregnant	32.4	30.6–33.8	
All groups	32.4 ± 1.38	29.6–36.0	$\overline{X}^a$ ± 2.7 (i.e., 29.7 −35.1)
Between cows within groups	$\overline{X}$ ± 0.47		$\overline{X}$ ± 0.94

[a] $\overline{X}$ = mean of appropriate category.

of variance, as shown in Table 15-5. Table 15-6 makes it evident that one protein level is all that is needed in the milking-cow ration, assuming that the ration is always the same. A deviation from the mean of ±2.7 g protein per 1,000 kcal ME, in a ration that has 1,000 kcal per pound, is equivalent to a difference of only $2.7 \times 100/454 = \pm0.6$ per cent of DP in the ration. In practical terms, it appears that 97.5 per cent of cows in milk can be adequately nourished by a ration supplying 32.4 g DP per 1,000 kcal ME, if it is consumed in amounts that meet their ME needs.

In general, it appears that three different meal mixtures will probably be required to adequately supply the protein requirements of the different feeding classes of the dairy herd. One will be about 20 per cent protein, for the calves and yearling heifers. The second would be about 18 per cent protein, for the two-year-old heifers and also for cows in milk who were for any reason on a somewhat restricted meal allowance rather than one that would meet their whole energy requirement, and for whom one might expect some loss in body weight because of too low an energy intake. The third meal mixture, which would cover most of the usual feeding of milking cows, should be about 16 per cent protein for the most economical results.

Other Nutrients of the Dairy-Cow Meal Mixture. The concentration of other nutrients needed in these meal mixtures will also be affected by the quantities of these nutrients that the forage will supply. The number of nutrients to be considered is not large, partly because the dairy cow (i.e., the microorganisms of her rumen) synthesizes all the *vitamin B-complex* required, and she normally gets in almost any kind of forage all the *vitamin A* she needs. Probably, also, the *vitamin D* requirements are met through her exposure to the sun in the summer during grazing. Of the minerals, we must consider *calcium* and *phosphorus,* and in many areas *cobalt, copper,* and *iodine* as well. In this country *salt* will also have to be supplied, but this is done as a matter of routine, regardless of any quantities of sodium or chlorine that may be in the forage. The problem of supplying cobalt, copper, and iodine is solved by specific additions to the meal ration rather than by choice of roughages.

We therefore need consider only the calcium and phosphorus that the roughage intake will supply to the animals, as compared with the quantities specified by feeding standards. As we saw in Chapter 14, when cattle are fed under conditions requiring no meal mixture, it will usually be necessary to supplement the forage ration with phosphorus but not with calcium. On the other hand, when meal mixtures are used in dairy-cattle feeding, how much calcium and phosphorus they should contain depends on the quantities and kinds of roughage also consumed. In order to get an idea of what supplementation is required for meal mixtures intended to be fed to cows in milk, we can use the same general scheme we used in determining the protein requirements.

Examination of the feeding standards indicates that differences in weight between adult cows do not affect the calcium and phosphorus requirement nearly as much as changes in the quantities of milk produced or in the stage of lactation do. Changes in the amount and kind of roughage also have a marked effect on the quantities of Ca and P that must eventually be provided in the meal mixture.

In estimating the probable quantities of calcium and phosphorus that would be furnished by the roughage allowance, we used figures that were 50 per cent of what would be supplied by *average* forage, and, because of the general tendency of feeders to restrict dairy cows in their meal allocations, we have based our calculations on a feeding program in which only three-quarters of the full allowance of meal was actually fed. We believe that by making these two modifications the calculations give the quantities of bone meal * that would be required under the most adverse conditions likely to be met in actual feeding practice. The assumption, that in any given case the roughage may furnish only half as much of either calcium or phosphorus as would be found in average samples, is predicated on the fact that soil phosphorus deficiency is widespread in farming areas, and that under such conditions the phosphorus content of the forage is appreciably lower than average. Similarly, there are many soils that have a high lime requirement, and such land tends to produce forage low in calcium.

The figures from these calculations show that the need for supplementary sources of calcium and phosphorus varies from none at all, where feeding of average roughage plus enough meal to fully meet energy requirements is followed, to amounts equivalent to including bone meal to the extent of 3 per cent in the meal allowance, where calcium- and phosphorus-poor roughage must be fed or where meager grain allowances are offered. The average requirement (where it is needed) of bone meal per 1,000 lb (or kg) of meal mixture is 18 lb (or kg) to meet calcium and 19 lb (or kg) to meet phosphorus deficits. We realize that individual cows for varying periods of time during a lactation are likely to be in one or another of all the different situations covered by the calculations, and that, because of the function of the bone trabeculae, fluctuations in intake of calcium and phosphorus above and below the exact requirement are not critical. We believe that the inclusion of supplements to provide 2,000 g of calcium plus 920 g of phosphorus per 1,000 lb (450 kg) of meal mixture should be routine practice. Twenty kg of feeding bone meal in 1,000 kg of an 18 per cent protein meal mixture will meet these levels. (Bone meal is merely a typical Ca–P supplement. We have no special preference for it as a source of these nutrients.)

We should also mention common salt. Standard dairy-cattle feeding for many years has called for the inclusion of 1 per cent common salt in the meal ration. This amount meets average conditions. In general, where normal amounts of meal are being fed, cattle show no requirement for salt beyond the 1 per cent included in their meal allowances. (Cows that are receiving no meal should have access to a mineral mixture that will provide salt and phosphorus.)

Wherever there is evidence of iodine deficiency, as indicated by the birth of goitrous calves, the meal should be fortified with iodine, which can be supplied most simply by using iodized salt. Five grams of potassium iodide

---

* Or other equivalent source of Ca and P.

per 50 kg of salt is probably adequate for this purpose. There are also geographic areas where cobalt deficiency is known to exist. In such areas the salt might also be cobaltized, using approximately 2.5 times as much cobalt sulfate per unit weight of salt as is recommended for iodine. (The problem of copper and molybdenum is discussed in Chapter 14.)

Beef-Cattle Meal Mixture Specifications

Meal mixtures for the several classes of beef cattle are similar in principle to those for dairy cattle, but have a different emphasis. Growth and fattening of steers and heifers become the major interest for concentrate feeding. This is reflected in the NRC feeding standard, where feeding categories are set out for ten separate groups of beef animals. These groups frequently differ more in management and in feeding practice than in the nature of the feed combination involved.

Forage plays a large part in the feeding program either as forage to be grazed or as a harvested crop. Indeed, for any beef stock over one year old that is not on a fattening program, roughage is likely to furnish the entire energy of the ration. A very general picture of the feeding groups and their requirements for feed is shown in Table 15-7.

If we accept the NRC feeding standard and make calculations on the basis of forage of 6 per cent DP, we find that four meal mixtures are recognized in terms of differing protein requirements (assuming that the use of a 14 per cent instead of 13 per cent mix is as satisfactory for the growing and breeding bulls as it is for the growing heifers). Thus, the 500-lb (230 kg) calves being finished as baby beef or as short yearlings should have a meal mixture with about 18 per cent total crude protein. The animals of this category, when they have reached 800 lb (360 kg), can use a 14 per cent protein mix. Growing heifers of 400 to 500 lb (or 180 to 230 kg) and growing bulls of 800 lb (360 kg) will also need the 14 per cent protein meal ration. It will be well to have about 12 per cent protein in the ration for the lighter yearlings when they start the feeding period. Finally, a meal carrying 7–9 per cent protein will meet the needs of all other groups to which feed other than roughage is to be given. This level of protein will be supplied by unsupplemented basal feeds and, consequently, is not a problem in meal mixture or protein supplement formulation.

It may be well to point out that *concentrate* and *meal* mixtures as here used are synonymous and refer to the total nonroughage feed combinations. For example, shelled corn, a protein supplement, and a mineral supplement may all be made available to the stock free-choice. If 10 lb (4.5 kg) of corn plus

TABLE 15-7 *Feeding Categories for Beef Cattle and Some Typical Feed Needs*

Feeding group	Av. wt of animal lb (kg)	Concentrate feed (lb)	Roughage feed, air-dry (lb)	DP needed daily (lb)	DP in roughage of 6 per cent DP (lb)	Protein needed in meal DP (lb)	CP (lb)	Per cent of CP needed in meal
Growing heifers	400 (180)	3.8	8.2	0.90	0.49	0.41	0.50	14
	600 (270)	2.0	14.0	0.90	0.84	0.06	0.08	4
Breeding bulls	800 (360)	10.2	6.8	1.40	0.40	1.00	1.25	13
	1,400 (640)	7.6	16.4	1.40	0.98	0.42	0.53	7
Wintering stock								
Calves	500 (230)	2.0	11.0	0.80	0.66	0.14	0.18	9
Yearlings	700 (320)	0	17.0	0.80	1.00	—	—	—
Pregnant heifers	800 (360)	0	20.0	0.90	1.20	—	—	—
Mature cows	1,100 (500)	0	18.0	0.80	1.20	—	—	—
Finishing stock								
Calves	500 (230)	4.5	9.5	1.20	0.57	0.63	0.80	18
	800 (360)	5.6	13.5	1.50	0.80	0.70	0.90	14
Yearlings	600 (270)	10.8	7.2	1.30	0.44	0.86	1.10	11
	1,000 (450)	15.6	10.4	1.70	0.62	1.10	1.38	8
Two-year-olds	800 (360)	11.5	12.5	1.50	0.75	0.75	0.94	8
	1,200 (550)	14.0	15.0	1.80	0.90	0.90	1.10	8

Source: Arranged from **Tables I** and **II** of the NRC *Nutrient Requirements for Beef Cattle,* 1950 revision. (Used as an example only.)

1.5 lb (681 g) of a 40 per cent protein mineral supplement are eaten, it is the equivalent of a *meal* mixture of 14 per cent protein.

The calculations of Table 15-7 were based on roughage with 6 per cent DP, which means a good-quality product that probably contains 50 per cent legumes. If roughage with less protein is used, the protein to be derived from the meal must increase. However, lower-protein roughage often also means lower TDN in the roughage and probably a tendency for reduced roughage consumption. This means more grain or meal must be fed to meet the energy needs. The end result is that the per cent of protein in the meal does not usually increase in direct inverse ratio to the decline in the per cent of protein in the roughage. When all factors are considered, the percentages of total crude protein in the concentrate feed that will adequately meet nutritional needs in the beef herd when fed good roughage appear to be:

Calves at start of feeding for baby beef	18%
Calves fed for normal growth	14%
Yearlings for fattening	12%
Two-year-olds for fattening	8%
Other classes	8%

It will be interesting to compare these percentages with those needed in the rations of dairy animals. For the young dairy calves, a meal mixture of 20 per cent protein was found satisfactory for average conditions. Beef calves (above) are found to need only 14 per cent protein in the meal. This difference is, in part, one of terminology. Beef calves are not fed man-made rations much before they have attained 30 per cent of their mature weight, and are nearly six months of age. They are still calves to the beef-cattle feeder, however. On the other hand, the dairy calf is weaned at one month, at a weight representing only 10 per cent or less of its adult size. The feeding for the next four months is one of replacing with a meal mixture the milk the beef calf gets during this period. Such meal mixtures must contain more protein than the calf will need from its sixth to its twelfth month. The situation is explained by a paragraph from the *Quebec Provincial Feeders Guide,* which states:

It is frequently assumed by feeders that once the calf reaches an age of six months and is ready to go out on regular herd rations, that little attention to feeding is needed until well along in pregnancy. Thus many heifers after the calf stage fail to grow and develop as fully as they should.

From this statement we can see that the calf stage in the dairy herd unofficially ends at about six months of age. The feeding until the calf is a year old is comparable in protein needs to that commonly referred to by the beef feeder as the "calf" period. When we recognize this difference in terminology the apparent discrepancy in protein recommendations largely disappears.

The more fundamental consideration arises from the difference in the objectives and hence in the feeding management of the beef and dairy herds. Other than for the young calves and for animals during final stages of finishing for market, the ration for all classes of beef animals is essentially roughage. Concentrate feeding in a commercial enterprise is always at a minimum consistent with health. Adult animals are at maintenance (or below) for most of the year and are bred at a time when the extra demands for reproduction and lactation coincide with the forage's being at its most abundant and nutritious stage.

When concentrate feeding becomes necessary it is usually because, relative to its useful energy, the content of the available forage is inadequate in protein and vitamin A. Occasionally there may also be a phosphorus shortage, especially if prairie hay or straws are to be the principal forage. Table 15-8 illustrates the general problem.

We can see from this table that for those classes of beef animals that are normally fed a ration with 50 per cent TDN (see the NRC standard) the so-called good roughages ordinarily form an adequate ration. Of the poorer quality roughages there are some whose energy value (TDN) is still essentially the same as that of the good-quality feeds. However, their protein may be inadequate, and they may also be short of phosphorus and carotene. Hence when we depend on these roughages, a protein and perhaps a mineral-vitamin supplement must be used for adequate nutrition.

We should point out that forages whose TDN is below 50 per cent will require extra energy from a grain allowance, and since any grain used will also provide protein, there may be no need for a high-protein supplement as well. Going back to the NRC standard we find given for mature cows of 1,100 lb, a requirement of 18 lb total feed with 4.5 per cent DP and 50 per cent TDN. If a forage with only 2.2 per cent DP and 40 per cent TDN were used, such as fully matured timothy, then grain would be needed to supplement the shortage of digestible energy. If 18 lb (8.2 kg) of total feed is still to be fed, then we would have to use 11 lb (5 kg) of the roughage plus about 7 lb (3.2 kg) of grain. This combination would also meet the protein needs if the grain were corn, and would provide a slight excess if barley, oats, or wheat were the grain. If, however, the cows were still to eat 18 lb (8.2 kg) of the 40 per cent TDN forage, it would require 2.2 lb (1 kg) of grain in addition to meet the full energy needs. This amount of corn grain would supply only about 0.15 lb (68 g) DP, leaving almost a third of a pound (150 g) to be supplied by some high-protein supplement. The solution there-

TABLE 15-8 *The Approximate Average Protein, Energy, Phosphorus, and Carotene Content of Some Common Beef-Cattle Forages*

Forage	Per cent of DP	ME (kcal/kg)	Per cent of TDN	Per cent of P	Carotene (mg/kg)	Supplement required [a]
Requirement	4.5	1808	50	0.19		
Alfalfa, hay	10.8	1816	50	0.23	54.8	None
Brome, hay	4.5	1427	39	0.28		Energy
Clover, hay	7.0	1880	52	0.19	18.0	None
Sorghum, kafir, fodder	3.1	1862	52	0.16	15.6	Protein, P
Lespedeza, hay	8.6	1809	50	0.17	42.8	P
Grass-legume, mixed, hay	4.5	1808	50	0.18		P
Oats, hay	3.9	1946	54	0.21	88.9	Protein
Soybean, hay	9.0	1677	46	0.20	31.8	Energy
Sorghum, Sudangrass, hay	4.9	1896	52	0.28		None
Corn, fodder	2.0	1647	46	0.07	3.1	Protein, energy, P, carotene
Grass, hay	3.5	1880	52	0.21		Protein
Lespedeza, hay	3.6	1446	40	0.15		Protein, energy, P
Prairie hay	1.5	1759	49	0.11	29.6	Protein, P
Prairie hay	1.2	1802	50	0.08	9.6	Protein, P, carotene
Oats, straw	1.3	1694	47	0.09		Protein, energy, P, carotene
Timothy, hay	3.1	1712	47	0.17	12.0	Protein, energy, P, carotene

[a] Based on all feeding groups whose total ration is shown in NRC standard as 50 per cent TDN.

fore might be a mixture of 1.2 lb (550 g) corn plus 1.0 lb (45 g) of soybean meal or cottonseed meal, which would meet quite nicely both the protein and energy demand. Such a mixture would be about 25 per cent protein, and would be referred to in the feed trade as a mixed-protein supplement.

Obviously there are innumerable combinations of forages, each of which yields a characteristic quantity of nutrients and energy. When these are matched against the need of some feeding groups of animals, we may find that we prefer in individual situations supplementary feeds, ranging all the way from simple basal feeds of 9 to 12 per cent protein (or even a single basal feed) to straight high-protein supplements such as cottonseed meal, furnishing 40 to 45 per cent protein (with combinations giving all inter-mediate protein levels).

There is no fixed pattern of "balanced rations" for beef cattle for yet an-other reason. Most of the situations that require protein supplements to the roughage call for relatively little extra energy. Consequently, these situations are met by the use of one high-protein feed or a mixture of several. Feeders who raise their own basal feed are interested in the formulation of protein or protein-mineral supplements. The nature of the basal feed is for them a constant. All of their final ready-to-feed rations will be combinations of their own grain (corn, barley, etc.) with varying proportions of protein, mineral, and vitamin supplements, according to the needs of the different feeding cate-gories.

Often the problem of meal mixtures for the beef herd is one of formulating (1) protein supplements, (2) protein-mineral supplements, and (3) mineral supplements, each with and without vitamin sources. Because one group of feeders is likely to employ corn or sorghum grain as the basal fed, while others may use coarse grains (barley, oats) or wheat, the basic differences in protein and in carotene between the two types will require due consideration in the formulations. And for those who use no grain, the mineral likely to be missing from the "poor forage" must be taken into account. In Chapter 16, flexible formulae for such supplements are given, together with charts for their use.

SUGGESTED READINGS

The series of individual pamphlets published by the NRC Committee in Animal Nutrition on the nutrient requirements of domestic animals: *Poultry* (1965); *Swine* (1964); *Dairy Cattle* (1966); *Beef Cattle* (1963); *Sheep* (1964); *Horses* (1966).

Guilbert, H. R., and J. K. Loosli, "Comparative Nutrition of Farm Animals," *J. Animal Sci.,* X, no. 1 (1951), 22.

CHAPTER 16

The Preparation of Flexible Formulae for Meal Mixtures

Ration formulation involves four distinct steps. The first is establishing the percentage of protein of the meal mixture, which is done with the help of feeding standards. The second is determining the proportions of the three main feed categories necessary to obtain that percentage of protein. The third is specifying the feeds to be used in each category, and stating the limits of substitution permissible within categories. Finally, the mixture must be checked against the appropriate feeding standard for adequacy of nutrients other than protein, and amendments specified where necessary.

It is as ingredients of meal mixtures that feeds other than roughages are put to work. However, there is no one best mixture. This is self-evident from the fact that the different feeding stuffs contain much the same kinds of nutrients, though often in widely different proportions and amounts. Some feeds are more nearly alike in proportions of nutrients than others and are thus practical substitutes for each other. Similar feeds may be grouped in a meal mixture formula according to the principal nutrient or nutrient combination they furnish in the final ration. In a working formula these groups may be:

1. Basal feeds (Energy Feeds).
2. Protein supplements.
    a. vegetable origin.
    b. animal origin.

3. Mineral supplements.
   a. macro-elements.
   b. micro-elements.
4. Vitamin supplements.

Because the average protein content of the common feeds of the basal group does not deviate widely from 12 per cent, and because the quantities in most animal meal rations of vitamin and mineral supplements are small, the proportions of these four feed groups are relatively fixed in the final mixture with a specific percentage of protein. Within the categories, however, considerable latitude in the selection of the individual feedstuffs is possible without appreciably altering the feeding values of the combinations. Thus, it becomes possible to prepare what might be called *pattern mixtures* for the various feeding categories of different classes of livestock. An example of such generalized formulae, shown in Table 16-1, was taken from the 1954 Quebec Feed Board bulletin, *Feeders Guide and Formulae for Meal Mixtures.**

In the absence of more detailed specifications, such general guides are enough to permit one familiar with feeding stuffs and feeding practice to formulate working combinations for balanced meal mixtures. It is preferable, however, to carry the pattern mixture idea further, to the preparation of a flexible formula designed to meet the specifications of adequate diets for described groups of animals. Our purpose in this chapter is to show how this is done.

The Pattern Meal Mixture

The step-by-step development of a pattern meal mixture for growing swine will serve as an example of the procedure. We note from Table 16-1 that a meal mixture suitable for growing market pigs should be 15 per cent total crude protein. We might also conclude from the innumerable formulae that have been proposed by experiment stations, extension services of colleges, and government departments of agriculture, or have found their way into textbooks on hog feeding, that such a meal mixture *could* be made up principally of farm-grown grains, linseed meal, meat meal, limestone, and salt. These same sources of information will also lead to the tentative conclusion that 1 per cent limestone or a comparable source of calcium is needed. Salt

---

* In this chapter the figures for the amounts of the several groups making up the batches are shown in pounds, but they can be considered to be in kilograms just as well.

TABLE 16-1 Generalized Formulae for Meal Mixtures for Farm Livestock and Poultry

Class of feed	Cattle[a] Cows in milk — Fed legume hay	Cattle[a] Cows in milk — Fed nonlegume hay	Cattle[a] Others excepting calves under six months	Hogs Early-weaned pigs to 60 lb[b]	Hogs Sows and market pigs to 100 lb[b]	Hogs All others[e]	Sheep Nursing ewes	Sheep Lambs	Chicks[f]	Chickens[e] Growing birds	Chickens[e] Layers and breeders	Turkeys[e] Poults	Turkeys[e] Growing birds	Turkeys[e] Fattening birds
Basal feeds														
Cereal grains and mill feeds	82	72	82	75	85	93	82	62	74	72	72	60	66	70
Protein supplements														
Vegetable	15	25	15	15	9	4	15	25	8	14	14	15	17	16
Animal	—	—	—	5	3	1	—	—	3	6	6	5	7	6
Minerals	3	3	3	5	3	2	3	3	3	2	2	3	3	2
Vitamins[d]	—	—	—	—	—	—	—	—	12	6	6	17	7	6

Note: In this table feedstuffs have been grouped into four main classes: (1) the basal feeds, which are represented by the farm grains and mill feeds, chiefly bran and shorts; (2) protein supplements, including subgroups of vegetable and animal sources; (3) minerals, and (4) vitamins. The livestock has been grouped in accordance with similarity of ration requirements. Thus, cows in milk will need two kinds of mixtures depending on the kind of roughage being fed. Fattening cattle may use the same feed as cows on legume hay, or, if desired, may be fed a mixture in which the mill feeds are replaced by basal feeds.

[a] Fattening rations should omit bran and shorts, and in others these feeds should be limited to 25% of the mixture.

[b] Not over 50% oats for sows or 30% for young pigs.

[c] Not over 50% corn or wheat. May contain 20% bran for production of lean bacon.

[d] Vitamin supplements: alfalfa, yeast, wheat germ, fish oil, etc.

[e] For poultry the scratch feed is considered a part of the formulae given.

[f] Mill feeds not to exceed 10%.

is universally added as 0.5 per cent of the ration. We could also calculate from feeding standards as in Table 16-2.

TABLE 16-2 *Mineral Supplements for Swine Rations*

Nutrient	Requirement for 100-lb pig per day	Normally in pig feeds	Needed in minerals	Equivalent in mineral carriers	Needed in 100 lb of feed mix[a]
Calcium	15.6 g	7.8 g	8 g	22 g limestone	400 g (~1%)
Phosphorus	10.8 g	10.7 g	0	—	—
Salt	20.0 g	0	12 g	12 g salt	240 g (~.5%)

[a] Based on 5.6 lb feed daily.

With respect to the linseed meal and meat meal proportions we may make use of a common rule of thumb that at least 10 per cent of the protein should come from an animal or marine source. Thus in a 1,000-lb batch of a 15 per cent protein meal mix, 15 lb of protein must be from meat meal, and this will call for 30 lb of this feed. With this information we can determine the quantities of each of the feeds to make a 1,000-lb batch.

	Feed	Protein
Total wanted	1,000 lb	150 lb
Meat meal	30	15
Limestone	10	0
Salt	5	0
Other feeds (by difference)	955	135

The per cent of protein to be in other feeds $= \dfrac{135 \times 100}{955} = 14$ per cent.

The proportions of the linseed meal and the farm grains (i.e., other feeds) needed can be easily calculated using the *cream-blenders* scheme. We need only to know that the linseed meal is 35 per cent protein and we shall figure in the example that the farm grains are 12 per cent. We have already determined that the mix must be 14 per cent protein.

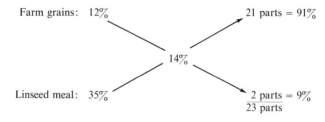

Farm grains: 12%   21 parts = 91%

14%

Linseed meal: 35%   2 parts = 9%
23 parts

Thus we require of linseed meal $955 \times 9\% = 87$ lb, and of farm grains $955 \times 91\%$ (or $955 - 87$) $= 868$ lb. Our final 15 per cent protein pattern mixture then becomes:

Farm grains	868
Linseed meal	87
Meat meal	30
Limestone	10
Salt	5
Total	1,000 lb

This is not yet the completed ration. It may require amendments to augment some of its minerals and vitamins, and we may wish to add an antibiotic. It does, however, form a pattern that establishes the proportions of key feeds needed to fix its protein level. From what we said in Chapter 11, it is evident that the particular feeds shown in this formula are merely key feeds, representing in each case feedstuffs of a particular category of the feed classification. Thus, linseed meal could as well be soybean meal, or peanut meal, or a mixture of two or more of the feeds in this category. Indeed, we might expand our simple formula to list some possible substitutes for each of the feeds of the simple formula (see Table 16-3).

Making the Pattern into a Flexible Formula

When the pattern is expanded as in Table 16-3, we have the start of a flexible formula that can serve as a guide to innumerable combinations of feedstuffs, any of which will result in a 15 per cent protein mixture. However, because of individual differences in feeds of the same category, care must be taken in the substitutes or alternative choices made. Often the question is not whether to use this or that feed, but how much of each. By making partial substitution we may be able to take advantage of desirable feeding properties or attractive prices and at the same time avoid undesirable consequences of an unrestricted use of some product. We can incorporate into our flexible formula a guide to cover these points. We have to add a column setting the maximum limit of use for each feed listed, and a further column in which a figure may be set to insure the minimum use of some feed.

The figures in these two added columns obviously should be based on sound evidence. Properly set they are insurance against inadvisable combinations that might be put together by one not fully aware of the peculiarities of certain feeds. The minimum figures are also of use in cases where, because of "protective" nutritional properties or because of price or availability, we want to be certain we have not overlooked a product.

TABLE 16-3 *Skeleton Formula for 15 Per Cent Protein Swine Ration*[a]

Main groups	Subgroups	Feeds	Recommended amounts
Basal feeds (Av. 12% protein) 868 lb		Barley Corn Oats Wheat	868
Protein supplements (Av. 40% protein) 117 lb	30–40% protein plant origin (87 lb)	Linseed meal Soybean meal Peanut meal Coconut meal	87
	35–70% protein animal origin (30 lb total to provide minimum 15 lb protein)	Meat meal Fish meal Milk powder Tankage	30
Mineral supplement, 15 lb		Bone meal Dicalcium phosphate Ground limestone Salt	10 5
Vitamin-antibiotic supplements[b]		Vit. A (IU) Vit. D (IU) B₁₂ (mg) Antibiotic (g)	800,000 90,000 5 5

[a] This is merely an example, not a complete ration (see Chapter 19 for hog-ration formulae).
[b] Values shown are for 1,000-lb batch of feed; for 1,000-kg batches multiply values by 2.2.

Such an oversight might well occur in commercial feed manufacture, where a by-product of a particular company is sold largely through the "balanced rations" they make. Thus a milling company that prepares table hominy might depend on including corn by-products in their balanced livestock rations for disposing of such feeds. The figures set for the minimum in this case might be based on anticipated supplies, other markets, company policy, etc., rather than entirely on nutritional considerations. In another case there might be a wish to ensure a low-energy ration for bacon production by the inclusion of a bulky feed, such as oats or bran.

The maximum and minimum limits are obviously subject to revision from time to time, and may also be widely different in different geographic areas. They can be, and very likely will be, tempered by prejudices and beliefs of the

author of the formula. Unless these are incompatible with nutritional considerations they are not fundamentally serious.

Subformulae

One useful feature of this type of formula (a consequence of the grouping
of feeds) is that it can be broken down into several subformulae. For example, we can derive recipes for a protein supplement and for a mineral-
vitamin supplement, as in Table 16-4.

TABLE 16-4 *Examples of Subformulae*

Feeds	Original amounts	On a 100 lb basis
For a protein supplement		
Linseed meal	87 lb	74
Meat meal	30 lb	26
For a mineral-vitamin supplement		
Limestone	10 lb	67
Salt	5 lb	33
Vitamin A	800,000 IU	5,000,000 IU[a]
Vitamin D	90,000 IU	600,000 IU[a]
Vitamin B_{12}	5 mg	33 mg[a]
Antibiotic	5 g	33 g[a]

[a] For 100 kg, multiply by 2.2.

Using the Supplement Formulae

To use these formulae to prepare mixtures with different percentages of
protein, we can easily calculate the quantities of protein supplement we need
per 1,000-lb batch. For an 18 per cent mixture we would figure:

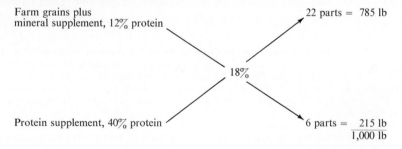

Farm grains plus mineral supplement, 12% protein   22 parts = 785 lb

18%

Protein supplement, 40% protein   6 parts =   215 lb
                                                            1,000 lb

Since we are to use 15 lb of minerals in our 1,000-lb batch (see Table 16-3), our proportions become:

Farm grains	770 lb
Protein supplement	215 lb
Mineral supplement	15 lb
Total	1,000 lb

Note that the protein and mineral supplements are treated separately. The latter is needed in proportion to the total mixture rather than to the protein supplement. The quantity of mineral supplement is so small that it can be neglected insofar as any practical effect on protein level is concerned. Hence in calculating proportionate parts we take the basal feeds plus minerals as 12 per cent protein.*

Here we see the origin of the mixed supplement. Such combinations are of interest to the feeder who wishes to use his home-grown grain to prepare the several rations needed in the feeding of his animals. Even though some ingredient of the supplement may not be needed for some one ration, its presence is not likely to be harmful. Such supplements can be made "flexible" merely by listing possible substitutes where they exist, and setting minimum and maximum limits for their use (see Chapter 17).

Precautions

We must remember that flexible formulae are guides and because of the latitude they permit in feed selection, are subject to abuse. Improper substitution can result in undesirable modifications of the protein in the final mix, although this modification will be minimized as we increase the number of feeds we use. Also, it may be possible to make mixtures that are of unsatisfactory palatability or are too heavy or too bulky. These limitations merely emphasize the fact that, in addition to whatever formulation guides may be available, some knowledge of feeds and some nutritional judgment are neces-

---

* In this example we have taken the farm grains as 12 per cent crude protein. This, of course, will be too high if the basal feeds to be used include a large proportion of corn or milo. The principle of the flexible formulae does not depend on any particular percentage of protein in the basal feeds or protein supplements. If, in a corn area, this feed is to make up 100 per cent of the farm grains, the only change from the example given above will be in the proportion of basal feeds to protein supplements. The steps in deriving the formulae are in every respect identical to those used in the example. Incidentally, the average protein of a mixture of equal parts of barley, oats, corn, and wheat is almost exactly 12 per cent.

sary qualifications of anyone who is to successfully compound nutritionally adequate livestock rations.

SUGGESTED READINGS ─────────────────────────

Crampton, E. W., "A Method for Determining the Composition of a Meal Mixture Suitable for a Given Purpose," *Sci. Agr.,* VIII (1927), 235.

Feeders Guide and Formulae for Meal Mixtures (Quebec Provincial Feed Board, 1962).

CHAPTER 17

The Preparation of Mixed Mineral Supplements

From the discussion in Chapter 16 it is evident that we must consider mineral supplements before we can formulate a balanced ration for any class of farm animal. It may be well, therefore, to deal with mixed mineral supplements before we go further with the preparation of complete meal rations.

Flexible Formula Mineral Supplements

When grain feeding is the normal practice, as it is with hogs and with most categories of dairy cattle, the problem of supplying any necessary mineral supplement is simply solved. But beef cattle and some classes of dairy cattle frequently receive no concentrate feeds, and the forage on which they subsist may be deficient in minerals.

In one sense the mineral supplement is a comparatively simple combination of materials, each often supplying fixed quantities of one or more needed elements. Bone meals are feedstuffs with 23 per cent calcium and 11 per cent phosphorus. They may deviate from these values, but only slightly, partly because in dry, fat-free bone all or part of the CO_3 in the $CaCO_3$ may be replaced by such radicals as F, OH, or Mn. The $Ca_3(PO_4)_2$ part of bone structure is for all practical purposes a constant. Furthermore, bone meal, as purchased, is a composite from thousands of animals, and so any variation between animals is averaged out in the commercial product.

There is more variation in the makeup of some of the phosphorus sources that have their origin in rock phosphate or in the softer deposits of high-phosphorus-containing bird skeletons, such as are found in the Curaçao Islands. Some of the variation is natural and some is a result of processing, such as the manufacture of phosphatic fertilizers or, incidentally, the removal of fluorine.

Mineral Supplements for Cattle or Sheep

Under practical cattle-feeding or sheep-feeding conditions, when normal amounts of roughage are fed, no intentional inclusion in a mineral supplement of specific sources of calcium is justified. The more common sources of feeding phosphorus are combinations that include as much calcium as phosphorus.

In more or less local areas there may be deficiencies in forage which make it desirable to include in the mineral mixture copper, cobalt, iodine, or manganese. Because they are needed in such small amounts relative to calcium, phosphorus, or common salt, they are sometimes referred to as trace minerals or microelements (see Chapter 8). Their function appears to be specific, but, except for iodine, their roles have not been fully elucidated. It is probable, however, that part of the function of at least some of them is related to the nourishment of the microflora of the rumen and caecum of the host animal, or that their use to the animal is mediated by the microflora. The details are less important here than the fact that some low-grade forages are readily eaten, and yield about as much available energy as the so-called good forages, if they are supplemented by vitamin A, *a small quantity of minerals,* and a source of nitrogen, which may be protein or a product such as urea. Corn cobs, for example, can furnish all of the roughage for fattening steers if vitamin A, plus a mineral mixture that is adequate qualitatively and quantitatively, and, of course, appropriate allowances of grain and protein supplement are fed. In experiments at the Ralston Purina Company Research Farm, cobs replaced pound for pound all of the hay, and since the amounts of other feeds were comparable, and equal gains were made by the steers (about 2.5 lb per day), it follows that the cobs furnished available energy to the animals equal to that from like weights of alfalfa hay.*

Information about the optimum composition of a perfect mineral supplement is lacking. We do know enough already to prepare combinations that will give acceptable results; and we have no reason to think that excesses of specific minerals, which undoubtedly exist in many of the rations giving

---

* H. B. Geurin, J. C. Thompson, H. L. Wilcke, and R. M. Bethke, "Cob Portion of Ground Ear Corn as Sole Roughage for Fattening Cattle," *J. Animal Sci.,* XIV (1955), 797.

excellent results by present standards, are nutritionally harmful. This thinking perhaps leads us to the use of "shotgun mixtures," which in the past have so often been decried by the animal husbandman. However, there are differences among shotgun mixtures, and perhaps the ones that are the result of educated guesses based on sound though limited experimental evidence can with advantage now be removed from the category of quack remedies. In any case we believe it to be amply proven that, because roughages are so highly variable in nutrient makeup from sample to sample, fortification with complex mineral mixtures may more often than not be sound feeding practice.

Flexible formulae for mineral mixtures that permit the local selection of carriers available are easily prepared. The chief item of choice will be the phosphorus carrier. The other variables are largely matters of inclusion or exclusion.

Supplementary Phosphorus Levels for Beef Cattle. Some idea of the supplementary phosphorus needs of beef animals, for example, can be had from feeding standards. Certain assumptions must be made concerning the kind and amounts of forage and meal (if any) that are fed.

If we assume that the average phosphorus content of poor roughage is 0.1 per cent, and that the animals of the several feeding groups are to receive rations of hay or hay plus grain to meet energy needs set out in the feeding standard, we can prepare a table showing the supplementary phosphorus that would be needed to meet probable shortages (see Table 17-1). From these calculations one could draw the conclusion that beef animals fed poor roughage (0.1 per cent P) might need up to 3 g of phosphorus per

TABLE 17-1 *Desirable Supplementary Phosphorus Levels for Beef Cattle (All figures are grams)*

Category	Daily P requirement	Phosphorus furnished daily in		Additional P needed per day
		Meal	Poor hay	
Growing heifers	15	8	4	3
Young bulls	18	12	8	—
Wintering				
Calves	12	3	6	3
Yearlings	12	—	9	3
Heifers and cows	16	—	10	6
Fattening				
Calves	16	9	5	2
Yearlings	20	15	4	1
Two-year-olds	20	15	6	—

day in addition to that in their normal feeds. It might be desirable to double this amount for pregnant heifers.

The ingredients of the mineral mixture, other than the phosphorus source, should include salt. Mineral mixtures without salt are not consumed uniformly well when offered as licks or in self-feeders.

Salt–Phosphorus Ratio. The principal ingredients contributing to the weight of the mineral mixture are the calcium-phosphorus carrier(s) and common salt. The proportions of these two are of some importance in the mixture if it is to be self-fed by itself. The ingredient which determines the quantity that will be eaten voluntarily is the salt. Hence, if the salt forms too large a portion, the consumption of the other ingredients will be inadequate.

Experiments indicate that voluntary salt intake by cattle on the range is from 1 to 2.5 lb (450 to 1,100 g) per month. Since we are concerned with adequate consumption of the mineral mixture, we are justified in figuring on the 1 pound per month salt intake, equivalent to 15 g per day. If, now, we use the figure of 3 g of phosphorus as the probable maximum daily requirement, then we need to provide about 30 g of a phosphorus carrier such as dicalcium phosphate or bone meal. From these two figures we arrive at the proportion of 1 part salt plus 2 of bone meal as the basic part of the mineral mixture, i.e., salt is 33 per cent by weight of the mineral mix. This ratio obviously will change if we use a more concentrated source of phosphorus, but since several products containing 12 per cent or less phosphorus are popular, the basic figure of 33 per cent salt in the mineral mix is a practical working figure. It also makes inclusion of any minor elements at the expense of the phosphorus carrier quite safe; and at the same time makes it possible to include the trace minerals in proportion to the salt, since the latter remains a constant.

There are also a few products with 20 to 24 per cent phosphorus that are now available for stock feeding. If these are used, calculations will show that to supply the phosphorus needed with the same (15 g) salt intake, the proportions of salt to phosphorus carrier will be about 1:1. To use the 1:2 ratio would be wasteful of phosphorus, though probably otherwise harmless. Consequently, it might be desirable to prepare two mineral formulae—one to be used where the source of phosphorus was a product analyzing from 10 to 14 per cent, and another for products of 20 to 24 per cent phosphorus.

Proportions of Other Elements. Of the other elements, the feed manufacturer will probably wish to include iodine, cobalt, iron, copper, and perhaps manganese, on the grounds that the mix may be fed in many different areas, in some of which some or all of these ingredients may be needed. For local use any one or more of these trace minerals can be omitted without changing the mixing formula appreciably, because of the small proportions in which they are used. *The more important consideration is the quantity that will be effective if they are to be included.* The use of too little results in an ineffective supplement, while excesses are uneconomical and nutritionally unnecessary.

TABLE 17-2   *Flexible Formulae for Mineral Supplements for Cattle and Sheep Feeding*

Source of mineral elements	Per cent of phosphorus in carrier	No. 1, using low P carriers	No. 2, using high P carriers	Salt licks
Monocalcium phosphate	24 ⎫		50 lb (23 kg)[a]	0
Dicalcium phosphate	20 ⎭			
Bone meal				
Steamed	14 ⎫			
Char	13	67 lb (30 kg)[a]		0
Raw	10			
Rock phosphate	9–13 ⎭			
Salt (NaCl)		33 lb (15 kg)	50 lb (23 kg)	100 lb (45 kg)
Trace elements (added at expense of P carrier)				
Copper sulfate		150 g	225 g	450 mg
Potassium iodide		2 g	2.8 g	5.6 mg
Cobalt chloride		5 g	8 g	16.0 mg
Manganese sulfate		140 g	210 g	420 mg
Daily amounts of mixture to supply salt needs of adult				
Beef cattle and sheep		40 g	30 g	15 g
Dairy cattle		85 g		

[a] Minus total of trace elements used.

Table 17-2 gives two flexible formulae for mineral mixtures for cattle and sheep, in which the quantities of the trace elements are adjusted to the salt level in accordance with experimental evidence and feeding standard recommendations. A salt lick is also included. These mixtures are flexible for the source of phosphorus and also in the sense that the trace elements may be omitted, but inflexible for the salt and the proportions of trace elements to salt where they are included. When the mixture is consumed in amounts to result in a daily intake of 15 g of salt, the intake of the other elements should be adequate. There is no evidence that larger consumption by individual animals will be harmful.

Requirements of Dairy Cattle for Mineral Supplements

The question now arises as to whether these mineral mixtures (Table 17-2) can also be used for producing dairy cows or for beef animals being fed grain, particularly where it is desirable for convenience and for the regulation of intake to incorporate the minerals in the meal mixture. Here our calculations must include the contribution of the meal allowance to the total requirement. Table 17-1 indicates that beef animals receiving liberal meal allowances to permit fattening may receive enough phosphorus without special mineral supplementation. However, much depends on the nature of the roughage. If, for example, the roughage consists largely of corn cobs, its phosphorus content may be negligible (corn-cob meal has only 0.02 per cent phosphorus), and so, although the meal may supply appreciable amounts of this element, it may have to be further fortified with phosphorus to make good the extreme deficiency in the roughage. Referring to Table 17-1 again, we can see that if the roughage cannot be depended on as a measurable source, a supplementary intake should be provided of perhaps 5 g of phosphorus daily.

The voluntary daily salt intake by beef cattle of from 15 to about 40 g appears to be normal. Thus if the mineral supplement is to be incorporated in the meal mixture, 25 g daily would not appear to be objectionable. On this basis the mineral supplement could be composed of 2 parts of phosphorus carrier to 1 of salt as shown in formula no. 1 of Table 17-2.

Lactation Demands. For the dairy cows, lactation requirements will constitute an additional factor. It is probable that the forage used for milking cows will be somewhat higher in phosphorus than for range beef cows or dry dairy stock. In addition, the meal ration will appreciably contribute to the phosphorus intake, even to providing a surplus over maintenance needs. The NRC feeding standard calls for 0.7 g of phosphorus for each pound of milk produced. Babcock long ago suggested that one gram of salt be added to the ration for each pound of milk. Feeding standards also recommend an extra 7 g of phosphorus daily during the last three months of pregnancy.

From this evidence we can calculate as an example in Table 17-3 the probable maximum need for supplemental phosphorus and salt of a 1,000-lb (450-kg) cow producing 30 lb (13.6 kg) of milk and receiving 20 lb (9 kg)

TABLE 17-3 *Mineral Requirements of a Dairy Cow*

Mineral requirement	Phosphorus	Salt
Requirement for:		
Maintenance	10 g	15 g
Pregnancy	7 g	0
Production	21 g	30 g
Total	38 g	45 g
Supplied in:		
Roughage	9 g	0
Meal	19 g	0
Supplement needed in 8 lb (3.7 kg) meal	10 g	45 g

of average roughage plus 8 lb (3.7 kg) of a 16 per cent protein meal mixture daily. If the phosphorus carrier is 10 to 12 per cent phosphorus, then this cow will require supplementary additions to her daily meal of 80 to 100 g of the phosphorus carrier plus 45 g salt. These proportions of phosphorus to salt are those of formula no. 1 in Table 17-2. If the daily meal is 8 lb (as above), then it will require about 36 lb (16.4 kg) of such a supplement per 1,000 lb (450 kg) of meal mixture, of which 12 lb (5.5 kg) would be salt. This is slightly more than the 1 per cent which the NRC standard indicates will usually meet requirements. Also, such a mineral supplement, if used at 3 per cent of the meal, will provide 15 per cent less phosphorus than is shown as needed in our calculation. We should note, however, that an allowance to cover pregnancy was included in the example. This extra is called for in feeding standards only during the last three months of pregnancy. Deleting this extra allowance of phosphorus would reduce the calculated daily requirement of bone meal some 25 per cent during most of the lactation period. Consequently, the inclusion in the milking-cow meal mixture of 3 per cent of formula no. 1, or of 2 per cent of formula no. 2 (Table 17-2), can be expected to meet the needs as nearly as is demanded in practical feeding practice.

Mixed Mineral Supplement for Swine

The problem of arriving at the basic makeup of the swine mineral supplement is not complicated by a variable roughage contribution to the elements needed. With swine the major mineral, other than common salt, which is likely to be needed in supplementary amounts, is calcium. The phosphorus requirement is usually met from grain intake.

Calcium Carriers. Calcium carriers can be roughly divided into two groups. The one includes products that do not also contain phosphorus. These furnish 36 to 40 per cent calcium, chiefly in the form of calcium carbonate. The other group is made up of calcium phosphates, and these run from about 23 per cent for the raw feeding bone meals to 30 per cent for edible steamed bone meal and dicalcium phosphate, with the various rock phosphates falling within this range. Thus, since the salt requirement is relatively constant, it may be desirable to consider two basic formulae differing according to which type of calcium carrier is to be used.

As with cattle mineral supplements the amounts of the trace minerals may conveniently be expressed in terms of the salt. Inclusion or omission can be at the expense of the calcium carrier without measurable harm to the usefulness of the mixture.

Calcium–Salt Ratio. The proportions of the calcium carriers to salt can be calculated directly from the NRC feeding standard. The quantities of the other ingredients have also been taken from this standard or, where necessary, from data in the more recent literature (see Table 17-4).

TABLE 17-4 *Flexible Formulae for Swine Mineral Supplements*

Source of mineral elements	Per cent of element in source	No. 1, using high Ca carriers	No. 2, using low Ca carriers
Calcium carbonate	40 Ca ⎫		
Oyster shells	38 Ca ⎬ 75 lb (34 kg)[a]		
Ground limestone (high Ca)	36 Ca ⎭		
Steamed bone meal (edible)	30 Ca ⎫		
Raw feeding bone meals	23 Ca ⎪		
Dicalcium phosphate	23 Ca ⎬		85 lb (39 kg)[a]
Rock phosphates	23–30 Ca ⎭		
Salt (NaCl)		25 lb (11 kg)	15 lb (7 kg)
Trace minerals (added at expense of Ca carrier)			
Ferrous sulfate	20 Fe	130 g	80 g
Copper sulfate	25 Cu	6 g	4 g
Potassium iodide	76 I	8 mg	5 mg
Zinc chloride	48 Zn		
Quantity of mixed supplement to include in each 1,000-lb batch of ration		20 lb (9 kg)	33 lb (15 kg)

[a] Minus total of trace minerals.

SUGGESTED READINGS————————————————

The series of individual pamphlets published by the NRC Committee on Animal Nutrition on the nutrient requirements of domestic animals.

Feeders Guide and Formulae for Meal Mixtures (Quebec Provincial Feed Board, 1962).

Geurin, H. B., J. C. Thompson, H. L. Wilcke, and R. M. Bethke, "Cob Portion of Ground Ear Corn as Sole Roughage for Fattening Cattle," *J. Animal Sci.,* XIV (1955), 797.

Stoddard, G. E., and C. H. Mickelsen, "Phosphorus Intake Depends on Taste," *Ut. Agr. Expt. Sta. Farm and Home Sci.,* XX (1961), 103.

CHAPTER 18

Flexible Formulae for Cattle
Meal Mixtures

To arrive at a basic formula* for cattle mixtures, the procedure outlined in Chapter 16 may in principle be followed. But there are two factors to be considered that were not dealt with in discussing the scheme of flexible ration formulation. The first is that quality of protein is a negligible problem in the choice of feedstuffs for herbivores; the second has to do with the percentages of protein in the basal feeds.

Corn vs. Barley as the Basis of Cattle Meal Rations

To be realistic, we must recognize that the basal-feed mixtures used in cattle feeding are essentially of two kinds. The one is based on corn (or milo), and the other is based on the coarse grains, barley and oats. Combinations in which corn predominates will not average 12 per cent protein, which figure was used for convenience in our example in Chapter 16, but will be closer to 10 per cent protein, unless more wheat bran is used than is usually desirable. Using the analyses given by Schneider, we find that corn grain varies from 7.8 to 11.7 per cent protein. The NRC committee gives, in its table of typical analyses for dairy-cow feeds as used, values for DP that are equivalent to protein of 8.5 per cent for dent and 9.5 per cent for flint corn of no. 2 grade; the beef committee figures on corn of only 8.5 per cent protein. Thus, if 3 parts of the basal feeds by weight were corn or milo, and the other seven parts were made up of course grains, the basal mixture would

* Formulae given in this chapter are examples for illustration of method only.

not exceed 10 per cent protein. If we also used molasses, it would be safer to calculate the total protein of the basal mixture as 9 per cent. On the other hand, when corn does not dominate the meal mixture, either barley or oats are likely to be emphasized, with the result that the basal-feed combination will usually be close to 12 per cent protein.

Thus it would seem worthwhile to present two formulae, one based on heavy corn feeding and the other on coarse grains. The difference between them will be in the proportions of basal feeds to protein supplement, the minerals remaining the same in each. This division will also *roughly* separate, for meal rations with the same percentage of protein, beef-cattle and dairy-cattle mixtures, in the sense that the former perhaps more than the latter will tend to emphasize heavy corn or milo feeding.

Percentage of Protein in the Formula

From previous discussions it seems evident that three dairy-cattle mixtures must be considered. One would be for young animals, and might well be called a calf meal; it should be at least 20 per cent protein. We have also shown that other classes of dairy cattle fed according to standard practice should have meal mixtures that are either 16 per cent or 18 per cent protein. In addition, a mixture of 14 per cent should be included for some classes of beef cattle. Of these three, the 16 per cent mixture is probably the most popular, so we will deal first with it.

In these formulae it is a good idea to subdivide the protein supplements; one group will include the oil meals, whose protein content ranges from 30 to 50 per cent, and the other will include the distillery and brewery by-products, whose protein ranges from 20 to 30 per cent.

Feedstuffs Included. The feedstuffs most often used in the preparation of cattle meal mixtures are included in the tables in this chapter. For simplicity the mineral component is shown only as a total weight of one of the supplements discussed in Chapter 16 and shown in Table 17-2.

In the column headed "Recommended formula" we show the quantities of the several feeds making up what might be considered a first-choice combination. The first choice means merely that if neither price, availability, nor any other feature (except nutritional usefulness) is a factor, these quantities give a mixture that has fewer peculiarities which limit its use than many other combinations we could make. It is a sort of guide in the use of the possible substitutions.

In the columns headed "Minimum limit" and "Maximum limit" are figures

TABLE 18-1 *Flexible Formula for 1,000 Pounds*[a] *of a 16 Per Cent Protein Cattle Meal Mixture Made of Basal Feeds Averaging 10 Per Cent Protein*

Main groups	Subgroups	Feedstuffs					
		Name	Per cent CP	Per cent TDN	Min. limit	Max. limit	Recommended formula
Basal feeds (av. 10% protein) 740 lb		Hominy feed	11	85	0	400	
		Wheat	15	83	0	240	
		Corn	9	80	500	600	500
		Milo	11	80	0	500	
		Barley	13	71	0	240	
		Oats	12	66	0	240	
		Wheat shorts	17	64	0	240	
		Wheat bran	16	57	0	240	160
		Molasses	3	72	0	80	80
Protein supplements (35% protein) 240 lb	(30–50% protein) 160 lb	Peanut meal	44	80	0	100	
		Gluten meal	41	76	0	100	
		Linseed meal	36	76	0	160	
		Soybean meal	44	73	0	160	60
		Cottonseed meal	41	79	0	100	100
	(20–30% protein) 80 lb	Corn dist. grains	27	81	0	80	80
		Gluten feed	25	75	0	80	
		Brewers' grains	22	60	0	80	
Minerals 20 lb		No. 2 formula from Table 17-2			20	20	20
Vitamins		Vit. A (as carotene)				10 g	10 g

[a] Figures, other than percentages, may be considered to be in kg if desired.

intended to aid in preparing satisfactory combinations where factors other than, or in addition to, nutritional properties must be considered. Table 18-1 presents a completed flexible formulae for a 16 per cent protein cattle meal mixture in which the basal-feed combination averages 10 per cent total protein.

Proportions of Groups. The proportions of the main groups may be arrived at as follows. It is convenient and reasonable to consider that the oil meals will average 40 per cent protein and the distillery and brewery by-products about 25 per cent, and that in practice we might use two or three parts of the former to one of the latter. The mixed-protein supplements would then, in practice, average about 35 per cent protein, and we could calculate for the 10 per cent protein basal-feed combinations the proportions of the formula in a 1,000-lb batch as:

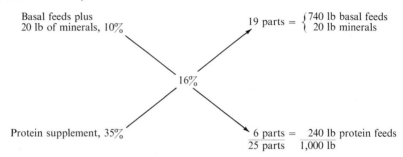

Table 18-2 presents another 16 per cent protein cattle meal mixture formula, based on a selection of basal feeds that averages 12 per cent protein. It will meet conditions where the coarse grains and perhaps also wheat will be more economical than corn or milo grain.

It may be well now to examine the makeup of these meal mixtures in some detail.

Basal Feeds. In these formulae nine commonly used low-protein feeds are included. They have been listed roughly in descending order of their TDN * contents. With the exception of molasses, differences in protein between these feeds are, for practical feeding, relatively minor. All the other feeds are between 9 and 16 per cent protein, and we know from experience that mixtures of them in practical rations usually work out to about 9 per cent protein in corn-feeding areas and to 12 per cent where coarse grains are emphasized.

Maximum Limits. The figures set in the maximum column are such that at least two feeds must be used from the basal category. We might argue that it doesn't matter whether there is more than one basal feed in the cattle meal mixture or not, and theoretically this statement is true. It is equally true that a mixture which contains more than one basal feed may be less restricted in its use, since the characteristics of any one feed do not predominate. Most feedstuffs have unique properties that limit their unrestricted

---

* If energy values are preferred, calculate each pound of TDN as equivalent to 2,000 kcal DE or 1,640 kcal ME.

TABLE 18-2 *Flexible Formula for 1,000 Pounds[a] of a 16 Per Cent Protein Cattle Meal Mixture Made of Basal Feeds Averaging 12 Per Cent Protein*

Main groups	Subgroups	Feedstuffs					
		Name	Per cent CP	Per cent TDN	Min. limit	Max. limit	Recom- mended formula
Basal feeds (av. 12% protein) 790 lb		Hominy feed	11	85		200	
		Wheat	15	83	0	300	
		Corn	9	80		200	
		Milo	11	80		200	
		Barley	13	71	300	500	400
		Oats	12	66	150	400	310
		Wheat shorts	17	64		250	
		Wheat bran	16	57	0	200	
		Molasses	3	72	0	80	80
Protein supplements (35% protein) 180 lb	(30–50% protein) 130 lb	Peanut meal	44	80	0	70	
		Gluten meal	41	76	0	70	
		Linseed meal	36	76	0	130	60
		Soybean meal	44	73	0	130	70
		Cottonseed meal	41	79	0	70	
	(20–30% protein) 50 lb	Corn dist. grains	27	81	0	50	
		Corn gluten feed	25	75	0	50	
		Brewers' grains	22	60	0	50	50
Minerals 30 lb		No. 1 formula (Table 17-2)					30
Vitamins		Vit. A (as carotene)				10 g	10 g

[a] Figures, other than percentages, may be considered to be in kg if desired.

use, especially in dairy-cow rations. Skilled feeders who prefer to employ amounts larger than the maximum set for any one of these products can do so, but for the less experienced, a mixture of basal feeds is partial insurance against a ration that might be somewhat undesirable in specific cases.

Specifically, the use of over 500 lb of *hominy* or *corn* in a 1,000-lb mix will tend to produce a heavy ration, which, if fed in large amounts, may tend to put animals "off feed" more than a bulkier ration might. More than 500 lb of *barley* will sometimes render a ration unpalatable, so that large allowances will not be readily eaten by high-producing animals requiring liberal feed. This unpalatability is more often due to the presence of weed seeds or other "dockage" than to the pure barley. With pure barley that is free from heat or weathering damage, the 500-lb maximum can safely be lifted. *Wheat,* if fed in quantities much over one-third of the ration, tends to produce a heavy, pasty, and somewhat unpalatable ration, especially if finely ground. *Oats,* on the other hand, if included as much more than 40 per cent, tends to result in a mixture of undesirably low TDN content, and of which a correspondingly larger allowance must be fed to meet specified energy requirements. *Wheat bran,* for the same reason, is limited to 25 per cent of the total ration. Even this amount of bran, unless counterbalanced by some of the heavier feeds, will result in too bulky a ration for normal feeding. Experimental evidence indicates that 8 per cent *molasses* in a meal mixture is about all that can safely be incorporated; more risks the danger of the feed becoming lumpy during storage and sometimes turning sour, especially during warm weather.

Minimum Limits. The minimum limits for the basal feeds present a somewhat different problem. One might put zeros for all of these feeds and justify it on the grounds that the maximum limits already require two feeds to be used, and that nutritionally the choice of feeds is probably almost immaterial. On the other hand, setting no minimums would make it possible for the basal portion of this ration to consist entirely of corn and wheat. Such a combination in the eyes of most practical dairy cow feeders would be undesirable, though beef feeders might not object to it. The feed would be heavy and tend to be pasty. Consequently, minimum limits have been set in the "coarse grain" formula for barley and oats. These two feeds together are satisfactory as basal feeds in a dairy-cow meal mixture, and when used in combinations with the heavier feeds will correct the undesirable physical properties of the latter. The maximum limit set for oats prevents its use beyond half the basal feeds. This is insurance that the ration will not be too light and bulky, even if the combination of barley and oats should be all the basal feeds in the mixture.

We should call attention to the fact that no minimum appears for *wheat bran,* in spite of the fact that in some districts feeders seem to think that it should not be omitted from a dairy ration under any circumstances. Bran has a particularly high phosphorus content, and if feeders do not wish to use phosphorus-containing minerals, they may use bran to help prevent phosphorus deficiency. However, the feed is exceedingly bulky and its inclusion in any appreciable proportion in the meal mixture increases unduly the quantity of feed required to meet given energy needs.

High-Protein Feeds. In these pattern mixtures we have indicated that four protein feeds are preferred. Here again, using more than one is not a nutritional necessity, especially in the preparation of a ready-to-feed mixture. Several feeds have been included, but for another purpose. From this 16 per cent formula can be derived a mixed mineral-protein supplement suitable for use in preparing cattle meal mixtures at home with farm-grown grains. Where such supplements contain a variety of feeds, they are less likely to have peculiarities that might limit their use with some combinations of farm grains. Note in connection with the protein supplements that the ratio of the oil meal group to the lower-protein group is fixed in order to give a final combination of 35 per cent protein. Finally, commercially prepared mixed-protein supplements normally involve a variety of feedstuffs, so that price fluctuations can be cushioned by adjusting the proportions of the feeds used.

Among the oil meal group linseed meal and soybean meal have equally good nutritional properties. But there may be reason to limit peanut meal and cottonseed meal somewhat, because these products have different qualities, especially the cottonseed meal. The amounts of gluten meal that can be used are limited because it is a heavy, concentrated feed, and if combined with corn or hominy will result in a mixture heavier than is desired for the best dairy-cattle feeding.

With the 20 to 30 per cent protein subgroups we have indicated no choice between the three feeds listed. We should point out, however, that whereas corn distillers' grains and gluten feed have around 80 per cent TDN, brewers' grains are decidedly light and bulky. Consequently, brewers' grains should not be combined with basal feeds that are also on the light side. Obviously there can be no minimum limit set within either subgroup for any of the common feeds.

Mineral and Vitamin Supplements. Of the 20 or 30 lb of mineral supplements that are called for in these formulae, the amount of salt is fixed at 10 lb per thousand of complete ration (see also Chapter 16). Whether it is necessary to add any vitamin A and D to a mixture of this kind will be questioned by different authorities. Primarily it will depend on what kind of roughage is being fed and on whether or not the cattle are exposed to ultraviolet rays. We do know that vitamin A, or its precursor, carotene, is critical during the last part of pregnancy for the normal development of the calf, but if requirements for the maintenance of the cow and her fetus are met, there is no evidence that additional amounts will be necessary to maintain maximum lactation. However, the concentration of vitamin A in the milk will depend to some extent on the amount of vitamin A in the cow's ration in excess of her maintenance needs. Consequently, in these formulae 10 grams of carotene or its equivalent in vitamin A is recommended as a minimum inclusion. This should be sufficient to supply the maintenance requirements plus the pregnancy requirements without any vitamin A coming from the roughage. Any quantities in excess of this figure as supplied by the roughage will contribute toward a high vitamin A content of the milk produced.

For vitamin D, there is no evidence on which to base a probable requirement. Vitamin D is probably necessary, but under normal conditions an adequate quantity is probably supplied by sun-cured roughage, particularly if cattle during a part of the year are on pasture and hence exposed to direct rays of the sun. Consequently, no figures are included in these formulae for vitamin D, nor are there any other vitamins which are needed as supplements in the preparation of meal mixtures for cattle.

A Protein Supplement for Use in Preparing Cattle Meal Mixtures

Referring to the 16 per cent protein flexible formulae on page 359, we can see that the ration consists of three major groups of feedstuffs. The first includes the basal feeds, the next the protein supplements, and the third the mineral supplement. Each of these may be considered as units, and if the protein group and the mineral group are expressed on the basis of a 1,000-lb batch and a 100-lb batch, respectively, they become useful in preparing

TABLE 18-3   *Flexible Formula for a 35 Per Cent Protein Supplement for a Balanced Cattle Meal Mixture*

Group totals (approx.)	Maximum limits as in 16 per cent protein formulae		Feedstuffs	Expressed on 1,000 lb basis	
	Table 18-1	*Table 18-2*		*Average Maximum*	*Recommended*
70% protein	100	70	Peanut meal	380	—
	100	70	Gluten meal	380	—
	160	130	Linseed meal	710	210
	160	130	Soybean meal	710	250
	100	70	Cottonseed meal	380	250
30% protein	80	50	Corn distillers' grains	290	—
	80	50	Corn gluten feed	290	290
	80	50	Brewers' grains	290	—

meal mixtures with differing percentages of protein. Tables 18-3 and 18-4 show these protein and mineral mixed supplements.

Flexible Formula for an 18 Per Cent Dairy-Cow Meal Mixture

One of the simplest ways to prepare an 18 per cent dairy-cow meal is to modify the proportions of protein and mineral supplements found in the 16 per cent formula. As we showed in Chapter 16, this is very simply done by considering the basal and mineral groups as one. The protein supplement is 35 per cent protein. Hence, for an 18 per cent protein mixture, we calculate (using basal feeds averaging 12 per cent protein):

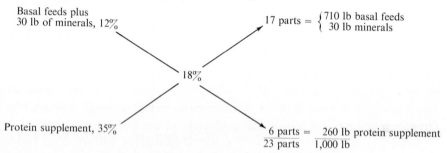

Such a mixture can be made by using the same substitutes among the basal feeds as with the original 16 per cent formula, and, of course, the formula for the mixed-protein supplement can also be modified by using alternative products, as in the flexible formula for the supplement.

Using the Protein and Mineral Supplement. If we wish to prepare a 14 per cent protein ration for cattle, such as might be useful for fattening beef stock, for dry dairy cows, or for cows on pasture, we may use the same procedure. These possibilities are shown in Table 18-4.

TABLE 18-4 *Proportions of Feed Groups to Prepare 1,000 Pounds of Cattle Meal Mixtures with Different Percentages of Protein*

Per cent of protein wanted in final mixture	Basal feeds of 12 per cent protein			Basal feeds with 10 per cent protein		
	Basal feeds	Protein suppl.	Mineral suppl. no. 1	Basal feeds	Protein suppl.	Mineral suppl. no. 2
14	880	90	30	820	160	20
16	790	180	30	740	240	20
18	710	260	30	640	340	20
20	620	350	30	560	420	20

Suggestions for Formulation of Cattle Meal Rations

Typical flexible formulae for meal mixtures suitable for feeding cattle have been given in the preceding pages. However, it is possible by selections within the permitted tolerances to compound meal mixtures having undesirable characteristics. Accordingly, the following suggestions may be useful in avoiding such situations.

The *fiber* content of the meal mixture should probably seldom exceed 12 per cent. Fiber content appreciably above this percentage is likely to lower the TDN of the mixture enough to affect the quantities of feed that have to be consumed to meet energy requirements. A fair guide here is to see that the meal mixture does not weigh less than 360 g per liter. When making such measurements pour the meal mixture into a measuring container until it overflows. Then with a small stick push off with one pass the feed that lies above the rim of the measure. *Do not pack the feed,* either by pressure or by agitation of the measure. It will also be desirable to take three or four tests on the same feed to get a reliable weight per unit volume. An erroneous measure will be misleading in terms of the TDN in the mixture.

Avoid the *fine grinding* of cattle feeds. Experiments indicate that grains that have been broken into three or more pieces are just as digestible as when finely ground. Coarsely ground feeds are more palatable to cattle than fine material. Grinding oat hulls or straw into a powder hides them in a meal mixture, but does not make them any more valuable for feeding. Some commercial mixtures for cows are now prepared by rolling or crushing a portion of the materials and pelleting the finer particles in order to avoid the dustiness of finely ground products.

Though useful in small amounts when available at economical prices, *wheat bran* is not an essential ingredient in a satisfactory dairy-cow meal ration. It should always be restricted to 25 per cent or less of the total mixture, regardless of any other consideration, because its low TDN will dilute the useful energy of the meal mixture.

Flexible Formula for a 22 Per Cent Protein Calf Meal (Calf Starter)

The flexible formulae we have presented so far in this chapter have been for meal mixtures intended for cattle stock over six months of age. More specifically, they are suitable for animals after an age at which the rumen has become fully functional. They are normally fed in conjunction with roughage, which represents from half to nearly all of the total ration.

TABLE 18-5 *Flexible Formula for a 1,000-pound Batch of 22 Per Cent Protein Calf Meal (Calf Starter)*

Major groups	Subgroups	Feedstuffs	Minimum limit	Maximum limit	Recom- mended
Basal feeds (12% protein) 590 lb		Oat groats	0	250	200
		Hominy feed	0	250	
		Corn	100	250	100
		Wheat middlings	50	115	115
		Oats	100	300	100
		Molasses	0	75	75
Protein supplement 380 lb	(43% protein) 320	Soybean meal	100	250	125
		Linseed meal	0	150	125
		Whitefish meal	25	50	30
		Meat meal	0	30	30
		Skim milk powder	5	50	10
	(24% protein) 60	Dried brewers' yeast	5	10	10
		Dried cereal grass	25	50	50
		Alfalfa leaf meal	0	50	
Minerals 30 lb		Bone meal	20	20	20
		Fine salt	10	10	10
		Potassium iodide	0	0.6 g	0.6 g
		Cobalt sulfate	0	1.5 g	1.5 g
Vitamins[a] (Added to 1000-lb batch)		Riboflavin	2,000,000 IU	?	2,000,000 IU
		Carotene	0	3 g	3 g
		Vitamin A	0	10,000,000 IU	
		Vitamin D	2,000,000 IU	?	2,000,000 IU
		Vitamin B_{12}	0	5 mg	
		Antibiotic	20 g	?	20 g

[a] For kilograms, multiply all values in this section by 2.2.

It now remains to present (in Table 18-5) a formula suitable for young dairy animals—animals from one month and perhaps up to six months of age. For the youngest calves, such a mixture will be a possible milk substitute, where animals are weaned from milk between 21 and 30 days of age. As such the mixture can be made into a water gruel for feeding, or it may be fed, preferably in pellet or crumble form, as a dry feed, if this is the calf-raising scheme being followed.

As with the previous flexible formulae, some latitude is permitted in the selection of the feedstuffs, but the substitutions are fewer and more restricted than with rations for older animals. This restriction is a reflection of more critical nutritional requirements plus much less factual information on the needs of young animals. The recommended formula has been found in practice to be satisfactory, but how far it can be modified by ingredient selection is not known. The maximum and minimum limits are therefore based on the general nutritional properties of the feeds rather than on any practical experience with the combinations that can be made.

Substitutions and Special Additives. Among the basal feeds there are few restrictions on substitution. The feeds shown are interchangeable within wide limits for all classes and ages of farm animals. In general, however, the *maximum of oats* should not be employed with the younger calves because of the fiber content.

It is with the protein supplements that more judgment must be exercised, since quality of protein is an important factor. Calves of this age group must get the essential amino acids from their rations. There is general agreement that of the plant sources *soybean* proteins are of higher biological value than the others, except for *wheat germ,* where only the embryo proteins are involved. Lysine or methionine will usually be the important limiting amino acid with most feeds of plant origin. To make up for this deficiency, low-fat *fish meal* and *meat meal* have been included, with the former having a minimum figure. As an added protection against low quality, a minimum of 5 per cent of *dried skim milk* is also called for.

Three potential natural sources of the vitamin B complex are included, two of which have minimum figures. Either the cereal grass or the alfalfa leaf meal will also provide carotene.

In addition to the natural feedstuffs, this formula is spiked with pure riboflavin, carotene (or vitamin A), concentrates of vitamin D, and an antibiotic. This group of additives is, in general, included without much basic knowledge of the quantities actually required, and the values given are therefore subject to revision (probably downward) as more information becomes available. We do not believe that any excesses the figures may represent are likely to be incompatible with the normal health or development of calves. There is some doubt whether B_{12} is needed. We have entered it in the formula only to indicate its maximum limit if it is to be included.

A calf starter meal would normally be discontinued for reasons of economy in favor of an 18 per cent (or, after an age of 12 months, a 16 per cent protein cattle meal mixture, though there is no nutritional reason for making such a change.

The Flexible Formulae in Practice

That the mixing formulae for commercially prepared balanced rations vary from time to time in accordance with fluctuations and availability of ingredients is no secret. Indeed, one of the older arguments for home mixing of feeds was that those made commercially were never twice the same, the implication being that such change in formulae was nutritionally undesirable. Such arguments are not often heard today, legitimate feed substitution being recognized as immaterial nutritionally and a practical necessity economically. Feeders generally, however, have little idea of how often or to what extent the mixture formulae are actually changed. The fact that the performance of the animals is not altered detectably by such changes is practical evidence that the feeder need not worry about them.

As a matter of interest, however, a tabulation of the actual formulae that one of the leading feed manufacturers in Canada used for their first-grade, 16 per cent protein dairy-cow meal mixture over a two-year period is shown in Table 18-6. The month when each change occurred is also shown. The small quantity of trace minerals in the formulae is omitted since it was constant. The quantities of bran and shorts were essentially constant, and reflect a constant supply of a by-product of the manufacturer's own flour-milling operations. Molasses, added largely for its favorable effect on the physical nature of the ration, fluctuated but little. Then, as we examine the formulae further, we can see that the minerals were constant except for a change from bone meal to dicalcium phosphate. The protein supplements, however, were exchanged to varying degrees. As a matter of insurance against local shortages of individual feeds, several protein feeds were included, and, consequently, the absolute amounts involved in an exchange of 50 per cent of the gluten for a like amount of distillers' grains would involve only 40 lb of either per ton of ration. Such changes could not be expected to be of nutritional significance in any one mixture. But in the overall feed-mill operations with a full line of mixed feeds, cutting in half the quantity of gluten used might easily involve several carloads of the ingredient over a relatively short time.

Among some of the feeds, however, there were more drastic substitutions. For example, in one case, 300 lb of screenings were replaced abruptly by 200 lb of barley plus 100 lb of wheat; in another, 100 lb of bran was replaced by 100 lb of oats. Also, 120 lb of linseed meal was replaced by 60 lb each of cottonseed and rapeseed meals.

Altogether twenty ration changes were made, or about one a month,

TABLE 18-6 *Formula Changes Over a Two-Year Period in a 16 Per Cent Dairy Meal Mixture (All figures are percentages)*

Ingredient	1953										1954						1955			
	Feb.	Feb.	Apr.	May	June	June	June	June	Aug.	Oct.	Jan.	Feb.	Mar.	July	Sept.	Oct.	Feb.	Mar.	Apr.	May
Molasses	10	10	10	10	10	9	7	7	7	7	7	7	7	7	9	9	9	9	9	9
Barley	15	5	5	5	5	5	15	15	15	15	15	15	15	15	15	15	20	20	20	20
Oats	15	15	15	15	15	20	20	20	20	20	20	20	20	20	18.5	18.5	13.5	18.5	18.5	16.5
No. 1 screenings	15	15	15	15	15	15														
Bran	15	15	15	15	15	15	15	15	15	15	15	15	15	15	15	15	15	10	10	12
Shorts	10	10	10	10	10	10	10	10	10	10	10	10	10	10	10	10	10	10	10	10
Linseed	4	4	6	5	6	8	7	9	5	3	3	3	5	5	6.5	3	3	3	3	3
Cottonseed									4	3	3	3								
Rapeseed	2	3	1	1	1	1					3	3	4	4	4	3	3	3	3	3
Gluten feed	3	3	2	3	3	4	4	7	7	7	5	5	5	5	4	4	4	4	6	4
Brewers' grains	2	2	3	3	3	3	3	3	3	3	3	5	5	4	4	4	4	4	4	4
Dist. grains	4	4	3	3	4	3	3	3	3	3	3	3	3	3	4	4	4	4	4	4
Malt sprouts	2	2	4	4	3	3	3	3	3	3	4	3	3	4	4	4	4	4	4	4
Clover scrngs.	10	9	10	8	7	5	7	5	5	5	5	5	5	5	3	7	7	7	5	7
Dical. phos.											1	1	1	1	1	1	1	1	1	1
Bone meal	1	1	1	1	1	1	1	1	1	1	1	1	1	1	1	1	1	1	1	1
Gr. limestone	1	1	1	1	1	1	1	1	1	1	1	1	1	1	1	1	1	1	1	1
Salt	1	1	1	1	1	1	1	1	1	1	1	1	1	1	1	1	1	1	1	1
Urea										0.5							0.5	0.5	0.5	0.5

not on any schedule, but rather largely according to ingredient price. By such procedures the selling price to the feeder can be kept reasonably stable. Maintenance of volume of sales and particularly maintenance of repeat sales is evidence that feed substitutions are in the best interests of all concerned.

Flexible Formulae for Range Cattle and Sheep Supplements

For adequate nutrition of range cows and ewes, the forage should contain an ample supply of protein, energy, minerals, and vitamins. Water, although not an important component of the forage, must also be available in some form to the grazing animals. Under favorable conditions and on well-managed ranges forages consisting of browse, grass, and forbs frequently supply all the nutrients necessary; however, on the winter range, or during unfavorable climatic conditions, or where there are nutrient deficiencies in the forage, it becomes necessary to supplement the forage diet.

Nutrient Deficiencies of Range Forage.   Much winter range forage is deficient in phosphorus, protein, and energy.* Calcium deficiency is rarely encountered. On some winter ranges there may be a deficiency of carotene (provitamin A), especially on a grass-type range, or where there is a late spring. However, sheep and cattle store a considerable amount of this vitamin in their liver during the summer, when they graze on succulent green forage, and if the winter is open and an early spring comes, animals in the Intermountain region will not suffer from a vitamin A shortage.

The Intermountain area is deficient in iodine, but it is not known whether other trace mineral deficiencies occur there. Since the cost of supplying supplements of iodine, cobalt, and copper are slight, it is good insurance to use trace-mineralized salt.

Formulation of a Supplement for Range Ewes or Cows.   In formulating supplements for range ewes and cows it is first necessary to determine the composition of the grazing animal's diet, and then to compute the nutrients needed to make up any deficiencies. A useful working calculation of the diet can be obtained by weighting the percentage of each floral component of the range forage by an index of the animal's preference for each species of forage present. The preferences that have been worked out for sheep are given in Table 18-7. The figures in this table together with those in Table 18-8 can be

---

* C. Wayne Cook, L. A. Stoddart, and Lorin E. Harris, *The Nutritive Value of Winter Range Plants in the Great Basin as Determined with Digestion Trials with Sheep* (Utah Agr. Expt. Sta. Bul. 372, 1954).

TABLE 18-7 *Average Degree of Utilization or Preference Indexes Used in the Calculation of a Diet for a Particular Sheep Allotment*

Plant	Plant composition (per cent)	Preference index (per cent)	Columns 1 × 2	Diet (per cent)
Black sage (*Artemisia nova*)	10	50	500	17
Bud sage (*Artemisia spinescens*)	5	40	200	7
Big sage (*Artemisia tridentata*)	11	15	165	5
Shadscale (*Artiplex confertifolia*)	13	20	260	9
Nuttall saltbush (*Atriplex nuttallii*)	8	35	280	9
Yellow brush (*Chrysothamnus stenophyllus*)	5	10	50	2
Winter fat (*Eurotia lanata*)	12	40	480	16
Desert molly (*Kochia vestita*)	10	15	150	5
Browse total	74			70
Western wheatgrass (*Agropyron smithii*)	3	30	90	3
Blue bunch wheatgrass (*Agropyron spicatum*)	3	40	120	4
Giant wildrye (*Elymus condensatus*)	1	10	10	0
Galleta curlygrass (*Hilaria jamesii*)	2	25	50	2
Indian ricegrass (*Oryzopsis hymenoides*)	6	45	270	9
Squirreltail grass (*Sitanion hystrix*)	3	50	150	5
Alkali sacaton (*Sporobolus airoides*)	1	10	10	0
Sand dropseed (*Sporobolus cryptandrus*)	2	10	20	1
Needleandthread grass (*Stipa comata*)	4	40	160	5
Grass total	25			29
Russian thistle (*Salsola tenuifolia*)	1	20	20	1
Grand total	100		2,865	100

used to calculate the nutrient make-up of the diet. Such calculations usually include only the amounts of eaten forage and its protein, energy, and phosphorus content.

Typical Requirement, Probable Intake, and Necessary Supplement for Ewes and Cows Grazing Intermountain Winter Range. The computations involved in arriving at the percentage of protein or of the energy needed in the supplements are shown in Table 18-9.

The percentage of protein needed in the supplement to the range forage is the weighted mean difference between the percentage of protein in the range forage and the percentage required in the daily ration of the ewe. An example is shown here.

	Forage dry matter	Dig protein	Protein
	(kg)	(per cent)	(kg)
Required per ewe	1.63	4.4	0.072
Eaten from range	1.50	2.6	0.039
Needed in supplement	0.13	24.9	0.033

Corresponding values for the concentration of the metabolizable energy and of phosphorus in the supplement are computed in the same manner and found in (Table 18-9).

Range grazing studies have shown that for both cows and ewes grazing in the Intermountain area, supplements containing about 16 percent crude protein are desirable for sagebrush type ranges, 28 per cent for mixed forage type, and 40 per cent for saltbush type and where grass predominates; the daily allowance per ewe is 100 g (0.22 lb),* and 450 g (1.0 lb) to cows.†

The supplements may be fed as pellets, blocks, or mixed with salt (to control intake). Economics should dictate which method to use. If salt is used, extra water must be provided.

Rate of Feeding.   The rate at which the supplement is fed is important. As the amount of feed is increased, the percentage of protein, phosphorus, and vitamin A can be decreased, and the percentage of less expensive energy feeds such as barley or corn can be increased. It will seldom be economical to feed the amounts of supplements necessary to keep ewes or cows gaining in weight or even to maintain their weight throughout the winter months if they are in good condition when they go to the winter range.

It is believed that it will, however, usually pay to feed supplements to ewes beginning about three weeks before the breeding season and continuing through it, in extremely cold weather, and for about thirty days before green feed is available in the spring. Small lambs, small ewes, old ewes with poor teeth, and thin ewes should be separated from the rest of the band and fed one of the necessary supplements from about the first of December until shearing time: old ewes, lambs, and yearlings from more than one band are often herded together in a "scad" herd.

When there is a deficiency of nutrients in the forage, or if the cows are thin,

---

* James, Lynn F., *Supplementing Cattle with Protein and Phosphorus on Desert Ranges* (Thesis, MS degree, Utah State Agr. College 1957).

† Lorin E. Harris, C. Wayne Cook, and L. A. Stoddart, *Feeding Phosphorus, Protein, and Energy Supplements to Ewes on Winter Ranges of Utah.* (Utah Agr. Expt. Sta. Bul. 398, 1956).

TABLE 18-8 Average Composition of Winter Range Plants (Moisture-Free Basis)

Plant	Gross energy (kcal/kg)	Metabolizable energy (kcal/kg)	Protein (per cent)	Digestible protein (per cent)	Phosphorus (per cent)
Bottlebrush, squirreltail (Sitanion hystrix)	3814	1414	4.5	1.1	0.07
Galleta grass (Hilaria jamesii)	3860	1312	5.5	1.4	0.07
Needleandthread grass (Stipa comata)	3915	1647	4.0	1.2	0.07
Rabbitbrush, small [yellow brush] (Chrysothamus stenophyllus)	4900	1321	6.6	3.4	0.10
Ricegrass, Indian (Oryzopsis hymenoides)	4281	280	3.5	0.5	0.06
Russian thistle, tumbling (Salsola tenuifolia)	3558	1780	14.2	9.7	0.16
Sacaton, alkali (Sporobolus airoides)	4195	999	3.4	0.0	0.08
Sagebrush, big (Artemisia tridentata)	5101	1268	9.4	5.4	0.18
Sagebrush, black (Artemisia nova)	5062	1124	8.5	4.4	0.16
Sagebrush, bud (Artemisia spinescens)	4239	2008	17.3	13.7	0.33
Saltbush, nuttall (Atriplex nuttallii)	3695	1321	7.2	3.4	0.19
Saltbush, shadscale (Atriplex confertifolia)	3633	880	7.7	4.3	0.09
Sand dropseed (Sporobolus cryptandrus)	4178	165	5.0	0.5	0.06
Summer cypress, gray [desert molly] (Kochia vestita)	3587	1903	9.0	5.5	0.12
Wheatgrass, beardless (Agropyron inerme)	4200	1991	3.1	0.0	0.06
Wheatgrass, western (Agropyron smithii)	4350	2469	2.4	0.2	0.06
Wildrye, giant (Elymus cinereus)	4094	311	3.2	0.3	0.06
Winterfat (Eurotia lanata)	3896	1310	11.0	6.9	0.12

TABLE 18-9 *Typical Requirement, Probable Intake, and Necessary Supplement for Ewes and Cows Grazing Intermountain Winter Range[a]*

Animal	Daily intake of dry matter		Per cent digestible protein (dry basis)	Metabolizable energy (dry basis)		Per cent phosphorus (dry basis)
	kg	lb		kcal/kg	kcal/lb	
Ewes (60 kg or 130 lb)						
Requirement	1.63	3.59	4.4	1466	665	0.18
Range forage	1.50	3.30	2.6	1410	640	0.09
Supplement	0.13	0.29	24.9	2092	949	1.08
Cows (370 kg or 800 lb)						
Requirement	8.61	19.0	4.4	1677	862	0.15
Range forage	8.16	18.0	2.6	1600	726	0.09
Supplement	0.45	1.0	36.0	3073	1394	1.24

[a] When these requirements are used it is assumed that the ewes and cows will graze on mountainous (National Forest) land during the summer.

TABLE 18-10 Flexible Formulae for Range Beef and Sheep Supplement Pellets (In Pounds per 1,000-pound Batch)

Main groups	Subgroups	Feedstuffs	Maximum proportion of mix to consist of the feed noted			Examples		
			16% protein	28% protein	40% protein	16% mix	28% mix	40% mix
Energy feeds (Av. 12% protein)	Grains	Barley, grain	700	90	100	500		60
		Corn, grain	500	150	100			
		Wheat, grain	500	250	100	50		
		Sorghum, milo, grain	250	150	100			
		Oats, grain	100	100	50			
		Wheat screenings	100	100	50	60	10	130[a]
	By-products	Wheat mill-run	100	50	50			
		Wheat middlings	100	100	50			
		Wheat shorts	100	100	50			
		Beet pulp, dried	100	50	50			
		Beet molasses	100	100	50	50		
		Sugarcane molasses	100	100	50		50	
		Total of energy feeds				660	60	190
Protein supplements (Av. 30% protein)	30–40% protein (except urea)	Cottonseed meal, mech-extd	15	600	850	50	500	440
		Linseed meal, mech-extd	10	250	250		250	250
		Soybean meal, mech-extd	15	450	850	50	130	
		Peanut meal, mech-extd	15	250	850			
		Urea	30	30	30			30
	20–30% protein	Corn gluten feed	10		5			
		Safflower meal, solv-extd	15		5			
		Bean, seeds	15		5			
		Corn distillers dried grains	10		5	70		
		Wheat distillers dried grains	10		5			
		Brewers dried grains	10		5	50		
		Total of protein supplement				220	880	750

Mineral supplements						
Dicalcium phosphate	5	20	4	5		
Bone meal, steamed	7	25	4		10	10
Defluorinated phosphate	7	25	4	5		
Monosodium phosphate	4	20	3			
Salt or trace mineralized salt	10	10	10	10	10	10
Total of mineral supplement				20	20	20
Vitamin supplements						
Alfalfa meal, sun-cured	150	100	50	100	40	40
Vitamin A and carotene supl.						
Alfalfa meal, dehydrated					40	40
Total vitamin supplement			100	100	40	40
Total batch			1000	1000	1000	1000

it will usually pay to feed a supplement during the winter months and up to calving time in the spring.

Flexible Formulae for Range Supplements. Examples of flexible formulae to meet protein requirements of 16, 28, and 40 per cent protein are shown in Table 18-10. The maximum levels for a 1,000-pound batch of individual feeds are given, as are the total amounts of the energy feeds, the protein supplements, the minerals, and the vitamin carriers. One example is given to illustrate each of the protein levels. The restrictions on the amounts of individual feeds are intended to ensure variety and to avoid the consequences of overuse of certain feeds that have peculiarities that may cause undesirable effects when fed to excess.

Water. It is usually assumed that water is readily available to range livestock, but on many ranges it is the limiting nutrient, being unpalatable because it contains dirt or filth, or has a high mineral content. In some areas or in some seasons, palatable water may be potentially available, but because it is frozen or because facilities are inadequate, animals may not obtain an ample supply.

For best production range livestock should be watered once each day, but the cost of supplying water sometimes makes it advisable to water them every other day.

SUGGESTED READING

Feeders Guide and Formulae for Meal Mixtures (Quebec Provincial Feed Board, 1962).

Harris, Lorin E. *Range nutrition in an arid region.* Honor Lecture Series, Utah State University, Logan, Utah. Jan., 1968.

CHAPTER 19

Flexible Formulae for Swine
Meal Mixtures

Data presented in Chapter 15 indicate that meal mixtures with three different percentages of protein are required for swine. Of the three, the 15 per cent mixture will normally be required in the largest quantities in any pig-feeding program, so we will deal with it first.

Two 15 Per Cent Protein Flexible Formulae for Swine Meal Mixtures

Table 19-1 gives a flexible formula for a swine meal mixture with 15 per cent protein made up of basal feeds with 12 per cent protein. A second formula, for a basal-feed combination with only 9 per cent protein, is shown in Table 19-2. A considerable selection of basal feeds has been included in the formula for reasons that will be evident shortly. Also, some of the animal-protein group are included, because with swine we have to select feeds that will assure the presence of essential amino acids. As with cattle rations, little choice is given with the minerals and none with the vitamin supplementation.

The maximum limits of individual basal feeds has been set so that at least two basal feeds must be employed. This variety among the basal feeds is not likely to be of any consequence in correcting low-quality protein, however, since all basal feeds are deficient in lysine.

The Protein Supplements. Only one high-protein feed of the plant group has been put into one of the final formulae. Many would question the

TABLE 19-1 *Flexible Formula for 15 Per Cent Protein Swine Meal Mixture Made of Basal Feeds Averaging 12 Per Cent Protein*

Main groups	Subgroups	Feedstuffs	Minimum limit	Maximum limit	Recommended
Basal feeds (av. 12% protein) 875 lb	Grains, 625 lb minimum	Oat groats	0	100	
		Rice polish	0	100	
		Wheat	0	300	
		Hominy feed	0	500	
		Corn	0	500	200
		Rye	0	200	
		Barley	200	500	300
		Feed flour	0	100	
		Oats	100	375	200
	Mill feeds, 250 lb maximum	Shorts (standard middlings)	0	250	175
		Wheat bran	0	200	
Protein supplements (av. 45% protein) 110 lb	Plant origin, 80 lb	Peanut meal	0	50	
		Soybean meal	0	80	
		Cottonseed meal	0	50	
		Linseed meal	0	80	80
		Brewers yeast	0	50	
	Animal origin, 30 lb	Fish meal	10	30	15
		Meat meal	0	30	15
		Skim milk powder	0	30	
		Buttermilk powder	0	30	
Minerals, 15 lb		Mixture no. 1 (Table 17-4)	15	15	15
Vitamin-antibiotic supplement[a]		Vitamin A	800,000 IU	1,000,000 IU	800,000 IU
		Vitamin D	900 IU	1,000 IU	900 IU
		Vitamin B_{12}	5 mg	10 mg	5 mg
		Antibiotic	5 mg	10 mg	5 mg

[a] If weight units in formulae are kilograms, multiply all values in this section by 2.2.

choice of *linseed meal* in this position. In some districts where *soybean meal* is an important and economical feed, it might well be preferred to linseed meal. Inasmuch as soybean meal may not fully correct the low biological value of the basal feeds, the choice of the high-protein feed from the plant group will not be critical.

The quantities of *meat* and *fish meal* are such that at least 10 per cent of the total protein of the ration will come from these two sources. There is

TABLE 19-2 *Flexible Formula for 15 Per Cent Protein Swine Meal Mixtures Made of Basal Feeds Averaging 9 Per Cent Protein*

Main groups	Subgroups	Feedstuffs	Minimum limit	Maximum limit	Recommended
Basal feeds (av. 9% protein) 790 lb	Grains, 540 lb minimum	Oat groats	0	100	
		Rice polish	0	100	
		Wheat	0	300	
		Hominy feed	0	300	
		Corn	490	790	790
		Rye	0	200	
		Barley	0	300	
		Feed flour	0	100	
		Oats	0	200	
	Mill feeds, 250 lb maximum	Shorts	0	200	
		Wheat bran	0	200	
Protein supplement (av. 45% protein) 195 lb	Plant origin, 140 lb	Peanut meal	0	90	
		Soybean meal	90	140	80
		Cottonseed meal	0	90	
		Linseed meal	0	90	60
		Brewers' yeast	0	50	
	Animal origin, 55 lb	Fish meal	20	55	25
		Meat meal	0	55	30
		Skim milk powder	0	55	
		Buttermilk powder	0	55	
Minerals, 15 lb		Mixture no. 1 (Table 17-4)	15	15	15
Vitamin-antibiotic supplement[a]		Vit. A	800,000 IU	1,000,000 IU	800,000 IU
		Vit. D	900 IU	1,000 IU	900 IU
		Vit. B_{12}	5 mg	10 mg	5 mg
		Antibiotic	5 mg	10 mg	5 mg

[a] If weight units in formula are kilograms, multiply all values in this section by 2.2.

a minimum limit set on fish meal, because experimental evidence indicates that when high-grade fish meal is included in the ration, the performance of the pigs tends to be somewhat better than when it is omitted, even though meat meal is also included.

Brewers' yeast has been included as a possible choice up to 5 per cent. Price will normally keep this feed at a minimum if it is used at all. Brewers' yeast is high in pantothenic acid, and wherever "goosestepping" or incoordination in the gait has been found with growing pigs, the use of some source of pantothenic acid has frequently corrected the trouble. Swine rations based largely on barley and oats tend to be borderline in this vitamin. Consequently, it is included as a possible selection. Since it furnishes about 40 per cent protein it is correctly classed in the protein group.

The Mineral and Vitamin Supplements. The mineral supplement is that discussed and formulated in Chapter 17 (see Table 17-4). The quantities of vitamin A and vitamin D we have recommended are based on the evidence that is available at present. These quantities are likely to be revised downward as further information becomes available. The vitamin A can, of course, be replaced by its equivalent in carotene.

The amounts of vitamin B_{12} which are included are essentially one-half the probable daily requirement. This figure has been set on the assumption that fish meal and meat meal will probably contain sufficient B_{12} to meet the remainder of the requirements. Somewhat the same line of reasoning has determined the level of antibiotic we have recommended.

The formula represented by the figures in the "recommended column" is essentially one that has been used in practice for many years and has given excellent results. It is, however, only a guide as to what type of selection should be preferred if no other considerations are involved.

TABLE 19-3 *Flexible Formula for a 45 Per Cent Protein Supplement for Swine Rations*

Group total (approx.)	Maximum limits as in 15 per cent protein formulae		Feedstuffs	Expressed on 1,000 lb basis	
	Table 19-1	Table 19-2		Average maximum	Recom-mended
72%	50	90	Peanut meal	480	
	80	140	Soybean meal	725	400
	50	90	Cottonseed meal	480	
	80	90	Linseed meal	480	225
	50	50	Brewers' yeast	250	100
28%	30	55	Fish meal	275	200
	30	55	Meat meal	275	
	30	55	Skim milk powder	275	75
	30	55	Buttermilk powder	275	

TABLE 19-4 *Proportions of Feed Groups Needed to Prepare 1,000 Pounds of Swine Meal Mixtures with Three Different Percentages of Protein*

Per cent of protein wanted	Using basal feeds averaging 12 per cent protein			Using basal feeds averaging 9 per cent protein		
	Basal feeds	Protein suppl.	Mineral suppl.	Basal feeds	Protein suppl.	Mineral suppl.
13	970	15	15	890	95	15
15	875	110	15	790	195	15
18	820	165	15	750	235	15

Protein-Mineral-Vitamin Supplement for Use with Home-Grown Feeds

The 15 per cent protein formulae can be separated into two parts. Deletion of the basal feeds, and restatement of the quantities of the other feeds in terms of 1,000-lb batches, will form the basis for flexible formulae for a 45 per cent protein supplement, as shown in Table 19-3. Table 19-4 gives the proportions of the basal feeds and the protein and mineral-vitamin supplements that should be combined to prepare mixtures with the three percentages of protein called for in hog feeding.

The maximum and minimum limits for the basal feeds shown in the original 15 per cent protein formula can be used in both the 18 per cent and the 13 per cent mixture. However, the heavier feeds should predominate in the 18 per cent ration; the minimum of *oats* would be used and *wheat bran* avoided. This ration should aim at 75 per cent or more TDN, and a low fiber level. On the other hand, the 13 per cent mixture, if intended for finishing bacon hogs or for feeding to dry sows and boars, will be more satisfactory if lighter feeds are emphasized, so that the TDN does not exceed 70 per cent; 67 per cent is not undesirably low. Such rations could carry the maximum of both *oats* and *wheat bran*.

The mineral-vitamin supplement is included in the same proportions in all mixtures, since these ingredients are needed more nearly in proportion to basal feeds or to total feed than to the protein supplements.

SUGGESTED READING

Feeder's Guide and Formulae for Meal Mixtures (Quebec Provincial Feed Board, 1962).

CHAPTER 20

Linear Programming of Meal Mixtures

In Chapters 16 to 19 we presented the idea of flexible formulae for meal mixtures intended for livestock rations, in contrast to the "one best formula" concept, where substitutions of ingredients are made on a second-choice or even third-choice basis. The flexible formula concept rests on the premise that the adequacy of the *nutrient* assembly of the ration is essentially independent of the *ingredient* assortment from which it is drawn. For example, the lysine deficiency of corn protein can be corrected by adding to corn any one of the many feeds that will supply supplemental amounts of lysine, such as fish meal, meat meal, milk, soybean meal, or even synthetic lysine itself. The animal to be nourished must receive in its ration a minimum of lysine, but nutritionally the dietary source contributing this amino acid is immaterial. The particular source used in a given ration is dictated largely by economic considerations.

In the flexible formula, insurance against nutrient deficiency in a chosen combination of feed ingredients is attained by setting minimum or maximum limits on the quantity to be used of each acceptable ingredient feed. These limits are based on general knowledge of the nutrient content of that feed. Thus the total of the energy and protein feeds is set to insure that the percentage of protein in the mixture will be maintained at the desired figure (16 per cent for growing pigs, for example); and the minimums and maximums for specific feed are stipulated to ensure that, for example, necessary minimum quantities of feeds with much lysine will be used to supplement those whose proteins are low in lysine. As an extra check, maximums would be set to prevent the use of undesirably large quantities of feeds that are low in lysine.

Where the principal quantitative concern in ration formulation is maintenance of adequate percentages of crude proteins with an acceptable amino-acid balance, flexible formulae can be prepared that will permit a wide latitude of choice among ingredient feeds in order to take advantage of local, current economic situations, but will at the same time restrict the proportions used in the ration mixtures to those which carry a nutritionally acceptable assortment of amino acids. For most home-mixed and farm-fed meal mixtures, where a large part of the ration consists of farm-grown grains, the flexible formula is all that is needed as a mixing guide.

As food technology produces more and more new feeds, and as feeder stock is more and more "fed out" in large commercial feeding units divorced from breeding herds, a rapidly increasing proportion of the nonroughage feedstuffs used for livestock and poultry is produced by feed manufacturers' operations, and reaches the feed lot or pen as a mixture blended to stated nutrient specifications. In such feed mixtures, not only is the percentage of protein stipulated, but energy, fat, fiber, minerals, vitamins, amino acids, and nonnutrient additives may also be specified. Thus instead of balancing the meal mixture quantitatively in terms of protein alone, it has become necessary to devise mixtures in which a dozen or more nutrients are called for in specific amounts or in a specified ratio to other nutrients. For example, the protein, some amino acids, phosphorus, and the B-complex vitamins are often adjusted to metabolizable energy in ration formulation.

Superimposed on this requirement about nutrient makeup is the demand that the particular combination of feedstuffs shall be the cheapest at current prices. Specifically, the question to be answered is: What is the least-cost formula that meets the stipulated nutrient makeup? The mathematical procedure by which this problem is solved is *linear programming*.

Linear Programming Defined

Linear programming is a mathematical technique for determining how best to choose among the available feedstuffs and ration ingredients, which have different nutrient makeups and prices, in order to get a mixture that has specified percentages of nutrients at the least possible cost. The results obtained from linear programming depend on the numerical values used for (1) the nutrient and other specifications demanded in the feed mixture, (2) the nutrient composition of the acceptable ingredients, and (3) the unit price of each ingredient feed or additive.

The Specifications Necessary

The specifications for the nutrients in the final mixture are usually based on an appropriate feeding standard for the specific animals to be fed. Other specifications include the quantity or weight of a batch, and the amounts of such nonnutrient ingredients as antibiotics and medicants.

As is suggested in the name "*linear* programming," all specified nutrients must be additive in their effects. That is, lysine or methionine or thiamine must be the same substance in all sources used, and regardless of source, the biological effects must always be additive. One unit weight of lysine from barley plus two from soybean meal must be biologically equivalent to three from any other sources, including synthetic 1-lysine monohydrochloride. The additive requirement is not a mathematical one, but a biological one. Within the effective range of the nutrient it is obviously necessary that the response of the animal shall conform to the percentage of the nutrient shown by chemical analysis to be present in the mixture.

Nonadditive Nutrients. There are some nutrients that are not always additive in response. Dietary phosphorus, for example, may be inorganic, organic, or both. Some acceptable ration ingredients supply only inorganic phosphorus, as in dicalcium phosphate, whereas in many plant seed products, such as wheat bran, a large part of the phosphorus is in an organic combination as phytin. The availability of phytin phosphorus to an animal depends on a phytase, which may itself be of dietary origin or may be a product of the metabolism of the microorganisms of the alimentary tract. Thus inorganic phosphorus is additive from different sources, but organic phosphorus is of variable availability, and hence is not necessarily additive biologically. It follows that phosphorus from mixtures of inorganic and organic forms may also not be additive in biological usefulness. Furthermore, there are species complications, in that ruminants appear to utilize phytin phosphorus satisfactorily, whereas with other species there is often a negligible, but also highly variable, absorbability of the organic forms. One solution to this problem is to specify one percentage for inorganic phosphorus and another for total phosphorus in the programmed ration.

Crude Fiber. Weende crude fiber from different sources may not be strictly additive in its effect in the ration, not only because it is a variable and unspecified mixture of substances that are individually insoluble in the acid-alkali digestion involved in the method (see Chapter 2), but also because the extent of its usefulness to the host depends initially on its digestion by rumen or caecal microflora. The extent of their activity in turn depends in part on how adequate the host's ration is in the nitrogen or phosphorus needed for the nourishment *of the microflora.* Also, this microbial breakdown of crude fiber is hampered by lignification of the cellulose. Thus the amount of crude fiber that disappears between its ingestion and the excretion of its residue is highly variable, both within and between species, as is indicated by the ranges of the digestion coefficients shown in Table 20-1. The variation in

TABLE 20-1 *Variations in Digestibility of Crude Fiber*

Species	Where digested	Per cent digested
Ruminants	Rumen, colon	50–90
Horse	Caecum, colon	13–40
Pig	Caecum, colon	3–25
Poultry	Caeca	20–30

digestion within species is principally a measure of differences between sources of the crude fiber. It is clear that crude fiber is not included in the specifications of a programmed ration mixture as a reliable guide to the useful energy of the mixture, but rather as an indicator of the bulkiness (or density) of the ration, which with nonherbivorous species may be of importance.

Some nutrients are partially convertible to others, as methionine is to cystine, and tryptophane is to niacin. Here again values for methionine cannot be interpreted unless the values for methionine plus cystine are also given, and those for tryptophane are meaningful only when those for niacin are also given.

Reliable Chemical Analyses Necessary

It may be appropriate also to point out that linear programming is not as useful as possible unless reliable analytical data on the content of acceptable mixture ingredients are available for each nutrient for which specifications are prescribed. For example, if the programmed mixture includes a requirement for lysine, and if there is a possible feedstuff for which no analytical value for lysine is given, none of that feed will be used in the mix, regardless of its price, until all the requirement for lysine has been met by feeds or products whose lysine values are given. If the feed thus discriminated against by the computer happens to be cheaper than those chosen, the mixture will be more expensive than it otherwise could have been. Thus we see why researchers are trying to complete as rapidly as possible the analytical data for the nutrients of feeds which present information indicates are useful as ingredients of balanced rations.

Incompatible Requirements

The nutritionist who prepares the information to be furnished to the computer must also be sure that the data are compatible with the values demanded. For example, if the composition of every permissible protein-containing ingredient is greater than 16 per cent, no possible combination of them can average 15 per cent. If the computer has been told to come up with a 15 per cent protein mix, it may come up with a 938-kg batch instead of the 1,000-kg batch called for ($150/938 \times 100 = 16$ per cent), and with a note that there is a "slack" variable value for weight of -62 kg. Sometimes it is easier to just multiply the quantities by a factor that will put them on the 1,000-unit basis wanted than to reprogram, or the slack variable may be so small that it can be disregarded in the mixing operations. More usually one gets such a formula when maximum tolerances for a mineral, such as calcium, cannot be met because of minimum levels set for an ingredient such as meat meal or bone meal.

Specifically, every restriction given in the specifications may cause such incompatibility, and there will be a "slack" variable for each such ingredient of the mix. In a possible solution, all slack variables become zero. Incompatible specifications and restrictions inevitably result in impossible solutions.

Actual Steps in the Linear Programming of Rations

It may now be helpful if we actually go through the steps in getting a programmed ration formula. A flexible formula for a 15 per cent protein swine meal mixture was presented as Table 19-1. Using this as a basis, a "stripped" version of the formula was prepared containing only three basal feeds, three protein supplements, three minerals sources, and, as a special case, one synthetic amino acid. This abbreviated formula does not represent a normal ration and is presented here only to illustrate the steps in the method.

Table 20-2 illustrates the information that must be assembled for programming a ration, except that in this table five sets of ingredient costs are shown, instead of the one set normally given, since five mixtures, all conforming to the same nutrient specifications and ingredient choices, were wanted to illustrate the programming scheme. We note first that specific limits of choice may be placed on each permitted ingredient. These restrictions

TABLE 20-2 Example of Acceptable Ingredients for 1,000 kg of Swine Meal Mixture

| Ingredient | Limits of use | | Composition of dry matter | | | | | | Cost of ingredient per kg dry weight (in dollars) | | | | |
	Min kg	Max kg	Per cent of protein	Kcal DE/kg	Per cent of Ca	Per cent of P	Per cent of lysine	Per cent of salt	1	2	3	4	5
Barley[a]	200	500	15.5	3,460	0.09	0.47	0.60	0	0.0660	0.0680	0.0660	0.0660	0.0660[a]
Corn		500	10.0	4,402	0.03	0.31	0.20	0	0.0680	0.0660	0.0600	0.0680	0.0680
Middlings		250	17.0	3,500	0.16	1.00	0.60	0	0.0590	0.0590	0.0590	0.0800	0.0800
Soybean meal	10	80	52.0	3,700	0.36	0.25	3.2	0	0.1180	0.1180	0.1180	0.1180	0.1181
Fish meal	10		70.0	3,344	3.00	2.20	6.5	0	0.1250	0.1250	0.1750	0.1750	0.1750
Meat meal		30	55.0	2,900	8.00	4.00	3.8	0	0.1290	0.1290	0.1290	0.1290	0.1290
Limestone		15	0	0	33.80	0.02	0	0	0.0064	0.0064	0.0064	0.0064	0.0064
Dicalcium phosphate		15	0	0	28.10	19.87	0	0	0.0945	0.0945	0.0945	0.0945	0.0945
Salt	56	56	0	0	0	0	0	100	0.0297	0.0297	0.0297	0.0297	0.0297
1-Lysine HCl		56	0	0	0	0	95.00	0	4.2500	4.2500	4.2500	4.2500	4.2500

[a] We are assuming this barley contains no lysine.

may be minimums, maximums, or both, or may be omitted where, except for cost, the quantities used are believed unimportant, as is true for lysine. If enough lysine can be obtained cheaply enough from other ingredients to meet the minimum requirement, none of the synthetic is needed. Conversely, no maximum is likely to be necessary in view of its cost and the known levels present in the ingredient feeds. The decision about whether or not restrictions are imposed on ingredient choice is entirely the responsibility of the nutritionist, and their validity will depend on how thoroughly he understands the nutritional consequences of their inclusion in the mix. Aside from, and often independent of, nutritional considerations that may call for limits on the choice of ingredients are such factors as local availability, or even general mill-management policy. Some feed manufacturers may be in the mixed-feed business as a means of marketing one of their own by-products. If so, they might be expected to stipulate minimum amounts which the machine must choose in the formulation.

TABLE 20-3 *Batch Weight and Percentages of Nutrients Required in the Dry Matter of the Mixture*

Nutrient specifications of mixtures	Weight in kg	Per cent of protein	Kcal DE/kg	Per cent of Ca	Per cent of P	Per cent of lysine	Per cent of salt
Minimum	1,000	18.2	3,700	0.72	0.55	0.82	0.56
Optimum	1,000	18.2	3,700	0.72	0.55	0.82	0.56
Maximum	1,000	18.5	3,800	0.72	0.65	0.82	0.56

In Table 20-3, the batch weight and the percentages of nutrients required in the formulated mixture are given. Except for the batch weight, the data in this table are usually taken from an appropriate feeding standard. Here we note that the analytical data for the ingredients (Table 20-2) and for the percentages of nutrients (Table 20-3) are and must be based on ingredients with the same per cent of dry matter (*in this example the basis is 100 per cent dry matter*). If the protein content given for barley is 15.5 per cent on a moisture-free basis, then we must stipulate the per cent of protein demanded in the mixture also on a moisture-free basis, even though the mixture *as fed* may be only 90 per cent dry matter. This requirement may necessitate converting the percentages of nutrients given in the standard on an "as fed" basis to a dry basis to match the analytical data, or vice versa.

Note also that minimum and optimum percentages are the same in Table 20-3, though this is not essential for the mechanics of formulation. In general, one might assume the optimum would be intermediate between the minimum and maximum points. Where the specification is relaxed to permit some leeway both above and below an optimum, it becomes more likely that an imbalance between energy and nutrients, or between nutrients, will be programmed. For example, given protein restrictions of 16 to 18 per cent, and energy of 3,600 to 3,800 kcal, where the extremes of one are acceptable only *if the other is at its mean value,* there is the possibility that, because of price, the cheapest formulation might be 16 per cent protein and an energy of 3,800 kcal. This ratio is far enough off the optimum to affect the ration efficiency. Consequently, since the minimum percentages of nutrients usually represent the minimum amounts of ingredients in a batch and hence minimum cost, minimum nutrient requirements have tended to be taken as the optimum ones. Once the specifications shown in Tables 20-2 and 20-3 have been established, the data are punched into data processing cards (IBM). The "punched deck" of cards is fed into the computer together with an appropriate program.

Program Defined

The program is a set of directions for the computer to follow in solving the mathematical problems involved in getting the least-cost formula for a ration mixture that meets the nutrient combination required. In effect, the part of the formulation that is actually done by the computer is finding the least-cost solution to a set of simultaneous algebraic equations.

Some of the Algebra of Linear Programming. Stated in algebraic terms our example problem may be represented as follows.

1. Let $x_1, x_2, x_3, \ldots, x_{10}$ represent the kg in the mixture to be formulated from: (1) barley grain, (2) corn grain, (3) middlings, (4) soybean meal, (5) fish meal, (6) meat meal, 7) limestone, (8) dicalcium phosphate, (9) salt, and (10) 1-lysine HCl.

2. Kg of barley *protein in mix* $= \dfrac{15.5\%}{100} x_1$; kg of corn protein $= \dfrac{10.0\%}{100}$ x_2; etc. Kg of barley protein plus corn protein in mix will be

$$0.155x_1 + 0.100x_2.$$

3. From Table 20-3 we note that the per cent of protein of the batch must lie between 18.2 and 18.5 per cent or between 182 and 185 kg in the 1,000-kg batch. A complete algebraic description of the protein requirement to be met is contained in the pair of *inequalities:*

(1) $0.155x_1 + 0.10x_2 + 0.17x_3 + 0.52x_4 + 0.70x_5 + 0.55x_6 \geq 182$ kg;

(2) $0.155x_1 + 0.10x_2 + 0.17x_3 + 0.52x_4 + 0.70x_5 + 0.55x_6 \leq 185$ kg.

Similar inequalities can be written for each of the other requirements, for energy, for example:

(1) $3.460x_1 + 4.402x_2 + 3.500x_3 + 3.700x_4 + 3.344x_5 + 2.900x_6$
$$\geq 3{,}700 \text{ kcal};$$

(2) $3.460x_1 + 4.402x_2 + 3.500x_3 + 3.700x_4 + 3.344x_5 + 2.900x_6$
$$\leq 3{,}800 \text{ kcal}.$$

4. The requirement from Table 20-3 that barley grain must be between 200 and 500 kg is expressed by two other inequalities, and that corn must not exceed 500 kg by a third:

$x_1 \geq 200$ kg; $x_1 \leq 500$ kg;
$x_2 \geq 500$ kg.

5. Where there is no range of choice (i.e., where the sum of the variables must exactly equal the quantity stipulated) we may write the *equations* directly, as for weight,

$$x_1 + x_2 + x_3 + x_4 + x_5 + x_6 + x_7 + x_8 + x_9 + x_{10} = 1{,}000 \text{ kg},$$

or for lysine,

$$0.006x_1 + 0.002x_2 + 0.096x_3 + 0.032x_4 + 0.065x_5 + 0.038x_6$$
$$+ 0.95x_{10} = 8.2$$

or for salt,

$1.0x_9 = 5.6$.

6. By writing as many inequalities and equalities as are necessary, we can make a mathematical model of the feed mixture, and the solution (i.e., the least-cost combination of the ingredients meeting the stipulation) will consist of numerical values of x_1, x_2, etc., all of which must be positive in sign. In other words, the solution will consist of values of x_1, x_2, etc., which satisfy the given inequalities, and at the same time make

$$c_1x_1 + c_1x_2 + c_1x_3 + \cdots + c_{10}x_{10}$$

a minimum, where c_1, c_2, etc., are the costs of x_1, x_2, etc.

7. Algebraically a set of inequalities has no simple solution. To overcome this difficulty, each inequality in the mathematical model is converted to an equation by the introduction of an "artificial" variable, U, which may be added to or subtracted from the left side of the inequality as required to make an equation. The two inequalities representing protein will illustrate.

(1) $0.155x_1 + 0.10x_2 + 0.17x_3 + 0.52x_4 + 0.70x_5 + 0.55x_6 \geq 182$ kg;
$$0.155x_1 + 0.10x_2 + 0.17x_3 + 0.52x_4 + 0.70x_5 + 0.55x_6 - U$$
$$= 182 \text{ kg}.$$

(2) $0.155x_1 + 0.10x_2 + 0.17x_3 + 0.52x_4 + 0.70x_5 + 0.55x_6 \leq 185$ kg;
$$0.155x_1 + 0.10x_2 + 0.17x_3 + 0.52x_4 + 0.70x_5 + 0.55x_6 + U$$
$$= 185 \text{ kg}.$$

By this device we arrive at a set of simultaneous equations whose solution is more straightforward than that for inequalities is.

Once the data of the equations and the program have been fed into the computer, the next thing the operator sees is the printout of the least-cost mixture formula, accompanied by a statement of the artificial variables. Such a printout for ration no. 1 is shown as Table 20-4. Table 20-5 gives the five

TABLE 20-4 *Example Printout of Programmed Ration No. 1*

Ingredient or nutrient	Amount in kg	Artificial variables (i.e., deviations from restrictions) in kg
Ingredient[a]		
Barley[b]	317.12	117.12 excess (from min)
		182.87 slack (from max)
Corn	305.15	194.84 slack
Middlings	250.00	0
Soybean meal	80.00	70.00 excess
Fish meal	22.08	12.08 excess
Meat meal	5.03	24.96 slack
Limestone	14.99	
	994.37	
Nutrient[c]		
Kcal/kg		99.99 slack
Protein		3.00 slack
Phosphorus		0.72 excess
		0.27 slack

[a] Ingredients not listed were not chosen for the mix.
[b] Where there is both a minimum and a maximum restriction on choice, there will be both an excess and a slack variable, though one may be zero and not shown. Together they define the amount of the ingredient to be used.
[c] Nutrients not listed are at exact percentages required (see Table 20-3).

computed ration formulae and their costs per 1,000 kg and per ton, as obtained from the example data of Tables 20-2 and 20-3.

The reader is reminded that the percentages of nutrients shown for the ingredients and required in the mix are based on the dry weight of the ma-

TABLE 20-5 *Programmed Formulae and Costs for Five Mixtures Using Composition and Prices of Ingredients as Shown in Table 20-2*

Ingredients	Ration formulae in kg				
	1	2	3	4	5
Barley	317	317	317	473	442[a]
Corn	305	305	305	310	434
Middlings	250	250	250	84	0
Soybean meal	80	80	80	80	10
Fish meal	22	22	22	10	97
Meat meal	5	5	5	27	0
Limestone	15	15	15	11	11
Dicalcium phosphate	0	0	0	0	0
Salt	6	6	6	6	6
1-Lysine HCl	0	0	0	0	730
Cost (in dollars) per 1,000 kg (2,200 lb) dry wt or per 1,100 kg (2,420 lb) 90% dry matter	69.60	69.63	68.27	73.80	68.04
Cost per 2,000 lb (90% dry matter)	57.50	57.55	56.40	61.00	56.15

[a] We are assuming this barley contains no lysine.

terial. Accordingly the requirement values in Table 20-2 (except the batch weight) should be reduced 10 per cent to have them correspond to the 1964 swine standard requirements, since the latter assume rations of about 90 per cent dry matter.

The feature of the five programmed formulae that might be unexpected is that the first three rations have identical ingredient formulae (to the nearest whole number). The only difference between rations 1 and 2 is an increase of $2.00 per 1,000 kg for barley and a similar decrease for corn, which is reflected in an increase of only five cents a ton for ration 2.

In ration 3, however, the cost of fish meal was increased to $175 from $125 per 1,000 kg, that of barley was reduced to $66 from $68 per 1,000 kg, and that of corn to $60 from $66. The net result was the same formulae as 1 and 2 but at a cost of $1.15 less per ton.

When we examine ration 4, however, we note an increase in the cost of middlings of $21 per 1,000 kg, and of $2 per 1,000 kg for corn, over their costs in ration 3, which has led to a marked change in the formula in order to maintain the same nutrient complex. The barley was increased some 160 kg per 1,000 kg of mix and the middlings reduced by 175 kg. The fish meal was reduced to its minimum choice level of 10 kg from the 22 kg previously used. To compensate, probably for loss of lysine from fish meal, the meat meal was increased to 27 kg from 5 kg. In addition, the cost per ton of this mixture was $61 as compared with an average of $57 for rations 1, 2, and 3. These changes in ingredients reflect the increased cost of middlings, which is an important source of lysine, but is not usually recognized as such, since it is by classification an energy feed.

In the fifth ration the ingredient costs were not changed from those of ration 4, but in its composition it was *assumed that no lysine was contained in barley.* Typical of what often happens when a feed with incomplete analytical data is used in a programmed ration, the analysis value for lysine was left out. This has the same effect in programming as putting a zero in the place of the missing data. The formula that the computer produced called for no middlings, the minimum 10 kg of soybean meal, and 97 kg of fish meal. The lysine was brought up to requirement by the inclusion of 730 grams of synthetic 1-lysine hydrochloride. And the price per ton of the mixture was $56.15, which was $5 cheaper than ration 4 at identical ingredient prices.

It is to be noted that none of the ingredient restrictions was violated in any ration, and all rations had the same percentages of nutrients. Once the computer program was prepared, it required about ten minutes to make the changes in price structure and recompute each of the successive rations.

SUMMARY OF SECTION **IV**

We have now completed, in this section of this book, integrating the facts and beliefs about the food needs of our animals with the makeup of the feeding stuffs that supply these needs. This integration has led us to express livestock feeding standards in terms of meal mixtures.

Theoretically, from an ideally balanced meal mixture ingredient formula, a suitable chemical analysis of the mixture, and a statement of the daily allowances to be given to an animal, we should be able to derive the corresponding feeding standard (if that meal mixture constituted the entire diet). By the same reasoning, we have attempted to derive ingredient-mixing formulae from feeding standards plus the analysis of the ingredients available for such mixtures.

In practice, we find it impossible to compound rations meeting exactly the quantities or proportions of nutrients set out in feeding standards, because we must deal with feeds which themselves are complex mixtures of nutrients. But we can prepare formulae for mixtures of feedstuffs whose nutrient combinations approximate the needs of specified animals closely enough so that the animal, by means of its own metabolic machinery, can make the final fitting, discarding the surpluses, and temporarily even making good minor shortages in the day's intake.

We have described one way of arriving at such formulae in this section. The flexibility of the scheme enhances its usefulness and its adaptability over a wide range of conditions under which livestock are fed; and its practicability is attested to by the success of the commercial mixed-feed industry.

Finally, we have attempted to give, albeit in an oversimplified and much abbreviated form, a picture of what is known as linear programming of meal mixtures, a method of deriving a least-cost formula that meets specified nutrient content. A linear-programmed formula differs from a flexible formula in three respects:

1. It adjusts more than the protein level to the desired concentration in the ration. It can produce not just a formula, but a least-cost formula, in which all of the required nutrients are at specified levels, within any imposed quantitative restrictions on ingredient choice other than cost.

2. It is not flexible in the sense that it is least-cost. It is based on the principle of flexibility in ingredient choice. Indeed it is by this feature that the least-cost combination of ingredients is arrived at.

3. It is arrived at by solving algebraic equations so extensive and complicated that the solutions are impractical except by electronic computation.

SECTION V NUTRIENT NEEDS OF ANIMALS

In the previous Sections we have discussed the nature of feeds, how they are named, and how they are used in the formulation of mixtures that will meet stipulated specifications. Here we consider specific feeding problems, present practical feeding guides for various categories of livestock, and indicate the general format of modern feeding standards for farm animals. We also discuss the legal regulation and control of the commercial feed industry, and reproduce a model feed bill designed by a committee of the Association of American Feed Control Officials.

CHAPTER 21

Livestock Feeding Guides

Daily Allowances of Feed for Animals

Successful livestock feeding is not learned out of a book, and there are many skilled feeders who know little or nothing of the science underlying the feeding practices which they successfully employ. There are, nevertheless, basic principles involved in livestock feeding, and some of these can be reduced to rules and guides that will be useful to feeders who lack the experience or apprenticeship otherwise necessary.

One of the ever-present problems is: how much meal should be fed per day to different animals? It is obvious that the answer depends on several factors. The first is the total available energy that must be provided to the animal in question. This amount is indicated by the TDN (or digestible energy, or metabolizable energy) shown in feeding standards. Such a figure, however, does not take into account the feeds other than the meal mixture that may also be fed to the livestock and which, of course, modifies the amount of meal that has to be fed. With some classes of animals whose increase in live-weight is itself the production, the feeder has a relatively simple guide in merely watching the gains of the animal. But with other classes of stock, particularly with producing dairy cattle, body-weight changes are not a direct index of whether or not the animal is receiving the correct quantities of meal. Consequently, with dairy-cattle feeding some guides other than the appearance or performance of the animal become of special importance.

Guides to Dairy-Cattle Feeding

Amounts of Meal Mixture for Dairy Cows in Milk. There have been many rules of thumb proposed as guides to the quantities of meal mixture that should be provided to cattle in the milking herd. These rules are based fundamentally on feeding standards, and on the roughage feeding practice. They are, nevertheless, guides and not fixed rules, and it may be well, therefore, to examine the requirements in some detail.

We might at first thought assume that the quantities of meal required by milking cows is influenced by the size of the cow. This, however, is not usually true, unless some unusual roughage feeding program is involved. As we shall see shortly, if the dairy cow is fed normal quantities of roughage, she will receive from it sufficient energy to fully meet her maintenance requirements. In excess of this requirement there are two things that need to be considered, the amount of milk being produced, and the stage of pregnancy. In Chapter 14 we discussed the effect of kind and quality of roughage on its voluntary intake. It will be evident from perusal of the data presented there that where high-quality roughage is used and fed to the limit of appetite, which is usually considered desirable practice, the energy thus obtained will meet maintenance requirements and be enough extra for almost 20 lb (nine kg) of milk of 4 per cent butterfat content. On the other hand, if poorer-quality roughage is fed, the available energy from the forage portion of the ration may be little more than enough to maintain the animal.

For purposes of determining guides to meal allowances, it is probably better to take average roughage containing about 50 per cent TDN (22 Mcal DE per kg or 18 Mcal ME per kg), and to make the assumption that approximately two kg of such roughage or its equivalent will be voluntarily consumed per cow per day for each 100 kg of liveweight. We can then quite simply calculate the amount of meal mixture that will supply the remaining amount of energy needed by cows at different levels of production according to the per cent of TDN in the meal used. The data have been calculated for a range of production levels and for meal of 70 and 75 per cent TDN. These data are presented in Fig. 21-1, from which we can see that whether a 70 or a 75 per cent TDN meal is fed, the place at which meal feeding should start is approximately after the first 10 lb (4.5 kg) of milk have been produced. That is, the roughage allowance should be sufficient for maintenance of the cow plus about one gallon* of milk. As the milk production increases, the

---

* The Imperial gallon used in Canada weighs 160 oz or 10 lb or 4.5 kg. The American gallon used in the United States weighs 128 oz or 8 lb or 3.7 kg.

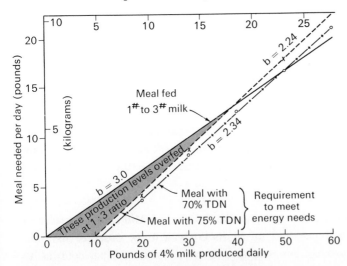

Figure 21-1. *Meal needed daily per pound of milk by a 1,000 pound cow consuming 20 pounds of dry forage of 50 per cent TDN.*

quantity of meal required increases in a linear fashion, and the rate is either one pound of meal for 2.34 lb of milk, or one pound of meal for 2.24 lb of milk, depending on whether a 75 or a 70 per cent TDN mixture is being used. This is just another way of saying that the allowances of meal to cows in milk should be at the rate of about one pound of meal to every 2.3 lb (or one kg to 2.3 kg) of milk produced in excess of the first gallon.

One of the commonest rules of thumb about meal allowances for dairy cattle calls for one pound of meal for every three lb of milk produced. The regression line for this ratio has been drawn into Fig. 21-1, and we can see that this ratio overfeeds all cows producing less than 40 lb of milk per day, but underfeeds cows producing over 40 or 50 lb of milk per day, depending on whether the meal used has 70 or 75 per cent TDN. Any feeding rate that calls for meal beginning immediately with any production above zero will obviously show this same defect.

Feeding During Late Lactation. There is another consideration, affecting cows producing relatively small amounts of milk during the last two months of pregnancy. These cows require more meal than their milk production indicates, and, consequently, a rule of thumb which calls for no meal allowance for productions of less than one gallon of milk is likely to underfeed them.

Probably the most satisfactory way to deal with this situation is to arrange to feed about five lb (2 kg) meal to such cows in excess of the quantities they require for milk production according to Fig. 21-1. Thus a cow that is within two months of term and producing only 10 lb (4.5 kg) of milk would receive 5 lb (2.3 kg) of a normal meal mixture, in spite of the fact that her production alone would not warrant any meal at all.

Rule of Thumb for Meal Feeding in Late Pregnancy. One pound of meal should be fed for each 2.3 lb (or 450 g of meal for each kg) of milk produced in excess of the first gallon; during the last 60 days of pregnancy cows should receive in addition to any such allowances an extra 5 lb (2.3 kg) of meal per day.

Adjustment of Meal Allowance for Fat Content of Milk. These rules are based on milk with 4 per cent fat or its equivalent. In order to use these feeding rules with productions that have more or less than 4 per cent fat, the actual production should be converted to its equivalent in 4 per cent milk. Table 21-1 gives approximate factors for converting milk produced to its equivalent in 4 per cent fat milk.

TABLE 21-1 *Converting Milk of a Given Fat Per Cent to Its Equivalent in 4 Per Cent Milk*

Per cent of fat in milk	Factor by which to multiply milk produced to obtain pounds of equivalent 4% milk	Factor to obtain kg of 4% milk
3.0	0.850	1.87
3.2	0.880	1.94
3.4	0.910	2.00
3.6	0.940	2.07
3.8	0.970	2.13
4.0	1.000	2.20
4.2	1.030	2.29
4.4	1.060	2.36
4.6	1.090	2.40
4.8	1.120	2.46
5.0	1.150	2.55

Roughage and Pasture Consumption by Adult Dairy Cattle. In general cattle should be full-fed on roughage if the most economical production is the objective. The quantity of roughage that will be consumed under such a program will be affected appreciably by its quality. Of roughage that has 50 to 55 per cent TDN, most cattle will consume daily 2.5 to 3.0 units per 100 units of liveweight. But as the TDN falls to 45 per cent, consumption is likely to be no more than 1.5 units per cow per day per 100 units liveweight. Probably the best general rule to follow is one calling for a consumption of 2 units of roughage or its equivalent per 100 units of liveweight daily. If better roughage is available and cows eat more of it, then meal allowances can be curtailed slightly below that given in the feeding chart, and vice versa.

When succulent roughages are also fed, a unit weight of dry roughage may be replaced by about three units of silage or by five units of roots, and one half the normal dry roughage allowance may be replaced on this basis.

Feeding a Meal Mixture to Cows on Pasture. Cows on abundant good pasture will eat sufficient herbage to produce 40 lb (18 kg) of 4 per cent milk or its equivalent without other food. Under these circumstances grain feeding at the rate of one pound for each 2.3 lb of milk in excess of 40 lb will be called for. For such feeding, use mixed farm grains without protein supplements, but be sure to include 1 per cent salt and 1 per cent feeding bone meal or dicalcium phosphate in the grain used. If farm grains are not available for such feeding, there will be no objection, nutritionally, to the use of a standard 16 per cent protein dairy-cow meal mixture.

For cows on limited or poor pasture, begin grain feeding after the first 20 lb (9 kg) of milk, using in this case 450 g of a 16 per cent protein meal mixture for each 2.3 lb (1 kg) of milk produced in excess of the first 20 lb (9 kg).

In general, mixed farm grains without protein supplement will be satisfactory as supplementary feed for spring pasture and for "aftermath," where meal feeding becomes necessary because of high milk production or because of shortage of sufficient pasturage. Midsummer pasture, on the other hand, or mature or dormant pasturage, is approximately the equivalent of timothy hay in protein content and should, therefore, be supplemented with a meal mixture of the sort used during winter feeding. It should be 16 or 18 per cent protein.

Salt Requirements of Cattle. All cows might well be tested weekly for salt hunger, and fed extra salt if they appear to be hungry for it. The usual inclusion of 1 per cent salt in the meal mixture will meet the needs of milking cows that are fed normal quantities of meal mixture. However, cows on pasture and cows that are fed very limited amounts of meal may need extra salt.

On the average a 1,000-lb (450-kg) dry cow should have 21 g of salt per day. In addition to this, cows in milk require about 9 g of salt for each 10 lb (4.5 kg) of milk produced.

Water. Cows in milk require from 4 to 5 lb (two kg) of water for each pound (450 g) of milk produced. On the average, from 12 to 15 gallons of water per day should be provided for every cow in the milking herd.

Rules of Thumb for Feeding Young Dairy Cattle

Where a system of milk feeding for young calves is followed, allow daily a quantity of milk equal to one-tenth of the calf's weight, up to a maximum of 20 lb (9 kg) of milk per day. For young cattle more than six months of age, feed 1 lb (450 g) daily of a normal milking-cow meal mixture of 16 to 18 per cent protein, together with 2 lb (1 kg) of good roughage for each 100 lb (45 kg) of the animal's live weight. For animals from the age of six months to breeding age, the grain allowances may be gradually reduced and roughage correspondingly increased so long as normal growth is maintained. When young heifers are turned to pasture, supplementary grain feeding may be necessary at first to keep them gaining at normal rates.

Fig. 21-2 has been prepared as a general guide to the amounts of hay

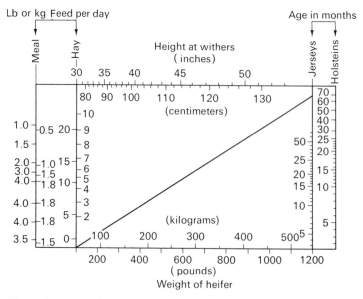

Figure 21-2. *Meal and hay allowances for growing dairy heifers according to age, height, and weight.*

and meal to be fed to growing dairy heifers according to their weight, age, and height.

Feeding Dairy Calves to Six Months of Age.   Beginning at two weeks of age, calves should have fine, leafy legume *hay* before them all the time. The idea that self-feeding hay leads to paunchy calves has been shown to be erroneous, providing the calves are fed normal amounts of grain and milk or its equivalent in gruels. Access to hay at all times has been found to help prevent the habit of sucking each other the calves may develop when penned together.

Roots and *silage* should not be given to the calf before it is three months of age. At this age the calf may be started on small quantities and the amounts fed gradually increased as indicated in the feeding tables following (see Table 21-2). If the calf does not show proper growth, succulent feeds should be decreased and concentrates increased. Fresh and pure *water* should be given to the calf. At from two to six months it can drink from 10 to 15 lb (4.5 to 7 kg) of water per day in addition to any other liquid such as milk or gruel.

Since calves are not able to subsist on roughage until the rumen has developed and become functional, they should not be turned out to *pasture* before they are four months old unless arrangements are made to feed them in the same way as if they were penned in the barn. Where satisfactory stabling facilities are not available for summer housing, the pasture chosen should provide shade and be well-drained.

Calf-Feeding Schedules.   For the first month, the feeding schedule for calves is the same, regardless of whether a method using milk, skim milk, or gruel is used, or whether a dry-meal feeding system is to be followed. Care should be taken that the calf, within the first 12 hours, obtains the colostrum of its mother. This first milk has special properties essential for the well-being of the calf. Antibodies that protect the calf from diseases in the early stages of its life and vitamin A are particularly important components of colostrum.

After the calf is a month old, its feeding on milk or milk substitutes will depend on the method chosen. Table 21-2 gives a schedule showing the amounts of the different kinds of feeds which would ordinarily be allowed to calves being raised on these methods, using either skim milk or calf-meal gruel.

Dry Calf-Meal Method of Feeding.   This method of calf feeding has much in its favor. Its saving in milk and labor, together with freedom from many troublesome digestive upsets often accompanying skim milk and gruel feeding, make it an attractive plan. Nevertheless, the system is not foolproof.

Most important is the fact that the success of this system depends in no

TABLE 21-2 *Quantities of Feeds to Be Allowed per Calf per Day According to Age and Calf-Raising Method*

| | Skim milk method | | | | Other feeds to be used with all methods | | | | | |
| | Whole milk | | Skim milk or gruel | | Dry-meal mixture[a] | | Hay | | Silage or roots | |
Age	lb	kg	lb	kg	lb	kg	lb	kg	lb	kg
1 week	6–8	3								
2 weeks	8–10	4								
3 weeks	10–12	5			0.25	0.10	0.25	0.10		
4 weeks	12–15	6			0.25 to 0.5	0.15	0.25 to 0.5	0.15		
5 weeks	14–18	7	1–8	2	0.5	0.25	0.5	0.25		
6 weeks	8–0	3	8–15	6	0.5 to 0.75	0.30	0.5 to 0.75	0.30		
7 weeks			15	7	0.75 to 1	0.40	0.75 to 1	0.40		
8 weeks			15	7	1	0.45	1	0.45		
3 months			15	7	1 to 2	0.70	2 to 3	1.15	1 to 2	0.70
4 months			15	7	2 to 2.5	1.00	3 to 4	1.60	2 to 4	1.40
5 months			15	7	2.5 to 3	1.25	4 to 5	2.00	4 to 5	2.00
6 months			15	7	3 to 3.5	1.50	5 to 6	2.50	5 to 6	2.50

[a] Use a 16 per cent protein milking-cow ration as the dry-meal mixture.

small measure on the use of a meal mixture designed for this plan of calf raising. A mixture that may be acceptable when used with whole or skim milk may prove entirely unsatisfactory with this plan, and unless feeders are prepared to provide the necessary kind of calf meal they should not consider this scheme of calf raising.

This method calls for the use of whole milk for the first month only. The amounts to be fed during this period are the same as shown in the calf-feeding schedule in Table 21-2. However, at the beginning of the fifth week the milk is abruptly discontinued. This sudden withdrawal of milk has the effect of requiring the calf to satisfy its hunger by eating other feeds that are provided. A *gradual* cutting down in the milk allowance, on the other hand, frequently results in the calf going partially hungry before it learns to eat the dry food.

From the time the calf is 2 weeks old a rack of high-grade, fine, dust-free legume hay, together with a continuous supply of fresh water, should be available at all times. In addition, a small quantity of calf meal, or better, calf-meal pellets, may be provided in a feed box to be eaten as desired. By the time milk is cut off the calf will have learned to eat hay and meal, and the removal of the milk will mean only that it will eat more meal.

We cannot too strongly urge that a calf meal especially prepared for this method of feeding be employed. Milk contains an abundance of riboflavin and of the minerals calcium and phosphorus, as well as proteins of high feeding value. These essentials must be provided in the calf meal if milk is to be replaced at this early age in the development of the calf.

Provision for drinking water is also important, and wherever it is possible free access to water is strongly recommended. The plan, in fact, is almost one of self-feeding; and calves raised on this system frequently make better progress than where hand-feeding is used, not necessarily because the ration is better, but because it is available when the calf wants it. A flexible formula for a calf meal suitable for this system of raising calves was discussed in Chapter 18 (see Table 18-5).

Feeding Pregnant Dairy Cows

About two-thirds the weight of the fetal calf is actually gained during the last 60 days of pregnancy. This is the usual dry period for the cow, and during this time liberal feed is needed for the calf growth, otherwise the nutrients required will be taken from the cow's own body reserves and, as a consequence, her subsequent lactation may suffer. Experiments with cattle show that the gain or loss in body weight by a cow during the last 60 days of pregnancy affects the quantity of milk produced in a lactation. In general, one series of studies found that for each pound (0.45 kg) of gain in body weight during her dry period, there was an increase during the lactation of 25 lb (11 kg) of milk for Holsteins, 20 lb (9 kg) of milk for Guernseys, and 15 lb (7 kg) of milk for Jerseys. Thus a gain of 50 lb (23 kg) in condition may mean from 750 to 1,250 lb (340–570 kg) more milk for the next lactation, which in itself is ample justification for attention to the dry-cow feeding.

During the dry period the cow should rebuild her reserves of calcium, phosphorus, and protein, which in high-producing animals have been partially depleted during the previous lactation. The feeding of farm grains alone to dry cows is not as satisfactory as the use of a ration properly balanced in minerals and protein during this period. Under most conditions a meal mixture with 14 per cent crude protein will be adequate, and while one can be produced by mixing a standard 16 per cent protein milking-cow meal mixture with an equal quantity of farm grains, such a practice will not provide adequate mineral supplementation. When such a practice is followed, add one unit weight of bone meal and 0.5 units of salt to each 1,000 units of the combination of grain and 16 per cent protein meal mixture.

Guides for Swine Feeding

The feeding of market hogs is largely a self-feeding program; hence, rules for daily allowances are not of primary importance. The gain made by self-fed market pigs is likely to be related to the TDN of the ration. When rapid gains are the objective, a mixture of 75 per cent TDN or more is normally fed during the finishing period, whereas the mixture better suited for finishing bacon hogs will have less than 70 per cent TDN. Of such a mixture, slightly larger amounts will often be eaten, but the gains will be slower than on the normal program. The protein of the rations to be fed will be 18 per cent,

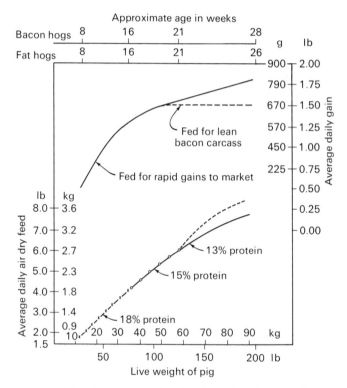

Figure 21-3. *Daily feed and gains of market pigs according to their attained weights or equivalent ages.*

15 per cent, and 13 per cent during the weaning, growing, and finishing periods, respectively. Some of the relations between the age and weight of pigs and their average feed intake and live weight gains are shown in Fig. 21-3.

Rules of Thumb for Feeding the Breeding Herd. The feeding of the breeding herd of swine differs from that of the market stock in that the former are treated largely as individual animals.

Sows

1. In general, feed to dry sows meal or its equivalent per day at 2 per cent of their live weight.

2. Pregnant sows require rations fortified with vitamins and minerals. A mineral mixture supplying at least calcium, salt, iron, and iodine should constitute about 3 per cent of the meal ration; and a daily allowance of one-half tablespoonful of a feeding oil to supply vitamins A and D will be profitable unless these are supplied in some other way.

3. Nursing sows should be fed what they will clean up promptly at three feeds daily. It will range between 3 and 4 per cent of their weight.

Boars

1. Allow meal daily equal to 2 per cent of live weight.

2. Feed enough to maintain a satisfactory condition of fleshing.

Self-Feeding of Brood Sows. Self-feeding of brood sows is entirely feasible if the grain ration is sufficiently bulky to prevent excessive fattening of the sows. A ration that may be used successfully for the self-feeding of brood sows consists of two parts of a normal meal mixture intended for growing market pigs diluted by one part of alfalfa meal or wheat bran. Fresh water may be conveniently supplied by an ordinary pressure water bowl of the type used in dairy stables.

Factors Affecting the Excellence of Hog Carcasses Intended for Bacon.

1. The greatest single factor causing an undesirable bacon carcass is excessive fat.

2. Male pigs normally yield fatter carcasses than females at equal marketing weight.

3. Pigs that have been fed heavily on wheat or corn during the last two months of feeding yield carcasses, within bacon weights, that are usually overfat.

4. Percentage of protein in the hog ration has no direct effect on carcass excellence, though it does affect the growth of the pig.

5. Restriction of rate of gain of pigs after a live weight of 125 lb (60 kg) tends to result in a leaner carcass. This is easily accomplished by using a lighter, bulkier fattening ration.

To produce the greatest number of "bonused" carcasses, feed young pigs liberally on a properly balanced low-fiber ration to about 125 lb (60 kg). Then change to a lighter ration for finishing. After pigs weigh 100 lb (45 kg), daily gains of 1.5 lb or 700 g are rapid enough. Faster gains than this are correlated with overfat carcasses.

As an example of this feeding practice, the following rations for the *growing* and *fattening* periods are given in Table 21-3. In this plan the

TABLE 21-3. *Growing and Fattening Swine Rations*

Feeds	Growing pigs [to 125 lb (60 kg)]	Fattening pigs [125 lb (60 kg) to market]
Barley (or comparable basal feed)	90	75
Wheat bran	0	20
Protein-mineral supplement	10	5

wheat bran may be replaced by alfalfa meal, or barley and bran may be replaced by oats. Note that oats are not used in the growing ration, but that barley here may be replaced in part by wheat, corn, or No. 1 feed screenings.

Horse Feeding

Horse feeding as such has not been considered before in this book. One of the reasons is that meal mixtures for horses are seldom compounded with the help of feeding standards. Adult work stock are adequately fed by a combination of oats (or its equivalent), nonlegume hay, and salt; and breeders usually require only iodine in addition to this ration. We do not mean that meal mixtures cannot be used. In fact horses do very well on any standard dairy-meal mixture, which is more acceptable if pelleted or crumbled to overcome the powdery condition of ground feed.

Rules of Thumb. The problems of horse feeding therefore are more with the quantities to be fed than with the makeup of meal mixture. Usually maintenance of body weight is a better guide than a formal feeding standard. Horses often seem to differ in their requirements more than cattle do, as is evident from the different quantities of feed needed by team mates to maintain steady body weight. Feeding standards, therefore, are no more than guides, and rules of thumb may be as practically accurate as is needed. One set of such rules is as follows: (1) Of roughage, feed a quantity daily equal to 1 or 1.5 per cent of the animal's weight. (2) Of grain, feed an allowance equal to 1 or 1.5 per cent of the animal's weight. (The smaller allowance of roughage and the larger one of grain is for horses at heavy work, and vice versa for those doing light work.)

Idle days. One of the problems of work-horse feeding arises when idle days occur. On such days feeding must be reduced to avoid a condition such as azoturea. The practical solution is to feed a bulky low-energy ration on the night before an idle day. This satisfies the appetite without the unwanted heavy energy intake at this time. A bran mash is made as follows:

To dry wheat bran, about equal by measure to the usual feed of grain, add a teaspoonful of salt and enough hot water to make a stiff mash. Cover and allow to steam. Feed when cool. If desired a cupful of molasses may be added to the bran before steaming.

Feeding Orphan Foals. Occasionally the horseman will have to raise an orphan foal. General directions that may be helpful are as follows.

To feed an orphan foal, secure milk from a fresh cow, preferably a low tester. To a pint of the cow's milk, add ¼ pint of limewater and a teaspoonful of sugar. This will make enough for two feedings at first.

Another formula that has been successfully used consists of:

Dried whole cow's milk	40 %	
Dried skim milk	30 % } 3 lb (1,370 g)	
Sugar	30 %	
Limewater	8 oz (1 cupful)	
Water	4 lb (2 Imperial quarts)	

Feed the orphan foal daily about four quarts of milk or the above fluid for each 100 lb (45 kg) of his live weight. Give in addition one tablespoonful of codliver oil per day. Warm the fluid to about 100°F. and feed from a nursing bottle. For the first few days feed every hour, using about half a pint of the prepared cow's milk at each feeding. Gradually lengthen the time between feedings and increase the allowance of milk. Care must be taken, especially during the first two weeks, not to overfeed the foal, or scours will result. (See Table 21-4.)

Teach foals to drink from a pail as soon as possible, so that the bottle feeding may be discontinued, thus eliminating the problem of sterilizing bottles and nipples. If a little grain is put into the bottom of the pail of milk, the foal soon learns to eat solid food. Once this occurs feed a little grain three times a day. Get the orphans on good pasture as soon as possible but continue to provide a little grain and hay. As a grain mixture, good results have been obtained from five parts ground oats, four parts wheat bran,

TABLE 21-4. *Schedule for Feeding Orphan Foals*

Age	Feed	Amount per feed	Times per day to feed
1 week	Cow's milk—1 pint Limewater—3 tablespoon- fuls Sugar—1 teaspoonful	¼ to ½ pint	Every hour
2 weeks	As above	½ to 1 pint	6 times
3 to 5 weeks	As above but omit the sugar	1 pint or more according to appetite and size of foal	6 times
5 to 8 weeks	Gradually substitute skim for whole milk Omit the lime water	As per appetite	4 times
8 to 12 weeks	Skim milk plus small grain allowance	As per appetite	3 times

and one part linseed meal. Good, bright clover hay, free from dust is to be preferred to timothy hay for foals. Good timothy, however, is far better than poor quality clover.

The Proximate Analysis as an Index of Feeding Value for Meal Mixtures Whose Formulae Are Unknown

The marked increase in the use of commercially prepared feed mixtures, whether they be complete ready-to-feed combinations or mixed supplements, has presented the problem of judging which one among the many available represents the preferred choice. With few exceptions little information is given to the feeder concerning such mixtures, other than the list of ingredients claimed to be present and the guarantee of minimum protein, minimum crude fat, and maximum crude-fiber content. For mineral supplements, and also for certain special types of mixtures, additional and sometimes alternate items are required to be guaranteed, such as the salt content or maximum fluorine content. But usually the mixtures will furnish protein and energy, and, consequently, will carry guarantees of protein, fat, and fiber, and it is on the basis of the figures for these three components that the feeder usually attempts to judge the quality and usefulness of the mixture. Unfortunately, figures for protein, fat, and fiber are not of much use as indices of the quality of such products, and feeding value cannot be so simply indicated.

For example, if the leaves of early-cut timothy hay are dried and ground, and compared with a sample of dried, ground feces produced by a herbivorous animal that has subsisted entirely on such feed, the two cannot be distinguished except by a strong magnifier. Their analysis might be:

	Ground timothy leaves (%)	*Feces from these leaves (%)*
Crude protein	7.9	8.2
Crude fat (ether extract)	1.3	2.4
Crude fiber	34.7	24.8
Nitrogen-free extract	50.9	59.2

Based on this chemical information, one might erroneously judge that the feces have the higher feeding value. Again, one might note that ground shoe leather is a high-protein, low-fiber product of animal origin, but of no feeding value. Weed seeds carry as much and often more protein, and sometimes less fiber, than cereal grains (Table 21-5). Furthermore, we should not

TABLE 21-5 *Proximate Analysis of a Few Weed Seeds*

Plant seed	*Per cent of water*	*Per cent of protein*	*Per cent of fat*	*Per cent of fiber*	*Per cent of N-Free extract*	*Per cent of ash*
Pigweed seed	9	19	7	11	50	4
Lamb's-quarters	10	15	5	20	46	4
Black bindweed	11	10	2	7	69	1
Wild radish seed	8	24	25	10	22	11
False flax	9	22	30	11	22	6

forget that many rations can be improved for some special purposes by *increasing* the fiber content.

In other words, where rations are compared in usefulness, the one with the higher protein or the lower fiber may not be the choice. It will depend upon what the ration is wanted for, what feedstuffs were used, and the content of essential nutrients, which are not indicated by either crude protein or crude fiber.

The Tag on the Bag. The list of ingredients on the tag of the commercially prepared feed and the integrity of the feed dealer (or manufacturer) are of much greater importance in indicating the usefulness of a mixed ration than are the figures for protein, fat, and fiber. Perhaps it may not be out of place to note that no reputable feed manufacturer or feed dealer will knowingly offer for sale or recommend a feed mixture that will not give satisfactory results when used as recommended. His business obviously depends on his continuing to sell to the same customers.

Are "Balanced" Rations Necessary?

All modern textbooks of feeding advocate the use of "balanced" rations. The reasons for doing so are not always prominently set down, and often the reasons advanced are not wholly sound in terms of economical feeding. Actually few humans, let alone animals, consume perfect rations. What, then, is the justification for the general advice to feed balanced rations to farm animals? The reason may be found in the definition: "A *balanced ration* is one in which the food requirements of the particular animal are completely met without excess of any nutrient."

The various nutrients required by an animal are numerous, and include not only the protein, fat, and carbohydrate of the older balanced ration, but a long list of minerals and vitamins. A few of these nutrients are interchangeable, completely or partially, but most of them are required in specific amounts as well as in definite proportions to other nutrients. In general, natural foodstuffs contain most of the nutrients needed by animals, but they are unbalanced in quantity. By mixing together foodstuffs of different nutrient composition, we can provide rations more nearly balanced to the needs of the animal than we can with single feedstuffs.

For example, let us suppose that a particular animal requires for its best growth or production 1.5 lb of protein and 8 lb of TDN. If a certain feed, such as corn, contains 10 per cent protein and 80 per cent TDN, then 10 lb of corn would provide this animal with exactly the amounts of TDN needed, but only one pound of protein would be provided. This shortage would limit full growth or production.

To get the extra half-pound of protein needed, we can do one of two things. Increasing the quantity of corn to 15 lb would provide the 1.5 lb of protein, but would also furnish 12 lb of carbohydrate, a surplus of four lb. This surplus could not be used, and accordingly would be burned in order to get rid of it. Alternatively, the balance of the corn ration (in protein and carbohydrate only) could be changed by mixing one pound of soybean meal with 6.5 lb of corn. The resulting combination would contain 15 per cent protein and 80 per cent TDN, and 10 lb of it would provide the 1.5 lb of protein and the 8 lb of TDN needed. Thus we add a protein supplement to basal feeds, not because the basal feeds contain no protein, but because a greater proportion of protein to TDN than is found in such feeds will better meet the animals' requirements in that respect.

The Economics of Balancing the Ration. What are the economics of such a balancing of a ration? In our example 8.66 lb of corn plus 1.34 lb of soybean meal have provided the 1.5 lb of protein needed by the animal. To provide the necessary protein with corn would have required 15 lb of corn. Thus, 1.34 lb of oil meal saved 6.34 lb of corn, or one pound of oil meal was worth about 4.7 lb of corn. Providing that the corn thus saved could be used to feed more animals, or put to some other equally profitable use, then if soybean meal could be bought for just 4.7 times the cost of corn, it would be immaterial whether the balanced ration were used or not (assuming for the sake of this example that no other factors were involved). If oil meal costs more than this, using the balanced ration would be uneconomical. If, however, oil meal could be bought for less than 4.7 times the price of corn, then balancing the ration would pay.

This principle applies to every nutrient of the ration; for the efficiency of the ration depends on how much it provides without waste the operating needs of the animals fed. Thus it may be that salt, calcium, or vitamin A may need to be increased in the ration to balance it properly. The perfectly balanced ration is always the most efficient ration, because it is the one of which the smallest quantity will be required to meet the food needs of the animal.

Evidently, then, we cannot say without qualification that it will necessarily pay to feed a balanced ration. If protein supplements are too costly in relation to basal feeds, it may not pay to completely balance the ration in terms of protein. We do not mean that the total intake of protein may be reduced without penalty; rather, we mean that it may pay better to feed extra amounts of lower-protein mixtures than to use the normal quantities of a balanced ration. Protein, however, is only one part of the ration, and one which may vary over relatively wide ranges with no perceptible effect on the health of the animal. Unfortunately, this is not true with minerals and vitamins. With these nutrients it is not only a matter of efficient rations in terms of maximum production; it is often a matter of the ability of the animal to produce or reproduce at all, if not of its ability to live. There can be no question, therefore, of the importance of balance in the ration in terms of the so-called protective nutrients. The greater the use of by-products, the greater the possibility that special supplementary sources of some nutrients will be required, since in the processing of natural feeds, losses and changes in the composition of the original material take place, and rations prepared from by-products become less well-balanced to the operating needs of the body.

The Misuse of Balanced Rations. In recent years there has been a marked increase in the number of farmers who have employed commercially mixed balanced rations in the feeding of livestock. However, these farmers have frequently diluted the purchased mixture with home-grown grains, either by deliberately mixing or by alternating the feeding of one with the other. This practice sometimes leads to difficulties.

For example, a 16 per cent protein dairy meal mixture diluted one-to-one with oats will have 14 per cent protein, and whereas the 16 per cent protein balanced ration normally would have about 2 per cent bone meal in it, the combination will have only half this quantity of mineral supplement. Thus when a balanced ration is used as only a part of the nonroughage feed, the protective nutrients may have been diluted in amount, so that the ultimate combination fed is inadequate nutritionally for the needs of the animals. If a commercial mixture is to be fed in combination with farm grains, a concentrate or supplement mixture should be chosen rather than a ready-to-feed balanced ration. The former is intentionally more heavily fortified with minerals and vitamins to cover the deficiency of the farm feeds that are to be fed with it.

Feeding Iodine

Fortification of the rations of pregnant females with iodine is indicated in all so-called "goitrous regions." The source of the iodine is largely unimportant. It is desirable, if inorganic sources are to be used (potassium iodide or sodium iodide), to insist on a stabilized material, since the iodine in unstabilized iodized salts will sublime and disappear when exposed to the air. There is so little danger from iodine poisoning that it may be disregarded in practical feeding.

The minimum quantities of iodine that should be provided to pregnant females have not been acceptably established. We should note, however, that salt that is 0.05 per cent potassium iodide has been effective in preventing clinical evidence of iodine deficiency. On this basis average daily intakes of potassium iodide of 25 mg for mares and cows, and of 15 mg for sows and ewes, appear to have been adequate. The iodine source may be incorporated in some part of the feed, such as the mineral supplement, or it may be given to the animals individually, in their feed or water.

When the iodine is to be fed to the animals directly, allowances can be conveniently measured out if the iodide is dissolved in water. For all classes of farm animals, dissolve 1.5 grams potassium iodide in 1 liter of water, and give daily to each pregnant female 1 tablespoonful of the solution. A solution seven times as strong may be made and a tablespoonful administered once a week. For beef cows having a common drinking trough, add to the water in the trough enough of the iodine solution for the whole herd.

Annual Requirements of Meal for Livestock

Feeders frequently must estimate the average quantities of grain or meal mixtures needed to feed different classes of livestock for some specified period of time. Table 21-6 has been prepared to meet average conditions.

TABLE 21-6 *Annual Meal Requirements of Livestock*

Class of animal	Special conditions	Amount of concentrates per head	
		lb	kg
Cattle			
Cows in milk[a]	4,000 lb (1,800 kg) milk	1,000	450
	6,000 lb (2,700 kg) milk	1,800	815
	8,000 lb (3,600 kg) milk	2,400	1,090
	10,000 lb (4,500 kg) milk	2,800	1,270
Breeding bulls		1,000	450
Young stock over 6 months		500	225
Calves under 6 months		450	205
Fattening cattle	Per 100 lb gain	300	135
Horses			
Work animals	1,400 lb weight[b]	2,500	1,140
Young stock over 1 year and stock at light work		1,000	450
Sheep			
Breeding flock		200	90
Fattening stock	90 days feeding	100	45
Swine[c]			
Sows	2 litters per year	2,000	910
	1 litter per year	1,500	670
Breeding boars		1,000	450
Market pigs	Weaning to 100 lb (45 kg) weight	250	115
	100 to 200 lb (45 to 90 kg) weight	400	180

[a] For spring calvers reduce these amounts 25%.
[b] For heavier animals increase amounts 200 lb for each 100 lb increased weight.
[c] Amount of concentrates is total meal or its equivalent.

The Format of Modern Feeding Standards

It was pointed out in Chapter 9 that the nutrient needs of many mammals, and probably of all, are the same if the nutrients are energy needs and if the latter are computed for animals of the same metabolic size (i.e., $W_{kg}^{.75}$) and *at equivalent performance* with respect to activity, (maintenance, work, play) growth, and reproductive status. It is interesting to note that all species are at the same physiological stage of development at the point of inflection on their growth curves, and that their growth curves from that point to maturity, coincide. Except for the exceptionally long juvenile stage of the human the growth curves of different species from birth to puberty also coincide.

It is obvious that the nutrient-caloric ratios will be affected if a nutrient (such as protein, calcium, or vitamin A) is intentionally fed in amounts that will induce body storage. Thus feeding standards must define separately the needs of each physiologically different group of animals of a herd. For example, the amount of protein per 1000 kcal required by growing animals declines continuously from the early juvenile period to the age of maturity. The rate of gain and hence the degree of protein storage may differ among different individuals and among different species. Thus the protein-kcal ratios for growing pigs differ for different age groups because the pigs are, in fact, different in "performance." To the degree that the energy (as kcal of DE, ME, or NE) is correlated with the weight of the ration the same is true of nutrient-feed ratios, and hence some standards give, in addition to daily amounts required, the per cent of nutrient in the ration.

Between species, equivalence in performance of growing stock can usually be assumed if comparisons between animals that have attained equivalent percentages of their expected adult weights show them to be in close agreement.

In general, it is implied that each animal grouping shown in a feeding standard defines a group of animals that can be fed identical rations, though some standards comprise more groups than necessary for rationing consideration.

Since it is necessary to use the appropriate feeding standard in its entirety to formulate a set of rations that will be adequate for the feeding of all different feeding groups in a breeding and producing herd, current NRC Feeding Standards are presented in the tables on the following pages.

TABLE 21-7 Daily Nutrient Requirements of Dairy Cattle

			Daily nutrients per animal[a]									
			Protein		Energy						Vit.	
Body weight (kg)	Daily gain (g)	Feed[b] (kg)	Total (g)	Digestible (g)	TDN (kg)	DE[a] (Mcal)	ME[c] (Mcal)	Ca (g)	P (g)	Carotene (mg)	A (1,000 IU)	Vit. D[d] (IU)
					Growth of heifers for herd replacement							
25	300	0.4	90	80	0.50	2.2	1.8	2.0	1.5	2.5	1.0	165
35	450	0.7	155	140	0.75	3.3	2.7	2.8	2.1	3.7	1.5	230
50	500	1.0	200	180	1.00	4.4	3.6	4.0	3.0	5.3	2.1	330
75	550	2.0	340	240	1.50	6.6	5.4	8.0	6.0	7.9	3.2	500
100	650	2.8	430	280	1.90	8.4	6.9	9.6	8.4	10.6	4.2	660
150	700	4.0	480	320	2.55	11.2	9.2	12	11	15.9	6.4	1,000
200	700	5.2	520	380	3.15	13.9	11.4	13	12	21.2	8.5	1,300
300	600	7.2	660	410	4.10	18.0	14.8	15	14	31.8	12.7	
400	600	8.8	700	420	4.60	20.2	16.7	16	15	42.4	17.0	
500	400	9.6	750	450	4.80	21.1	17.3	16	15	53.0	21.2	
600	200	10.0	780	470	5.00	22.0	18.0	16	15	63.5	25.4	
					Maintenance of mature cows							
400		5.8	417	250	2.95	13.0	10.7	11	11	42	16.8	
500		7.0	500	300	3.45	15.2	12.5	14	14	53	21.2	
600		8.0	567	340	3.95	17.4	14.3	16	16	64	25.6	
700		9.2	650	390	4.40	19.4	15.9	18	18	74	29.6	
800		10.4	733	440	4.90	21.6	17.7	22	22	85	34.0	
					Reproduction (add to maintenance during last 2 or 3 months of gestation)							
400		4.0	400	240	2.4	10.6	8.7	10	8	22	8.8	
550		5.0	460	275	3.0	13.2	10.8	13	11	30	12.0	
700		6.0	550	330	3.6	15.8	13.0	16	14	38	15.2	
					Growth of dairy bulls							
500	800	10.4	890	580	6.2	27.2	22.3	18	17	53.0	21.2	
700	600	12.0	1,050	650	7.0	30.8	25.2	19	18	74	29.6	
900		13.6	1,150	710	8.0	35.2	28.9	22	20	95	38.0	
					Maintenance of mature breeding bulls							
700		10.0	810	540	5.7	25.1	20.5	15	15	74	29.6	
900		12.2	960	640	7.0	30.8	25.2	20	20	95	38.0	

[a] Thiamine, riboflavin, niacin, pyridoxine, pantothenic acid, folic acid, vitamin B_{12}, and vitamin K are synthesized by bacteria in the rumen. Trace minerals are discussed in the text.
[b] Based on air-dry feed containing 90 per cent dry matter. These figures are only rough estimations.
[c] ME has been estimated on the basis that 1 g TDN has 4.4 kcal of DE (4.4 Mcal per kg) and that 82 per cent of the DE is available as ME. The ME values can be converted to DE by multiplying them by 1.22.
[d] Quantitative data are not available for growing animals above 200 kg weight. Animals exposed to direct sunlight or fed sun-cured forages do not need supplemental vitamin D.

TABLE 21-8 Nutrient Requirements of Beef Cattle in Percentage or Amount Per Kilogram of Total Ration (Based on Air-Dry Feed Containing 90 Per Cent Dry Matter)

Body weight (kg)	Average daily gain[a] (kg)	Daily feed per animal (kg)	Total protein (%)	Digestible protein (%)	DE (kcal/kg)	TDN (%)	Ca (%)	P (%)	Carotene[b] (mg/kg)	Vitamin A (IU/kg)
									Percentage of ration or amount per kilogram of feed	
Calves finished as short yearlings										
181	1.04	5.35	11.0	8.2	2,921	66	0.37	0.28	4.2	1,650
272	1.09	7.44	11.0	8.2	2,921	66	0.27	0.23	4.2	1,650
363	1.00	8.80	10.0	7.5	2,921	66	0.23	0.21	4.2	1,650
454	1.00	10.43	10.0	7.5	2,921	66	0.20	0.20	4.2	1,650
Finishing yearling cattle										
272	1.18	7.94	10.0	7.5	2,866	65	0.25	0.21	4.2	1,650
363	1.22	10.12	10.0	7.5	2,866	65	0.20	0.20	4.2	1,650
454	1.18	11.70	10.0	7.5	2,866	65	0.20	0.20	4.2	1,650
499	1.04	11.70	10.0	7.5	2,866	65	0.20	0.20	4.2	1,650
Finishing two-year-old cattle										
363	1.27	10.57	10.0	7.5	2,811	64	0.21	0.21	4.2	1,650
454	1.32	12.79	10.0	7.5	2,811	64	0.20	0.20	4.2	1,650
544	1.22	14.06	10.0	7.5	2,811	64	0.20	0.20	4.2	1,650
Normal growth, heifers and steers										
181	.72	5.53	11.7	7.0	2,315	53	0.29	0.21	4.2	1,650
272	.64	7.44	9.3	5.6	2,205	50	0.21	0.16	4.2	1,650
363	.54	8.66	7.8	4.7	2,205	50	0.18	0.15	4.2	1,650
454	.45	9.57	7.8	4.7	2,205	50	0.15	0.15	4.2	1,650

Wintering weanling calves										
181	.45	4.76	10.3	6.2	2,205	50	0.27	0.21	4.2	1,650
227	.45	5.72	10.3	6.2	2,205	50	0.23	0.18	4.2	1,650
272	.45	6.49	9.1	5.5	2,205	50	0.20	0.16	4.2	1,650
Wintering yearling cattle										
272	.45	6.49	8.3	5.0	2,205	50	0.20	0.17	4.2	1,650
363	.32	7.17	7.5	4.5	2,205	50	0.18	0.17	4.2	1,650
408	.23	7.17	7.5	4.5	2,205	50	0.18	0.17	4.2	1,650
Wintering pregnant heifers										
318	.68	9.07	7.5	4.5	2,205	50	0.16	0.15	5.5	2,200
408	.36	8.16	7.5	4.5	2,205	50	0.16	0.15	5.5	2,200
454	.23	8.16	7.5	4.5	1,874	43	0.16	0.15	5.5	2,200
Wintering mature pregnant cows										
363	.68	9.98	7.5	4.5	2,205	50	0.16	0.15	5.5	2,200
454	.18	8.16	7.5	4.5	2,205	50	0.16	0.15	5.5	2,200
544	.00	8.16	7.5	4.5	2,205	50	0.16	0.15	5.5	2,200
544	−.23	7.98	7.5	4.5	1,874	43	0.16	0.15	5.5	2,200
Cows nursing calves, first 3–4 months postpartum										
408–499	.00	12.70	8.3	5.0	2,646	60	0.24	0.18	8.4	3,310
Bulls, growth and maintenance (moderate activity)										
272	1.04	7.35	12.5	7.5	2,756	63	0.29	0.21	8.4	3,310
454	.72	9.07	12.0	7.2	2,646	60	0.21	0.17	8.4	3,310
635	.45	11.20	10.0	6.0	2,535	58	0.16	0.15	8.4	3,310
816	.00	11.57	9.3	5.6	2,425	55	0.15	0.15	8.4	3,310

[a] Average daily gain for finishing cattle is based upon cattle receiving stilbestrol; those not receiving stilbestrol gain from 10 to 20% slower.

[b] Cattle can use carotene to satisfy their vitamin A requirement. Figures given assume 400 IU of vitamin A per mg of carotene.

TABLE 21-9 *Daily Nutrients Required per Kilogram of Milk[a]*
(To be Added to Requirements for Growth or Maintenance)

Fat content of milk (%)	Protein (g)	Digestible protein (g)	TDN (g)	DE (Mcal)	ME (Mcal)	Ca (g)	P (g)
		For cows producing more than 35 kg of milk daily					
3.0	78	50	360	1.59	1.30	2.8	2.0
3.5	83	53	390	1.72	1.41	2.8	2.0
4.0	88	56	420	1.85	1.52	2.8	2.0
4.5	93	59	450	1.98	1.62	2.8	2.0
5.0	98	62	480	2.12	1.74	2.8	2.0
5.5	103	66	510	2.25	1.84	2.8	2.0
6.0	108	70	540	2.38	1.95	2.8	2.0
		For cows producing 20 to 35 kg of milk daily					
3.0	70	45	320	1.41	1.16	2.4	1.8
3.5	74	48	345	1.52	1.25	2.4	1.8
4.0	78	51	370	1.63	1.34	2.4	1.8
4.5	82	54	395	1.74	1.43	2.4	1.8
5.0	86	56	420	1.85	1.52	2.4	1.8
5.5	90	58	445	1.96	1.61	2.4	1.8
6.0	94	60	470	2.07	1.70	2.4	1.8
		For cows producing less than 20 kg of milk daily					
3.0	62	40	280	1.23	1.01	2.2	1.6
3.5	66	43	305	1.34	1.10	2.2	1.6
4.0	70	46	330	1.46	1.20	2.2	1.6
4.5	74	48	355	1.57	1.29	2.2	1.6
5.0	78	50	380	1.68	1.38	2.2	1.6
5.5	82	53	405	1.79	1.47	2.2	1.6
6.0	86	56	430	1.90	1.56	2.2	1.6

[a] See the footnotes to Table 21-7. When calculating the intakes for lactating heifers that are still growing, the figures for growth rather than maintenance should be used. When adequate amounts of carotene, vitamin A, and vitamin D are fed for growth and reproduction, extra amounts will not stimulate milk production. For pasture levels of vitamin A activity of the milk, cows should be fed 300 mg of carotene or 36 mg (120,000 IU) of true vitamin A daily.

TABLE 21-10 *Nutrient Requirements of Dairy Cattle in Percentage or Amount Per Kilogram of Total Ration (Based on Air-Dry Feed Containing 90 Per Cent Dry Matter)*

Body weight (kg)	Daily feed Total (kg)	Per cent of weight (%)	Digestible protein (%)	TDN (%)	DE (Mcal/kg)	ME (Mcal/kg)	Ca (%)	P (%)	Carotene (mg/kg)	Vitamin A (1,000 IU/kg)	Vitamin D (IU/kg)
				Growth of heifers							
25	0.4	1.6	20.0	125	5.5	4.5	0.50	0.38	6.5	2.5	410
35	0.7	2.0	20.0	107	4.7	3.9	0.40	0.30	5.3	2.0	330
50	1.0	2.0	18.0	100	4.4	3.6	0.40	0.30	5.3	2.0	330
75	2.0	2.7	12.0	75	3.3	2.7	0.40	0.30	4.0	1.6	250
100	2.8	2.8	10.0	68	3.0	2.5	0.34	0.30	3.8	1.5	250
150	4.0	2.7	8.0	64	2.8	2.3	0.30	0.28	4.0	1.6	250
200	5.2	2.6	7.3	61	2.7	2.2	0.25	0.23	4.0	1.6	250
250	6.2	2.6	6.5	58	2.5	2.1	0.22	0.21	4.3	1.7	
300	7.2	2.4	5.7	57	2.5	2.1	0.21	0.19	4.4	1.8	
350	8.0	2.3	5.2	56	2.5	2.0	0.20	0.19	4.6	1.8	
400	8.8	2.2	4.8	52	2.3	1.9	0.18	0.17	4.8	1.9	
450	9.0	2.0	4.8	51	2.3	1.9	0.18	0.17	5.3	2.0	
500	9.6	1.9	4.7	50	2.2	1.8	0.17	0.16	5.5	2.1	
550	9.8	1.8	4.7	50	2.2	1.8	0.16	0.15	5.9	2.4	
600	10.0	1.7	4.7	50	2.2	1.8	0.16	0.15	6.4	2.5	

(continued)

TABLE 21-10 (continued)

Body weight (kg)	Daily feed		Percentage or amount per kg of feed								
	Total (kg)	Per cent of weight (%)	Digestible protein (%)	TDN (%)	DE (Mcal/kg)	ME (Mcal/kg)	Ca (%)	P (%)	Carotene (mg/kg)	Vitamin A (1,000 IU/kg)	Vitamin D (IU/kg)
				Energy							
Growth of veal calves for slaughter											
35	0.7	2.0	20.0	115	5.0	4.1	0.40	0.30	5.3	2.1	330
50	1.2	2.4	20.0	115	5.0	4.1	0.40	0.30	4.4	1.8	275
75	2.0	2.7	18.0	115	5.0	4.1	0.40	0.30	4.0	1.6	250
100	2.8	2.8	16.0	110	4.7	3.9	0.34	0.30	3.8	1.5	250
150	3.2	2.1	15.0	100	4.4	3.6	0.38	0.36	5.0	1.9	310
Maintenance of mature cows											
350	5.2	1.5	4.3	54	2.4	1.9	0.19	0.19	7.1	2.7	
400	5.8	1.5	4.3	51	2.2	1.9	0.19	0.19	7.2	2.9	
450	6.2	1.4	4.3	52	2.3	1.9	0.19	0.19	7.7	3.0	
500	7.0	1.4	4.3	49	2.2	1.9	0.20	0.20	7.6	3.0	
550	7.8	1.4	4.2	49	2.2	1.8	0.20	0.20	7.5	3.0	
600	8.0	1.3	4.2	49	2.2	1.8	0.20	0.20	8.0	3.2	
650	8.6	1.3	4.2	49	2.2	1.8	0.20	0.20	8.0	3.2	
700	9.2	1.3	4.2	48	2.1	1.8	0.20	0.20	8.0	3.2	
750	9.8	1.3	4.2	47	2.1	1.7	0.20	0.20	8.1	3.3	
800	10.4	1.3	4.2	47	2.1	1.7	0.21	0.21	8.1	3.3	

Reproduction (add to maintenance)

400	4.0		6.0	60	2.6	2.2	0.25	0.20	5.5	2.1
550	5.0		5.5	60	2.6	2.2	0.26	0.22	6.0	2.4
700	6.0		5.5	60	2.6	2.2	0.27	0.23	6.3	2.5

Growth of dairy bulls (25 to 150: Use table for heifers)

200	5.8	2.9	7.3	60	2.7	2.2	0.25	0.23	3.7	1.5
250	6.7	2.7	6.5	60	2.6	2.2	0.22	0.21	3.9	1.6
300	8.0	2.7	6.0	60	2.6	2.2	0.21	0.19	4.0	1.6
400	9.6	2.4	5.8	60	2.6	2.2	0.18	0.17	4.4	1.8
500	10.4	2.1	5.6	60	2.6	2.1	0.17	0.16	5.1	2.0
600	11.2	1.9	5.5	58	2.6	2.1	0.16	0.15	5.6	2.2
700	12.0	1.7	5.4	58	2.6	2.1	0.16	0.15	6.2	2.5
800	13.0	1.6	5.3	58	2.6	2.1	0.16	0.15	6.5	2.6
900	13.6	1.5	5.2	58	2.6	2.1	0.16	0.15	7.0	2.8
1,000	14.5	1.4	5.1	58	2.6	2.1	0.16	0.15	7.3	2.9

Maintenance of mature breeding bulls

500	7.8	1.6	5.8	56	2.5	2.0	0.14	0.14	6.8	2.7
600	8.8	1.5	5.6	57	2.5	2.0	0.14	0.14	7.3	2.9
700	10.0	1.4	5.4	57	2.5	2.1	0.15	0.15	7.4	3.0
800	11.0	1.4	5.4	58	2.6	2.1	0.15	0.15	7.7	3.1
900	12.2	1.4	5.2	58	2.5	2.1	0.16	0.16	7.8	3.1
1,000	13.5	1.4	5.1	58	2.5	2.1	0.16	0.16	7.9	3.1
1,100	14.5	1.3	5.1	58	2.5	2.1	0.16	0.16	8.1	3.2
1,200	15.5	1.3	5.1	58	2.5	2.1	0.16	0.16	8.2	3.3

TABLE 21-11 *Nutrient Requirements of Foxes in Percentage or Amount Per Kilogram of Total Ration (Based on Feed Containing 100 Per Cent Dry Matter)*

Nutrient	Growth		Mainte-nance	Preg-nancy	Lac-tation
	7 to 23 weeks	23 weeks to maturity			
Energy (kcal GE)	?	?	3,227	?	?
Protein, per cent	25	19	?	?	?
Fat soluble vitamins					
Vitamin A, IU	2,410.	2,410.	?	?	?
Water soluble vitamins					
Folic acid, mg	0.2	0.2	?	?	?
Niacin, mg	10.0	10.0	?	?	?
Pantothenic acid, mg	8.0	8.0	?	?	?
Pyridoxine, mg	2.0	2.0	?	?	?
Riboflavin, mg	2.6	2.6	?	4.0	4.0
Thiamine, mg	1.0	1.0	1.0	0.5	?
Minerals					
Salt, per cent	0.5	0.5	0.5	0.5	0.5
Calcium, per cent	0.6	0.6	0.6	?	?
Phosphorus, per cent	0.6	0.6	0.4	?	?
Ca:P ratio	1.0:1.0 to 1.7:1.0	1.0:1.0 to 1.7:1.0	1.0:1.0 to 1.7:1.0	?	?

TABLE 21-12 *Nutrient Requirements of Horses in Percentage or Amount Per Kilogram of Total Ration (Based on Air-Dry Feed Containing 90 Per Cent Dry Matter)*

Body weight (kg)	Average daily gain (kg)	Daily feed (kg)	Total protein (%)	Digestible protein (%)	TDN (%)	DE (Mcal/kg)	Ca (%)	P (%)	Carotene (mg/kg)	Vitamin A (IU/kg)
										Percentage, or amount per kg, of feed
Growing horses: 270-kg mature weight										
90	0.41	2.77	13.1	8.7	63	2.75	0.40	0.36	0.9	550
185	0.18	2.68	10.1	7.3	63	2.75	0.41	0.41	1.5	840
270	0	3.40	8.0	5.3	63	2.75	0.18	0.18	2.6	1,480
Growing horses: 365-kg mature weight										
90	0.64	3.04	16.4	11.2	63	2.75	0.46	0.36	0.9	550
185	0.41	4.26	9.6	6.6	63	2.75	0.40	0.30	1.3	770
270	0.23	4.72	7.7	5.7	63	2.75	0.28	0.28	2.0	1,060
365	0	4.22	7.5	5.5	63	2.75	0.21	0.21	2.9	1,590
Growing horses: 455-kg mature weight										
90	0.73	3.04	17.9	12.5	63	2.75	0.52	0.36	0.9	550
185	0.54	4.49	11.1	7.6	63	2.75	0.33	0.27	1.3	730
270	0.36	5.17	8.8	5.9	63	2.75	0.27	0.23	1.8	970
365	0.23	5.58	7.3	5.3	63	2.75	0.23	0.22	2.2	1,190
455	0	4.94	8.2	5.5	63	2.75	0.22	0.22	3.1	1,680

(continued)

TABLE 21-12 (*continued*)

Body weight (kg)	Average daily gain (kg)	Daily feed (kg)	Total protein (%)	Digestible protein (%)	TDN (%)	DE (Mcal/kg)	Ca (%)	P (%)	Carotene (mg/kg)	Vitamin A (IU/kg)
									Percentage, or amount per kg, of feed	
				Growing horses: 545-kg mature weight						
90	1.00	3.40	21.3	14.8	63	2.75	0.56	0.47	0.9	550
185	0.82	5.08	12.5	8.9	63	2.75	0.35	0.33	1.3	750
270	0.59	5.94	9.2	6.7	63	2.75	0.30	0.29	1.5	880
365	0.36	6.08	8.2	5.7	63	2.75	0.30	0.27	2.0	1,100
455	0.18	6.08	7.4	5.3	63	2.75	0.20	0.20	2.4	1,370
545	0	5.67	8.0	5.5	63	2.75	0.21	0.21	3.1	1,760
				Growing horses: 635-kg mature weight						
90	1.22	3.63	23.8	16.5	63	2.75	0.66	0.47	0.9	550
185	1.00	5.31	14.5	10.0	63	2.75	0.40	0.32	1.1	640
270	0.82	6.53	10.4	7.4	63	2.75	0.29	0.26	1.5	820
365	0.59	6.99	9.0	6.2	63	2.75	0.26	0.24	1.8	950
455	0.36	6.99	7.8	5.5	63	2.75	0.20	0.20	2.2	1,190
545	0.18	6.80	7.3	5.3	63	2.75	0.20	0.20	2.6	1,480
635	0	6.35	7.8	5.5	63	2.75	0.20	0.20	3.3	1,850
				Mature horses at light work						
185		3.76	5.2	3.6	63	2.75	0.16	0.16	1.5	880
270		5.08	5.2	3.7	63	2.75	0.16	0.16	1.8	990
365		6.26	5.3	3.7	63	2.75	0.16	0.16	2.0	1,080
455		7.39	5.3	3.7	63	2.75	0.16	0.16	2.0	1,130
545		8.48	5.3	3.7	63	2.75	0.16	0.16	2.2	1,170
635		9.53	5.2	3.7	63	2.75	0.16	0.16	2.2	1,240

Mature horses at medium work

185	4.35	4.4	3.1	63	2.75	0.18	0.18	1.3	750
270	5.90	4.5	3.2	63	2.75	0.17	0.17	1.5	840
365	7.35	4.5	3.2	63	2.75	0.16	0.16	1.5	900
455	8.62	4.5	3.2	63	2.75	0.16	0.16	1.8	970
545	9.93	4.5	3.2	63	2.75	0.16	0.16	1.8	1,020
635	11.11	4.5	3.2	63	2.75	0.16	0.16	2.0	1,060

Mares, last quarter of pregnancy

185	2.63	9.8	6.9	63	2.75	0.34	0.30	10.6	3,530
270	3.63	9.9	6.9	63	2.75	0.33	0.30	11.7	3,860
365	4.45	9.8	6.8	63	2.75	0.31	0.29	12.6	4,210
455	5.31	9.7	6.8	63	2.75	0.30	0.28	13.2	4,390
545	6.08	9.8	6.9	63	2.75	0.30	0.28	13.9	4,610
635	6.80	9.8	6.9	63	2.75	0.29	0.28	14.3	4,180

Mares, peak of lactation

185	6.99	11.3	7.9	63	2.75	0.26	0.19	4.0	1,320
270	7.98	11.2	7.8	63	2.75	0.29	0.23	5.3	1,760
365	9.43	11.3	7.9	63	2.75	0.29	0.23	6.0	1,980
455	10.43	11.1	7.8	63	2.75	0.29	0.23	6.6	2,230
545	11.52	11.3	7.9	63	2.75	0.29	0.23	7.3	2,430
635	13.15	10.7	7.5	63	2.75	0.29	0.23	7.5	2,490

TABLE 21-13 *Nutrient Requirements of Mink in Percentage or Amount Per Kilogram of Total Ration (Based on Feed Containing 100 Per Cent Dry Matter)*

Nutrient	Growth (*weaning to pelting*)	Maintenance (*mature*)	Pregnancy	Lactation
Energy (kcal GE)	5,300	4,250	5,300	a
Protein, per cent	25	?	?	?
Fat soluble vitamins				
Vitamin A, IU	3,500	?	?	?
Vitamin E, mg	25	?	?	?
Water soluble vitamins				
Thiamine, mg	1.2	1.1	?	?
Riboflavin, mg	1.5	?	?	?
Pantothenic acid, mg	6.0	?	?	?
Niacin, mg	20	?	?	?
Pyridoxine, mg	1.1	?	?	?
Folic acid, mg	0.5	?	?	?
Minerals				
Salt (NaCl), per cent	0.5	0.5	0.5	0.5
Calcium, per cent	0.4	0.3	0.4	0.6
Phosphorus, per cent	0.4	0.3	0.4	0.6
Ca:P ratio	1:1 to 2:1	1:1 to 2:1	1:1 to 2:1	1:1 to 2:1

[a] Energy requirements for lactation increase sharply with number of young produced and growth of the young. The recommended level for growth may be taken as basal and increased according to the above criteria.

TABLE 21-14 *Nutrient Requirements of Chickens in Percentage or Amount Per Kilogram of Total Ration*[a]
(*Based on Air-Dry Feed Containing 90 Per Cent Dry Matter*)

Nutrient	Starting chickens (0–8 weeks)	Growing chickens (8–18 weeks)	Laying hens	Breeding hens
Total protein, per cent	20	16	15	15
Vitamins				
Vitamin A activity (IU)[b]	2,000	2,000	4,000	4,000
Vitamin D (ICU)	200	200	500	500
Vitamin E	see text			
Vitamin K$_1$, mg	0.53	?	?	?
Thiamine, mg	1.8	?	?	*0.8*
Riboflavin, mg	3.6	1.8	2.2	3.8
Pantothenic acid, mg	10	10	2.2	10
Niacin, mg	27	*11*	?	?
Pyridoxine, mg	3	?	3	4.5
Biotin, mg	0.09	?	?	*0.15*
Choline, mg	1,300	?	?	?
Folacin, mg	*1.2*	?	0.25	0.35
Vitamin B$_{12}$, mg	0.009	?	?	*0.003*
Minerals				
Calcium, per cent	1.0	1.0	2.75[c]	2.75[c]
Phosphorus, per cent[d]	0.7	0.6	0.6	0.6
Sodium, per cent[e]	0.15	0.15	0.15	0.15
Potassium, per cent	0.2	0.16	?	?
Manganese, mg	55	?	?	33
Iodine, mg	0.35	0.35	0.30	0.30
Magnesium, mg	500	?	?	?
Iron, mg	40	?	?	?
Copper, mg	4	?	?	?
Zinc, mg	35	?	?	?

[a] These figures are estimates of requirements and include no margins of safety. (See text, page 2–5.) Italicized figures are tentative.
[b] May be vitamin A or pro-vitamin A.
[c] This amount of calcium need not be incorporated in the mixed feed, inasmuch as calcium supplements fed free-choice are considered as part of the ration.
[d] At least 0.5% of the total feed of starting chickens should be inorganic phosphorus. All the phosphorus of non-plant-feed ingredients is considered to be inorganic. Approximately 30% of the phosphorus of plant products is non-Phytin phosphorus and may be considered as part of the inorganic phosphorus required. A portion of the phosphorus requirement of growing chickens and laying and breeding hens must also be supplied in inorganic form. For birds in these categories the requirement for inorganic phosphorus is lower and not as well defined as for starting chickens.
[e] Equivalent to 0.37% of sodium chloride.

TABLE 21-15 *Nutrient Requirements of Turkeys in Percentage or Amount Per Kilogram of Total Ration*[a] *(Based on Air Dry Feed Containing 90 Per Cent Dry Matter)*

Nutrient	Starting poults (0–8 weeks)	Growing turkeys (8–16 weeks)	Breeding turkeys
Total protein, per cent[b]	28	20	15
Vitamins			
Vitamin A activity (IU)[c]	4,000	4,000	4,000
Vitamin D (ICU)	900	900	*900*
Vitamin E	see text		
Vitamin K₁, mg	*0.7*	?	?
Thiamine, mg	*2*	?	?
Riboflavin, mg	3.6	?	*3.8*
Pantothenic acid, mg	11	?	*16*
Niacin, mg	*70*	?	?
Pyridoxine, mg	*3*	?	?
Choline, mg	1,900	?	?
Folacin, mg	0.9	?	*0.8*
Vitamin B₁₂, mg	*0.003*	?	?
Minerals			
Calcium, per cent	1.2	1.2	2.25[d]
Phosphorus, per cent[e]	0.8	0.8	0.75
Sodium, per cent[f]	0.15	0.15	0.15
Potassium, per cent	*0.4*	?	?
Manganese, mg	55	?	33
Iron, mg	*60*	?	?
Copper, mg	*6*	?	?
Zinc, mg	*70*	?	?

[a] These figures are estimates of requirements and include no margins of safety. (See text, page 2–5.) Italicized figures are tentative.
[b] The protein content of rations for growing turkeys from 16 weeks to market weight may be reduced to 16%.
[c] May be vitamin A or pro-vitamin A.
[d] This amount of calcium need not be incorporated in the mixed feed, inasmuch as calcium supplements fed free-choice are considered as part of the ration.
[e] At least 0.5% of the total feed of starting poults should be inorganic phosphorus. All the phosphorus of non-plant-feed ingredients is considered to be inorganic. Approximately 30% of the phosphorus of plant products is non-phytin phosphorus and may be considered as part of the inorganic phosphorus required. Presumably a portion of the requirement of growing and breeding turkeys must also be furnished in inorganic form.
[f] Equivalent to 0.39% of sodium chloride.

TABLE 21-16 *Essential Amino Acid Requirements of Chickens and Turkeys*[a]
(*Based on Air-Dry Feed Containing 90 Per Cent Dry Matter*)

Amino acid	Starting chicks percentage of diet	Starting poults percentage of diet	Laying chickens percentage of diet
Arginine	1.2	1.6	0.8
Lysine	1.1	1.5	0.5
Histidine	0.4	?	?
Methionine	0.75	0.87	0.53
or			
{ Methionine	0.4	0.52	0.28
{ Cystine	0.35	0.35	0.25
Tryptophan	0.2	0.26	0.15
Glycine[b]	1.0	1.0	?
Phenylalanine	1.3	?	?
or			
{ Phenylalanine	0.7	?	?
{ Tyrosine	0.6	?	?
Leucine	1.4	?	1.2
Isoleucine	0.75	0.84	0.5
Threonine	0.7	?	0.4
Valine	0.85	?	?
Protein	20.0	28.0	15.0
Metabolizable energy, kcal/kg	2,750	2,450	2,850

[a] These figures are estimates of requirements.
[b] The chick can synthesize glycine but the synthesis does not proceed at a rate sufficient for maximum growth.

TABLE 21-17 Nutrient Requirements of Sheep in Percentage or Amount Per Kilogram of Total Ration (Based on Air-Dry Feed Containing 90 Per Cent Dry Matter)

Body weight (kg)	Daily gain or loss (kg)	Daily feed Per animal (kg)	Daily feed Per cent of live weight (%)	TDN (%)	DE (Mcal)	Protein (%)	DP (%)	Percentage of ration or amount per kg of feed Ca (%)	P (%)	Salt (%)	Caro-tene (mg)	Vit. A (µg)[a]	Vit. A (IU)	Vit. D (IU)
Ewes: nonlactating and first 15 weeks of gestation
45.4	0.03	1.18	2.6	50	2.20	8.0	4.4	.27	.21	0.8	1.5	238	794	212
54.4	0.03	1.36	2.5	50	2.20	8.0	4.4	.24	.19	0.7	1.5	242	809	220
63.5	0.03	1.54	2.4	50	2.20	8.0	4.4	.22	.17	0.7	1.5	256	855	227
72.6	0.03	1.72	2.4	50	2.20	8.0	4.4	.20	.16	0.7	1.5	258	862	231
Ewes: last 6 weeks of gestation
45.4	0.17	1.72	3.8	52	2.29	8.4	4.6	.24	.18	0.6	3.3	403	1,344	146
54.4	0.17	1.90	3.5	52	2.29	8.2	4.5	.23	.17	0.6	3.5	428	1,428	156
63.5	0.17	2.09	3.3	52	2.29	8.0	4.4	.22	.16	0.6	3.7	454	1,514	168
72.6	0.17	2.18	3.0	52	2.29	7.8	4.3	.22	.16	0.6	4.0	503	1,671	183
Ewes: first 8 to 10 weeks of lactation
45.4	−0.04	2.09	4.6	59	2.60	8.7	4.8	.30	.22	0.5	2.9	333	1,111	119
54.4	−0.04	2.27	4.2	58	2.56	8.4	4.6	.28	.21	0.5	3.1	359	1,199	132
63.5	−0.04	2.49	3.9	56	2.47	8.0	4.4	.27	.20	0.5	3.3	379	1,265	141
72.6	−0.04	2.58	3.6	55	2.42	8.0	4.4	.27	.20	0.5	3.5	423	1,406	154

Ewes: last 12 to 14 weeks of lactation

45.4	0.03	1.72	3.8	52	2.29	8.4	4.6	.26	.20	0.6	3.3	403	1,344	146
54.4	0.03	1.90	3.5	52	2.29	8.2	4.5	.25	.19	0.6	3.5	428	1,428	156
63.5	0.03	2.09	3.3	52	2.29	8.0	4.4	.24	.18	0.6	3.7	454	1,514	168
72.6	0.03	2.18	3.0	52	2.29	7.8	4.3	.24	.18	0.6	4.2	503	1,671	183

Ewes: replacement lambs and yearlings

27.2	0.14	1.22	4.5	55	2.42	11.0	6.0	.21	.19	0.6	1.3	187	624	110
36.3	0.09	1.45	4.0	50	2.20	8.7	4.8	.20	.18	0.6	1.5	214	712	137
45.4	0.06	1.54	3.4	50	2.20	7.6	4.2	.20	.18	0.6	1.8	243	816	163
54.4	0.03	1.54	2.8	50	2.20	7.0	3.9	.20	.18	0.7	2.2	298	992	194

Rams: lambs and yearlings

36.3	0.18	1.45	4.0	62	2.73	10.0	5.5	.20	.18	0.6	1.5	214	712	137
45.4	0.14	1.68	3.7	57	2.51	8.6	4.7	.18	.16	0.6	1.8	225	750	150
54.4	0.09	1.90	3.5	50	2.29	7.6	4.2	.17	.15	0.6	1.8	240	802	156
63.5	0.05	2.09	3.3	50	2.29	6.9	3.8	.16	.14	0.5	2.0	258	862	168
72.6	0.05	2.18	3.0	50	2.29	6.6	3.6	.15	.14	0.5	2.0	280	930	183

Lambs: fattening

27.2	0.16	1.22	4.5	55	2.42	12.0	6.6	.23	.21	0.6	0.9	134	450	123
31.8	0.18	1.41	4.4	58	2.56	11.0	6.1	.21	.18	0.6	0.9	141	470	126
36.3	0.20	1.54	4.3	62	2.73	10.7	5.9	.19	.18	0.6	0.9	150	498	130
40.8	0.20	1.68	4.2	62	2.73	9.5	5.3	.18	.16	0.6	0.9	154	507	134
45.4	0.18	1.77	3.9	62	2.73	9.4	5.2	.18	.16	0.6	0.9	159	529	141

a Vitamin A alcohol, 0.3 µg is equivalent to IU of vitamin A activity. If vitamin A acetate is used, the vitamin A alcohol should be multiplied by 1.15 to obtain equivalent vitamin A activity. The comparable multiplier for vitamin A palmitate is 1.83.

TABLE 21-18 *Nutrient Requirements of Pheasants and Quail in Percentage or Amount Per Kilogram of Total Ration*[a] *(Based on Air-Dry Feed Containing 90 Per Cent Dry Matter)*

Nutrient	Starting and growing pheasants	Starting and growing quail	Breeding quail
Total protein, per cent	30[b]	28	?
Vitamins			
Vitamin A activity (IU)[c]	?	*13,000*	?
Vitamin D (ICU)	*1,200*	?	?
Riboflavin, mg	*3.5*	?	?
Niacin, mg	*60*	?	?
Minerals			
Calcium, per cent	?	?	*2.3*
Phosphorus, per cent	*1.0*	?	*1.0*
Sodium, per cent[d]	*0.085*	*0.085*	?
Chlorine, per cent	*0.11*	*0.11*	?
Iodine, mg	*0.30*	*0.30*	?

[a] These figures are estimates of requirements. Italicized figures are tentative.
[b] At energy level of 2,300 kcal of metabolizable energy per kg of feed.
[c] May be vitamin A or pro-vitamin A.
[d] Equivalent to 0.21% of sodium chloride.

TABLE 21-19 *Nutrient Requirements of Growing and Finishing Swine in Percentage or Amount Per Kilogram of Total Ration (Based on Air-Dry Feed Containing 90 Per Cent Dry Matter)*

Nutrients	Growing pigs			Finishing pigs			
				Corn diet		Small grain diet	
	5–10	10–20	20–35	35–60	60–100	35–60	60–100
Live weight range, kg							
Expected daily gain, kg	0.30	0.50	0.60	0.75	0.90	0.70	0.80
Protein and energy							
Crude protein, per cent	22	18	16	14	13	15	14
Digestible energy, kcal	3,500	3,500	3,300	3,300	3,300	3,100	3,100
Inorganic Nutrients							
Calcium, per cent	0.80	0.65	0.65	0.50	0.50	0.50	0.50
Phosphorus, per cent	0.60	0.50	0.50	0.40	0.40	0.40	0.40
Sodium, per cent	—	0.10	0.10	—	—	—	—
Chlorine, per cent	—	0.13	0.13	—	—	—	—
Vitamins							
β-Carotene, mg	4.4	3.5	2.6	2.6	2.6	2.6	2.6
Vitamin A, IU	2,200	1,750	1,300	1,300	1,300	1,300	1,300
Vitamin D, IU	220	200	200	125	125	125	125
Thiamine, mg	1.3	1.1	1.1	1.1	1.1	1.1	1.1
Riboflavin, mg	3.0	3.0	2.6	2.2	2.2	2.2	2.2
Niacin, mg[a]	22.0	18.0	14.0	10.0	10.0	10.0	10.0
Pantothenic acid, mg	13.0	11.0	11.0	11.0	11.0	11.0	11.0
Vitamin B_6, mg	1.5	1.5	1.1	—	—	—	—
Choline, mg	1,100	900	—	—	—	—	—
Vitamin B_{12}, μg	22.0	15.0	11.0	11.0	11.0	11.0	11.0

[a] The niacin requirement assumes that all of the niacin in the cereal grains and their by-products is in a bound form and thus is largely unavailable.

TABLE 21-20 *Nutrient Requirements of Breeding Swine in Percentage
or Amount Per Kilogram of Total Ration
(Based on Air-Dry Feed Containing 90 Per Cent Dry Matter)*

Nutrient	Breeding swine		
	Bred gilts and sows	*Lactating gilts and sows*	*Boars young and adult*
Live weight range, kg	110–160	140–200	110–180
Protein and energy			
Crude protein, per cent	14	15	14
Digestible energy, kcal	3,300	3,300	3,300
Inorganic nutrients			
Calcium, per cent	0.75	0.6	0.75
Phosphorus, per cent	0.50	0.4	0.50
NaCl (salt), per cent	0.5	0.5	0.5
Vitamins			
β-Carotene, mg	8.2	6.6	8.2
Vitamin A, IU	4,100	3,300	4,100
Vitamin D, IU	275	220	275
Thiamine, mg	1.4	1.1	1.4
Riboflavin, mg	4.1	3.3	4.1
Niacin, mg	22.0	17.6	22.0
Pantothenic acid, mg	16.5	13.2	16.5
Vitamin B_{12}, μg	13.8	11.0	13.8

TABLE 21-21 *Essential Amino Acid Requirements of Swine in Percentage or Amount Per Kilogram of Total Ration (Based on Air-Dry Feed Containing 90 Per Cent Dry Matter)*

Variable	Growing pigs		Finishing pig	Bred sows and gilts
Live weight range, kg	5–10	20–35	60–100	100+
Protein, per cent	22	16	13	14
Dig. energy, kcal/kg	3,500	3,300	3,300	3,300
Amino acid				
Arginine	—	0.20[a]	—	—
Histidine	0.27	0.18	—	0.20[a]
Isoleucine	0.76	0.50	0.35	0.43
Leucine	0.90	0.60	—	0.66[a]
Lysine	1.20	0.70	0.50	0.49
Methionine[b]	0.80	0.50	—	0.35
Phenylalanine[c]	—	0.50	—	0.52[a]
Threonine	0.70	0.45	—	0.42
Tryptophan	0.18	0.12	0.09[a]	0.08[a]
Valine	0.65	0.50	—	0.446

[a] This level is adequate but the minimum requirement has not been established.
[b] Cystine can satisfy 40% of the need for methionine.
[c] Tyrosine can satisfy 30% of the need for phenylalanine.

TABLE 21-22 *Trace Minerals for Swine in Percentage or Amount Per Kilogram of Total Ration (Based on Air-Dry Feed Containing 90 Per Cent Dry Matter)*

Mineral element	Requirement (mg/kg diet)	Toxic level (mg/kg diet)
Copper	6[a]	250[b]
Iron	80[a]	5,000
Iodine	0.2	—
Magnesium	400[a]	—
Manganese	12	4,000
Zinc	50[c]	2,000
Selenium	0.1	5

[a] Baby pig requirement.
[b] Toxic symptoms have been obtained on a few occasions.
[c] Higher levels may be needed if excess calcium is fed.

CHAPTER 22

Feedstuffs Control and Legislation

In Canada and in most of the United States, feeds passing through commercial channels are subject to registration and to some measure of legal regulation. Such regulation is for the protection of the consumers and the manufacturers alike. For the consumer it is intended to insure that the product offered is properly labeled (in the broad sense) and that it is wholesome as a food for his livestock. For the honest manufacturer it is an attempt to protect him from unscrupulous competitors who might misrepresent their products to the potential consumer.

Canadian Legislation

In Canada, feeding stuffs control is vested in a branch of the Federal Department of Agriculture, with the Canadian Feed Act being the enabling legislation. Canadian feed legislation did not suddenly appear full-blown. Rather, it developed step by step in response to changing conditions of the industry. Many specific sections of the Act were incorporated at the request of the feed manufacturers themselves, and the more recent modifications were presented to the feed trade for study and suggestion before official adoption. With the increasing use of commercially mixed rations the importance of the Feed Act as a guide has also increased, and this has naturally been paralleled by an increased interest in the legislation both by the trade and by the feeder. It seems appropriate, therefore, to include in a text of this nature a brief discussion of the background of feed control and of typical present-day feed legislation. Since pertinent information is not available elsewhere in as precise form, we shall use as an example the situation in Canada. In general principles and philosophy it will not be unlike that in force elsewhere in America.

The Canadian Act originated as a consequence of the development of the commerce of feeding stuffs, which brought about problems of weed-seed distribution. The seriousness of this problem was evident from an investigation of the weed-seed content of the feedstuffs that were being shipped from the grain-growing areas and from flour-milling centers to the livestock farms, a study which revealed a situation indicated by the figures in Table 22-1.

TABLE 22-1   *Weed Seeds in Commercial Feeding Stuffs*

Feed	Whole weed seeds per pound		
	Maximum	Minimum	Average
Bran, shorts, middlings	4,704	0	246
Crushed grain	2,248	8	677
Meals (various)	18,768	16	1,802
Feed grain (unground)	8,888	908	4,022

The weed seeds included those from about 50 species of plants troublesome on barren lands, and their hidden presence in ground feeds was becoming a menace, introducing noxious weeds on farms previously free of them. The situation resulted in amendments to the Adulteration Act requiring that bran, shorts, middlings, and chop feeds be free of vital seeds of any noxious weeds defined under the Seed Control Act.

Investigations revealed that much of the weed-seed problem arose from the disposition of screenings cleaned from western-grown grains at terminal elevators at the head of the Great Lakes. Elevator screenings have potential feeding value, since they include broken and shrunken wheat, wild buckwheat, wild oats, and some other cultivated coarse grains. Screenings, however, were foul with fine weed seeds, such as lamb's-quarters, stinkweed, mustards, etc., some of which are injurious to livestock. Equipment for the removal of these weed seeds was not generally available at the time.

These investigations revealed another problem. Microscopic examinations revealed that finely ground oat hulls were being added to shorts. This addition resulted in a lighter colored and more floury product, and led farmers to believe they were getting a better-quality feed.

The one most common adulteration, and most often complained about, was the addition to mill feeds of ground grain screenings. This was, in effect, returning to the bran and shorts the screenings cleaned out of the wheat before the flour was milled. Had this practice not included the harmful weed seeds it might not have given rise to so much complaint. The petitions from farmers individually and through their livestock organizations led to new legislation in 1920 to regulate the sale and inspection of feeding stuffs. This Canada Feed Act of 1920 was designed to make it possible for feeders to know what they were buying and at the same time to protect manufacturers and dealers from unscrupulous competition. Certain materials, including hays, straws, and the common whole grains, as well as roots, wet brewers' grains, and other watery materials, which are often perishable, were exempted from the provisions of the law. The materials that are covered are classified into three main groups, chop feeds, flour-mill by-products, and commercial feeding stuffs. These each have different standards of quality and demand different labeling, but all carry a uniform restriction against the presence of mustards, purple cockle, ergotized grains, and other seeds or materials regarded as injurious to the health of livestock or poultry.

The large number and great variety of materials that enter into the composition of mixed feeds makes it impossible to judge quality from simple inspection. The 1920 law required every bag (or parcel) of such feeds to be labeled by the manufacturer with the minimum percentage of protein, and maximum percentage of fat and crude fiber, as well as with the specific name of every ingredient contained in the feed. (This provision has been modified in some respects in the later revision.) The administration of the last part of this requirement made it necessary to establish official names and definitions of feeding stuffs involved, and this part of the Act has proved to be most useful to all who have occasion to use feeds.

Feed Legislation in the United States

In the United States there is an active organization, the Association of American Feed Control Officials (AAFCO), consisting of (1) officers charged with the execution of state and federal laws regulating the sale of commercial feedstuffs, (2) the heads of experiment stations and laboratories responsible for the inspection of such products, and (3) research workers of state or federal agencies engaged in the investigations of feedstuffs. It was organized in 1909. The object of this group is "to promote uniformity in legislation, definitions, and rulings, and the enforcement of laws relating to the manufacture, sale, and distribution of feedstuffs on the continent of North America."

This association issues annually its official publication, in which is given a wealth of information about the control of the commerce of feedstuffs. It includes such items as official regulations concerning naming of ingredients, registration, labeling, weights of packages, and use of preservatives, as well as resolutions expressing the official attitude on a wide variety of matters incidental to feeding stuffs legislation and regulation. There is also a section giving official definitions of feeding stuffs. Another section presents a considerable number of analytical procedures of the Association of Official Agricultural Chemists "to be used as a guide and a source of ready reference in the analysis of feedstuffs." The list is not exhaustive but covers those techniques necessary for the more fundamental examination of feedstuffs. To one wishing an insight into the problems and the workings of the machinery of feeding stuffs control, these publications are invaluable.

Actual feed legislation in the United States is entirely a state responsibility. As a consequence, feed legislation is not the same in all states. One of the objectives of AAFCO has been the unification of feed legislation, and in this connection they have, through a standing committee, designed a model feed bill as a guide to what they believe such legislation should embrace. It is reproduced on the following pages.

UNIFORM STATE FEED BILL

Prepared and approved by
ASSOCIATION OF AMERICAN FEED CONTROL OFFICIALS
and the
AMERICAN FEED MANUFACTURERS ASSOCIATION

NOTE—Although this Bill and the Regulations have not been passed into law in all the states the subject matter covered herein does represent the official policy of this association.

AN ACT

To regulate the distribution of commercial feeds and customer-formula feeds in the State of . . . . BE IT ENACTED by the Legislature of the State of . . . .

Section 1. Title

This Act shall be known as the " . . . . Commercial Feed Law of 19 . . . ."

Section 2. Enforcing Official

This Act shall be administered by the . . . . of the State of . . . . , hereinafter referred to as the " . . . ."

Section 3. Definitions of Words and Terms

When used in this Act:

- (a) The term "person" includes individual, partnership, corporation and association.
- (b) The term "distribute" means to offer for sale, sell or barter, commercial feed or customer-formula feed; or to supply, furnish or otherwise provide commercial feed or customer-formula feed to a contract feeder;

  The term "distributor" means any person who distributes.
- (c) The term "sell" or "sale" includes exchange.
- (d) The term "commercial feed" means all materials which are distributed for use as feed or for mixing in feed, for animals other than man except;
    - (1) Option A—Unmixed seed, whole or processed, made directly from the entire seed. Option B—Unmixed or unprocessed whole seeds.
    - (2) Hay, straw, stover, silage, cobs, husks, and hulls
        - (i) when unground, and
        - (ii) when unmixed with other materials.
    - (3) Individual chemical compounds when not mixed with other materials.
- (e) The term "feed ingredient" means each of the constituent materials making up a commercial feed.
- (f) The term "mineral feed" shall mean a substance or mixture of substances designed or intended to supply primarily mineral elements or inorganic nutrients.
- (g) The term "customer-formula feed" means a mixture of commercial feeds and/or materials each batch of which mixture is mixed according to the specific instructions of the final purchaser, or contract feeder.
- (h) The term "brand name" means any word, name, symbol or device, or any combination thereof, identifying the commercial feed of a distributor and distinguishing it from that of others.
- (i) The term "product name" means the name of the commercial feed which identifies it as to kind, class, or specific use.
- (j) The term "label" means a display of written, printed, or graphic matter upon or affixed to the container in which a commercial feed

is distributed, or on the invoice or delivery slip with which a commercial feed or customer-formula feed is distributed.

(k) The term "ton" means a net weight of two thousand pounds avoirdupois.

(l) The terms "per cent" or "percentage" mean percentage by weight.

(m) The term "official sample" means any sample of feed taken by the . . . . or his agent and designated as "official" by the . . . .

(n) The term "contract feeder" means a person who, as an independent contractor, feeds commercial feed to animals pursuant to a contract whereby such commercial feed is supplied, furnished, or otherwise provided to such person and whereby such person's remuneration is determined all or in part by feed consumption, mortality, profits, or amount or quality of product.

Section 4. Registration

(a) Each commercial feed shall be registered before being distributed in this State; provided, however, that customer-formula feeds are exempt from registration. The application for registration shall be submitted on forms furnished by the . . . . , and, if the . . . . so requests, shall also be accompanied by a label or other printed matter describing the product. Upon approval by the . . . . a copy of the registration shall be furnished to the applicant. All registrations are considered permanent unless new registrations are called for by the . . . . or unless cancelled by the registrant. (Option: All registrations expire on . . . . of each year.) The application shall include the information required by subparagraphs (2), (3), (4), and (5) of paragraph (a) of Section 5. The . . . . may by regulation permit on the registration the alternative listing of ingredients of comparable feeding value, provided that the label for each package shall state the specific ingredients which are in such package.

(b) A distributor shall not be required to register any brand of commercial feed which is already registered under this Act by another person.

(c) Changes in the guarantee of either chemical or ingredient composition of a registered commercial feed may be permitted provided there is satisfactory evidence that such changes would not result in a lowering of the feeding value of the product for the purpose for which designed.

(d) The . . . . is empowered to refuse registration of any application not in compliance with the provisions of this Act and to cancel any registration subsequently found not to be in compliance with any provision of this Act; provided, however, that no registration shall be refused or cancelled until the registrant shall have been given opportunity to be heard before the . . . . and to amend his application in order to comply with the requirements of this Act.

Section 5. Labeling

(a) Any commercial feed distributed in this State shall be accompanied by a legible label bearing the following information:

(1) The net weight.

(2) The product name; and brand name, if any, under which the commercial feed is distributed.

(3) The guaranteed analysis of the commercial feed, listing the minimum percentage of crude protein, minimum percentage of crude fat, and maximum percentage of crude fiber; additional guarantees required to be or intentionally shown, shall appear only in the guaranteed analysis section of the label after the guarantee for maximum crude fiber. For all mineral feeds and for those commercial feeds containing a level of added mineral ingredients established by regulation, the list shall include the following, if added: minimum and maximum percentages of calcium (Ca), minimum percentage of phosphorus (P), minimum percentage of iodine (I), and minimum and maximum percentages of salt (NaCl). Other substances or elements, determinable by laboratory methods, may be guaranteed by permission of the . . . . When any items are guaranteed, they shall be subject to inspection and analysis in accordance with the methods and regulations that may be prescribed by the . . . . The . . . . may be regulation designate certain commercial feeds which need not be labeled to show guarantees for crude protein, crude fat and crude fiber.

(4) The common or usual name of each ingredient used in the manufacture of the commercial feed, except as the . . . . may, by regulation, permit the use of a collective term for a group of ingredients all of which perform the same function. An ingredient statement is not required for single standardized ingredient feeds which are officially defined.

(5) The name and principal address of the person responsible for distributing the commercial feed.

(b) When a commercial feed is distributed in this State in bags or other containers, the label shall be placed on or affixed to the container; when a commercial feed is distributed in bulk, the label shall accompany delivery and be furnished to the purchaser at time of delivery.

(c) A customer-formula feed shall be labeled by invoice. The invoice, which is to accompany delivery and be supplied to the purchaser at the time of delivery, shall bear the following information:

(1) Name and address of the mixer.

(2) Name and address of the purchaser.

(3) Date of sale.

(4) The product name and brand name, if any, and number of pounds of each registered commercial feed used in the mixture and the name and number of pounds of each other feed ingredient added.

(d) If a commercial feed or a customer-formula feed contains a non-nutritive substance which is intended for use in the diagnosis, cure, mitigation, treatment, or prevention of disease or which is intended to affect the structure or any function of the animal body, the........ may require the label to show the amount present, directions for use, and/or warnings against misuse of the feed.

Section 6. Inspection Fees

(a) There shall be paid to the.........for all commercial feeds distributed in this State an inspection fee at the rate of........... cents per ton; provided, however, that customer-formula feeds are hereby exempted if the inspection fee is paid on the commercial feeds which they contain; and provided, further, that distribution of commercial feeds to manufacturers are hereby exempted if the commercial feeds so distributed are used solely in manufacture of feeds which are registered; and provided, further, that any distributor shall pay an annual registration fee of twenty-five dollars ($25.00) for each brand of commercial feed distributed only in individual packages of 10 pounds or less, and the distributor of such brand shall not be required to pay the inspection fee on such packages of the brand so registered. Fees so collected shall constitute a fund for the payment of the costs of inspection, sampling, and analysis, and other expenses necessary for the administration of this Act.

(b) Every person, except as hereinafter provided, who distributes commercial feed in this State shall:

(1) File, not later than the last day of January, April, July, and October of each year, a quarterly statement under oath, setting forth the number of net tons of commercial feeds distributed in this State during the preceding calendar quarter; and upon filing

such statement shall pay the inspection fee at the rate stated in paragraph (a) of this Section. When more than one person is involved in the distribution of a commercial feed, the person who distributes to the consumer is responsible for reporting the tonnage and paying the inspection fee unless the report and payment have been made by a prior distributor of the feed.

(2) Keep such records as may be necessary or required by the . . . . to indicate accurately the tonnage of commercial feed distributed in this State, and the . . . . shall have the right to examine such records to verify statements of tonnage.

Failure to make an accurate statement of tonnage or to pay the inspection fee or comply as provided herein shall constitute sufficient cause for the cancellation of all registrations on file for the distributor.

Section 7. Adulteration

No person shall distribute an adulterated feed. A commercial feed or customer-formula feed shall be deemed to be adulterated:

(a) If any poisonous, deleterious, or non-nutritive ingredient has been found in sufficient amount to render it injurious to health when fed in accordance with directions for use on the label.

(b) If any valuable constituent has been in whole or in part omitted or abstracted therefrom or any less valuable substance substituted therefor.

(c) If its composition or quality falls below or differs from that which it is purported or is represented to possess by its labeling.

(d) If it contains added hulls, screenings, straw, cobs, or other high fiber material unless the name of each such material is stated on the label.

(e) If it contains viable weed seeds in amounts exceeding the limits which the . . . . shall establish by rule or regulation.

Section 8. Misbranding

No person shall distribute misbranded feed. A commercial feed or customer-formula feed shall be deemed to be misbranded:

(a) If its labeling is false or misleading in any particular.

(b) If it is distributed under the name of another feed.

(c) If it is not labeled as required in Section 5 of this Act and in regulations prescribed under this Act.

(d) If it purports to be or is represented as a feed ingredient, or if it purports to contain or is represented as containing a feed ingredient, unless such feed ingredient conforms to the definition of identity, if any, prescribed by regulation of the . . . . ; in the adopting of such regulations the . . . . shall give due regard to commonly accepted definitions and official feed terms such as those issued by the Association of American Feed Control Officials.

(e) If any word, statement, or other information required by or under authority of this Act to appear on the label or labeling is not prominently placed thereon with such conspicuousness (as compared with other words, statements, designs, or devices, in the labeling) and in such terms as to render it likely to be read and understood by the ordinary individual under customary conditions of purchase and use.

Section 9. Inspection, Sampling, Analysis

(a) It shall be the duty of the . . . . , who may act through his authorized agent, to sample, inspect, make analyses of, and test commercial feeds and customer-formula feeds distributed within this State at such time and place to such an extent as he may deem necessary to determine whether such feeds are in compliance with the provisions of this Act. The . . . . individually or through his agent, is authorized to enter upon any public or private premises including any vehicle of transport during regular business hours in order to have access to commercial feeds and customer-formula feeds and to records relating to their distribution.

(b) The methods of sampling and analysis shall be those adopted by the . . . . from sources such as the Journal of the Association of Official Analytical Chemists.

(c) The . . . . , in determining for administrative purposes whether a commercial feed is deficient in any component, shall be guided solely by the official sample as defined in paragraph (m) of Section 3 and obtained and analyzed as provided for in paragraph (b) of Section 9.

(d) When the inspection and analysis of an official sample indicates a commercial feed has been adulterated or misbranded, the results of analysis shall be forwarded by the . . . . to the distributor and the purchaser. Upon request within thirty days the . . . . shall furnish to the distributor a portion of the sample concerned.

Section 10. Rules and Regulations

The . . . . is hereby charged with the enforcement of this Act, and after due publicity as prescribed in the provisions of Chapter . . . . . . of the State Law is empowered to promulgate and adopt such reasonable rules and regulations as may be necessary in order to secure the efficient administration of this Act. Publicity concerning the proposed regulations shall be reasonably calculated to give interested parties adequate notice and adequate opportunity to be heard. A public hearing shall be provided when requested in writing from any interested person.

Section 11. Detained Commercial Feeds

(a) "Withdrawal from distribution" orders. When the . . . . or his authorized agent has reasonable cause to believe any lot of commercial feed is being distributed in violation of any of the provisions of this Act or of any of the prescribed regulations under this Act, he may issue and enforce a written or printed "withdrawal from distribution" order, warning the distributor not to dispose of the lot of feed in any manner until written permission is given by the . . . . or the Court. The . . . . . . shall release the lot of commercial feed so withdrawn when said provisions and regulations have been complied with. If compliance is not obtained within 30 days, the . . . . may begin, or upon request of the distributor shall begin, proceedings for condemnation.

(b) Condemnation and Confiscation. Any lot of commercial feed not in compliance with said provisions and regulations shall be subject to seizure on complaint of the . . . . to a court of competent jurisdiction in the area in which said commercial feed is located. In the event the court finds the said commercial feed to be in violation of this Act and orders the condemnation of said commercial feed, it shall be disposed of in any manner consistent with the quality of the commercial feed and the laws of the State: Provided, that in no instance shall the disposition of said commercial feed be ordered by the court without first giving the claimant an opportunity to apply to the court for release of said commercial feed or for permission to process or re-label said commercial feed to bring it into compliance with this Act.

Section 12. Penalties

(a) Any person convicted of violating any of the provisions of this Act or the rules and regulations issued thereunder or who shall impede, obstruct, hinder, or otherwise prevent or attempt to prevent said . . . . or his duly authorized agent in performance of his duty in connection with the provisions of this Act, shall be adjudged guilty of a misdemeanor and shall be fined not less than . . . . or more than . . . . for the first violation, and not less than . . . . or more than . . . . for a subsequent violation. In all prosecutions under this Act involving the composition of a lot of commercial feed, a certified copy of the official analysis signed by the . . . . shall be accepted as prima facie evidence of the composition.

(b) Nothing in this Act shall be construed as requiring the . . . . or his representative to report for prosecution or for the institution of seizure proceedings as a result of minor violations of the Act when he believes that the public interest will be best served by a suitable notice of warning in writing.

(c) It shall be the duty of each . . . . attorney to whom any violation is reported to cause appropriate proceedings to be instituted and prosecuted in a court of competent jurisdiction without delay. Before the . . . . reports a violation for such prosecution, an opportunity shall be given the distributor to present his view to the . . . .

(d) The . . . . is hereby authorized to apply for and the court to grant a temporary or permanent injunction restraining any person from violating or continuing to violate any of the provisions of this Act or any rule or regulation promulgated under the Act notwithstanding the existence of other remedies at law. Said injunction to be issued without bond.

(e) Any person adversely affected by an act, order, or ruling made pursuant to the provisions of this Act may within 45 days thereafter bring action in the (here name the particular Court in the County where the enforcement official has his office) for new trial of the issues bearing upon such act, order, or ruling, and upon such trial the Court may issue and enforce such orders, judgments, or decrees as the Court may deem proper, just, and equitable.

Section 13. Publications

The . . . . shall publish at least annually, in such forms as he may deem proper, information concerning the sales of commercial feeds, together with such data on their production and use as he may consider advisable, and a report of the results of the analyses of official samples of commercial feeds sold within the State as compared with the analyses guaranteed in the registration and on the label; provided, however, that the information concerning production and use of commercial feeds shall not disclose the operations of any person.

Section 14. Constitutionality

If any clause, sentence, paragraph, or part of this Act shall for any reason be judged invalid by any court of competent jurisdiction, such judgment shall not affect, impair, or invalidate the remainder thereof but shall be confined in its operation to the clause, sentence, paragraph, or part thereof directly involved in the controversy in which such judgment shall have been rendered.

Section 15. Repeal

All laws and parts of laws in conflict with or inconsistent with the provisions of this Act are hereby repealed. (The specific statute and specific code sections to be repealed may have to be stated.)

Section 16. Effective Date

This Act shall take effect and be in force from and after the first day of . . . .

RULES AND REGULATIONS
under the
UNIFORM STATE FEED BILL
by the
. . . . of the State of . . . .

Pursuant to due publication and public hearing required by the provisions of Chapter . . . . of the State laws, the . . . . has adopted the following Rules and Regulations.

1. Brand and Production Names

(a) The brand or product name must not be misleading. If the name indicates the feed is made for a specific use, the character of the feed must conform therewith. A mixture labeled "dairy feed," for example, must be suitable for that purpose.

(b) Single ingredient feeds shall have a product name in accordance with the designated definitions of feed ingredients as recognized by the Association of American Feed Control Officials unless the . . . . designates otherwise.

(c) A name of a commercial feed shall not be derived from one or more ingredients of a mixture to the exclusion of other ingredients and shall not be one representing any component of a mixture unless all components are included in the name. Provided, however, that if any ingredient or combination of ingredients is intended to impart a distinctive characteristic to the product which is of significance to the purchaser, the name of that ingredient or combination of ingredients may be used as a part of the brand name or product name if, in the opinion of the control official, the ingredient or combination of ingredients is present in sufficient quantity to impart a distinctive characteristic to the product, that it does not constitute a representation that the ingredient or combination of ingredients is present to the exclusion of other ingredients, and that is not otherwise false or misleading.

(d) The word vitamin, or a contraction thereof, or any word suggesting vitamin can be used only in the name of a feed which is represented to be a vitamin supplement, and which is labeled with the minimum content of each vitamin declared, as specified in Regulation 2(c).

(e) The term "mineralized" shall not be used in the name of a feed except "Trace Mineralized Salt". When so used, the product must contain significant amounts of trace minerals which are recognized as essential for animal nutrition.

(f) The term "meat" and "meat by-products" when applied to the corresponding portions of the animals other than cattle, swine, sheep, and goats shall be used in qualified form, as, for example, "horse meat by-products," "reindeer meat by-products," etc.

2. Expression of Guarantees

(a) The sliding-scale method of expressing guarantees (for example, "Protein 15–18%") is prohibited, except as specifically provided by the law or by regulation.

(b) Drugs in commercial feeds shall be guaranteed in terms of percentage by weight, except that antibiotics present at less than 2000 grams (total) of antibiotics per ton of feed shall be guaranteed in terms of grams per ton of feed or when present at 2000 grams per ton or more they shall be guaranteed in terms of grams per pound of feed. The term "milligrams per pound" may be used, in lieu of percentage declarations for drugs, including antibiotics, on the label only in those cases where a dosage is given as "milligrams per day" in the feeding directions.

(c) Guarantees of minimum vitamin content of feeds and feed supplements shall be stated in units or milligrams per pound as provided herein: vitamin E in USP or International Units, vitamin A, other than precursors of vitamin A, in USP Units, vitamin D in products offered for poultry feeding in International Chick Units, vitamin D for other uses in USP Units, all other vitamins as true vitamins, not compounds, excepting only pyridoxine hydrochloride, choline chloride, and thiamine; oils and concentrates containing vitamin A or vitamin

D or both may be additionally labeled to show vitamin content in units per gram; and providing that the term "d-pantothenic acid" be used in stating the pantothenic acid guarantee.

(d) Pursuant to Section 5(a)(3) of the law, commercial feeds containing 5% or more mineral ingredients, shall include in the guaranteed analysis the minimum and maximum percentages of calcium (Ca) and salt (NaCl) and the minimum percentages of phosphorus (P) and iodine (I), if added. Minerals, except salt (NaCl), when quantitatively guaranteed, shall be stated in terms of percentage of the element.

(e) Products which need not be labeled to show guarantees for crude protein, crude fat, and crude fiber are:
1. Products distributed solely as mineral and/or vitamin supplements.
2. Molasses.
3. Drug compounds.

3. Definitions, Sampling and Analysis

Except as the . . . . designates otherwise in specific cases, the names and definitions for commercial feeds shall be those adopted by the Association of American Feed Control Officials, and the methods of sampling and analysis shall be the official methods of the Association of Official Analytical Chemists.

4. Ingredient Statement

(a) Each ingredient must be specifically named. The names and definitions adopted by the Association of American Feed Control Officials are to be used as the common or usual names unless the . . . . designates otherwise.

(b) When water is added in the preparation of canned foods for animals, water must be listed as an ingredient.

(c) The term "dehydrated" may precede the name of any product that has been artificially dried.

(d) No reference to quality or grade of an ingredient shall appear in the ingredient statement of a feed.

(e) Pursuant to Section 4(a) of the law, alternative listing of ingredients within the following groups may be shown on the registration certificate:
(1) Corn, hominy feed, wheat, barley, and grain sorghums (as presently accepted).
(2) Cottonseed meal, soybean meal, peanut meal, linseed meal, and corn gluten meal.

(3) Fish meal, meat and bone meal, tankage, and poultry by-product meal.

(4) Beet molasses, corn sugar molasses, citrus molasses, and cane molasses.

(5) Wheat bran, wheat mill run, and wheat middlings.

(6) Wheat shorts, wheat red dog, corn germ meal, corn gluten feed, and grain sorghum gluten feed.

5. Labeling

(a) The information required in Section 5(a) of the law must appear in its entirety on one side of a label or on one side of the container; this information shall not be subordinated or obscured by other statements and designs.

(b) The names of all ingredients must be shown in letters or type of the same size.

6. Minerals

(a) When the word "iodized" is used in connection with a feed ingredient, the ingredient shall not contain less than 0.007% iodine, uniformly distributed.

(b) Mineral phosphatic materials for feeding purposes shall be labeled with a guarantee for the minimum percentages of calcium and phosphorus, and the maximum percentage of fluorine.

(c) The fluorine content of any mineral or mineral mixture which is to be used directly for the feeding of domestic animals shall not exceed 0.30% for cattle; 0.35% for sheep; 0.45% for swine; and 0.60% for poultry.

Soft rock phosphate, rock phosphates, or other fluorine-bearing ingredients may be used only in such amounts that they will not raise the fluorine concentration of the total (grain) ration above the following amounts: 0.009% for cattle; 0.01% for sheep; 0.014% for swine; and 0.035% for poultry.

7. Non-Protein Nitrogen

Urea and the ammonium compounds and ammoniated products defined in this Official Publication are acceptable ingredients in commercial feeds for ruminant animals and shall not be used in commercial feeds for other animals and birds as sources of equivalent crude protein.

The maximum percentage of equivalent crude protein from added non-protein nitrogen shall appear immediately below the guarantee for total crude protein.

If the commercial feed contains more than 8.75% of equivalent crude protein from all forms of non-protein nitrogen, added as such, or if the equivalent crude protein from all forms of non-protein nitrogen, added as such, exceeds one third of the total crude protein, the label shall bear adequate directions for the safe use of such feeds and the statement:

"CAUTION: Use as directed."

The directions for use and the caution statement shall be in type of such size so that they will be read and understood by ordinary persons under customary conditions of purchase and use.

On labels such as those for medicated feeds which bear adequate feeding directions and/or warning statements, the presence of added non-protein nitrogen products shall not require a duplication of feeding directions or the warning statements.

8. Artificial Color

An artificial color may be used in feeds only if it has been shown to be harmless to animals. No material shall be used to enhance the natural color of a feed or feed ingredient whereby inferiority would be concealed.

9. Drugs, etc.

Before a registration is accepted for a commercial feed which contains drugs or other ingredients which are potentially harmful to animals, the distributor may be required: (1) to submit evidence to show the safety of the feed when used according to the directions which the distributor furnishes with the feed; (2) to furnish a written statement that adequate written or printed warnings and feeding directions will accompany each delivery of the feed; (3) to state the percentage of the drug, or other ingredients in a prominent place on the label of the feed.

APPENDICES

The items we have included in the three Appendices are intended chiefly as reference material to furnish more detailed information in certain areas than seemed practical or necessary in the text proper.

In Appendix 1, we have given a table to be used to calculate metabolic size for a given live body weight. Also included are conversion tables to convert data from the United States system of measurements to the metric system, or from the metric system to the United States system.

Appendix 2 contains three glossaries: (1) the abbreviations of terms used in the NRC names of feeds; (2) the definitions of the part and process terms used in the NRC feed names; and (3) the terms for stages of maturity of forages.

Appendix 3 is a table of feed composition. The choice, from the 5,500 feeds known, of the 1,000 or so feeds listed in it was made by integrating the lists of feeds requested by each of the NRC Feeding Standard Committees for inclusion in the published standards for the species of animals for which they were responsible. This composite list was then augmented by a few other feeds to complete certain series, such as maturities, grades, or processes. Some duplication was caused by also including some feeds of Canadian origin. The format of the table and the use in it of the recently adopted NRC feed nomenclature are innovations which will enhance its usefulness both for feed manufacturers and for students of applied animal nutrition. It is preceded by detailed notes explaining its use.

Tables for Metabolic Size and Numerical Conversion

TABLE A-1-1 *Metabolic Size for Live Body Weight* ($W_{kg}{}^{.75}$)

W	$W^{.75}$	W	$W^{.75}$	W	$W^{.75}$
0.5	0.60	130	38.50	480	102.55
1.0	1.00	140	40.70	490	104.15
1.5	1.36	150	42.86	500	105.74
2.0	1.68	160	44.99	510	107.32
2.5	1.99	170	47.08	520	108.89
3.0	2.28	180	49.14	530	110.47
3.5	2.56	190	51.17	540	112.02
4.0	2.83	200	53.18	550	113.57
4.5	3.09	210	55.16	560	115.12
5.0	3.34	220	57.12	570	116.65
5.5	3.59	230	59.06	580	118.19
6.0	3.83	240	60.98	590	119.71
6.5	4.07	250	62.87	600	121.23
7.0	4.30	260	64.75	620	124.2
7.5	4.53	270	66.61	640	127.2
8.0	4.76	280	68.45	660	130.2
8.5	4.98	290	70.28	680	133.2
9.0	5.20	300	72.08	700	136.1
9.5	5.41	310	73.88	720	139.0
10.0	5.62	320	75.66	740	141.9
15.0	7.62	330	77.42	760	144.7
20.0	9.46	340	79.18	780	147.6
25.0	11.18	350	80.92	800	150.4
30.0	12.82	360	82.65	820	153.2
35.0	14.39	370	84.36	840	156.0
40.0	15.91	380	86.07	860	158.8
45.0	17.37	390	87.76	880	161.6
50.0	18.80	400	89.44	900	164.3
60.0	21.56	410	91.11	920	167.0
70.0	24.20	420	92.78	940	169.8
80.0	26.75	430	94.43	960	172.5
90.0	29.22	440	96.07	980	175.2
100	31.62	450	97.70	1,000	177.8
110	33.97	460	99.33		
120	36.26	470	100.94		

TABLE A-1-2 *Weight Unit Conversion Factors*

Units given	Units wanted	For conversion multiply by	Units given	Units wanted	For conversion multiply by
lb	g	453.6	μg/kg	μg/lb	0.4536
lb	kg	0.4536	kcal/kg	kcal/lb	0.4536
oz	g	28.35	kcal/lb	kcal/kg	2.2046
kg	lb	2.2046	ppm	μg/g	1.
kg	mg	1,000,000.	ppm	mg/kg	1.
kg	g	1,000.	ppm	mg/lb	0.4536
g	mg	1,000.	mg/kg	%	0.0001
g	μg	1,000,000.	ppm	%	0.0001
mg	μg	1,000.	mg/g	%	0.1
mg/g	mg/lb	453.6	g/kg	%	0.1
mg/kg	mg/lb	0.4536			

TABLE A-1-3 *British or United States System, Length. This system is used in the United States and most of the British Commonwealth Countries.*

Inches (in)	Feet (ft)	Yards (yd)	Rods (rd)	Miles (mi)	Metric equivalent
1	0.08333	0.2778	0.005051	0.00001578	2.5400 cm
12	1	0.3333	0.06061	0.0001894	0.3048 m
36	3	1	0.1818	0.0005682	0.9144 m
198	16.5	5.5	1	0.003125	5.0292 m
63,360	5,280	1,760	320	1	1.6094 km

TABLE A-1-4 *Metric System, Length*

Millimeters (mm)	Centimeters (cm)	Decimeters (dm)	Meters (m)	British or U.S. equivalent
1	0.1	0.01	0.001	0.03937 inch
10	1	0.1	0.01	0.3937 inch
100	10	1	0.1	3.9370 inches
100	10	1	0.1	0.3281 foot
1,000	100	10	1	39.370 inches
1,000	100	10	1	3.2808 feet

ABLE A-1-5 *British or United States System, Area. This system is used in the United ates and most of the British Commonwealth Countries.*

inches q in)	Sq feet (sq ft)	Sq yards (sq yd)	Sq rods (sq rd)	Acres (A)	Sq miles (sq mi)	Metric equivalent
1	0.006944	—	—	—	—	6.4516 sq cm
144	1	0.1111	—	—	—	0.0929 sq m
,296	9	1	0.0331	—	—	0.8361 sq m
	272.25	30.25	1	0.00625	—	25.2930 sq m
	43,560	4,840	160	1	0.001563	40.4687 sq dkm
	27,878,400	3,097,600	102,400	640	1	2.5900 sq km

TABLE A-1-6 *Metric System, Area*

Sq meters or centares (m², ca)	Sq dekameters or ares (dkm², a)	Sq hectometers or hectares (hm², ha)	Sq kilometers (km²)	British or U.S. equivalent
1	0.01	0.0001	0.000001	0.3954 sq rod
100	1	0.01	0.0001	0.02471 acre
10,000	100	1	0.01	2.4710 acres
1,000,000	10,000	100	1	0.3861 sq mile

TABLE A-1-7 *British or United States System, Capacity (Liquid Measure)*

Gills (gi)	Pints (pt)	Quarts (qt)	Gallons (gal)	Metric equivalent	
				U.S.	British
1	0.25	0.125	0.03125	118.292 ml	142.06 ml
4	1	0.5	0.125	0.4732 l	0.5682 l
8	2	1	0.25	0.9463 l	1.1365 l
32	8	4	1	3.7853 l	4.5460 l

TABLE A-1-8 *Metric System, Capacity. One liter is the volume of pure water at 4° C and 760 mm pressure which weighs 1 kilogram. 1 liter = 1.000027 cubic decimeter = 1000.027 cubic centimeters.*

Milliliters (ml)	Centiliters (cl)	Deciliters (dl)	Liters (l)	U.S. equivalent
1	0.1	0.01	0.001	16.2311 minims
10	1	0.1	0.01	2.7052 fl drams
100	10	1	0.1	3.3815 fl ounces
1,000	100	10	1	270.518 fl drams 33.815 fl ounces

Liters (l)	Dekaliters (dkl)	Hectoliters (hl)	Kiloliters (kl)	U.S. equivalent
1	0.1	0.01	0.001	1.05671 liq quarts 0.264178 gallon 1.81620 dry pints 0.908102 dry quart
10	1	0.1	0.01	18.1620 dry pints 9.08102 dry quarts 1.13513 pecks
100	10	1	0.1	2.83782 bushels
1,000	100	10	1	(no equivalent)

The Chemical and Biological Composition
of Feedstuffs

The analytical data of a feed is tabulated only under its National Research Council (NRC) name (see Chapter 1). The name is designed to give—to the extent that the information is available or applicable—a qualitative description of the product as to: origin or parent material; species, variety, or kind; the part actually eaten; the process(es) and treatment(s) to which it has been subjected; the stage of maturity (primarily for forages or roughages); cutting or crop (forages or roughages); grade or quality designations; and class.

In the tabulation, feeds of the same origin—and of the same species, variety, or kind, if designated—have been subgrouped into eight feed classes: (1) dry forages and roughages, (2) pasture, range plants, and forages fed green, (3) silages, (4) energy feeds, (5) protein supplements, (6) minerals, (7) vitamins, and (8) additives. The class to which a feed belongs is indicated in the NRC name by including in parentheses the class number as its final term. Within each origin (and species, variety, or kind, if given) there may be feeds that belong to several classes. These classes are grouped in ascending numerical order of their class code number within origin or within origin and species. The scientific name precedes each group with the same scientific name.

In the NRC nomenclature, feeds which in the dry state contain more than 18 per cent crude fiber are classified as forages and roughages. Products that contain 20 per cent or more of protein are classified as protein supplements. Products with less than these two minima are classified as energy feeds. (These guide lines are approximate and there is some overlapping.)

In order that the names may be used in the limited space available in printed tables or on punch-cards, standard abbreviations have been devised to cover most of the terms included in the eight components of the name.

GLOSSARY A-2-1 *Abbreviations*

AAFCO	Association of American Feed Control Officials	F	fluorine
acid-precip	acid precipitated	fbr	fiber
antox	antioxidant	ferm	fermentation, fermented
As	arsenic	f-grnd	fine ground
asp	aspiration	fm	foreign material
blm	bloom	f-scr	fine screened
by-prod	by-product	f-sift	fine sifted
c	coarse	g	gram
Ca	calcium	gel	gelatin
Can	Canadian	gland	glandular
c-bolt	coarse bolted	gr	grade
CE	Canadian Eastern	grav-sep	gravity separation
CFA	Canada Feeds Act	grnd	ground
CGA	Canada Grain Act	heat-proc	heat processed
c-grnd	coarse ground	heat-rend	heat rendered
CHO	carbohydrate	homog	homogenized
chop	chopped	hydro	hydrolyzed
c-scr	coarse screened	I	iodine
c-sift	coarse sifted	ICU	International Chick Unit
coag	coagulated	insol	insoluble
comm	commercial	IU	international units
cond	condensed	K	potassium
CP	chemically pure	kcal	kilocalorie
crumbl	crumbled	kg	kilogram
cult	cultured	lb	pound
CW	Canadian Western	m	million
dehy	dehydrated	mech	mechanical
dig	digestibility, digestible	mech-expr	mechanically expressed
distil	distillation, distillers	mech-extd	mechanically extracted, expeller extracted, hydraulic extracted, or old process
dl	dextro levo		
dry-mil	dry milled		
dry-rend	dry rendered	μg	microgram
equiv	equivalent	mg	milligram
evap	evaporated	mil-rn	mill run
exprd	expressed	mm	millimeter
extd	extracted	mn	minimum
extn	extraction	m-sacch	monosaccharide(s)
extn-unspec	extraction unspecified	mx	maximum
extru	extruded	N	nitrogen
f	fine	Na	sodium

No	Northern	sol	solubles
NRC	National Research Council	solv-extd	solvent extracted
orig	original	sp	species
P	phosphorus	substand	substandard
Pb	lead	supl	supplement
pltd	pelleted	tech	technical
ppm	parts per million	trim	trimmings
precip	precipitated	unscr	unscreened
pres	preservative	unspec	unspecified
proc	processed, processing	US	United States
prod	product	USP	United States Pharmacopeia
prot	protein	vac	vacuum
pt	part(s)	vac-dehy	vacuum dehydrated
qual	quality	veg	vegetable
rend	rendered	vit	vitamin
res	residue	w	with
s-c	sun cured	wet-mil	wet milled
scr	screened	wet-rend	wet rendered
sep	separation	wo	without
shred	shredded	wt	weight
sift	sifted		
skim	skimmed		

GLOSSARY A-2-2 *Definitions of Parts and Processes Used in the NRC Feed Names*

Term used	Definitions
Aerial parts	The above-ground parts of a plant.
Air-ashed	Reduced by combustion in air to a mineral residue.
Ammoniated	Combined or impregnated with ammonia or an ammonium compound.
Bolls	The pods or capsules of certain plants, especially flax or cotton.
Bone glue residue	Part of bone remaining after removal of that portion used in the manufacture of bone glue.
Bran	Pericarp of grain.
Brewers' grains	The coarse, insoluble residue from brewed malt.
Browse	Small stems, leaves and/or flowers and fruits of shrubs, trees, or woody vines.
Buds	Small lateral or terminal protuberances on the stems of a plant consisting of an undeveloped shoot made up of rudimentary foliage or floral leaves or both, overarching a growing point.
Bulbs	Masses of overlapping membranous or fleshy leaves on a short stem-base enclosing one or more buds that may develop into a new plant.
Buttermilk	Liquid residue from churning cream.
Cannery residue	Residue from a canning process suitable for feeding.

GLOSSARY A-2-2 (*continued*)

Term used	Definitions
Carcass residue	Residues from animal tissues exclusive of hair, hoofs, horns, and contents of the digestive tract.
Casein	The precipitate resulting from treatment of skim milk with acid or rennet.
Cereal by-product	Product produced in addition to the principal product during the manufacture of table cereal.
Chaff	Glumes, husks, or other seed covering, together with other plant parts separated from seed in threshing or processing.
Charcoal	Dark-colored porous form of carbon made from the organic parts of vegetable or animal substances by their incomplete combustion.
Cheese trimmings	Extraneous matter, such as the rind, removed from cheese.
Chopped	Reduced in particle size by cutting with knives or other sharp instruments.
Coarse-bolted	Separated from its parent material by means of a coarsely woven bolting cloth.
Coarse-sifted	Passed through coarsely woven wire sieves for the separation of particles of different sizes.
Cobs	The fibrous inner portion of the ear of maize from which the kernels have been removed.
Cob fractions	Cob butts and beeswings.
Condensed	Reduced to denser form by removal of moisture.
Cooked	Heated in the presence of moisture to alter chemical or physical characteristics or to sterilize.
Cracked	Particle size reduced by a combined breaking and crushing action.
Cracklings	Residue after removal of fat from adipose tissue or skin of animals by dry heat.
Crown	The head of foliage of a shrub; the region in seed plants, usually at ground level, at which stems and roots merge.
D-activated	Plant or animal sterol fractions which have been vitamin D activated by ultraviolet light or by other means.
Debittered	The condition following removal of bitter substances.
Dehydrated	Having been freed of moisture by heat.
Deribbed	Having had the primary veins of the leaves removed.
Distillation grains	Grains from which alcohol or alcoholic beverages have been distilled.
Distillation residue	*See* Stillage.
Distillation solubles	Stillage filtrate.
Distillation stillage	*See* Stillage.
Dry-milled	Ground into flour, meal, or powder in the absence of moisture.
Dry-rendered	Residues of animal tissues cooked in open steam-jacketed vessels until the water has evaporated. Fat is removed by draining and pressing the solid residue.
Ears	Entire fruiting heads of *Zea mays*, including only the cob and grain.
Elevator chaff and dust	Fine particles that accumulate in cyclones above elevators that store grain.

GLOSSARY A-2-2 (*continued*)

Term used	Definitions
Endosperm	Starchy portion of seed.
Endosperm oil	Oil obtained from the endosperm of the seed.
Ensiled	Aerial parts of plants which have been preserved by putting into a container or pit or trench in the ground. Normally the original material is finely cut and blown into an air-tight chamber, as a silo, where it is compressed to exclude air and where it undergoes an acid fermentation that retards spoilage.
Entire plant	The whole or complete plant. This term has been used with water plants such as seaweed.
Etiolated	Material grown in the absence of sunlight; blanched; bleached; colorless; pale.
Fan air-dried	Having had moisture removed by means of a device producing an artificial current of air.
Fermentation product	Product formed as a result of an enzymatic transformation of organic substrates.
Fermentation solubles	That portion of the stillage which passes through screens, being composed mostly of water, water-soluble substances, and fine particles from the fermentation process.
Fermented	Acted upon by yeasts, molds, or bacteria in a controlled aerobic or anaerobic process in the manufacture of such products as alcohols, acids, vitamins of the B-complex group, or antibiotics.
Fiber	An elongate tapering cell that has at maturity no protoplasm content, is found in many plant organs, is well developed in the xylem and phloem of the vascular system, and imparts elasticity, flexibility, and tensile strength to the plant.
Fiber by-product	A secondary product obtained during the manufacture of a fiber plant product.
Fine-ground	Reduced to very small particles size by impact, shearing, or attrition.
Fines (in crumbles or pellets)	Any material which will pass through a screen whose openings are immediately smaller than the specified minimum crumble size or pellet diameter.
Fine-screened	Particles separated according to size by passage over or through wire screens.
Fine-sifted	Separated by passage through a finely woven meshed material.
Flaked	*See* Rolled.
Flour	Soft, finely ground, and bolted meal obtained from the milling of cereal grains, other seeds, or products. It consists essentially of the starch and gluten of the endosperm.
Flour by-product	A secondary product obtained during the milling of bread flour from grain.
Fresh	Newly produced or gathered; not stored, cured, or preserved.
Fruit	The edible, more or less succulent product of a perennial or woody plant, consisting of the ripened seeds and adjacent or surrounding tissues, or the latter alone.
Gland tissue	An aggregate of cells with their intercellular contents that form secreted substances.

GLOSSARY A-2-2 *(continued)*

Term used	Definitions
Gluten	The tough, viscid, nitrogenous substance remaining when the flour of wheat or other grain is washed to remove the starch.
Graham flour	Whole wheat flour.
Grain	Seed from cereal plants.
Grain clippings	The hulls, fragments or groats, light immature grains, and chaffy material resulting from the dehulling of cereal grains, principally oats.
Grain screenings	The small imperfect grains, weed seeds, and other foreign material having feeding value that is separated in cleaning grain by a screen.
Grits	Coarsely ground grain from which the bran and germ have been removed, usually screened to uniform particle size.
Groats	Grain from which the hulls have been removed.
Hatchery by-product	A mixture of egg shells, infertile and unhatched eggs, and culled chicks, which has been cooked, dried and ground, with or without partial removal of fat.
Hay	The aerial portion of grass or herbage especially cut and cured for animal feeding.
Heat-processed	Subjected to a method of preparation involving the use of elevated temperatures, with or without pressure.
Heat-rendered	Melted, extracted, or clarified through use of heat. Usually water and fat are removed.
Hulls	Outer covering of grain or other seed.
Husks	Leaves enveloping an ear of maize, or the outer coverings of kernels or seeds, especially when dry and membranous.
Joints	Nodes of stems of plants.
Juice	The aqueous substance obtainable from biological tissue by pressing or filtering, with or without addition of water.
Kernel	A whole grain. For other species, dehulled seed.
Kibbled	Cracked or crushed baked dough, or extruded feed that has been cooked prior to or during the extrusion process.
Leaves	Lateral outgrowths of stems that constitute part of the foliage of a plant, typically a flattened green blade, and primarily functions in photosynthesis.
Malt hulls	The product consisting almost entirely of hulls as obtained in the cleaning of malting barley.
Meat	The clean, wholesome flesh derived from slaughtered mammals and limited to skeletal muscle, including that which is found in the tongue, diaphragm, or esophagus, and to the cardiac muscle of the heart. It may be with or without the accompanying and overlaying fat, skin, sinews, nerves, and blood vessels. It does not include tissues found in the lips, snout, or ears.
Meats	The edible parts of some nuts, such as coconut.
Middlings	A by-product of flour milling comprising several grades of granular particles containing different proportions of endosperm, bran, and germ, each of which contains different percentages of crude fiber.

GLOSSARY A-2-2 *(continued)*

Term used	*Definitions*
Milk	Total lacteal secretion from the mammary gland.
Milk albumin	The coagulated protein fraction from whey.
Mill dust	Feed or food particles which escape from mill equipment.
Mill residue	Part remaining after a milling process.
Mill-run	The state in which a material comes from the mill, ungraded and usually uninspected.
Molasses	The thick, viscous by-product resulting from refined sugar production.
Molasses distillation solubles	Liquid containing dissolved substances obtained from stillage made with molasses.
Molasses fermentation solubles	That portion of molasses stillage which passes through screens, being composed mostly of water, water-soluble substances, and fine particles.
Nut	Dry indehiscent fruit from either a simple or compound ovary, and characterized by a hard, bony ovary wall.
Offal	Material left as a by-product from the preparation of some specific product; less valuable portions; the by-products of milling.
Oil refinery liquid	By-product produced from the refining of an edible oil.
Pearl by-product	Additional product obtained during the grinding of barley into small round pellets, known as pearl barley.
Pearled	Dehulled grains reduced by machine brushing into smaller smooth particles.
Pectin	Any of the group of colorless amorphous methylated pectic substances occurring in plant tissues or obtained by restricted treatment of protopectin that are obtained from fruits or succulent vegetables, that yield viscous solutions with water and when combined with acid and sugar yield a gel.
Peelings	Outer layers or coverings which have been removed by stripping or tearing.
Pelleted	Having agglomerated feed by compacting and forcing through die openings.
Pith	Continuous central strand of parenchymatous tissue occurring in the stems of most vascular plants.
Pits	Stones of drupaceous fruits.
Pods	A dehiscent seed vessel or fruit.
Polishings	A by-product of rice consisting of the fine residue resulting from the brushing of the grain to polish the kernel after the hulls and bran have been removed.
Precipitated	Separated from suspension or a solution as a result of some chemical or physical change brought about by a chemical reagent, by cold, or by any other means.
Pressed	Compacted or molded by pressure; having fat, oil, or juice extracted under pressure.
Pulp	The solid residue remaining after extraction of juices from fruits, roots, or stems. Similar terms, bagasse and pomace.
Retort-charred	Material partially burned in a closed retort, as is done in the manufacture of bone black.

GLOSSARY A-2-2 *(continued)*

Term used	Definitions
Rolled	Having changed the shape and/or size of particles by compressing between rollers. It may entail tempering or conditioning.
Roots	Subterranean parts of plants.
Scoured	Cleansed from natural impurities.
Screened	Having separated various sized particles by passing over or through screens.
Screenings	By-product obtained from screening grains and other seeds.
Seedballs	Rounded and usually dry or capsular fruits.
Shells	The hard, fibrous, or calcareous covering of a plant or animal product; i.e., nut, egg, oyster.
Shredded	Cut into long narrow pieces.
Sifted	Materials that have been passed through wire sieves to separate particles of different sizes. The common connotation is the separation of finer material than would be done by screening.
Skimmed	Material from which floating solid material has been removed. It is also applied to milk from which fat has been removed by centrifuging.
Skin	Outer coverings of fruits or seeds, as the rinds, husks, or peels; dermal tissue of animals.
Solubles	Liquid containing dissolved substances obtained from processing animal or plant materials. It may contain some fine suspended solids.
Solvent-extracted	Having removed fat or oil from materials by organic solvents. Similar term, "new process."
Spent	Exhausted of active or effective properties, i.e., absorbing activity.
Spines	Stiff, sharp-pointed processes; a specialized stiff, sharp-pointed leaf form.
Spray-dehydrated	Material which has been dried by spraying onto the surface of a heated drum. It is recovered by scraping from the drum.
Stalk(s)	The main stem of an herbaceous plant; often with its dependent parts, as leaves, twigs, and fruit.
Steffen's filtrate	The filtrate obtained from the precipitation of calcium sucrate in the Steffen process and used chiefly as a source of amino acids.
Stems	The coarse, aerial parts of plants which serve as supporting structures for leaves, buds, flowers, and fruit.
Stickwater solubles	Water-soluble fraction from fish from which the liquid has been removed, originally obtained by steam-cooking and pressing the fish.
Stillage	The mash from fermentation of grains after removal of alcohol by distillation.
Straw	The plant residue remaining after separation of the seeds in threshing. It includes chaff.
Straw pulp	A moist, slightly cohering mass of ground straw.
Sugar	A sweet, crystallizable substance that consists essentially of sucrose (as used in describing feeds), and that

GLOSSARY A-2-2 *(continued)*

Term used	Definitions
	occurs naturally in the most readily available amounts in sugar cane, sugar beet, sugar maple, sorghum, and sugar palm.
Sun-cured	Material dried by exposure in open air to the direct rays of the sun.
Syrup	Concentrated juice of a fruit or plant.
Toasted	Brown, dried, or parched by exposure to a fire, or to gas or electric heat.
Tubers	Short, thickened, fleshy stems or terminal portions of stems or rhizomes that are usually formed underground, bear minute leaves, each with a bud capable under suitable conditions of developing into a new plant, and constitute the resting stage of various plants.
Vacuum-dehydrated	Freed of moisture after removal of surrounding air while in an air-tight enclosure.
Vines	The aerial parts of a plant whose stems require support or lie on the ground.
Wet-milled	Subjected to a milling process while containing moisture.
Whey	The watery part of milk separated from the curd.
Whey low lactose	The product resulting from partial removal of milk sugar from whey.
Wilted	A product without turgor because of water loss.
Yeast fermentation grains	Grains used as the substrate by yeast, providing a source of carbohydrate which is transformed by them into some other product.

GLOSSARY A-2-3 *Terms for Stage of Maturity Used in NRC Feed Names*

Preferred maturity term	Definition	Comparable term
Germinated	Resumption of growth by the embryo in a seed after a period of dormancy.	Sprouted.
Early leaf	Stage at which the plant reaches 1/3 of its growth before blooming.	Fresh new growth, very immature.
Immature	Period between 1/3 and 2/3 of its growth before blooming (this may include fall aftermath).	Pre-bud stage, young before boot, before heading out.
Pre-bloom	Stage including the last third of growth before blooming.	Bud, bud stage, budding plants, in bud, pre-flowering, before bloom, heading to in bloom, boot, heads just showing.
Early bloom	Period between initiation of bloom up to stage at which 1/10 of the plants are in bloom.	Up to 1/10 bloom, initial bloom, headed out, in head.
Mid-bloom	Period during which 1/10 to 2/3 of the plants are in bloom.	Bloom, flowering plants, flowering, half bloom, in bloom.
Full bloom	When 2/3 or more of the plants are in bloom.	3/4 to full bloom.
Late bloom	When blossoms begin to dry and fall.	15 days after silking, before milk, in bloom to early pod.
Milk stage	Period after bloom when the seeds begin to form.	Early seed, post-bloom, soft immature, in milk, in tassel, after bloom, fruiting, pod stage, late bloom to early seed, seed forming, seed developing.
Dough stage	Stage at which the seeds are soft and immature.	Seeds dough, seed well-developed, nearly mature.
Mature	State at which the plant would normally be harvested for seed.	Fruiting plants, fruiting, in seed, well-matured, dough to glazing, kernels ripe.
Overripe	Stage after the plant is mature and the seeds are ripe (applies mostly to range plants).	Late seed, ripe, very mature, well-matured.
Dormant	Plants cured on the stem; applied to most nongrowing range plants after the seeds have been cast (applies mostly to range plants).	Seeds cast, mature and weathered.

Table of Feed Composition

The System of Naming

The system of naming can be illustrated by listing, for a specific feed, the appropriate designation applicable to each of the eight terms of the NRC name.

Name Components	NRC Name
Origin	Alfalfa
Variety	Ranger
Part eaten	aerial pt
Process	dehy
Maturity*	pre-blm
Cutting	cut 1
Grade	mn 17 prot mx 27 fbr
Classification	(1)

The NRC name is normally written in linear form, with the components of the name separated by commas but without other punctuation. The feed listed above would appear as:

Alfalfa, Ranger, aerial pt, dehy, pre-blm, cut 1, mn 17 prot mx 27 fbr, (1)

Locating Name and Data for a Specific Feed

To find in Table A–3–1 the NRC name of a specific feed, one must know its origin or the name of the parent material (and usually also the variety or kind). The first word of the NRC name is invariably the name of the

* See Glossary A-2-3.

parent material of the product fed. This word (origin-term) is, for feeds of vegetable origin, the *name of the plant* rather than the word "plant." For non-vegetable feeds it is the *name of the species of animal or fowl,* Cattle, Horse, Whale, Crab, Chicken, Turkey. "Fish" is the origin term for all fishes, followed by species or variety: Fish, cod; Fish, perch. Where *species or kind is not specified,* the terms "animal," "fish," "plant," or "poultry" serve to indicate the origin.

If the reader is uncertain about the NRC origin-term, he may determine it from any of its common names, since they are included in the primary (origin-term) alphabetical listing. Thus, in the H section one finds:

Hominy feed, *see* Corn, grits by-prod

or under M,

Molasses, *see* Beet, sugar, molasses; *see* Citrus, syrup; *see* Sugarcane, molasses

or under T,

Tankage, *see* Animal, carcass res w blood; *see* Animal–Poultry, carcass res mx 35 blood; *see* Animal–Poultry

Tallow, see Cattle

Feeds listed under the same origin-term, but *without designation of species, variety or kind,* are alphabetized by origin, subgrouped by class within origin, and listed alphabetically within classes by the part eaten. If stage of maturity is included in the name, the listing is in order of increasing maturity.

Feeds of the same origin-term that are *designated by species, variety, and/or kind* are alphabetized by origin and species, subgrouped by class within origin and species, and listed alphabetically within class by part eaten. These feeds are listed following those unspecified as to kind. For example:

Corn, ——, aerial part, s-c. (1)
, ——, cobs, grnd, (1)
, ——, aerial part, ensiled, (3)
, ——, aerial part, ensiled, milk stage, (3)
, ——, ears w husks, ensiled, (3)
, ——, ears, grnd, (4)
, ——, grains, flaked, (4)
, ——, distillers grains, dehy, (5)
, ——, gluten, (5)
Corn, dent, grain, (4)
, flint, grain, (4)
, sweet, cannery res, ensiled, (3)
, white, grits by-prod, (4)

Tabulated under each NRC name are: the name assigned to the product by the Association of American Feed Control Officials (AAFCO) and by the Canada Feed Act (CFA), as well as any regional or local names by which it is, or has been, known; these are followed by its six-digit reference number, the first digit of which indicates the class of the feed. Thus one may wish to ascertain the composition of a feed he calls "Linseed oil meal"—a name that he will find in its alphabetical place, with a cross-reference to its NRC name,

> Linseed oil meal, *see* Flax.

Under "Flax" he then finds,

> Flax, seed, mech-extd grnd, mx 0.5 acid insol ash, (5)
> Linseed meal (AAFCO)
> Linseed meal (CFA)
> Linseed oil meal, expeller extracted
> Linseed oil meal, hydraulic extracted
> Linseed meal, old process
> Ref No 5-02-045

> Flax, seed, solv-extd grnd, mx 0.5 acid insol ash, (5)
> Solvent extracted linseed meal (AAFCO)
> Linseed oil meal, solvent extracted
> Ref No 5-02-048

If the NRC name of a listed feed omits one or more of the descriptive terms (such as variety, process, maturity, cutting, grade) the analytical data represent the average of available analyses undifferentiated with respect to the missing terms, usually referred to as "all analyses." Thus the name "Corn, grain, (4)" includes all varieties, and grades of corn. The name "Clover, red, aerial parts, ensiled, (3)" includes all maturities and cuttings, and indicates that maturity or cutting was not specified by the analysts reporting the data. The variability among the samples from which the average composition of such feeds is determined will tend to be greater than among samples of feed that have been more precisely defined. In general, the more complete the name is—with respect to the eight component parts—the less variable will be the average compositional data. Consequently, the average composition of the more completely described feeds will tend better to indicate the probable composition of a particular sample of that feed that may be included in a diet or feed-mixture formulation.

If the variability in average composition of a given feed is high, the probability that a sample of a feed of that name will have a composition similar to that listed in the table is relatively low, even though the average value reported for a particular nutrient that it contains may be exactly the same as that reported for a more precisely described feed.

The analytical data are expressed in the metric system (with the exception of the bushel weights of the cereal grains) and are shown on an "as fed" as well as a "dry" basis. The coefficient of variation is shown where there are a sufficient number of analyses to warrant the calculation.

The NRC feed reference number may be used as an identification on electronic computers for linear programing.

It has not been possible to obtain values for the apparent digestible energy (DE) and metabolizable energy (ME) of all feedstuffs. For some of these, values have been estimated, using the following formulae:

$$\text{DE (kcal/kg)} = \frac{\text{TDN \%}}{100} \times 4409$$

$$\text{ME (kcal/kg) for ruminants} = \text{DE (kcal)} \times 0.82$$

$$\text{ME (kcal/kg) for swine} = \text{DE (kcal/kg)} \left[\frac{(96 - 0.202) \times \text{protein \%}}{100} \right].$$

Net energy values for some cattle feeds including net energy for maintenance (NE_m) and net energy for gain (NE_{gain}) have been taken from Lofgreen's[*] data or calculated from the following formulae:

$$\text{Log F} = 2.2577 - 0.2213 \text{ ME}$$

$$NE_m = 77/F$$

$$NE_{gain} = 2.54 - 0.0314 \text{ F}$$

The terms used in the formulae are those defined by Harris[†] and are on a dry matter basis (moisture free).

1. ME is the metabolizable energy in kcal/g of dry matter (DM) (or Mcal/kg DM).
2. F is the grams of dry matter per unit of $W^{.75}$ required to maintain energy equilibrium.
3. NE_m is the net energy for maintenance in kcal/g DM (Mcal/kg DM).
4. NE_{gain} is the net energy for gain in weight in kcal/g DM (Mcal/kg DM).

To convert NE_m and NE_{gain} to the standardized units in Table A–3–1 (kcal/kg) the values were multiplied by 1,000.

If digestion coefficients were not available and could not be reliably esti-

---

[*] G. P. Lofgreen and W. N. Garrett, "A System for Expressing Net Energy Requirements and Feed Values for Growing and Finishing Beef Cattle," *J. Animal Sci.*, XXVII (1968), 793.

[†] L. E. Harris, *Biological Energy Relationships and Glossary of Energy Terms*, (Natl. Acad. Sci., Natl. Res. Council, Pub. 1411). First revised edition, 1966.

mated from data for closely related feeds, digestible protein and total digestible nutrients (TDN) were estimated by use of the regression equations presented by Schneider *et al.**

The International Standards for vitamin A activity based on vitamin A and beta-carotene are as follows: One International Unit (IU) of vitamin A = one USP unit = vitamin A activity of 0.300 μg of crystalline vitamin A alchohol, corresponding to 0.344 μg of vitamin A acetate or 0.550 μg of vitamin A palmitate. Beta-carotene is the standard for pro-vitamin A. One IU of vitamin A activity is equivalent to 0.6 μg of beta-carotene or 1 mg of beta-carotene = 1,667 IU of vitamin A. International Standards for vitamin A are based on the utilization by rats of vitamin A and/or beta-carotene.

The vitamin A equivalent for carotene was calculated by assuming that 0.6 μg of beta-carotene = one IU of vitamin A. Because the various species do not convert carotene to vitamin A in the same ratio as rats, it is suggested that the conversion rates given in Table 13-1 be used.

For more details about the nomenclature and methods of summarization, see Harris *et al.*†

* B. H. Schneider, H. L. Lucas, M. A. Cipolloni, and H. M. Pavlech, "The Prediction of Digestibility for Feeds for Which There Are Only Proximate Data," *J. Animal Sci.,* XI (1952), 77.
† Harris, L. E., J. Malcolm Asplund, and Earle W. Crampton. *Feed Nomenclature and Methods for Summarizing Feed Data.* (Utah Agr. Expt. Sta. Bul. 479, 1968.)

TABLE A-3-1 Composition of Feeds

Feed name or analyses		Mean		C.V.
		As fed	Dry	± %

Active dry yeast - see Yeast, active

ALFALFA. Medicago sativa

Alfalfa, aerial pt, dehy grnd, mn 15 prot, (1)

Ref no 1-00-022

		As fed	Dry	
Dry matter	%	93.1	100.0	
Ash	%	8.4	9.0	
Crude fiber	%	26.4	28.4	
Ether extract	%	2.3	2.5	
N-free extract	%	40.8	43.8	
Protein (N x 6.25)	%	15.2	16.3	
Cattle	dig prot %	11.8	12.7	
Sheep	dig prot %	10.4	11.2	
Swine	dig prot %	7.0	7.5	
Energy				
Cattle	DE kcal/kg	2504.	2690.	
Sheep	DE kcal/kg	2258.	2425.	
Swine	DE kcal/kg	1436.	1543.	
Cattle	ME kcal/kg	2054.	2206.	
Chickens	MEn kcal/kg	661.	710.	
Sheep	ME kcal/kg	1851.	1988.	
Swine	ME kcal/kg	1331.	1430.	
Cattle	TDN %	57.	61.	
Sheep	TDN %	51.	55.	
Swine	TDN %	32.	35.	
Calcium	%	1.23	1.32	
Chlorine	%	.44	.47	
Iron	%	.031	.033	
Magnesium	%	.29	.31	
Phosphorus	%	.22	.24	
Potassium	%	2.33	2.50	
Sodium	%	.07	.08	
Cobalt	mg/kg	.180	.190	
Copper	mg/kg	10.4	11.2	
Iodine	mg/kg	.120	.129	
Manganese	mg/kg	29.0	31.1	
Selenium	mg/kg	.500	.540	
Zinc	mg/kg	20.0	21.5	
Carotene	mg/kg	102.0	109.5	
Choline	mg/kg	1550.	1665.	
Folic acid	mg/kg	1.54	1.65	
Niacin	mg/kg	41.9	45.0	
Pantothenic acid	mg/kg	20.9	22.4	
Riboflavin	mg/kg	10.6	11.4	
Thiamine	mg/kg	3.0	3.2	
a-tocopherol	mg/kg	98.0	105.2	
Vitamin B6	mg/kg	6.50	7.00	
Vitamin K	mg/kg	9.90	10.60	

Continued

Feed name or analyses		Mean		C.V.
		As fed	Dry	± %
Vitamin A equiv	IU/g	170.0	182.5	
Alanine	%	.80	.86	
Arginine	%	.60	.64	
Aspartic acid	%	1.70	1.82	
Glutamic acid	%	1.50	1.61	
Glycine	%	.70	.75	
Histidine	%	.30	.32	
Hydroxyproline	%	.60	.64	
Leucine	%	1.10	1.18	
Lysine	%	.60	.64	
Methionine	%	.20	.21	
Phenylalanine	%	.80	.86	
Proline	%	.80	.86	
Serine	%	.70	.75	
Threonine	%	.60	.64	
Tryptophan	%	.40	.43	
Tyrosine	%	.40	.43	
Valine	%	.70	.75	

Alfalfa, aerial pt, dehy grnd, mn 17 prot, (1)

Ref no 1-00-023

		As fed	Dry	
Dry matter	%	93.0	100.0	
Ash	%	9.0	9.7	
Crude fiber	%	24.3	26.1	
Ether extract	%	3.0	3.2	
N-free extract	%	38.9	41.8	
Protein (N x 6.25)	%	17.9	19.2	
Cattle	dig prot %	14.0	15.0	
Sheep	dig prot %	12.3	13.2	
Swine	dig prot %	8.3	8.9	
Energy				
Cattle	DE kcal/kg	2543.	2734.	
Sheep	DE kcal/kg	2255.	2425.	
Swine	DE kcal/kg	1435.	1543.	
Cattle	ME kcal/kg	2085.	2242.	
Chickens	MEn kcal/kg	1212.	1303.	
Sheep	ME kcal/kg	1849.	1988.	
Swine	ME kcal/kg	1322.	1421.	
Cattle	NEm kcal/kg	1218.	1310.	
Cattle	NEgain kcal/kg	642.	690.	
Cattle	TDN %	58.	62.	
Sheep	TDN %	51.	55.	
Swine	TDN %	32.	35.	
Calcium	%	1.33	1.43	
Chlorine	%	.46	.49	
Iron	%	.046	.049	
Magnesium	%	.29	.31	
Phosphorus	%	.24	.26	
Potassium	%	2.49	2.68	
Sodium	%	.09	.10	
Cobalt	mg/kg	.360	.390	

Continued

(1) dry forages and roughages
(2) pasture, range plants, and forages fed green
(3) silages
(4) energy feeds

(5) protein supplements
(6) minerals
(7) vitamins
(8) additives

Feed name or analyses		Mean		C.V. ± %
		As fed	Dry	
Copper	mg/kg	9.9	10.6	
Iodine	mg/kg	.150	.161	
Manganese	mg/kg	29.0	31.2	
Selenium	mg/kg	.600	.645	
Zinc	mg/kg	16.0	17.2	
Carotene	mg/kg	161.2	173.3	
Choline	mg/kg	1518.	1632.	
Folic acid	mg/kg	2.10	2.26	
Niacin	mg/kg	45.8	49.2	
Pantothenic acid	mg/kg	30.0	32.2	
Riboflavin	mg/kg	12.3	13.2	
Thiamine	mg/kg	3.5	3.8	
a-tocopherol	mg/kg	128.0	137.6	
Vitamin B_6	mg/kg	6.30	6.77	
Vitamin K	mg/kg	8.70	9.35	
Vitamin A equiv	IU/g	268.7	288.9	
Alanine	%	.90	.97	
Arginine	%	.70	.75	
Aspartic acid	%	1.90	2.04	
Glutamic acid	%	1.70	1.83	
Glycine	%	.90	.97	
Histidine	%	.40	.43	
Isoleucine	%	.70	.75	
Leucine	%	1.30	1.40	
Lysine	%	.80	.86	
Methionine	%	.20	.22	
Phenylalanine	%	.80	.86	
Proline	%	.90	.97	
Serine	%	.70	.75	
Threonine	%	.80	.86	
Tryptophan	%	.40	.43	
Tyrosine	%	.50	.54	
Valine	%	.90	.97	

Alfalfa, aerial pt, dehy grnd, mn 20 prot, (1)

Ref no 1-00-024

Dry matter	%	93.1	100.0	
Ash	%	10.3	11.1	
Crude fiber	%	20.2	21.7	
Ether extract	%	3.6	3.9	
N-free extract	%	38.4	41.2	
Protein (N x 6.25)	%	20.6	22.1	
Cattle	dig prot %	16.0	17.2	
Sheep	dig prot %	14.2	15.2	
Swine	dig prot %	12.6	13.5	
Energy				
Cattle	DE kcal/kg	2545.	2734.	
Sheep	DE kcal/kg	2258.	2425.	
Swine	DE kcal/kg	2217.	2381.	
Cattle	ME kcal/kg	2087.	2242.	
Chickens	MEn kcal/kg	1587.	1705.	
Sheep	ME kcal/kg	1851.	1988.	
Swine	ME kcal/kg	2029.	2179.	

Continued

Feed name or analyses		Mean		C.V. ± %
		As fed	Dry	
Cattle	NEm kcal/kg	1238.	1330.	
Cattle	NE gain kcal/kg	680.	730.	
Cattle	TDN %	58.	62.	
Sheep	TDN %	51.	55.	
Swine	TDN %	50.	54.	
Calcium	%	1.52	1.63	
Chlorine	%	.58	.62	
Iron	%	.040	.043	
Magnesium	%	.35	.38	
Phosphorus	%	.27	.29	
Potassium	%	2.52	2.71	
Sodium	%	.86	.92	
Cobalt	mg/kg	.320	.344	
Copper	mg/kg	10.6	11.4	
Iodine	mg/kg	.140	.150	
Manganese	mg/kg	34.0	36.5	
Selenium	mg/kg	.500	.537	
Zinc	mg/kg	18.0	19.3	
Carotene	mg/kg	216.4	232.4	
Choline	mg/kg	1618.	1738.	
Folic acid	mg/kg	2.67	2.87	
Niacin	mg/kg	54.7	58.7	
Pantothenic acid	mg/kg	32.8	35.2	
Riboflavin	mg/kg	15.5	16.6	
Thiamine	mg/kg	3.9	4.2	
a-tocopherol	mg/kg	147.0	157.9	
Vitamin B_6	mg/kg	7.90	8.48	
Vitamin K	mg/kg	14.70	15.79	
Vitamin A equiv	IU/g	360.7	387.4	
Alanine	%	1.10	1.18	
Arginine	%	.90	.97	
Aspartic acid	%	2.10	2.26	
Glutamic acid	%	2.10	2.26	
Glycine	%	1.00	1.07	
Histidine	%	.40	.43	
Isoleucine	%	.80	.86	
Leucine	%	1.50	1.61	
Lysine	%	.90	.97	
Methionine	%	.30	.32	
Phenylalanine	%	1.10	1.18	
Proline	%	1.00	1.07	
Serine	%	.90	.97	
Threonine	%	.90	.97	
Tryptophan	%	.50	.54	
Tyrosine	%	.70	.75	
Valine	%	.10	.11	

Alfalfa, aerial pt, dehy grnd, mn 22 prot, (1)

Ref no 1-07-851

Dry matter	%	92.9	100.0	
Ash	%	10.3	11.1	
Crude fiber	%	18.5	19.9	
Ether extract	%	3.7	4.0	

Continued

(1) dry forages and roughages
(2) pasture, range plants, and forages fed green

(3) silages
(4) energy feeds

Feed name or analyses		Mean		C.V. ± %
		As fed	Dry	
N-free extract	%	37.9	40.8	
Protein (N x 6.25)	%	22.5	24.2	
Swine	dig prot %	13.7	14.8	
Energy				
Swine	DE kcal/kg	2253.	2425.	
Swine	ME kcal/kg	2052.	2209.	
Swine	TDN %	51.	55.	
Calcium	%	1.48	1.59	
Chlorine	%	.52	.56	
Iron	%	.045	.048	
Magnesium	%	.34	.36	
Phosphorus	%	.28	.30	
Potassium	%	2.51	2.70	
Sodium	%	.11	.12	
Cobalt	mg/kg	.300	.323	
Copper	mg/kg	11.1	11.9	
Iodine	mg/kg	.200	.215	
Manganese	mg/kg	37.0	39.8	
Selenium	mg/kg	.540	.581	
Zinc	mg/kg	20.0	21.5	
Carotene	mg/kg	252.5	271.7	
Choline	mg/kg	1853.	1994.	
Folic acid	mg/kg	3.00	3.23	
Niacin	mg/kg	58.8	63.3	
Pantothenic acid	mg/kg	33.0	35.5	
Riboflavin	mg/kg	17.4	18.7	
Thiamine	mg/kg	4.2	4.5	
a-tocopherol	mg/kg	151.0	162.5	
Vitamin B6	mg/kg	7.80	8.39	
Vitamin K	mg/kg	8.50	9.15	
Vitamin A equiv	IU/g	422.0	454.1	
Alanine	%	1.30	1.40	
Arginine	%	1.00	1.08	
Aspartic acid	%	2.30	2.47	
Glutamic acid	%	2.30	2.47	
Glycine	%	1.10	1.18	
Histidine	%	.50	.54	
Isoleucine	%	.90	.97	
Leucine	%	1.70	1.83	
Lysine	%	1.00	1.08	
Methionine	%	.40	.43	
Phenylalanine	%	1.20	1.29	
Proline	%	1.10	1.18	
Serine	%	.90	.97	
Threonine	%	1.00	1.08	
Tryptophan	%	.60	.64	
Tyrosine	%	.80	.86	
Valine	%	1.20	1.29	

Alfalfa, aerial pt, dehy grnd, (1)
  Dehydrated alfalfa meal (AAFCO)

Ref no 1-00-025

Feed name or analyses		Mean		C.V. ± %
		As fed	Dry	
Dry matter	%	92.5	100.0	2
Ash	%	9.6	10.4	24
Crude fiber	%	24.8	26.8	13
Ether extract	%	2.6	2.8	27
N-free extract	%	38.0	41.1	
Protein (N x 6.25)	%	17.4	18.9	17
Cattle	dig prot %	13.6	14.7	
Sheep	dig prot %	12.0	13.0	
Swine	dig prot %	8.2	8.9	
Cellulose	%	19.5	21.1	
Energy				
Cattle	DE kcal/kg	2488.	2690.	
Sheep	DE kcal/kg	2202.	2381.	
Swine	DE kcal/kg	1468.	1587.	
Cattle	ME kcal/kg	2040.	2206.	
Sheep	ME kcal/kg	1806.	1952.	
Swine	ME kcal/kg	1353.	1463.	
Cattle	TDN %	56.	61.	
Sheep	TDN %	50.	54.	
Swine	TDN %	33.	36.	
Calcium	%	1.32	1.43	12
Chlorine	%	.67	.72	
Phosphorus	%	.32	.35	61
Sodium	%	.18	.19	26
Cobalt	mg/kg	.180	.200	99
Copper	mg/kg	19.6	21.2	99
Manganese	mg/kg	43.8	47.4	49
Zinc	mg/kg	32.5	35.1	49
Carotene	mg/kg	113.4	122.6	87
Vitamin A equiv	IU/g	189.0	204.4	

Alfalfa, aerial pt, dehy, immature, (1)

Ref no 1-00-041

Feed name or analyses		Mean		C.V. ± %
		As fed	Dry	
Dry matter	%	91.1	100.0	2
Ash	%	10.9	12.0	13
Crude fiber	%	25.6	28.1	16
Ether extract	%	2.1	2.3	19
N-free extract	%	33.3	36.6	
Protein (N x 6.25)	%	19.1	21.0	17
Cattle	dig prot %	13.4	14.7	
Sheep	dig prot %	13.8	15.1	
Swine	dig prot %	11.7	12.8	
Energy				
Cattle	DE kcal/kg	2410.	2646.	
Sheep	DE kcal/kg	2249.	2469.	
Swine	DE kcal/kg	2009.	2205.	
Cattle	ME kcal/kg	1977.	2170.	

Continued

(5) protein supplements
(6) minerals

(7) vitamins
(8) additives

Feed name or analyses		Mean		C.V. ± %
		As fed	Dry	
Sheep	ME kcal/kg	1844.	2024.	
Swine	ME kcal/kg	1844.	2024.	
Cattle	TDN %	55.	60.	
Sheep	TDN %	51.	56.	
Swine	TDN %	46.	50.	

Alfalfa, hay, s-c, immature, (1)

Ref no 1-00-050

		As fed	Dry	C.V.
Dry matter	%	89.1	100.0	4
Ash	%	8.6	9.7	18
Crude fiber	%	23.4	26.3	15
Ether extract	%	2.4	2.7	29
N-free extract	%	35.5	39.8	
Protein (N x 6.25)	%	19.1	21.5	14
Cattle	dig prot %	13.4	15.0	
Sheep	dig prot %	13.8	15.5	
Energy				
Cattle	DE kcal/kg	2239.	2513.	
Sheep	DE kcal/kg	2239.	2513.	
Cattle	ME kcal/kg	1836.	2061.	
Sheep	ME kcal/kg	1836.	2061.	
Cattle	NEm kcal/kg	1212.	1360.	
Cattle	NEgain kcal/kg	507.	569.	
Cattle	TDN %	51.	57.	
Sheep	TDN %	51.	57.	
Calcium	%	1.89	2.12	14
Chlorine	%	.30	.34	20
Iron	%	.020	.020	30
Magnesium	%	.23	.26	15
Phosphorus	%	.27	.30	14
Potassium	%	2.01	2.26	16
Sodium	%	.20	.22	28
Sulfur	%	.56	.63	12
Manganese	mg/kg	34.4	38.6	36
Carotene	mg/kg	446.6	501.2	33
Vitamin A equiv	IU/g	744.5	835.5	

Alfalfa, hay, s-c, pre-blm, (1)

Ref no 1-00-054

		As fed	Dry	
Dry matter	%	84.5	100.0	
Ash	%	6.0	7.1	
Crude fiber	%	24.1	28.5	
Ether extract	%	2.7	3.2	
N-free extract	%	35.3	41.8	
Protein (N x 6.25)	%	16.4	19.4	
Cattle	dig prot %	9.9	11.7	
Sheep	dig prot %	7.5	8.9	
Energy				
Cattle	DE kcal/kg	2347.	2778.	
Sheep	DE kcal/kg	1938.	2293.	

Continued

Feed name or analyses		Mean		C.V. ± %
		As fed	Dry	
Cattle	ME kcal/kg	1925.	2278.	
Sheep	ME kcal/kg	1589.	1880.	
Cattle	TDN %	53.	63.	
Sheep	TDN %	44.	52.	
Calcium	%	1.06	1.25	
Phosphorus	%	.19	.23	

Alfalfa, hay, s-c, early blm, (1)

Ref no 1-00-059

		As fed	Dry	C.V.
Dry matter	%	90.0	100.0	4
Ash	%	8.5	9.4	12
Crude fiber	%	26.8	29.8	12
Ether extract	%	2.0	2.2	20
N-free extract	%	36.2	40.2	
Protein (N x 6.25)	%	16.6	18.4	11
Cattle	dig prot %	11.4	12.7	
Sheep	dig prot %	13.0	14.4	
Cellulose	%	1.8	2.0	8
Lignin	%	.8	.9	9
Energy	GE kcal/kg	4050.	4500.	
Cattle	DE kcal/kg	2262.	2513.	
Sheep	DE kcal/kg	2301.	2557.	
Cattle	ME kcal/kg	1855.	2061.	
Sheep	ME kcal/kg	1887.	2097.	
Cattle	NEm kcal/kg	1212.	1347.	
Cattle	NEgain kcal/kg	441.	490.	
Cattle	TDN %	51.	57.	
Sheep	TDN %	52.	58.	
Calcium	%	1.12	1.25	36
Chlorine	%	.34	.38	15
Iron	%	.020	.020	
Magnesium	%	.27	.30	16
Phosphorus	%	.21	.23	16
Potassium	%	1.87	2.08	11
Sodium	%	.14	.15	17
Sulfur	%	.27	.30	14
Cobalt	mg/kg	.080	.090	
Copper	mg/kg	12.1	13.4	
Manganese	mg/kg	28.4	31.5	28
Carotene	mg/kg	114.5	127.2	41
Vitamin A equiv	IU/g	190.9	212.0	

Alfalfa, hay, s-c, mid-blm, (1)

Ref no 1-00-063

		As fed	Dry	C.V.
Dry matter	%	89.2	100.0	3
Ash	%	7.6	8.5	16
Crude fiber	%	27.6	30.9	15
Ether extract	%	1.8	2.0	23
N-free extract	%	37.0	41.5	
Protein (N x 6.25)	%	15.2	17.1	10

Continued

(1) dry forages and roughages
(2) pasture, range plants, and forages fed green

(3) silages
(4) energy feeds

TABLE A-3-1 Composition of Feeds (Continued)

Feed name or analyses		Mean As fed	Mean Dry	C.V. ± %
Cattle	dig prot %	10.8	12.1	
Sheep	dig prot %	11.8	13.2	
Cellulose	%	2.4	2.7	
Lignin	%	.7	.8	
Energy				
Cattle	DE kcal/kg	2281.	2557.	
Sheep	DE kcal/kg	2202.	2469.	
Cattle	ME kcal/kg	1870.	2097.	
Sheep	ME kcal/kg	1805.	2024.	
Cattle	TDN %	52.	58.	
Sheep	TDN %	50.	56.	
Calcium	%	1.20	1.35	25
Iron	%	.010	.010	
Magnesium	%	.31	.35	
Phosphorus	%	.20	.22	31
Potassium	%	1.30	1.46	
Copper	mg/kg	13.7	15.4	
Manganese	mg/kg	14.7	16.5	
Carotene	mg/kg	29.7	33.3	73
Vitamin A equiv	IU/g	49.5	55.5	

Alfalfa, hay, s-c, full blm, (1)

Ref no 1-00-068

Feed name or analyses		Mean As fed	Mean Dry	C.V. ± %
Dry matter	%	87.7	100.0	4
Ash	%	7.8	8.9	12
Crude fiber	%	29.7	33.9	7
Ether extract	%	1.6	1.8	24
N-free extract	%	34.6	39.5	
Protein (N x 6.25)	%	14.0	15.9	9
Cattle	dig prot %	10.0	11.4	
Sheep	dig prot %	10.4	11.9	
Energy				
Cattle	DE kcal/kg	2204.	2513.	
Sheep	DE kcal/kg	2011.	2293.	
Cattle	ME kcal/kg	1807.	2061.	
Sheep	ME kcal/kg	1649.	1880.	
Cattle	TDN %	50.	57.	
Sheep	TDN %	46.	52.	
Calcium	%	1.13	1.28	6
Iron	%	.010	.020	33
Magnesium	%	.31	.35	23
Phosphorus	%	.18	.20	22
Potassium	%	.48	.55	62
Copper	mg/kg	11.8	13.4	
Manganese	mg/kg	29.6	33.7	38
Carotene	mg/kg	32.4	37.0	45
Vitamin A equiv	IU/g	54.0	61.7	

Alfalfa, hay, s-c, mature, (1)

Ref no 1-00-071

Feed name or analyses		Mean As fed	Mean Dry	C.V. ± %
Dry matter	%	91.2	100.0	4
Ash	%	7.1	7.8	18
Crude fiber	%	34.2	37.5	11
Ether extract	%	1.5	1.7	28
N-free extract	%	35.9	39.4	
Protein (N x 6.25)	%	12.4	13.6	10
Cattle	dig prot %	8.7	9.5	
Sheep	dig prot %	8.9	9.8	
Energy				
Cattle	DE kcal/kg	2212.	2425.	
Sheep	DE kcal/kg	2212.	2425.	
Cattle	ME kcal/kg	1813.	1988.	
Sheep	ME kcal/kg	1813.	1988.	
Cattle	TDN %	50.	55.	
Sheep	TDN %	50.	55.	

Alfalfa, hay, s-c, over ripe, (1)

Ref no 1-00-072

Feed name or analyses		Mean As fed	Mean Dry	C.V. ± %
Dry matter	%	91.3	100.0	3
Ash	%	6.8	7.5	6
Crude fiber	%	36.5	40.0	15
Ether extract	%	1.5	1.6	29
N-free extract	%	36.4	39.9	
Protein (N x 6.25)	%	10.0	11.0	12
Cattle	dig prot %	7.0	7.7	
Sheep	dig prot %	7.1	7.8	
Energy				
Cattle	DE kcal/kg	2214.	2425.	
Sheep	DE kcal/kg	2334.	2557.	
Cattle	ME kcal/kg	1815.	1988.	
Sheep	ME kcal/kg	1914.	2097.	
Cattle	TDN %	50.	55.	
Sheep	TDN %	53.	58.	
Calcium	%	.37	.41	
Phosphorus	%	.17	.19	
Carotene	mg/kg	7.2	7.9	
Vitamin A equiv	IU/g	12.0	13.2	

Alfalfa, hay, s-c, cut 1, (1)

Ref no 1-00-073

Feed name or analyses		Mean As fed	Mean Dry	C.V. ± %
Dry matter	%	89.1	100.0	3
Ash	%	7.9	8.9	20
Crude fiber	%	30.1	33.8	15
Ether extract	%	1.7	1.9	35
N-free extract	%	34.9	39.2	

Continued

(5) protein supplements
(6) minerals

(7) vitamins
(8) additives

Feed name or analyses		Mean		C.V. ± %
		As fed	Dry	
Protein (N x 6.25)	%	14.4	16.2	17
Cattle	dig prot %	9.4	10.5	
Sheep	dig prot %	10.2	11.5	
Lignin	%	10.3	11.6	16
Energy	GE kcal/kg	3864.	4337.	5
Cattle	DE kcal/kg	2121.	2381.	
Sheep	DE kcal/kg	2121.	2381.	
Cattle	ME kcal/kg	1739.	1952.	
Sheep	ME kcal/kg	1739.	1952.	
Cattle	TDN %	48.	54.	
Sheep	TDN %	48.	54.	
Calcium	%	1.35	1.52	35
Iron	%	.020	.020	75
Magnesium	%	.27	.30	20
Phosphorus	%	.21	.24	39
Potassium	%	1.66	1.87	29
Sulfur	%	.34	.38	
Copper	mg/kg	5.1	5.7	
Manganese	mg/kg	27.7	31.1	
Carotene	mg/kg	22.4	25.1	99
Vitamin A equiv	IU/g	37.3	41.8	
Vitamin D$_2$	IU/g	1.2	1.4	

Alfalfa, hay, s-c, cut 2, (1)

Ref no 1-00-075

		As fed	Dry	C.V. ± %
Dry matter	%	89.2	100.0	4
Ash	%	7.4	8.3	20
Crude fiber	%	27.8	31.2	20
Ether extract	%	1.9	2.1	30
N-free extract	%	35.9	40.3	
Protein (N x 6.25)	%	16.1	18.1	17
Cattle	dig prot %	11.3	12.7	
Sheep	dig prot %	12.1	13.6	
Lignin	%	10.3	11.6	23
Energy	GE kcal/kg	3761.	4216.	2
Cattle	DE kcal/kg	2242.	2513.	
Sheep	DE kcal/kg	2281.	2557.	
Cattle	ME kcal/kg	1838.	2061.	
Sheep	ME kcal/kg	1870.	2097.	
Cattle	TDN %	51.	57.	
Sheep	TDN %	52.	58.	
Calcium	%	1.28	1.44	32
Iron	%	.020	.020	99
Magnesium	%	.28	.31	23
Phosphorus	%	.23	.26	47
Potassium	%	1.64	1.84	30
Cobalt	mg/kg	.100	.110	10
Copper	mg/kg	12.2	13.7	10
Manganese	mg/kg	27.1	30.4	32
Carotene	mg/kg	27.3	30.6	99
Vitamin A equiv	IU/g	45.5	51.0	
Vitamin D$_2$	IU/g	.9	1.1	

Alfalfa, hay, s-c, cut 3, (1)

Ref no 1-00-076

Feed name or analyses		As fed	Dry	C.V. ± %
Dry matter	%	89.0	100.0	3
Ash	%	7.8	8.8	15
Crude fiber	%	26.9	30.2	14
Ether extract	%	1.9	2.1	21
N-free extract	%	36.2	40.7	
Protein (N x 6.25)	%	16.2	18.2	16
Cattle	dig prot %	11.3	12.7	
Sheep	dig prot %	11.6	13.1	
Cellulose	%	18.0	20.2	10
Lignin	%	9.3	10.4	21
Energy	GE kcal/kg	3749.	4212.	
Cattle	DE kcal/kg	2119.	2381.	
Sheep	DE kcal/kg	2236.	2513.	
Cattle	ME kcal/kg	1737.	1952.	
Sheep	ME kcal/kg	1834.	2061.	
Cattle	TDN %	48.	54.	
Sheep	TDN %	51.	57.	
Calcium	%	1.21	1.36	35
Iron	%	.020	.020	99
Magnesium	%	.27	.30	20
Phosphorus	%	.21	.24	28
Potassium	%	1.73	1.94	31
Carotene	mg/kg	118.7	133.4	50
Vitamin A equiv	IU/g	197.9	222.4	
Vitamin D$_2$	IU/g	1.2	1.4	

Alfalfa, hay, s-c, (1)

Ref no 1-00-078

		As fed	Dry	C.V. ± %
Dry matter	%	89.7	100.0	4
Ash	%	8.0	8.9	22
Crude fiber	%	28.2	31.4	18
Ether extract	%	1.9	2.1	32
N-free extract	%	36.1	40.3	
Protein (N x 6.25)	%	15.6	17.3	21
Cattle	dig prot %	10.8	12.1	
Horses	dig prot %	11.5	12.8	
Sheep	dig prot %	11.1	12.4	
Swine	dig prot %	7.3	8.1	
Cellulose	%	20.8	23.2	23
Lignin	%	9.5	10.6	25
Energy	GE kcal/kg	3964.	4419.	6
Cattle	DE kcal/kg	2215.	2469.	
Sheep	DE kcal/kg	2215.	2469.	
Swine	DE kcal/kg	1384.	1543.	
Cattle	ME kcal/kg	1816.	2024.	
Sheep	ME kcal/kg	1816.	2024.	
Swine	ME kcal/kg	1280.	1427.	
Cattle	NE$_m$ kcal/kg	996.	1110.	

Continued

(1) dry forages and roughages
(2) pasture, range plants, and forages fed green

(3) silages
(4) energy feeds

Feed name or analyses		Mean		C.V.
		As fed	Dry	± %
Cattle	NE$_{gain}$ kcal/kg	332.	370.	
Cattle	TDN %	50.	56.	
Horses	TDN %	48.	53.	
Sheep	TDN %	50.	56.	
Swine	TDN %	31.	35.	
Calcium	%	1.48	1.64	27
Chlorine	%	.25	.28	34
Iron	%	.020	.020	99
Magnesium	%	.29	.32	43
Phosphorus	%	.23	.26	58
Potassium	%	1.59	1.77	30
Sodium	%	.14	.16	46
Sulfur	%	.32	.36	31
Cobalt	mg/kg	.120	.130	43
Copper	mg/kg	12.3	13.7	24
Manganese	mg/kg	46.5	51.8	32
Zinc	mg/kg	15.2	17.0	24
Carotene	mg/kg	54.8	61.1	99
Vitamin A equiv	IU/g	91.4	101.8	
Vitamin D$_2$	IU/g	1.2	1.3	99

Alfalfa, hay, s-c grnd, early blm, (1)

Ref no 1-00-108

		As fed	Dry	± %
Dry matter	%	91.0	100.0	1
Ash	%	9.8	10.8	5
Crude fiber	%	23.3	25.6	12
Ether extract	%	2.6	2.9	20
N-free extract	%	37.6	41.3	
Protein (N x 6.25)	%	17.7	19.4	8
Cattle	dig prot %	13.7	15.1	
Sheep	dig prot %	12.7	14.0	
Energy				
Cattle	DE kcal/kg	2448.	2690.	
Sheep	DE kcal/kg	2327.	2557.	
Cattle	ME kcal/kg	2007.	2206.	
Sheep	ME kcal/kg	1908.	2097.	
Cattle	TDN %	56.	61.	
Sheep	TDN %	53.	58.	
Carotene	mg/kg	163.3	179.5	
Vitamin A equiv	IU/g	272.2	299.2	

Alfalfa, hay, s-c grnd, stemmy, (1)

Ref no 1-00-118

		As fed	Dry	± %
Dry matter	%	92.7	100.0	2
Ash	%	7.2	7.8	12
Crude fiber	%	34.8	37.5	7
Ether extract	%	1.6	1.7	9
N-free extract	%	35.5	38.3	
Protein (N x 6.25)	%	13.6	14.7	13
Cattle	dig prot %	10.7	11.5	

Continued

Feed name or analyses		Mean		C.V.
		As fed	Dry	± %
Sheep	dig prot %	9.8	10.6	
Energy				
Cattle	DE kcal/kg	2453.	2646.	
Sheep	DE kcal/kg	2330.	2513.	
Cattle	ME kcal/kg	2012.	2170.	
Sheep	ME kcal/kg	1910.	2061.	
Cattle	TDN %	56.	60.	
Sheep	TDN %	53.	57.	

Alfalfa, hay, s-c grnd, (1)
Sun-cured alfalfa meal (AAFCO)
Ground alfalfa hay (AAFCO)

Ref no 1-00-111

		As fed	Dry	± %
Dry matter	%	92.2	100.0	3
Ash	%	9.5	10.3	22
Crude fiber	%	25.8	28.0	15
Ether extract	%	2.3	2.5	31
N-free extract	%	37.8	41.0	
Protein (N x 6.25)	%	16.7	18.2	16
Cattle	dig prot %	13.1	14.2	
Sheep	dig prot %	12.1	13.1	
Swine	dig prot %	7.7	8.4	
Cellulose	%	22.1	24.0	25
Lignin	%	9.8	10.7	
Energy	GE kcal/kg	4204.	4560.	
Cattle	DE kcal/kg	2440.	2646.	
Sheep	DE kcal/kg	2358.	2557.	
Swine	DE kcal/kg	1382.	1499.	
Cattle	ME kcal/kg	2001.	2170.	
Sheep	ME kcal/kg	1933.	2097.	
Swine	ME kcal/kg	1276.	1384.	
Cattle	TDN %	55.	60.	
Sheep	TDN %	53.	58.	
Swine	TDN %	31.	34.	
Calcium	%	1.24	1.35	42
Chlorine	%	.43	.47	33
Iron	%	.040	.050	56
Magnesium	%	.31	.34	68
Phosphorus	%	.28	.30	41
Potassium	%	2.27	2.46	8
Sodium	%	.18	.19	26
Cobalt	mg/kg	.220	.240	86
Copper	mg/kg	17.2	18.7	99
Manganese	mg/kg	42.9	46.5	99
Zinc	mg/kg	32.4	35.1	49
Carotene	mg/kg	96.3	104.5	92
Vitamin A equiv	IU/g	160.5	174.2	

(5) protein supplements
(6) minerals

(7) vitamins
(8) additives

Feed name or analyses		Mean		C.V.
		As fed	Dry	± %

Alfalfa, hay, s-c pltd, (1)

Ref no 1-00-128

Dry matter	%	91.6	100.0	1
Ash	%	10.4	11.3	9
Crude fiber	%	22.6	24.7	9
Ether extract	%	3.2	3.5	16
N-free extract	%	35.8	39.1	
Protein (N x 6.25)	%	19.6	21.4	10
Cattle	dig prot %	13.7	15.0	
Sheep	dig prot %	14.1	15.4	
Energy				
Cattle	DE kcal/kg	2262.	2469.	
Sheep	DE kcal/kg	2262.	2469.	
Cattle	ME kcal/kg	1854.	2024.	
Sheep	ME kcal/kg	1854.	2024.	
Cattle	TDN %	51.	56.	
Sheep	TDN %	51.	56.	
Calcium	%	1.44	1.56	
Chlorine	%	.24	.26	
Iron	%	.040	.040	
Magnesium	%	.33	.36	
Phosphorus	%	.32	.35	
Potassium	%	2.14	2.33	
Sodium	%	.04	.04	
Sulfur	%	.28	.31	
Cobalt	mg/kg	.220	.240	
Copper	mg/kg	31.1	34.0	
Manganese	mg/kg	47.6	52.0	
Carotene	mg/kg	113.3	123.7	36
Vitamin A equiv	IU/g	188.9	206.2	

Alfalfa, leaves, dehy grnd, (1)
    Alfalfa leaf meal, dehydrated (AAFCO)

Ref no 1-00-137

Dry matter	%	92.2	100.0	1
Ash	%	11.0	11.9	16
Crude fiber	%	19.6	21.3	18
Ether extract	%	3.0	3.2	14
N-free extract	%	38.0	41.2	
Protein (N x 6.25)	%	20.6	22.4	12
Cattle	dig prot %	15.3	16.6	
Sheep	dig prot %	16.5	17.9	
Energy				
Cattle	DE kcal/kg	2276.	2469.	
Sheep	DE kcal/kg	2602.	2822.	
Cattle	ME kcal/kg	1866.	2024.	
Sheep	ME kcal/kg	2134.	2314.	
Cattle	TDN %	52.	56.	
Sheep	TDN %	59.	64.	
Calcium	%	1.64	1.78	10

Continued

Feed name or analyses		Mean		C.V.
		As fed	Dry	± %

Chlorine	%	.31	.34	
Iron	%	.036	.039	10
Phosphorus	%	.23	.25	16
Potassium	%	2.07	2.25	
Cobalt	mg/kg	.199	.216	8
Copper	mg/kg	10.6	11.5	8
Manganese	mg/kg	36.8	39.9	16
Carotene	mg/kg	149.0	161.6	36
Niacin	mg/kg	36.4	39.5	14
Pantothenic acid	mg/kg	32.9	35.7	23
Riboflavin	mg/kg	18.1	19.6	14
Thiamine	mg/kg	5.5	6.0	18
Vitamin D₂	IU/g	.4	.4	59

Alfalfa, leaves, s-c, (1)

Ref no 1-00-146

Dry matter	%	88.8	100.0	2
Ash	%	9.5	10.7	8
Crude fiber	%	14.6	16.4	13
Ether extract	%	2.8	3.1	15
N-free extract	%	40.7	45.8	
Protein (N x 6.25)	%	21.3	24.0	7
Cattle	dig prot %	14.9	16.8	
Sheep	dig prot %	16.2	18.2	
Energy	GE kcal/kg	3802.	4282.	2
Cattle	DE kcal/kg	2310.	2601.	
Sheep	DE kcal/kg	2428.	2734.	
Cattle	ME kcal/kg	1894.	2133.	
Sheep	ME kcal/kg	1991.	2242.	
Cattle	NEm kcal/kg	1235.	1391.	
Cattle	NE gain kcal/kg	639.	720.	
Cattle	TDN %	52.	59.	
Sheep	TDN %	55.	62.	
Calcium	%	2.11	2.38	21
Iron	%	.030	.030	19
Magnesium	%	.36	.41	7
Phosphorus	%	.26	.29	18
Potassium	%	1.80	2.02	27
Sulfur	%	.54	.61	
Manganese	mg/kg	42.9	48.3	42
Carotene	mg/kg	62.4	70.3	31
Vitamin A equiv	IU/g	104.0	117.2	

Alfalfa. leaves, s-c grnd, mn 20 prot mx 18 fbr, (1)
    Alfalfa leaf meal, sun-cured (AAFCO)

Ref no 1-00-246

Dry matter	%	88.8	100.0	2
Ash	%	9.5	10.7	8
Crude fiber	%	14.6	16.4	13

Continued

(1) dry forages and roughages

(2) pasture, range plants, and forages fed green

(3) silages

(4) energy feeds

TABLE A-3-1 Composition of Feeds (Continued)

Feed name or analyses		Mean		C.V.
		As fed	Dry	± %
Ether extract	%	2.8	3.1	15
N-free extract	%	40.7	45.8	
Protein (N x 6.25)	%	21.3	24.0	7
Cattle	dig prot %	15.8	17.8	
Sheep	dig prot %	16.2	18.2	
Swine	dig prot %	13.0	14.6	
Energy	GE kcal/kg	3802.	4282.	2
Cattle	DE kcal/kg	2310.	2601.	
Sheep	DE kcal/kg	2428.	2734.	
Swine	DE kcal/kg	2192.	2469.	
Cattle	ME kcal/kg	1894.	2133.	
Sheep	ME kcal/kg	1991.	2242.	
Swine	ME kcal/kg	2000.	2252.	
Cattle	TDN %	52.	59.	
Sheep	TDN %	55.	62.	
Swine	TDN %	50.	56.	
Calcium	%	2.11	2.38	21
Iron	%	.030	.034	19
Magnesium	%	.36	.41	7
Phosphorus	%	.26	.29	18
Potassium	%	1.80	2.02	27
Sulfur	%	.54	.61	
Manganese	mg/kg	42.9	48.3	42
Carotene	mg/kg	62.4	70.3	31
Vitamin A equiv	IU/g	104.0	117.2	

Alfalfa, stems, s-c, (1)

Ref no 1-00-164

		As fed	Dry	C.V.
Dry matter	%	89.9	100.0	2
Ash	%	5.7	6.3	12
Crude fiber	%	39.9	44.4	7
Ether extract	%	1.2	1.3	12
N-free extract	%	33.5	37.3	
Protein (N x 6.25)	%	9.6	10.7	16
Cattle	dig prot %	6.7	7.5	
Sheep	dig prot %	4.8	5.4	
Energy	GE kcal/kg	3770.	4194.	1
Cattle	DE kcal/kg	2140.	2381.	
Sheep	DE kcal/kg	1744.	1940.	
Cattle	ME kcal/kg	1755.	1952.	
Sheep	ME kcal/kg	1430.	1591.	
Cattle	TDN %	48.	54.	
Sheep	TDN %	40.	44.	
Calcium	%	.80	.89	20
Iron	%	.010	.020	31
Magnesium	%	.24	.27	25
Phosphorus	%	.20	.22	26
Potassium	%	1.91	2.13	23
Manganese	mg/kg	16.6	18.5	37

Alfalfa, aerial pt, fresh, immature, (2)

Ref no 2-00-177

Feed name or analyses		As fed	Dry	C.V.
Dry matter	%	25.8	100.0	47
Ash	%	2.6	10.1	15
Crude fiber	%	5.4	21.0	16
Ether extract	%	.9	3.4	18
N-free extract	%	10.6	41.0	
Protein (N x 6.25)	%	6.3	24.5	11
Cattle	dig prot %	4.9	19.1	
Sheep	dig prot %	4.7	18.4	
Energy				
Cattle	DE kcal/kg	705.	2734.	
Sheep	DE kcal/kg	683.	2646.	
Cattle	ME kcal/kg	578.	2242.	
Sheep	ME kcal/kg	560.	2170.	
Cattle	TDN %	16.	62.	
Sheep	TDN %	15.	60.	
Calcium	%	.60	2.33	16
Chlorine	%	.09	.36	13
Iron	%	.010	.020	
Magnesium	%	.06	.22	27
Phosphorus	%	.10	.40	12
Potassium	%	.58	2.23	22
Sodium	%	.05	.20	13
Sulfur	%	.16	.61	12
Manganese	mg/kg	8.1	31.3	40
Carotene	mg/kg	62.4	241.7	47
Vitamin A equiv	IU/g	104.0	402.9	

Alfalfa, aerial pt, fresh, pre-blm, (2)

Ref no 2-00-181

		As fed	Dry	C.V.
Dry matter	%	21.1	100.0	
Ash	%	2.0	9.6	
Crude fiber	%	5.5	26.0	
Ether extract	%	.5	2.2	
N-free extract	%	8.8	41.7	
Protein (N x 6.25)	%	4.3	20.5	
Cattle	dig prot %	3.4	16.0	
Sheep	dig prot %	3.4	16.0	
Energy				
Cattle	DE kcal/kg	558.	2646.	
Sheep	DE kcal/kg	568.	2690.	
Cattle	ME kcal/kg	458.	2170.	
Sheep	ME kcal/kg	465.	2206.	
Cattle	TDN %	13.	60.	
Sheep	TDN %	13.	61.	
Calcium	%	.48	2.30	
Magnesium	%	.01	.03	
Phosphorus	%	.06	.31	

Continued

(5) protein supplements

(6) minerals

(7) vitamins

(8) additives

Feed name or analyses		Mean As fed	Dry	C.V. ± %
Potassium	%	.40	1.92	
Carotene	mg/kg	23.1	109.6	
Pantothenic acid	mg/kg	4.0	18.9	

Feed name or analyses		Mean As fed	Dry	C.V. ± %
Sulfur	%	.07	.26	
Zinc	mg/kg	6.0	23.6	
Carotene	mg/kg	36.3	141.8	15
Vitamin A equiv	IU/g	60.5	236.4	

Alfalfa, aerial pt, fresh, early blm, (2)

Ref no 2-00-184

		As fed	Dry	C.V.
Dry matter	%	25.2	100.0	9
Ash	%	2.3	9.2	8
Crude fiber	%	6.9	27.3	4
Ether extract	%	.7	2.9	17
N-free extract	%	10.4	41.3	
Protein (N x 6.25)	%	4.9	19.3	8
Cattle	dig prot %	3.8	15.0	
Sheep	dig prot %	3.8	15.2	
Energy				
Cattle	DE kcal/kg	678.	2690.	
Sheep	DE kcal/kg	700.	2778.	
Cattle	ME kcal/kg	556.	2206.	
Sheep	ME kcal/kg	574.	2278.	
Cattle	TDN %	15.	61.	
Sheep	TDN %	16.	63.	
Calcium	%	.58	2.30	
Magnesium	%	.01	.03	
Phosphorus	%	.08	.31	
Potassium	%	.48	1.92	
Carotene	mg/kg	44.0	174.6	42
Vitamin A equiv	IU/g	73.3	291.0	

Alfalfa, aerial pt, fresh, mid-blm, (2)

Ref no 2-00-185

		As fed	Dry	C.V.
Dry matter	%	25.6	100.0	14
Ash	%	2.1	8.3	13
Crude fiber	%	7.9	30.9	8
Ether extract	%	.7	2.6	14
N-free extract	%	10.3	40.4	
Protein (N x 6.25)	%	4.6	17.8	8
Cattle	dig prot %	3.6	13.9	
Sheep	dig prot %	3.4	13.2	
Energy				
Cattle	DE kcal/kg	677.	2646.	
Sheep	DE kcal/kg	666.	2601.	
Cattle	ME kcal/kg	556.	2170.	
Sheep	ME kcal/kg	546.	2133.	
Cattle	TDN %	15.	60.	
Sheep	TDN %	15.	59.	
Calcium	%	.51	2.01	
Chlorine	%	.13	.51	
Magnesium	%	.06	.25	
Phosphorus	%	.07	.28	7
Potassium	%	.50	1.95	31

Continued

Alfalfa, aerial pt, fresh, full blm, (2)

Ref no 2-00-188

		As fed	Dry	C.V.
Dry matter	%	25.3	100.0	4
Ash	%	2.1	8.4	11
Crude fiber	%	8.0	31.7	7
Ether extract	%	.8	3.0	10
N-free extract	%	10.1	40.0	
Protein (N x 6.25)	%	4.3	16.9	6
Cattle	dig prot %	3.3	13.2	
Sheep	dig prot %	3.2	12.7	
Energy				
Cattle	DE kcal/kg	669.	2646.	
Sheep	DE kcal/kg	625.	2469.	
Cattle	ME kcal/kg	549.	2170.	
Sheep	ME kcal/kg	512.	2024.	
Cattle	TDN %	15.	60.	
Sheep	TDN %	14.	56.	
Calcium	%	.39	1.53	14
Chlorine	%	.11	.43	9
Iron	%	.010	.040	
Magnesium	%	.07	.27	16
Phosphorus	%	.07	.27	13
Potassium	%	.54	2.13	13
Sodium	%	.04	.15	26
Sulfur	%	.08	.31	9
Manganese	mg/kg	39.3	155.2	
Zinc	mg/kg	3.6	14.1	

Alfalfa, aerial pt, fresh, cut 1, (2)

Ref no 2-00-191

		As fed	Dry	C.V.
Dry matter	%	22.2	100.0	15
Ash	%	1.9	8.7	11
Crude fiber	%	6.5	29.4	19
Ether extract	%	.7	3.2	17
N-free extract	%	8.8	39.6	
Protein (N x 6.25)	%	4.2	19.1	24
Cattle	dig prot %	3.3	14.9	
Sheep	dig prot %	3.1	13.8	
Energy				
Cattle	DE kcal/kg	587.	2646.	
Sheep	DE kcal/kg	538.	2425.	
Cattle	ME kcal/kg	482.	2170.	
Sheep	ME kcal/kg	441.	1988.	
Cattle	TDN %	13.	.60.	
Sheep	TDN %	12.	55.	

Continued

(1) dry forages and roughages
(2) pasture, range plants, and forages fed green

(3) silages
(4) energy feeds

Feed name or analyses		Mean As fed	Dry	C.V. ± %
Calcium	%	.42	1.87	
Chlorine	%	.16	.73	
Iron	%	.010	.030	
Magnesium	%	.01	.06	
Phosphorus	%	.08	.36	38
Potassium	%	.49	2.21	35
Sodium	%	.05	.24	
Sulfur	%	.08	.36	
Zinc	mg/kg	4.4	19.8	
Carotene	mg/kg	34.2	153.9	18
Vitamin A equiv	IU/g	57.0	256.6	

Alfalfa, aerial pt, fresh, cut 2, (2)

Ref no 2-00-193

Feed name or analyses		As fed	Dry	C.V. ± %
Dry matter	%	30.1	100.0	20
Ash	%	2.6	8.6	13
Crude fiber	%	8.3	27.6	17
Ether extract	%	.8	2.8	22
N-free extract	%	12.8	42.4	
Protein (N x 6.25)	%	5.6	18.6	19
Cattle	dig prot %	4.4	14.5	
Sheep	dig prot %	4.2	13.8	
Energy				
Cattle	DE kcal/kg	810.	2690.	
Sheep	DE kcal/kg	770.	2557.	
Cattle	ME kcal/kg	664.	2206.	
Sheep	ME kcal/kg	631.	2097.	
Cattle	TDN %	18.	61.	
Sheep	TDN %	17.	58.	
Calcium	%	.71	2.37	
Iron	%	.060	.180	
Magnesium	%	.06	.19	
Phosphorus	%	.11	.37	
Potassium	%	.67	2.22	
Sodium	%	.12	.39	
Sulfur	%	.09	.31	
Zinc	mg/kg	9.3	30.9	
Carotene	mg/kg	55.5	184.3	28
Vitamin A equiv	IU/g	92.5	307.2	

Alfalfa, aerial pt, fresh, cut 3, (2)

Ref no 2-00-194

Feed name or analyses		As fed	Dry	C.V. ± %
Dry matter	%	27.4	100.0	43
Ash	%	2.7	9.9	10
Crude fiber	%	6.5	23.7	13
Ether extract	%	.7	2.5	20
N-free extract	%	11.4	41.6	
Protein (N x 6.25)	%	6.1	22.3	18
Cattle	dig prot %	4.5	16.5	

Continued

Feed name or analyses		Mean As fed	Dry	C.V. ± %
Energy				
Cattle	DE kcal/kg	713.	2601.	
Cattle	ME kcal/kg	584.	2133.	
Cattle	TDN %	16.	59.	
Calcium	%	.48	1.76	43
Chlorine	%	.16	.60	
Iron	%	.040	.140	
Magnesium	%	.02	.09	
Phosphorus	%	.10	.35	50
Potassium	%	.60	2.21	
Sodium	%	.05	.19	
Sulfur	%	.09	.33	

Alfalfa, aerial pt, fresh, cut 4, (2)

Ref no 2-00-195

Feed name or analyses		As fed	Dry	C.V. ± %
Dry matter	%	63.4	100.0	22
Ash	%	5.6	8.9	10
Crude fiber	%	14.1	22.2	19
Ether extract	%	1.3	2.0	36
N-free extract	%	27.6	43.5	
Protein (N x 6.25)	%	14.8	23.4	8
Cattle	dig prot %	11.5	18.2	
Sheep	dig prot %	11.2	17.6	
Energy				
Cattle	DE kcal/kg	1733.	2734.	
Sheep	DE kcal/kg	1678.	2646.	
Cattle	ME kcal/kg	1421.	2242.	
Sheep	ME kcal/kg	1376.	2170.	
Cattle	TDN %	39.	62.	
Sheep	TDN %	38.	60.	
Calcium	%	1.58	2.49	
Chlorine	%	.43	.68	
Iron	%	.020	.030	
Magnesium	%	.35	.55	
Phosphorus	%	.28	.45	
Potassium	%	1.74	2.75	
Sodium	%	.14	.22	
Sulfur	%	.26	.41	

Alfalfa, aerial pt, fresh, (2)

Ref no 2-00-196

Feed name or analyses		As fed	Dry	C.V. ± %
Dry matter	%	27.2	100.0	37
Ash	%	2.4	9.0	18
Crude fiber	%	7.4	27.4	18
Ether extract	%	.8	3.0	17
N-free extract	%	11.2	41.3	
Protein (N x 6.25)	%	5.2	19.3	20
Cattle	dig prot %	4.1	15.0	
Sheep	dig prot %	3.9	14.5	

Continued

(5) protein supplements

(6) minerals

(7) vitamins

(8) additives

Feed name or analyses		Mean		C.V.
		As fed	Dry	± %
Energy				
Cattle	DE kcal/kg	732.	2690.	
Sheep	DE kcal/kg	707.	2601.	
Cattle	ME kcal/kg	600.	2206.	
Sheep	ME kcal/kg	580.	2133.	
Cattle	NEm kcal/kg	359.	1320.	
Cattle	NE gain kcal/kg	193.	710.	
Cattle	TDN %	16.	61.	
Sheep	TDN %	16.	59.	
Calcium	%	.47	1.72	23
Chlorine	%	.13	.47	22
Iron	%	.010	.030	25
Magnesium	%	.07	.27	30
Phosphorus	%	.08	.31	28
Potassium	%	.55	2.03	25
Sodium	%	.05	.20	67
Sulfur	%	.11	.39	27
Cobalt	mg/kg	.020	.090	8
Copper	mg/kg	2.7	9.9	14
Manganese	mg/kg	13.7	50.5	60
Zinc	mg/kg	4.8	17.6	65
Carotene	mg/kg	54.1	198.9	33
Vitamin A equiv	IU/g	90.2	331.6	
Vitamin D2	IU/g	.	.2	

Alfalfa, aerial pt, ensiled, immature, (3)

Ref no 3-00-203

		As fed	Dry	C.V. ± %
Dry matter	%	31.1	100.0	19
Ash	%	2.9	9.2	16
Crude fiber	%	8.8	28.4	12
Ether extract	%	1.2	3.9	21
N-free extract	%	11.7	37.6	
Protein (N x 6.25)	%	6.5	20.9	12
Cattle	dig prot %	4.4	14.0	
Sheep	dig prot %	4.2	13.4	
Energy				
Cattle	DE kcal/kg	782.	2513.	
Sheep	DE kcal/kg	740.	2381.	
Cattle	ME kcal/kg	641.	2061.	
Sheep	ME kcal/kg	607.	1952.	
Cattle	TDN %	18.	57.	
Sheep	TDN %	17.	54.	
Calcium	%	.55	1.77	14
Phosphorus	%	.15	.49	32
Carotene	mg/kg	53.0	170.4	28
Vitamin A equiv	IU/g	88.4	284.0	

Alfalfa, aerial pt, ensiled, early blm, mn 50 dry matter, (3)

Ref no 3-08-151

		As fed	Dry	C.V. ± %
Dry matter	%	55.0	100.0	
Ash	%	5.0	9.0	
Crude fiber	%	17.8	32.4	
Ether extract	%	2.0	3.6	
N-free extract	%	20.4	37.1	
Protein (N x 6.25)	%	9.8	17.9	
Cattle	dig prot %	5.9	10.7	
Sheep	dig prot %	6.3	11.4	
Energy				
Cattle	DE kcal/kg	1261.	2293.	
Sheep	DE kcal/kg	1358.	2469.	
Cattle	ME kcal/kg	1034.	1880.	
Sheep	ME kcal/kg	1113.	2024.	
Cattle	TDN %	29.	52.	
Sheep	TDN %	31.	56.	
Calcium	%	.88	1.61	
Phosphorus	%	.21	.38	

Alfalfa, aerial pt, ensiled, early blm, mn 30 mx 50 dry matter, (3)

Ref no 3-08-150

		As fed	Dry	C.V. ± %
Dry matter	%	38.5	100.0	
Ash	%	3.0	7.8	
Crude fiber	%	11.0	28.6	
Ether extract	%	1.2	3.1	
N-free extract	%	16.5	42.9	
Protein (N x 6.25)	%	6.8	17.6	
Cattle	dig prot %	4.1	10.6	
Sheep	dig prot %	4.4	11.3	
Energy				
Cattle	DE kcal/kg	900.	2337.	
Sheep	DE kcal/kg	934.	2425.	
Cattle	ME kcal/kg	738.	1916.	
Sheep	ME kcal/kg	765.	1988.	
Cattle	TDN %	20.	53.	
Sheep	TDN %	21.	55.	
Calcium	%	.58	1.52	
Phosphorus	%	.12	.30	

(1) dry forages and roughages

(2) pasture, range plants, and forages fed green

(3) silages

(4) energy feeds

Feed name or analyses		Mean		C.V. ± %
		As fed	Dry	

Alfalfa, aerial pt, ensiled, early blm, mx 30 dry matter, (3)

Ref no 3-08-149

		As fed	Dry	
Dry matter	%	28.3	100.0	
Ash	%	2.5	8.8	
Crude fiber	%	8.2	28.9	
Ether extract	%	1.1	4.0	
N-free extract	%	11.2	39.7	
Protein (N x 6.25)	%	5.3	18.6	
Cattle	dig prot %	3.2	11.2	
Sheep	dig prot %	3.4	11.9	
Energy				
Cattle	DE kcal/kg	661.	2337.	
Sheep	DE kcal/kg	674.	2381.	
Cattle	ME kcal/kg	542.	1916.	
Sheep	ME kcal/kg	552.	1952.	
Cattle	TDN %	15.	53.	
Sheep	TDN %	15.	54.	
Calcium	%	.40	1.40	
Magnesium	%	.10	.36	
Phosphorus	%	.09	.32	
Potassium	%	.67	2.36	

Alfalfa, aerial pt, ensiled, early blm, (3)

Ref no 3-00-205

		As fed	Dry	
Dry matter	%	33.0	100.0	15
Ash	%	2.6	8.0	16
Crude fiber	%	9.7	29.3	6
Ether extract	%	1.1	3.3	21
N-free extract	%	13.7	41.5	
Protein (N x 6.25)	%	5.9	17.9	7
Cattle	dig prot %	3.7	11.1	
Sheep	dig prot %	3.8	11.4	
Energy				
Cattle	DE kcal/kg	786.	2381.	
Sheep	DE kcal/kg	786.	2381.	
Cattle	ME kcal/kg	644.	1952.	
Sheep	ME kcal/kg	644.	1952.	
Cattle	TDN %	18.	54.	
Sheep	TDN %	18.	54.	
Calcium	%	.50	1.52	
Phosphorus	%	.10	.30	

Alfalfa, aerial pt, ensiled, (3)

Ref no 3-00-212

Feed name or analyses		As fed	Dry	C.V. ± %
Dry matter	%	30.4	100.0	18
Ash	%	2.8	9.2	31
Crude fiber	%	9.2	30.4	12
Ether extract	%	1.1	3.5	36
N-free extract	%	11.9	39.1	
Protein (N x 6.25)	%	5.4	17.8	18
Cattle	dig prot %	3.6	11.9	
Sheep	dig prot %	3.5	11.4	
Energy	GE kcal/kg	1344.	4421.	4
Cattle	DE kcal/kg	750.	2469.	
Sheep	DE kcal/kg	724.	2381.	
Cattle	ME kcal/kg	615.	2024.	
Sheep	ME kcal/kg	593.	1952.	
Cattle	TDN %	17.	56.	
Sheep	TDN %	16.	54.	
Calcium	%	.49	1.61	17
Chlorine	%	.15	.50	33
Iron	%	.010	.030	17
Magnesium	%	.10	.34	19
Phosphorus	%	.12	.38	57
Potassium	%	.73	2.40	4
Sodium	%	.05	.16	24
Sulfur	%	.11	.36	19
Cobalt	mg/kg	.050	.150	
Copper	mg/kg	2.9	9.7	16
Manganese	mg/kg	15.3	50.3	29
Carotene	mg/kg	27.3	89.7	62
Vitamin A equiv	IU/g	45.5	149.5	

Alfalfa, aerial pt, wilted ensiled, (3)

Ref no 3-00-221

		As fed	Dry	
Dry matter	%	36.2	100.0	10
Ash	%	3.1	8.6	9
Crude fiber	%	10.9	30.2	13
Ether extract	%	1.2	3.3	15
N-free extract	%	14.5	40.1	
Protein (N x 6.25)	%	6.4	17.8	9
Cattle	dig prot %	4.3	11.9	
Sheep	dig prot %	4.1	11.4	
Energy				
Cattle	DE kcal/kg	926.	2557.	
Sheep	DE kcal/kg	862.	2381.	
Cattle	ME kcal/kg	759.	2097.	
Sheep	ME kcal/kg	707.	1952.	
Cattle	NE$_m$ kcal/kg	474.	1310.	
Cattle	NE$_{gain}$ kcal/kg	250.	690.	
Cattle	TDN %	21.	58.	
Sheep	TDN %	20.	54.	

Continued

(5) protein supplements

(6) minerals

(7) vitamins

(8) additives

Feed name or analyses		Mean As fed	Mean Dry	C.V. ± %
Calcium	%	.51	1.40	8
Iron	%	.010	.030	11
Magnesium	%	.12	.33	14
Phosphorus	%	.12	.32	13
Potassium	%	.85	2.36	2
Copper	mg/kg	3.4	9.3	14
Manganese	mg/kg	18.8	52.0	18
Carotene	mg/kg	18.7	51.6	34
Vitamin A equiv	IU/g	31.2	86.0	
Vitamin D₂	IU/g	.2	.6	

Feed name or analyses		Mean As fed	Mean Dry	C.V. ± %
Magnesium	%	.11	.34	
Phosphorus	%	.10	.31	34
Potassium	%	.82	2.56	7
Copper	mg/kg	4.1	12.6	
Manganese	mg/kg	13.7	42.6	
Carotene	mg/kg	31.3	97.2	56
Vitamin A equiv	IU/g	52.2	162.0	

Alfalfa, aerial pt w H₃PO₄ pres added, ensiled, (3)

Ref no 3-00-231

		As fed	Dry	
Dry matter	%	29.0	100.0	6
Ash	%	2.7	9.2	12
Crude fiber	%	9.2	31.7	6
Ether extract	%	1.1	3.7	18
N-free extract	%	10.9	37.5	
Protein (N x 6.25)	%	5.2	17.9	9
Cattle	dig prot %	3.5	12.2	
Sheep	dig prot %	3.3	11.4	
Energy				
Cattle	DE kcal/kg	690.	2381.	
Sheep	DE kcal/kg	690.	2381.	
Cattle	ME kcal/kg	566.	1952.	
Sheep	ME kcal/kg	566.	1952.	
Cattle	TDN %	16.	54.	
Sheep	TDN %	16.	54.	
Carotene	mg/kg	30.0	103.6	72
Vitamin A equiv	IU/g	50.0	172.7	

Alfalfa, aerial pt w molasses added, ensiled, (3)

Ref no 3-00-238

		As fed	Dry	
Dry matter	%	32.2	100.0	16
Ash	%	2.8	8.8	33
Crude fiber	%	9.3	28.8	8
Ether extract	%	1.1	3.5	31
N-free extract	%	13.3	41.4	
Protein (N x 6.25)	%	5.6	17.5	14
Cattle	dig prot %	3.9	12.2	
Sheep	dig prot %	3.6	11.2	
Energy				
Cattle	DE kcal/kg	838.	2601.	
Sheep	DE kcal/kg	767.	2381.	
Cattle	ME kcal/kg	687.	2133.	
Sheep	ME kcal/kg	628.	1952.	
Cattle	TDN %	19.	59.	
Sheep	TDN %	17.	54.	
Calcium	%	.56	1.74	13
Iron	%	.010	.030	

Continued

ALFALFA—BROME, SMOOTH. Medicago sativa, Bromus inermis

Alfalfa-Brome, smooth, aerial pt, fresh, early blm, (2)

Ref no 2-00-261

		As fed	Dry	
Dry matter	%	21.6	100.0	
Ash	%	2.1	9.8	
Crude fiber	%	5.5	25.3	
Ether extract	%	.8	3.6	
N-free extract	%	9.0	41.7	
Protein (N x 6.25)	%	4.2	19.6	
Cattle	dig prot %	3.1	14.4	
Sheep	dig prot %	3.0	14.1	
Energy				
Cattle	DE kcal/kg	590.	2734.	
Sheep	DE kcal/kg	600.	2778.	
Cattle	ME kcal/kg	484.	2242.	
Sheep	ME kcal/kg	492.	2278.	
Cattle	TDN %	13.	62.	
Sheep	TDN %	14.	63.	
Calcium	%	.33	1.52	
Magnesium	%	.08	.35	
Phosphorus	%	.08	.37	
Potassium	%	.84	3.87	

Alfalfa-brome, smooth, aerial pt, fresh, (2)

Ref no 2-00-262

		As fed	Dry	
Dry matter	%	21.6	100.0	
Ash	%	2.1	9.8	
Crude fiber	%	5.5	25.3	
Ether extract	%	.8	3.6	
N-free extract	%	9.0	41.7	
Protein (N x 6.25)	%	4.2	19.6	
Cattle	dig prot %	3.2	14.6	
Sheep	dig prot %	3.3	15.2	
Energy				
Cattle	DE kcal/kg	600.	2778.	
Sheep	DE kcal/kg	600.	2778.	
Cattle	ME kcal/kg	492.	2278.	
Sheep	ME kcal/kg	492.	2278.	
Cattle	TDN %	14.	63.	

Continued

(1) dry forages and roughages

(2) pasture, range plants, and forages fed green

(3) silages

(4) energy feeds

TABLE A-3-1 Composition of Feeds (Continued)

Feed name or analyses		Mean		C.V.
		As fed	Dry	± %
Sheep	TDN %	14.	63.	
Calcium	%	.33	1.52	69
Magnesium	%	.08	.35	48
Phosphorus	%	.08	.37	86
Potassium	%	.84	3.87	31

Alfalfa-Brome, smooth, aerial pt, ensiled, mn 50 dry matter, (3)

Ref no 3-08-148

		As fed	Dry	
Dry matter	%	55.0	100.0	
Ash	%	4.8	8.7	
Crude fiber	%	18.5	33.7	
Ether extract	%	1.9	3.4	
N-free extract	%	21.9	39.8	
Protein (N x 6.25)	%	7.9	14.4	
Cattle	dig prot %	4.6	8.4	
Sheep	dig prot %	4.1	7.5	
Energy				
Cattle	DE kcal/kg	1310.	2381.	
Sheep	DE kcal/kg	1358.	2469.	
Cattle	ME kcal/kg	1074.	1952.	
Sheep	ME kcal/kg	1113.	2024.	
Cattle	TDN %	30.	54.	
Sheep	TDN %	31.	56.	

Alfalfa-Brome, smooth, aerial pt, ensiled, mn 30 mx 50 dry matter, (3)

Ref no 3-08-147

		As fed	Dry	
Dry matter	%	46.5	100.0	
Ash	%	4.1	8.8	
Crude fiber	%	15.3	33.0	
Ether extract	%	1.1	2.3	
N-free extract	%	18.8	40.4	
Protein (N x 6.25)	%	7.2	15.5	
Cattle	dig prot %	4.2	9.0	
Sheep	dig prot %	3.8	8.1	
Energy				
Cattle	DE kcal/kg	1087.	2337.	
Sheep	DE kcal/kg	1128.	2425.	
Cattle	ME kcal/kg	891.	1916.	
Sheep	ME kcal/kg	924.	1988.	
Cattle	TDN %	25.	53.	
Sheep	TDN %	26.	55.	

Alfalfa-Brome, smooth, aerial pt, ensiled, mx 30 dry matter, (3)

Ref no 3-08-146

		As fed	Dry	
Dry matter	%	25.0	100.0	
Ash	%	2.1	8.4	
Crude fiber	%	7.7	30.8	
Ether extract	%	1.1	4.4	
N-free extract	%	10.3	41.2	
Protein (N x 6.25)	%	3.8	15.2	
Cattle	dig prot %	2.2	8.8	
Sheep	dig prot %	2.0	7.9	
Energy				
Cattle	DE kcal/kg	606.	2425.	
Sheep	DE kcal/kg	628.	2513.	
Cattle	ME kcal/kg	497.	1988.	
Sheep	ME kcal/kg	515.	2061.	
Cattle	TDN %	14.	55.	
Sheep	TDN %	14.	57.	

ALFALFA-ORCHARDGRASS. Medicago sativa, Dactylis glomerata

Alfalfa-Orchardgrass, aerial pt, ensiled, mn 50 dry matter, (3)

Ref no 3-08-143

		As fed	Dry	
Dry matter	%	61.0	100.0	
Ash	%	6.2	10.1	
Crude fiber	%	18.6	30.5	
Ether extract	%	2.6	4.3	
N-free extract	%	23.7	38.9	
Protein (N x 6.25)	%	9.9	16.2	
Cattle	dig prot %	5.7	9.4	
Sheep	dig prot %	5.1	8.4	
Energy				
Cattle	DE kcal/kg	1452.	2381.	
Sheep	DE kcal/kg	1506.	2469.	
Cattle	ME kcal/kg	1190.	1952.	
Sheep	ME kcal/kg	1235.	2024.	
Cattle	TDN %	33.	54.	
Sheep	TDN %	34.	56.	

(5) protein supplements
(6) minerals

(7) vitamins
(8) additives

Feed name or analyses		Mean		C.V
		As fed	Dry	± %

Alfalfa-Orchardgrass, aerial pt, ensiled, mn 30 mx 50 dry matter, (3)

Ref no 3-08-144

Dry matter	%	40.0	100.0	
Ash	%	3.6	9.1	
Crude fiber	%	12.6	31.6	
Ether extract	%	1.6	4.0	
N-free extract	%	15.2	38.1	
Protein (N x 6.25)	%	6.9	17.2	
Cattle	dig prot %	4.0	10.0	
Sheep	dig prot %	3.6	8.9	
Energy				
Cattle	DE kcal/kg	952.	2381.	
Sheep	DE kcal/kg	988.	2469.	
Cattle	ME kcal/kg	781.	1952.	
Sheep	ME kcal/kg	810.	2024.	
Cattle	TDN %	22.	54.	
Sheep	TDN %	22.	56.	

Alfalfa-Orchardgrass, aerial pt, ensiled, mx 30 dry matter, (3)

Ref no 3-08-145

Dry matter	%	28.0	100.0	
Ash	%	2.5	9.0	
Crude fiber	%	8.8	31.4	
Ether extract	%	1.1	4.1	
N-free extract	%	10.8	38.4	
Protein (N x 6.25)	%	4.8	17.1	
Cattle	dig prot %	2.8	9.9	
Sheep	dig prot %	2.5	8.9	
Energy				
Cattle	DE kcal/kg	667.	2381.	
Sheep	DE kcal/kg	691.	2469.	
Cattle	ME kcal/kg	546.	1952.	
Sheep	ME kcal/kg	567.	2024.	
Cattle	TDN %	15.	54.	
Sheep	TDN %	16.	56.	

ALKALI SACATON. Sporobolus airiodes

Alkali sacaton, aerial pt, fresh, dormant, (2)

Ref no 2-05-599

Dry matter	%	86.0	100.0	
Ash	%	10.8	12.6	
Ether extract	%	1.9	2.2	
Protein (N x 6.25)	%	2.9	3.4	

Continued

Feed name or analyses		Mean		C.V
		As fed	Dry	± %

Sheep	dig prot %	.0	.0	
Cellulose	%	28.4	33.0	
Lignin	%	8.6	10.0	
Energy	GE kcal/kg	3608.	4195.	
Sheep	DE kcal/kg	1689.	1964.	
Sheep	ME kcal/kg	1422.	1653.	
Sheep	TDN %	30.	35.	
Calcium	%	.58	.67	
Phosphorus	%	.07	.08	
Carotene	mg/kg	.3	.4	

ANIMAL. Scientific name not used

Animal, blood, dehy grnd, (5)
Blood meal (AAFCO)
Blood meal (CFA)

Ref no 5-00-380

Dry matter	%	91.0	100.0	4
Ash	%	5.6	6.2	31
Crude fiber	%	1.0	1.1	51
Ether extract	%	1.6	1.8	87
N-free extract	%	2.8	3.1	
Protein (N x 6.25)	%	79.9	87.8	6
Sheep	dig prot %	56.7	62.3	
Swine	dig prot %	62.3	68.5	
Energy	GE kcal/kg	5210.	5726.	
Sheep	DE kcal/kg	2608.	2866.	
Swine	DE kcal/kg	2684.	2949.	
Chickens	MEn kcal/kg	2844.	3125.	
Sheep	ME kcal/kg	2138.	2350.	
Swine	ME kcal/kg	2101.	2309.	
Sheep	TDN %	59.	65.	
Swine	TDN %	61.	67.	
Calcium	%	.28	.31	50
Iron	%	.376	.413	60
Magnesium	%	.22	.24	
Phosphorus	%	.22	.24	44
Potassium	%	.09	.10	
Sodium	%	.32	.35	
Copper	mg/kg	9.9	10.9	17
Manganese	mg/kg	5.3	5.8	82
Choline	mg/kg	757.	832.	
Niacin	mg/kg	31.5	34.6	
Pantothenic acid	mg/kg	1.1	1.2	
Riboflavin	mg/kg	1.5	1.6	94
Arginine	%	3.50	3.85	
Cystine	%	1.40	1.54	
Histidine	%	4.20	4.62	
Isoleucine	%	1.00	1.10	
Leucine	%	10.30	11.32	
Lysine	%	6.90	7.58	
Methionine	%	.90	.99	

Continued

(1) dry forages and roughages
(2) pasture, range plants, and forages fed green

(3) silages
(4) energy feeds

TABLE A-3-1 Composition of Feeds (Continued)

Feed name or analyses		Mean		C.V.
		As fed	Dry	± %
Phenylalanine	%	6.10	6.70	
Threonine	%	3.70	4.07	
Tryptophan	%	1.10	1.21	
Tyrosine	%	1.80	1.98	
Valine	%	6.50	7.14	

Animal, blood, spray dehy, (5)
   Blood flour

Ref no 5-00-381

Feed name or analyses		Mean		C.V.
		As fed	Dry	± %
Dry matter	%	91.0	100.0	2
Ash	%	4.8	5.3	43
Crude fiber	%	1.0	1.1	99
Ether extract	%	1.0	1.1	99
N-free extract	%	2.0	2.2	
Protein (N x 6.25)	%	82.2	90.3	5
Sheep	dig prot %	58.3	64.1	
Swine	dig prot %	64.1	70.4	
Energy				
Sheep	DE kcal/kg	2648.	2910.	
Swine	DE kcal/kg	2608.	2866.	
Sheep	ME kcal/kg	2171.	2386.	
Swine	ME kcal/kg	2029.	2230.	
Sheep	TDN %	60.	66.	
Swine	TDN %	59.	65.	
Calcium	%	.45	.49	56
Iron	%	.300	.330	69
Magnesium	%	.04	.04	52
Phosphorus	%	.37	.41	39
Potassium	%	.41	.45	
Sodium	%	.33	.36	
Cobalt	mg/kg	.100	.100	99
Copper	mg/kg	8.1	8.9	52
Manganese	mg/kg	6.4	7.0	46
Choline	mg/kg	279.	307.	
Niacin	mg/kg	28.6	31.4	23
Pantothenic acid	mg/kg	5.3	5.8	58
Riboflavin	mg/kg	4.2	4.6	46
Thiamine	mg/kg	.4	.4	29
Arginine	%	3.30	3.63	8
Histidine	%	4.80	5.28	10
Isoleucine	%	1.10	1.21	19
Leucine	%	10.60	11.65	19
Lysine	%	8.20	9.01	11
Methionine	%	1.00	1.10	7
Phenylalanine	%	5.60	6.15	6
Threonine	%	3.60	3.96	7
Tryptophan	%	1.00	1.10	11
Tyrosine	%	2.00	2.20	15
Valine	%	7.20	7.91	7

Animal carcass res, dry-rend dehy grnd,
   mx 4.4 P, (5)
   Meat meal (AAFCO)
   Meat scrap

Ref no 5-00-385

Feed name or analyses		Mean		C.V.
		As fed	Dry	± %
Dry matter	%	93.5	100.0	2
Ash	%	25.2	27.0	22
Crude fiber	%	2.4	2.5	40
Ether extract	%	9.9	10.6	24
N-free extract	%	2.6	2.8	
Protein (N x 6.25)	%	53.4	57.1	6
Cattle	dig prot %	48.6	52.0	
Sheep	dig prot %	43.8	46.8	
Swine	dig prot %	47.5	50.8	
Energy	GE kcal/kg	3181.	3402.	
Cattle	DE kcal/kg	3133.	3351.	
Sheep	DE kcal/kg	2885.	3086.	
Swine	DE kcal/kg	3010.	3219.	
Cattle	ME kcal/kg	2569.	2748.	
Chickens	MEn kcal/kg	1984.	2122.	
Sheep	ME kcal/kg	2366.	2530.	
Swine	ME kcal/kg	2543.	2720.	
Cattle	TDN %	71.	76.	
Sheep	TDN %	65.	70.	
Swine	TDN %	68.	73.	
Calcium	%	7.94	8.49	33
Chlorine	%	1.31	1.40	
Iron	%	.044	.047	43
Magnesium	%	.27	.29	
Phosphorus	%	4.03	4.31	43
Potassium	%	.55	.59	
Sodium	%	1.68	1.80	
Sulfur	%	.50	.53	
Cobalt	mg/kg	.128	.137	55
Copper	mg/kg	9.7	10.4	30
Manganese	mg/kg	9.5	10.2	51
Biotin	mg/kg	.09	.10	
Choline	mg/kg	1955.	2091.	33
Niacin	mg/kg	56.9	60.8	20
Pantothenic acid	mg/kg	4.8	5.1	41
Riboflavin	mg/kg	5.3	5.7	42
Thiamine	mg/kg	.2	.2	
Vitamin B_{12}	mcg/kg	51.1	54.6	
Arginine	%	3.70	3.96	27
Cystine	%	.60	.64	35
Cysteine	%	.80	.86	
Glutamic acid	%	8.10	8.66	23
Glycine	%	2.20	2.35	
Histidine	%	1.10	1.18	35
Isoleucine	%	1.90	2.03	22
Leucine	%	3.50	3.74	17
Lysine	%	3.80	4.06	39
Methionine	%	.80	.86	25

Continued

(5) protein supplements
(6) minerals

(7) vitamins
(8) additives

Feed name or analyses		Mean		C.V.
		As fed	Dry	± %
Phenylalanine	%	1.90	2.03	23
Serine	%	2.10	2.24	
Threonine	%	1.80	1.92	14
Tryptophan	%	.30	.32	25
Tyrosine	%	.90	.96	20
Valine	%	2.60	2.78	17

Animal, carcass res w blood, dry- or wet-rend dehy grnd, mx 4.4 P, (5)

Meat meal tankage (AAFCO)

Digester tankage

Ref no 5-00-386

		As fed	Dry	C.V. ± %
Dry matter	%	92.0	100.0	3
Ash	%	21.4	23.3	20
Crude fiber	%	2.0	2.2	63
Ether extract	%	8.1	8.8	29
N-free extract	%	.6	.7	
Protein (N x 6.25)	%	59.8	65.0	10
Cattle	dig prot %	54.5	59.2	
Sheep	dig prot %	49.7	54.0	
Swine	dig prot %	37.1	40.3	
Lignin	%	3.0	3.3	
Energy				
Cattle	DE kcal/kg	3204.	3483.	
Sheep	DE kcal/kg	2839.	3086.	
Swine	DE kcal/kg	2475.	2690.	
Cattle	ME kcal/kg	2628.	2856.	
Chickens	MEn kcal/kg	2646.	2876.	
Sheep	ME kcal/kg	2328.	2530.	
Swine	ME kcal/kg	2052.	2230.	
Cattle	TDN %	73.	79.	
Sheep	TDN %	64.	70.	
Swine	TDN %	56.	61.	
Calcium	%	5.94	6.46	36
Magnesium	%	.16	.17	
Phosphorus	%	3.17	3.45	24
Potassium	%	.56	.61	
Sodium	%	1.67	1.82	
Cobalt	mg/kg	.200	.200	53
Copper	mg/kg	38.7	42.1	33
Manganese	mg/kg	19.1	20.8	57
Choline	mg/kg	2169.	2358.	
Folic acid	mg/kg	1.50	1.60	
Niacin	mg/kg	39.2	42.6	20
Pantothenic acid	mg/kg	2.4	2.6	32
Riboflavin	mg/kg	2.4	2.6	61
Arginine	%	3.60	3.91	23
Histidine	%	1.90	2.07	42
Isoleucine	%	1.90	2.07	40
Leucine	%	5.10	5.54	25
Lysine	%	4.00	4.35	23
Methionine	%	.80	.87	11
Phenylalanine	%	2.70	2.93	37

Continued

Feed name or analyses		Mean		C.V.
		As fed	Dry	± %
Threonine	%	2.40	2.61	25
Tryptophan	%	.70	.76	34
Valine	%	4.20	4.57	29

Animal, carcass res w bone, dry-rend dehy grnd, mn 4.4 P, (5)

Meat and bone meal (AAFCO)

Meat and bone scrap

Ref no 5-00-388

		As fed	Dry	C.V. ± %
Dry matter	%	94.0	100.0	2
Ash	%	29.1	31.0	15
Crude fiber	%	2.2	2.3	52
Ether extract	%	9.5	10.1	31
N-free extract	%	2.6	2.8	
Protein (N x 6.25)	%	50.6	53.8	6
Cattle	dig prot %	46.1	49.0	
Sheep	dig prot %	41.4	44.1	
Swine	dig prot %	45.0	47.9	
Energy				
Cattle	DE kcal/kg	2984.	3175.	
Sheep	DE kcal/kg	2735.	2910.	
Swine	DE kcal/kg	2859.	3042.	
Cattle	ME kcal/kg	2448.	2604.	
Chickens	MEn kcal/kg	1984.	2111.	
Sheep	ME kcal/kg	2243.	2386.	
Swine	ME kcal/kg	2434.	2589.	
Cattle	TDN %	68.	72.	
Sheep	TDN %	62.	66.	
Swine	TDN %	65.	69.	
Calcium	%	10.57	11.25	13
Chlorine	%	.35	.37	
Iron	%	.050	.053	
Magnesium	%	1.13	1.20	
Phosphorus	%	5.07	5.39	20
Potassium	%	1.46	1.55	
Sodium	%	.73	.78	
Cobalt	mg/kg	.183	.195	
Copper	mg/kg	1.5	1.6	
Manganese	mg/kg	12.3	13.1	
Choline	mg/kg	2189.	2329.	30
Niacin	mg/kg	47.8	50.8	46
Pantothenic acid	mg/kg	3.7	3.9	30
Riboflavin	mg/kg	4.4	4.7	31
Thiamine	mg/kg	1.1	1.2	
Vitamin B12	mcg/kg	44.8	47.7	
Arginine	%	4.00	4.26	18
Cystine	%	.60	.64	
Glutamic acid	%	11.00	11.70	
Glycine	%	6.60	7.02	
Histidine	%	.90	.96	36
Isoleucine	%	1.70	1.81	20
Leucine	%	3.10	3.30	6
Lysine	%	3.50	3.72	45

Continued

(1) dry forages and roughages

(2) pasture, range plants, and forages fed green

(3) silages

(4) energy feeds

Feed name or analyses		Mean		C.V. ± %
		As fed	Dry	
Methionine	%	.70	.74	11
Phenylalanine	%	1.80	1.92	19
Threonine	%	1.80	1.92	24
Tryptophan	%	.20	.21	25
Valine	%	2.40	2.55	30

Animal fat - see Animal-Poultry, fat; see Swine, lard

Animal, liver, dehy grnd, (5)
Animal liver meal (AAFCO)
Animal liver meal (CFA)
Liver meal

Ref no 5-00-389

		As fed	Dry	
Dry matter	%	92.6	100.0	2
Ash	%	6.0	6.5	27
Crude fiber	%	1.3	1.4	40
Ether extract	%	15.1	16.3	18
N-free extract	%	3.7	4.0	
Protein (N x 6.25)	%	66.5	71.8	7
Swine	dig prot %	64.4	69.6	
Energy				
Swine	DE kcal/kg	3920.	4233.	
Swine	ME kcal/kg	3195.	3450.	
Swine	TDN %	89.	96.	
Calcium	%	.50	.54	52
Iron	%	.063	.068	32
Phosphorus	%	1.25	1.35	23
Cobalt	mg/kg	.134	.144	31
Copper	mg/kg	89.3	96.4	63
Manganese	mg/kg	8.8	9.5	62
Biotin	mg/kg	.02	.02	
Folic acid	mg/kg	5.56	6.00	
Niacin	mg/kg	203.7	220.0	44
Pantothenic acid	mg/kg	45.2	48.8	62
Riboflavin	mg/kg	46.3	50.0	32
Thiamine	mg/kg	.2	.2	80
Vitamin B$_{12}$	mcg/kg	501.5	541.6	
Arginine	%	4.10	4.43	10
Cystine	%	.90	.97	
Glutamic acid	%	8.10	8.75	
Glycine	%	5.60	6.05	
Histidine	%	1.50	1.62	12
Isoleucine	%	3.40	3.67	29
Leucine	%	5.40	5.83	10
Lysine	%	4.80	5.18	21
Methionine	%	1.30	1.40	28
Phenylalanine	%	2.90	3.13	13
Threonine	%	2.60	2.81	13
Tryptophan	%	.60	.65	30
Tyrosine	%	1.70	1.84	7
Valine	%	4.20	4.54	14

Animal, bone, cooked dehy grnd, mn 10 P, (6)
Feeding bone meal (CFA)

Ref no 6-00-397

		As fed	Dry	
Dry matter	%	94.5	100.0	
Ether extract	%	9.6	10.2	
Protein (N x 6.25)	%	17.8	18.8	
Calcium	%	25.82	27.32	
Phosphorus	%	12.35	13.07	
Fluorine	mg/kg	2000.00	2116.40	

Animal, bone, steamed dehy grnd, (6)
Bone meal, steamed (AAFCO)

Ref no 6-00-400

		As fed	Dry	
Dry matter	%	95.0	100.0	2
Ash	%	71.8	75.6	12
Crude fiber	%	2.0	2.1	58
Ether extract	%	3.2	3.4	94
Protein (N x 6.25)	%	12.1	12.7	53
Cattle	dig prot %	8.2	8.6	
Sheep	dig prot %	8.2	8.6	
Swine	dig prot %	9.4	9.9	
Energy				
Cattle	DE kcal/kg	670.	705.	
Sheep	DE kcal/kg	670.	705.	
Swine	DE kcal/kg	660.	695.	
Cattle	ME kcal/kg	549.	578.	
Chickens	ME$_n$ kcal/kg	992.	1044.	
Sheep	ME kcal/kg	549.	578.	
Swine	ME kcal/kg	616.	649.	
Cattle	TDN %	15.	16.	
Sheep	TDN %	15.	16.	
Swine	TDN %	15.	16.	
Calcium	%	28.98	30.51	9
Iron	%	.084	.088	71
Magnesium	%	.64	.67	41
Phosphorus	%	13.59	14.31	14
Sodium	%	.46	.48	9
Cobalt	mg/kg	.100	.100	29
Copper	mg/kg	16.3	17.2	44
Manganese	mg/kg	30.4	32.0	66
Zinc	mg/kg	424.6	447.1	
Niacin	mg/kg	4.2	4.4	58
Pantothenic acid	mg/kg	2.4	2.5	53
Riboflavin	mg/kg	.9	.9	33
Thiamine	mg/kg	.4	.4	99

(5) protein supplements
(6) minerals

(7) vitamins
(8) additives

Feed name or analyses		Mean		C.V.
		As fed	Dry	± %

Animal, bone charcoal, retort-charred grnd, (6)
  Bone black (CFA)
  Bone char (CFA)
  Spent bone black

Ref no 6-00-403

		As fed	Dry	
Dry matter	%	90.0	100.0	
Protein (N x 6.25)	%	8.5	9.4	
Calcium	%	27.10	30.11	
Magnesium	%	.53	.59	
Phosphorus	%	12.73	14.14	6
Potassium	%	.14	.16	
Arginine	%	1.80	2.00	
Histidine	%	.20	.22	
Isoleucine	%	.60	.67	
Leucine	%	.80	.89	
Lysine	%	1.00	1.11	
Methionine	%	.20	.22	
Phenylalanine	%	.50	.56	
Threonine	%	.50	.56	
Valine	%	.70	.78	

Animal, bone phosphate, precip dehy, mn 17 P, (6)
  Bone phosphate (AAFCO)

Ref no 6-00-406

		As fed	Dry	
Dry matter	%	99.0	100.0	
Ash	%	86.4	87.3	
Ether extract	%	.3	.3	
Protein (N x 6.25)	%	.4	.4	
Calcium	%	28.00	28.28	
Phosphorus	%	11.20	11.31	

ANIMAL—POULTRY. Scientific name not used

Animal-poultry, fat, heat-rend, mn 90 fatty
  acids mx 2.5 unsaponifiable matter mx
  1 insol matter, (4)
  Animal fat (AAFCO)

Ref no 4-00-409

		As fed	Dry	
Dry matter	%	99.5	100.0	
Ether extract	%	99.4	99.9	
Energy	GE kcal/kg	9450.	9497.	
Chickens	ME$_n$ kcal/kg	7090.	7126.	
Cattle	NE$_m$ kcal/kg	4552.	4575.	
Cattle	NE$_{gain}$ kcal/kg	2603.	2616.	

Animal-poultry, carcass res mx 35 blood, dry- or
  wet-rend dehy grnd, mn 50 prot, (5)
  Feeding tankage (CFA)

Ref no 5-00-410

		As fed	Dry	
Dry matter	%	93.1	100.0	
Ash	%	8.5	9.1	
Ether extract	%	13.6	14.6	
Protein (N x 6.25)	%	52.1	55.9	
Cattle	dig prot %	47.4	50.9	
Sheep	dig prot %	43.2	46.4	
Swine	dig prot %	42.6	45.8	
Energy				
Cattle	DE kcal/kg	3448.	3704.	
Sheep	DE kcal/kg	2956.	3175.	
Swine	DE kcal/kg	3161.	3395.	
Cattle	ME kcal/kg	2827.	3037.	
Sheep	ME kcal/kg	2424.	2604.	
Swine	ME kcal/kg	2678.	2876.	
Cattle	TDN %	78.	84.	
Sheep	TDN %	67.	72.	
Swine	TDN %	72.	77.	
Calcium	%	11.56	12.42	
Phosphorus	%	5.37	5.77	

Animal-poultry, carcass res w bone mx 35 blood,
  dry- or wet-rend dehy grnd, mn 40 prot, (5)
  Feeding meat and bone tankage (CFA)

Ref no 5-00-413

		As fed	Dry	
Dry matter	%	94.0	100.0	1
Ash	%	26.3	28.0	17
Crude fiber	%	3.0	3.2	20
Ether extract	%	11.9	12.7	26
N-free extract	%	3.1	3.3	
Protein (N x 6.25)	%	49.6	52.8	10
Cattle	dig prot %	45.1	48.0	
Sheep	dig prot %	41.2	43.8	
Swine	dig prot %	40.7	43.3	
Energy				
Cattle	DE kcal/kg	3191.	3395.	
Sheep	DE kcal/kg	2735.	2910.	
Swine	DE kcal/kg	2984.	3175.	
Cattle	ME kcal/kg	2617.	2784.	
Chickens	ME$_n$ kcal/kg	1874.	1994.	
Sheep	ME kcal/kg	2243.	2386.	
Swine	ME kcal/kg	2546.	2708.	
Cattle	TDN %	72.	77.	
Sheep	TDN %	62.	66.	
Swine	TDN %	68.	72.	
Calcium	%	10.97	11.67	
Phosphorus	%	5.14	5.47	

(1) dry forages and roughages
(2) pasture, range plants, and forages fed green

(3) silages
(4) energy feeds

Feed name or analyses		Mean		C.V.
		As fed	Dry	± %

APPLES. Malus spp

Apples, pulp, dehy grnd, (4)
Dried apple pomace (AAFCO)

Ref no 4-00-423

Dry matter	%	91.0	100.0	3
Ash	%	1.9	2.1	28
Crude fiber	%	16.0	17.6	17
Ether extract	%	4.7	5.2	16
N-free extract	%	63.0	69.2	
Protein (N x 6.25)	%	5.4	5.9	20
Cattle	dig prot %	1.7	1.9	
Sheep	dig prot %	.4	.5	
Energy	GE kcal/kg	4165.	4577.	
Cattle	DE kcal/kg	3049.	3351.	
Sheep	DE kcal/kg	2528.	2778.	
Cattle	ME kcal/kg	2501.	2748.	
Sheep	ME kcal/kg	2073.	2278.	
Cattle	TDN %	69.	76.	
Sheep	TDN %	57.	63.	
Calcium	%	.13	.14	32
Iron	%	.030	.030	
Magnesium	%	.06	.07	42
Phosphorus	%	.12	.13	55
Potassium	%	.45	.49	
Sodium	%	.03	.03	
Manganese	mg/kg	7.3	8.0	

Atlas - see Sorghum, atlas

Bagasse - see Sugarcane, pulp

BAKERY REFUSE. Scientific name not used

Bakery, refuse, dehy grnd, mx salt declared above 3, (4)
Dried bakery product (AAFCO)

Ref no 4-00-466

Dry matter	%	91.6	100.0	
Ash	%	1.6	1.7	
Crude fiber	%	.7	.8	
Ether extract	%	13.7	15.0	
N-free extract	%	64.7	70.6	
Protein (N x 6.25)	%	10.9	11.9	
Cattle	dig prot %	6.3	6.9	
Sheep	dig prot %	7.4	8.1	

Continued

Feed name or analyses		Mean		C.V.
		As fed	Dry	± %
Energy				
Cattle	DE kcal/kg	3594.	3924.	
Sheep	DE kcal/kg	3635.	3968.	
Cattle	ME kcal/kg	2948.	3218.	
Sheep	ME kcal/kg	2981.	3254.	
Cattle	TDN %	82.	89.	
Sheep	TDN %	82.	90.	

Baking refuse - see Bread, dehy

BARLEY. Hordeum vulgare

Barley, hay, s-c, (1)

Ref no 1-00-495

Dry matter	%	87.3	100.0	2
Ash	%	6.8	7.8	9
Crude fiber	%	23.0	26.4	11
Ether extract	%	1.8	2.2	17
N-free extract	%	47.8	54.7	
Protein (N x 6.25)	%	7.8	8.9	19
Cattle	dig prot %	4.4	5.0	
Sheep	dig prot %	4.4	5.0	
Energy				
Cattle	DE kcal/kg	2194.	2513.	
Sheep	DE kcal/kg	2194.	2513.	
Cattle	ME kcal/kg	1799.	2061.	
Sheep	ME kcal/kg	1799.	2061.	
Cattle	NE_m kcal/kg	1074.	1230.	
Cattle	NE_gain kcal/kg	498.	570.	
Cattle	TDN %	50.	57.	
Sheep	TDN %	50.	57.	
Calcium	%	.18	.21	20
Phosphorus	%	.26	.30	23
Manganese	mg/kg	34.2	39.2	18

Barley, straw, (1)

Ref no 1-00-498

Dry matter	%	88.2	100.0	3
Ash	%	5.8	6.6	10
Crude fiber	%	37.4	42.4	6
Ether extract	%	1.6	1.8	12
N-free extract	%	39.8	45.1	
Protein (N x 6.25)	%	3.6	4.1	9
Cattle	dig prot %	.4	.5	
Sheep	dig prot %	.6	.7	
Energy				
Cattle	DE kcal/kg	1595.	1808.	
Sheep	DE kcal/kg	1905.	2160.	

Continued

Feed name or analyses		Mean		C.V.
		As fed	Dry	± %
Cattle	ME kcal/kg	1307.	1482.	
Sheep	ME kcal/kg	1562.	1771.	
Cattle	NE$_m$ kcal/kg	891.	1010.	
Cattle	NE$_{gain}$ kcal/kg	123.	140.	
Cattle	TDN %	36.	41.	
Sheep	TDN %	43.	49.	
Calcium	%	.30	.34	19
Iron	%	.030	.030	21
Magnesium	%	.17	.19	17
Phosphorus	%	.08	.09	35
Potassium	%	2.01	2.28	10
Sodium	%	.12	.14	
Manganese	mg/kg	15.2	17.2	51

Barley, aerial pt, fresh, immature, (2)

Ref no 2-00-499

		As fed	Dry	
Dry matter	%	23.1	100.0	17
Ash	%	3.3	14.2	25
Crude fiber	%	3.8	16.5	13
Ether extract	%	1.0	4.5	24
N-free extract	%	8.4	36.5	
Protein (N x 6.25)	%	6.5	28.3	15
Cattle	dig prot %	5.0	21.9	
Sheep	dig prot %	4.6	20.1	
Cellulose	%	4.4	19.0	
Lignin	%	.4	1.8	
Energy				
Cattle	DE kcal/kg	733.	3175.	
Sheep	DE kcal/kg	632.	2734.	
Cattle	ME kcal/kg	602.	2604.	
Sheep	ME kcal/kg	518.	2242.	
Cattle	TDN %	17.	72.	
Sheep	TDN %	14.	62.	
Calcium	%	.17	.74	29
Phosphorus	%	.13	.56	16
Potassium	%	.71	3.06	32
Carotene	mg/kg	127.3	551.2	22
Vitamin A equiv	IU/g	212.2	918.8	

Barley, aerial pt, fresh, early blm, (2)

Ref no 2-00-500

		As fed	Dry	
Dry matter	%	11.7	100.0	
Ash	%	1.9	16.5	9
Crude fiber	%	2.3	19.4	5
Ether extract	%	.5	4.0	22
N-free extract	%	4.3	36.4	
Protein (N x 6.25)	%	2.8	23.7	18
Cattle	dig prot %	2.1	18.0	
Sheep	dig prot %	2.0	16.8	
Cellulose	%	2.4	20.1	

Continued

Feed name or analyses		Mean		C.V.
		As fed	Dry	± %
Lignin	%	.3	2.5	
Energy				
Cattle	DE kcal/kg	330.	2822.	
Sheep	DE kcal/kg	304.	2601.	
Cattle	ME kcal/kg	271.	2314.	
Sheep	ME kcal/kg	250.	2133.	
Cattle	TDN %	7.	64.	
Sheep	TDN %	7.	59.	

Barley, aerial pt, fresh, mid-blm, (2)

Ref no 2-00-501

		As fed	Dry	
Dry matter	%	16.7	100.0	
Ash	%	1.5	8.9	
Crude fiber	%	3.6	21.5	
Ether extract	%	.7	4.3	
N-free extract	%	8.1	48.3	
Protein (N x 6.25)	%	2.8	17.0	
Cattle	dig prot %	2.0	12.3	
Sheep	dig prot %	2.1	12.4	
Cellulose	%	3.6	21.9	
Lignin	%	.5	2.9	
Energy				
Cattle	DE kcal/kg	501.	2998.	
Sheep	DE kcal/kg	486.	2910.	
Cattle	ME kcal/kg	410.	2458.	
Sheep	ME kcal/kg	398.	2386.	
Cattle	TDN %	11.	68.	
Sheep	TDN %	11.	66.	

Barley, aerial pt, fresh, (2)

Ref no 2-00-511

		As fed	Dry	
Dry matter	%	21.1	100.0	19
Ash	%	2.7	12.9	27
Crude fiber	%	4.4	20.8	26
Ether extract	%	.8	3.9	23
N-free extract	%	8.7	41.2	
Protein (N x 6.25)	%	4.5	21.2	25
Cattle	dig prot %	3.4	15.9	
Sheep	dig prot %	3.2	15.0	
Cellulose	%	5.7	27.0	15
Lignin	%	1.0	4.7	39
Energy				
Cattle	DE kcal/kg	623.	2954.	
Sheep	DE kcal/kg	577.	2734.	
Cattle	ME kcal/kg	511.	2422.	
Sheep	ME kcal/kg	473.	2242.	
Cattle	NE$_m$ kcal/kg	283.	1340.	
Cattle	NE$_{gain}$ kcal/kg	156.	740.	
Cattle	TDN %	14.	67.	
Sheep	TDN %	13.	62.	

Continued

(1) dry forages and roughages
(2) pasture, range plants, and forages fed green

(3) silages
(4) energy feeds

Feed name or analyses		Mean		C.V. ± %
		As fed	Dry	
Calcium	%	.10	.48	43
Phosphorus	%	.08	.37	31
Potassium	%	.64	3.06	32
Carotene	mg/kg	82.2	389.6	32
Vitamin A equiv	IU/g	137.0	649.5	

Barley, grain, grnd, (4)

Ref no 4-00-526

		As fed	Dry	
Dry matter	%	91.0	100.0	
Ash	%	5.5	6.0	
Crude fiber	%	2.0	2.2	
Ether extract	%	1.0	1.1	
N-free extract	%	67.8	74.5	
Protein (N x 6.25)	%	14.7	16.2	
Cattle	dig prot %	11.1	12.2	
Sheep	dig prot %	11.6	12.8	
Swine	dig prot %	11.4	12.5	
Energy				
Cattle	DE kcal/kg	3290.	3616.	
Sheep	DE kcal/kg	3371.	3704.	
Swine	DE kcal/kg	3210.	3527.	
Cattle	ME kcal/kg	2698.	2965.	
Sheep	ME kcal/kg	2764.	3037.	
Swine	ME kcal/kg	2976.	3270.	
Cattle	TDN %	75.	82.	
Sheep	TDN %	76.	84.	
Swine	TDN %	73.	80.	
Calcium	%	.05	.05	
Iron	%	.010	.010	
Phosphorus	%	.70	.77	
Copper	mg/kg	9.5	10.4	
Manganese	mg/kg	16.9	18.6	
Niacin	mg/kg	36.5	40.1	
Pantothenic acid	mg/kg	14.5	15.9	
Riboflavin	mg/kg	4.0	4.4	
Thiamine	mg/kg	4.6	5.1	

Barley, grain, thresher-run, mn 48 wt mn 10 mx 20 fm, (4)

Ref no 4-08-159 Canada

		As fed	Dry	
Dry matter	%	90.0	100.0	
Ash	%	2.6	2.9	
Crude fiber	%	4.5	5.0	
Ether extract	%	.9	1.0	
N-free extract	%	70.0	77.8	
Protein (N x 6.25)	%	11.8	13.1	
Cattle	dig prot %	8.8	9.8	
Sheep	dig prot %	9.3	10.3	
Swine	dig prot %	9.1	10.1	
Energy	GE kcal/kg	4208.	4675.	

Continued

Feed name or analyses		Mean		C.V. ± %
		As fed	Dry	
Cattle	DE kcal/kg	3254.	3616.	
Sheep	DE kcal/kg	3373.	3748.	
Swine	DE kcal/kg	3214.	3571.	
Cattle	ME kcal/kg	2668.	2965.	
Sheep	ME kcal/kg	2766.	3073.	
Swine	ME kcal/kg	3002.	3335.	
Cattle	TDN %	74.	82.	
Sheep	TDN %	76.	85.	
Swine	TDN %	73.	81.	

Barley, grain, thresher-run, mn 48 wt mx 10 fm, (4)

Ref no 4-08-158 Canada

		As fed	Dry	
Dry matter	%	90.0	100.0	
Ash	%	2.8	3.1	
Crude fiber	%	4.5	5.0	
Ether extract	%	1.0	1.1	
N-free extract	%	70.4	78.2	
Protein (N x 6.25)	%	11.3	12.6	
Cattle	dig prot %	8.5	9.4	
Sheep	dig prot %	9.0	10.0	
Swine	dig prot %	8.7	9.7	
Energy	GE kcal/kg	4140.	4600.	
Cattle	DE kcal/kg	3254.	3616.	
Sheep	DE kcal/kg	3373.	3748.	
Swine	DE kcal/kg	3214.	3571.	
Cattle	ME kcal/kg	2668.	2965.	
Sheep	ME kcal/kg	2766.	3073.	
Swine	ME kcal/kg	3024.	3360.	
Cattle	TDN %	74.	82.	
Sheep	TDN %	76.	85.	
Swine	TDN %	73.	81.	

Barley, grain, thresher-run, mx 48 wt mn 10 mx 20 fm, (4)

Ref no 4-08-156 Canada

		As fed	Dry	
Dry matter	%	90.0	100.0	
Ash	%	2.8	3.1	
Crude fiber	%	5.9	6.6	
Ether extract	%	1.1	1.2	
N-free extract	%	68.5	76.1	
Protein (N x 6.25)	%	11.7	13.0	
Cattle	dig prot %	8.8	9.8	
Sheep	dig prot %	9.3	10.3	
Swine	dig prot %	9.0	10.0	
Energy				
Cattle	DE kcal/kg	3214.	3571.	
Sheep	DE kcal/kg	3334.	3704.	
Swine	DE kcal/kg	3174.	3527.	
Cattle	ME kcal/kg	2635.	2928.	

Continued

Feed name or analyses		Mean		C.V.
		As fed	Dry	± %
Sheep	ME kcal/kg	2733.	3037.	
Swine	ME kcal/kg	2965.	3294.	
Cattle	TDN %	73.	81.	
Sheep	TDN %	76.	84.	
Swine	TDN %	72.	80.	

Barley, grain, thresher-run, mx 48 wt mx 10 fm, (4)

Ref no 4-08-155 Canada

Dry matter	%	90.0	100.0	
Ash	%	2.9	3.2	
Crude fiber	%	5.1	5.7	
Ether extract	%	.8	.9	
N-free extract	%	69.4	77.1	
Protein (N x 6.25)	%	11.8	13.1	
Cattle	dig prot %	8.8	9.8	
Sheep	dig prot %	9.3	10.3	
Swine	dig prot %	9.1	10.1	
Energy	GE kcal/kg	4140.	4600.	
Cattle	DE kcal/kg	3254.	3616.	
Sheep	DE kcal/kg	3373.	3748.	
Swine	DE kcal/kg	3174.	3527.	
Cattle	ME kcal/kg	2668.	2965.	
Sheep	ME kcal/kg	2766.	3073.	
Swine	ME kcal/kg	2952.	3280.	
Cattle	TDN %	74.	82.	
Sheep	TDN %	76.	85.	
Swine	TDN %	72.	80.	

Barley, grain, (4)

Ref no 4-00-530

Dry matter	%	89.0	100.0	3
Ash	%	2.4	2.7	60
Crude fiber	%	5.0	5.6	36
Ether extract	%	1.9	2.1	4
N-free extract	%	68.2	76.6	
Protein (N x 6.25)	%	11.6	13.0	15
Cattle	dig prot %	8.7	9.8	
Sheep	dig prot %	9.2	10.3	
Swine	dig prot %	8.2	9.2	
Energy	GE kcal/kg	4084.	4589.	
Cattle	DE kcal/kg	3257.	3660.	
Sheep	DE kcal/kg	3375.	3792.	
Swine	DE kcal/kg	3080.	3461.	
Cattle	ME kcal/kg	2671.	3001.	
Chickens	MEn kcal/kg	2646.	2973.	
Sheep	ME kcal/kg	2767.	3109.	
Swine	ME kcal/kg	2876.	3232.	
Cattle	NEm kcal/kg	1896.	2130.	
Cattle	NEgain kcal/kg	1246.	1400.	
Cattle	TDN %	74.	83.	

Continued

Feed name or analyses		Mean		C.V.
		As fed	Dry	± %
Sheep	TDN %	76.	86.	
Swine	TDN %	70.	79.	
Calcium	%	.08	.09	66
Iron	%	.005	.006	25
Magnesium	%	.12	.14	11
Phosphorus	%	.42	.47	13
Potassium	%	.56	.63	19
Sodium	%	.02	.02	
Cobalt	mg/kg	.100	.100	
Copper	mg/kg	7.6	8.6	62
Manganese	mg/kg	16.3	18.3	33
Zinc	mg/kg	15.3	17.2	
Biotin	mg/kg	.20	.20	99
Choline	mg/kg	1030.	1157.	
Folic acid	mg/kg	.50	.60	
Niacin	mg/kg	57.4	64.5	31
Pantothenic acid	mg/kg	6.5	7.3	26
Riboflavin	mg/kg	2.0	2.2	70
Thiamine	mg/kg	5.1	5.7	24
a-tocopherol	mg/kg	6.1	6.8	50
Vitamin B6	mg/kg	2.90	3.30	99
Arginine	%	.53	.60	16
Cystine	%	.18	.20	20
Glycine	%	.36	.40	
Histidine	%	.27	.30	23
Isoleucine	%	.53	.60	22
Leucine	%	.80	.90	26
Lysine	%	.53	.60	45
Methionine	%	.18	.20	50
Phenylalanine	%	.62	.70	20
Threonine	%	.36	.40	17
Tryptophan	%	.18	.20	35
Tyrosine	%	.36	.40	40
Valine	%	.62	.70	14

Barley, grain, Can 1 feed mn 46 wt mx 4 fm, (4)

Ref no 4-00-531

Dry matter	%	86.5	100.0	
Ash	%	2.3	2.6	20
Crude fiber	%	4.8	5.6	62
Ether extract	%	1.9	2.2	52
N-free extract	%	66.1	76.4	
Protein (N x 6.25)	%	11.4	13.2	10
Cattle	dig prot %	8.6	9.9	
Sheep	dig prot %	9.0	10.4	
Swine	dig prot %	8.9	10.3	
Energy	GE kcal/kg	4599.	5317.	
Cattle	DE kcal/kg	3166.	3660.	
Sheep	DE kcal/kg	3280.	3792.	
Swine	DE kcal/kg	3128.	3616.	
Cattle	ME kcal/kg	2596.	3001.	
Sheep	ME kcal/kg	2689.	3109.	
Swine	ME kcal/kg	2918.	3374.	

Continued

(1) dry forages and roughages
(2) pasture, range plants, and forages fed green

(3) silages
(4) energy feeds

Feed name or analyses		Mean		C.V.
		As fed	Dry	± %
Cattle	TDN %	72.	83.	
Sheep	TDN %	74.	86.	
Swine	TDN %	71.	82.	
Alanine	%	.41	.47	
Arginine	%	.46	.53	
Aspartic acid	%	.55	.64	
Glutamic acid	%	2.37	2.74	
Glycine	%	.42	.49	
Histidine	%	.21	.24	
Isoleucine	%	.32	.37	
Leucine	%	.67	.78	
Lysine	%	.35	.40	
Methionine	%	.10	.12	
Phenylalanine	%	.47	.54	
Proline	%	.99	1.14	
Serine	%	.41	.48	
Threonine	%	.32	.37	
Tyrosine	%	.26	.30	
Valine	%	.42	.49	

Barley, grain, Can 2 feed mn 43 wt mx 10 fm, (4)

Ref no 4-00-532

Feed name or analyses		As fed	Dry	C.V. ± %
Dry matter	%	86.5	100.0	
Ash	%	2.2	2.6	
Crude fiber	%	5.0	5.7	
Ether extract	%	1.9	2.2	
N-free extract	%	66.1	76.4	
Protein (N x 6.25)	%	11.4	13.1	
Cattle	dig prot %	8.5	9.8	
Sheep	dig prot %	8.9	10.3	
Swine	dig prot %	8.8	10.2	
Energy	GE kcal/kg	4599.	5317.	
Cattle	DE kcal/kg	3166.	3660.	
Sheep	DE kcal/kg	3280.	3792.	
Swine	DE kcal/kg	3128.	3616.	
Cattle	ME kcal/kg	2596.	3001.	
Sheep	ME kcal/kg	2689.	3109.	
Swine	ME kcal/kg	2921.	3377.	
Cattle	TDN %	72.	83.	
Sheep	TDN %	74.	86.	
Swine	TDN %	71.	82.	
Alanine	%	.40	.46	
Arginine	%	.46	.53	
Aspartic acid	%	.64	.74	
Glutamic acid	%	2.40	2.78	
Glycine	%	.40	.46	
Histidine	%	.19	.22	
Isoleucine	%	.36	.42	
Leucine	%	.67	.78	
Lysine	%	.37	.43	
Methionine	%	.10	.12	
Phenylalanine	%	.50	.58	
Proline	%	1.05	1.21	

Continued

(5) protein supplements

(6) minerals

Feed name or analyses		Mean		C.V.
		As fed	Dry	± %
Serine	%	.40	.46	
Threonine	%	.34	.39	
Tyrosine	%	.24	.28	
Valine	%	.50	.58	

Barley, grain, Can 3 feed mx 20 fm, (4)

Ref no 4-00-533

Feed name or analyses		As fed	Dry	C.V. ± %
Dry matter	%	86.5	100.0	
Ash	%	2.4	2.7	27
Crude fiber	%	5.4	6.3	60
Ether extract	%	2.1	2.4	48
N-free extract	%	65.2	75.4	
Protein (N x 6.25)	%	11.4	13.2	11
Cattle	dig prot %	8.6	9.9	
Sheep	dig prot %	9.0	10.4	
Swine	dig prot %	8.7	10.1	
Energy	GE kcal/kg	4599.	5317.	
Cattle	DE kcal/kg	3128.	3616.	
Sheep	DE kcal/kg	3280.	3792.	
Swine	DE kcal/kg	3051.	3527.	
Cattle	ME kcal/kg	2565.	2965.	
Sheep	ME kcal/kg	2689.	3109.	
Swine	ME kcal/kg	2847.	3291.	
Cattle	TDN %	71.	82.	
Sheep	TDN %	74.	86.	
Swine	TDN %	69.	80.	
Alanine	%	.40	.46	
Arginine	%	.45	.52	
Aspartic acid	%	.63	.73	
Glutamic acid	%	2.45	2.83	
Glycine	%	.41	.47	
Histidine	%	.21	.24	
Isoleucine	%	.40	.46	
Leucine	%	.72	.83	
Lysine	%	.37	.43	
Methionine	%	.09	.10	
Proline	%	1.05	1.21	
Serine	%	.42	.48	
Threonine	%	.35	.40	
Valine	%	.51	.59	

Barley, grain, gr 1 US mn 47 wt mx 1 fm, (4)

Ref no 4-00-535

Feed name or analyses		As fed	Dry	C.V. ± %
Dry matter	%	89.0	100.0	
Ash	%	2.7	3.0	
Crude fiber	%	6.0	6.7	
Ether extract	%	1.9	2.1	
N-free extract	%	66.4	74.6	
Protein (N x 6.25)	%	12.1	13.6	
Cattle	dig prot %	9.1	10.2	

Continued

(7) vitamins

(8) additives

Feed name or analyses		As fed (Mean)	Dry (Mean)	C.V. ± %
Sheep	dig prot %	9.5	10.7	
Swine	dig prot %	9.4	10.6	
Energy				
Cattle	DE kcal/kg	3218.	3616.	
Sheep	DE kcal/kg	3336.	3748.	
Swine	DE kcal/kg	3178.	3571.	
Cattle	ME kcal/kg	2639.	2965.	
Sheep	ME kcal/kg	2735.	3073.	
Swine	ME kcal/kg	2962.	3328.	
Cattle	TDN %	73.	82.	
Sheep	TDN %	76.	85.	
Swine	TDN %	72.	81.	
Calcium	%	.24	.27	
Phosphorus	%	.36	.41	
Thiamine	mg/kg	4.3	4.8	36

Feed name or analyses		As fed (Mean)	Dry (Mean)	C.V. ± %
Cattle	dig prot %	8.8	10.0	
Sheep	dig prot %	9.3	10.6	
Swine	dig prot %	9.1	10.3	
Energy				
Cattle	DE kcal/kg	3182.	3616.	
Sheep	DE kcal/kg	3337.	3792.	
Swine	DE kcal/kg	3104.	3527.	
Cattle	ME kcal/kg	2609.	2965.	
Sheep	ME kcal/kg	2736.	3109.	
Swine	ME kcal/kg	2896.	3291.	
Cattle	TDN %	72.	82.	
Sheep	TDN %	76.	86.	
Swine	TDN %	70.	80.	
Calcium	%	.05	.06	88
Phosphorus	%	.34	.39	83

Barley, grain, gr 2 US mn 45 wt mx 2 fm, (4)

Ref no 4-00-536

		Mean (As fed)	Mean (Dry)	C.V.
Dry matter	%	88.1	100.0	2
Ash	%	2.4	2.7	
Crude fiber	%	5.6	6.3	
Ether extract	%	1.8	2.0	
N-free extract	%	66.7	75.7	
Protein (N x 6.25)	%	11.7	13.3	
Cattle	dig prot %	8.8	10.0	
Sheep	dig prot %	9.2	10.5	
Swine	dig prot %	9.2	10.4	
Energy				
Cattle	DE kcal/kg	3186.	3616.	
Sheep	DE kcal/kg	3341.	3792.	
Swine	DE kcal/kg	3146.	3171.	
Cattle	ME kcal/kg	2612.	2965.	
Sheep	ME kcal/kg	2739.	3109.	
Swine	ME kcal/kg	2935.	3332.	
Cattle	TDN %	72.	82.	
Sheep	TDN %	76.	86.	
Swine	TDN %	71.	81.	
Calcium	%	.05	.06	
Phosphorus	%	.31	.35	
Riboflavin	mg/kg	1.3	1.5	
Thiamine	mg/kg	4.7	5.3	6

Barley, grain, gr 5 US mn 36 wt mx 6 fm, (4)
Barley grain, light

Ref no 4-00-540

		Mean (As fed)	Mean (Dry)	C.V.
Dry matter	%	88.0	100.0	2
Ash	%	3.0	3.4	6
Crude fiber	%	10.0	11.4	7
Ether extract	%	2.4	2.7	6
N-free extract	%	62.3	70.8	
Protein (N x 6.25)	%	10.3	11.7	9
Cattle	dig prot %	7.7	8.8	
Sheep	dig prot %	8.1	9.2	
Swine	dig prot %	7.6	8.6	
Energy				
Cattle	DE kcal/kg	3026.	3439.	
Sheep	DE kcal/kg	3260.	3704.	
Swine	DE kcal/kg	3142.	3571.	
Cattle	ME kcal/kg	2482.	2820.	
Sheep	ME kcal/kg	2672.	3037.	
Swine	ME kcal/kg	2941.	3342.	
Cattle	NE_m kcal/kg	1558.	1770.	
Cattle	NE_{gain} kcal/kg	1030.	1170.	
Cattle	TDN %	69.	78.	
Sheep	TDN %	74.	84.	
Swine	TDN %	71.	81.	

Barley, grain, gr 3 US mn 43 wt mx 3 fm, (4)

Ref no 4-00-537

		Mean (As fed)	Mean (Dry)	
Dry matter	%	88.0	100.0	
Ash	%	1.8	2.1	
Crude fiber	%	6.0	6.8	
Ether extract	%	1.7	1.9	
N-free extract	%	66.7	75.8	
Protein (N x 6.25)	%	11.8	13.4	

Continued

Barley, grain, Pacific coast, (4)

Ref no 4-07-939

		Mean (As fed)	Mean (Dry)	
Dry matter	%	89.0	100.0	
Ash	%	2.3	2.6	
Crude fiber	%	6.2	7.0	
Ether extract	%	2.2	2.5	
N-free extract	%	68.5	77.0	
Protein (N x 6.25)	%	9.7	10.9	
Cattle	dig prot %	7.3	8.2	

Continued

(1) dry forages and roughages
(2) pasture, range plants, and forages fed green

(3) silages
(4) energy feeds

Feed name or analyses		Mean		C.V.
		As fed	Dry	± %
Sheep	dig prot %	6.9	7.8	
Swine	dig prot %	7.5	8.4	
Energy				
Cattle	DE kcal/kg	3218.	3616.	
Sheep	DE kcal/kg	3100.	3483.	
Swine	DE kcal/kg	3139.	3527.	
Cattle	ME kcal/kg	2639.	2965.	
Sheep	ME kcal/kg	2542.	2856.	
Swine	ME kcal/kg	2944.	3308.	
Cattle	TDN %	73.	82.	
Sheep	TDN %	70.	79.	
Swine	TDN %	71.	80.	
Calcium	%	.06	.07	
Phosphorus	%	.40	.45	
Choline	mg/kg	937.	1054.	
Niacin	mg/kg	44.1	49.6	
Pantothenic acid	mg/kg	7.3	8.2	
Riboflavin	mg/kg	1.3	1.5	
Thiamine	mg/kg	4.0	4.5	

Feed name or analyses		Mean		C.V.
		As fed	Dry	± %
Cattle	dig prot %	9.7	10.9	
Sheep	dig prot %	10.1	11.4	
Swine	dig prot %	9.2	10.4	
Energy				
Cattle	DE kcal/kg	3375.	3792.	
Sheep	DE kcal/kg	3453.	3880.	
Swine	DE kcal/kg	3139.	3527.	
Cattle	ME kcal/kg	2767.	3109.	
Sheep	ME kcal/kg	2832.	3182.	
Swine	ME kcal/kg	2923.	3284.	
Cattle	TDN %	76.	86.	
Sheep	TDN %	78.	88.	
Swine	TDN %	72.	80.	
Calcium	%	.07	.08	
Magnesium	%	.12	.13	
Phosphorus	%	.36	.40	
Potassium	%	.53	.60	

Barley, grain screenings, (4)

Ref no 4-00-542

Dry matter	%	89.0	100.0	
Ash	%	2.9	3.3	
Crude fiber	%	8.0	9.0	
Ether extract	%	2.3	2.6	
N-free extract	%	63.7	71.6	
Protein (N x 6.25)	%	12.0	13.5	
Cattle	dig prot %	9.0	10.1	
Sheep	dig prot %	9.6	10.8	
Swine	dig prot %	9.2	10.4	
Energy				
Cattle	DE kcal/kg	3139.	3527.	
Sheep	DE kcal/kg	3296.	3704.	
Swine	DE kcal/kg	3061.	3439.	
Cattle	ME kcal/kg	2574.	2892.	
Sheep	ME kcal/kg	2703.	3037.	
Swine	ME kcal/kg	2855.	3208.	
Cattle	TDN %	71.	80.	
Sheep	TDN %	75.	84.	
Swine	TDN %	69.	78.	

Barley, groats, (4)

Ref no 4-00-543

Dry matter	%	89.0	100.0	2
Ash	%	1.8	2.0	16
Crude fiber	%	2.0	2.2	25
Ether extract	%	1.9	2.1	19
N-free extract	%	70.5	79.2	
Protein (N x 6.25)	%	12.9	14.5	17

Continued

Barley, pearl by-prod, (4)
Pearl barley by-product (AAFCO)
Barley feed (CFA)

Ref no 4-00-548

Dry matter	%	90.0	100.0	2
Ash	%	3.6	4.0	22
Crude fiber	%	8.0	8.9	46
Ether extract	%	3.0	3.3	33
N-free extract	%	62.8	69.8	
Protein (N x 6.25)	%	12.6	14.0	17
Cattle	dig prot %	9.4	10.5	
Sheep	dig prot %	10.2	11.3	
Cellulose	%	5.0	5.6	
Lignin	%	3.0	3.3	
Energy				
Cattle	DE kcal/kg	3135.	3483.	
Sheep	DE kcal/kg	2817.	3130.	
Cattle	ME kcal/kg	2570.	2856.	
Sheep	ME kcal/kg	2310.	2567.	
Cattle	TDN %	71.	79.	
Sheep	TDN %	64.	71.	
Calcium	%	.06	.07	51
Iron	%	.010	.010	25
Phosphorus	%	.43	.48	25
Manganese	mg/kg	30.6	34.0	16
Choline	mg/kg	1201.	1335.	
Folic acid	mg/kg	.80	.90	
Niacin	mg/kg	63.6	70.7	
Pantothenic acid	mg/kg	7.7	8.6	
Riboflavin	mg/kg	2.2	2.4	
Thiamine	mg/kg	5.9	6.6	41

(5) protein supplements
(6) minerals

(7) vitamins
(8) additives

Feed name or analyses		Mean As fed	Dry	C.V. ± %

Barley, distil grains, dehy, (5)
  Barley distillers dried grains (AAFCO)

Ref no 5-00-518

		As fed	Dry	
Dry matter	%	92.0	100.0	
Ash	%	1.8	2.0	
Crude fiber	%	10.0	10.9	
Ether extract	%	11.6	12.6	
N-free extract	%	40.8	44.4	
Protein (N x 6.25)	%	27.7	30.1	
Cattle	dig prot %	21.7	23.6	
Sheep	dig prot %	22.8	24.8	
Energy				
Cattle	DE kcal/kg	3650.	3968.	
Sheep	DE kcal/kg	3042.	3307.	
Cattle	ME kcal/kg	2994.	3254.	
Sheep	ME kcal/kg	2495.	2712.	
Cattle	TDN %	83.	90.	
Sheep	TDN %	69.	75.	
Niacin	mg/kg	57.6	62.6	
Riboflavin	mg/kg	10.3	11.2	
Thiamine	mg/kg	9.2	10.0	

Barley, malt sprout cleanings w hulls, dehy, mx 24 prot, (5)
  Malt cleanings (AAFCO)

Ref no 5-00-544

		As fed	Dry	C.V. ± %
Dry matter	%	92.0	100.0	2
Ash	%	5.3	5.8	13
Crude fiber	%	15.0	16.3	17
Ether extract	%	1.7	1.8	21
N-free extract	%	51.2	55.6	
Protein (N x 6.25)	%	18.9	20.5	16
Cattle	dig prot %	13.6	14.8	
Sheep	dig prot %	14.7	16.0	
Energy				
Cattle	DE kcal/kg	2839.	3086.	
Sheep	DE kcal/kg	2839.	3086.	
Cattle	ME kcal/kg	2328.	2530.	
Sheep	ME kcal/kg	2328.	2530.	
Cattle	TDN %	64.	70.	
Sheep	TDN %	64.	70.	

Barley, malt sprouts w hulls, dehy, mn 24 prot, (5)
  Malt sprouts (AAFCO)

Ref no 5-00-545

		As fed	Dry	C.V. ± %
Dry matter	%	93.0	100.0	2
Ash	%	6.4	6.9	8
Crude fiber	%	14.0	15.1	11
Ether extract	%	1.4	1.5	29
N-free extract	%	44.9	48.3	
Protein (N x 6.25)	%	26.2	28.2	7
Cattle	dig prot %	20.4	21.9	
Sheep	dig prot %	23.9	25.7	
Swine	dig prot %	20.7	22.3	
Energy				
Cattle	DE kcal/kg	2829.	3042.	
Sheep	DE kcal/kg	2829.	3042.	
Swine	DE kcal/kg	1558.	1675.	
Cattle	ME kcal/kg	2319.	2494.	
Chickens	ME n kcal/kg	1411.	1517.	
Sheep	ME kcal/kg	2319.	2494.	
Swine	ME kcal/kg	1406.	1512.	
Cattle	TDN %	64.	69.	
Sheep	TDN %	64.	69.	
Swine	TDN %	35.	38.	
Calcium	%	.22	.24	22
Magnesium	%	.18	.19	
Phosphorus	%	.73	.78	9
Potassium	%	.21	.23	
Manganese	mg/kg	31.7	34.1	34
Choline	mg/kg	1584.	1703.	
Folic acid	mg/kg	.20	.20	
Niacin	mg/kg	43.3	46.5	
Pantothenic acid	mg/kg	8.6	9.2	
Riboflavin	mg/kg	1.5	1.6	9
Thiamine	mg/kg	.7	.8	

BARLEY, WESTERN. Hordeum vulgare

Barley, western, grain, gr 5 US mn 36 wt mx 4 fm, (4)
  Barley, western, grain, light

Ref no 4-00-566

		As fed	Dry	C.V. ± %
Dry matter	%	88.6	100.0	1
Ash	%	3.3	3.7	10
Crude fiber	%	7.4	8.4	26
Ether extract	%	2.1	2.4	8
N-free extract	%	63.8	72.0	
Protein (N x 6.25)	%	12.0	13.5	6
Cattle	dig prot %	8.9	10.1	
Sheep	dig prot %	9.5	10.7	

Continued

(1) dry forages and roughages
(2) pasture, range plants, and forages fed green
(3) silages
(4) energy feeds

Feed name or analyses		Mean		C.V.
		As fed	Dry	± %
Swine	dig prot %	8.9	10.0	
Energy				
Cattle	DE kcal/kg	3125.	3527.	
Sheep	DE kcal/kg	3282.	3704.	
Swine	DE kcal/kg	2970.	3351.	
Cattle	ME kcal/kg	2562.	2892.	
Sheep	ME kcal/kg	2691.	3037.	
Swine	ME kcal/kg	2770.	3126.	
Cattle	TDN %	71.	80.	
Sheep	TDN %	74.	84.	
Swine	TDN %	67.	76.	

Feed name or analyses		Mean		C.V.
		As fed	Dry	± %
Sheep	DE kcal/kg	3242.	3748.	
Swine	DE kcal/kg	3051.	3527.	
Cattle	ME kcal/kg	2533.	2928.	
Sheep	ME kcal/kg	2658.	3073.	
Swine	ME kcal/kg	2837.	3280.	
Cattle	TDN %	70.	81.	
Sheep	TDN %	74.	85.	
Swine	TDN %	69.	80.	

BARLEY, YELLOW. Hordeum vulgare

Barley, yellow, grain, Can 2 CW, (4)

Ref no 4-00-570

Dry matter	%	86.5	100.0	
Ash	%	2.2	2.5	15
Crude fiber	%	5.9	6.8	9
Ether extract	%	1.5	1.7	27
N-free extract	%	65.4	75.6	
Protein (N x 6.25)	%	11.6	13.4	7
Cattle	dig prot %	8.6	10.0	
Sheep	dig prot %	9.2	10.6	
Swine	dig prot %	8.9	10.3	
Energy	GE kcal/kg	4599.	5317.	
Cattle	DE kcal/kg	3128.	3616.	
Sheep	DE kcal/kg	3280.	3792.	
Swine	DE kcal/kg	3051.	3527.	
Cattle	ME kcal/kg	2565.	2965.	
Sheep	ME kcal/kg	2689.	3109.	
Swine	ME kcal/kg	2847.	3291.	
Cattle	TDN %	71.	82.	
Sheep	TDN %	74.	86.	
Swine	TDN %	69.	80.	

Barley, yellow, grain, Can 3 CW, (4)

Ref no 4-00-571

Dry matter	%	86.5	100.0	
Ash	%	2.2	2.5	
Crude fiber	%	6.1	7.1	3
Ether extract	%	1.5	1.7	19
N-free extract	%	64.0	74.0	
Protein (N x 6.25)	%	12.7	14.7	5
Cattle	dig prot %	9.5	11.0	
Sheep	dig prot %	10.0	11.6	
Swine	dig prot %	9.8	11.3	
Energy				
Cattle	DE kcal/kg	3089.	3571.	

Continued

BARLEY, 2-ROW. Hordeum distichon

Barley, 2-row, grain, Can 2 CW mn 49 wt mn 90 purity mx 1.5 fm, (4)

Ref no 4-00-572

Dry matter	%	86.5	100.0	
Ash	%	2.1	2.4	33
Crude fiber	%	4.0	4.6	73
Ether extract	%	2.0	2.3	63
N-free extract	%	66.7	77.1	
Protein (N x 6.25)	%	11.7	13.6	
Cattle	dig prot %	8.8	10.2	
Sheep	dig prot %	9.2	10.7	
Swine	dig prot %	9.1	10.5	
Energy	GE kcal/kg	4599.	5317.	
Cattle	DE kcal/kg	3204.	3704.	
Sheep	DE kcal/kg	3318.	3836.	
Swine	DE kcal/kg	3128.	3616.	
Cattle	ME kcal/kg	2627.	3037.	
Sheep	ME kcal/kg	2721.	3146.	
Swine	ME kcal/kg	2915.	3370.	
Cattle	TDN %	73.	84.	
Sheep	TDN %	75.	87.	
Swine	TDN %	71.	82.	

Barley, 2-row, grain, Can 3 CW, (4)

Ref no 4-00-573

Dry matter	%	86.5	100.0	
Ash	%	2.1	2.4	33
Crude fiber	%	4.1	4.7	88
Ether extract	%	2.0	2.3	39
N-free extract	%	66.7	77.1	
Protein (N x 6.25)	%	11.7	13.5	18
Cattle	dig prot %	8.8	10.2	
Sheep	dig prot %	9.2	10.7	
Swine	dig prot %	9.0	10.4	
Energy				
Cattle	DE kcal/kg	3204.	3704.	
Sheep	DE kcal/kg	3318.	3836.	
Swine	DE kcal/kg	3128.	3616.	

Continued

(5) protein supplements

(6) minerals

(7) vitamins

(8) additives

Feed name or analyses		Mean As fed	Mean Dry	C.V. ± %
Cattle	ME kcal/kg	2627.	3037.	
Sheep	ME kcal/kg	2721.	3146.	
Swine	ME kcal/kg	2915.	3370.	
Cattle	TDN %	73.	84.	
Sheep	TDN %	75.	87.	
Swine	TDN %	71.	82.	

Feed name or analyses		Mean As fed	Mean Dry	C.V. ± %
Methionine	%	.12	.14	
Phenylalanine	%	.42	.48	
Proline	%	1.01	1.17	
Serine	%	.38	.44	
Threonine	%	.29	.34	
Tyrosine	%	.27	.31	
Valine	%	.33	.38	

BARLEY, 6-ROW. Hordeum vulgare

Barley, 6-row, grain, Can 1 CW mn 50 wt mn 95 purity mx 1 fm, (4)

Ref no 4-00-574

		As fed	Dry	C.V.
Dry matter	%	86.5	100.0	
Ash	%	2.2	2.5	
Crude fiber	%	4.6	5.3	
Ether extract	%	2.0	2.3	
Energy	GE kcal/kg	4599.	5317.	

Barley, 6-row, grain, Can 2 CW mn 48 wt mn 90 purity mx 1.5 fm, (4)

Ref no 4-00-575

		As fed	Dry	C.V.
Dry matter	%	86.5	100.0	
Ash	%	2.3	2.6	5
Crude fiber	%	5.0	5.8	89
Ether extract	%	1.9	2.2	60
N-free extract	%	66.6	77.0	
Protein (N x 6.25)	%	10.7	12.4	14
Cattle	dig prot %	8.0	9.3	
Sheep	dig prot %	8.5	9.8	
Swine	dig prot %	8.2	9.5	
Energy	GE kcal/kg	4599.	5317.	
Cattle	DE kcal/kg	3166.	3660.	
Sheep	DE kcal/kg	3280.	3792.	
Swine	DE kcal/kg	3128.	3616.	
Cattle	ME kcal/kg	2596.	3001.	
Sheep	ME kcal/kg	2689.	3109.	
Swine	ME kcal/kg	2915.	3370.	
Cattle	TDN %	72.	83.	
Sheep	TDN %	74.	86.	
Swine	TDN %	71.	82.	
Alanine	%	.40	.46	
Arginine	%	.38	.44	
Aspartic acid	%	.55	.64	
Glutamic acid	%	2.08	2.40	
Glycine	%	.38	.44	
Histidine	%	.16	.18	
Isoleucine	%	.26	.30	
Leucine	%	.57	.66	
Lysine	%	.28	.33	

Continued

Barley, 6-row, grain, Can 3 CW mn 46 wt mn 85 purity mx 4 fm, (4)

Ref no 4-00-576

		As fed	Dry	C.V.
Dry matter	%	86.5	100.0	
Ash	%	2.3	2.7	28
Crude fiber	%	4.9	5.7	67
Ether extract	%	1.9	2.1	55
N-free extract	%	66.4	76.8	
Protein (N x 6.25)	%	11.0	12.7	13
Cattle	dig prot %	8.2	9.5	
Sheep	dig prot %	8.6	10.0	
Swine	dig prot %	8.5	9.8	
Energy				
Cattle	DE kcal/kg	3166.	3660.	
Sheep	DE kcal/kg	3280.	3792.	
Swine	DE kcal/kg	3128.	3616.	
Cattle	ME kcal/kg	2596.	3001.	
Sheep	ME kcal/kg	2689.	3109.	
Swine	ME kcal/kg	2915.	3370.	
Cattle	TDN %	72.	83.	
Sheep	TDN %	74.	86.	
Swine	TDN %	71.	82.	

Barley, 6-row, grain, Can 4 CW, (4)

Ref no 4-00-577

		As fed	Dry	C.V.
Dry matter	%	86.5	100.0	
Ash	%	2.3	2.6	25
Crude fiber	%	4.9	5.6	56
Ether extract	%	1.9	2.2	41
N-free extract	%	66.7	77.1	
Protein (N x 6.25)	%	10.9	12.5	29
Cattle	dig prot %	8.1	9.4	
Sheep	dig prot %	8.6	9.9	
Swine	dig prot %	8.3	9.6	
Energy	GE kcal/kg	4599.	5317.	
Cattle	DE kcal/kg	3166.	3660.	
Sheep	DE kcal/kg	3318.	3836.	
Swine	DE kcal/kg	3128.	3616.	
Cattle	ME kcal/kg	2596.	3001.	
Sheep	ME kcal/kg	2721.	3146.	
Swine	ME kcal/kg	2915.	3370.	

Continued

(1) dry forages and roughages
(2) pasture, range plants, and forages fed green
(3) silages
(4) energy feeds

TABLE A-3-1 Composition of Feeds (Continued)

Feed name or analyses		Mean		C.V.
		As fed	Dry	± %
Cattle	TDN %	72.	83.	
Sheep	TDN %	75.	87.	
Swine	TDN %	71.	82.	

BEAN. Phaseolus spp

Bean, kidney, seed, (5)

Ref no 5-00-600

		As fed	Dry	
Dry matter	%	89.0	100.0	
Ash	%	3.9	4.4	
Crude fiber	%	4.1	4.6	
Ether extract	%	1.2	1.3	
N-free extract	%	56.9	63.9	
Protein (N x 6.25)	%	23.0	25.8	
Cattle	dig prot %	15.4	17.3	
Sheep	dig prot %	15.4	17.3	
Energy				
Cattle	DE kcal/kg	3256.	3659.	
Sheep	DE kcal/kg	3022.	3395.	
Cattle	ME kcal/kg	2670.	3000.	
Sheep	ME kcal/kg	2478.	2784.	
Cattle	TDN %	74.	83.	
Sheep	TDN %	68.	77.	
Calcium	%	.13	.15	
Phosphorus	%	.51	.57	

Bean, navy, seed, (5)

Ref no 5-00-623

		As fed	Dry	
Dry matter	%	90.0	100.0	
Ash	%	4.2	4.7	
Crude fiber	%	4.2	4.7	
Ether extract	%	1.4	1.6	
N-free extract	%	57.2	63.6	
Protein (N x 6.25)	%	22.9	25.4	
Cattle	dig prot %	20.2	22.4	
Sheep	dig prot %	20.2	22.4	
Energy				
Cattle	DE kcal/kg	3293.	3659.	
Sheep	DE kcal/kg	3452.	3836.	
Cattle	ME kcal/kg	2700.	3000.	
Sheep	ME kcal/kg	2831.	3146.	
Cattle	TDN %	75.	83.	
Sheep	TDN %	78.	87.	
Calcium	%	.15	.17	
Phosphorus	%	.57	.63	
Potassium	%	1.70	1.89	

Beef - see Cattle

BEET, MANGELS. Beta spp

Beet, mangels, roots, (4)
Mangel, roots

Ref no 4-00-637

		As fed	Dry	
Dry matter	%	10.6	100.0	
Ash	%	1.1	10.8	
Crude fiber	%	.9	8.3	
Ether extract	%	.1	.7	
N-free extract	%	7.1	67.0	
Protein (N x 6.25)	%	1.4	13.2	
Cattle	dig prot %	1.0	9.8	
Sheep	dig prot %	.5	5.1	
Energy				
Cattle	DE kcal/kg	364.	3439.	
Sheep	DE kcal/kg	402.	3792.	
Cattle	ME kcal/kg	299.	2820.	
Sheep	ME kcal/kg	330.	3109.	
Cattle	TDN %	8.	78.	
Sheep	TDN %	9.	86.	
Calcium	%	.02	.19	
Chlorine	%	.13	1.23	
Iron	%	.002	.019	
Magnesium	%	.02	.19	
Phosphorus	%	.02	.19	
Potassium	%	.21	1.98	
Sodium	%	.07	.66	
Sulfur	%	.02	.19	

BEET, SUGAR. Beta saccharifera

Beet, sugar, aerial pt w crowns, dehy, (1)

Ref no 1-00-640

		As fed	Dry	
Dry matter	%	82.6	100.0	2
Ash	%	18.0	21.8	8
Crude fiber	%	10.2	12.4	16
Ether extract	%	.9	1.1	5
N-free extract	%	43.6	52.8	
Protein (N x 6.25)	%	9.8	11.9	6
Cattle	dig prot %	6.3	7.6	
Sheep	dig prot %	6.3	7.6	
Energy				
Cattle	DE kcal/kg	2076.	2513.	
Sheep	DE kcal/kg	2076.	2513.	
Cattle	ME kcal/kg	1702.	2061.	

Continued

(5) protein supplements

(6) minerals

(7) vitamins

(8) additives

Feed name or analyses		Mean		C.V.
		As fed	Dry	± %
Sheep	ME kcal/kg	1702.	2061.	
Cattle	TDN %	47.	57.	
Sheep	TDN %	47.	57.	

Beet, sugar, aerial pt w crowns, fresh, (2)

Ref no 2-00-649

Dry matter	%	16.7	100.0	7
Ash	%	3.6	21.5	14
Crude fiber	%	1.8	10.9	2
Ether extract	%	.4	2.7	25
N-free extract	%	8.2	48.9	
Protein (N x 6.25)	%	2.7	16.0	4
Cattle	dig prot %	2.0	12.0	
Sheep	dig prot %	2.0	12.0	
Energy				
Cattle	DE kcal/kg	471.	2822.	
Sheep	DE kcal/kg	471.	2822.	
Cattle	ME kcal/kg	386.	2314.	
Sheep	ME kcal/kg	386.	2314.	
Cattle	NE_m kcal/kg	205.	1230.	
Cattle	NE_gain kcal/kg	95.	570.	
Cattle	TDN %	11.	64.	
Sheep	TDN %	11.	64.	
Calcium	%	.17	1.02	
Phosphorus	%	.04	.24	

Beet, sugar, aerial pt w crowns, ensiled, (3)

Ref no 3-00-660

Dry matter	%	20.7	100.0	16
Ash	%	7.5	36.3	20
Crude fiber	%	2.8	13.3	18
Ether extract	%	.4	2.2	27
N-free extract	%	7.3	35.5	
Protein (N x 6.25)	%	2.6	12.7	11
Cattle	dig prot %	2.1	10.0	
Sheep	dig prot %	2.1	10.0	
Energy				
Cattle	DE kcal/kg	493.	2381.	
Sheep	DE kcal/kg	493.	2381.	
Cattle	ME kcal/kg	404.	1952.	
Sheep	ME kcal/kg	404.	1952.	
Cattle	TDN %	11.	54.	
Sheep	TDN %	11.	54.	
Calcium	%	.48	2.32	
Phosphorus	%	.04	.20	

Beet, sugar, molasses, mn 48 invert sugar min 79.5 degrees brix, (4)
Beet molasses (AAFCO)
Molasses (CFA)

Ref no 4-00-668

Dry matter	%	77.0	100.0	7
Ash	%	8.2	10.6	19
Ether extract	%	.2	.3	99
N-free extract	%	61.9	80.4	
Protein (N x 6.25)	%	6.7	8.7	35
Cattle	dig prot %	3.8	5.0	
Sheep	dig prot %	-2.3	-3.0	
Energy				
Cattle	DE kcal/kg	3021.	3924.	
Sheep	DE kcal/kg	2546.	3307.	
Cattle	ME kcal/kg	2478.	3218.	
Chickens	MEn kcal/kg	1962.	2548.	
Sheep	ME kcal/kg	2088.	2712.	
Cattle	TDN %	68.	89.	
Sheep	TDN %	58.	75.	
Calcium	%	.16	.21	80
Iron	%	.010	.010	
Magnesium	%	.23	.30	
Phosphorus	%	.03	.04	89
Potassium	%	4.77	6.20	
Sodium	%	1.17	1.52	
Cobalt	mg/kg	.400	.500	
Copper	mg/kg	17.6	22.9	
Manganese	mg/kg	4.6	6.0	
Niacin	mg/kg	42.2	54.8	15
Pantothenic acid	mg/kg	4.6	6.0	
Riboflavin	mg/kg	2.4	3.1	

Beet, sugar, pulp, dehy, (4)
Dried beet pulp (AAFCO)
Dried beet pulp (CFA)

Ref no 4-00-669

Dry matter	%	91.0	100.0	4
Ash	%	3.6	3.9	28
Crude fiber	%	19.0	20.9	11
Ether extract	%	.6	.7	68
N-free extract	%	58.7	64.5	
Protein (N x 6.25)	%	9.1	10.0	10
Cattle	dig prot %	4.1	4.5	
Horses	dig prot %	6.3	6.9	
Sheep	dig prot %	4.6	5.0	
Swine	dig prot %	3.7	4.1	
Lignin	%	8.0	8.8	
Energy	GE kcal/kg	3837.	4217.	
Cattle	DE kcal/kg	2889.	3175.	

Continued

(1) dry forages and roughages
(2) pasture, range plants, and forages fed green
(3) silages
(4) energy feeds

Feed name or analyses		Mean		C.V.
		As fed	Dry	± %
Sheep	DE kcal/kg	2889.	3175.	
Swine	DE kcal/kg	2860.	3143.	
Cattle	ME kcal/kg	2370.	2604.	
Chickens	MEn kcal/kg	617.	678.	
Sheep	ME kcal/kg	2370.	2604.	
Swine	ME kcal/kg	2688.	2954.	
Cattle	TDN %	66.	72.	
Horses	TDN %	76.	83.	
Sheep	TDN %	66.	72.	
Swine	TDN %	65.	71.	
Calcium	%	.68	.75	22
Iron	%	.030	.033	44
Magnesium	%	.27	.30	42
Phosphorus	%	.10	.11	27
Potassium	%	.21	.23	41
Cobalt	mg/kg	.100	.100	
Copper	mg/kg	12.5	13.7	29
Manganese	mg/kg	35.0	38.5	45
Zinc	mg/kg	.7	.8	
Choline	mg/kg	829.	912.	24
Niacin	mg/kg	16.3	17.9	55
Pantothenic acid	mg/kg	1.5	1.6	38
Riboflavin	mg/kg	.7	.8	51
Thiamine	mg/kg	.4	.4	38
Vitamin D_3	ICU/g	1.0	1.0	
Arginine	%	.30	.33	
Histidine	%	.20	.22	
Isoleucine	%	.30	.33	
Leucine	%	.60	.66	
Lysine	%	.60	.66	
Phenylalanine	%	.30	.33	
Threonine	%	.40	.44	
Tryptophan	%	.10	.11	
Tyrosine	%	.40	.44	
Valine	%	.40	.44	

Beet, sugar, pulp, wet, (4)

Ref no 4-00-671

		As fed	Dry	
Dry matter	%	10.0	100.0	
Ash	%	.4	4.0	
Crude fiber	%	2.0	20.0	
Ether extract	%	.2	2.0	
N-free extract	%	6.5	65.0	
Protein (N x 6.25)	%	.9	9.0	
Cattle	dig prot %	.4	4.0	
Sheep	dig prot %	.5	5.0	
Energy				
Cattle	DE kcal/kg	300.	2998.	
Sheep	DE kcal/kg	331.	3307.	
Cattle	ME kcal/kg	246.	2458.	
Sheep	ME kcal/kg	271.	2712.	
Cattle	TDN %	7.	68.	
Sheep	TDN %	8.	75.	

Continued

(5) protein supplements
(6) minerals

Feed name or analyses		Mean		C.V.
		As fed	Dry	± %
Calcium	%	.09	.90	
Phosphorus	%	.01	.10	
Potassium	%	.02	.20	

Beet, sugar, pulp w molasses, dehy, (4)

Ref no 4-00-672

		As fed	Dry	
Dry matter	%	92.0	100.0	2
Ash	%	5.7	6.2	18
Crude fiber	%	16.0	17.4	13
Ether extract	%	.5	.5	34
N-free extract	%	60.7	66.0	
Protein (N x 6.25)	%	9.1	9.9	16
Cattle	dig prot %	6.0	6.5	
Sheep	dig prot %	6.0	6.5	
Swine	dig prot %	2.3	2.5	
Energy				
Cattle	DE kcal/kg	3002.	3263.	
Sheep	DE kcal/kg	3204.	3483.	
Swine	DE kcal/kg	2992.	3252.	
Cattle	ME kcal/kg	2462.	2676.	
Chickens	MEn kcal/kg	661.	718.	
Sheep	ME kcal/kg	2628.	2856.	
Swine	ME kcal/kg	2812.	3057.	
Cattle	NEm kcal/kg	1868.	2030.	
Cattle	NEgain kcal/kg	1233.	1340.	
Cattle	TDN %	68.	74.	
Sheep	TDN %	73.	79.	
Swine	TDN %	68.	74.	
Calcium	%	.56	.61	
Magnesium	%	.13	.14	
Phosphorus	%	.08	.11	
Potassium	%	1.64	1.78	

Beet, sugar, pulp w Steffens filtrate, dehy, (4)
Dried beet product (AAFCO)

Ref no 4-00-675

		As fed	Dry	
Dry matter	%	91.5	100.0	
Ash	%	5.6	6.1	
Crude fiber	%	15.2	16.6	
Ether extract	%	.3	.3	
N-free extract	%	59.3	64.8	
Protein (N x 6.25)	%	11.1	12.1	
Cattle	dig prot %	7.3	8.0	
Sheep	dig prot %	7.3	8.0	
Energy				
Cattle	DE kcal/kg	2986.	3263.	
Sheep	DE kcal/kg	3187.	3483.	
Cattle	ME kcal/kg	2448.	2676.	

Continued

(7) vitamins
(8) additives

Feed name or analyses		Mean		C.V.
		As fed	Dry	± %
Sheep	ME kcal/kg	2613.	2856.	
Cattle	TDN %	68.	74.	
Sheep	TDN %	72.	79.	

Feed name or analyses		Mean		C.V.
		As fed	Dry	± %
Sheep	ME kcal/kg	1612.	1735.	
Cattle	TDN %	44.	48.	
Sheep	TDN %	44.	48.	

Beet, sugar, sol w low K salts and glutamic acid, cond, (4)

Condensed beet solubles product (AAFCO)

Ref no 4-00-679

		As fed	Dry	
Dry matter	%	62.9	100.0	
Ash	%	12.3	19.6	
Protein (N x 6.25)	%	14.3	22.7	

BERMUDAGRASS. Cynodon dactylon

Bermudagrass, hay, s-c, immature, (1)

Ref no 1-00-699

		As fed	Dry	C.V.
Dry matter	%	90.6	100.0	1
Ash	%	8.2	9.0	39
Crude fiber	%	22.5	24.8	13
Ether extract	%	1.6	1.8	18
N-free extract	%	45.8	50.6	
Protein (N x 6.25)	%	12.5	13.8	13
Cattle	dig prot %	6.3	7.0	
Sheep	dig prot %	6.3	7.0	
Energy				
Cattle	DE kcal/kg	1917.	2116.	
Sheep	DE kcal/kg	1917.	2116.	
Cattle	ME kcal/kg	1572.	1735.	
Sheep	ME kcal/kg	1572.	1735.	
Cattle	TDN %	43.	48.	
Sheep	TDN %	43.	48.	

Bermudagrass, hay, s-c, mid-blm, (1)

Ref no 1-00-700

		As fed	Dry	
Dry matter	%	92.9	100.0	
Ash	%	9.5	10.2	
Crude fiber	%	25.9	27.9	
Ether extract	%	2.0	2.2	
N-free extract	%	46.6	50.2	
Protein (N x 6.25)	%	8.8	9.5	
Cattle	dig prot %	4.4	4.8	
Sheep	dig prot %	4.4	4.8	
Energy				
Cattle	DE kcal/kg	1966.	2116.	
Sheep	DE kcal/kg	1966.	2116.	
Cattle	ME kcal/kg	1612.	1735.	

Continued

Bermudagrass, hay, s-c, full blm, (1)

Ref no 1-00-701

		As fed	Dry	
Dry matter	%	92.2	100.0	
Ash	%	9.9	10.7	
Crude fiber	%	26.3	28.5	
Ether extract	%	1.7	1.8	
N-free extract	%	46.9	50.9	
Protein (N x 5.25)	%	7.5	8.1	
Cattle	dig prot %	3.8	4.1	
Sheep	dig prot %	3.8	4.1	
Energy				
Cattle	DE kcal/kg	1910.	2072.	
Sheep	DE kcal/kg	1910.	2072.	
Cattle	ME kcal/kg	1566.	1699.	
Sheep	ME kcal/kg	1566.	1699.	
Cattle	TDN %	43.	47.	
Sheep	TDN %	43.	47.	

Bermudagrass, hay, s-c, mature, (1)

Ref no 1-00-702

		As fed	Dry	C.V.
Dry matter	%	92.2	100.0	3
Ash	%	6.4	7.0	35
Crude fiber	%	27.6	29.9	6
Ether extract	%	1.5	1.6	25
N-free extract	%	51.2	55.5	
Protein (N x 6.25)	%	5.5	6.0	6
Cattle	dig prot %	2.8	3.1	
Sheep	dig prot %	2.8	3.1	
Energy				
Cattle	DE kcal/kg	1992.	2160.	
Sheep	DE kcal/kg	1992.	2160.	
Cattle	ME kcal/kg	1633.	1771.	
Sheep	ME kcal/kg	1633.	1771.	
Cattle	TDN %	45.	49.	
Sheep	TDN %	45.	49.	

Bermudagrass, hay, s-c, (1)

Ref no 1-00-703

		As fed	Dry	C.V.
Dry matter	%	91.1	100.0	2
Ash	%	6.1	6.7	37
Crude fiber	%	27.0	29.6	16
Ether extract	%	1.8	2.0	18
N-free extract	%	48.1	52.8	

Continued

(1) dry forages and roughages

(2) pasture, range plants, and forages fed green

(3) silages

(4) energy feeds

Feed name or analyses		As fed	Dry	C.V. ± %
Protein (N x 6.25)	%	8.1	8.9	24
Cattle	dig prot %	4.4	4.8	
Sheep	dig prot %	4.1	4.5	
Energy				
Cattle	DE kcal/kg	1727.	1896.	
Sheep	DE kcal/kg	1968.	2160.	
Cattle	ME kcal/kg	1417.	1555.	
Sheep	ME kcal/kg	1613.	1771.	
Cattle	NE_m kcal/kg	966.	1060.	
Cattle	NE_{gain} kcal/kg	228.	250.	
Cattle	TDN %	39.	43.	
Sheep	TDN %	45.	49.	
Calcium	%	.42	.46	36
Iron	%	.026	.029	
Magnesium	%	.15	.17	58
Phosphorus	%	.18	.20	30
Potassium	%	1.34	1.47	13
Iodine	mg/kg	.105	.115	
Carotene	mg/kg	117.2	128.7	17

Bermudagrass, aerial pt, fresh, immature, (2)

Ref no 2-00-706

		As fed	Dry	C.V. ± %
Dry matter	%	46.7	100.0	6
Ash	%	4.4	9.4	30
Crude fiber	%	11.6	24.9	12
Ether extract	%	1.0	2.2	18
N-free extract	%	21.7	46.4	
Protein (N x 6.25)	%	8.0	17.1	19
Cattle	dig prot %	5.8	12.4	
Sheep	dig prot %	6.0	12.9	
Energy				
Cattle	DE kcal/kg	1462.	3130.	
Sheep	DE kcal/kg	1297.	2778.	
Cattle	ME kcal/kg	1199.	2567.	
Sheep	ME kcal/kg	1064.	2278.	
Cattle	TDN %	33.	71.	
Sheep	TDN %	29.	63.	
Calcium	%	.42	.89	30
Magnesium	%	.14	.30	39
Phosphorus	%	.15	.32	23
Potassium	%	.71	1.52	58

Bermudagrass, aerial pt, fresh, early blm, (2)

Ref no 2-00-707

		As fed	Dry	C.V. ± %
Dry matter	%	29.1	100.0	46
Ash	%	2.7	9.3	11
Crude fiber	%	7.0	23.9	10
Ether extract	%	1.2	4.0	34
N-free extract	%	13.2	45.4	
Protein (N x 6.25)	%	5.1	17.4	16

Continued

Feed name or analyses		As fed	Dry	C.V. ± %
Cattle	dig prot %	3.7	12.7	
Sheep	dig prot %	3.8	13.2	
Energy				
Cattle	DE kcal/kg	885.	3042.	
Sheep	DE kcal/kg	796.	2734.	
Cattle	ME kcal/kg	726.	2494.	
Sheep	ME kcal/kg	652.	2242.	
Cattle	TDN %	20.	69.	
Sheep	TDN %	18.	62.	
Calcium	%	.17	.58	
Magnesium	%	.04	.14	
Phosphorus	%	.07	.23	
Potassium	%	.62	2.13	

Bermudagrass, aerial pt, fresh, full blm, (2)

Ref no 2-00-708

		As fed	Dry	C.V. ± %
Dry matter	%	35.0	100.0	
Ash	%	3.5	10.0	5
Crude fiber	%	9.1	26.1	5
Ether extract	%	.7	2.0	8
N-free extract	%	18.1	51.6	
Protein (N x 6.25)	%	3.6	10.3	12
Cattle	dig prot %	2.3	6.6	
Sheep	dig prot %	2.3	6.6	
Energy				
Cattle	DE kcal/kg	972.	2778.	
Sheep	DE kcal/kg	1034.	2954.	
Cattle	ME kcal/kg	797.	2278.	
Sheep	ME kcal/kg	848.	2422.	
Cattle	TDN %	22.	63.	
Sheep	TDN %	23.	67.	
Calcium	%	.19	.54	8
Magnesium	%	.06	.17	38
Phosphorus	%	.07	.20	21
Potassium	%	.55	1.57	11

Bermudagrass, aerial pt, fresh, mature, (2)

Ref no 2-00-711

		As fed	Dry	C.V. ± %
Dry matter	%	40.0	100.0	
Ash	%	3.1	7.8	24
Crude fiber	%	11.4	28.5	6
Ether extract	%	.8	2.0	22
N-free extract	%	22.4	55.9	
Protein (N x 6.25)	%	2.3	5.8	17
Cattle	dig prot %	1.1	2.8	
Sheep	dig prot %	1.0	2.4	
Cellulose	%	13.5	33.7	
Lignin	%	4.0	9.9	
Energy				
Cattle	DE kcal/kg	1094.	2734.	

Continued

(5) protein supplements
(6) minerals

(7) vitamins
(8) additives

Feed name or analyses		Mean		C.V.
		As fed	Dry	± %
Sheep	DE kcal/kg	1234.	3086.	
Cattle	ME kcal/kg	897.	2242.	
Sheep	ME kcal/kg	1012.	2530.	
Cattle	TDN %	25.	62.	
Sheep	TDN %	28.	70.	
Calcium	%	.16	.40	25
Magnesium	%	.06	.15	
Phosphorus	%	.07	.18	25
Potassium	%	.40	1.01	

Bermudagrass, aerial pt, fresh, (2)

Ref no 2-00-712

Dry matter	%	36.7	100.0	33
Ash	%	3.8	10.4	31
Crude fiber	%	9.5	25.9	10
Ether extract	%	.8	2.1	38
N-free extract	%	18.4	50.0	
Protein (N x 6.25)	%	4.2	11.6	29
Cattle	dig prot %	2.9	7.8	
Sheep	dig prot %	2.9	7.8	
Energy				
Cattle	DE kcal/kg	1036.	2822.	
Sheep	DE kcal/kg	1068.	2910.	
Cattle	ME kcal/kg	849.	2314.	
Sheep	ME kcal/kg	876.	2386.	
Cattle	TDN %	23.	64.	
Sheep	TDN %	24.	66.	
Calcium	%	.19	.53	41
Iron	%	.040	.110	
Magnesium	%	.08	.23	41
Phosphorus	%	.08	.22	28
Potassium	%	.60	1.63	36
Sodium	%	.16	.44	
Cobalt	mg/kg	.020	.070	80
Copper	mg/kg	2.1	5.7	24
Manganese	mg/kg	36.7	100.1	
Carotene	mg/kg	103.2	281.1	30
Vitamin A equiv	IU/g	172.0	468.6	

BERMUDAGRASS, COASTAL. Cynodon dactylon

Bermudagrass, coastal, hay, s-c, (1)

Ref no 1-00-716

Dry matter	%	91.5	100.0	1
Ash	%	4.7	5.1	26
Crude fiber	%	27.9	30.5	6
Ether extract	%	2.0	2.2	11
N-free extract	%	48.2	52.7	
Protein (N x 6.25)	%	8.7	9.5	25

Continued

(1) dry forages and roughages
(2) pasture, range plants, and forages fed green

Feed name or analyses		Mean		C.V.
		As fed	Dry	± %
Cattle	dig prot %	4.7	5.1	
Sheep	dig prot %	4.4	4.8	
Energy				
Cattle	DE kcal/kg	1775.	1940.	
Sheep	DE kcal/kg	2018.	2205.	
Cattle	ME kcal/kg	1456.	1591.	
Sheep	ME kcal/kg	1654.	1808.	
Cattle	TDN %	40.	44.	
Sheep	TDN %	46.	50.	
Calcium	%	.42	.46	46
Magnesium	%	.16	.17	76
Phosphorus	%	.16	.18	44

Birdsfoot trefoil - see Trefoil, birdsfoot

Blood flour - see Animal blood, spray dehy

Blood meal - see Animal, blood, dehy grnd

BLUEGRASS. Poa spp

Bluegrass, hay, s-c, (1)

Ref no 1-00-744

Dry matter	%	90.6	100.0	3
Ash	%	7.3	8.1	13
Crude fiber	%	27.4	30.3	13
Ether extract	%	2.7	3.0	31
N-free extract	%	42.6	47.0	
Protein (N x 6.25)	%	10.5	11.6	30
Cattle	dig prot %	6.3	7.0	
Sheep	dig prot %	6.6	7.3	
Energy				
Cattle	DE kcal/kg	2517.	2778.	
Sheep	DE kcal/kg	2517.	2778.	
Cattle	ME kcal/kg	2064.	2278.	
Sheep	ME kcal/kg	2064.	2278.	
Cattle	TDN %	57.	63.	
Sheep	TDN %	57.	63.	
Calcium	%	.35	.39	21
Iron	%	.020	.030	45
Magnesium	%	.19	.21	31
Phosphorus	%	.24	.27	24
Potassium	%	1.56	1.72	17
Copper	mg/kg	9.0	9.9	17
Manganese	mg/kg	83.9	92.6	57
Carotene	mg/kg	224.8	248.1	39
Vitamin A equiv	IU/g	374.7	413.6	

(3) silages
(4) energy feeds

Feed name or analyses		Mean		C.V.
		As fed	Dry	± %

Bluegrass, aerial pt, fresh, immature, (2)

Ref no 2-00-747

		As fed	Dry	± %
Dry matter	%	30.9	100.0	29
Ash	%	2.4	7.9	13
Crude fiber	%	7.8	25.3	15
Ether extract	%	1.1	3.6	20
N-free extract	%	14.1	45.6	
Protein (N x 6.25)	%	5.4	17.6	19
Cattle	dig prot %	4.0	12.8	
Sheep	dig prot %	4.1	13.4	
Energy				
Cattle	DE kcal/kg	981.	3175.	
Sheep	DE kcal/kg	845.	2734.	
Cattle	ME kcal/kg	805.	2604.	
Sheep	ME kcal/kg	693.	2242.	
Cattle	TDN %	22.	72.	
Sheep	TDN %	19.	62.	
Calcium	%	.17	.55	36
Iron	%	.010	.030	
Magnesium	%	.06	.18	9
Phosphorus	%	.13	.42	46
Potassium	%	.77	2.48	17
Copper	mg/kg	4.4	14.1	
Manganese	mg/kg	24.8	80.3	
Carotene	mg/kg	123.9	401.1	20
Vitamin A equiv	IU/g	206.5	668.6	

Bluegrass, aerial pt, fresh, early blm, (2)

Ref no 2-00-749

		As fed	Dry	± %
Dry matter	%	35.7	100.0	7
Ash	%	2.8	7.9	12
Crude fiber	%	10.1	28.2	10
Ether extract	%	1.4	3.9	24
N-free extract	%	16.2	45.3	
Protein (N x 6.25)	%	5.2	14.7	18
Cattle	dig prot %	3.7	10.4	
Sheep	dig prot %	3.8	10.7	
Energy				
Cattle	DE kcal/kg	1086.	3042.	
Sheep	DE kcal/kg	992.	2778.	
Cattle	ME kcal/kg	890.	2494.	
Sheep	ME kcal/kg	813.	2278.	
Cattle	TDN %	25.	69.	
Sheep	TDN %	22.	63.	
Calcium	%	.15	.42	22
Magnesium	%	.04	.11	29
Phosphorus	%	.10	.29	22
Potassium	%	.76	2.14	19
Carotene	mg/kg	99.6	278.9	28
Vitamin A equiv	IU/g	166.0	464.9	

Bluegrass, aerial pt, fresh, mid-blm, (2)

Ref no 2-00-750

		As fed	Dry	± %
Dry matter	%	29.3	100.0	
Ash	%	2.2	7.5	7
Crude fiber	%	8.6	29.3	8
Ether extract	%	1.2	4.0	22
N-free extract	%	13.7	46.7	
Protein (N x 6.25)	%	3.7	12.5	14
Cattle	dig prot %	2.5	8.5	
Sheep	dig prot %	2.5	8.6	
Energy				
Cattle	DE kcal/kg	866.	2954.	
Sheep	DE kcal/kg	827.	2822.	
Cattle	ME kcal/kg	710.	2422.	
Sheep	ME kcal/kg	678.	2314.	
Cattle	TDN %	20.	67.	
Sheep	TDN %	19.	64.	
Calcium	%	.12	.40	
Phosphorus	%	.09	.31	
Potassium	%	.58	1.97	

Bluegrass, aerial pt, fresh, full blm, (2)

Ref no 2-00-752

		As fed	Dry	± %
Dry matter	%	40.0	100.0	
Ash	%	1.8	4.6	
Crude fiber	%	13.8	34.6	
Protein (N x 6.25)	%	2.9	7.3	
Calcium	%	.12	.31	25
Phosphorus	%	.11	.27	27
Potassium	%	.55	1.37	
Carotene	mg/kg	57.8	144.6	44
Vitamin A equiv	IU/g	96.4	241.0	

Bluegrass, aerial pt, fresh, mature, (2)

Ref no 2-00-754

		As fed	Dry	± %
Dry matter	%	41.6	100.0	
Ash	%	2.6	6.2	26
Crude fiber	%	13.7	32.9	12
Ether extract	%	1.3	3.1	30
N-free extract	%	21.2	50.9	
Protein (N x 6.25)	%	2.9	6.9	22
Cattle	dig prot %	1.6	3.8	
Sheep	dig prot %	1.4	3.4	
Energy				
Cattle	DE kcal/kg	1174.	2822.	
Sheep	DE kcal/kg	1247.	2998.	
Cattle	ME kcal/kg	963.	2314.	

Continued

(5) protein supplements

(6) minerals

(7) vitamins

(8) additives

Feed name or analyses		Mean		C.V.
		As fed	Dry	± %
Sheep	ME kcal/kg	1022.	2458.	
Cattle	TDN %	27.	64.	
Sheep	TDN %	28.	68.	
Carotene	mg/kg	30.4	73.2	26
Vitamin A equiv	IU/g	50.7	122.0	

Feed name or analyses		Mean		C.V.
		As fed	Dry	± %
Potassium	%	.63	1.96	27
Copper	mg/kg	4.5	13.9	54
Manganese	mg/kg	18.6	58.0	30
Carotene	mg/kg	65.7	204.8	45
Vitamin A equiv	IU/g	109.5	341.4	

Bluegrass, aerial pt, fresh, over ripe, (2)

Ref no 2-00-755

		As fed	Dry	C.V.
Dry matter	%	85.6	100.0	
Ash	%	5.0	5.8	39
Crude fiber	%	35.3	41.2	6
Ether extract	%	2.5	2.9	71
N-free extract	%	39.6	46.3	
Protein (N x 6.25)	%	3.2	3.8	31
Cattle	dig prot %	.9	1.1	
Sheep	dig prot %	.4	.5	
Energy				
Cattle	DE kcal/kg	2416.	2822.	
Sheep	DE kcal/kg	2491.	2910.	
Cattle	ME kcal/kg	1981.	2314.	
Sheep	ME kcal/kg	2042.	2386.	
Cattle	TDN %	55.	64.	
Sheep	TDN %	56.	66.	
Calcium	%	.25	.29	26
Phosphorus	%	.19	.22	75
Potassium	%	.66	.77	16
Carotene	mg/kg	17.5	20.5	
Vitamin A equiv	IU/g	29.2	34.2	

BLUEGRASS, CANADA. Poa compressa

Bluegrass, Canada, hay, s-c, immature, (1)

Ref no 1-00-760

		As fed	Dry	C.V.
Dry matter	%	96.7	100.0	
Ash	%	9.0	9.3	4
Crude fiber	%	24.9	25.8	1
Ether extract	%	3.0	3.1	2
N-free extract	%	43.0	44.5	
Protein (N x 6.25)	%	16.7	17.3	15
Cattle	dig prot %	11.7	12.1	
Sheep	dig prot %	10.5	10.9	
Energy				
Cattle	DE kcal/kg	3027.	3130.	
Sheep	DE kcal/kg	2644.	2734.	
Cattle	ME kcal/kg	2482.	2567.	
Sheep	ME kcal/kg	2168.	2242.	
Cattle	TDN %	69.	71.	
Sheep	TDN %	60.	62.	

Bluegrass, aerial pt, fresh, (2)

Ref no 2-00-756

		As fed	Dry	C.V.
Dry matter	%	32.1	100.0	39
Ash	%	2.6	8.0	23
Crude fiber	%	8.8	27.3	19
Ether extract	%	1.2	3.6	26
N-free extract	%	14.7	45.8	
Protein (N x 6.25)	%	4.9	15.3	31
Cattle	dig prot %	3.5	10.9	
Sheep	dig prot %	3.6	11.2	
Energy				
Cattle	DE kcal/kg	962.	2998.	
Sheep	DE kcal/kg	892.	2778.	
Cattle	ME kcal/kg	789.	2458.	
Sheep	ME kcal/kg	731.	2278.	
Cattle	TDN %	22.	68.	
Sheep	TDN %	20.	63.	
Calcium	%	.14	.43	43
Iron	%	.010	.020	68
Magnesium	%	.06	.20	17
Phosphorus	%	.12	.38	41

Continued

Bluegrass, Canada, hay, s-c, (1)

Ref no 1-00-762

		As fed	Dry	C.V.
Dry matter	%	93.4	100.0	2
Ash	%	7.4	7.9	11
Crude fiber	%	27.0	28.9	7
Ether extract	%	2.5	2.7	10
N-free extract	%	45.7	48.9	
Protein (N x 6.25)	%	10.8	11.6	30
Cattle	dig prot %	6.5	7.0	
Sheep	dig prot %	6.8	7.3	
Energy				
Cattle	DE kcal/kg	2677.	2866.	
Sheep	DE kcal/kg	2595.	2778.	
Cattle	ME kcal/kg	2195.	2350.	
Sheep	ME kcal/kg	2128.	2278.	
Cattle	TDN %	61.	65.	
Sheep	TDN %	59.	63.	
Calcium	%	.28	.30	19
Magnesium	%	.31	.33	
Phosphorus	%	.27	.29	16
Potassium	%	1.48	1.59	
Manganese	mg/kg	86.5	92.6	

(1) dry forages and roughages
(2) pasture, range plants, and forages fed green

(3) silages
(4) energy feeds

Feed name or analyses		Mean		C.V.
		As fed	Dry	± %

Bluegrass, Canada, aerial pt, fresh, immature, (2)

Ref no 2-00-763

Dry matter	%	25.9	100.0	7
Ash	%	2.4	9.1	7
Crude fiber	%	6.6	25.5	16
Ether extract	%	1.0	3.7	10
N-free extract	%	11.1	43.0	
Protein (N x 6.25)	%	4.8	18.7	7
Cattle	dig prot %	3.6	13.8	
Sheep	dig prot %	3.7	14.4	
Energy				
Cattle	DE kcal/kg	811.	3130.	
Sheep	DE kcal/kg	720.	2778.	
Cattle	ME kcal/kg	665.	2567.	
Sheep	ME kcal/kg	590.	2278.	
Cattle	TDN %	18.	71.	
Sheep	TDN %	16.	63.	

Bluegrass, Canada, aerial pt, fresh, (2)

Ref no 2-00-764

Dry matter	%	30.6	100.0	11
Ash	%	2.7	8.9	7
Crude fiber	%	8.1	26.4	15
Ether extract	%	1.1	3.7	10
N-free extract	%	13.5	44.0	
Protein (N x 6.25)	%	5.2	17.0	17
Cattle	dig prot %	3.8	12.3	
Sheep	dig prot %	3.9	12.8	
Energy				
Cattle	DE kcal/kg	944.	3086.	
Sheep	DE kcal/kg	837.	2734.	
Cattle	ME kcal/kg	774.	2530.	
Sheep	ME kcal/kg	686.	2242.	
Cattle	TDN %	21.	70.	
Sheep	TDN %	19.	62.	
Calcium	%	.12	.39	9
Magnesium	%	.05	.16	
Phosphorus	%	.12	.39	18
Potassium	%	.62	2.04	
Manganese	mg/kg	24.2	79.2	6

BLUEGRASS, KENTUCKY. Poa pratensis

Bluegrass, Kentucky, hay, s-c, immature, (1)

Ref no 1-00-770

Dry matter	%	91.6	100.0	1
Ash	%	8.0	8.7	13
Crude fiber	%	24.6	26.9	6
Ether extract	%	2.9	3.2	31
N-free extract	%	40.6	44.3	
Protein (N x 6.25)	%	15.5	16.9	6
Cattle	dig prot %	11.2	12.2	
Sheep	dig prot %	9.0	9.8	
Energy				
Cattle	DE kcal/kg	2746.	2998.	
Sheep	DE kcal/kg	2424.	2646.	
Cattle	ME kcal/kg	2252.	2458.	
Sheep	ME kcal/kg	1988.	2170.	
Cattle	TDN %	62.	68.	
Sheep	TDN %	55.	60.	
Calcium	%	.45	.49	
Phosphorus	%	.35	.38	
Carotene	mg/kg	309.4	337.8	13
Vitamin A equiv	IU/g	515.8	563.1	

Bluegrass, Kentucky, hay, s-c, mid-blm, (1)

Ref no 1-00-771

Dry matter	%	88.1	100.0	
Ash	%	7.0	7.9	
Crude fiber	%	27.9	31.7	
Ether extract	%	3.4	3.9	
N-free extract	%	40.4	45.9	
Protein (N x 6.25)	%	9.3	10.6	
Cattle	dig prot %	5.4	6.1	
Sheep	dig prot %	5.4	6.1	
Energy				
Cattle	DE kcal/kg	2409.	2734.	
Sheep	DE kcal/kg	2370.	2690.	
Cattle	ME kcal/kg	1975.	2242.	
Sheep	ME kcal/kg	1943.	2206.	
Cattle	TDN %	55.	62.	
Sheep	TDN %	54.	61.	
Calcium	%	.30	.34	
Phosphorus	%	.21	.24	

(5) protein supplements
(6) minerals

(7) vitamins
(8) additives

Feed name or analyses		Mean		C.V.
		As fed	Dry	± %

Bluegrass, Kentucky, hay, s-c, full blm, (1)

Ref no 1-00-772

Dry matter	%	92.1	100.0	
Ash	%	5.4	5.9	
Crude fiber	%	29.9	32.5	
Ether extract	%	3.0	3.3	
N-free extract	%	45.5	49.4	
Protein (N x 6.25)	%	8.2	8.9	
Cattle	dig prot %	4.2	4.6	
Sheep	dig prot %	4.8	5.2	
Energy				
Cattle	DE kcal/kg	2355.	2557.	
Sheep	DE kcal/kg	2518.	2734.	
Cattle	ME kcal/kg	1931.	2097.	
Sheep	ME kcal/kg	2065.	2242.	
Cattle	TDN %	53.	58.	
Sheep	TDN %	57.	62.	
Calcium	%	.24	.26	
Phosphorus	%	.25	.27	
Potassium	%	1.40	1.52	

Bluegrass, Kentucky, hay, s-c, mature, (1)

Ref no 1-00-774

Dry matter	%	93.0	100.0	
Ash	%	6.7	7.2	
Crude fiber	%	28.9	31.1	
Ether extract	%	2.7	2.9	
N-free extract	%	48.9	52.6	
Protein (N x 6.25)	%	5.8	6.2	
Cattle	dig prot %	2.1	2.3	
Sheep	dig prot %	3.3	3.6	
Energy				
Cattle	DE kcal/kg	2543.	2734.	
Sheep	DE kcal/kg	2502.	2690.	
Cattle	ME kcal/kg	2085.	2242.	
Sheep	ME kcal/kg	2052.	2206.	
Cattle	TDN %	58.	62.	
Sheep	TDN %	57.	61.	

Bluegrass, Kentucky, hay, s-c, (1)

Ref no 1-00-776

Dry matter	%	88.8	100.0	3
Ash	%	7.3	8.2	13
Crude fiber	%	26.6	29.9	11
Ether extract	%	2.8	3.1	27
N-free extract	%	41.6	46.8	
Protein (N x 6.25)	%	10.7	12.0	25

Continued

Cattle	dig prot %	6.5	7.3	
Sheep	dig prot %	6.2	7.0	
Energy				
Cattle	DE kcal/kg	2506.	2822.	
Sheep	DE kcal/kg	2350.	2646.	
Cattle	ME kcal/kg	2055.	2314.	
Sheep	ME kcal/kg	1927.	2170.	
Cattle	TDN %	57.	64.	
Sheep	TDN %	53.	60.	
Calcium	%	.36	.40	20
Chlorine	%	.41	.46	32
Iron	%	.020	.030	45
Magnesium	%	.19	.21	31
Phosphorus	%	.23	.26	25
Potassium	%	1.53	1.72	17
Sodium	%	.12	.13	24
Sulfur	%	.20	.23	23
Copper	mg/kg	8.8	9.9	17
Manganese	mg/kg	82.2	92.6	58

Bluegrass, Kentucky, aerial pt, fresh, immature, (2)

Ref no 2-00-778

Dry matter	%	30.5	100.0	22
Ash	%	2.2	7.4	15
Crude fiber	%	7.6	25.1	14
Ether extract	%	1.1	3.6	21
N-free extract	%	14.2	46.6	
Protein (N x 6.25)	%	5.3	17.3	20
Cattle	dig prot %	3.8	12.6	
Sheep	dig prot %	4.0	13.1	
Cellulose	%	6.1	20.0	8
Lignin	%	1.2	3.8	21
Energy				
Cattle	DE kcal/kg	968.	3175.	
Sheep	DE kcal/kg	847.	2778.	
Cattle	ME kcal/kg	794.	2604.	
Sheep	ME kcal/kg	695.	2278.	
Cattle	TDN %	22.	72.	
Sheep	TDN %	19.	63.	
Calcium	%	.17	.56	37
Iron	%	.010	.030	
Magnesium	%	.05	.18	11
Phosphorus	%	.14	.47	34
Potassium	%	.70	2.28	14
Copper	mg/kg	4.3	14.1	
Manganese	mg/kg	24.5	80.3	
Carotene	mg/kg	116.8	383.0	24
Vitamin A equiv	IU/g	194.7	638.5	

(1) dry forages and roughages
(2) pasture, range plants, and forages fed green

(3) silages
(4) energy feeds

TABLE A-3-1 Composition of Feeds (Continued)

Feed name or analyses		Mean As fed	Dry	C.V. ± %

Bluegrass, Kentucky, aerial pt, fresh, early blm, (2)

Ref no 2-00-779

Feed name or analyses		As fed	Dry	C.V. ± %
Dry matter	%	35.7	100.0	7
Ash	%	2.8	7.9	12
Crude fiber	%	9.9	27.8	10
Ether extract	%	1.4	3.9	23
N-free extract	%	16.3	45.6	
Protein (N x 6.25)	%	5.3	14.8	13
Cattle	dig prot %	3.7	10.5	
Sheep	dig prot %	3.8	10.8	
Cellulose	%	10.1	28.3	9
Lignin	%	1.6	4.6	8
Energy				
Cattle	DE kcal/kg	1086.	3042.	
Sheep	DE kcal/kg	992.	2778.	
Cattle	ME kcal/kg	890.	2494.	
Sheep	ME kcal/kg	813.	2278.	
Cattle	TDN %	25.	69.	
Sheep	TDN %	22.	63.	
Calcium	%	.16	.46	14
Magnesium	%	.04	.11	29
Phosphorus	%	.14	.39	15
Potassium	%	.72	2.01	10

Bluegrass, Kentucky, aerial pt, fresh, mid-blm, (2)

Ref no 2-00-780

Feed name or analyses		As fed	Dry	C.V. ± %
Dry matter	%	29.3	100.0	
Ash	%	2.2	7.5	7
Crude fiber	%	8.6	29.3	8
Ether extract	%	1.2	4.0	22
N-free extract	%	13.5	46.0	
Protein (N x 6.25)	%	3.9	13.2	5
Cattle	dig prot %	2.7	9.1	
Sheep	dig prot %	2.7	9.3	
Cellulose	%	9.0	30.8	9
Lignin	%	1.7	5.8	9
Energy				
Cattle	DE kcal/kg	878.	2998.	
Sheep	DE kcal/kg	827.	2822.	
Cattle	ME kcal/kg	720.	2458.	
Sheep	ME kcal/kg	678.	2314.	
Cattle	TDN %	20.	68.	
Sheep	TDN %	19.	64.	
Calcium	%	.11	.38	
Phosphorus	%	.11	.38	
Potassium	%	.58	1.97	

Bluegrass, Kentucky, aerial pt, fresh, milk stage, (2)

Ref no 2-00-782

Feed name or analyses		As fed	Dry	C.V. ± %
Dry matter	%	35.0	100.0	
Ash	%	2.6	7.3	
Crude fiber	%	10.6	30.3	
Ether extract	%	1.3	3.6	
N-free extract	%	16.5	47.2	
Protein (N x 6.25)	%	4.1	11.6	
Cattle	dig prot %	2.7	7.8	
Sheep	dig prot %	2.7	7.8	
Energy				
Cattle	DE kcal/kg	1034.	2954.	
Sheep	DE kcal/kg	988.	2822.	
Cattle	ME kcal/kg	848.	2422.	
Sheep	ME kcal/kg	810.	2314.	
Cattle	TDN %	23.	67.	
Sheep	TDN %	22.	64.	
Calcium	%	.07	.19	
Phosphorus	%	.09	.27	

Bluegrass, Kentucky, aerial pt, fresh, mature, (2)

Ref no 2-00-784

Feed name or analyses		As fed	Dry	C.V. ± %
Dry matter	%	41.6	100.0	
Ash	%	2.6	6.2	15
Crude fiber	%	13.4	32.2	12
Ether extract	%	1.3	3.1	30
N-free extract	%	20.4	49.0	
Protein (N x 6.25)	%	4.0	9.5	15
Cattle	dig prot %	2.5	6.0	
Sheep	dig prot %	2.4	5.8	
Cellulose	%	16.3	39.1	11
Lignin	%	3.2	7.7	12
Energy				
Cattle	DE kcal/kg	1229.	2954.	
Sheep	DE kcal/kg	1210.	2910.	
Cattle	ME kcal/kg	1008.	2422.	
Sheep	ME kcal/kg	992.	2386.	
Cattle	TDN %	28.	67.	
Sheep	TDN %	27.	66.	
Calcium	%	.08	.19	28
Iron	%	.010	.030	26
Magnesium	%	.05	.12	25
Phosphorus	%	.11	.27	24
Potassium	%	.77	1.86	9
Manganese	mg/kg	12.1	29.1	
Carotene	mg/kg	38.1	91.7	
Vitamin A equiv	IU/g	63.5	152.9	

Feed name or analyses		Mean		C.V.
		As fed	Dry	± %

Bluegrass, Kentucky, aerial pt, fresh, over ripe, (2)

Ref no 2-00-785

		As fed	Dry	
Dry matter	%	45.0	100.0	
Ash	%	2.8	6.3	
Crude fiber	%	18.9	42.1	
Ether extract	%	.6	1.3	
N-free extract	%	21.2	47.0	
Protein (N x 6.25)	%	1.5	3.3	
Cattle	dig prot %	.3	.7	
Sheep	dig prot %	.05	.1	
Energy				
Cattle	DE kcal/kg	1290.	2866.	
Sheep	DE kcal/kg	1310.	2910.	
Cattle	ME kcal/kg	1058.	2350.	
Sheep	ME kcal/kg	1074.	2386.	
Cattle	TDN %	29.	65.	
Sheep	TDN %	30.	66.	
Carotene	mg/kg	18.4	40.8	
Vitamin A equiv	IU/g	30.7	68.0	

Bluegrass, Kentucky, aerial pt, fresh, (2)

Ref no 2-00-786

		As fed	Dry	C.V.
Dry matter	%	31.5	100.0	19
Ash	%	2.5	7.9	16
Crude fiber	%	8.3	26.5	20
Ether extract	%	1.1	3.6	27
N-free extract	%	14.6	46.2	
Protein (N x 6.25)	%	5.0	15.8	30
Cattle	dig prot %	3.6	11.3	
Sheep	dig prot %	3.7	11.7	
Cellulose	%	9.5	30.2	18
Lignin	%	1.9	6.0	34
Energy	GE kcal/kg	1448.	4597.	
Cattle	DE kcal/kg	972.	3086.	
Sheep	DE kcal/kg	875.	2778.	
Cattle	ME kcal/kg	797.	2530.	
Sheep	ME kcal/kg	718.	2278.	
Cattle	TDN %	22.	70.	
Sheep	TDN %	20.	63.	
Calcium	%	.14	.45	43
Chlorine	%	.13	.40	
Iron	%	.010	.020	68
Magnesium	%	.06	.20	17
Phosphorus	%	.12	.39	37
Potassium	%	.65	2.06	12
Sodium	%	.09	.30	50
Sulfur	%	.14	.46	42
Cobalt	mg/kg	.030	.090	
Copper	mg/kg	4.4	13.9	54
Manganese	mg/kg	7.3	23.1	89

Continued

Feed name or analyses		Mean		C.V.
		As fed	Dry	± %
Zinc	mg/kg	5.4	17.0	
Carotene	mg/kg	63.1	200.2	47
Vitamin A equiv	IU/g	105.2	333.7	

BLUESTEM. Andropogon spp

Bluestem, aerial pt, fresh, immature, (2)

Ref no 2-00-821

		As fed	Dry	C.V.
Dry matter	%	31.6	100.0	11
Ash	%	3.1	9.7	16
Crude fiber	%	9.1	28.9	13
Ether extract	%	.8	2.4	16
N-free extract	%	15.2	48.0	
Protein (N x 6.25)	%	3.5	11.0	21
Cattle	dig prot %	2.3	7.2	
Sheep	dig prot %	2.3	7.2	
Energy				
Cattle	DE kcal/kg	892.	2822.	
Sheep	DE kcal/kg	920.	2910.	
Cattle	ME kcal/kg	731.	2314.	
Sheep	ME kcal/kg	754.	2386.	
Cattle	TDN %	20.	64.	
Sheep	TDN %	21.	66.	
Calcium	%	.20	.63	19
Iron	%	.020	.070	
Phosphorus	%	.05	.17	56
Potassium	%	.43	1.35	
Copper	mg/kg	11.6	36.8	
Manganese	mg/kg	26.3	83.3	
Carotene	mg/kg	69.3	219.2	35
Vitamin A equiv	IU/g	115.5	365.4	

Bluestem, aerial pt, fresh, early blm, (2)

Ref no 2-00-822

		As fed	Dry	C.V.
Dry matter	%	29.4	100.0	
Ash	%	2.8	9.6	19
Crude fiber	%	8.9	30.2	7
Ether extract	%	.8	2.8	30
N-free extract	%	13.8	46.9	
Protein (N x 6.25)	%	3.1	10.5	21
Cattle	dig prot %	2.0	6.8	
Sheep	dig prot %	2.0	6.8	
Energy				
Cattle	DE kcal/kg	817.	2778.	
Sheep	DE kcal/kg	843.	2866.	
Cattle	ME kcal/kg	670.	2278.	
Sheep	ME kcal/kg	691.	2350.	
Cattle	TDN %	18.	63.	

Continued

(1) dry forages and roughages

(2) pasture, range plants, and forages fed green

(3) silages

(4) energy feeds

Feed name or analyses		As fed	Dry	C.V. ± %
Sheep	TDN %	19.	65.	
Carotene	mg/kg	55.6	189.0	
Vitamin A equiv	IU/g	92.7	315.1	

Bluestem, aerial pt, fresh, mid-blm, (2)

Ref no 2-00-823

		As fed	Dry	C.V. ± %
Dry matter	%	50.0	100.0	
Ash	%	5.0	10.1	12
Crude fiber	%	15.5	31.0	12
Ether extract	%	1.1	2.2	20
N-free extract	%	24.6	49.3	
Protein (N x 6.25)	%	3.7	7.4	16
Cattle	dig prot %	2.1	4.2	
Sheep	dig prot %	2.0	3.9	
Energy				
Cattle	DE kcal/kg	1323.	2646.	
Sheep	DE kcal/kg	1477.	2954.	
Cattle	ME kcal/kg	1085.	2170.	
Sheep	ME kcal/kg	1211.	2422.	
Cattle	TDN %	30.	60.	
Sheep	TDN %	34.	67.	

Bluestem, aerial pt, fresh, full blm, (2)

Ref no 2-00-824

		As fed	Dry	C.V. ± %
Dry matter	%	70.4	100.0	
Ash	%	5.7	8.1	17
Crude fiber	%	24.5	34.8	7
Ether extract	%	1.5	2.1	14
N-free extract	%	34.6	49.2	
Protein (N x 6.25)	%	4.1	5.8	15
Cattle	dig prot %	2.0	2.8	
Sheep	dig prot %	1.7	2.4	
Energy				
Cattle	DE kcal/kg	1925.	2734.	
Sheep	DE kcal/kg	2080.	2954.	
Cattle	ME kcal/kg	1578.	2242.	
Sheep	ME kcal/kg	1705.	2422.	
Cattle	TDN %	44.	62.	
Sheep	TDN %	47.	67.	
Calcium	%	.31	.44	28
Phosphorus	%	.09	.13	77
Carotene	mg/kg	55.8	79.2	15
Vitamin A equiv	IU/g	93.0	132.0	

Bluestem, aerial pt, fresh, mature, (2)

Ref no 2-00-825

		As fed	Dry	C.V. ± %
Dry matter	%	71.3	100.0	17
Ash	%	5.2	7.3	28
Crude fiber	%	24.2	34.0	10
Ether extract	%	1.5	2.1	26
N-free extract	%	37.1	52.1	
Protein (N x 6.25)	%	3.2	4.5	21
Cattle	dig prot %	1.2	1.7	
Sheep	dig prot %	.8	1.2	
Energy				
Cattle	DE kcal/kg	1949.	2734.	
Sheep	DE kcal/kg	2169.	3042.	
Cattle	ME kcal/kg	1598.	2242.	
Sheep	ME kcal/kg	1778.	2494.	
Cattle	TDN %	44.	62.	
Sheep	TDN %	49.	69.	
Calcium	%	.28	.40	15
Iron	%	.050	.060	
Magnesium	%	.04	.06	
Phosphorus	%	.08	.11	50
Potassium	%	.36	.51	
Copper	mg/kg	11.5	16.1	
Manganese	mg/kg	26.2	36.8	

Bluestem, aerial pt, fresh, over ripe, (2)

Ref no 2-00-826

		As fed	Dry	C.V. ± %
Dry matter	%	84.7	100.0	
Ash	%	6.7	7.9	35
Crude fiber	%	30.2	35.6	9
Ether extract	%	1.6	1.9	23
N-free extract	%	43.6	51.5	
Protein (N x 6.25)	%	2.6	3.1	29
Cattle	dig prot %	.4	.5	
Sheep	dig prot %	.0	.0	
Energy				
Cattle	DE kcal/kg	2241.	2646.	
Sheep	DE kcal/kg	2576.	3042.	
Cattle	ME kcal/kg	1838.	2170.	
Sheep	ME kcal/kg	2112.	2494.	
Cattle	TDN %	51.	60.	
Sheep	TDN %	58.	69.	
Calcium	%	.25	.30	21
Iron	%	.030	.040	
Phosphorus	%	.07	.08	74
Copper	mg/kg	18.6	22.0	
Manganese	mg/kg	40.0	46.7	
Carotene	mg/kg	2.2	2.6	40
Vitamin A equiv	IU/g	3.7	4.3	

(5) protein supplements

(6) minerals

(7) vitamins

(8) additives

Feed name or analyses		Mean		C.V.
		As fed	Dry	± %

Bluestem, aerial pt, fresh, (2)

Ref no 2-00-827

Dry matter	%	54.4	100.0	24
Ash	%	4.5	8.3	27
Crude fiber	%	17.8	32.8	16
Ether extract	%	1.2	2.2	26
N-free extract	%	27.3	50.1	
Protein (N x 6.25)	%	3.6	6.6	45
Cattle	dig prot %	1.9	3.5	
Sheep	dig prot %	1.7	3.1	
Lignin	%	6.4	11.8	
Energy				
Cattle	DE kcal/kg	1487.	2734.	
Sheep	DE kcal/kg	1631.	2998.	
Cattle	ME kcal/kg	1220.	2242.	
Sheep	ME kcal/kg	1337.	2458.	
Cattle	TDN %	34.	62.	
Sheep	TDN %	37.	68.	
Calcium	%	.22	.40	34
Iron	%	.020	.040	55
Magnesium	%	.10	.18	39
Phosphorus	%	.08	.15	62
Potassium	%	.65	1.19	32
Copper	mg/kg	15.0	27.6	30
Manganese	mg/kg	33.6	61.7	34
Carotene	mg/kg	64.4	118.4	68
Vitamin A equiv	IU/g	107.4	197.4	

Bone - see Animal

Bone black - see Animal, bone charcoal

Bone char - see Animal, bone charcoal

Bone charcoal - see Animal

Bone phosphate - see Animal, bone phosphate

Bran - see Wheat

BREAD. Scientific name not used

Bread, dehy, (4)

Ref no 4-07-944

Dry matter	%	95.0	100.0	
Ash	%	1.9	2.0	
Crude fiber	%	.5	.5	
Ether extract	%	1.0	1.1	
N-free extract	%	80.6	84.8	
Protein (N x 6.25)	%	11.0	11.6	
Cattle	dig prot %	7.8	8.2	
Sheep	dig prot %	7.8	8.2	
Swine	dig prot %	9.8	10.3	
Energy				
Cattle	DE kcal/kg	3644.	3836.	
Sheep	DE kcal/kg	3644.	3836.	
Swine	DE kcal/kg	3686.	3880.	
Cattle	ME kcal/kg	2989.	3146.	
Sheep	ME kcal/kg	2989.	3146.	
Swine	ME kcal/kg	3454.	3636.	
Cattle	TDN %	83.	87.	
Sheep	TDN %	83.	87.	
Swine	TDN %	84.	88.	
Calcium	%	.03	.03	
Phosphorus	%	.10	.10	

Brewers dried grains - see Grains

Brewers dried yeast - see Yeast, brewers

Brewers rice - see Rice, groats, polished and broken

BROME. Bromus spp

Brome, hay, s-c, (1)

Ref no 1-00-890

Dry matter	%	89.7	100.0	2
Ash	%	7.7	8.6	16
Crude fiber	%	28.7	32.0	16
Ether extract	%	2.3	2.6	39
N-free extract	%	40.4	45.0	
Protein (N x 6.25)	%	10.6	11.8	36
Cattle	dig prot %	4.5	5.0	
Sheep	dig prot %	6.4	7.2	

Continued

(1) dry forages and roughages
(2) pasture, range plants, and forages fed green

(3) silages
(4) energy feeds

TABLE A-3-1 Composition of Feeds (Continued)

Feed name or analyses		Mean		C.V.
		As fed	Dry	± %
Energy				
Cattle	DE kcal/kg	1740.	1940.	
Sheep	DE kcal/kg	2254.	2513.	
Cattle	ME kcal/kg	1427.	1591.	
Sheep	ME kcal/kg	1849.	2061.	
Cattle	TDN %	39.	44.	
Sheep	TDN %	51.	57.	

Feed name or analyses		Mean		C.V.
		As fed	Dry	± %
Magnesium	%	.03	.11	
Phosphorus	%	.12	.39	22
Potassium	%	.80	2.67	32
Carotene	mg/kg	55.2	183.9	
Vitamin A equiv	IU/g	92.0	306.6	

Brome, aerial pt, fresh, immature, (2)

Ref no 2-00-892

		As fed	Dry	C.V. ± %
Dry matter	%	32.5	100.0	19
Ash	%	3.9	12.0	27
Crude fiber	%	7.8	23.9	15
Ether extract	%	1.2	3.6	28
N-free extract	%	13.1	40.2	
Protein (N x 6.25)	%	6.6	20.3	32
Cattle	dig prot %	4.9	15.1	
Sheep	dig prot %	5.4	16.6	
Energy				
Cattle	DE kcal/kg	974.	2998.	
Sheep	DE kcal/kg	1118.	3439.	
Cattle	ME kcal/kg	799.	2458.	
Sheep	ME kcal/kg	916.	2820.	
Cattle	TDN %	22.	68.	
Sheep	TDN %	25.	78.	
Calcium	%	.19	.59	29
Magnesium	%	.06	.18	
Phosphorus	%	.12	.37	35
Potassium	%	1.40	4.30	22
Carotene	mg/kg	149.3	459.5	19
Vitamin A equiv	IU/g	248.9	766.0	

Brome, aerial pt, fresh, early blm, (2)

Ref no 2-00-893

		As fed	Dry	C.V. ± %
Dry matter	%	30.0	100.0	11
Ash	%	2.4	8.1	17
Crude fiber	%	8.5	28.3	8
Ether extract	%	1.1	3.7	22
N-free extract	%	13.8	46.0	
Protein (N x 6.25)	%	4.2	13.9	20
Cattle	dig prot %	2.9	9.7	
Sheep	dig prot %	3.0	9.9	
Energy				
Cattle	DE kcal/kg	899.	2998.	
Sheep	DE kcal/kg	847.	2822.	
Cattle	ME kcal/kg	737.	2458.	
Sheep	ME kcal/kg	694.	2314.	
Cattle	TDN %	20.	68.	
Sheep	TDN %	19.	64.	
Calcium	%	.12	.41	25

Continued

Brome, aerial pt, fresh, mid-blm, (2)

Ref no 2-00-894

		As fed	Dry	C.V. ± %
Dry matter	%	30.0	100.0	
Ash	%	2.0	6.7	13
Crude fiber	%	9.1	30.4	6
Ether extract	%	.9	3.0	6
N-free extract	%	14.8	49.2	
Protein (N x 6.25)	%	3.2	10.7	15
Cattle	dig prot %	2.1	7.0	
Sheep	dig prot %	2.1	7.0	
Energy				
Cattle	DE kcal/kg	899.	2998.	
Sheep	DE kcal/kg	873.	2910.	
Cattle	ME kcal/kg	737.	2458.	
Sheep	ME kcal/kg	716.	2386.	
Cattle	TDN %	20.	68.	
Sheep	TDN %	20.	66.	

Brome, aerial pt, fresh, full blm, (2)

Ref no 2-00-895

		As fed	Dry	C.V. ± %
Dry matter	%	27.1	100.0	
Ash	%	2.0	7.5	33
Crude fiber	%	8.2	30.3	19
Ether extract	%	.9	3.3	79
N-free extract	%	13.1	48.3	
Protein (N x 6.25)	%	2.9	10.6	23
Cattle	dig prot %	1.9	6.9	
Sheep	dig prot %	1.9	6.9	
Energy				
Cattle	DE kcal/kg	789.	2910.	
Sheep	DE kcal/kg	789.	2910.	
Cattle	ME kcal/kg	647.	2386.	
Sheep	ME kcal/kg	647.	2386.	
Cattle	TDN %	18.	66.	
Sheep	TDN %	18.	66.	
Calcium	%	.09	.35	40
Phosphorus	%	.08	.29	25
Potassium	%	.55	2.04	29
Carotene	mg/kg	22.4	82.7	
Vitamin A equiv	IU/g	37.3	137.9	

(5) protein supplements

(6) minerals

(7) vitamins

(8) additives

Feed name or analyses	Mean As fed	Dry	C.V. ± %

Brome, aerial pt, fresh, mature, (2)

Ref no 2-00-898

		Mean As fed	Dry	C.V. ± %
Dry matter	%	56.1	100.0	8
Ash	%	3.7	6.6	28
Crude fiber	%	18.5	33.0	13
Ether extract	%	1.2	2.2	27
N-free extract	%	29.1	51.8	
Protein (N x 6.25)	%	3.6	6.4	33
Cattle	dig prot %	1.8	3.3	
Sheep	dig prot %	1.7	3.0	
Energy				
Cattle	DE kcal/kg	1608.	2866.	
Sheep	DE kcal/kg	1682.	2998.	
Cattle	ME kcal/kg	1318.	2350.	
Sheep	ME kcal/kg	1379.	2458.	
Cattle	TDN %	36.	65.	
Sheep	TDN %	38.	68.	
Calcium	%	.17	.30	30
Phosphorus	%	.14	.26	24
Potassium	%	.70	1.25	28

Brome, aerial pt, fresh, (2)

Ref no 2-00-900

		Mean As fed	Dry	C.V. ± %
Dry matter	%	35.9	100.0	34
Ash	%	3.2	8.9	41
Crude fiber	%	10.4	29.0	21
Ether extract	%	1.0	2.9	49
N-free extract	%	17.2	47.8	
Protein (N x 6.25)	%	4.1	11.4	55
Cattle	dig prot %	2.7	7.6	
Sheep	dig prot %	2.7	7.6	
Energy				
Cattle	DE kcal/kg	1029.	2866.	
Sheep	DE kcal/kg	1029.	2866.	
Cattle	ME kcal/kg	844.	2350.	
Sheep	ME kcal/kg	844.	2350.	
Cattle	TDN %	23.	65.	
Sheep	TDN %	23.	65.	
Calcium	%	.16	.44	33
Magnesium	%	.09	.26	35
Phosphorus	%	.11	.30	41
Potassium	%	.83	2.31	51
Copper	mg/kg	.8	2.2	
Carotene	mg/kg	111.6	310.9	39
Vitamin A equiv	IU/g	186.0	518.3	

Brome, aerial pt, fresh, over ripe, (2)

Ref no 2-00-899

		Mean As fed	Dry	C.V. ± %
Dry matter	%	75.0	100.0	11
Ash	%	4.5	6.0	33
Crude fiber	%	28.0	37.4	11
Ether extract	%	1.0	1.4	21
N-free extract	%	38.3	51.1	
Protein (N x 6.25)	%	3.1	4.1	25
Cattle	dig prot %	1.0	1.4	
Sheep	dig prot %	.6	.8	
Energy				
Cattle	DE kcal/kg	2150.	2866.	
Sheep	DE kcal/kg	2248.	2998.	
Cattle	ME kcal/kg	1762.	2350.	
Sheep	ME kcal/kg	1844.	2458.	
Cattle	TDN %	49.	65.	
Sheep	TDN %	51.	68.	
Calcium	%	.21	.28	44
Magnesium	%	.10	.14	
Phosphorus	%	.14	.18	38
Potassium	%	.92	1.23	16

BROME, CHEATGRASS. Bromus tectorum

Brome, cheatgrass, aerial pt, fresh, immature, (2)

Ref no 2-00-908

		Mean As fed	Dry	C.V. ± %
Dry matter	%	21.0	100.0	
Ash	%	2.0	9.6	
Crude fiber	%	4.8	22.9	
Ether extract	%	.6	2.7	
N-free extract	%	10.3	49.0	
Protein (N x 6.25)	%	3.3	15.8	
Cattle	dig prot %	2.4	11.3	
Sheep	dig prot %	2.4	11.7	
Energy				
Cattle	DE kcal/kg	630.	2998.	
Sheep	DE kcal/kg	602.	2866.	
Cattle	ME kcal/kg	516.	2458.	
Sheep	ME kcal/kg	494.	2350.	
Cattle	TDN %	14.	68.	
Sheep	TDN %	14.	65.	
Calcium	%	.13	.64	
Phosphorus	%	.06	.28	

(1) dry forages and roughages
(2) pasture, range plants, and forages fed green

(3) silages
(4) energy feeds

Feed name or analyses		Mean		C.V.
		As fed	Dry	± %

Brome, cheatgrass, aerial pt, fresh, dough stage, (2)

Ref no 2-00-910

		As fed	Dry	
Dry matter	%	30.0	100.0	
Ash	%	2.6	8.7	
Crude fiber	%	10.4	34.8	
Ether extract	%	.4	1.3	
N-free extract	%	15.0	49.9	
Protein (N x 6.25)	%	1.6	5.3	
Cattle	dig prot %	.7	2.4	
Sheep	dig prot %	.6	1.9	
Energy				
Cattle	DE kcal/kg	820.	2734.	
Sheep	DE kcal/kg	899.	2998.	
Cattle	ME kcal/kg	672.	2241.	
Sheep	ME kcal/kg	737.	2458.	
Cattle	TDN %	19.	62.	
Sheep	TDN %	20.	68.	
Calcium	%	.11	.38	
Phosphorus	%	.08	.27	

BROME, SMOOTH. Bromus inermis

Brome, smooth, hay, s-c, immature, (1)

Ref no 1-00-940

		As fed	Dry	
Dry matter	%	90.3	100.0	3
Ash	%	8.1	9.0	15
Crude fiber	%	25.8	28.6	15
Ether extract	%	2.5	2.8	39
N-free extract	%	38.8	43.0	
Protein (N x 6.25)	%	15.0	16.6	23
Cattle	dig prot %	11.1	12.3	
Sheep	dig prot %	6.9	7.6	
Energy				
Cattle	DE kcal/kg	2628.	2910.	
Sheep	DE kcal/kg	2070.	2293.	
Cattle	ME kcal/kg	2154.	2386.	
Sheep	ME kcal/kg	1698.	1880.	
Cattle	TDN %	60.	66.	
Sheep	TDN %	47.	52.	
Calcium	%	.59	.65	24
Iron	%	.020	.020	41
Magnesium	%	.28	.31	8
Phosphorus	%	.33	.37	43
Potassium	%	2.16	2.39	6
Cobalt	mg/kg	.080	.090	58
Copper	mg/kg	13.5	15.0	33
Manganese	mg/kg	32.0	35.5	10
Carotene	mg/kg	58.5	64.8	76
Vitamin A equiv	IU/g	97.5	108.0	

Brome, smooth, hay, s-c, early blm, (1)

Ref no 1-00-941

		As fed	Dry	
Dry matter	%	90.3	100.0	
Ash	%	7.8	8.7	
Crude fiber	%	28.2	31.2	
Ether extract	%	2.2	2.4	
N-free extract	%	41.2	45.6	
Protein (N x 6.25)	%	10.9	12.1	
Cattle	dig prot %	6.9	7.4	
Sheep	dig prot %	5.0	5.6	
Energy				
Cattle	DE kcal/kg	2469.	2734.	
Sheep	DE kcal/kg	2110.	2337.	
Cattle	ME kcal/kg	2024.	2242.	
Sheep	ME kcal/kg	1730.	1916.	
Cattle	TDN %	56.	62.	
Sheep	TDN %	48.	53.	

Brome, smooth, hay, s-c, full blm, (1)

Ref no 1-00-942

		As fed	Dry	
Dry matter	%	88.9	100.0	2
Ash	%	6.6	7.4	20
Crude fiber	%	31.9	35.9	3
Ether extract	%	1.8	2.0	13
N-free extract	%	39.5	44.4	
Protein (N x 6.25)	%	9.2	10.3	10
Cattle	dig prot %	5.2	5.8	
Sheep	dig prot %	4.2	4.7	
Energy				
Cattle	DE kcal/kg	2078.	2337.	
Sheep	DE kcal/kg	2117.	2381.	
Cattle	ME kcal/kg	1703.	1916.	
Sheep	ME kcal/kg	1735.	1952.	
Cattle	TDN %	47.	53.	
Sheep	TDN %	48.	54.	

Brome, smooth, hay, s-c, mature, (1)

Ref no 1-00-944

		As fed	Dry	
Dry matter	%	92.8	100.0	2
Ash	%	7.6	8.2	22
Crude fiber	%	31.7	34.2	11
Ether extract	%	2.8	3.0	27
N-free extract	%	45.3	48.8	
Protein (N x 6.25)	%	5.4	5.8	32
Cattle	dig prot %	1.8	2.0	
Sheep	dig prot %	2.5	2.7	
Lignin	%	13.2	14.2	

Continued

(5) protein supplements
(6) minerals

(7) vitamins
(8) additives

Feed name or analyses		Mean		C.V.
		As fed	Dry	± %
Energy				
Cattle	DE kcal/kg	2373.	2557.	
Sheep	DE kcal/kg	2210.	2381.	
Cattle	ME kcal/kg	1946.	2097.	
Sheep	ME kcal/kg	1811.	1952.	
Cattle	TDN %	54.	58.	
Sheep	TDN %	50.	54.	
Calcium	%	.40	.43	20
Chlorine	%	.12	.13	
Iron	%	.010	.010	
Magnesium	%	.18	.19	
Phosphorus	%	.20	.22	64
Potassium	%	2.56	2.76	
Cobalt	mg/kg	.120	.130	
Copper	mg/kg	6.3	6.8	
Manganese	mg/kg	98.2	105.8	
Carotene	mg/kg	4.4	4.8	
Vitamin A equiv	IU/g	7.3	8.0	

Brome, smooth, hay, s-c, (1)

Ref no 1-00-947

		As fed	Dry	C.V. ± %
Dry matter	%	89.7	100.0	2
Ash	%	7.9	8.8	16
Crude fiber	%	28.4	31.7	17
Ether extract	%	2.3	2.6	38
N-free extract	%	40.0	44.6	
Protein (N x 6.25)	%	11.0	12.3	36
Cattle	dig prot %	6.8	7.6	
Sheep	dig prot %	5.0	5.6	
Cellulose	%	27.8	31.0	
Lignin	%	8.1	9.0	34
Energy	GE kcal/kg	4010.	4470.	2
Cattle	DE kcal/kg	2413.	2690.	
Sheep	DE kcal/kg	2096.	2337.	
Cattle	ME kcal/kg	1979.	2206.	
Sheep	ME kcal/kg	1719.	1916.	
Cattle	TDN %	55.	61.	
Sheep	TDN %	48.	53.	
Calcium	%	.39	.43	35
Chlorine	%	.40	.45	35
Iron	%	.010	.010	61
Magnesium	%	.19	.21	32
Phosphorus	%	.25	.28	48
Potassium	%	2.12	2.36	17
Sodium	%	.45	.50	41
Sulfur	%	.18	.20	33
Cobalt	mg/kg	.100	.110	99
Copper	mg/kg	10.3	11.5	34
Manganese	mg/kg	46.9	52.3	57
Zinc	mg/kg	15.6	17.4	12
Carotene	mg/kg	33.0	36.8	99
Vitamin A equiv	IU/g	55.0	61.3	

Brome, smooth, aerial pt, fresh, immature, (2)

Ref no 2-00-956

		As fed	Dry	C.V. ± %
Dry matter	%	32.5	100.0	19
Ash	%	3.4	10.6	15
Crude fiber	%	7.3	22.4	18
Ether extract	%	1.4	4.3	22
N-free extract	%	13.2	40.6	
Protein (N x 6.25)	%	7.2	22.1	30
Cattle	dig prot %	5.4	16.7	
Sheep	dig prot %	5.9	18.1	
Energy				
Cattle	DE kcal/kg	1017.	3130.	
Sheep	DE kcal/kg	1146.	3527.	
Cattle	ME kcal/kg	834.	2567.	
Sheep	ME kcal/kg	940.	2892.	
Cattle	TDN %	23.	71.	
Sheep	TDN %	26.	80.	
Calcium	%	.20	.62	
Phosphorus	%	.18	.57	
Carotene	mg/kg	189.4	582.9	

Brome, smooth, aerial pt, fresh, early blm, (2)

Ref no 2-00-957

		As fed	Dry	C.V. ± %
Dry matter	%	30.0	100.0	11
Ash	%	2.3	7.7	15
Crude fiber	%	8.5	28.4	8
Ether extract	%	1.2	3.9	11
N-free extract	%	13.8	46.0	
Protein (N x 6.25)	%	4.2	14.0	21
Cattle	dig prot %	2.9	9.8	
Sheep	dig prot %	3.0	10.0	
Energy				
Cattle	DE kcal/kg	899.	2998.	
Sheep	DE kcal/kg	807.	2690.	
Cattle	ME kcal/kg	737.	2458.	
Sheep	ME kcal/kg	662.	2206.	
Cattle	TDN %	20.	68.	
Sheep	TDN %	18.	61.	
Calcium	%	.13	.44	
Phosphorus	%	.08	.25	

Brome, smooth, aerial pt, fresh, over ripe, (2)

Ref no 2-00-961

		As fed	Dry	C.V. ± %
Dry matter	%	75.0	100.0	10
Ash	%	4.2	5.6	22
Crude fiber	%	28.8	38.4	7
Ether extract	%	1.0	1.4	22

Continued

(1) dry forages and roughages
(2) pasture, range plants, and forages fed green

(3) silages
(4) energy feeds

TABLE A-3-1 Composition of Feeds (Continued)

Feed name or analyses		Mean		C.V.
		As fed	Dry	± %
N-free extract	%	37.9	50.5	
Protein (N x 6.25)	%	3.1	4.1	8
Cattle	dig prot %	1.3	1.7	
Sheep	dig prot %	1.3	1.7	
Energy				
Cattle	DE kcal/kg	1753.	2337.	
Sheep	DE kcal/kg	1753.	2337.	
Cattle	ME kcal/kg	1437.	1916.	
Sheep	ME kcal/kg	1437.	1916.	
Cattle	TDN %	40.	53.	
Sheep	TDN %	40.	53.	
Calcium	%	.19	.25	
Phosphorus	%	.14	.19	

Broomcorn - see Sorghum, broomcorn

Broomcorn millet - see Millet, proso

Brown rice - see Rice, groats

BUCKWHEAT. Fagopyrum spp

Buckwheat, grain w added hulls, high gr, (1)
   Buckwheat feed, high grade

   Ref no 1-08-003

		As fed	Dry	
Dry matter	%	89.3	100.0	
Ash	%	4.2	4.7	
Crude fiber	%	18.2	20.4	
Ether extract	%	4.9	5.5	
N-free extract	%	43.5	48.7	
Protein (N x 6.25)	%	18.5	20.7	
Cattle	dig prot %	11.7	13.1	
Sheep	dig prot %	11.7	13.1	
Energy				
Cattle	DE kcal/kg	2283.	2557.	
Sheep	DE kcal/kg	2283.	2557.	
Cattle	ME kcal/kg	1873.	2097.	
Sheep	ME kcal/kg	1873.	2097.	
Cattle	TDN %	52.	58.	
Sheep	TDN %	52.	58.	
Phosphorus	%	.48	.54	
Potassium	%	.66	.74	

Buckwheat, grain w added hulls, low gr, (1)
   Buckwheat feed, low grade

   Ref no 1-08-002

		As fed	Dry	
Dry matter	%	88.3	100.0	
Ash	%	3.2	3.6	
Crude fiber	%	28.6	32.4	
Ether extract	%	3.4	3.8	
N-free extract	%	39.8	45.1	
Protein (N x 6.25)	%	13.3	15.1	
Cattle	dig prot %	5.2	5.9	
Sheep	dig prot %	5.2	5.9	
Energy				
Cattle	DE kcal/kg	1440.	1631.	
Sheep	DE kcal/kg	1440.	1631.	
Cattle	ME kcal/kg	1180.	1337.	
Sheep	ME kcal/kg	1180.	1337.	
Cattle	TDN %	33.	37.	
Sheep	TDN %	33.	37.	
Phosphorus	%	.37	.42	
Potassium	%	.68	.77	

Buckwheat, hulls, (1)
   Buckwheat hulls (AAFCO)

   Ref no 1-00-987

		As fed	Dry	
Dry matter	%	88.6	100.0	1
Ash	%	1.6	1.8	20
Crude fiber	%	42.7	48.2	8
Ether extract	%	1.0	1.1	13
N-free extract	%	40.3	45.5	
Protein (N x 6.25)	%	3.0	3.4	13
Cattle	dig prot %	.2	.2	
Sheep	dig prot %	.2	.2	
Cellulose	%	39.3	44.4	
Lignin	%	28.0	31.5	
Energy				
Cattle	DE kcal/kg	625.	705.	
Sheep	DE kcal/kg	625.	705.	
Cattle	ME kcal/kg	512.	578.	
Sheep	ME kcal/kg	512.	578.	
Cattle	TDN %	14.	16.	
Sheep	TDN %	14.	16.	
Calcium	%	.26	.29	
Phosphorus	%	.02	.02	
Potassium	%	.27	.31	

(5) protein supplements
(6) minerals

(7) vitamins
(8) additives

Feed name or analyses		Mean		C.V.
		As fed	Dry	± %

Buckwheat, straw, (1)

Ref no 1-00-988

Dry matter	%	88.7	100.0	1
Ash	%	8.7	9.8	19
Crude fiber	%	35.6	40.2	10
Ether extract	%	1.0	1.1	15
N-free extract	%	39.1	44.1	
Protein (N x 6.25)	%	4.2	4.8	14
Cattle	dig prot %	1.0	1.1	
Sheep	dig prot %	.8	.9	
Energy				
Cattle	DE kcal/kg	1916.	2160.	
Sheep	DE kcal/kg	1995.	2249.	
Cattle	ME kcal/kg	1571.	1771.	
Sheep	ME kcal/kg	1636.	1844.	
Cattle	TDN %	43.	49.	
Sheep	TDN %	45.	51.	
Calcium	%	1.26	1.42	22
Phosphorus	%	.18	.20	72
Potassium	%	2.86	3.22	28

Buckwheat, aerial pt, fresh, (2)

Ref no 2-00-989

Dry matter	%	36.6	100.0	
Ash	%	3.6	9.8	
Crude fiber	%	8.0	21.9	
Ether extract	%	.9	2.5	
N-free extract	%	19.5	53.2	
Protein (N x 6.25)	%	4.6	12.6	
Cattle	dig prot %	3.1	8.6	
Sheep	dig prot %	3.2	8.7	
Energy				
Cattle	DE kcal/kg	1033.	2822.	
Sheep	DE kcal/kg	1097.	2998.	
Cattle	ME kcal/kg	847.	2314.	
Sheep	ME kcal/kg	900.	2458.	
Cattle	TDN %	23.	64.	
Sheep	TDN %	25.	68.	

Buckwheat, grain, (4)

Ref no 4-00-994

Dry matter	%	88.0	100.0	3
Ash	%	1.8	2.0	13
Crude fiber	%	9.0	10.2	24
Ether extract	%	2.5	2.8	14
N-free extract	%	63.7	72.4	
Protein (N x 6.25)	%	11.1	12.6	12

Continued

Feed name or analyses		Mean		C.V.
		As fed	Dry	± %
Cattle	dig prot %	6.7	7.6	
Horses	dig prot %	7.2	8.2	
Sheep	dig prot %	8.0	9.1	
Swine	dig prot %	8.0	9.1	
Energy	GE kcal/kg	3967.	4508.	
Cattle	DE kcal/kg	3026.	3439.	
Sheep	DE kcal/kg	2794.	3175.	
Swine	DE kcal/kg	3026.	3439.	
Cattle	ME kcal/kg	2482.	2820.	
Chickens	MEn kcal/kg	2712.	3082.	
Sheep	ME kcal/kg	2292.	2604.	
Swine	ME kcal/kg	2829.	3215.	
Cattle	TDN %	69.	78.	
Horses	TDN %	62.	70.	
Sheep	TDN %	63.	72.	
Swine	TDN %	69.	78.	
Calcium	%	.11	.13	90
Phosphorus	%	.33	.38	21
Potassium	%	.45	.51	12
Cobalt	mg/kg	.060	.070	26
Copper	mg/kg	9.5	10.8	8
Manganese	mg/kg	33.7	38.3	
Zinc	mg/kg	8.7	9.9	
Niacin	mg/kg	17.8	20.2	71
Riboflavin	mg/kg	10.6	12.1	19
Thiamine	mg/kg	3.3	3.7	50
Arginine	%	.97	1.10	
Histidine	%	.26	.30	
Isoleucine	%	.35	.40	
Leucine	%	.53	.60	
Lysine	%	.62	.70	
Methionine	%	.18	.20	
Phenylalanine	%	.44	.50	
Threonine	%	.44	.50	
Tryptophan	%	.18	.20	
Valine	%	.53	.60	

Buckwheat, flour by-prod wo hulls, c-sift, mx 10 fbr, (5)
  Buckwheat middlings, (AAFCO)

Ref no 5-00-991

Dry matter	%	89.0	100.0	
Ash	%	4.7	5.3	
Crude fiber	%	6.0	6.7	
Ether extract	%	6.9	7.8	
N-free extract	%	43.0	48.3	
Protein (N x 6.25)	%	28.4	31.9	
Cattle	dig prot %	25.8	29.0	
Sheep	dig prot %	24.1	27.1	
Swine	dig prot %	22.7	25.5	
Energy				
Cattle	DE kcal/kg	3336.	3748.	
Sheep	DE kcal/kg	3257.	3660.	

Continued

(1) dry forages and roughages
(2) pasture, range plants, and forages fed green

(3) silages
(4) energy feeds

TABLE A-3-1 Composition of Feeds (Continued)

Feed name or analyses		Mean As fed	Mean Dry	C.V. ± %
Swine	DE kcal/kg	3178.	3571.	
Cattle	ME kcal/kg	2735.	3073.	
Sheep	ME kcal/kg	2671.	3001.	
Swine	ME kcal/kg	2848.	3200.	
Cattle	TDN %	76.	85.	
Sheep	TDN %	74.	83.	
Swine	TDN %	72.	81.	
Phosphorus	%	1.02	1.15	
Potassium	%	.98	1.10	

Feed name or analyses		Mean As fed	Mean Dry	C.V. ± %
Cattle	ME kcal/kg	2482.	2820.	
Sheep	ME kcal/kg	2672.	3037.	
Swine	ME kcal/kg	2844.	3232.	
Cattle	TDN %	69.	78.	
Sheep	TDN %	74.	84.	
Swine	TDN %	70.	79.	

BUFFALOGRASS. Buchloe dactyloides

Buffalograss, hay, s-c, immature, (1)

Ref no 1-01-000

BUCKWHEAT, TARTARY. Fagopyrum tataricum

Buckwheat, tartary, grain, (4)

Ref no 4-00-996

		As fed	Dry	C.V.
Dry matter	%	88.0	100.0	2
Ash	%	1.8	2.1	8
Crude fiber	%	8.0	9.1	34
Ether extract	%	2.9	3.3	9
N-free extract	%	63.3	71.9	
Protein (N x 6.25)	%	12.0	13.6	13
Cattle	dig prot %	7.5	8.5	
Sheep	dig prot %	8.5	9.7	
Swine	dig prot %	8.6	9.8	
Energy				
Cattle	DE kcal/kg	2871.	3263.	
Sheep	DE kcal/kg	3104.	3527.	
Swine	DE kcal/kg	3026.	3439.	
Cattle	ME kcal/kg	2355.	2676.	
Sheep	ME kcal/kg	2545.	2892.	
Swine	ME kcal/kg	2820.	3205.	
Cattle	TDN %	65.	74.	
Sheep	TDN %	70.	80.	
Swine	TDN %	69.	78.	

		As fed	Dry	C.V.
Dry matter	%	92.0	100.0	
Ash	%	11.3	12.3	
Crude fiber	%	24.1	26.2	
Ether extract	%	2.0	2.2	
N-free extract	%	43.9	47.7	
Protein (N x 6.25)	%	10.7	11.6	
Cattle	dig prot %	5.8	6.3	
Sheep	dig prot %	5.8	6.3	
Energy				
Cattle	DE kcal/kg	2271.	2469.	
Sheep	DE kcal/kg	2110.	2293.	
Cattle	ME kcal/kg	1862.	2024.	
Sheep	ME kcal/kg	1730.	1880.	
Cattle	TDN %	52.	56.	
Sheep	TDN %	48.	52.	

Buckwheat, tartary, groats, (4)

Ref no 4-00-997

		As fed	Dry	C.V.
Dry matter	%	88.0	100.0	2
Ash	%	1.9	2.2	9
Crude fiber	%	2.0	2.3	14
Ether extract	%	3.4	3.9	7
N-free extract	%	66.5	75.6	
Protein (N x 6.25)	%	14.1	16.0	10
Cattle	dig prot %	9.4	10.7	
Sheep	dig prot %	10.5	11.9	
Swine	dig prot %	10.1	11.5	
Energy				
Cattle	DE kcal/kg	3026.	3439.	
Sheep	DE kcal/kg	3260.	3704.	
Swine	DE kcal/kg	3065.	3483.	

Buffalograss, hay, s-c, full blm, (1)

Ref no 1-01-001

		As fed	Dry	C.V.
Dry matter	%	88.9	100.0	
Ash	%	13.8	15.5	
Crude fiber	%	23.3	26.2	
Ether extract	%	1.9	2.1	
N-free extract	%	41.2	46.4	
Protein (N x 6.25)	%	8.7	9.8	
Cattle	dig prot %	4.7	5.3	
Sheep	dig prot %	4.7	5.3	
Energy				
Cattle	DE kcal/kg	2117.	2381.	
Sheep	DE kcal/kg	1960.	2205.	
Cattle	ME kcal/kg	1735.	1952.	
Sheep	ME kcal/kg	1607.	1808.	
Cattle	TDN %	48.	54.	
Sheep	TDN %	44.	50.	

Continued

(5) protein supplements

(6) minerals

(7) vitamins

(8) additives

Feed name or analyses		Mean		C.V. ± %
		As fed	Dry	

Buffalograss, hay, s-c, mature, (1)

Ref no 1-01-002

		As fed	Dry	
Dry matter	%	92.9	100.0	
Ash	%	13.4	14.4	
Crude fiber	%	25.5	27.5	
Ether extract	%	1.6	1.7	
N-free extract	%	47.1	50.7	
Protein (N x 6.25)	%	5.3	5.7	
Cattle	dig prot %	2.9	3.1	
Sheep	dig prot %	2.9	3.1	
Energy				
Cattle	DE kcal/kg	2253.	2425.	
Sheep	DE kcal/kg	2089.	2249.	
Cattle	ME kcal/kg	1847.	1988.	
Sheep	ME kcal/kg	1713.	1844.	
Cattle	TDN %	51.	55.	
Sheep	TDN %	47.	51.	

Buffalograss, hay, s-c, (1)

Ref no 1-01-003

		As fed	Dry	
Dry matter	%	89.9	100.0	2
Ash	%	12.0	13.4	22
Crude fiber	%	24.1	26.8	6
Ether extract	%	1.7	1.9	16
N-free extract	%	45.0	50.0	
Protein (N x 6.25)	%	7.1	7.9	28
Cattle	dig prot %	3.9	4.3	
Sheep	dig prot %	3.9	4.3	
Energy				
Cattle	DE kcal/kg	2180.	2425.	
Sheep	DE kcal/kg	2061.	2293.	
Cattle	ME kcal/kg	1787.	1988.	
Sheep	ME kcal/kg	1690.	1880.	
Cattle	TDN %	49.	55.	
Sheep	TDN %	47.	52.	
Calcium	%	.40	.45	28
Phosphorus	%	.11	.12	20

Buffalograss, aerial pt, fresh, immature, (2)

Ref no 2-01-004

		As fed	Dry	
Dry matter	%	50.0	100.0	
Ash	%	5.2	10.4	8
Crude fiber	%	13.4	26.8	6
Ether extract	%	1.2	2.3	23
N-free extract	%	14.0	48.1	
Protein (N x 6.25)	%	6.2	12.4	19
Cattle	dig prot %	4.2	8.4	

Continued

Feed name or analyses		Mean		C.V. ± %
		As fed	Dry	
Sheep	dig prot %	4.2	8.5	
Energy				
Cattle	DE kcal/kg	1411.	2822.	
Sheep	DE kcal/kg	1433.	2866.	
Cattle	ME kcal/kg	1157.	2314.	
Sheep	ME kcal/kg	1175.	2350.	
Cattle	TDN %	32.	64.	
Sheep	TDN %	32.	65.	
Calcium	%	.28	.56	12
Phosphorus	%	.12	.23	13
Carotene	mg/kg	62.6	125.2	43
Vitamin A equiv	IU/g	104.4	208.7	

Buffalograss, aerial pt, fresh, full blm, (2)

Ref no 2-01-006

		As fed	Dry	
Dry matter	%	54.5	100.0	
Ash	%	6.3	11.5	13
Crude fiber	%	15.8	29.0	7
Ether extract	%	.6	1.1	48
N-free extract	%	26.5	48.6	
Protein (N x 6.25)	%	5.3	9.8	19
Cattle	dig prot %	3.4	6.2	
Sheep	dig prot %	3.3	6.1	
Energy				
Cattle	DE kcal/kg	1490.	2734.	
Sheep	DE kcal/kg	1586.	2910.	
Cattle	ME kcal/kg	1222.	2242.	
Sheep	ME kcal/kg	1300.	2386.	
Cattle	TDN %	34.	62.	
Sheep	TDN %	36.	66.	
Carotene	mg/kg	40.4	74.1	36
Vitamin A equiv	IU/g	67.3	123.5	

Buffalograss, aerial pt, fresh, mature, (2)

Ref no 2-01-008

		As fed	Dry	
Dry matter	%	72.3	100.0	
Ash	%	11.0	15.2	21
Crude fiber	%	22.2	30.7	4
Ether extract	%	1.2	1.7	36
N-free extract	%	33.6	46.5	
Protein (N x 6.25)	%	4.3	5.9	14
Cattle	dig prot %	2.1	2.9	
Sheep	dig prot %	1.8	2.5	
Energy				
Cattle	DE kcal/kg	1658.	2293.	
Sheep	DE kcal/kg	2136.	2954.	
Cattle	ME kcal/kg	1359.	1880.	
Sheep	ME kcal/kg	1751.	2422.	
Cattle	TDN %	38.	52.	
Sheep	TDN %	48.	67.	

Continued

(1) dry forages and roughages
(2) pasture, range plants, and forages fed green

(3) silages
(4) energy feeds

Feed name or analyses		Mean		C.V.
		As fed	Dry	± %
Calcium	%	.29	.40	27
Magnesium	%	.09	.12	
Phosphorus	%	.12	.16	44
Potassium	%	.26	.36	
Carotene	mg/kg	46.4	64.2	9
Vitamin A equiv	IU/g	77.3	107.0	

Buffalograss, aerial pt, fresh, over ripe, (2)

Ref no 2-01-009

Cattle	dig prot %	1.8	2.6	
Dry matter	%	69.8	100.0	
Ash	%	10.5	15.0	28
Crude fiber	%	21.9	31.4	9
Ether extract	%	1.0	1.4	35
N-free extract	%	32.6	46.7	
Protein (N x 6.25)	%	3.8	5.5	14
Sheep	dig prot %	1.5	2.1	
Energy				
Cattle	DE kcal/kg	1600.	2293.	
Sheep	DE kcal/kg	2062.	2954.	
Cattle	ME kcal/kg	1312.	1880.	
Sheep	ME kcal/kg	1690.	2422.	
Cattle	TDN %	36.	52.	
Sheep	TDN %	47.	67.	
Carotene	mg/kg	31.5	45.0	34
Vitamin A equiv	IU/g	52.5	75.0	

Buffalograss, aerial pt, fresh, (2)

Ref no 2-01-010

Dry matter	%	47.7	100.0	24
Ash	%	5.9	12.4	23
Crude fiber	%	13.2	27.7	9
Ether extract	%	.8	1.8	36
N-free extract	%	23.3	48.9	
Protein (N x 6.25)	%	4.4	9.2	30
Cattle	dig prot %	2.7	5.7	
Sheep	dig prot %	2.7	5.6	
Energy				
Cattle	DE kcal/kg	1241.	2601.	
Sheep	DE kcal/kg	1409.	2954.	
Cattle	ME kcal/kg	1017.	2133.	
Sheep	ME kcal/kg	1155.	2422.	
Cattle	TDN %	28.	59.	
Sheep	TDN %	32.	67.	
Calcium	%	.25	.52	18
Magnesium	%	.07	.14	22
Phosphorus	%	.08	.16	32
Potassium	%	.34	.71	34
Carotene	mg/kg	44.7	93.7	55
Vitamin A equiv	IU/g	74.5	156.2	

BURCLOVER, CALIFORNIA. Medicago hispida

Burclover, California, hay, s-c, immature, (1)

Ref no 1-01-028

Dry matter	%	90.3	100.0	
Ash	%	12.1	13.4	
Crude fiber	%	20.9	23.1	
Ether extract	%	2.1	2.3	
N-free extract	%	31.8	35.2	
Protein (N x 6.25)	%	23.5	26.0	
Cattle	dig prot %	17.5	19.4	
Sheep	dig prot %	16.4	18.2	
Energy				
Cattle	DE kcal/kg	2628.	2910.	
Sheep	DE kcal/kg	2230.	2469.	
Cattle	ME kcal/kg	2154.	2386.	
Sheep	ME kcal/kg	1828.	2024.	
Cattle	TDN %	60.	66.	
Sheep	TDN %	50.	56.	

Burclover, California, hay, s-c, full blm, (1)

Ref no 1-01-029

Dry matter	%	88.4	100.0	1
Ash	%	7.6	8.6	13
Crude fiber	%	22.3	25.2	9
Ether extract	%	2.4	2.7	7
N-free extract	%	40.5	45.8	
Protein (N x 6.25)	%	15.6	17.7	10
Cattle	dig prot %	10.9	12.3	
Sheep	dig prot %	11.0	12.4	
Energy				
Cattle	DE kcal/kg	2378.	2690.	
Sheep	DE kcal/kg	2260.	2557.	
Cattle	ME kcal/kg	1950.	2206.	
Sheep	ME kcal/kg	1854.	2097.	
Cattle	TDN %	54.	61.	
Sheep	TDN %	51.	58.	

Burclover, California, hay, s-c, (1)

Ref no 1-01-030

Dry matter	%	89.7	100.0	1
Ash	%	8.7	9.7	17
Crude fiber	%	22.5	25.1	9
Ether extract	%	2.5	2.8	10
N-free extract	%	38.9	43.4	
Protein (N x 6.25)	%	17.0	19.0	16

Continued

(5) protein supplements

(6) minerals

(7) vitamins

(8) additives

Feed name or analyses		Mean		C.V.
		As fed	Dry	± %
Cattle	dig prot %	12.0	13.4	
Sheep	dig prot %	11.9	13.3	
Energy				
Cattle	DE kcal/kg	2452.	2734.	
Sheep	DE kcal/kg	2294.	2557.	
Cattle	ME kcal/kg	2011.	2242.	
Sheep	ME kcal/kg	1881.	2097.	
Cattle	TDN %	56.	62.	
Sheep	TDN %	52.	58.	

Burclover, California, aerial pt, fresh, immature, (2)

Ref no 2-01-032

Dry matter	%	20.0	100.0	
Ash	%	1.9	9.6	
Crude fiber	%	3.6	17.9	
Ether extract	%	.6	2.9	27
N-free extract	%	8.3	41.6	
Protein (N x 6.25)	%	5.6	28.0	
Cattle	dig prot %	4.6	23.1	
Sheep	dig prot %	4.6	23.1	
Energy				
Cattle	DE kcal/kg	582.	2910.	
Sheep	DE kcal/kg	582.	2910.	
Cattle	ME kcal/kg	477.	2386.	
Sheep	ME kcal/kg	477.	2386.	
Cattle	TDN %	13.	66.	
Sheep	TDN %	13.	66.	

Burclover, California, aerial pt, fresh, (2)

Ref no 2-01-035

Dry matter	%	20.8	100.0	
Ash	%	1.9	9.3	19
Crude fiber	%	4.2	20.4	31
Ether extract	%	1.0	4.9	41
N-free extract	%	8.6	41.3	
Protein (N x 6.25)	%	5.0	24.1	24
Cattle	dig prot %	4.0	19.4	
Sheep	dig prot %	4.0	19.4	
Energy				
Cattle	DE kcal/kg	614.	2954.	
Sheep	DE kcal/kg	614.	2954.	
Cattle	ME kcal/kg	504.	2422.	
Sheep	ME kcal/kg	504.	2422.	
Cattle	TDN %	14.	67.	
Sheep	TDN %	14.	67.	

Buttermilk - see Cattle

CABBAGE. Brassica oleracea capitata

Cabbage, aerial pt, fresh, (2)

Ref no 2-01-046

Dry matter	%	11.7	100.0	
Ash	%	1.4	12.2	
Crude fiber	%	1.2	10.3	
Ether extract	%	.2	1.9	
N-free extract	%	6.3	53.8	
Protein (N x 6.25)	%	2.6	21.8	
Cattle	dig prot %	2.2	18.7	
Sheep	dig prot %	2.2	18.7	
Energy				
Cattle	DE kcal/kg	428.	3660.	
Sheep	DE kcal/kg	428.	3660.	
Cattle	ME kcal/kg	351.	3001.	
Sheep	ME kcal/kg	351.	3001.	
Cattle	TDN %	10.	83.	
Sheep	TDN %	10.	83.	
Calcium	%	.06	.51	
Chlorine	%	.05	.43	
Iron	%	.001	.008	
Magnesium	%	.02	.17	
Phosphorus	%	.03	.26	
Potassium	%	.24	2.05	
Sodium	%	.01	.08	
Sulfur	%	.11	.94	

CACTUS, PRICKLYPEAR. Opuntia spp

Cactus, pricklypear, aerial pt, fresh, (2)

Ref no 2-01-061

Dry matter	%	17.1	100.0	30
Ash	%	3.4	20.0	18
Crude fiber	%	2.3	13.3	13
Ether extract	%	.3	2.0	33
N-free extract	%	10.2	59.7	
Protein (N x 6.25)	%	.8	5.0	25
Cattle	dig prot %	.5	2.8	
Sheep	dig prot %	.4	2.5	
Energy				
Cattle	DE kcal/kg	445.	2601.	
Sheep	DE kcal/kg	400.	2337.	
Cattle	ME kcal/kg	365.	2133.	
Sheep	ME kcal/kg	328.	1916.	

Continued

(1) dry forages and roughages
(2) pasture, range plants, and forages fed green

(3) silages
(4) energy feeds

Feed name or analyses		Mean		C.V.
		As fed	Dry	± %
Cattle	TDN %	10.	59.	
Sheep	TDN %	9.	53.	
Calcium	%	1.08	6.29	74
Chlorine	%	.04	.21	99
Iron	%	.015	.090	99
Magnesium	%	.28	1.65	94
Phosphorus	%	.01	.08	38
Potassium	%	.21	1.21	78
Sodium	%	.05	.30	7
Sulfur	%	.04	.23	17

CALCIUM PHOSPHATE, DIBASIC

Calcium phosphate, dibasic, comm, (6)
Dicalcium phosphate (AAFCO)

Ref no 6-01-080

Dry matter	%	96.0	100.0	
Calcium	%	22.20	23.13	
Phosphorus	%	17.90	18.65	
Fluorine	mg/kg	768.00	800.00	

CANARYGRASS, REED. Phalaris arundinacea

Canarygrass, reed, hay, s-c, immature, (1)

Ref no 1-01-097

Dry matter	%	91.7	100.0	
Ash	%	6.7	7.3	12
Crude fiber	%	27.8	30.3	3
Ether extract	%	2.7	2.9	37
N-free extract	%	40.9	44.6	
Protein (N x 6.25)	%	13.7	14.9	8
Cattle	dig prot %	9.0	9.8	
Sheep	dig prot %	8.6	9.4	
Energy				
Cattle	DE kcal/kg	2507.	2734.	
Sheep	DE kcal/kg	2022.	2205.	
Cattle	ME kcal/kg	2056.	2242.	
Sheep	ME kcal/kg	1658.	1808.	
Cattle	TDN %	57.	62.	
Sheep	TDN %	46.	50.	
Calcium	%	.32	.35	
Phosphorus	%	.26	.28	
Carotene	mg/kg	188.8	205.9	
Vitamin A equiv	IU/g	314.7	343.2	

Canarygrass, reed, hay, s-c, mature, (1)

Ref no 1-01-101

Dry matter	%	92.0	100.0	2
Ash	%	5.9	6.4	20
Crude fiber	%	33.6	36.5	7
Ether extract	%	1.7	1.9	21
N-free extract	%	43.9	47.7	
Protein (N x 6.25)	%	6.9	7.5	18
Cattle	dig prot %	3.1	3.4	
Sheep	dig prot %	4.3	4.7	
Energy				
Cattle	DE kcal/kg	2069.	2249.	
Sheep	DE kcal/kg	2069.	2249.	
Cattle	ME kcal/kg	1696.	1844.	
Sheep	ME kcal/kg	1696.	1844.	
Cattle	TDN %	47.	51.	
Sheep	TDN %	47.	51.	
Calcium	%	.28	.31	
Phosphorus	%	.16	.17	
Carotene	mg/kg	82.8	90.0	
Vitamin A equiv	IU/g	138.0	150.0	

Canarygrass, reed, hay, s-c, (1)

Ref no 1-01-104

Dry matter	%	91.3	100.0	2
Ash	%	6.6	7.2	15
Crude fiber	%	31.3	34.3	8
Ether extract	%	2.0	2.2	25
N-free extract	%	43.4	47.5	
Protein (N x 6.25)	%	8.0	8.8	28
Cattle	dig prot %	4.2	4.6	
Sheep	dig prot %	5.0	5.5	
Energy				
Cattle	DE kcal/kg	2254.	2469.	
Sheep	DE kcal/kg	2013.	2205.	
Cattle	ME kcal/kg	1848.	2024.	
Sheep	ME kcal/kg	1651.	1808.	
Cattle	TDN %	51.	56.	
Sheep	TDN %	46.	50.	
Calcium	%	.31	.34	27
Iron	%	.010	.020	
Magnesium	%	.24	.26	
Phosphorus	%	.23	.25	37
Potassium	%	2.14	2.35	
Copper	mg/kg	10.9	11.9	
Manganese	mg/kg	84.4	92.4	39
Carotene	mg/kg	82.5	90.4	76
Vitamin A equiv	IU/g	137.5	150.7	

(5) protein supplements
(6) minerals

(7) vitamins
(8) additives

Feed name or analyses		Mean		C.V.
		As fed	Dry	± %

Canarygrass, reed, aerial pt, fresh, immature, (2)

Ref no 2-01-105

		As fed	Dry	C.V.
Dry matter	%	24.5	100.0	8
Ash	%	2.5	10.2	20
Crude fiber	%	5.2	21.3	10
Ether extract	%	1.2	4.7	22
N-free extract	%	11.0	44.7	
Protein (N x 6.25)	%	4.7	19.1	15
Cattle	dig prot %	3.4	14.1	
Sheep	dig prot %	3.6	14.8	
Cellulose	%	5.3	21.8	
Lignin	%	.7	3.0	
Energy				
Cattle	DE kcal/kg	734.	2998.	
Sheep	DE kcal/kg	734.	2998.	
Cattle	ME kcal/kg	602.	2458.	
Sheep	ME kcal/kg	602.	2458.	
Cattle	TDN %	17.	68.	
Sheep	TDN %	17.	68.	
Calcium	%	.14	.57	
Phosphorus	%	.12	.49	
Potassium	%	.89	3.64	

Canarygrass, reed, aerial pt, fresh, mid-blm, (2)

Ref no 2-01-107

		As fed	Dry	C.V.
Dry matter	%	27.7	100.0	
Ash	%	2.3	8.2	
Crude fiber	%	8.1	29.4	
Ether extract	%	.9	3.3	
N-free extract	%	12.9	46.7	
Protein (N x 6.25)	%	3.4	12.4	
Cattle	dig prot %	2.3	8.4	
Sheep	dig prot %	2.4	8.5	
Energy				
Cattle	DE kcal/kg	818.	2954.	
Sheep	DE kcal/kg	782.	2822.	
Cattle	ME kcal/kg	671.	2422.	
Sheep	ME kcal/kg	641.	2314.	
Cattle	TDN %	18.	67.	
Sheep	TDN %	18.	64.	

Canarygrass, reed, aerial pt, fresh, full blm, (2)

Ref no 2-01-108

		As fed	Dry	C.V.
Dry matter	%	30.0	100.0	
Ash	%	2.5	8.2	5
Crude fiber	%	9.3	31.0	3
Ether extract	%	1.0	3.4	8

Continued

		As fed	Dry	C.V.
N-free extract	%	14.3	47.7	
Protein (N x 6.25)	%	2.9	9.7	15
Cattle	dig prot %	1.8	6.1	
Sheep	dig prot %	1.8	6.0	
Cellulose	%	9.9	32.9	10
Lignin	%	1.9	6.2	6
Energy				
Cattle	DE kcal/kg	847.	2822.	
Sheep	DE kcal/kg	873.	2910.	
Cattle	ME kcal/kg	694.	2314.	
Sheep	ME kcal/kg	716.	2386.	
Cattle	TDN %	19.	64.	
Sheep	TDN %	20.	66.	

Canarygrass, reed, aerial pt, fresh, mature, (2)

Ref no 2-01-111

		As fed	Dry	C.V.
Dry matter	%	30.0	100.0	
Ash	%	2.1	7.0	12
Crude fiber	%	9.9	33.1	8
Ether extract	%	.9	2.9	5
N-free extract	%	14.8	49.2	
Protein (N x 6.25)	%	2.3	7.8	9
Cattle	dig prot %	1.4	4.5	
Sheep	dig prot %	1.3	4.2	
Cellulose	%	11.9	39.6	6
Lignin	%	2.0	6.8	
Energy				
Cattle	DE kcal/kg	860.	2866.	
Sheep	DE kcal/kg	873.	2910.	
Cattle	ME kcal/kg	705.	2350.	
Sheep	ME kcal/kg	716.	2386.	
Cattle	TDN %	20.	65.	
Sheep	TDN %	20.	66.	
Calcium	%	.08	.27	
Phosphorus	%	.06	.20	

Canarygrass, reed, aerial pt, fresh, (2)

Ref no 2-01-113

		As fed	Dry	C.V.
Dry matter	%	25.8	100.0	21
Ash	%	2.2	8.4	22
Crude fiber	%	6.9	26.8	15
Ether extract	%	1.0	3.7	30
N-free extract	%	12.4	47.9	
Protein (N x 6.25)	%	3.4	13.2	31
Cattle	dig prot %	2.3	9.1	
Sheep	dig prot %	2.4	9.3	
Cellulose	%	7.5	29.0	16
Lignin	%	1.2	4.7	21
Energy				
Cattle	DE kcal/kg	751.	2910.	

Continued

(1) dry forages and roughages
(2) pasture, range plants, and forages fed green

(3) silages
(4) energy feeds

Feed name or analyses		Mean		C.V.
		As fed	Dry	± %
Sheep	DE kcal/kg	739.	2866.	
Cattle	ME kcal/kg	616.	2386.	
Sheep	ME kcal/kg	606.	2350.	
Cattle	TDN %	17.	66.	
Sheep	TDN %	17.	65.	
Calcium	%	.10	.40	42
Phosphorus	%	.08	.30	47
Potassium	%	.94	3.64	

Cane molasses - see Sugarcane

CARPETGRASS. Axonopus spp

Carpetgrass, hay, s-c, mid-blm, (1)

Ref no 1-01-135

Dry matter	%	93.0	100.0
Ash	%	10.7	11.5
Crude fiber	%	26.3	28.3
Ether extract	%	1.8	1.9
N-free extract	%	46.5	50.0
Protein (N x 6.25)	%	7.7	8.3
Cattle	dig prot %	3.8	4.1
Sheep	dig prot %	3.7	4.0
Energy			
Cattle	DE kcal/kg	2870.	3086.
Sheep	DE kcal/kg	2337.	2513.
Cattle	ME kcal/kg	2353.	2530.
Sheep	ME kcal/kg	1917.	2061.
Cattle	TDN %	65.	70.
Sheep	TDN %	53.	57.
Calcium	%	.34	.37
Phosphorus	%	.15	.16

Carpetgrass, hay, s-c, full blm, (1)

Ref no 1-01-136

Dry matter	%	92.4	100.0
Ash	%	8.4	9.1
Crude fiber	%	30.7	33.2
Ether extract	%	1.3	1.4
N-free extract	%	45.1	48.8
Protein (N x 6.25)	%	6.9	7.5
Cattle	dig prot %	3.1	3.4
Sheep	dig prot %	3.0	3.3
Energy			
Cattle	DE kcal/kg	2403.	2601.
Sheep	DE kcal/kg	2241.	2425.
Cattle	ME kcal/kg	1971.	2133.
Sheep	ME kcal/kg	1837.	1988.

Continued

(5) protein supplements
(6) minerals

Feed name or analyses		Mean		C.V.
		As fed	Dry	± %
Cattle	TDN %	54.	59.	
Sheep	TDN %	51.	55.	
Calcium	%	.37	.40	
Phosphorus	%	.08	.09	

Carpetgrass, hay, s-c, mature, (1)

Ref no 1-01-137

Dry matter	%	91.0	100.0
Ash	%	7.3	8.0
Crude fiber	%	31.0	34.1
Ether extract	%	1.3	1.4
N-free extract	%	46.0	50.6
Protein (N x 6.25)	%	5.4	5.9
Cattle	dig prot %	1.8	2.0
Sheep	dig prot %	1.7	1.9
Energy			
Cattle	DE kcal/kg	2287.	2513.
Sheep	DE kcal/kg	2167.	2381.
Cattle	ME kcal/kg	1876.	2061.
Sheep	ME kcal/kg	1776.	1952.
Cattle	TDN %	52.	57.
Sheep	TDN %	49.	54.
Calcium	%	.31	.34
Phosphorus	%	.10	.11

Carpetgrass, hay, s-c, (1)

Ref no 1-01-138

Dry matter	%	91.8	100.0	1
Ash	%	8.7	9.5	18
Crude fiber	%	28.2	30.7	6
Ether extract	%	1.6	1.7	17
N-free extract	%	45.8	49.9	
Protein (N x 6.25)	%	7.5	8.2	19
Cattle	dig prot %	3.7	4.0	
Sheep	dig prot %	3.6	3.9	
Energy				
Cattle	DE kcal/kg	2550.	2778.	
Sheep	DE kcal/kg	2307.	2513.	
Cattle	ME kcal/kg	2091.	2278.	
Sheep	ME kcal/kg	1892.	2061.	
Cattle	TDN %	58.	63.	
Sheep	TDN %	52.	57.	
Calcium	%	.37	.40	17
Phosphorus	%	.12	.13	23

(7) vitamins
(8) additives

Feed name or analyses	Mean As fed	Mean Dry	C.V. ± %

Carpetgrass, aerial pt, fresh, (2)

Ref no 2-01-140

Feed name or analyses		As fed	Dry	C.V. ± %
Dry matter	%	29.3	100.0	25
Ash	%	3.2	11.1	12
Crude fiber	%	8.1	27.7	6
Ether extract	%	.5	1.7	21
N-free extract	%	14.7	50.2	
Protein (N x 6.25)	%	2.7	9.3	12
Cattle	dig prot %	1.7	5.8	
Sheep	dig prot %	1.6	5.6	
Cellulose	%	10.1	34.5	8
Lignin	%	3.6	12.4	16
Energy				
Cattle	DE kcal/kg	788.	2690.	
Sheep	DE kcal/kg	866.	2954.	
Cattle	ME kcal/kg	646.	2206.	
Sheep	ME kcal/kg	710.	2422.	
Cattle	TDN %	18.	61.	
Sheep	TDN %	20.	67.	
Calcium	%	.13	.46	24
Chlorine	%	.12	.42	15
Iron	%	.010	.040	32
Magnesium	%	.06	.21	52
Phosphorus	%	.06	.19	42
Potassium	%	.25	.85	13
Sulfur	%	.03	.09	42
Cobalt	mg/kg	.030	.090	74
Manganese	mg/kg	141.6	483.3	21
Carotene	mg/kg	40.9	139.6	42
Vitamin A equiv	IU/g	68.2	232.7	

CARROT. Daucus spp

Carrot, roots, fresh, (4)

Ref no 4-01-145

Feed name or analyses		As fed	Dry	C.V. ± %
Dry matter	%	11.9	100.0	
Ash	%	1.2	10.1	
Crude fiber	%	1.1	9.2	
Ether extract	%	.2	1.6	
N-free extract	%	8.2	69.0	
Protein (N x 6.25)	%	1.2	10.1	
Cattle	dig prot %	.6	5.0	
Sheep	dig prot %	.9	7.7	
Swine	dig prot %	.9	7.2	
Energy				
Cattle	DE kcal/kg	430.	3616.	
Sheep	DE kcal/kg	456.	3836.	
Swine	DE kcal/kg	430.	3616.	
Cattle	ME kcal/kg	353.	2965.	

Continued

Feed name or analyses		As fed	Dry	C.V. ± %
Sheep	ME kcal/kg	374.	3146.	
Swine	ME kcal/kg	404.	3399.	
Cattle	NEm kcal/kg	245.	2060.	
Cattle	NEgain kcal/kg	163.	1370.	
Cattle	TDN %	10.	82.	
Sheep	TDN %	10.	87.	
Swine	TDN %	10.	82.	
Calcium	%	.05	.42	
Chlorine	%	.06	.50	
Iron	%	.002	.017	
Magnesium	%	.02	.17	
Phosphorus	%	.04	.34	
Potassium	%	.25	2.10	
Sodium	%	.19	1.60	
Sulfur	%	.02	.17	
Copper	mg/kg	1.3	10.9	
Manganese	mg/kg	3.7	31.1	
Carotene	mg/kg	106.0	890.8	
Niacin	mg/kg	14.8	124.4	
Pantothenic acid	mg/kg	2.0	16.8	
Riboflavin	mg/kg	.7	5.9	
Thiamine	mg/kg	.7	5.9	

Casein - see Cattle

Cattail millet - see Millet, pearl

CATTLE. Bos spp

Cattle, whey, cond, mn solids declared, (4)
  Condensed whey (AAFCO)
  Whey, condensed
  Whey, evaporated
  Whey, semisolid

Ref no 4-01-180

Feed name or analyses		As fed	Dry	C.V. ± %
Dry matter	%	68.0	100.0	6
Ash	%	6.3	9.3	35
Crude fiber	%	.1	.1	99
Ether extract	%	.6	.9	39
N-free extract	%	52.8	77.6	
Protein (N x 6.25)	%	8.2	12.1	33
Calcium	%	.26	.38	23
Phosphorus	%	.40	.59	83
Pantothenic acid	mg/kg	15.6	22.9	
Riboflavin	mg/kg	18.7	27.5	
Thiamine	mg/kg	3.5	5.1	

(1) dry forages and roughages
(2) pasture, range plants, and forages fed green

(3) silages
(4) energy feeds

Feed name or analyses		Mean		C.V. ± %
		As fed	Dry	

Cattle, whey, dehy, mn 65 lactose, (4)
Dried whey (AAFCO)
Whey, dried

Ref no 4-01-182

Feed name or analyses		As fed	Dry	C.V. ± %
Dry matter	%	94.0	100.0	3
Ash	%	9.7	10.3	16
Ether extract	%	.8	.9	84
N-free extract	%	69.6	74.1	
Protein (N x 6.25)	%	13.8	14.7	10
Swine	dig prot %	12.6	13.4	
Energy				
Swine	DE kcal/kg	3432.	3651.	
Chickens	MEn kcal/kg	1852.	1970.	
Swine	ME kcal/kg	3191.	3395.	
Cattle	NEm kcal/kg	1880.	2000.	
Cattle	NEgain kcal/kg	1250.	1330.	
Swine	TDN %	78.	83.	
Calcium	%	.87	.93	31
Iron	%	.016	.017	5
Magnesium	%	.13	.14	
Phosphorus	%	.79	.84	19
Cobalt	mg/kg	.094	.100	51
Copper	mg/kg	43.1	45.9	44
Manganese	mg/kg	4.6	4.9	98
Biotin	mg/kg	.40	.40	41
Choline	mg/kg	20.	21.	25
Folic acid	mg/kg	.90	1.00	
Niacin	mg/kg	11.2	11.9	51
Pantothenic acid	mg/kg	47.7	50.8	29
Riboflavin	mg/kg	29.9	31.8	34
Thiamine	mg/kg	3.7	3.9	26
Arginine	%	.40	.43	18
Cystine	%	.30	.32	
Histidine	%	.20	.21	
Isoleucine	%	.90	.96	13
Leucine	%	1.40	1.49	29
Lysine	%	1.10	1.17	26
Methionine	%	.20	.21	34
Phenylalanine	%	.40	.43	10
Threonine	%	.80	.85	15
Tryptophan	%	.20	.21	36
Tyrosine	%	.30	.32	40
Valine	%	.70	.74	21

Cattle, whey low lactose, cond, mn solids declared, (4)
Condensed whey-product (AAFCO)
Whey-product, condensed
Whey-product, evaporated

Ref no 4-01-185

		As fed	Dry	
Dry matter	%	62.0	100.0	
Ash	%	5.0	8.1	
Crude fiber	%	1.0	1.6	
Ether extract	%	.8	1.3	
N-free extract	%	46.3	74.6	
Protein (N x 6.25)	%	8.9	14.4	
Riboflavin	mg/kg	14.1	22.7	

Cattle, whey wo albumin low lactose, dehy, (4)
Dried whey solubles (AAFCO)
Whey solubles, dried

Ref no 4-01-189

		As fed	Dry	
Dry matter	%	96.0	100.0	
Ash	%	13.7	14.3	
Ether extract	%	1.9	2.0	
N-free extract	%	63.3	65.9	
Protein (N x 6.25)	%	17.1	17.8	
Energy				
Chickens	MEn kcal/kg	1653.	1722.	
Arginine	%	1.00	1.04	
Histidine	%	.10	.10	
Isoleucine	%	.30	.31	
Leucine	%	.20	.21	
Methionine	%	.10	.10	
Phenylalanine	%	.10	.10	
Threonine	%	.50	.52	
Tryptophan	%	.10	.10	
Tyrosine	%	.20	.21	
Valine	%	.30	.31	

Cattle, buttermilk, cond, mn 27 total solids 0.055 fat mx 0.14 ash per 1 solids, (5)
Condensed buttermilk (AAFCO)
Buttermilk, concentrated
Buttermilk, condensed
Buttermilk, evaporated

Ref no 5-01-159

		As fed	Dry	
Dry matter	%	29.0	100.0	13
Ash	%	3.4	11.9	70
Ether extract	%	2.5	8.6	99
N-free extract	%	12.4	42.6	

Continued

(5) protein supplements
(6) minerals

(7) vitamins
(8) additives

Feed name or analyses		Mean		C.V.
		As fed	Dry	± %
Protein (N x 6.25)	%	10.7	36.9	27
Sheep	dig prot %	9.6	33.2	
Swine	dig prot %	9.9	34.3	
Energy				
Sheep	DE kcal/kg	1176.	4056.	
Swine	DE kcal/kg	959.	3307.	
Sheep	ME kcal/kg	965.	3326.	
Swine	ME kcal/kg	850.	2930.	
Sheep	TDN %	27.	92.	
Swine	TDN %	22.	75.	
Calcium	%	.44	1.52	
Magnesium	%	.19	.66	
Phosphorus	%	.26	.90	
Potassium	%	.23	.79	
Sodium	%	.31	1.07	
Riboflavin	mg/kg	14.3	49.3	

Feed name or analyses		Mean		C.V.
		As fed	Dry	± %
Isoleucine	%	2.70	2.90	
Leucine	%	3.40	3.66	
Lysine	%	2.40	2.58	
Methionine	%	.70	.75	
Phenylalanine	%	1.50	1.61	
Threonine	%	1.60	1.72	
Tryptophan	%	.50	.54	
Tyrosine	%	1.00	1.08	
Valine	%	2.80	3.01	

Cattle, buttermilk, dehy, feed gr mx 8 moisture mx 13 ash mn 5 fat, (5)
   Dried buttermilk, feed grade (AAFCO)
   Buttermilk, dried

   Ref no 5-01-160

Dry matter	%	93.0	100.0	2
Ash	%	9.6	10.8	25
Ether extract	%	5.8	6.2	41
N-free extract	%	45.2	48.6	
Protein (N x 6.25)	%	32.0	34.4	7
Sheep	dig prot %	28.8	31.0	
Swine	dig prot %	29.8	32.0	
Energy				
Sheep	DE kcal/kg	3690.	3968.	
Swine	DE kcal/kg	3388.	3643.	
Chickens	MEn kcal/kg	2756.	2963.	
Sheep	ME kcal/kg	3026.	3254.	
Swine	ME kcal/kg	3015.	3242.	
Sheep	TDN %	84.	90.	
Swine	TDN %	77.	83.	
Calcium	%	1.34	1.44	27
Magnesium	%	.48	.52	
Phosphorus	%	.94	1.01	9
Potassium	%	.71	.76	
Sodium	%	.95	1.02	
Manganese	mg/kg	3.5	3.8	78
Biotin	mg/kg	.30	.30	26
Choline	mg/kg	1808.	1944.	35
Folic acid	mg/kg	.40	.40	
Niacin	mg/kg	8.6	9.2	57
Pantothenic acid	mg/kg	30.1	32.4	36
Riboflavin	mg/kg	31.0	33.3	31
Thiamine	mg/kg	3.5	3.8	18
Vitamin B6	mg/kg	2.40	2.60	23
Arginine	%	1.10	1.18	
Histidine	%	.90	.97	

Continued

Cattle, casein, milk acid-precip dehy, mn 80 prot, (5)
   Casein (AAFCO)
   Casein, dried

   Ref no 5-01-162

Dry matter	%	90.0	100.0	2
Ash	%	3.3	3.7	33
Ether extract	%	.5	.6	93
N-free extract	%	4.3	4.8	
Protein (N x 6.25)	%	81.8	90.9	5
Sheep	dig prot %	79.4	88.2	
Swine	dig prot %	76.0	84.5	
Energy				
Sheep	DE kcal/kg	3730.	4145.	
Swine	DE kcal/kg	3532.	3924.	
Sheep	ME kcal/kg	3059.	3399.	
Swine	ME kcal/kg	2740.	3045.	
Sheep	TDN %	85.	94.	
Swine	TDN %	80.	89.	
Calcium	%	.61	.68	29
Phosphorus	%	.99	1.10	33
Manganese	mg/kg	4.4	4.9	12
Choline	mg/kg	209.	232.	
Folic acid	mg/kg	.40	.40	
Niacin	mg/kg	1.3	1.4	
Pantothenic acid	mg/kg	2.6	2.9	
Riboflavin	mg/kg	1.5	1.7	
Thiamine	mg/kg	.4	.4	
Vitamin B6	mg/kg	.40	.50	
Arginine	%	3.40	3.78	8
Cystine	%	.30	.33	
Glycine	%	1.50	1.67	
Histidine	%	2.50	2.78	11
Isoleucine	%	5.70	6.33	15
Leucine	%	8.60	9.55	14
Lysine	%	7.00	7.78	11
Methionine	%	2.70	3.00	9
Phenylalanine	%	4.60	5.11	8
Threonine	%	3.80	4.22	11
Tryptophan	%	1.00	1.11	12
Tyrosine	%	4.70	5.22	16
Valine	%	6.80	7.55	9

(1) dry forages and roughages
(2) pasture, range plants, and forages fed green

(3) silages
(4) energy feeds

Feed name or analyses		Mean		C.V.
		As fed	Dry	± %

Cattle, cheese, cottage wet, (5)
 Cottage cheese

Ref no 5-08-001

		As fed	Dry	
Dry matter	%	21.0	100.0	
Ash	%	1.0	4.8	
Crude fiber	%	.0	.0	
Ether extract	%	.3	1.4	
N-free extract	%	2.7	12.8	
Protein (N x 6.25)	%	17.0	81.0	
Swine	dig prot %	15.3	72.9	
Energy				
Swine	DE kcal/kg	889.	4233.	
Swine	ME kcal/kg	707.	3369.	
Swine	TDN %	20.	96.	
Calcium	%	.09	.43	
Iron	%	.004	.019	
Phosphorus	%	.18	.86	
Potassium	%	.07	.33	
Sodium	%	.29	1.38	
Niacin	mg/kg	1.0	4.8	
Riboflavin	mg/kg	2.8	13.3	
Thiamine	mg/kg	.3	1.4	

Cattle, cheese trim, (5)
 Cheese rind (AAFCO)

Ref no 5-01-163

		As fed	Dry	
Dry matter	%	75.0	100.0	9
Ash	%	4.0	5.3	55
Ether extract	%	27.4	36.5	26
N-free extract	%	12.4	16.5	
Protein (N x 6.25)	%	31.3	41.7	36
Calcium	%	.86	1.15	31
Magnesium	%	.02	.03	21
Phosphorus	%	.49	.65	22
Potassium	%	.24	.32	99
Sodium	%	.71	.95	27

Cattle, lips, raw, (5)

Ref no 5-07-940

		As fed	Dry	
Dry matter	%	30.0	100.0	
Ether extract	%	7.0	23.3	
Protein (N x 6.25)	%	18.0	60.0	

Cattle, liver, raw, (5)
 Beef liver

Ref no 5-01-166

		As fed	Dry	
Dry matter	%	26.0	100.0	
Crude fiber	%	.0	.0	
Ether extract	%	3.2	12.3	
Protein (N x 6.25)	%	17.3	66.7	
Calcium	%	.01	.04	
Phosphorus	%	.23	.88	

Cattle, lungs, raw, (5)

Ref no 5-07-941

		As fed	Dry	
Dry matter	%	20.0	100.0	
Ether extract	%	3.0	15.0	
Protein (N x 6.25)	%	16.0	80.0	

Cattle, milk, dehy, feed gr mx 8 moisture mn 26 fat, (5)
 Dried whole milk (AAFCO)
 Milk, whole, dried

Ref no 5-01-167

		As fed	Dry	
Dry matter	%	93.7	100.0	1
Ash	%	5.4	5.8	
Crude fiber	%	.2	.2	
Ether extract	%	26.4	28.2	
N-free extract	%	36.4	38.9	
Protein (N x 6.25)	%	25.2	26.9	
Swine	dig prot %	24.4	26.1	
Energy				
Swine	DE kcal/kg	5165.	5512.	
Swine	ME kcal/kg	4679.	4994.	
Swine	TDN %	117.	125.	
Calcium	%	.89	.95	
Chlorine	%	1.45	1.55	
Iron	%	.017	.018	
Phosphorus	%	.68	.72	
Potassium	%	1.01	1.08	
Sodium	%	.36	.38	
Manganese	mg/kg	.4	.4	
Biotin	mg/kg	.37	.39	
Carotene	mg/kg	7.0	7.5	
Niacin	mg/kg	8.4	9.0	
Pantothenic acid	mg/kg	22.7	24.2	
Riboflavin	mg/kg	19.6	20.9	
Thiamine	mg/kg	3.7	3.9	
Vitamin B6	mg/kg	4.63	4.94	
Vitamin A equiv	IU/g	11.7	12.5	

Continued

(5) protein supplements
(6) minerals

(7) vitamins
(8) additives

Feed name or analyses		Mean		C.V.
		As fed	Dry	± %
Vitamin D₂	IU/g	.3	.3	
Arginine	%	.90	.96	
Histidine	%	.70	.75	
Isoleucine	%	1.30	1.39	
Leucine	%	2.50	2.67	
Lysine	%	2.20	2.35	
Methionine	%	.60	.64	
Phenylalanine	%	1.30	1.39	
Threonine	%	1.00	1.07	
Tryptophan	%	.40	.43	
Tyrosine	%	1.30	1.39	
Valine	%	1.70	1.81	

Cattle, milk, fresh, (5)
Milk, cattle, fresh

Ref no 5-01-168

		As fed	Dry	
Dry matter	%	12.0	100.0	
Ash	%	.8	6.7	
Ether extract	%	3.7	30.8	
N-free extract	%	4.4	36.7	
Protein (N x 6.25)	%	3.1	25.8	
Cattle	dig prot %	3.0	24.8	
Swine	dig prot %	3.0	25.0	
Energy				
Cattle	DE kcal/kg	688.	5732.	
Swine	DE kcal/kg	660.	5500.	
Cattle	ME kcal/kg	564.	4700.	
Swine	ME kcal/kg	599.	4994.	
Cattle	TDN %	16.	130.	
Swine	TDN %	15.	125.	
Choline	mg/kg	876.	7296.	
Niacin	mg/kg	1.8	15.0	
Pantothenic acid	mg/kg	8.1	67.5	
Riboflavin	mg/kg	1.8	15.0	
Thiamine	mg/kg	0.4	3.3	
Arginine	%	.10	.83	
Histidine	%	.10	.83	
Isoleucine	%	.20	1.67	
Leucine	%	.30	2.50	
Lysine	%	.30	2.50	
Methionine	%	.10	.83	
Phenylalanine	%	.10	.83	
Threonine	%	.10	.83	
Tyrosine	%	.20	1.67	
Valine	%	.20	1.67	

Cattle, milk, skim centrifugal, (5)

Ref no 5-01-170

		As fed	Dry	
Dry matter	%	9.6	100.0	
Ash	%	.6	6.1	
Crude fiber	%	.0	.0	
Ether extract	%	.1	1.5	
N-free extract	%	6.1	63.9	
Protein (N x 6.25)	%	2.7	28.5	
Cattle	dig prot %	2.6	27.4	
Sheep	dig prot %	2.6	26.8	
Swine	dig prot %	2.7	27.9	
Energy				
Cattle	DE kcal/kg	394.	4100.	
Sheep	DE kcal/kg	402.	4189.	
Swine	DE kcal/kg	415.	4321.	
Cattle	ME kcal/kg	323.	3362.	
Sheep	ME kcal/kg	330.	3435.	
Swine	ME kcal/kg	375.	3910.	
Cattle	TDN %	9.	93.	
Sheep	TDN %	9.	95.	
Swine	TDN %	9.	98.	
Calcium	%	.12	1.26	
Iron	%	.002	.017	
Phosphorus	%	.10	1.03	
Potassium	%	.10	1.01	
Sulfur	%	.03	.32	
Cobalt	mg/kg	.010	.110	
Copper	mg/kg	.1	.9	
Manganese	mg/kg	.	.4	
Niacin	mg/kg	1.1	11.5	
Pantothenic acid	mg/kg	3.5	36.8	
Riboflavin	mg/kg	2.0	20.7	
Thiamine	mg/kg	.4	4.6	
Arginine	%	1.20	12.50	
Histidine	%	.90	9.38	
Isoleucine	%	2.30	23.96	
Leucine	%	3.30	34.38	
Lysine	%	2.80	29.17	
Phenylalanine	%	1.50	15.62	
Serine	%	1.60	16.67	
Threonine	%	1.40	14.58	

Cattle, milk, skim dehy, mx 8 moisture, (5)
Dried skimmed milk, feed grade (AAFCO)
Milk, skimmed, dried

Ref no 5-01-175

		As fed	Dry	
Dry matter	%	94.0	100.0	2
Ash	%	7.6	8.1	10
Crude fiber	%	.2	.2	90
Ether extract	%	.9	1.0	98

Continued

(1) dry forages and roughages
(2) pasture, range plants, and forages fed green

(3) silages
(4) energy feeds

Feed name or analyses		Mean		C.V.
		As fed	Dry	± %
N-free extract	%	51.8	55.1	
Protein (N x 6.25)	%	33.5	35.6	5
Sheep	dig prot %	30.1	32.0	
Swine	dig prot %	32.8	34.9	
Energy	GE kcal/kg	3456.	3677.	
Sheep	DE kcal/kg	3564.	3792.	
Swine	DE kcal/kg	3784.	4026.	
Chickens	MEn kcal/kg	2513.	2673.	
Sheep	ME kcal/kg	2922.	3109.	
Swine	ME kcal/kg	3360.	3575.	
Sheep	TDN %	81.	86.	
Swine	TDN %	86.	92.	
Calcium	%	1.26	1.34	10
Iron	%	.005	.005	60
Magnesium	%	.11	.12	
Phosphorus	%	1.03	1.10	9
Potassium	%	1.67	1.78	
Cobalt	mg/kg	.110	.117	78
Copper	mg/kg	11.5	12.2	55
Manganese	mg/kg	2.2	2.3	58
Biotin	mg/kg	.33	.35	27
Choline	mg/kg	1426.	1517.	37
Folic acid	mg/kg	.62	.66	17
Niacin	mg/kg	11.5	12.2	26
Pantothenic acid	mg/kg	33.7	35.8	23
Riboflavin	mg/kg	20.1	21.4	28
Thiamine	mg/kg	3.5	3.7	50
a-tocopherol	mg/kg	9.2	9.8	
Vitamin B_6	mg/kg	3.97	4.22	14
Vitamin B_{12}	mcg/kg	41.9	44.6	
Vitamin D_2	IU/g	.4	.4	
Arginine	%	1.20	1.28	17
Cystine	%	.50	.53	
Glutamic acid	%	6.80	7.24	
Glycine	%	.20	.21	
Histidine	%	.90	.96	20
Isoleucine	%	2.30	2.45	19
Leucine	%	3.30	3.51	5
Lysine	%	2.80	2.98	5
Methionine	%	.80	.85	21
Phenylalanine	%	1.50	1.60	16
Threonine	%	1.40	1.49	8
Tryptophan	%	.40	.42	18
Tyrosine	%	1.30	1.38	31
Valine	%	2.20	2.34	4

Cattle, whey albumin, heat and acid-precip dehy,
  mn 75 prot, (5)
  Dried milk albumin (AAFCO)
  Milk, albumin, dried

Ref no 5-01-177

Dry matter	%	92.3	100.0	
Ash	%	30.4	32.9	

Continued

(5) protein supplements
(6) minerals

Feed name or analyses		Mean		C.V.
		As fed	Dry	± %
Crude fiber	%	.7	.8	
Ether extract	%	1.1	1.2	
N-free extract	%	12.8	13.9	
Protein (N x 6.25)	%	47.3	51.2	
Swine	dig prot %	43.0	46.6	
Energy				
Swine	DE kcal/kg	2035.	2205.	
Swine	ME kcal/kg	1744.	1890.	
Swine	TDN %	46.	50.	

Cattle, spleen, raw, (5)
  Cattle, melts, raw

Ref no 5-07-942

Dry matter	%	25.0	100.0	
Ether extract	%	4.0	16.0	
Protein (N x 6.25)	%	18.0	72.0	

Cattle, udders, raw, (5)

Ref no 5-07-943

Dry matter	%	25.0	100.0	
Ether extract	%	12.0	48.0	
Protein (N x 6.25)	%	12.0	48.0	

Charcoal - see Plant, see Animal, bone charcoal

Cheese - see Cattle

Chicken - see also Poultry, see Turkey

CHICKEN. Gallus domesticus

Chicken, broilers, whole, raw, (5)

Ref no 5-07-945

Dry matter	%	68.0	100.0	
Ether extract	%	7.8	11.5	
Protein (N x 6.25)	%	17.6	25.9	

(7) vitamins
(8) additives

Feed name or analyses		Mean		C.V.
		As fed	Dry	± %

Chicken, cull hens, whole, raw, (5)

Ref no 5-07-950

Dry matter	%	70.0	100.0	
Ether extract	%	8.0	11.5	
Protein (N x 6.25)	%	18.1	25.9	

Chicken, day-old chicks, whole, raw, (5)

Ref no 5-07-946

Dry matter	%	24.4	100.0	
Crude fiber	%	.9	3.6	
Ether extract	%	5.7	23.5	
Protein (N x 6.25)	%	13.9	57.0	

Chicken, eggs w shells, raw, (5)

Ref no 5-01-213

Dry matter	%	34.1	100.0	
Ash	%	10.7	31.4	
Crude fiber	%	.0	.0	
Ether extract	%	10.6	31.1	
N-free extract	%	.0	.0	
Protein (N x 6.25)	%	12.8	37.5	
Calcium	%	1.50	4.40	

Chicken, feet, raw, (5)

Ref no 5-07-947

Dry matter	%	47.0	100.0	
Ether extract	%	11.0	23.4	
Protein (N x 6.25)	%	25.0	53.2	

Chicken, gizzards, raw, (5)

Ref no 5-07-948

Dry matter	%	69.0	100.0	
Ether extract	%	6.2	9.0	
Protein (N x 6.25)	%	20.3	29.4	

Feed name or analyses		Mean		C.V.
		As fed	Dry	± %

Chicken, heads, raw, (5)

Ref no 5-07-949

Dry matter	%	33.0	100.0	
Ether extract	%	6.0	18.2	
Protein (N x 6.25)	%	19.0	57.6	

Chicken, offal w feet, raw, (5)

Ref no 5-07-951

Dry matter	%	31.0	100.0	
Ether extract	%	12.9	41.6	
Protein (N x 6.25)	%	13.1	42.3	
Calcium	%	.82	2.64	
Phosphorus	%	.42	1.35	

Chicken, offal wo feet, raw, (5)

Ref no 5-07-952

Dry matter	%	27.0	100.0	
Crude fiber	%	.2	.7	
Ether extract	%	11.4	42.2	
Protein (N x 6.25)	%	11.8	43.7	
Calcium	%	.27	1.00	
Phosphorus	%	.19	.70	

Chipped rice - see Rice, groats, polished and broken

CITRUS. Citrus spp

Citrus, pulp, ensiled, (3)

Ref no 3-01-234

Dry matter	%	19.5	100.0	
Ash	%	1.0	5.3	
Crude fiber	%	3.1	15.9	
Ether extract	%	1.7	8.8	
Protein (N x 6.25)	%	1.4	7.1	
Cattle	dig prot %	.4	1.8	
Sheep	dig prot %	.4	1.8	
Energy				
Cattle	DE kcal/kg	757.	3880.	
Sheep	DE kcal/kg	757.	3880.	
Cattle	ME kcal/kg	620.	3182.	
Sheep	ME kcal/kg	620.	3182.	
Cattle	TDN %	17.	88.	

Continued

(1) dry forages and roughages
(2) pasture, range plants, and forages fed green

(3) silages
(4) energy feeds

TABLE A-3-1 Composition of Feeds (Continued)

Feed name or analyses		Mean		C.V.
		As fed	Dry	± %
Sheep	TDN %	17.	88.	
Calcium	%	.40	2.04	
Iron	%	.003	.016	
Magnesium	%	.03	.16	
Phosphorus	%	.03	.15	
Potassium	%	.12	.62	

Citrus, pulp wo fines, shred dehy, (4)
Dried citrus pulp (AAFCO)
Citrus pulp, dried

Ref no 4-01-237

Dry matter	%	90.0	100.0	3
Ash	%	6.0	6.7	15
Crude fiber	%	13.0	14.4	13
Ether extract	%	4.6	5.1	45
N-free extract	%	59.8	66.5	
Protein (N x 6.25)	%	6.6	7.3	17
Cattle	dig prot %	3.5	3.9	
Sheep	dig prot %	3.5	3.9	
Swine	dig prot %	2.7	3.0	
Energy				
Cattle	DE kcal/kg	3056.	3395.	
Sheep	DE kcal/kg	3056.	3395.	
Swine	DE kcal/kg	1984.	2205.	
Cattle	ME kcal/kg	2506.	2784.	
Sheep	ME kcal/kg	2506.	2784.	
Swine	ME kcal/kg	1876.	2084.	
Cattle	NE_m kcal/kg	1773.	1970.	
Cattle	NE_{gain} kcal/kg	1188.	1320.	
Cattle	TDN %	69.	77.	
Sheep	TDN %	69.	77.	
Swine	TDN %	45.	50.	
Calcium	%	1.96	2.18	28
Iron	%	.016	.018	38
Magnesium	%	.16	.18	83
Phosphorus	%	.12	.13	30
Potassium	%	.62	.69	
Copper	mg/kg	5.7	6.3	78
Manganese	mg/kg	6.8	7.6	79
Zinc	mg/kg	14.5	16.1	
Choline	mg/kg	845.	939.	
Niacin	mg/kg	21.6	24.0	39
Pantothenic acid	mg/kg	13.0	14.4	20
Riboflavin	mg/kg	2.4	2.7	47
Thiamine	mg/kg	1.5	1.7	32

Feed name or analyses		Mean		C.V.
		As fed	Dry	± %

Citrus, pulp wo fines, ammoniated shred dehy, (4)

Ref no 4-01-238

Dry matter	%	86.0	100.0	
Ash	%	4.9	5.7	
Crude fiber	%	14.0	16.3	
Ether extract	%	5.1	5.9	
N-free extract	%	48.6	56.5	
Protein (N x 6.25)	%	13.4	15.6	
Cattle	dig prot %	9.9	11.5	
Sheep	dig prot %	9.9	11.5	
Energy				
Cattle	DE kcal/kg	2768.	3219.	
Sheep	DE kcal/kg	2768.	3219.	
Cattle	ME kcal/kg	2270.	2640.	
Sheep	ME kcal/kg	2270.	2640.	
Cattle	TDN %	63.	73.	
Sheep	TDN %	63.	73.	
Calcium	%	1.64	1.91	
Magnesium	%	.07	.08	
Phosphorus	%	.12	.14	

Citrus, syrup, mn 45 invert sugar mn 71 degrees
brix, (4)
Citrus molasses (AAFCO)

Ref no 4-01-241

Dry matter	%	65.0	100.0	7
Ash	%	6.1	9.4	40
Ether extract	%	.2	.3	99
N-free extract	%	51.6	79.4	
Protein (N x 6.25)	%	7.1	10.9	53
Cattle	dig prot %	3.6	5.6	
Sheep	dig prot %	3.6	5.6	
Energy				
Cattle	DE kcal/kg	2207.	3395.	
Sheep	DE kcal/kg	2207.	3395.	
Cattle	ME kcal/kg	1810.	2784.	
Sheep	ME kcal/kg	1810.	2784.	
Cattle	TDN %	50.	77.	
Sheep	TDN %	50.	77.	
Calcium	%	1.31	2.01	43
Iron	%	.030	.050	99
Magnesium	%	.14	.22	99
Phosphorus	%	.16	.25	99
Potassium	%	.09	.14	
Copper	mg/kg	72.8	112.0	99
Manganese	mg/kg	26.0	40.0	65
Zinc	mg/kg	88.9	136.7	
Niacin	mg/kg	26.6	40.9	24
Pantothenic acid	mg/kg	12.5	19.2	
Riboflavin	mg/kg	6.2	9.5	

(5) protein supplements

(6) minerals

(7) vitamins

(8) additives

Feed name or analyses		Mean		C.V.
		As fed	Dry	± %

Feed name or analyses		Mean		C.V.
		As fed	Dry	± %
Calcium	%	1.30	1.48	66
Magnesium	%	.38	.43	51
Phosphorus	%	.16	.18	27

Citrus, seed, mech-extd grnd, (5)
Citrus seed meal, mechanical extracted (AAFCO)

Ref no 5-01-239

		As fed	Dry	± %
Dry matter	%	88.0	100.0	4
Ash	%	6.5	7.4	15
Crude fiber	%	10.0	11.4	17
Ether extract	%	6.6	7.5	73
N-free extract	%	29.6	33.6	
Protein (N x 6.25)	%	35.3	40.1	25
Cattle	dig prot %	28.9	32.8	
Sheep	dig prot %	29.9	34.0	
Energy				
Cattle	DE kcal/kg	2988.	3395.	
Sheep	DE kcal/kg	2638.	2998.	
Cattle	ME kcal/kg	2450.	2784.	
Sheep	ME kcal/kg	2163.	2458.	
Cattle	TDN %	68.	77.	
Sheep	TDN %	60.	68.	
Calcium	%	1.20	1.36	25
Iron	%	.030	.030	
Magnesium	%	.60	.68	35
Phosphorus	%	.69	.78	68
Potassium	%	1.31	1.49	
Copper	mg/kg	6.6	7.5	
Manganese	mg/kg	7.5	8.5	
Zinc	mg/kg	7.5	8.5	

CITRUS, LEMON. Citrus limon

Citrus, lemon, pulp wo fines, shred dehy, (4)
Lemon pulp, dried

Ref no 4-01-247

		As fed	Dry	
Dry matter	%	93.0	100.0	
Ash	%	5.8	6.2	
Crude fiber	%	14.0	15.1	
Ether extract	%	1.7	1.8	
N-free extract	%	64.9	69.8	
Protein (N x 6.25)	%	6.6	7.1	
Cattle	dig prot %	3.1	3.3	
Sheep	dig prot %	3.1	3.3	
Energy				
Cattle	DE kcal/kg	3198.	3439.	
Sheep	DE kcal/kg	3198.	3439.	
Cattle	ME kcal/kg	2623.	2820.	
Sheep	ME kcal/kg	2623.	2820.	
Cattle	TDN %	73.	78.	
Sheep	TDN %	73.	78.	

CITRUS, GRAPEFRUIT. Citrus paradisi

Citrus, grapefruit, pulp wo fines, shred dehy, (4)
Grapefruit pulp dried

Ref no 4-01-244

		As fed	Dry	
Dry matter	%	88.0	100.0	3
Ash	%	4.9	5.6	33
Crude fiber	%	13.0	14.8	15
Ether extract	%	4.8	5.5	46
N-free extract	%	57.6	65.5	
Protein (N x 6.25)	%	7.6	8.6	21
Cattle	dig prot %	4.5	5.1	
Sheep	dig prot %	4.5	5.1	
Energy				
Cattle	DE kcal/kg	2910.	3307.	
Sheep	DE kcal/kg	2910.	3307.	
Cattle	ME kcal/kg	2386.	2712.	
Sheep	ME kcal/kg	2386.	2712.	
Cattle	TDN %	66.	75.	
Sheep	TDN %	66.	75.	

CITRUS, LIME. Citrus aurantifolia

Citrus, lime, pulp wo fines, shred dehy, (4)
Lime pulp, dried

Ref no 4-01-249

		As fed	Dry	
Dry matter	%	89.0	100.0	
Ash	%	5.1	5.7	
Crude fiber	%	15.0	16.9	
Ether extract	%	2.9	3.2	
N-free extract	%	58.4	65.6	
Protein (N x 6.25)	%	7.7	8.6	
Cattle	dig prot %	4.5	5.1	
Sheep	dig prot %	4.5	5.1	
Energy				
Cattle	DE kcal/kg	2943.	3307.	
Sheep	DE kcal/kg	2943.	3307.	
Cattle	ME kcal/kg	2414.	2712.	
Sheep	ME kcal/kg	2414.	2712.	
Cattle	TDN %	67.	75.	
Sheep	TDN %	67.	75.	

Continued

(1) dry forages and roughages
(2) pasture, range plants, and forages fed green

(3) silages
(4) energy feeds

Feed name or analyses		Mean		C.V. ± %
		As fed	Dry	

CITRUS, ORANGE. Citrus sinensis

Citrus, orange, pulp wo fines, shred dehy, (4)
Orange pulp, dried

Ref no 4-01-254

Dry matter	%	89.0	100.0	2
Ash	%	4.4	4.9	40
Crude fiber	%	10.0	11.2	25
Ether extract	%	1.8	2.0	43
N-free extract	%	65.9	74.0	
Protein (N x 6.25)	%	7.0	7.9	15
Cattle	dig prot %	5.5	6.2	
Sheep	dig prot %	5.5	6.2	
Energy				
Cattle	DE kcal/kg	3453.	3880.	
Sheep	DE kcal/kg	3453.	3880.	
Cattle	ME kcal/kg	2832.	3182.	
Sheep	ME kcal/kg	2832.	3182.	
Cattle	TDN %	78.	88.	
Sheep	TDN %	78.	88.	
Calcium	%	.63	.71	
Phosphorus	%	.10	.11	

CLAMS. Mercenaria mercenaria, Mya arenaria, Siliqua patula

Clams, shells, grnd, (6)

Ref no 6-01-259

Dry matter	%	99.0	100.0	
Ash	%	66.9	67.6	
Protein (N x 6.25)	%	1.4	1.4	
Calcium	%	36.39	36.75	
Iron	%	.460	.460	
Magnesium	%	.27	.27	
Phosphorus	%	.03	.03	
Potassium	%	.16	.16	
Sodium	%	.51	.52	
Manganese	mg/kg	335.7	339.1	

Feed name or analyses		Mean		C.V. ± %
		As fed	Dry	

CLOVER, ALSIKE. Trifolium hybridum

Clover, alsike, hay, s-c, immature, (1)

Ref no 1-01-307

Dry matter	%	92.2	100.0	
Ash	%	7.5	8.1	
Crude fiber	%	17.8	19.3	
Ether extract	%	2.3	2.5	
N-free extract	%	39.3	42.7	
Protein (N x 6.25)	%	25.3	27.4	
Cattle	dig prot %	16.0	17.3	
Sheep	dig prot %	17.0	18.4	
Energy				
Cattle	DE kcal/kg	2480.	2690.	
Sheep	DE kcal/kg	2358.	2557.	
Cattle	ME kcal/kg	2034.	2206.	
Sheep	ME kcal/kg	1933.	2097.	
Cattle	TDN %	56.	61.	
Sheep	TDN %	53.	58.	
Calcium	%	1.35	1.46	11
Phosphorus	%	.38	.41	73
Carotene	mg/kg	237.4	257.5	15
Vitamin A equiv	IU/g	395.7	429.3	

Clover, alsike, hay, s-c, early blm, (1)

Ref no 1-01-308

Dry matter	%	83.5	100.0	
Ash	%	7.8	9.4	
Crude fiber	%	23.7	28.4	
Ether extract	%	2.3	2.8	
N-free extract	%	35.2	42.1	
Protein (N x 6.25)	%	14.4	17.3	
Cattle	dig prot %	9.1	10.9	
Sheep	dig prot %	9.5	11.4	
Energy				
Cattle	DE kcal/kg	2172.	2601.	
Sheep	DE kcal/kg	1988.	2381.	
Cattle	ME kcal/kg	1781.	2133.	
Sheep	ME kcal/kg	1630.	1952.	
Cattle	TDN %	49.	59.	
Sheep	TDN %	45.	54.	

(5) protein supplements
(6) minerals

(7) vitamins
(8) additives

Feed name or analyses		Mean		C.V. ± %
		As fed	Dry	

Clover, alsike, hay, s-c, full blm, (1)

Ref no 1-01-309

Feed name or analyses		As fed	Dry	C.V. ± %
Dry matter	%	90.4	100.0	2
Ash	%	8.2	9.1	26
Crude fiber	%	27.4	30.3	9
Ether extract	%	3.2	3.5	26
N-free extract	%	38.0	42.0	
Protein (N x 6.25)	%	13.6	15.1	16
Cattle	dig prot %	8.6	9.5	
Sheep	dig prot %	9.1	10.1	
Energy				
Cattle	DE kcal/kg	2392.	2646.	
Sheep	DE kcal/kg	2392.	2646.	
Cattle	ME kcal/kg	1962.	2170.	
Sheep	ME kcal/kg	1962.	2170.	
Cattle	TDN %	54.	60.	
Sheep	TDN %	54.	60.	

Clover, alsike, hay, s-c, milk stage, (1)

Ref no 1-01-311

		As fed	Dry	
Dry matter	%	82.3	100.0	
Ash	%	6.9	8.4	
Crude fiber	%	25.9	31.5	
Ether extract	%	2.4	2.9	
N-free extract	%	35.1	42.6	
Protein (N x 6.25)	%	12.0	14.6	
Cattle	dig prot %	7.6	9.2	
Sheep	dig prot %	8.1	9.8	
Energy				
Cattle	DE kcal/kg	2178.	2646.	
Sheep	DE kcal/kg	2032.	2469.	
Cattle	ME kcal/kg	1786.	2170.	
Sheep	ME kcal/kg	1666.	2024.	
Cattle	TDN %	49.	60.	
Sheep	TDN %	46.	56.	

Clover, alsike, hay, s-c, (1)

Ref no 1-01-313

		As fed	Dry	
Dry matter	%	87.9	100.0	3
Ash	%	7.6	8.7	23
Crude fiber	%	25.8	29.4	10
Ether extract	%	2.5	2.9	31
N-free extract	%	38.9	44.3	
Protein (N x 6.25)	%	12.9	14.7	22
Cattle	dig prot %	8.2	9.3	
Sheep	dig prot %	8.6	9.8	
Energy	GE kcal/kg	3890.	4425.	

Continued

Feed name or analyses		Mean		C.V. ± %
		As fed	Dry	
Cattle	DE kcal/kg	2326.	2646.	
Sheep	DE kcal/kg	2170.	2469.	
Cattle	ME kcal/kg	1907.	2170.	
Sheep	ME kcal/kg	1779.	2024.	
Cattle	TDN %	53.	60.	
Sheep	TDN %	49.	56.	
Calcium	%	1.15	1.31	14
Chlorine	%	.69	.78	7
Iron	%	.020	.030	33
Magnesium	%	.40	.45	20
Phosphorus	%	.22	.25	75
Potassium	%	1.50	1.70	26
Sodium	%	.40	.46	6
Sulfur	%	.18	.21	12
Copper	mg/kg	5.3	6.0	25
Manganese	mg/kg	60.7	69.0	29
Carotene	mg/kg	164.4	187.0	40
Vitamin A equiv	IU/g	274.0	311.7	

Clover, alsike, aerial pt, fresh, immature, (2)

Ref no 2-01-314

		As fed	Dry	
Dry matter	%	18.8	100.0	10
Ash	%	2.4	12.8	10
Crude fiber	%	3.3	17.5	11
Ether extract	%	.6	3.2	5
N-free extract	%	8.0	42.4	
Protein (N x 6.25)	%	4.5	24.1	8
Cattle	dig prot %	3.2	16.9	
Sheep	dig prot %	3.0	16.1	
Energy				
Cattle	DE kcal/kg	547.	2910.	
Sheep	DE kcal/kg	497.	2646.	
Cattle	ME kcal/kg	448.	2386.	
Sheep	ME kcal/kg	408.	2170.	
Cattle	TDN %	12.	66.	
Sheep	TDN %	11.	60.	
Calcium	%	.22	1.19	13
Magnesium	%	.06	.34	
Phosphorus	%	.08	.42	17
Potassium	%	.43	2.31	

Clover, alsike, aerial pt, fresh, full blm, (2)

Ref no 2-01-315

		As fed	Dry	
Dry matter	%	23.6	100.0	6
Ash	%	2.3	9.7	10
Crude fiber	%	6.4	27.1	6
Ether extract	%	.7	3.0	7
N-free extract	%	10.5	44.5	
Protein (N x 6.25)	%	3.7	15.7	8
Cattle	dig prot %	2.6	11.0	

Continued

(1) dry forages and roughages

(2) pasture, range plants, and forages fed green

(3) silages

(4) energy feeds

Feed name or analyses		Mean		C.V.
		As fed	Dry	± %
Sheep	dig prot %	2.5	10.5	
Energy				
Cattle	DE kcal/kg	697.	2954.	
Sheep	DE kcal/kg	635.	2690.	
Cattle	ME kcal/kg	572.	2422.	
Sheep	ME kcal/kg	521.	2206.	
Cattle	TDN %	16.	67.	
Sheep	TDN %	14.	61.	
Calcium	%	.31	1.30	10
Iron	%	.010	.050	
Phosphorus	%	.07	.29	22
Potassium	%	.61	2.58	
Manganese	mg/kg	27.6	116.9	

Clover, alsike, aerial pt, fresh, (2)

Ref no 2-01-316

		As fed	Dry	± %
Dry matter	%	22.6	100.0	8
Ash	%	2.3	10.0	13
Crude fiber	%	5.7	25.3	11
Ether extract	%	.7	3.1	10
N-free extract	%	9.9	44.1	
Protein (N x 6.25)	%	4.0	17.5	15
Cattle	dig prot %	2.8	12.2	
Sheep	dig prot %	2.6	11.7	
Energy				
Cattle	DE kcal/kg	668.	2954.	
Sheep	DE kcal/kg	608.	2690.	
Cattle	ME kcal/kg	547.	2422.	
Sheep	ME kcal/kg	498.	2206.	
Cattle	TDN %	15.	67.	
Sheep	TDN %	14.	61.	
Calcium	%	.29	1.27	13
Chlorine	%	.17	.77	6
Iron	%	.010	.040	24
Magnesium	%	.07	.32	18
Phosphorus	%	.07	.30	20
Potassium	%	.61	2.70	5
Sodium	%	.10	.45	7
Sulfur	%	.05	.22	21
Copper	mg/kg	1.4	6.0	9
Manganese	mg/kg	26.5	117.1	12
Zinc	mg/kg	13.6	60.2	

CLOVER, CRIMSON. Trifolium incarnatum

Clover, crimson, hay, s-c, early blm, (1)

Ref no 1-01-325

		As fed	Dry	
Dry matter	%	84.9	100.0	
Ash	%	14.2	16.7	
Crude fiber	%	19.5	23.0	

Continued

Feed name or analyses		Mean		C.V.
		As fed	Dry	± %
Ether extract	%	2.4	2.8	
N-free extract	%	28.8	33.9	
Protein (N x 6.25)	%	20.0	23.6	
Cattle	dig prot %	14.0	16.5	
Sheep	dig prot %	14.6	17.2	
Energy				
Cattle	DE kcal/kg	2096.	2469.	
Sheep	DE kcal/kg	2358.	2778.	
Cattle	ME kcal/kg	1718.	2024.	
Sheep	ME kcal/kg	1934.	2278.	
Cattle	TDN %	48.	56.	
Sheep	TDN %	53.	63.	

Clover, crimson, hay, s-c, milk stage, (1)

Ref no 1-01-327

		As fed	Dry	± %
Dry matter	%	86.1	100.0	2
Ash	%	7.2	8.4	18
Crude fiber	%	31.3	36.4	5
Ether extract	%	1.8	2.1	13
N-free extract	%	31.6	36.7	
Protein (N x 6.25)	%	14.1	16.4	4
Cattle	dig prot %	9.9	11.5	
Sheep	dig prot %	9.5	11.0	
Energy				
Cattle	DE kcal/kg	2239.	2601.	
Sheep	DE kcal/kg	1974.	2293.	
Cattle	ME kcal/kg	1836.	2133.	
Sheep	ME kcal/kg	1619.	1880.	
Cattle	TDN %	51.	59.	
Sheep	TDN %	45.	52.	

Clover, crimson, hay, s-c, (1)

Ref no 1-01-328

		As fed	Dry	± %
Dry matter	%	87.4	100.0	3
Ash	%	8.3	9.4	23
Crude fiber	%	28.1	32.2	10
Ether extract	%	2.0	2.3	16
N-free extract	%	34.2	39.2	
Protein (N x 6.25)	%	14.8	16.9	10
Cattle	dig prot %	10.3	11.8	
Sheep	dig prot %	9.9	11.3	
Energy				
Cattle	DE kcal/kg	2313.	2646.	
Sheep	DE kcal/kg	2004.	2293.	
Cattle	ME kcal/kg	1896.	2170.	
Sheep	ME kcal/kg	1643.	1880.	
Cattle	TDN %	52.	60.	
Sheep	TDN %	45.	52.	
Calcium	%	1.24	1.42	3
Chlorine	%	.55	.63	13

Continued

Feed name or analyses		Mean		C.V.
		As fed	Dry	± %
Iron	%	.060	.070	49
Magnesium	%	.24	.27	11
Phosphorus	%	.16	.18	14
Potassium	%	1.35	1.54	10
Sodium	%	.34	.39	16
Sulfur	%	.24	.28	9
Manganese	mg/kg	149.7	171.3	34

Feed name or analyses		Mean		C.V.
		As fed	Dry	± %
Chlorine	%	.10	.58	19
Iron	%	.010	.070	21
Magnesium	%	.07	.41	21
Phosphorus	%	.06	.35	29
Potassium	%	.43	2.41	21
Sodium	%	.07	.40	
Sulfur	%	.05	.28	14
Manganese	mg/kg	51.4	290.6	36

Clover, crimson, leaves, s-c, (1)

Ref no 1-01-329

		As fed	Dry	
Dry matter	%	90.0	100.0	
Ash	%	8.9	9.9	
Crude fiber	%	23.4	26.0	
Ether extract	%	1.9	2.1	
N-free extract	%	37.4	41.6	
Protein (N x 6.25)	%	18.4	20.4	
Cattle	dig prot %	12.9	14.3	
Sheep	dig prot %	12.3	13.7	
Energy				
Cattle	DE kcal/kg	2381.	2646.	
Sheep	DE kcal/kg	2381.	2646.	
Cattle	ME kcal/kg	1953.	2170.	
Sheep	ME kcal/kg	1953.	2170.	
Cattle	TDN %	54.	60.	
Sheep	TDN %	54.	60.	
Calcium	%	1.09	1.21	34
Iron	%	.020	.020	20
Magnesium	%	.19	.21	11
Phosphorus	%	.18	.20	24
Potassium	%	1.72	1.91	19
Manganese	mg/kg	39.5	43.9	36

CLOVER, EGYPTIAN. Trifolium alexandrinum

Clover, Egyptian, aerial pt, fresh, immature, (2)

Ref no 2-01-341

		As fed	Dry	
Dry matter	%	11.7	100.0	10
Ash	%	2.3	19.8	3
Crude fiber	%	2.5	21.6	1
Ether extract	%	.4	3.8	7
N-free extract	%	4.0	34.5	
Protein (N x 6.25)	%	2.4	20.3	4
Cattle	dig prot %	1.7	14.2	
Sheep	dig prot %	1.8	15.6	
Energy				
Cattle	DE kcal/kg	278.	2381.	
Sheep	DE kcal/kg	315.	2690.	
Cattle	ME kcal/kg	228.	1952.	
Sheep	ME kcal/kg	258.	2206.	
Cattle	TDN %	6.	54.	
Sheep	TDN %	7.	61.	

Clover, crimson, aerial pt, fresh, (2)

Ref no 2-01-336

		As fed	Dry	
Dry matter	%	17.7	100.0	4
Ash	%	1.6	9.3	4
Crude fiber	%	5.0	28.3	7
Ether extract	%	.5	3.1	22
N-free extract	%	7.5	42.6	
Protein (N x 6.25)	%	3.0	16.7	2
Cattle	dig prot %	2.1	11.7	
Sheep	dig prot %	2.0	11.2	
Energy				
Cattle	DE kcal/kg	523.	2954.	
Sheep	DE kcal/kg	476.	2690.	
Cattle	ME kcal/kg	429.	2422.	
Sheep	ME kcal/kg	390.	2206.	
Cattle	TDN %	12.	67.	
Sheep	TDN %	11.	61.	
Calcium	%	.29	1.62	19

Continued

Clover, Egyptian, aerial pt, fresh, early blm, cut 3, (2)

Ref no 2-01-345

		As fed	Dry	
Dry matter	%	17.2	100.0	10
Ash	%	1.9	10.8	7
Crude fiber	%	4.2	24.2	10
Ether extract	%	.6	3.4	7
N-free extract	%	7.1	41.8	
Protein (N x 6.25)	%	3.4	19.8	8
Cattle	dig prot %	2.4	13.9	
Sheep	dig prot %	2.8	16.6	
Energy				
Cattle	DE kcal/kg	455.	2646.	
Sheep	DE kcal/kg	538.	3130.	
Cattle	ME kcal/kg	373.	2170.	
Sheep	ME kcal/kg	442.	2567.	
Cattle	TDN %	10.	60.	
Sheep	TDN %	12.	71.	

(1) dry forages and roughages
(2) pasture, range plants, and forages fed green

(3) silages
(4) energy feeds

TABLE A-3-1 Composition of Feeds (Continued)

Feed name or analyses		As fed (Mean)	Dry (Mean)	C.V. ± %

Clover, Egyptian, aerial pt, fresh, (2)

Ref no 2-01-349

Feed name or analyses		As fed	Dry	C.V. ± %
Dry matter	%	14.4	100.0	20
Ash	%	2.4	16.9	16
Crude fiber	%	3.2	22.1	13
Ether extract	%	.5	3.7	13
N-free extract	%	5.5	38.1	
Protein (N x 6.25)	%	2.8	19.2	9
Cattle	dig prot %	1.9	13.4	
Sheep	dig prot %	2.1	14.8	
Energy				
Cattle	DE kcal/kg	356.	2469.	
Sheep	DE kcal/kg	400.	2778.	
Cattle	ME kcal/kg	291.	2024.	
Sheep	ME kcal/kg	328.	2278.	
Cattle	TDN %	8.	56.	
Sheep	TDN %	9.	63.	
Calcium	%	.51	3.56	
Phosphorus	%	.05	.32	
Potassium	%	.34	2.39	

CLOVER, LADINO. Trifolium repens

Clover, ladino, hay, s-c, immature, (1)

Ref no 1-01-368

Feed name or analyses		As fed	Dry	C.V. ± %
Dry matter	%	92.2	100.0	3
Ash	%	8.8	9.6	15
Crude fiber	%	16.4	17.8	24
Ether extract	%	2.6	2.8	41
N-free extract	%	40.0	43.4	
Protein (N x 6.25)	%	24.3	26.4	15
Cattle	dig prot %	15.3	16.6	
Sheep	dig prot %	15.1	16.4	
Energy				
Cattle	DE kcal/kg	2440.	2646.	
Sheep	DE kcal/kg	2317.	2513.	
Cattle	ME kcal/kg	2001.	2170.	
Sheep	ME kcal/kg	1900.	2061.	
Cattle	TDN %	55.	60.	
Sheep	TDN %	52.	57.	
Calcium	%	1.29	1.40	4
Phosphorus	%	.32	.35	52
Carotene	mg/kg	199.1	215.9	7
Vitamin A equiv	IU/g	331.9	359.9	

Clover, ladino, hay, s-c, early blm, (1)

Ref no 1-01-369

Feed name or analyses		As fed	Dry	C.V. ± %
Dry matter	%	91.5	100.0	3
Ash	%	11.3	12.3	
Crude fiber	%	15.9	17.4	19
Ether extract	%	3.6	3.9	34
N-free extract	%	39.9	43.6	
Protein (N x 6.25)	%	20.8	22.8	6
Cattle	dig prot %	13.2	14.4	
Sheep	dig prot %	12.9	14.1	
Energy				
Cattle	DE kcal/kg	2380.	2601.	
Sheep	DE kcal/kg	2259.	2469.	
Cattle	ME kcal/kg	1952.	2133.	
Sheep	ME kcal/kg	1852.	2024.	
Cattle	TDN %	54.	59.	
Sheep	TDN %	51.	56.	
Calcium	%	1.14	1.25	
Phosphorus	%	.30	.33	

Clover, ladino, hay, s-c, (1)

Ref no 1-01-378

Feed name or analyses		As fed	Dry	C.V. ± %
Dry matter	%	91.2	100.0	3
Ash	%	8.7	9.5	24
Crude fiber	%	17.5	19.2	25
Ether extract	%	3.1	3.4	46
N-free extract	%	40.9	44.9	
Protein (N x 6.25)	%	21.0	23.0	18
Cattle	dig prot %	13.2	14.5	
Sheep	dig prot %	13.0	14.3	
Lignin	%	10.6	11.7	23
Energy				
Cattle	DE kcal/kg	2453.	2690.	
Sheep	DE kcal/kg	2292.	2513.	
Cattle	ME kcal/kg	2012.	2206.	
Sheep	ME kcal/kg	1880.	2061.	
Cattle	TDN %	56.	61.	
Sheep	TDN %	52.	57.	
Calcium	%	1.26	1.38	19
Chlorine	%	.26	.28	26
Iron	%	.060	.060	99
Magnesium	%	.46	.50	47
Phosphorus	%	.36	.40	36
Potassium	%	1.97	2.17	22
Sodium	%	.12	.13	
Sulfur	%	.20	.22	20
Cobalt	mg/kg	.140	.150	28
Copper	mg/kg	8.0	8.8	11
Manganese	mg/kg	120.8	132.5	31

Continued

(5) protein supplements
(6) minerals

(7) vitamins
(8) additives

Feed name or analyses		Mean		C.V.
		As fed	Dry	± %
Zinc	mg/kg	15.5	17.0	
Carotene	mg/kg	147.0	161.2	38
Vitamin A equiv	IU/g	245.0	268.7	

Clover, ladino, aerial pt, fresh, immature, (2)

Ref no 2-01-380

Feed name or analyses		As fed	Dry	C.V. ± %
Dry matter	%	20.2	100.0	29
Ash	%	2.3	11.2	20
Crude fiber	%	2.9	14.3	17
Ether extract	%	1.0	5.2	25
N-free extract	%	8.9	44.2	
Protein (N x 6.25)	%	5.1	25.1	10
Cattle	dig prot %	3.6	17.6	
Sheep	dig prot %	3.4	16.8	
Energy	GE kcal/kg	971.	4809.	
Cattle	DE kcal/kg	614.	3042.	
Sheep	DE kcal/kg	561.	2778.	
Cattle	ME kcal/kg	504.	2494.	
Sheep	ME kcal/kg	460.	2278.	
Cattle	TDN %	14.	69.	
Sheep	TDN %	13.	63.	
Calcium	%	.32	1.58	22
Iron	%	.010	.040	
Magnesium	%	.08	.40	21
Phosphorus	%	.07	.35	17
Potassium	%	.52	2.56	24
Sodium	%	.02	.12	45
Sulfur	%	.03	.16	
Manganese	mg/kg	15.0	74.3	38
Carotene	mg/kg	71.2	352.6	25
Vitamin A equiv	IU/g	118.7	587.8	

Clover, ladino, aerial pt, fresh, (2)

Ref no 2-01-383

		As fed	Dry	C.V.
Dry matter	%	18.7	100.0	29
Ash	%	2.1	11.4	19
Crude fiber	%	2.7	14.5	16
Ether extract	%	.9	4.8	25
N-free extract	%	8.3	44.2	
Protein (N x 6.25)	%	4.7	25.1	10
Cattle	dig prot %	3.3	17.6	
Sheep	dig prot %	3.1	16.8	
Energy				
Cattle	DE kcal/kg	569.	3042.	
Sheep	DE kcal/kg	519.	2778.	
Cattle	ME kcal/kg	466.	2494.	
Sheep	ME kcal/kg	426.	2278.	
Cattle	TDN %	13.	69.	
Sheep	TDN %	12.	63.	
Calcium	%	.25	1.33	28

Continued

Feed name or analyses		Mean		C.V.
		As fed	Dry	± %
Iron	%	.010	.040	17
Magnesium	%	.08	.41	19
Phosphorus	%	.07	.37	15
Potassium	%	.41	2.19	23
Sodium	%	.02	.10	63
Sulfur	%	.02	.13	46
Cobalt	mg/kg	.020	.130	86
Copper	mg/kg	2.1	11.0	20
Manganese	mg/kg	14.3	76.3	69
Zinc	mg/kg	7.2	38.6	53
Carotene	mg/kg	59.7	319.5	29
Vitamin A equiv	IU/g	99.5	532.6	

CLOVER, MAMMOTH RED. Trifolium pratense

Clover, mammoth red, hay, s-c, (1)

Ref no 1-01-386

		As fed	Dry	C.V.
Dry matter	%	89.2	100.0	3
Ash	%	7.0	7.9	12
Crude fiber	%	28.6	32.1	10
Ether extract	%	3.2	3.6	18
N-free extract	%	38.0	42.6	
Protein (N x 6.25)	%	12.3	13.8	9
Cattle	dig prot %	7.8	8.7	
Sheep	dig prot %	7.7	8.6	
Energy				
Cattle	DE kcal/kg	2399.	2690.	
Sheep	DE kcal/kg	2242.	2513.	
Cattle	ME kcal/kg	1968.	2206.	
Sheep	ME kcal/kg	1838.	2061.	
Cattle	TDN %	54.	61.	
Sheep	TDN %	51.	57.	
Calcium	%	1.69	1.90	4
Phosphorus	%	.26	.29	14

Clover, mammoth red, aerial pt, fresh, (2)

Ref no 2-01-387

		As fed	Dry	C.V.
Dry matter	%	24.7	100.0	7
Ash	%	2.3	9.4	15
Crude fiber	%	7.3	29.7	12
Ether extract	%	.5	2.2	9
N-free extract	%	10.7	43.4	
Protein (N x 6.25)	%	3.8	15.3	17
Cattle	dig prot %	2.6	10.7	
Sheep	dig prot %	2.5	10.2	
Energy				
Cattle	DE kcal/kg	719.	2910.	
Sheep	DE kcal/kg	654.	2646.	
Cattle	ME kcal/kg	589.	2386.	

Continued

(1) dry forages and roughages
(2) pasture, range plants, and forages fed green
(3) silages
(4) energy feeds

Feed name or analyses		Mean		C.V. ± %
		As fed	Dry	
Sheep	ME kcal/kg	536.	2170.	
Cattle	TDN %	16.	66.	
Sheep	TDN %	15.	60.	

CLOVER, RED. Trifolium pratense

Clover, red, hay, s-c, immature, (1)

Ref no 1-01-394

Feed name or analyses		Mean		C.V. ± %
		As fed	Dry	
Dry matter	%	87.3	100.0	4
Ash	%	8.5	9.7	23
Crude fiber	%	17.8	20.4	23
Ether extract	%	3.4	3.9	30
N-free extract	%	38.9	44.6	
Protein (N x 6.25)	%	18.7	21.4	9
Cattle	dig prot %	11.2	12.8	
Sheep	dig prot %	11.6	13.3	
Energy				
Cattle	DE kcal/kg	2271.	2601.	
Sheep	DE kcal/kg	2579.	2954.	
Cattle	ME kcal/kg	1862.	2133.	
Sheep	ME kcal/kg	2114.	2422.	
Cattle	TDN %	52.	59.	
Sheep	TDN %	58.	67.	
Calcium	%	1.54	1.77	13
Magnesium	%	.45	.51	23
Phosphorus	%	.27	.31	19
Potassium	%	2.24	2.57	4
Carotene	mg/kg	216.9	248.5	20
Vitamin A equiv	IU/g	361.6	414.2	

Clover, red, hay, s-c, early blm, (1)

Ref no 1-01-400

Dry matter	%	86.4	100.0	3
Ash	%	8.5	9.8	10
Crude fiber	%	23.7	27.4	5
Ether extract	%	1.9	2.2	9
N-free extract	%	38.6	44.7	
Protein (N x 6.25)	%	13.7	15.9	13
Cattle	dig prot %	8.6	10.0	
Sheep	dig prot %	8.5	9.8	
Energy				
Cattle	DE kcal/kg	2362.	2734.	
Sheep	DE kcal/kg	2095.	2425.	
Cattle	ME kcal/kg	1937.	2242.	
Sheep	ME kcal/kg	1718.	1988.	
Cattle	TDN %	54.	62.	
Sheep	TDN %	48.	55.	
Calcium	%	1.44	1.67	
Phosphorus	%	.32	.37	

Clover, red, hay, s-c, mid-blm, (1)

Ref no 1-01-401

Dry matter	%	88.2	100.0	2
Ash	%	6.8	7.7	22
Crude fiber	%	25.3	28.7	11
Ether extract	%	3.5	4.0	21
N-free extract	%	39.3	44.6	
Protein (N x 6.25)	%	13.2	15.0	14
Cattle	dig prot %	9.3	10.5	
Sheep	dig prot %	8.1	9.2	
Energy				
Cattle	DE kcal/kg	2528.	2866.	
Sheep	DE kcal/kg	2294.	2601.	
Cattle	ME kcal/kg	2073.	2350.	
Sheep	ME kcal/kg	1881.	2133.	
Cattle	TDN %	57.	65.	
Sheep	TDN %	52.	59.	
Calcium	%	1.81	2.05	
Phosphorus	%	.30	.34	
Potassium	%	1.53	1.74	

Clover, red, hay, s-c, full blm, (1)

Ref no 1-01-403

Dry matter	%	86.0	100.0	6
Ash	%	7.1	8.3	10
Crude fiber	%	25.9	30.1	16
Ether extract	%	3.1	3.6	36
N-free extract	%	37.4	43.5	
Protein (N x 6.25)	%	12.5	14.5	7
Cattle	dig prot %	7.5	8.7	
Sheep	dig prot %	8.1	9.4	
Energy				
Cattle	DE kcal/kg	2237.	2601.	
Sheep	DE kcal/kg	2351.	2734.	
Cattle	ME kcal/kg	1834.	2133.	
Sheep	ME kcal/kg	1928.	2242.	
Cattle	TDN %	51.	59.	
Sheep	TDN %	53.	62.	
Calcium	%	1.43	1.66	10
Magnesium	%	.44	.51	16
Phosphorus	%	.22	.25	13
Potassium	%	1.59	1.85	7
Sulfur	%	.16	.19	

(5) protein supplements
(6) minerals

(7) vitamins
(8) additives

Feed name or analyses		Mean		C.V.
		As fed	Dry	± %

Clover, red, hay, s-c, mature, (1)

Ref no 1-01-405

		As fed	Dry	± %
Dry matter	%	88.9	100.0	3
Ash	%	6.0	6.7	11
Crude fiber	%	30.6	34.4	13
Ether extract	%	2.1	2.4	36
N-free extract	%	40.9	46.0	
Protein (N x 6.25)	%	9.3	10.5	17
Cattle	dig prot %	5.6	6.3	
Sheep	dig prot %	5.7	6.4	
Lignin	%	16.0	18.0	
Energy				
Cattle	DE kcal/kg	2312.	2601.	
Sheep	DE kcal/kg	2273.	2557.	
Cattle	ME kcal/kg	1896.	2133.	
Sheep	ME kcal/kg	1864.	2097.	
Cattle	TDN %	52.	59.	
Sheep	TDN %	52.	58.	
Calcium	%	.97	1.09	
Iron	%	.030	.030	
Magnesium	%	.31	.35	
Phosphorus	%	.19	.21	
Potassium	%	1.51	1.70	
Carotene	mg/kg	2.1	2.4	
Vitamin A equiv	IU/g	3.5	4.0	

Clover, red, hay, s-c, cut 1, (1)

Ref no 1-01-406

		As fed	Dry	± %
Dry matter	%	87.6	100.0	4
Ash	%	6.9	7.9	16
Crude fiber	%	25.8	29.4	13
Ether extract	%	2.4	2.7	31
N-free extract	%	39.4	45.0	
Protein (N x 6.25)	%	13.1	15.0	17
Cattle	dig prot %	8.2	9.4	
Sheep	dig prot %	8.4	9.6	
Lignin	%	13.8	15.8	21
Energy				
Cattle	DE kcal/kg	2395.	2734.	
Sheep	DE kcal/kg	2318.	2646.	
Cattle	ME kcal/kg	1964.	2242.	
Sheep	ME kcal/kg	1901.	2170.	
Cattle	TDN %	54.	62.	
Sheep	TDN %	53.	60.	
Calcium	%	1.12	1.28	22
Iron	%	.020	.020	99
Magnesium	%	.32	.36	20
Phosphorus	%	.17	.19	28
Potassium	%	1.37	1.56	27
Carotene	mg/kg	9.6	11.0	99
Vitamin A equiv	IU/g	16.0	18.3	

Clover, red, hay, s-c, cut 2, (1)

Ref no 1-01-407

		As fed	Dry	± %
Dry matter	%	87.3	100.0	2
Ash	%	7.5	8.6	20
Crude fiber	%	24.4	27.9	13
Ether extract	%	2.4	2.8	17
N-free extract	%	38.0	43.5	
Protein (N x 6.25)	%	15.0	17.2	9
Cattle	dig prot %	9.0	10.3	
Sheep	dig prot %	9.2	10.5	
Lignin	%	12.3	14.1	24
Energy				
Cattle	DE kcal/kg	2271.	2601.	
Sheep	DE kcal/kg	2232.	2557.	
Cattle	ME kcal/kg	1862.	2133.	
Sheep	ME kcal/kg	1831.	2097.	
Cattle	TDN %	52.	59.	
Sheep	TDN %	51.	58.	
Calcium	%	1.26	1.44	17
Iron	%	.010	.010	35
Magnesium	%	.36	.41	30
Phosphorus	%	.18	.21	23
Potassium	%	1.26	1.44	24
Carotene	mg/kg	21.4	24.5	66
Vitamin A equiv	IU/g	35.7	40.8	

Clover, red, hay, s-c, cut 3, (1)

Ref no 1-01-408

		As fed	Dry	± %
Dry matter	%	87.9	100.0	4
Ash	%	8.1	9.2	29
Crude fiber	%	24.9	28.3	14
Ether extract	%	2.0	2.3	20
N-free extract	%	37.4	42.6	
Protein (N x 6.25)	%	15.5	17.6	30
Cattle	dig prot %	9.3	10.6	
Sheep	dig prot %	9.4	10.7	
Lignin	%	13.4	15.2	
Energy				
Cattle	DE kcal/kg	2248.	2557.	
Sheep	DE kcal/kg	2209.	2513.	
Cattle	ME kcal/kg	1843.	2097.	
Sheep	ME kcal/kg	1812.	2061.	
Cattle	TDN %	51.	58.	
Sheep	TDN %	50.	57.	
Calcium	%	1.23	1.40	
Iron	%	.020	.020	
Magnesium	%	.21	.24	
Phosphorus	%	.11	.12	
Potassium	%	1.22	1.39	
Carotene	mg/kg	10.1	11.5	
Vitamin A equiv	IU/g	16.8	19.2	

(1) dry forages and roughages
(2) pasture, range plants, and forages fed green

(3) silages
(4) energy feeds

Feed name or analyses		Mean		C.V. ± %
		As fed	Dry	

Clover, red, hay, s-c, (1)

Ref no 1-01-415

Feed name or analyses		As fed	Dry	C.V. ± %
Dry matter	%	87.7	100.0	3
Ash	%	6.9	7.9	23
Crude fiber	%	26.4	30.1	15
Ether extract	%	2.5	2.9	32
N-free extract	%	38.8	44.2	
Protein (N x 6.25)	%	13.1	14.9	18
Cattle	dig prot %	7.8	8.9	
Horses	dig prot %	7.0	8.0	
Sheep	dig prot %	8.0	9.1	
Cellulose	%	22.9	26.1	
Lignin	%	12.8	14.6	20
Energy	GE kcal/kg	3900.	4447.	4
Cattle	DE kcal/kg	2281.	2601.	
Horses	DE kcal/kg	1817.	2072.	
Sheep	DE kcal/kg	2242.	2557.	
Cattle	ME kcal/kg	1871.	2133.	
Horses	ME kcal/kg	1490.	1699.	
Sheep	ME kcal/kg	1839.	2097.	
Cattle	TDN %	52.	59.	
Horses	TDN %	41.	47.	
Sheep	TDN %	51.	58.	
Calcium	%	1.41	1.61	14
Chlorine	%	.23	.26	62
Iron	%	.010	.010	99
Magnesium	%	.39	.45	23
Phosphorus	%	.19	.22	29
Potassium	%	1.54	1.76	22
Sodium	%	.13	.15	44
Sulfur	%	.11	.17	17
Cobalt	mg/kg	.130	.150	22
Copper	mg/kg	9.8	11.2	17
Manganese	mg/kg	57.6	65.7	30
Zinc	mg/kg	15.1	17.2	30
Carotene	mg/kg	32.3	36.8	99
Vitamin A equiv	IU/g	53.8	61.3	

Clover, red, aerial pt, fresh, immature, (2)

Ref no 2-01-427

		As fed	Dry	
Dry matter	%	18.4	100.0	14
Ash	%	2.0	10.6	12
Crude fiber	%	2.7	14.5	7
Ether extract	%	.8	4.4	17
N-free extract	%	8.0	43.5	
Protein (N x 6.25)	%	5.0	27.0	12
Cattle	dig prot %	3.6	19.4	
Sheep	dig prot %	3.3	18.1	
Energy				
Cattle	DE kcal/kg	568.	3086.	

Continued

Feed name or analyses		Mean		C.V. ± %
		As fed	Dry	
Sheep	DE kcal/kg	511.	2778.	
Cattle	ME kcal/kg	466.	2530.	
Sheep	ME kcal/kg	419.	2278.	
Cattle	TDN %	13.	70.	
Sheep	TDN %	12.	63.	
Calcium	%	.48	2.63	19
Magnesium	%	.10	.53	6
Phosphorus	%	.07	.36	16
Potassium	%	.43	2.32	17

Clover, red, aerial pt, fresh, early blm, (2)

Ref no 2-01-428

		As fed	Dry	
Dry matter	%	19.6	100.0	14
Ash	%	2.1	10.6	6
Crude fiber	%	3.7	19.0	27
Ether extract	%	1.0	5.0	10
N-free extract	%	8.7	44.3	
Protein (N x 6.25)	%	4.1	21.1	23
Cattle	dig prot %	3.0	15.2	
Sheep	dig prot %	2.8	14.1	
Energy				
Cattle	DE kcal/kg	605.	3086.	
Sheep	DE kcal/kg	544.	2778.	
Cattle	ME kcal/kg	496.	2530.	
Sheep	ME kcal/kg	446.	2278.	
Cattle	TDN %	14.	70.	
Sheep	TDN %	12.	63.	
Calcium	%	.44	2.26	
Phosphorus	%	.07	.38	
Potassium	%	.49	2.49	

Clover, red, aerial pt, fresh, full blm, (2)

Ref no 2-01-429

		As fed	Dry	
Dry matter	%	27.7	100.0	8
Ash	%	2.0	7.3	5
Crude fiber	%	8.2	29.6	4
Ether extract	%	1.1	4.0	8
N-free extract	%	12.2	44.2	
Protein (N x 6.25)	%	4.1	14.9	7
Cattle	dig prot %	2.7	9.7	
Sheep	dig prot %	2.8	10.0	
Energy				
Cattle	DE kcal/kg	782.	2822.	
Sheep	DE kcal/kg	770.	2778.	
Cattle	ME kcal/kg	641.	2314.	
Sheep	ME kcal/kg	631.	2278.	
Cattle	TDN %	18.	64.	
Sheep	TDN %	17.	63.	
Calcium	%	.28	1.01	33

Continued

(5) protein supplements

(6) minerals

(7) vitamins

(8) additives

Feed name or analyses		Mean		C.V.
		As fed	Dry	± %
Magnesium	%	.14	.51	4
Phosphorus	%	.07	.27	9
Potassium	%	.54	1.96	4

Clover, red, aerial pt, fresh, cut 2, (2)

Ref no 2-01-432

		As fed	Dry	C.V. ± %
Dry matter	%	27.1	100.0	16
Ash	%	2.2	8.3	10
Crude fiber	%	6.8	25.1	17
Ether extract	%	1.0	3.6	3
N-free extract	%	12.4	45.7	
Protein (N x 6.25)	%	4.7	17.3	24
Cattle	dig prot %	3.0	11.2	
Sheep	dig prot %	3.0	11.2	
Energy				
Cattle	DE kcal/kg	765.	2822.	
Sheep	DE kcal/kg	765.	2822.	
Cattle	ME kcal/kg	627.	2314.	
Sheep	ME kcal/kg	627.	2314.	
Cattle	TDN %	17.	64.	
Sheep	TDN %	17.	64.	
Calcium	%	.51	1.88	
Phosphorus	%	.12	.46	
Potassium	%	.70	2.59	

Clover, red, aerial pt, fresh, (2)

Ref no 2-01-434

		As fed	Dry	C.V. ± %
Dry matter	%	23.6	100.0	18
Ash	%	2.1	8.8	15
Crude fiber	%	5.7	24.2	15
Ether extract	%	.9	4.0	13
N-free extract	%	10.6	44.8	
Protein (N x 6.25)	%	4.3	18.2	24
Cattle	dig prot %	2.8	11.8	
Sheep	dig prot %	2.9	12.2	
Energy				
Cattle	DE kcal/kg	666.	2822.	
Sheep	DE kcal/kg	656.	2778.	
Cattle	ME kcal/kg	546.	2314.	
Sheep	ME kcal/kg	538.	2278.	
Cattle	TDN %	15.	64.	
Sheep	TDN %	15.	63.	
Calcium	%	.42	1.76	32
Chlorine	%	.17	.72	27
Iron	%	.010	.030	18
Magnesium	%	.11	.45	13
Phosphorus	%	.07	.29	21
Potassium	%	.50	2.10	16
Sodium	%	.05	.20	21
Sulfur	%	.08	.17	31

Continued

Feed name or analyses		Mean		C.V.
		As fed	Dry	± %
Cobalt	mg/kg	.030	.130	23
Copper	mg/kg	2.1	8.8	15
Manganese	mg/kg	37.6	159.2	73
Carotene	mg/kg	43.5	184.3	30
Vitamin A equiv	IU/g	72.5	307.2	

Clover, red, aerial pt, ensiled, early blm, (3)

Ref no 3-01-436

		As fed	Dry	C.V. ± %
Dry matter	%	30.1	100.0	21
Ash	%	3.0	9.8	14
Crude fiber	%	8.9	29.7	6
Ether extract	%	1.0	3.2	8
N-free extract	%	12.6	42.1	
Protein (N x 6.25)	%	4.6	15.2	12
Cattle	dig prot %	2.6	8.5	
Sheep	dig prot %	3.1	10.3	
Energy				
Cattle	DE kcal/kg	743.	2469.	
Sheep	DE kcal/kg	690.	2293.	
Cattle	ME kcal/kg	609.	2024.	
Sheep	ME kcal/kg	566.	1880.	
Cattle	TDN %	17.	56.	
Sheep	TDN %	16.	52.	

Clover, red, aerial pt, ensiled, mid-blm, (3)

Ref no 3-01-437

		As fed	Dry	C.V. ± %
Dry matter	%	31.4	100.0	6
Ash	%	2.5	7.8	9
Crude fiber	%	10.5	33.3	5
Ether extract	%	.9	2.8	13
N-free extract	%	13.7	43.6	
Protein (N x 6.25)	%	3.9	12.5	4
Cattle	dig prot %	2.3	7.2	
Sheep	dig prot %	2.6	8.4	
Energy				
Cattle	DE kcal/kg	803.	2557.	
Sheep	DE kcal/kg	775.	2469.	
Cattle	ME kcal/kg	658.	2097.	
Sheep	ME kcal/kg	636.	2024.	
Cattle	TDN %	18.	58.	
Sheep	TDN %	18.	56.	

Clover, red, aerial pt, ensiled, (3)

Ref no 3-01-441

		As fed	Dry	C.V. ± %
Dry matter	%	27.7	100.0	17
Ash	%	2.6	9.3	14
Crude fiber	%	8.3	30.0	8

Continued

(1) dry forages and roughages
(2) pasture, range plants, and forages fed green

(3) silages
(4) energy feeds

Feed name or analyses		Mean		C.V.
		As fed	Dry	± %
Ether extract	%	1.1	3.8	30
N-free extract	%	11.5	41.8	
Protein (N x 6.25)	%	4.2	15.1	13
Cattle	dig prot %	2.4	8.8	
Sheep	dig prot %	2.8	10.1	
Energy				
Cattle	DE kcal/kg	708.	2557.	
Sheep	DE kcal/kg	684.	2469.	
Cattle	ME kcal/kg	581.	2097.	
Sheep	ME kcal/kg	561.	2024.	
Cattle	TDN %	16.	58.	
Sheep	TDN %	16.	56.	
Calcium	%	.43	1.53	21
Iron	%	.010	.030	17
Magnesium	%	.11	.39	15
Phosphorus	%	.06	.22	27
Potassium	%	.48	1.72	5
Sodium	%	.06	.23	
Carotene	mg/kg	57.2	206.4	81
Vitamin A equiv	IU/g	95.4	344.1	

Clover, red, aerial pt w AIV pres added, ensiled, (3)

Ref no 3-01-446

		As fed	Dry	
Dry matter	%	24.5	100.0	
Ash	%	2.1	8.7	
Crude fiber	%	7.6	30.8	
Ether extract	%	1.3	5.1	
N-free extract	%	9.7	39.8	
Protein (N x 6.25)	%	3.8	15.6	
Cattle	dig prot %	2.9	11.7	
Sheep	dig prot %	2.5	10.4	
Energy				
Cattle	DE kcal/kg	767.	3130.	
Sheep	DE kcal/kg	616.	2513.	
Cattle	ME kcal/kg	629.	2567.	
Sheep	ME kcal/kg	505.	2061.	
Cattle	TDN %	17.	71.	
Sheep	TDN %	14.	57.	

Clover, red, aerial pt w H3 PO4 pres added, ensiled, (3)

Ref no 3-01-448

		As fed	Dry	
Dry matter	%	34.1	100.0	4
Ash	%	3.2	9.4	10
Crude fiber	%	10.2	30.0	9
Ether extract	%	1.0	3.0	9
N-free extract	%	15.0	44.0	
Protein (N x 6.25)	%	4.6	13.6	10
Cattle	dig prot %	2.4	7.2	
Sheep	dig prot %	3.1	9.1	

Continued

Feed name or analyses		Mean		C.V.
		As fed	Dry	± %
Energy				
Cattle	DE kcal/kg	842.	2469.	
Sheep	DE kcal/kg	827.	2425.	
Cattle	ME kcal/kg	690.	2024.	
Sheep	ME kcal/kg	678.	1988.	
Cattle	TDN %	19.	56.	
Sheep	TDN %	19.	55.	

Clover, red, aerial pt w molasses added, ensiled, (3)

Ref no 3-01-451

		As fed	Dry	
Dry matter	%	31.4	100.0	15
Ash	%	2.7	8.6	12
Crude fiber	%	9.3	29.6	10
Ether extract	%	1.0	3.1	21
N-free extract	%	14.0	44.6	
Protein (N x 6.25)	%	4.4	14.1	14
Cattle	dig prot %	2.5	8.0	
Sheep	dig prot %	3.2	10.2	
Energy				
Cattle	DE kcal/kg	803.	2557.	
Sheep	DE kcal/kg	872.	2778.	
Cattle	ME kcal/kg	658.	2097.	
Sheep	ME kcal/kg	715.	2278.	
Cattle	TDN %	18.	58.	
Sheep	TDN %	20.	63.	

Clover, red, seed, (5)

Ref no 5-08-004

		As fed	Dry	
Dry matter	%	87.5	100.0	
Ash	%	6.7	7.6	
Crude fiber	%	9.2	10.5	
Ether extract	%	7.8	8.9	
N-free extract	%	31.3	35.8	
Protein (N x 6.25)	%	32.6	37.2	
Cattle	dig prot %	24.4	27.9	
Sheep	dig prot %	24.4	27.9	
Energy				
Cattle	DE kcal/kg	3280.	3748.	
Sheep	DE kcal/kg	3280.	3748.	
Cattle	ME kcal/kg	2689.	3073.	
Sheep	ME kcal/kg	2689.	3073.	
Cattle	TDN %	74.	85.	
Sheep	TDN %	74.	85.	

(5) protein supplements

(6) minerals

(7) vitamins

(8) additives

Feed name or analyses		Mean		C.V.
		As fed	Dry	± %

Clover, red, seed screenings, (5)

Ref no 5-08-005

Dry matter	%	90.5	100.0	
Ash	%	5.9	6.5	
Crude fiber	%	10.2	11.3	
Ether extract	%	5.9	6.5	
N-free extract	%	40.3	44.5	
Protein (N x 6.25)	%	28.2	31.2	
Cattle	dig prot %	22.0	24.3	
Sheep	dig prot %	22.9	25.3	
Energy				
Cattle	DE kcal/kg	2913.	3219.	
Sheep	DE kcal/kg	3112.	3439.	
Cattle	ME kcal/kg	2389.	2640.	
Sheep	ME kcal/kg	2552.	2820.	
Cattle	TDN %	66.	73.	
Sheep	TDN %	70.	78.	

Clover, sweet - see Sweetclover

CLOVER, WHITE. Trifolium repens

Clover, white, hay, s-c, immature, (1)

Ref no 1-01-459

Dry matter	%	88.1	100.0	2
Ash	%	8.3	9.4	15
Crude fiber	%	16.1	18.3	16
Ether extract	%	2.9	3.3	18
N-free extract	%	37.1	42.1	
Protein (N x 6.25)	%	23.7	26.9	6
Cattle	dig prot %	14.9	16.9	
Sheep	dig prot %	14.7	16.7	
Energy				
Cattle	DE kcal/kg	2331.	2646.	
Sheep	DE kcal/kg	2214.	2513.	
Cattle	ME kcal/kg	1912.	2170.	
Sheep	ME kcal/kg	1816.	2061.	
Cattle	TDN %	53.	60.	
Sheep	TDN %	50.	57.	

Clover, white, hay, s-c, mature, (1)

Ref no 1-01-463

Dry matter	%	92.2	100.0	
Ash	%	4.5	4.9	
Crude fiber	%	33.3	36.1	

Continued

Feed name or analyses		Mean		C.V.
		As fed	Dry	± %

Ether extract	%	1.7	1.8	
N-free extract	%	39.7	43.1	
Protein (N x 6.25)	%	13.0	14.1	
Cattle	dig prot %	8.2	8.9	
Sheep	dig prot %	8.0	8.7	
Energy				
Cattle	DE kcal/kg	2480.	2690.	
Sheep	DE kcal/kg	2358.	2557.	
Cattle	ME kcal/kg	2034.	2206.	
Sheep	ME kcal/kg	1933.	2097.	
Cattle	TDN %	56.	61.	
Sheep	TDN %	53.	58.	

Clover, white, hay, s-c, (1)

Ref no 1-01-464

Dry matter	%	90.9	100.0	2
Ash	%	9.2	10.1	22
Crude fiber	%	21.2	23.3	22
Ether extract	%	2.4	2.6	27
N-free extract	%	38.1	42.0	
Protein (N x 6.25)	%	20.0	22.0	17
Cattle	dig prot %	12.6	13.9	
Sheep	dig prot %	12.4	13.6	
Energy				
Cattle	DE kcal/kg	2364.	2601.	
Sheep	DE kcal/kg	2244.	2469.	
Cattle	ME kcal/kg	1939.	2133.	
Sheep	ME kcal/kg	1840.	2024.	
Cattle	TDN %	54.	59.	
Sheep	TDN %	51.	56.	
Calcium	%	1.13	1.24	16
Copper	mg/kg	10.8	11.9	
Manganese	mg/kg	129.7	142.7	

Clover, white, aerial pt, fresh, immature, (2)

Ref no 2-01-465

Dry matter	%	21.0	100.0	39
Ash	%	2.7	12.9	40
Crude fiber	%	3.3	15.8	11
Ether extract	%	.7	3.4	29
N-free extract	%	8.5	40.5	
Protein (N x 6.25)	%	5.8	27.4	10
Cattle	dig prot %	4.0	19.2	
Sheep	dig prot %	3.9	18.4	
Energy				
Cattle	DE kcal/kg	611.	2910.	
Sheep	DE kcal/kg	556.	2646.	
Cattle	ME kcal/kg	501.	2386.	

Continued

(1) dry forages and roughages
(2) pasture, range plants, and forages fed green

(3) silages
(4) energy feeds

TABLE A-3-1 Composition of Feeds (Continued)

Feed name or analyses		Mean		C.V.
		As fed	Dry	± %
Sheep	ME kcal/kg	456.	2170.	
Cattle	TDN %	14.	66.	
Sheep	TDN %	13.	60.	

Clover, white, aerial pt, fresh, (2)

Ref no 2-01-468

		As fed	Dry	± %
Dry matter	%	18.5	100.0	36
Ash	%	2.2	12.1	36
Crude fiber	%	2.9	15.6	20
Ether extract	%	.6	3.2	24
N-free extract	%	7.6	41.3	
Protein (N x 6.25)	%	5.1	27.8	13
Cattle	dig prot %	3.6	19.5	
Sheep	dig prot %	3.4	18.6	
Energy				
Cattle	DE kcal/kg	538.	2910.	
Sheep	DE kcal/kg	498.	2690.	
Cattle	ME kcal/kg	441.	2386.	
Sheep	ME kcal/kg	408.	2206.	
Cattle	TDN %	12.	66.	
Sheep	TDN %	11.	61.	
Calcium	%	.27	1.48	19
Iron	%	.010	.030	15
Magnesium	%	.06	.32	16
Phosphorus	%	.09	.46	11
Potassium	%	.41	2.20	21
Sodium	%	.07	.39	15
Manganese	mg/kg	68.5	370.4	52

COCONUT. Cocos nucifera

Coconut, meats, fresh, (4)

Ref no 4-01-574

		As fed	Dry	
Dry matter	%	49.1	100.0	
Ash	%	.9	1.8	
Crude fiber	%	4.0	8.1	
Ether extract	%	35.3	71.9	
N-free extract	%	5.4	11.1	
Protein (N x 6.25)	%	3.5	7.1	
Calcium	%	.13	.26	
Iron	%	.017	.035	
Phosphorus	%	.95	1.93	
Potassium	%	2.56	5.21	
Sodium	%	.23	.47	
Niacin	mg/kg	5.0	10.2	
Riboflavin	mg/kg	.2	.4	
Thiamine	mg/kg	.5	1.0	

Coconut, meats, mech-extd grnd, (5)
Coconut meal, expeller (AAFCO)
Copra meal, expeller (AAFCO)
Coconut meal, hydraulic (AAFCO)
Copra meal, hydraulic (AAFCO)

Ref no 5-01-572

Feed name or analyses		As fed	Dry	± %
Dry matter	%	93.0	100.0	2
Ash	%	6.9	7.4	18
Crude fiber	%	12.0	12.9	14
Ether extract	%	6.6	7.1	32
N-free extract	%	47.2	50.7	
Protein (N x 6.25)	%	20.4	21.9	5
Cattle	dig prot %	16.5	17.7	
Sheep	dig prot %	17.1	18.4	
Swine	dig prot %	14.9	16.0	
Energy				
Cattle	DE kcal/kg	3321.	3571.	
Sheep	DE kcal/kg	3404.	3660.	
Swine	DE kcal/kg	3363.	3616.	
Cattle	ME kcal/kg	2723.	2928.	
Chickens	MEn kcal/kg	1764.	1897.	
Sheep	ME kcal/kg	2791.	3001.	
Swine	ME kcal/kg	3080.	3312.	
Cattle	NEm kcal/kg	1748.	1880.	
Cattle	NEgain kcal/kg	1162.	1250.	
Cattle	TDN %	75.	81.	
Sheep	TDN %	77.	83.	
Swine	TDN %	76.	82.	
Calcium	%	.21	.23	23
Iron	%	.196	.211	
Magnesium	%	.26	.28	
Phosphorus	%	.61	.66	10
Potassium	%	1.12	1.20	
Sodium	%	.04	.04	
Cobalt	mg/kg	2.300	2.500	
Copper	mg/kg	18.7	20.1	
Manganese	mg/kg	55.4	59.6	
Choline	mg/kg	920.	989.	
Folic acid	mg/kg	1.30	1.40	
Niacin	mg/kg	24.9	26.8	
Pantothenic acid	mg/kg	6.6	7.1	
Riboflavin	mg/kg	3.1	3.3	
Thiamine	mg/kg	.7	.8	

Coconut, meats, solv-extd grnd, (5)
Solvent extracted coconut meal (AAFCO)
Solvent extracted copra meal (AAFCO)

Ref no 5-01-573

		As fed	Dry	
Dry matter	%	92.0	100.0	
Ash	%	5.6	6.1	
Crude fiber	%	15.0	16.3	

Continued

(5) protein supplements
(6) minerals

(7) vitamins
(8) additives

Feed name or analyses		Mean		C.V.
		As fed	Dry	± %
Ether extract	%	1.8	2.0	
N-free extract	%	48.3	52.5	
Protein (N x 6.25)	%	21.3	23.1	
Cattle	dig prot %	17.2	18.7	
Sheep	dig prot %	17.8	19.4	
Swine	dig prot %	15.5	16.9	
Lignin	%	1.0	1.1	
Energy				
Cattle	DE kcal/kg	3002.	3263.	
Sheep	DE kcal/kg	3042.	3307.	
Swine	DE kcal/kg	3123.	3395.	
Cattle	ME kcal/kg	2462.	2676.	
Sheep	ME kcal/kg	2495.	2712.	
Swine	ME kcal/kg	2852.	3100.	
Cattle	NE$_m$ kcal/kg	1527.	1660.	
Cattle	NE$_{gain}$ kcal/kg	994.	1080.	
Cattle	TDN %	68.	74.	
Sheep	TDN %	69.	75.	
Swine	TDN %	71.	77.	
Calcium	%	.17	.18	
Chlorine	%	.03	.03	
Phosphorus	%	.61	.66	
Riboflavin	mg/kg	13.2	14.3	
Thiamine	mg/kg	.9	1.0	

Copra - see Coconut

CORN. Zea mays

Corn, aerial pt, s-c, dough stage, (1)
Corn fodder, sun-cured, dough stage

Ref no 1-02-774

		As fed	Dry	
Dry matter	%	76.4	100.0	6
Ash	%	5.0	6.6	11
Crude fiber	%	21.6	28.3	1
Ether extract	%	2.1	2.8	8
N-free extract	%	40.0	52.3	
Protein (N x 6.25)	%	7.6	10.0	15
Cattle	dig prot %	4.4	5.7	
Sheep	dig prot %	4.0	5.3	
Energy				
Cattle	DE kcal/kg	2426.	3175.	
Sheep	DE kcal/kg	2156.	2822.	
Cattle	ME kcal/kg	1989.	2604.	
Sheep	ME kcal/kg	1768.	2314.	
Cattle	TDN %	55.	72.	
Sheep	TDN %	49.	64.	

Corn, aerial pt, s-c, mature, (1)
Corn fodder, sun-cured, mature

Ref no 1-02-772

Feed name or analyses		As fed	Dry	C.V. ± %
Dry matter	%	82.4	100.0	8
Ash	%	5.5	6.7	19
Crude fiber	%	21.3	25.9	18
Ether extract	%	2.0	2.4	18
N-free extract	%	46.2	56.1	
Protein (N x 6.25)	%	7.3	8.9	20
Cattle	dig prot %	3.4	4.1	
Sheep	dig prot %	3.9	4.7	
Energy				
Cattle	DE kcal/kg	2362.	2866.	
Sheep	DE kcal/kg	2289.	2778.	
Cattle	ME kcal/kg	1936.	2350.	
Sheep	ME kcal/kg	1877.	2278.	
Cattle	NE$_m$ kcal/kg	1145.	1390.	
Cattle	NE$_{gain}$ kcal/kg	667.	810.	
Cattle	TDN %	54.	65.	
Sheep	TDN %	52.	63.	
Calcium	%	.59	.72	19
Magnesium	%	.27	.33	9
Phosphorus	%	.25	.30	23

Corn, aerial pt, s-c, (1)
Corn fodder, sun-cured

Ref no 1-02-775

		As fed	Dry	
Dry matter	%	82.4	100.0	8
Ash	%	5.5	6.7	19
Crude fiber	%	21.3	25.9	18
Ether extract	%	2.0	2.4	20
N-free extract	%	46.2	56.1	
Protein (N x 6.25)	%	7.3	8.9	21
Cattle	dig prot %	3.4	4.1	
Sheep	dig prot %	4.2	5.1	
Energy				
Cattle	DE kcal/kg	2362.	2866.	
Sheep	DE kcal/kg	2325.	2822.	
Cattle	ME kcal/kg	1936.	2350.	
Sheep	ME kcal/kg	1907.	2314.	
Cattle	NE$_m$ kcal/kg	1190.	1444.	
Cattle	NE$_{gain}$ kcal/kg	507.	615.	
Cattle	TDN %	54.	65.	
Sheep	TDN %	53.	64.	
Calcium	%	.41	.50	29
Chlorine	%	.16	.19	14
Iron	%	.010	.010	19
Magnesium	%	.24	.29	22
Phosphorus	%	.21	.25	27
Potassium	%	.77	.93	9

Continued

(1) dry forages and roughages
(2) pasture, range plants, and forages fed green

(3) silages
(4) energy feeds

TABLE A-3-1 Composition of Feeds (Continued)

Feed name or analyses		Mean		C.V.
		As fed	Dry	± %
Sodium	%	.02	.03	28
Sulfur	%	.12	.14	18
Copper	mg/kg	4.0	4.8	15
Manganese	mg/kg	56.1	68.1	11
Carotene	mg/kg	3.6	4.4	
Vitamin A equiv	IU/g	6.0	7.3	

Corn, aerial pt wo ears wo husks, s-c, mature, (1)
Corn stover, sun-cured, mature

Ref no 1-02-776

Dry matter	%	87.2	100.0	6
Ash	%	6.2	7.1	26
Crude fiber	%	32.4	37.1	14
Ether extract	%	1.0	1.2	29
N-free extract	%	42.5	48.7	
Protein (N x 6.25)	%	5.1	5.9	21
Cattle	dig prot %	1.9	2.2	
Sheep	dig prot %	2.7	3.1	
Energy				
Cattle	DE kcal/kg	2268.	2601.	
Sheep	DE kcal/kg	2384.	2734.	
Cattle	ME kcal/kg	1860.	2133.	
Sheep	ME kcal/kg	1955.	2242.	
Cattle	NE$_m$ kcal/kg	1055.	1210.	
Cattle	NE$_{gain}$ kcal/kg	480.	550.	
Cattle	TDN %	51.	59.	
Sheep	TDN %	54.	62.	
Calcium	%	.43	.49	
Phosphorus	%	.08	.09	

Corn, aerial pt wo ears wo husks, s-c, (1)
Corn stover, sun-cured

Ref no 1-02-778

Dry matter	%	78.5	100.0	9
Ash	%	5.3	6.8	20
Crude fiber	%	26.8	34.2	15
Ether extract	%	1.3	1.6	19
N-free extract	%	40.0	51.0	
Protein (N x 6.25)	%	5.0	6.4	26
Cattle	dig prot %	2.0	2.6	
Sheep	dig prot %	1.6	2.0	
Energy				
Cattle	DE kcal/kg	2007.	2557.	
Sheep	DE kcal/kg	1904.	2425.	
Cattle	ME kcal/kg	1647.	2098.	
Sheep	ME kcal/kg	1560.	1988.	
Cattle	NE$_m$ kcal/kg	860.	1096.	
Cattle	NE$_{gain}$ kcal/kg	309.	394.	
Cattle	TDN %	46.	58.	
Sheep	TDN %	43.	55.	

Continued

Feed name or analyses		Mean		C.V.
		As fed	Dry	± %
Calcium	%	.38	.49	13
Chlorine	%	.24	.31	11
Iron	%	.020	.020	34
Magnesium	%	.24	.31	20
Phosphorus	%	.07	.09	56
Potassium	%	.72	.92	41
Sodium	%	.05	.07	25
Sulfur	%	.13	.17	15
Copper	mg/kg	3.8	4.8	14
Manganese	mg/kg	119.2	151.9	32
Carotene	mg/kg	3.1	4.0	34
Vitamin A equiv	IU/g	5.2	6.7	

Corn, cobs, grnd, (1)
Ground corn cob (AAFCO)

Ref no 1-02-782

Dry matter	%	90.4	100.0	7
Ash	%	1.5	1.7	23
Crude fiber	%	32.4	35.8	11
Ether extract	%	.5	.5	34
N-free extract	%	53.5	59.2	
Protein (N x 6.25)	%	2.5	2.8	32
Cattle	dig prot %	.0	.0	
Horses	dig prot %	.4	.4	
Sheep	dig prot %	.6	.7	
Swine	dig prot %	.0	.0	
Energy	GE kcal/kg	3998.	4423.	
Cattle	DE kcal/kg	1873.	2072.	
Horses	DE kcal/kg	1156.	1279.	
Sheep	DE kcal/kg	1953.	2160.	
Swine	DE kcal/kg	319.	353.	
Cattle	ME kcal/kg	1536.	1699.	
Horses	ME kcal/kg	948.	1049.	
Sheep	ME kcal/kg	1601.	1771.	
Swine	ME kcal/kg	305.	337.	
Cattle	NE$_m$ kcal/kg	958.	1060.	
Cattle	NE$_{gain}$ kcal/kg	226.	250.	
Cattle	TDN %	42.	47.	
Horses	TDN %	26.	29.	
Sheep	TDN %	44.	49.	
Swine	TDN %	7.	8.	
Calcium	%	.11	.12	63
Iron	%	.021	.023	
Magnesium	%	.06	.07	19
Phosphorus	%	.04	.04	54
Potassium	%	.76	.84	21
Sulfur	%	.42	.47	
Cobalt	mg/kg	.120	.130	
Copper	mg/kg	6.6	7.3	
Manganese	mg/kg	5.6	6.2	
Carotene	mg/kg	.6	.7	
Vitamin A equiv	IU/g	1.0	1.2	

(5) protein supplements

(6) minerals

(7) vitamins

(8) additives

Feed name or analyses		Mean		C.V. ± %
		As fed	Dry	

Corn, cobs, (1)

Ref no 1-02-783

		As fed	Dry	C.V. ± %
Dry matter	%	90.4	100.0	7
Ash	%	1.5	1.7	23
Crude fiber	%	32.4	35.8	11
Ether extract	%	.5	.5	34
N-free extract	%	53.5	59.2	
Protein (N x 6.25)	%	2.5	2.8	32
Cattle	dig prot %	.0	.0	
Sheep	dig prot %	.6	.7	
Energy	GE kcal/kg	3998.	4423.	
Cattle	DE kcal/kg	1873.	2072.	
Sheep	DE kcal/kg	1953.	2160.	
Cattle	ME kcal/kg	1536.	1699.	
Sheep	ME kcal/kg	1601.	1771.	
Cattle	TDN %	42.	47.	
Sheep	TDN %	44.	49.	
Calcium	%	.11	.12	63
Iron	%	.020	.020	
Magnesium	%	.06	.07	19
Phosphorus	%	.04	.04	54
Potassium	%	.76	.84	21
Sulfur	%	.42	.47	
Cobalt	mg/kg	.120	.130	
Copper	mg/kg	6.6	7.3	
Manganese	mg/kg	5.6	6.2	
Carotene	mg/kg	.6	.7	
Vitamin A equiv	IU/g	1.0	1.2	

Corn, husks, (1)

Ref no 1-02-786

		As fed	Dry	C.V. ± %
Dry matter	%	88.9	100.0	2
Ash	%	2.7	3.0	18
Crude fiber	%	30.7	34.5	3
Ether extract	%	.7	.8	29
N-free extract	%	51.7	58.2	
Protein (N x 6.25)	%	3.1	3.5	8
Cattle	dig prot %	.9	1.0	
Sheep	dig prot %	.4	.4	
Energy				
Cattle	DE kcal/kg	2862.	3219.	
Sheep	DE kcal/kg	2352.	2646.	
Cattle	ME kcal/kg	2347.	2640.	
Sheep	ME kcal/kg	1929.	2170.	
Cattle	TDN %	65.	73.	
Sheep	TDN %	53.	60.	
Carotene	mg/kg	.6	.7	
Vitamin A equiv	IU/g	1.0	1.2	

Corn, aerial pt, fresh, (2)
Corn fodder, fresh

Ref no 2-02-806

		As fed	Dry	C.V. ± %
Dry matter	%	30.6	100.0	31
Ash	%	1.7	5.7	18
Crude fiber	%	7.7	25.3	12
Ether extract	%	.8	2.7	22
N-free extract	%	17.8	58.1	
Protein (N x 6.25)	%	2.5	8.2	31
Cattle	dig prot %	1.2	3.8	
Sheep	dig prot %	1.5	4.8	
Energy				
Cattle	DE kcal/kg	890.	2910.	
Sheep	DE kcal/kg	917.	2998.	
Cattle	ME kcal/kg	730.	2386.	
Sheep	ME kcal/kg	752.	2458.	
Cattle	TDN %	20.	66.	
Sheep	TDN %	21.	68.	
Calcium	%	.08	.27	18
Chlorine	%	.06	.21	20
Iron	%	.	.010	28
Magnesium	%	.05	.15	28
Phosphorus	%	.07	.22	19
Potassium	%	.39	1.26	14
Sodium	%	.01	.04	43
Sulfur	%	.05	.17	34
Cobalt	mg/kg	.030	.090	
Copper	mg/kg	1.6	5.3	38
Manganese	mg/kg	19.8	64.6	22
Carotene	mg/kg	19.4	63.5	42
Vitamin A equiv	IU/g	32.3	105.8	

Corn, aerial pt, ensiled, milk stage, (3)
Corn fodder silage, milk stage

Ref no 3-02-818

		As fed	Dry	C.V. ± %
Dry matter	%	23.1	100.0	
Ash	%	1.4	6.0	
Crude fiber	%	6.3	27.4	
Ether extract	%	.8	3.5	
N-free extract	%	12.8	55.6	
Protein (N x 6.25)	%	1.7	7.5	
Cattle	dig prot %	.8	3.3	
Sheep	dig prot %	1.0	4.4	
Energy				
Cattle	DE kcal/kg	703.	3042.	
Sheep	DE kcal/kg	733.	3175.	
Cattle	ME kcal/kg	576.	2494.	
Sheep	ME kcal/kg	602.	2604.	
Cattle	TDN %	16.	69.	
Sheep	TDN %	17.	72.	

Continued

(1) dry forages and roughages
(2) pasture, range plants, and forages fed green

(3) silages
(4) energy feeds

TABLE A-3-1 Composition of Feeds (Continued)

Feed name or analyses		Mean		C.V.
		As fed	Dry	± %
Calcium	%	.06	.28	
Magnesium	%	.09	.41	
Phosphorus	%	.06	.24	
Potassium	%	.36	1.57	

Corn, aerial pt, ensiled, dough stage, (3)
Corn fodder silage, dough stage

Ref no 3-02-819

Feed name or analyses		Mean		C.V.
		As fed	Dry	± %
Dry matter	%	25.6	100.0	12
Ash	%	1.5	6.0	13
Crude fiber	%	6.6	26.0	13
Ether extract	%	.8	3.3	24
N-free extract	%	14.4	56.3	
Protein (N x 6.25)	%	2.2	8.4	11
Cattle	dig prot %	1.0	3.9	
Sheep	dig prot %	1.2	4.7	
Energy				
Cattle	DE kcal/kg	745.	2910.	
Sheep	DE kcal/kg	779.	3042.	
Cattle	ME kcal/kg	611.	2386.	
Sheep	ME kcal/kg	638.	2494.	
Cattle	TDN %	17.	66.	
Sheep	TDN %	18.	69.	
Calcium	%	.04	.17	
Phosphorus	%	.04	.14	

Corn, aerial pt, ensiled, mature, well-eared mn 50 dry matter, (3)

Ref no 3-08-152

Feed name or analyses		Mean		C.V.
		As fed	Dry	± %
Dry matter	%	55.0	100.0	
Ash	%	3.0	5.4	
Crude fiber	%	12.6	23.0	
Ether extract	%	1.6	2.9	
N-free extract	%	33.5	60.9	
Protein (N x 6.25)	%	4.3	7.8	
Cattle	dig prot %	2.5	4.5	
Sheep	dig prot %	2.4	4.3	
Energy				
Cattle	DE kcal/kg	1722.	3130.	
Sheep	DE kcal/kg	1697.	3086.	
Cattle	ME kcal/kg	1412.	2567.	
Sheep	ME kcal/kg	1392.	2530.	
Cattle	TDN %	39.	71.	
Sheep	TDN %	38.	70.	
Calcium	%	.15	.27	
Phosphorus	%	.10	.19	

Corn, aerial pt, ensiled, mature, well-eared mn 30 mx 50 dry matter, (3)

Ref no 3-08-153

Feed name or analyses		Mean		C.V.
		As fed	Dry	± %
Dry matter	%	40.0	100.0	
Ash	%	2.4	6.0	
Crude fiber	%	9.8	24.4	
Ether extract	%	1.2	2.9	
N-free extract	%	23.4	58.6	
Protein (N x 6.25)	%	3.2	8.1	
Cattle	dig prot %	1.9	4.7	
Sheep	dig prot %	1.8	4.4	
Energy				
Cattle	DE kcal/kg	1234.	3086.	
Sheep	DE kcal/kg	1217.	3042.	
Cattle	ME kcal/kg	1012.	2530.	
Sheep	ME kcal/kg	998.	2494.	
Cattle	TDN %	28.	70.	
Sheep	TDN %	28.	69.	
Calcium	%	.11	.27	
Phosphorus	%	.08	.20	
Potassium	%	.42	1.05	

Corn, aerial pt, ensiled, mature, well-eared mx 30 dry matter, (3)

Ref no 3-08-154

Feed name or analyses		Mean		C.V.
		As fed	Dry	± %
Dry matter	%	27.9	100.0	
Ash	%	1.7	6.2	
Crude fiber	%	7.3	26.3	
Ether extract	%	.8	2.7	
N-free extract	%	15.7	56.4	
Protein (N x 6.25)	%	2.3	8.4	
Cattle	dig prot %	1.4	4.9	
Sheep	dig prot %	1.3	4.6	
Energy				
Cattle	DE kcal/kg	861.	3086.	
Sheep	DE kcal/kg	836.	2998.	
Cattle	ME kcal/kg	706.	2530.	
Sheep	ME kcal/kg	686.	2458.	
Cattle	TDN %	20.	70.	
Sheep	TDN %	19.	68.	
Calcium	%	.08	.28	
Magnesium	%	.05	.18	
Phosphorus	%	.06	.21	
Potassium	%	.26	.95	

(5) protein supplements
(6) minerals

(7) vitamins
(8) additives

Feed name or analyses		Mean		C.V.
		As fed	Dry	± %

Corn, aerial pt, ensiled, mature, (3)
Corn fodder silage, mature

Ref no 3-02-820

		As fed	Dry	± %
Dry matter	%	26.7	100.0	15
Ash	%	1.6	6.0	20
Crude fiber	%	6.5	24.4	11
Ether extract	%	.8	2.9	32
N-free extract	%	15.6	58.6	
Protein (N x 6.25)	%	2.2	8.1	22
Cattle	dig prot %	1.4	5.3	
Sheep	dig prot %	1.2	4.4	
Energy				
Cattle	DE kcal/kg	836.	3130.	
Sheep	DE kcal/kg	812.	3042.	
Cattle	ME kcal/kg	685.	2567.	
Sheep	ME kcal/kg	666.	2494.	
Cattle	TDN %	19.	71.	
Sheep	TDN %	18.	69.	
Cobalt	mg/kg	.050	.180	41
Carotene	mg/kg	4.2	15.6	42
Vitamin A equiv	IU/g	7.0	26.0	

Corn, aerial pt, ensiled, (3)
Corn fodder silage

Ref no 3-02-822

		As fed	Dry	± %
Dry matter	%	25.6	100.0	21
Ash	%	1.5	6.0	24
Crude fiber	%	6.4	25.1	14
Ether extract	%	.8	3.0	31
N-free extract	%	14.7	57.6	
Protein (N x 6.25)	%	2.1	8.3	23
Cattle	dig prot %	.9	3.7	
Sheep	dig prot %	1.0	3.9	
Energy				
Cattle	DE kcal/kg	722.	2822.	
Sheep	DE kcal/kg	722.	2822.	
Cattle	ME kcal/kg	592.	2314.	
Sheep	ME kcal/kg	592.	2314.	
Cattle	NEm kcal/kg	529.	2066.	
Cattle	NE gain kcal/kg	242.	945.	
Cattle	TDN %	16.	64.	
Sheep	TDN %	16.	64.	
Calcium	%	.08	.33	41
Chlorine	%	.05	.18	70
Iron	%	.010	.020	66
Magnesium	%	.06	.24	46
Phosphorus	%	.06	.23	39
Potassium	%	.29	1.15	32
Sodium	%	.01	.03	63
Sulfur	%	.03	.11	20

Continued

		As fed	Dry	± %
Cobalt	mg/kg	.020	.090	64
Copper	mg/kg	2.6	10.1	99
Manganese	mg/kg	12.5	49.0	38
Zinc	mg/kg	5.4	20.9	
Carotene	mg/kg	11.7	45.6	70
Vitamin A equiv	IU/g	19.5	76.0	
Vitamin D2	IU/g	.1	.4	15

Corn, aerial pt wo ears wo husks, ensiled, (3)
Corn stover silage

Ref no 3-02-836

		As fed	Dry	± %
Dry matter	%	27.2	100.0	13
Ash	%	2.1	7.6	10
Crude fiber	%	8.7	32.1	8
Ether extract	%	.7	2.4	20
N-free extract	%	13.8	50.7	
Protein (N x 6.25)	%	2.0	7.2	10
Cattle	dig prot %	.8	2.9	
Sheep	dig prot %	.7	2.7	
Energy				
Cattle	DE kcal/kg	696.	2557.	
Sheep	DE kcal/kg	648.	2381.	
Cattle	ME kcal/kg	570.	2097.	
Sheep	ME kcal/kg	531.	1952.	
Cattle	TDN %	16.	58.	
Sheep	TDN %	15.	54.	
Calcium	%	.10	.38	51
Magnesium	%	.08	.31	57
Phosphorus	%	.05	.19	16
Potassium	%	.39	1.43	38

Corn, ears w husks, ensiled, (3)

Ref no 3-02-839

		As fed	Dry	± %
Dry matter	%	43.4	100.0	
Ash	%	2.1	4.9	
Crude fiber	%	5.1	11.8	
Ether extract	%	1.6	3.8	
N-free extract	%	30.7	70.7	
Protein (N x 6.25)	%	3.8	8.8	
Cattle	dig prot %	2.1	4.8	
Sheep	dig prot %	2.1	4.8	
Energy				
Cattle	DE kcal/kg	1378.	3175.	
Sheep	DE kcal/kg	1378.	3175.	
Cattle	ME kcal/kg	1130.	2604.	
Sheep	ME kcal/kg	1130.	2604.	
Cattle	TDN %	31.	72.	
Sheep	TDN %	31.	72.	
Calcium	%	.03	.06	
Phosphorus	%	.12	.27	

(1) dry forages and roughages
(2) pasture, range plants, and forages fed green

(3) silages
(4) energy feeds

Feed name or analyses		Mean		C.V.
		As fed	Dry	± %
Corn, bran, wet- or dry-mil dehy, (4)				
Corn bran (AAFCO)				
Corn bran (CFA)				
Ref no 4-02-841				
Dry matter	%	89.0	100.0	2
Ash	%	2.1	2.4	34
Crude fiber	%	10.0	11.2	17
Ether extract	%	4.4	4.9	64
N-free extract	%	65.0	73.1	
Protein (N x 6.25)	%	7.5	8.4	37
Cattle	dig prot %	3.9	4.4	
Horses	dig prot %	3.9	4.4	
Sheep	dig prot %	4.3	4.8	
Energy	GE kcal/kg	4008.	4505.	
Cattle	DE kcal/kg	2512.	2822.	
Horses	DE kcal/kg	2904.	3263.	
Sheep	DE kcal/kg	2904.	3263.	
Cattle	ME kcal/kg	2059.	2314.	
Horses	ME kcal/kg	2463.	2676.	
Sheep	ME kcal/kg	2463.	2676.	
Cattle	TDN %	57.	64.	
Horses	TDN %	66.	74.	
Sheep	TDN %	66.	74.	
Calcium	%	.03	.03	56
Magnesium	%	.26	.29	
Phosphorus	%	.19	.21	
Potassium	%	.73	.82	
Manganese	mg/kg	16.1	18.1	
Biotin	mg/kg	.10	.10	
Niacin	mg/kg	42.0	47.2	
Pantothenic acid	mg/kg	5.3	5.9	
Riboflavin	mg/kg	1.5	1.7	
Thiamine	mg/kg	4.4	4.9	2

Feed name or analyses		Mean		C.V.
		As fed	Dry	± %
Sheep	DE kcal/kg	2992.	3439.	
Swine	DE kcal/kg	3107.	3571.	
Cattle	ME kcal/kg	2831.	3254.	
Chickens	MEn kcal/kg	2822.	3244.	
Sheep	ME kcal/kg	2453.	2820.	
Swine	ME kcal/kg	2923.	3360.	
Cattle	NEm kcal/kg	1940.	2230.	
Cattle	NEgain kcal/kg	1212.	1393.	
Cattle	TDN %	78.	90.	
Sheep	TDN %	68.	78.	
Swine	TDN %	70.	81.	
Calcium	%	.04	.05	
Iron	%	.007	.008	
Magnesium	%	.15	.17	
Phosphorus	%	.27	.31	
Potassium	%	.53	.61	
Cobalt	mg/kg	.300	.300	
Copper	mg/kg	7.7	8.8	
Manganese	mg/kg	13.0	15.0	

Corn, ears w husks, grnd, (4)
   Corn and cob meal with husks (AAFCO)
   Ear corn chop with husks (AAFCO)
   Ground ear corn with husks (AAFCO)
   Ground snapped corn

Ref no 4-02-850

Dry matter	%	89.0	100.0	8
Ash	%	2.6	2.9	25
Crude fiber	%	11.0	12.4	12
Ether extract	%	2.9	3.3	15
N-free extract	%	64.4	72.4	
Protein (N x 6.25)	%	8.0	9.0	8
Cattle	dig prot %	3.8	4.3	
Sheep	dig prot %	4.9	5.5	
Energy				
Cattle	DE kcal/kg	3532.	3968.	
Sheep	DE kcal/kg	3100.	3483.	
Cattle	ME kcal/kg	2896.	3254.	
Sheep	ME kcal/kg	2542.	2856.	
Cattle	NEm kcal/kg	1807.	2030.	
Cattle	NEgain kcal/kg	1202.	1350.	
Cattle	TDN %	80.	90.	
Sheep	TDN %	70.	79.	

Corn, ears, grnd, (4)
   Corn and cob meal (AAFCO)
   Ear corn chop (AAFCO)
   Ground ear corn (AAFCO)

Ref no 4-02-849

Dry matter	%	87.0	100.0	4
Ash	%	1.6	1.8	29
Crude fiber	%	8.0	9.2	23
Ether extract	%	3.2	3.7	13
N-free extract	%	66.1	76.0	
Protein (N x 6.25)	%	8.1	9.3	10
Cattle	dig prot %	4.0	4.6	
Sheep	dig prot %	4.4	5.1	
Swine	dig prot %	5.8	6.7	
Energy				
Cattle	DE kcal/kg	3452.	3968.	

Continued

Feed name or analyses		Mean		C.V.
		As fed	Dry	± %

Corn fodder - see Corn, aerial pt

Corn, grain, flaked, (4)
Flaked corn (AAFCO)
Corn grain, flaked

Ref no 4-02-859

		As fed	Dry	
Dry matter	%	97.0	100.0	
Crude fiber	%	.4	.4	
Ether extract	%	.3	.3	
Protein (N x 6.25)	%	7.8	8.0	
Calcium	%	.01	.01	
Phosphorus	%	.04	.04	
Niacin	mg/kg	21.0	21.6	
Riboflavin	mg/kg	1.3	1.3	
Thiamine	mg/kg	4.1	4.2	

Corn grain - see Corn, dent; see Corn, dent white; see Corn, dent yellow; see Corn, flint; see Corn, pop; see Corn, sweet; see Corn, white; see Corn, yellow

Corn, grain, Can 1 CE mn 56 wt mx 2 fm, (4)

Ref no 4-02-870

		As fed	Dry	
Dry matter	%	86.5	100.0	
Ash	%	1.2	1.4	18
Crude fiber	%	1.7	2.0	12
Ether extract	%	3.9	4.5	17
N-free extract	%	70.9	82.0	
Protein (N x 6.25)	%	8.7	10.1	11
Cattle	dig prot %	6.6	7.6	
Sheep	dig prot %	6.8	7.9	
Swine	dig prot %	7.0	8.1	
Energy	GE kcal/kg	4804.	5553.	
Cattle	DE kcal/kg	3470.	4012.	
Sheep	DE kcal/kg	3738.	4321.	
Swine	DE kcal/kg	3508.	4056.	
Cattle	ME kcal/kg	2846.	3290.	
Sheep	ME kcal/kg	3065.	3543.	
Swine	ME kcal/kg	3298.	3813.	
Cattle	TDN %	79.	91.	
Sheep	TDN %	85.	98.	
Swine	TDN %	80.	92.	

Corn, grain, Can 1 CW mn 56 wt mx 2 fm, (4)

Ref no 4-02-871

		As fed	Dry	
Dry matter	%	86.5	100.0	
Ash	%	1.2	1.4	18
Crude fiber	%	1.8	2.1	29
Ether extract	%	3.9	4.5	18
N-free extract	%	70.6	81.6	
Protein (N x 6.25)	%	9.0	10.4	10
Cattle	dig prot %	6.7	7.8	
Sheep	dig prot %	7.0	8.1	
Swine	dig prot %	7.2	8.3	
Energy	GE kcal/kg	4804.	5553.	
Cattle	DE kcal/kg	3470.	4012.	
Sheep	DE kcal/kg	3738.	4321.	
Swine	DE kcal/kg	3508.	4056.	
Cattle	ME kcal/kg	2846.	3290.	
Sheep	ME kcal/kg	3065.	3543.	
Swine	ME kcal/kg	3298.	3813.	
Cattle	TDN %	79.	91.	
Sheep	TDN %	85.	98.	
Swine	TDN %	80.	92.	

Corn, grain, Can 2 CE mn 54 wt mx 3 fm, (4)

Ref no 4-02-872

		As fed	Dry	
Dry matter	%	86.5	100.0	
Ash	%	1.2	1.4	22
Crude fiber	%	1.8	2.1	20
Ether extract	%	3.9	4.6	27
N-free extract	%	70.9	82.0	
Protein (N x 6.25)	%	8.6	9.9	12
Cattle	dig prot %	6.4	7.4	
Sheep	dig prot %	6.7	7.7	
Swine	dig prot %	6.8	7.9	
Energy	GE kcal/kg	4804.	5553.	
Cattle	DE kcal/kg	3470.	4012.	
Sheep	DE kcal/kg	3738.	4321.	
Swine	DE kcal/kg	3508.	4056.	
Cattle	ME kcal/kg	2846.	3290.	
Sheep	ME kcal/kg	3065.	3543.	
Swine	ME kcal/kg	3298.	3813.	
Cattle	TDN %	79.	91.	
Sheep	TDN %	85.	98.	
Swine	TDN %	80.	92.	

(1) dry forages and roughages
(2) pasture, range plants, and forages fed green
(3) silages
(4) energy feeds

TABLE A-3-1 Composition of Feeds (Continued)

Feed name or analyses		Mean As fed	Mean Dry	C.V. ± %

Corn, grain, Can 2 CW mn 54 wt mx 3 fm, (4)

Ref no 4-02-873

Feed name or analyses		As fed	Dry	C.V. ± %
Dry matter	%	86.5	100.0	
Ash	%	1.2	1.4	27
Crude fiber	%	1.8	2.1	33
Ether extract	%	4.0	4.6	20
N-free extract	%	70.3	81.3	
Protein (N x 6.25)	%	9.1	10.6	6
Cattle	dig prot %	6.9	8.0	
Sheep	dig prot %	7.2	8.3	
Swine	dig prot %	7.4	8.5	
Energy	GE kcal/kg	4804.	5553.	
Cattle	DE kcal/kg	3470.	4012.	
Sheep	DE kcal/kg	3738.	4321.	
Swine	DE kcal/kg	3508.	4056.	
Cattle	ME kcal/kg	2846.	3290.	
Sheep	ME kcal/kg	3065.	3543.	
Swine	ME kcal/kg	3298.	3813.	
Cattle	TDN %	79.	91.	
Sheep	TDN %	85.	98.	
Swine	TDN %	80.	92.	

Corn, grain, Can 3 CW mn 52 wt mx 5 fm, (4)

Ref no 4-02-875

Feed name or analyses		As fed	Dry	C.V. ± %
Dry matter	%	86.5	100.0	
Ash	%	1.3	1.5	21
Crude fiber	%	1.8	2.0	35
Ether extract	%	3.8	4.4	13
N-free extract	%	70.5	81.5	
Protein (N x 6.25)	%	9.2	10.6	14
Cattle	dig prot %	6.9	8.0	
Sheep	dig prot %	7.2	8.3	
Swine	dig prot %	7.4	8.5	
Energy	GE kcal/kg	4804.	5553.	
Cattle	DE kcal/kg	3470.	4012.	
Sheep	DE kcal/kg	3738.	4321.	
Swine	DE kcal/kg	3508.	4056.	
Cattle	ME kcal/kg	2846.	3290.	
Sheep	ME kcal/kg	3065.	3543.	
Swine	ME kcal/kg	3298.	3813.	
Cattle	TDN %	79.	91.	
Sheep	TDN %	85.	98.	
Swine	TDN %	80.	92.	

Corn, grain, Can 3 CE mn 52 wt mx 5 fm, (4)

Ref no 4-02-874

Feed name or analyses		As fed	Dry	C.V. ± %
Dry matter	%	86.5	100.0	
Ash	%	1.2	1.4	20
Crude fiber	%	1.8	2.1	22
Ether extract	%	4.1	4.7	31
N-free extract	%	70.8	81.9	
Protein (N x 6.25)	%	8.6	9.9	10
Cattle	dig prot %	6.4	7.4	
Sheep	dig prot %	6.7	7.7	
Swine	dig prot %	6.8	7.9	
Energy	GE kcal/kg	4804.	5553.	
Cattle	DE kcal/kg	3470.	4012.	
Sheep	DE kcal/kg	3776.	4365.	
Swine	DE kcal/kg	3508.	4056.	
Cattle	ME kcal/kg	2846.	3290.	
Sheep	ME kcal/kg	3096.	3579.	
Swine	ME kcal/kg	3298.	3813.	
Cattle	TDN %	79.	91.	
Sheep	TDN %	86.	99.	
Swine	TDN %	80.	92.	

Corn, grain, Can 4 CE mn 50 wt mx 7 fm, (4)

Ref no 4-02-876

Feed name or analyses		As fed	Dry	C.V. ± %
Dry matter	%	86.5	100.0	
Ash	%	1.2	1.4	
Crude fiber	%	1.9	2.2	
Ether extract	%	3.2	3.7	
N-free extract	%	71.6	82.8	
Protein (N x 6.25)	%	8.6	9.9	
Cattle	dig prot %	6.4	7.4	
Sheep	dig prot %	6.7	7.7	
Swine	dig prot %	6.8	7.9	
Energy	GE kcal/kg	4804.	5553.	
Cattle	DE kcal/kg	3432.	3968.	
Sheep	DE kcal/kg	3738.	4321.	
Swine	DE kcal/kg	3508.	4056.	
Cattle	ME kcal/kg	2815.	3254.	
Sheep	ME kcal/kg	3065.	3543.	
Swine	ME kcal/kg	3298.	3813.	
Cattle	TDN %	78.	90.	
Sheep	TDN %	85.	98.	
Swine	TDN %	80.	92.	

(5) protein supplements
(6) minerals
(7) vitamins
(8) additives

Feed name or analyses		Mean		C.V.
		As fed	Dry	± %

Corn, grain, Can 4 CW mn 50 wt mx 7 fm, (4)

Ref no 4-02-877

		As fed	Dry	± %
Dry matter	%	86.5	100.0	
Ash	%	1.3	1.4	34
Crude fiber	%	1.8	2.1	24
Ether extract	%	3.8	4.4	23
N-free extract	%	70.5	81.5	
Protein (N x 6.25)	%	9.2	10.6	11
Cattle	dig prot %	6.9	8.0	
Sheep	dig prot %	7.2	8.3	
Swine	dig prot %	7.4	8.5	
Energy	GE kcal/kg	4804.	5553.	
Cattle	DE kcal/kg	3470.	4012.	
Sheep	DE kcal/kg	3738.	4321.	
Swine	DE kcal/kg	3508.	4056.	
Cattle	ME kcal/kg	2846.	3290.	
Sheep	ME kcal/kg	3065.	3543.	
Swine	ME kcal/kg	3298.	3813.	
Cattle	TDN %	79.	91.	
Sheep	TDN %	85.	98.	
Swine	TDN %	80.	92.	

Corn, grain, Can 5 CW mn 47 wt mx 12 fm, (4)

Ref no 4-02-878

		As fed	Dry	± %
Dry matter	%	86.5	100.0	
Ash	%	1.3	1.5	29
Crude fiber	%	1.8	2.1	29
Ether extract	%	4.1	4.7	35
N-free extract	%	69.9	80.9	
Protein (N x 6.25)	%	9.4	10.8	10
Cattle	dig prot %	7.0	8.1	
Sheep	dig prot %	7.3	8.4	
Swine	dig prot %	7.4	8.6	
Energy	GE kcal/kg	4804.	5553.	
Cattle	DE kcal/kg	3470.	4012.	
Sheep	DE kcal/kg	3738.	4321.	
Swine	DE kcal/kg	3508.	4056.	
Cattle	ME kcal/kg	2846.	3290.	
Sheep	ME kcal/kg	3065.	3543.	
Swine	ME kcal/kg	3298.	3813.	
Cattle	TDN %	79.	91.	
Sheep	TDN %	85.	98.	
Swine	TDN %	80.	92.	

Corn, grits by-prod, mn 5 fat, (4)
Hominy feed (AAFCO)
Hominy feed (CFA)

Ref no 4-02-887

		As fed	Dry	± %
Dry matter	%	90.6	100.0	2
Ash	%	2.5	2.8	17
Crude fiber	%	5.0	5.5	21
Ether extract	%	6.5	7.2	24
N-free extract	%	65.9	72.7	
Protein (N x 6.25)	%	10.7	11.8	96
Cattle	dig prot %	7.2	7.9	
Sheep	dig prot %	6.9	7.6	
Swine	dig prot %	8.5	9.4	
Energy	GE kcal/kg	4275.	4702.	
Cattle	DE kcal/kg	3795.	4189.	
Sheep	DE kcal/kg	3555.	3924.	
Swine	DE kcal/kg	3595.	3968.	
Cattle	ME kcal/kg	3112.	3435.	
Chickens	MEn kcal/kg	2866.	3163.	
Sheep	ME kcal/kg	2916.	3218.	
Swine	ME kcal/kg	3365.	3714.	
Cattle	NEm kcal/kg	2147.	2370.	
Cattle	NE gain kcal/kg	1477.	1630.	
Cattle	TDN %	86.	95.	
Sheep	TDN %	81.	89.	
Swine	TDN %	82.	90.	
Calcium	%	.05	.06	40
Iron	%	.006	.007	50
Magnesium	%	.24	.26	4
Phosphorus	%	.53	.58	18
Potassium	%	.67	.74	9
Sulfur	%	.03	.03	99
Cobalt	mg/kg	.060	.066	
Copper	mg/kg	14.6	16.1	52
Manganese	mg/kg	14.6	16.1	16
Carotene	mg/kg	9.2	10.1	
Niacin	mg/kg	51.1	56.2	38
Pantothenic acid	mg/kg	7.5	8.2	12
Riboflavin	mg/kg	2.0	2.2	11
Thiamine	mg/kg	7.9	8.7	64
Vitamin A equiv	IU/g	15.3	16.8	

Corn, molasses, mn 43 dextrose-equiv 50 total dextrose mn 78 degrees brix, (4)
Corn sugar molasses (AAFCO)

Ref no 4-02-888

		As fed	Dry	± %
Dry matter	%	73.0	100.0	6
Ash	%	8.0	11.0	21
Protein (N x 6.25)	%	.3	.4	27
Sodium	%	3.06	4.19	

(1) dry forages and roughages
(2) pasture, range plants, and forages fed green

(3) silages
(4) energy feeds

TABLE A-3-1 Composition of Feeds (Continued)

Feed name or analyses		Mean		C.V.
		As fed	Dry	± %

Corn, distil grains, dehy, (5)
    Corn distillers dried grains (AAFCO)
    Corn distillers dried grains (CFA)

Ref no 5-02-842

Feed name or analyses		As fed	Dry	C.V. ± %
Dry matter	%	92.0	100.0	3
Ash	%	2.6	2.8	43
Crude fiber	%	12.0	13.0	17
Ether extract	%	9.3	10.1	17
N-free extract	%	41.0	44.6	
Protein (N x 6.25)	%	27.1	29.5	11
Cattle	dig prot %	21.2	23.1	
Sheep	dig prot %	19.5	21.2	
Energy	GE kcal/kg	4990.	5424.	
Cattle	DE kcal/kg	3408.	3704.	
Sheep	DE kcal/kg	3327.	3616.	
Cattle	ME kcal/kg	2794.	3037.	
Chickens	MEn kcal/kg	1631.	1773.	
Sheep	ME kcal/kg	2728.	2965.	
Cattle	TDN %	77.	84.	
Sheep	TDN %	75.	82.	
Calcium	%	.09	.10	71
Iron	%	.020	.020	
Magnesium	%	.06	.07	
Phosphorus	%	.37	.40	51
Potassium	%	.09	.10	
Sodium	%	.90	.98	
Cobalt	mg/kg	.100	.100	
Copper	mg/kg	44.7	48.6	
Manganese	mg/kg	18.9	20.5	
Biotin	mg/kg	.40	.40	
Choline	mg/kg	1859.	2021.	77
Niacin	mg/kg	42.2	45.9	43
Pantothenic acid	mg/kg	5.9	6.4	51
Riboflavin	mg/kg	3.1	3.4	71
Thiamine	mg/kg	1.8	2.0	69
Arginine	%	1.00	1.09	
Histidine	%	.60	.65	
Isoleucine	%	1.00	1.09	
Leucine	%	3.60	3.91	
Lysine	%	.90	.98	
Methionine	%	.40	.43	
Phenylalanine	%	.60	.65	
Threonine	%	.30	.33	
Tryptophan	%	.20	.22	
Tyrosine	%	.90	.98	
Valine	%	1.20	1.30	

Corn, distil grains w sol, dehy, mn 75 orig solids, (5)
    Corn distillers dried grains with solubles (AAFCO)

Ref no 5-02-843

Feed name or analyses		As fed	Dry	C.V. ± %
Dry matter	%	92.0	100.0	2
Ash	%	2.6	2.8	24
Crude fiber	%	9.0	9.8	19
Ether extract	%	9.3	10.1	21
N-free extract	%	43.7	47.5	
Protein (N x 6.25)	%	27.4	29.8	12
Cattle	dig prot %	21.5	23.4	
Sheep	dig prot %	13.4	14.6	
Lignin	%	6.0	6.5	
Energy	GE kcal/kg	4528.	4922.	
Cattle	DE kcal/kg	3570.	3880.	
Sheep	DE kcal/kg	3042.	3307.	
Cattle	ME kcal/kg	2927.	3182.	
Chickens	MEn kcal/kg	2425.	2636.	
Sheep	ME kcal/kg	2495.	2712.	
Cattle	TDN %	81.	88.	
Sheep	TDN %	69.	75.	
Calcium	%	.09	.10	27
Iron	%	.020	.020	47
Magnesium	%	.06	.07	36
Phosphorus	%	.37	.40	24
Potassium	%	.09	.10	49
Sodium	%	.90	.98	
Cobalt	mg/kg	.100	.100	
Copper	mg/kg	44.7	48.6	
Manganese	mg/kg	18.9	20.5	61
Biotin	mg/kg	.70	.80	60
Carotene	mg/kg	3.7	4.0	82
Choline	mg/kg	2471.	2686.	49
Niacin	mg/kg	66.9	72.7	21
Pantothenic acid	mg/kg	11.0	11.9	57
Riboflavin	mg/kg	8.6	9.3	39
Thiamine	mg/kg	2.9	3.2	52
Vitamin A equiv	IU/g	6.2	6.7	
Vitamin D3	ICU/g	1.0	1.0	
Glycine	%	.50	.54	
Leucine	%	2.20	2.39	15
Lysine	%	.70	.76	14
Methionine	%	.50	.54	22
Phenylalanine	%	1.70	1.85	16
Threonine	%	1.00	1.09	
Tryptophan	%	.10	.11	36
Tyrosine	%	.60	.65	11
Valine	%	1.60	1.74	6

(5) protein supplements
(6) minerals

(7) vitamins
(8) additives

Feed name or analyses		Mean		C.V.
		As fed	Dry	± %

Corn, distil sol, dehy, (5)
   Corn distillers dried solubles (AAFCO)

Ref no 5-02-844

Feed name or analyses		As fed	Dry	C.V. ± %
Dry matter	%	93.0	100.0	2
Ash	%	8.0	8.6	18
Crude fiber	%	4.0	4.3	74
Ether extract	%	9.1	9.8	68
N-free extract	%	45.0	48.4	
Protein (N x 6.25)	%	26.9	28.9	28
Cattle	dig prot %	21.0	22.6	
Sheep	dig prot %	22.0	23.7	
Swine	dig prot %	16.1	17.3	
Lignin	%	2.0	2.2	
Energy				
Cattle	DE kcal/kg	3608.	3880.	
Sheep	DE kcal/kg	3690.	3968.	
Swine	DE kcal/kg	3300.	3548.	
Cattle	ME kcal/kg	2959.	3182.	
Chickens	MEn kcal/kg	2932.	3153.	
Sheep	ME kcal/kg	3026.	3254.	
Swine	ME kcal/kg	2976.	3200.	
Cattle	TDN %	82.	88.	
Sheep	TDN %	84.	90.	
Swine	TDN %	75.	81.	
Calcium	%	.35	.38	34
Iron	%	.055	.059	30
Magnesium	%	.64	.69	19
Phosphorus	%	1.37	1.47	28
Potassium	%	1.74	1.87	5
Copper	mg/kg	82.7	88.9	4
Manganese	mg/kg	73.5	79.0	27
Biotin	mg/kg	1.50	1.60	56
Carotene	mg/kg	.7	.8	
Choline	mg/kg	4818.	5179.	28
Folic acid	mg/kg	1.10	1.20	
Niacin	mg/kg	115.3	123.9	20
Pantothenic acid	mg/kg	20.9	22.5	32
Riboflavin	mg/kg	16.9	18.2	41
Thiamine	mg/kg	6.8	7.3	53
Vitamin A equiv	IU/g	1.2	1.3	
Cystine	%	.60	.65	
Glycine	%	1.10	1.18	
Leucine	%	2.10	2.26	46
Lysine	%	.90	.97	49
Methionine	%	.60	.65	
Phenylalanine	%	1.50	1.61	10
Threonine	%	1.00	1.07	20
Tryptophan	%	.20	.22	40
Tyrosine	%	.70	.75	34
Valine	%	1.50	1.61	11

Corn, germ, solv-extd grnd, (5)

Ref no 5-02-895

Feed name or analyses		As fed	Dry	C.V. ± %
Dry matter	%	90.0	100.0	1
Ash	%	1.7	1.9	33
Crude fiber	%	11.0	12.2	12
Ether extract	%	1.4	1.6	39
N-free extract	%	54.3	60.3	
Protein (N x 6.25)	%	21.6	24.0	4
Cattle	dig prot %	16.2	18.0	
Sheep	dig prot %	16.2	18.0	
Energy				
Cattle	DE kcal/kg	3095.	3439.	
Sheep	DE kcal/kg	3413.	3792.	
Cattle	ME kcal/kg	2538.	2820.	
Sheep	ME kcal/kg	2798.	3109.	
Cattle	TDN %	70.	78.	
Sheep	TDN %	77.	86.	
Calcium	%	.05	.06	42
Iron	%	.030	.030	
Phosphorus	%	.50	.56	
Cobalt	mg/kg	.100	.100	
Copper	mg/kg	7.0	7.8	
Manganese	mg/kg	17.4	19.3	31
Niacin	mg/kg	42.0	46.7	
Pantothenic acid	mg/kg	3.3	3.7	
Riboflavin	mg/kg	3.7	4.1	
a-tocopherol	mg/kg	1.1	1.2	

Corn, germ wo sol, wet-mil mech-extd dehy grnd, (5)
   Corn germ meal, mechanical extracted, (wet-
      milled) (AAFCO)

Ref no 5-02-897

Feed name or analyses		As fed	Dry	C.V. ± %
Dry matter	%	93.0	100.0	2
Ash	%	2.9	3.1	46
Crude fiber	%	9.0	9.7	22
Ether extract	%	7.6	8.2	35
N-free extract	%	53.3	57.3	
Protein (N x 6.25)	%	20.2	21.7	19
Cattle	dig prot %	14.8	15.9	
Sheep	dig prot %	15.2	16.3	
Energy	GE kcal/kg	4710.	5063.	
Cattle	DE kcal/kg	3567.	3836.	
Sheep	DE kcal/kg	3363.	3616.	
Cattle	ME kcal/kg	2926.	3146.	
Sheep	ME kcal/kg	2757.	2965.	
Cattle	TDN %	81.	87.	
Sheep	TDN %	76.	82.	
Calcium	%	.05	.05	25
Iron	%	.040	.040	21
Magnesium	%	.63	.68	

Continued

(1) dry forages and roughages
(2) pasture, range plants, and forages fed green

(3) silages
(4) energy feeds

Feed name or analyses		Mean		C.V.
		As fed	Dry	± %
Phosphorus	%	.51	.55	27
Potassium	%	.21	.23	
Copper	mg/kg	10.1	10.9	62
Manganese	mg/kg	15.4	16.6	58
Choline	mg/kg	1586.	1705.	
Niacin	mg/kg	40.7	43.8	25
Pantothenic acid	mg/kg	5.1	5.5	64
Riboflavin	mg/kg	3.1	3.3	26
Thiamine	mg/kg	19.8	21.3	

Corn, germ wo sol, wet-mil solv-extd dehy grnd, (5)
Corn germ meal, solvent extracted, (wet-milled) (AAFCO)

Ref no 5-02-898

Feed name or analyses		As fed	Dry	
Dry matter	%	93.0	100.0	
Crude fiber	%	12.0	12.9	
Ether extract	%	2.0	2.2	
Protein (N x 6.25)	%	18.0	19.4	
Calcium	%	.10	.11	
Phosphorus	%	.40	.43	
Choline	mg/kg	1800.	1936.	
Folic acid	mg/kg	.20	.21	
Niacin	mg/kg	35.1	37.7	
Pantothenic acid	mg/kg	4.1	4.4	
Riboflavin	mg/kg	4.1	4.4	
Thiamine	mg/kg	1.0	1.1	

Corn, gluten, wet-mil dehy, (5)
Corn gluten meal (AAFCO)
Corn gluten meal (CFA)

Ref no 5-02-900

Feed name or analyses		As fed	Dry	C.V. ± %
Dry matter	%	91.0	100.0	2
Ash	%	2.4	2.6	34
Crude fiber	%	4.0	4.4	31
Ether extract	%	2.3	2.5	28
N-free extract	%	39.5	43.4	
Protein (N x 6.25)	%	42.9	47.1	7
Cattle	dig prot %	35.7	39.2	
Sheep	dig prot %	36.8	40.5	
Energy				
Cattle	DE kcal/kg	3371.	3704.	
Sheep	DE kcal/kg	3451.	3792.	
Cattle	ME kcal/kg	2764.	3037.	
Chickens	MEn kcal/kg	3307.	3634.	
Sheep	ME kcal/kg	2829.	3109.	
Cattle	TDN %	76.	84.	
Sheep	TDN %	78.	86.	
Calcium	%	.16	.18	69
Iron	%	.040	.040	40
Magnesium	%	.05	.05	98

Continued

Feed name or analyses		Mean		C.V.
		As fed	Dry	± %
Phosphorus	%	.40	.44	33
Potassium	%	.03	.03	87
Sodium	%	.10	.10	
Cobalt	mg/kg	.100	.100	50
Copper	mg/kg	28.2	31.0	40
Manganese	mg/kg	7.3	8.0	52
Choline	mg/kg	330.	363.	23
Folic acid	mg/kg	.20	.20	
Niacin	mg/kg	49.9	54.8	21
Pantothenic acid	mg/kg	10.3	11.3	39
Riboflavin	mg/kg	1.5	1.6	53
Thiamine	mg/kg	.2	.2	97
Arginine	%	1.40	1.54	20
Cystine	%	.60	.66	
Glycine	%	1.50	1.65	
Histidine	%	1.00	1.10	
Isoleucine	%	2.30	2.53	
Leucine	%	7.60	8.35	
Lysine	%	.80	.88	10
Methionine	%	1.00	1.10	32
Phenylalanine	%	2.90	3.19	
Threonine	%	1.40	1.54	
Tryptophan	%	.20	.22	
Tyrosine	%	1.00	1.10	
Valine	%	2.20	2.42	

Corn, gluten w bran, wet-mil dehy, (5)
Corn gluten feed (AAFCO)
Corn gluten feed (CFA)

Ref no 5-02-903

Feed name or analyses		As fed	Dry	C.V. ± %
Dry matter	%	90.0	100.0	2
Ash	%	6.3	7.0	25
Crude fiber	%	8.0	8.9	17
Ether extract	%	2.4	2.7	46
N-free extract	%	48.1	53.4	
Protein (N x 6.25)	%	25.3	28.1	9
Cattle	dig prot %	21.8	24.2	
Sheep	dig prot %	21.8	24.2	
Energy	GE kcal/kg	4041.	4490.	
Cattle	DE kcal/kg	3254.	3616.	
Sheep	DE kcal/kg	3334.	3704.	
Cattle	ME kcal/kg	2668.	2965.	
Chickens	MEn kcal/kg	1675.	1861.	
Sheep	ME kcal/kg	2733.	3037.	
Cattle	TDN %	74.	82.	
Sheep	TDN %	76.	84.	
Calcium	%	.46	.51	50
Iron	%	.050	.060	38
Magnesium	%	.29	.32	52
Phosphorus	%	.77	.86	22
Potassium	%	.60	.67	22
Sodium	%	.95	1.06	
Cobalt	mg/kg	.090	.100	79

Continued

(5) protein supplements
(6) minerals

(7) vitamins
(8) additives

Feed name or analyses		Mean		C.V.
		As fed	Dry	± %
Copper	mg/kg	47.7	53.0	40
Manganese	mg/kg	23.8	26.4	33
Biotin	mg/kg	.30	.30	
Choline	mg/kg	1516.	1684.	56
Folic acid	mg/kg	.20	.20	
Niacin	mg/kg	71.9	79.9	22
Pantothenic acid	mg/kg	17.2	19.1	31
Riboflavin	mg/kg	2.4	2.7	66
Thiamine	mg/kg	2.0	2.2	84
Arginine	%	.80	.89	20
Histidine	%	.60	.67	13
Isoleucine	%	1.20	1.33	27
Leucine	%	2.60	2.89	20
Lysine	%	.80	.89	10
Methionine	%	.30	.33	13
Phenylalanine	%	.90	1.00	22
Threonine	%	.80	.89	10
Tryptophan	%	.20	.22	20
Tyrosine	%	.90	1.00	31
Valine	%	1.30	1.44	25

CORN, DENT. Zea mays indentata

Corn, dent, grain gr 1 US mn 56 wt, (4)

Ref no 4-02-914

Feed name or analyses		As fed	Dry	C.V. ± %
Dry matter	%	86.0	100.0	2
Ash	%	1.0	1.2	52
Crude fiber	%	2.0	2.3	7
Ether extract	%	3.7	4.3	8
N-free extract	%	70.7	82.2	
Protein (N x 6.25)	%	8.6	10.0	5
Cattle	dig prot %	6.4	7.5	
Sheep	dig prot %	6.7	7.8	
Swine	dig prot %	6.9	8.0	
Energy				
Cattle	DE kcal/kg	3450.	4012.	
Sheep	DE kcal/kg	3716.	4321.	
Swine	DE kcal/kg	3488.	4056.	
Cattle	ME kcal/kg	2829.	3290.	
Sheep	ME kcal/kg	3047.	3543.	
Swine	ME kcal/kg	3279.	3813.	
Cattle	TDN %	78.	91.	
Sheep	TDN %	84.	98.	
Swine	TDN %	79.	92.	
Calcium	%	.02	.02	75
Magnesium	%	.10	.11	15
Phosphorus	%	.28	.32	13
Potassium	%	.28	.32	12
Copper	mg/kg	4.0	4.6	13
Manganese	mg/kg	5.7	6.6	22

Corn, dent, grain, gr 2 US mn 54 wt, (4)

Ref no 4-02-915

Feed name or analyses		As fed	Dry	C.V. ± %
Dry matter	%	88.0	100.0	2
Ash	%	1.1	1.3	26
Crude fiber	%	2.0	2.3	19
Ether extract	%	4.3	4.9	13
N-free extract	%	71.2	80.9	
Protein (N x 6.25)	%	9.3	10.6	9
Cattle	dig prot %	7.0	8.0	
Sheep	dig prot %	7.3	8.3	
Swine	dig prot %	7.5	8.5	
Energy	GE kcal/kg	3874.	4402.	
Cattle	DE kcal/kg	3569.	4056.	
Sheep	DE kcal/kg	3841.	4365.	
Swine	DE kcal/kg	3569.	4056.	
Cattle	ME kcal/kg	2927.	3326.	
Sheep	ME kcal/kg	3150.	3579.	
Swine	ME kcal/kg	3351.	3808.	
Cattle	NE$_m$ kcal/kg	1998.	2270.	
Cattle	NE$_{gain}$ kcal/kg	1302.	1480.	
Cattle	TDN %	81.	92.	
Sheep	TDN %	87.	99.	
Swine	TDN %	81.	92.	
Calcium	%	.02	.02	28
Magnesium	%	.11	.12	14
Phosphorus	%	.29	.33	10
Potassium	%	.29	.33	16
Copper	mg/kg	4.0	4.6	
Manganese	mg/kg	5.6	6.4	17

Corn, dent, grain, gr 3 US mn 52 wt, (4)

Ref no 4-02-916

Feed name or analyses		As fed	Dry	C.V. ± %
Dry matter	%	85.0	100.0	
Ash	%	1.0	1.2	21
Crude fiber	%	2.0	2.4	
Ether extract	%	3.6	4.2	11
N-free extract	%	69.8	82.1	
Protein (N x 6.25)	%	8.6	10.1	7
Cattle	dig prot %	6.5	7.6	
Sheep	dig prot %	6.7	7.9	
Swine	dig prot %	7.0	8.2	
Energy				
Cattle	DE kcal/kg	3410.	4012.	
Sheep	DE kcal/kg	3673.	4321.	
Swine	DE kcal/kg	3485.	4100.	
Cattle	ME kcal/kg	2796.	3290.	
Sheep	ME kcal/kg	3012.	3543.	
Swine	ME kcal/kg	3276.	3854.	
Cattle	TDN %	77.	91.	
Sheep	TDN %	83.	98.	

Continued

(1) dry forages and roughages
(2) pasture, range plants, and forages fed green

(3) silages
(4) energy feeds

TABLE A-3-1 Composition of Feeds (Continued)

Feed name or analyses		Mean		C.V.
		As fed	Dry	± %
Swine	TDN %	79.	93.	
Calcium	%	.02	.02	63
Magnesium	%	.10	.12	99
Copper	mg/kg	3.6	4.2	
Manganese	mg/kg	5.4	6.4	16

Corn, dent, grain, gr 4 US mn 49 wt, (4)

Ref no 4-02-917

		As fed	Dry	± %
Dry matter	%	83.0	100.0	
Ash	%	1.0	1.2	16
Crude fiber	%	2.0	2.4	
Ether extract	%	3.6	4.3	8
N-free extract	%	68.0	81.9	
Protein (N x 6.25)	%	8.5	10.2	5
Cattle	dig prot %	6.3	7.6	
Sheep	dig prot %	6.6	8.0	
Swine	dig prot %	6.5	7.8	
Energy				
Cattle	DE kcal/kg	3330.	4012.	
Sheep	DE kcal/kg	3586.	4321.	
Swine	DE kcal/kg	3293.	3968.	
Cattle	ME kcal/kg	2731.	3290.	
Sheep	ME kcal/kg	2941.	3543.	
Swine	ME kcal/kg	3170.	3730.	
Cattle	TDN %	76.	91.	
Sheep	TDN %	81.	98.	
Swine	TDN %	75.	90.	
Calcium	%	.02	.03	42
Magnesium	%	.11	.13	28
Phosphorus	%	.26	.31	17
Potassium	%	.25	.30	
Copper	mg/kg	3.1	3.7	
Manganese	mg/kg	4.9	5.9	27

Corn, dent, grain, soft, (4)

Ref no 4-02-919

		As fed	Dry	± %
Dry matter	%	54.0	100.0	15
Ash	%	.4	.8	26
Crude fiber	%	1.0	1.9	41
Ether extract	%	2.5	4.6	12
N-free extract	%	44.2	81.8	
Protein (N x 6.25)	%	5.9	10.9	17
Cattle	dig prot %	4.4	8.2	
Sheep	dig prot %	4.6	8.5	
Swine	dig prot %	4.7	8.7	
Energy				
Cattle	DE kcal/kg	2190.	4056.	
Sheep	DE kcal/kg	2357.	4365.	
Swine	DE kcal/kg	2214.	4100.	
Cattle	ME kcal/kg	1796.	3326.	

Continued

Feed name or analyses		Mean		C.V.
		As fed	Dry	± %
Sheep	ME kcal/kg	1933.	3579.	
Swine	ME kcal/kg	2077.	3846.	
Cattle	TDN %	50.	92.	
Sheep	TDN %	53.	99.	
Swine	TDN %	50.	93.	
Calcium	%	.02	.04	99
Magnesium	%	.07	.13	25
Phosphorus	%	.17	.31	32
Potassium	%	.17	.31	19
Copper	mg/kg	1.1	2.0	44
Manganese	mg/kg	3.1	5.7	45
Folic acid	mg/kg	.10	.20	21
Niacin	mg/kg	13.7	25.3	29
Pantothenic acid	mg/kg	4.2	7.7	12
Riboflavin	mg/kg	.8	1.5	40
Thiamine	mg/kg	2.3	4.2	19

Corn, dent, grain, (4)

Ref no 4-02-920

		As fed	Dry	± %
Dry matter	%	86.0	100.0	8
Ash	%	1.1	1.3	42
Crude fiber	%	2.0	2.3	22
Ether extract	%	3.9	4.5	15
N-free extract	%	70.0	81.4	
Protein (N x 6.25)	%	9.0	10.5	13
Cattle	dig prot %	6.8	7.9	
Sheep	dig prot %	7.0	8.2	
Swine	dig prot %	7.2	8.4	
Energy	GE kcal/kg	3786.	4402.	
Cattle	DE kcal/kg	3450.	4012.	
Sheep	DE kcal/kg	3716.	4321.	
Swine	DE kcal/kg	3488.	4056.	
Cattle	ME kcal/kg	2829.	3290.	
Sheep	ME kcal/kg	3047.	3543.	
Swine	ME kcal/kg	3275.	3808.	
Cattle	TDN %	78.	91.	
Sheep	TDN %	84.	98.	
Swine	TDN %	79.	92.	
Calcium	%	.03	.03	99
Magnesium	%	.12	.14	99
Phosphorus	%	.27	.31	34
Potassium	%	.28	.33	32
Copper	mg/kg	2.1	2.4	56
Manganese	mg/kg	5.1	5.9	99
Zinc	mg/kg	16.9	19.6	
Biotin	mg/kg	.10	.10	
Choline	mg/kg	537.	625.	49
Folic acid	mg/kg	.20	.20	49
Niacin	mg/kg	21.4	24.9	40
Pantothenic acid	mg/kg	5.3	6.2	29
Riboflavin	mg/kg	1.3	1.5	52
Thiamine	mg/kg	4.0	4.6	19
Vitamin B₆	mg/kg	7.20	8.40	

(5) protein supplements
(6) minerals

(7) vitamins
(8) additives

Feed name or analyses		Mean		C.V.
		As fed	Dry	± %

CORN, DENT WHITE. Zea mays indentata

Corn, dent white, grain, (4)

Ref no 4-02-928

Dry matter	%	88.0	100.0	2
Ash	%	1.1	1.2	13
Crude fiber	%	2.0	2.3	23
Ether extract	%	3.7	4.2	15
N-free extract	%	72.6	82.5	
Protein (N x 6.25)	%	8.6	9.8	7
Cattle	dig prot %	6.5	7.4	
Sheep	dig prot %	6.7	7.6	
Swine	dig prot %	6.9	7.8	
Energy				
Cattle	DE kcal/kg	3530.	4012.	
Sheep	DE kcal/kg	3802.	4321.	
Swine	DE kcal/kg	3569.	4056.	
Cattle	ME kcal/kg	2895.	3290.	
Sheep	ME kcal/kg	3118.	3543.	
Swine	ME kcal/kg	3354.	3812.	
Cattle	TDN %	80.	91.	
Sheep	TDN %	86.	98.	
Swine	TDN %	81.	92.	
Calcium	%	.04	.04	99
Phosphorus	%	.27	.31	13
Cobalt	mg/kg	.100	.100	
Copper	mg/kg	5.8	6.6	
Manganese	mg/kg	8.5	9.7	48
Zinc	mg/kg	23.6	26.8	
Biotin	mg/kg	.10	.10	13
Carotene	mg/kg	.4	.4	
Niacin	mg/kg	15.1	17.2	31
Pantothenic acid	mg/kg	3.9	4.4	29
Riboflavin	mg/kg	1.3	1.5	20
Thiamine	mg/kg	4.5	5.1	24
Vitamin A equiv	IU/g	.7	.7	
Arginine	%	.26	.30	
Cystine	%	.09	.10	
Histidine	%	.18	.20	
Isoleucine	%	.44	.50	
Leucine	%	.88	1.00	
Lysine	%	.26	.30	
Methionine	%	.09	.10	
Phenylalanine	%	.35	.40	
Threonine	%	.35	.40	
Tryptophan	%	.09	.10	
Tyrosine	%	.44	.50	
Valine	%	.35	.40	

CORN, DENT YELLOW. Zea mays indentata

Corn, dent yellow, grain, Can 2 CW mn 54 wt mx 3 fm, (4)

Ref no 4-02-955

Dry matter	%	86.5	100.0
Ash	%	1.2	1.4
Crude fiber	%	2.3	2.7
Ether extract	%	3.8	4.4
N-free extract	%	68.9	79.7
Protein (N x 6.25)	%	10.2	11.8
Cattle	dig prot %	7.6	8.8
Sheep	dig prot %	8.0	9.2
Swine	dig prot %	8.1	9.4
Energy	GE kcal/kg	4804.	5553.
Cattle	DE kcal/kg	3470.	4012.
Sheep	DE kcal/kg	3738.	4321.
Swine	DE kcal/kg	3508.	4056.
Cattle	ME kcal/kg	2846.	3290.
Sheep	ME kcal/kg	3065.	3543.
Swine	ME kcal/kg	3284.	3796.
Cattle	TDN %	79.	91.
Sheep	TDN %	85.	98.
Swine	TDN %	80.	92.

Corn, dent yellow, grain, Can 3 CW mn 52 wt mx 5 fm, (4)

Ref no 4-02-956

Dry matter	%	86.5	100.0
Ash	%	1.6	1.9
Crude fiber	%	2.3	2.7
Ether extract	%	3.9	4.5
N-free extract	%	68.0	78.6
Protein (N x 6.25)	%	10.6	12.3
Cattle	dig prot %	8.0	9.2
Sheep	dig prot %	8.3	9.6
Swine	dig prot %	8.5	9.8
Energy	GE kcal/kg	4804.	5553.
Cattle	DE kcal/kg	3432.	3968.
Sheep	DE kcal/kg	3700.	4277.
Swine	DE kcal/kg	3470.	4012.
Cattle	ME kcal/kg	2815.	3254.
Sheep	ME kcal/kg	3034.	3507.
Swine	ME kcal/kg	3245.	3751.
Cattle	TDN %	78.	90.
Sheep	TDN %	84.	97.
Swine	TDN %	79.	91.

(1) dry forages and roughages
(2) pasture, range plants, and forages fed green
(3) silages
(4) energy feeds

Feed name or analyses		Mean		C.V. ± %
		As fed	Dry	

Corn, dent yellow, grain, gr 1 US mn 56 wt, (4)

Ref no 4-02-930

Feed name or analyses		As fed	Dry	C.V. ± %
Dry matter	%	87.0	100.0	1
Ash	%	1.4	1.6	58
Crude fiber	%	2.0	2.3	35
Ether extract	%	3.8	4.4	8
N-free extract	%	70.9	81.5	
Protein (N x 6.25)	%	8.9	10.2	5
Cattle	dig prot %	6.6	7.6	
Sheep	dig prot %	7.0	8.0	
Swine	dig prot %	7.1	8.2	
Energy				
Cattle	DE kcal/kg	3490.	4012.	
Sheep	DE kcal/kg	3759.	4321.	
Swine	DE kcal/kg	3529.	4056.	
Cattle	ME kcal/kg	2862.	3290.	
Sheep	ME kcal/kg	3082.	3543.	
Swine	ME kcal/kg	3317.	3813.	
Cattle	TDN %	79.	91.	
Sheep	TDN %	85.	98.	
Swine	TDN %	80.	92.	
Iron	%	.002	.002	
Copper	mg/kg	4.2	4.8	
Manganese	mg/kg	5.6	6.4	

Corn, dent yellow, grain, gr 2 US mn 54 wt, (4)

Ref no 4-02-931

Feed name or analyses		As fed	Dry	C.V. ± %
Dry matter	%	89.0	100.0	2
Ash	%	1.1	1.2	25
Crude fiber	%	2.0	2.2	30
Ether extract	%	3.9	4.4	9
N-free extract	%	73.1	82.2	
Protein (N x 6.25)	%	8.9	10.0	7
Cattle	dig prot %	6.7	7.5	
Sheep	dig prot %	6.9	7.8	
Swine	dig prot %	7.1	8.0	
Energy	GE kcal/kg	3918.	4402.	
Cattle	DE kcal/kg	3571.	4012.	
Sheep	DE kcal/kg	3846.	4321.	
Swine	DE kcal/kg	3610.	4056.	
Cattle	ME kcal/kg	2928.	3290.	
Sheep	ME kcal/kg	3153.	3543.	
Swine	ME kcal/kg	3394.	3813.	
Cattle	TDN %	81.	91.	
Sheep	TDN %	87.	98.	
Swine	TDN %	82.	92.	
Calcium	%	.02	.02	
Phosphorus	%	.31	.35	
Carotene	mg/kg	1.8	2.0	
Niacin	mg/kg	26.3	29.5	

Continued

Feed name or analyses		As fed	Dry	C.V. ± %
Pantothenic acid	mg/kg	3.9	4.4	
Riboflavin	mg/kg	1.3	1.5	
Thiamine	mg/kg	3.6	4.0	
Vitamin A equiv	IU/g	3.0	3.3	
Arginine	%	.45	.51	37
Cystine	%	.09	.10	
Histidine	%	.18	.20	40
Isoleucine	%	.45	.51	
Leucine	%	.99	1.11	62
Lysine	%	.18	.20	37
Methionine	%	.09	.10	
Phenylalanine	%	.45	.51	
Threonine	%	.36	.40	21
Tryptophan	%	.09	.10	
Valine	%	.36	.40	

Corn, dent yellow, grain, gr 3 US mn 52 wt, (4)

Ref no 4-02-932

Feed name or analyses		As fed	Dry	C.V. ± %
Dry matter	%	86.0	100.0	
Ash	%	1.0	1.2	14
Crude fiber	%	2.0	2.3	
Ether extract	%	3.7	4.3	8
N-free extract	%	70.6	82.1	
Protein (N x 6.25)	%	8.7	10.1	4
Cattle	dig prot %	6.5	7.6	
Sheep	dig prot %	6.8	7.9	
Swine	dig prot %	7.0	8.2	
Energy				
Cattle	DE kcal/kg	3450.	4012.	
Sheep	DE kcal/kg	3716.	4321.	
Swine	DE kcal/kg	3526.	4100.	
Cattle	ME kcal/kg	2829.	3290.	
Sheep	ME kcal/kg	3047.	3543.	
Swine	ME kcal/kg	3314.	3854.	
Cattle	TDN %	78.	91.	
Sheep	TDN %	84.	98.	
Swine	TDN %	80.	93.	
Calcium	%	.02	.02	
Iron	%	.002	.002	
Phosphorus	%	.25	.29	
Manganese	mg/kg	5.5	6.4	

Corn, dent yellow, grain, gr 4 US mn 49 wt, (4)

Ref no 4-02-933

Feed name or analyses		As fed	Dry	C.V. ± %
Dry matter	%	87.0	100.0	
Ash	%	1.0	1.2	16
Crude fiber	%	2.0	2.3	
Ether extract	%	3.8	4.4	7
N-free extract	%	71.2	81.9	
Protein (N x 6.25)	%	8.9	10.2	6

Continued

Feed name or analyses		Mean		C.V.
		As fed	Dry	± %
Cattle	dig prot %	6.6	7.6	
Sheep	dig prot %	7.0	8.0	
Swine	dig prot %	7.1	8.2	
Energy				
Cattle	DE kcal/kg	3490.	4012.	
Sheep	DE kcal/kg	3759.	4321.	
Swine	DE kcal/kg	3529.	4056.	
Cattle	ME kcal/kg	2862.	3290.	
Sheep	ME kcal/kg	3082.	3543.	
Swine	ME kcal/kg	3313.	3808.	
Cattle	TDN %	79.	91.	
Sheep	TDN %	85.	98.	
Swine	TDN %	80.	92.	
Calcium	%	.04	.05	
Magnesium	%	.17	.20	
Phosphorus	%	.30	.34	

Corn, dent yellow, grain, gr 5 US mn 46 wt, (4)

Ref no 4-02-934

Dry matter	%	79.0	100.0	
Ash	%	.9	1.1	26
Crude fiber	%	2.0	2.5	
Ether extract	%	3.3	4.2	11
N-free extract	%	64.9	82.2	
Protein (N x 6.25)	%	7.9	10.0	8
Cattle	dig prot %	5.9	7.5	
Sheep	dig prot %	6.2	7.8	
Swine	dig prot %	6.1	7.7	
Energy				
Cattle	DE kcal/kg	3169.	4012.	
Sheep	DE kcal/kg	3414.	4321.	
Swine	DE kcal/kg	3135.	3968.	
Cattle	ME kcal/kg	2599.	3290.	
Sheep	ME kcal/kg	2799.	3543.	
Swine	ME kcal/kg	2947.	3730.	
Cattle	TDN %	72.	91.	
Sheep	TDN %	77.	98.	
Swine	TDN %	71.	90.	

Corn, dent yellow, grain, grnd cooked, (4)

Ref no 4-07-953

Dry matter	%	88.0	100.0	
Crude fiber	%	2.1	2.4	
Ether extract	%	4.0	4.5	
Protein (N x 6.25)	%	9.2	10.5	
Calcium	%	.02	.02	
Phosphorus	%	.26	.30	

Corn, dent yellow, grain, (4)

Ref no 4-02-935

Feed name or analyses		Mean		C.V.
		As fed	Dry	± %
Dry matter	%	86.0	100.0	3
Ash	%	1.1	1.3	45
Crude fiber	%	2.0	2.3	23
Ether extract	%	3.8	4.4	99
N-free extract	%	70.3	81.8	
Protein (N x 6.25)	%	8.8	10.2	9
Cattle	dig prot %	6.5	7.6	
Sheep	dig prot %	6.9	8.0	
Swine	dig prot %	7.0	8.2	
Energy	GE kcal/kg	3786.	4402.	
Cattle	DE kcal/kg	3450.	4012.	
Sheep	DE kcal/kg	3716.	4321.	
Swine	DE kcal/kg	3488.	4056.	
Cattle	ME kcal/kg	2829.	3290.	
Chickens	MEn kcal/kg	3417.	3973.	
Sheep	ME kcal/kg	3047.	3543.	
Swine	ME kcal/kg	3275.	3808.	
Cattle	TDN %	78.	91.	
Sheep	TDN %	84.	98.	
Swine	TDN %	79.	92.	
Calcium	%	.03	.03	65
Chlorine	%	.03	.04	25
Iron	%	.003	.003	67
Magnesium	%	.15	.17	99
Phosphorus	%	.27	.31	27
Potassium	%	.33	.38	29
Sodium	%	.01	.01	99
Sulfur	%	.12	.14	14
Cobalt	mg/kg	.100	.100	60
Copper	mg/kg	3.4	4.0	48
Manganese	mg/kg	4.1	4.8	99
Zinc	mg/kg	10.4	12.1	
Biotin	mg/kg	.06	.07	21
Carotene	mg/kg	4.1	4.8	52
Choline	mg/kg	537.	625.	48
Folic acid	mg/kg	.20	.20	99
Niacin	mg/kg	22.9	26.6	41
Pantothenic acid	mg/kg	5.0	5.9	32
Riboflavin	mg/kg	1.1	1.3	69
Thiamine	mg/kg	4.0	4.6	16
Vitamin B6	mg/kg	7.20	8.40	
Vitamin A equiv	IU/g	6.8	8.0	
Cystine	%	.09	.10	42
Glycine	%	.43	.50	
Methionine	%	.17	.20	50

(1) dry forages and roughages

(2) pasture, range plants, and forages fed green

(3) silages

(4) energy feeds

Feed name or analyses		Mean		C.V.
		As fed	Dry	± %

CORN, FLINT. Zea mays indurata

Corn, flint, grain, (4)

Ref no 4-02-948

		As fed	Dry	± %
Dry matter	%	89.0	100.0	3
Ash	%	1.3	1.5	14
Crude fiber	%	2.0	2.3	25
Ether extract	%	4.3	4.8	19
N-free extract	%	71.5	80.3	
Protein (N x 6.25)	%	9.9	11.1	15
Cattle	dig prot %	7.4	8.3	
Sheep	dig prot %	8.4	9.5	
Swine	dig prot %	7.9	8.9	
Energy				
Cattle	DE kcal/kg	3571.	4012.	
Sheep	DE kcal/kg	3924.	4409.	
Swine	DE kcal/kg	3610.	4056.	
Cattle	ME kcal/kg	2928.	3290.	
Sheep	ME kcal/kg	3217.	3615.	
Swine	ME kcal/kg	3386.	3804.	
Cattle	TDN %	81.	91.	
Sheep	TDN %	89.	100.	
Swine	TDN %	82.	92.	
Iron	%	.003	.003	
Phosphorus	%	.21	.24	
Copper	mg/kg	11.6	13.0	
Manganese	mg/kg	7.0	7.9	
Niacin	mg/kg	15.8	17.8	
Lysine	%	.27	.30	
Methionine	%	.18	.20	
Tryptophan	%	.09	.10	

CORN, POP. Zea mays everta

Corn, pop, grain, (4)

Ref no 4-02-964

		As fed	Dry	± %
Dry matter	%	90.0	100.0	1
Ash	%	1.3	1.4	21
Crude fiber	%	2.0	2.2	21
Ether extract	%	5.0	5.6	8
N-free extract	%	70.8	78.7	
Protein (N x 6.25)	%	10.9	12.1	10
Cattle	dig prot %	8.2	9.1	
Sheep	dig prot %	8.5	9.4	
Swine	dig prot %	8.7	9.7	
Energy				
Cattle	DE kcal/kg	3650.	4056.	
Sheep	DE kcal/kg	3928.	4365.	

		As fed	Dry	± %
Swine	DE kcal/kg	3690.	4100.	
Cattle	ME kcal/kg	2993.	3326.	
Sheep	ME kcal/kg	3221.	3579.	
Swine	ME kcal/kg	3451.	3834.	
Cattle	TDN %	83.	92.	
Sheep	TDN %	89.	99.	
Swine	TDN %	84.	93.	
Phosphorus	%	.31	.34	11
Copper	mg/kg	2.3	2.6	43
Biotin	mg/kg	.08	.09	
Niacin	mg/kg	17.2	19.1	9
Pantothenic acid	mg/kg	3.3	3.7	19
Riboflavin	mg/kg	1.2	1.3	38
Vitamin B_6	mg/kg	4.3	4.8	

Corn stover - see Corn, aerial pt wo ears wo husks

CORN, SWEET. Zea mays saccharata

Corn, sweet, cannery res, fresh, (2)
Corn, sweet, cannery refuse

Ref no 2-02-975

		As fed	Dry	± %
Dry matter	%	77.0	100.0	
Ash	%	2.5	3.2	
Crude fiber	%	17.0	22.1	
Ether extract	%	1.8	2.3	
N-free extract	%	49.0	63.6	
Protein (N x 6.25)	%	6.8	8.8	
Cattle	dig prot %	3.8	4.9	
Sheep	dig prot %	4.1	5.3	
Energy				
Cattle	DE kcal/kg	2376.	3086.	
Sheep	DE kcal/kg	2445.	3175.	
Cattle	ME kcal/kg	1948.	2530.	
Sheep	ME kcal/kg	2005.	2604.	
Cattle	TDN %	54.	70.	
Sheep	TDN %	55.	72.	
Phosphorus	%	.69	.90	

Corn, sweet, cannery res, ensiled, (3)

Ref no 3-07-955

		As fed	Dry	± %
Dry matter	%	29.4	100.0	
Ash	%	1.4	4.9	
Crude fiber	%	7.9	26.8	
Ether extract	%	1.5	5.1	
N-free extract	%	16.0	54.4	
Protein (N x 6.25)	%	2.6	8.8	
Cattle	dig prot %	1.4	4.9	

Continued

(5) protein supplements
(6) minerals

(7) vitamins
(8) additives

Continued

Feed name or analyses		Mean As fed	Mean Dry	C.V. ± %
Sheep	dig prot %	1.4	4.8	
Energy				
Cattle	DE kcal/kg	933.	3175.	
Sheep	DE kcal/kg	933.	3175.	
Cattle	ME kcal/kg	766.	2604.	
Sheep	ME kcal/kg	766.	2604.	
Cattle	TDN %	21.	72.	
Sheep	TDN %	21.	72.	

Corn, sweet, grain, (4)

Ref no 4-02-977

		As fed	Dry	C.V.
Dry matter	%	91.0	100.0	1
Ash	%	1.7	1.9	18
Crude fiber	%	3.0	3.3	9
Ether extract	%	7.9	8.7	25
N-free extract	%	66.8	73.4	
Protein (N x 6.25)	%	11.6	12.7	13
Cattle	dig prot %	8.6	9.5	
Sheep	dig prot %	9.0	9.9	
Swine	dig prot %	9.3	10.2	
Energy				
Cattle	DE kcal/kg	3772.	4145.	
Sheep	DE kcal/kg	4052.	4453.	
Swine	DE kcal/kg	3772.	4145.	
Cattle	ME kcal/kg	3093.	3399.	
Sheep	ME kcal/kg	3322.	3651.	
Swine	ME kcal/kg	3523.	3871.	
Cattle	TDN %	86.	94.	
Sheep	TDN %	92.	101.	
Swine	TDN %	86.	94.	
Calcium	%	.01	.01	
Iron	%	.01	.01	
Phosphorus	%	.41	.45	
Copper	mg/kg	5.4	5.9	
Niacin	mg/kg	46.2	50.8	37
Pantothenic acid	mg/kg	15.0	16.5	58
Riboflavin	mg/kg	1.8	2.0	

CORN, WHITE. Zea mays

Corn, white, grain, grnd, (4)
Corn white, meal

Ref no 4-02-979

		As fed	Dry	
Dry matter	%	88.0	100.0	
Ash	%	.6	.7	
Crude fiber	%	1.0	1.1	20
Ether extract	%	1.2	1.4	24
N-free extract	%	76.6	87.0	
Protein (N x 6.25)	%	8.6	9.8	8

Continued

Feed name or analyses		Mean As fed	Mean Dry	C.V. ± %
Cattle	dig prot %	6.5	7.4	
Sheep	dig prot %	6.7	7.6	
Swine	dig prot %	6.9	7.8	
Energy				
Cattle	DE kcal/kg	3492.	3968.	
Sheep	DE kcal/kg	3764.	4277.	
Swine	DE kcal/kg	3530.	4012.	
Cattle	ME kcal/kg	2864.	3254.	
Sheep	ME kcal/kg	3086.	3507.	
Swine	ME kcal/kg	3318.	3771.	
Cattle	TDN %	79.	90.	
Sheep	TDN %	85.	97.	
Swine	TDN %	80.	91.	
Calcium	%	.02	.02	
Phosphorus	%	.16	.18	
Niacin	mg/kg	17.6	20.0	
Riboflavin	mg/kg	.9	1.0	
Thiamine	mg/kg	1.5	1.7	
Tyrosine	%	.40	.45	

Corn, white, grain, Can 1 CE mn 56 wt mx 2 fm, (4)

Ref no 4-02-980

		As fed	Dry	
Dry matter	%	86.5	100.0	
Ash	%	1.1	1.3	
Crude fiber	%	2.4	2.8	
Ether extract	%	3.8	4.4	
N-free extract	%	71.4	82.5	
Protein (N x 6.25)	%	7.8	9.0	
Cattle	dig prot %	5.9	6.8	
Sheep	dig prot %	6.0	7.0	
Swine	dig prot %	6.2	7.2	
Energy	GE kcal/kg	4804.	5553.	
Cattle	DE kcal/kg	3470.	4012.	
Sheep	DE kcal/kg	3738.	4321.	
Swine	DE kcal/kg	3508.	4056.	
Cattle	ME kcal/kg	2846.	3290.	
Sheep	ME kcal/kg	3065.	3543.	
Swine	ME kcal/kg	3302.	3817.	
Cattle	TDN %	79.	91.	
Sheep	TDN %	85.	98.	
Swine	TDN %	80.	92.	

Corn, white, grain, Can 2 CE mn 54 wt mx 3 fm, (4)

Ref no 4-02-981

		As fed	Dry	
Dry matter	%	86.5	100.0	
Ash	%	1.2	1.4	17
Crude fiber	%	2.8	3.2	43
Ether extract	%	3.7	4.3	
N-free extract	%	70.7	81.7	
Protein (N x 6.25)	%	8.2	9.4	4

Continued

(1) dry forages and roughages
(2) pasture, range plants, and forages fed green
(3) silages
(4) energy feeds

Feed name or analyses		Mean		C.V.
		As fed	Dry	± %
Cattle	dig prot %	6.0	7.0	
Sheep	dig prot %	6.3	7.3	
Swine	dig prot %	6.5	7.5	
Energy	GE kcal/kg	4804.	5553.	
Cattle	DE kcal/kg	3470.	4012.	
Sheep	DE kcal/kg	3738.	4321.	
Swine	DE kcal/kg	3508.	4056.	
Cattle	ME kcal/kg	2846.	3290.	
Sheep	ME kcal/kg	3065.	3543.	
Swine	ME kcal/kg	3302.	3817.	
Cattle	TDN %	79.	91.	
Sheep	TDN %	85.	98.	
Swine	TDN %	80.	92.	

Corn, white, grain wo germ, grnd, (4)
Degermed white corn meal

Ref no 4-02-988

Feed name or analyses		As fed	Dry	C.V. ± %
Dry matter	%	89.0	100.0	
Ash	%	.4	.4	
Crude fiber	%	1.0	1.1	
Ether extract	%	1.1	1.2	
N-free extract	%	77.4	87.0	
Protein (N x 6.25)	%	9.2	10.3	
Cattle	dig prot %	6.8	7.7	
Sheep	dig prot %	7.1	8.0	
Energy				
Cattle	DE kcal/kg	3492.	3924.	
Sheep	DE kcal/kg	3806.	4277.	
Cattle	ME kcal/kg	2864.	3218.	
Sheep	ME kcal/kg	3121.	3507.	
Cattle	TDN %	79.	89.	
Sheep	TDN %	86.	97.	
Calcium	%	.01	.01	
Phosphorus	%	.17	.19	
Niacin	mg/kg	10.3	11.6	
Pantothenic acid	mg/kg	3.1	3.5	
Riboflavin	mg/kg	.7	.8	
Thiamine	mg/kg	1.1	1.2	
Arginine	%	.40	.45	
Histidine	%	.20	.22	
Isoleucine	%	.30	.34	
Leucine	%	1.00	1.12	
Lysine	%	.30	.34	
Methionine	%	.20	.22	
Phenylalanine	%	.40	.45	
Threonine	%	.30	.34	
Tryptophan	%	.10	.11	
Valine	%	.50	.56	

Corn, white, grits by-prod, mn 5 fat, (4)
White hominy feed (AAFCO)
White hominy feed (CFA)
Hominy, white corn, feed
Corn, white, hominy feed

Ref no 4-02-990

Feed name or analyses		As fed	Dry	C.V. ± %
Dry matter	%	89.9	100.0	
Crude fiber	%	4.7	5.2	
Ether extract	%	5.7	6.3	
Protein (N x 6.25)	%	10.8	12.0	
Calcium	%	.05	.06	
Phosphorus	%	1.00	1.10	
Niacin	mg/kg	55.3	61.5	
Pantothenic acid	mg/kg	6.7	7.5	
Riboflavin	mg/kg	2.2	2.4	
Thiamine	mg/kg	13.1	14.6	

CORN, YELLOW. Zea mays

Corn, yellow, grain, grnd, (4)

Ref no 4-02-992

Feed name or analyses		As fed	Dry	C.V. ± %
Dry matter	%	88.0	100.0	1
Ash	%	1.9	2.2	49
Crude fiber	%	2.0	2.3	51
Ether extract	%	3.4	3.9	51
N-free extract	%	71.7	81.5	
Protein (N x 6.25)	%	8.9	10.1	9
Cattle	dig prot %	6.7	7.6	
Sheep	dig prot %	7.0	7.9	
Swine	dig prot %	7.2	8.1	
Energy				
Cattle	DE kcal/kg	3530.	4012.	
Sheep	DE kcal/kg	3802.	4321.	
Swine	DE kcal/kg	3530.	4012.	
Cattle	ME kcal/kg	2895.	3290.	
Sheep	ME kcal/kg	3118.	3543.	
Swine	ME kcal/kg	3318.	3771.	
Cattle	TDN %	80.	91.	
Sheep	TDN %	86.	98.	
Swine	TDN %	80.	91.	
Choline	mg/kg	592.	672.	
Niacin	mg/kg	10.3	11.7	
Pantothenic acid	mg/kg	6.8	7.7	
Riboflavin	mg/kg	.7	.8	53
Thiamine	mg/kg	1.5	1.7	

(5) protein supplements
(6) minerals

(7) vitamins
(8) additives

Feed name or analyses		Mean		C.V.
		As fed	Dry	± %

Corn, yellow, grain, Can 1 CE mn 56 wt mx 2 fm, (4)

Ref no 4-02-993

		As fed	Dry	± %
Dry matter	%	86.5	100.0	
Ash	%	1.2	1.4	13
Crude fiber	%	2.6	3.0	64
Ether extract	%	4.0	4.6	20
N-free extract	%	70.3	81.3	
Protein (N x 6.25)	%	8.4	9.7	6
Cattle	dig prot %	6.3	7.3	
Sheep	dig prot %	6.6	7.6	
Swine	dig prot %	6.7	7.8	
Energy	GE kcal/kg	4804.	5553.	
Cattle	DE kcal/kg	3470.	4012.	
Sheep	DE kcal/kg	3738.	4321.	
Swine	DE kcal/kg	3508.	4056.	
Cattle	ME kcal/kg	2846.	3290.	
Sheep	ME kcal/kg	3065.	3543.	
Swine	ME kcal/kg	3298.	3813.	
Cattle	TDN %	79.	91.	
Sheep	TDN %	85.	98.	
Swine	TDN %	80.	92.	
Alanine	%	.62	.72	
Arginine	%	.34	.39	
Aspartic acid	%	.56	.65	
Glutamic acid	%	1.51	1.75	
Glycine	%	.32	.37	
Histidine	%	.19	.22	
Isoleucine	%	.19	.22	
Leucine	%	.94	1.09	
Lysine	%	.22	.26	
Methionine	%	.14	.16	
Phenylalanine	%	.36	.42	
Proline	%	.71	.82	
Serine	%	.42	.48	
Threonine	%	.29	.34	
Tyrosine	%	.28	.33	
Valine	%	.28	.33	

Corn, yellow, grain, Can 1 CW mn 56 wt mx 2 fm, (4)

Ref no 4-02-994

		As fed	Dry	± %
Dry matter	%	86.5	100.0	
Ash	%	1.2	1.4	10
Crude fiber	%	3.1	3.6	51
Ether extract	%	3.8	4.4	13
N-free extract	%	69.0	79.8	
Protein (N x 6.25)	%	9.3	10.8	2
Cattle	dig prot %	7.0	8.1	
Sheep	dig prot %	7.3	8.4	

Continued

Feed name or analyses		Mean		C.V.
		As fed	Dry	± %
Swine	dig prot %	7.4	8.6	
Energy	GE kcal/kg	4804.	5553.	
Cattle	DE kcal/kg	3432.	3968.	
Sheep	DE kcal/kg	3700.	4277.	
Swine	DE kcal/kg	3470.	4012.	
Cattle	ME kcal/kg	2815.	3254.	
Sheep	ME kcal/kg	3034.	3507.	
Swine	ME kcal/kg	3255.	3763.	
Cattle	TDN %	78.	90.	
Sheep	TDN %	84.	97.	
Swine	TDN %	79.	91.	

Corn, yellow, grain, Can 2 CE mn 54 wt mx 3 fm, (4)

Ref no 4-02-995

		As fed	Dry	± %
Dry matter	%	86.5	100.0	
Ash	%	1.2	1.4	9
Crude fiber	%	2.7	3.1	71
Ether extract	%	4.1	4.7	19
N-free extract	%	70.1	81.0	
Protein (N x 6.25)	%	8.4	9.8	3
Cattle	dig prot %	6.4	7.4	
Sheep	dig prot %	6.6	7.6	
Swine	dig prot %	6.8	7.9	
Energy	GE kcal/kg	4804.	5553.	
Cattle	DE kcal/kg	3470.	4012.	
Sheep	DE kcal/kg	3700.	4277.	
Swine	DE kcal/kg	3508.	4056.	
Cattle	ME kcal/kg	2846.	3290.	
Sheep	ME kcal/kg	3034.	3507.	
Swine	ME kcal/kg	3298.	3813.	
Cattle	TDN %	79.	91.	
Sheep	TDN %	84.	97.	
Swine	TDN %	80.	92.	
Alanine	%	.61	.71	
Arginine	%	.32	.37	
Aspartic acid	%	.54	.62	
Glutamic acid	%	1.54	1.78	
Glycine	%	.29	.34	
Histidine	%	.19	.22	
Isoleucine	%	.18	.21	
Leucine	%	.94	1.09	
Lysine	%	.21	.24	
Methionine	%	.13	.15	
Phenylalanine	%	.36	.42	
Proline	%	.69	.80	
Serine	%	.41	.47	
Threonine	%	.28	.32	
Tyrosine	%	.28	.32	
Valine	%	.27	.31	

(1) dry forages and roughages

(2) pasture, range plants, and forages fed green

(3) silages

(4) energy feeds

Feed name or analyses		Mean		C.V.
		As fed	Dry	± %

**Corn, yellow, grain, Can 2 CW mn 54 wt
mx 3 fm, (4)**

Ref no 4-02-996

Dry matter	%	86.5	100.0	
Ash	%	1.2	1.4	9
Crude fiber	%	2.9	3.3	67
Ether extract	%	4.0	4.6	17
N-free extract	%	69.2	80.0	
Protein (N x 6.25)	%	9.3	10.7	7
Cattle	dig prot %	6.9	8.0	
Sheep	dig prot %	7.2	8.3	
Swine	dig prot %	7.3	8.5	
Energy	GE kcal/kg	4804.	5553.	
Cattle	DE kcal/kg	3432.	3968.	
Sheep	DE kcal/kg	3700.	4277.	
Swine	DE kcal/kg	3470.	4012.	
Cattle	ME kcal/kg	2815.	3254.	
Sheep	ME kcal/kg	3034.	3507.	
Swine	ME kcal/kg	3255.	3763.	
Cattle	TDN %	78.	90.	
Sheep	TDN %	84.	97.	
Swine	TDN %	79.	91.	

**Corn, yellow, grain, Can 3 CE mn 52 wt
mx 5 fm, (4)**

Ref no 4-02-997

Dry matter	%	86.5	100.0	
Ash	%	1.3	1.5	22
Crude fiber	%	2.7	3.1	56
Ether extract	%	4.3	5.0	26
N-free extract	%	69.8	80.7	
Protein (N x 6.25)	%	8.4	9.7	6
Cattle	dig prot %	6.3	7.3	
Sheep	dig prot %	6.6	7.6	
Swine	dig prot %	6.7	7.8	
Energy	GE kcal/kg	4804.	5553.	
Cattle	DE kcal/kg	3470.	4012.	
Sheep	DE kcal/kg	3738.	4321.	
Swine	DE kcal/kg	3508.	4056.	
Cattle	ME kcal/kg	2846.	3290.	
Sheep	ME kcal/kg	3065.	3543.	
Swine	ME kcal/kg	3298.	3813.	
Cattle	TDN %	79.	91.	
Sheep	TDN %	85.	98.	
Swine	TDN %	80.	92.	
Alanine	%	.60	.70	
Arginine	%	.32	.37	
Aspartic acid	%	.54	.62	
Glutamic acid	%	1.52	1.76	
Glycine	%	.30	.35	

Feed name or analyses		Mean		C.V.
		As fed	Dry	± %

Histidine	%	.20	.23	
Isoleucine	%	.18	.21	
Leucine	%	.93	1.08	
Lysine	%	.22	.25	
Methionine	%	.13	.15	
Phenylalanine	%	.35	.41	
Proline	%	.69	.80	
Serine	%	.42	.48	
Threonine	%	.27	.31	
Tyrosine	%	.27	.31	
Valine	%	.28	.32	

**Corn, yellow, grain, Can 3 CW mn 52 wt
mx 5 fm, (4)**

Ref no 4-02-998

Dry matter	%	86.5	100.0	
Ash	%	1.3	1.5	13
Crude fiber	%	2.9	3.3	61
Ether extract	%	3.9	4.5	22
N-free extract	%	68.9	79.7	
Protein (N x 6.25)	%	9.6	11.0	7
Cattle	dig prot %	7.1	8.2	
Sheep	dig prot %	7.4	8.6	
Swine	dig prot %	7.6	8.8	
Energy	GE kcal/kg	4804.	5553.	
Cattle	DE kcal/kg	3432.	3968.	
Sheep	DE kcal/kg	3700.	4277.	
Swine	DE kcal/kg	3470.	4012.	
Cattle	ME kcal/kg	2815.	3254.	
Sheep	ME kcal/kg	3034.	3507.	
Swine	ME kcal/kg	3255.	3763.	
Cattle	TDN %	78.	90.	
Sheep	TDN %	84.	97.	
Swine	TDN %	79.	91.	

**Corn, yellow, grain, Can 4 CE mn 50 wt
mx 7 fm, (4)**

Ref no 4-02-999

Dry matter	%	86.5	100.0	
Ash	%	1.2	1.4	
Crude fiber	%	2.6	3.0	54
Ether extract	%	3.6	4.2	5
N-free extract	%	70.7	81.7	
Protein (N x 6.25)	%	8.4	9.7	4
Cattle	dig prot %	6.3	7.3	
Sheep	dig prot %	6.6	7.6	
Swine	dig prot %	6.7	7.8	
Energy	GE kcal/kg	4804.	5553.	
Cattle	DE kcal/kg	3470.	4012.	
Sheep	DE kcal/kg	3738.	4321.	

Continued   *Continued*

(5) protein supplements

(6) minerals

(7) vitamins

(8) additives

Feed name or analyses		Mean		C.V. ± %
		As fed	Dry	
Swine	DE kcal/kg	3508.	4056.	
Cattle	ME kcal/kg	2846.	3290.	
Sheep	ME kcal/kg	3065.	3543.	
Swine	ME kcal/kg	3298.	3813.	
Cattle	TDN %	79.	91.	
Sheep	TDN %	85.	98.	
Swine	TDN %	80.	92.	
Alanine	%	.55	.64	
Arginine	%	.30	.35	
Aspartic acid	%	.51	.59	
Glutamic acid	%	1.41	1.63	
Glycine	%	.28	.33	
Histidine	%	.19	.22	
Isoleucine	%	.15	.17	
Leucine	%	.84	.97	
Lysine	%	.21	.24	
Methionine	%	.12	.14	
Phenylalanine	%	.34	.39	
Proline	%	.64	.74	
Serine	%	.40	.46	
Threonine	%	.27	.31	
Tyrosine	%	.26	.30	
Valine	%	.25	.29	

Corn, yellow, grain, Can 5 CE mn 47 wt mx 12 fm, (4)

Ref no 4-03-001

Feed name or analyses		Mean		C.V. ± %
		As fed	Dry	
Dry matter	%	86.5	100.0	
Ash	%	1.3	1.4	25
Crude fiber	%	3.3	3.8	59
Ether extract	%	3.7	4.3	5
N-free extract	%	69.5	80.4	
Protein (N x 6.25)	%	8.7	10.1	10
Cattle	dig prot %	6.6	7.6	
Sheep	dig prot %	6.8	7.9	
Swine	dig prot %	7.0	8.1	
Energy	GE kcal/kg	4804.	5553.	
Cattle	DE kcal/kg	3432.	3968.	
Sheep	DE kcal/kg	3700.	4277.	
Swine	DE kcal/kg	3470.	4012.	
Cattle	ME kcal/kg	2815.	3254.	
Sheep	ME kcal/kg	3034.	3507.	
Swine	ME kcal/kg	3262.	3771.	
Cattle	TDN %	78.	90.	
Sheep	TDN %	84.	97.	
Swine	TDN %	79.	91.	

Corn, yellow, grain, Can 4 CW mn 50 wt mx 7 fm, (4)

Ref no 4-03-000

Feed name or analyses		Mean		C.V. ± %
		As fed	Dry	
Dry matter	%	86.5	100.0	
Ash	%	1.4	1.6	8
Crude fiber	%	2.6	3.0	70
Ether extract	%	3.9	4.5	18
N-free extract	%	68.9	79.6	
Protein (N x 6.25)	%	9.8	11.3	10
Cattle	dig prot %	7.4	8.5	
Sheep	dig prot %	7.6	8.8	
Swine	dig prot %	7.8	9.0	
Energy	GE kcal/kg	4804.	5553.	
Cattle	DE kcal/kg	3432.	3968.	
Sheep	DE kcal/kg	3700.	4277.	
Swine	DE kcal/kg	3470.	4012.	
Cattle	ME kcal/kg	2815.	3254.	
Sheep	ME kcal/kg	3034.	3507.	
Swine	ME kcal/kg	3252.	3759.	
Cattle	TDN %	78.	90.	
Sheep	TDN %	84.	97.	
Swine	TDN %	79.	91.	

Corn, yellow, grain, Can 5 CW mn 47 wt mx 12 fm, (4)

Ref no 4-03-002

Feed name or analyses		Mean		C.V. ± %
		As fed	Dry	
Dry matter	%	86.5	100.0	
Ash	%	1.4	1.7	19
Crude fiber	%	3.2	3.7	46
Ether extract	%	3.4	4.0	4
N-free extract	%	68.5	79.2	
Protein (N x 6.25)	%	9.9	11.4	11
Cattle	dig prot %	7.4	8.6	
Sheep	dig prot %	7.7	8.9	
Swine	dig prot %	7.9	9.1	
Energy	GE kcal/kg	4804.	5553.	
Cattle	DE kcal/kg	3432.	3968.	
Sheep	DE kcal/kg	3700.	4277.	
Swine	DE kcal/kg	3470.	4012.	
Cattle	ME kcal/kg	2815.	3254.	
Sheep	ME kcal/kg	3034.	3507.	
Swine	ME kcal/kg	3252.	3759.	
Cattle	TDN %	78.	90.	
Sheep	TDN %	84.	97.	
Swine	TDN %	79.	91.	

(1) dry forages and roughages
(2) pasture, range plants, and forages fed green

(3) silages
(4) energy feeds

Feed name or analyses		Mean		C.V.
		As fed	Dry	± %

Corn, yellow, grain wo germ, grnd, (4)
Degermed yellow corn meal

Ref no 4-03-009

Feed name or analyses		As fed	Dry	
Dry matter	%	89.0	100.0	
Ash	%	.4	.4	
Crude fiber	%	1.0	1.1	
Ether extract	%	1.2	1.3	
N-free extract	%	77.6	87.2	
Protein (N x 6.25)	%	8.9	10.0	
Cattle	dig prot %	6.7	7.5	
Sheep	dig prot %	6.9	7.8	
Energy				
Cattle	DE kcal/kg	3532.	3968.	
Sheep	DE kcal/kg	3806.	4277.	
Cattle	ME kcal/kg	2896.	3254.	
Sheep	ME kcal/kg	3121.	3507.	
Cattle	TDN %	80.	90.	
Sheep	TDN %	86.	97.	
Calcium	%	.01	.01	
Phosphorus	%	.17	.19	
Copper	mg/kg	.9	1.0	
Riboflavin	mg/kg	.4	.4	
Thiamine	mg/kg	.7	.8	

CORN—SOYBEAN. Zea mays, glycine max

Corn-soybean, aerial pt, ensiled, (3)

Ref no 3-03-015

		As fed	Dry	
Dry matter	%	26.1	100.0	
Ash	%	1.8	6.9	
Crude fiber	%	7.0	26.9	
Ether extract	%	1.0	3.8	
N-free extract	%	13.7	52.4	
Protein (N x 6.25)	%	2.6	10.0	
Cattle	dig prot %	1.6	6.2	
Sheep	dig prot %	1.8	6.8	
Energy				
Cattle	DE kcal/kg	829.	3175.	
Sheep	DE kcal/kg	829.	3175.	
Cattle	ME kcal/kg	680.	2604.	
Sheep	ME kcal/kg	680.	2604.	
Cattle	TDN %	19.	72.	
Sheep	TDN %	19.	72.	
Calcium	%	.28	1.09	
Phosphorus	%	.12	.46	

(5) protein supplements

(6) minerals

Cottage cheese - see Cattle, Cheese, Cottage

COTTON. Gossypium spp

Cotton, bolls, s-c, (1)

Ref no 1-01-596

		As fed	Dry	C.V. ± %
Dry matter	%	91.0	100.0	1
Ash	%	6.6	7.3	15
Crude fiber	%	30.6	33.6	27
Ether extract	%	2.4	2.7	20
N-free extract	%	42.2	46.4	
Protein (N x 6.25)	%	9.1	10.0	21
Cattle	dig prot %	.5	.6	
Sheep	dig prot %	2.4	2.6	
Energy				
Cattle	DE kcal/kg	1685.	1852.	
Sheep	DE kcal/kg	1886.	2072.	
Cattle	ME kcal/kg	1382.	1519.	
Sheep	ME kcal/kg	1546.	1699.	
Cattle	TDN %	38.	42.	
Sheep	TDN %	43.	47.	
Calcium	%	1.03	1.13	
Magnesium	%	.25	.28	
Phosphorus	%	.11	.12	
Potassium	%	2.80	3.08	

Cotton, seed hulls, (1)
Cottonseed hulls (AAFCO)

Ref no 1-01-599

		As fed	Dry	C.V. ± %
Dry matter	%	90.3	100.0	3
Ash	%	2.5	2.8	14
Crude fiber	%	42.9	47.5	7
Ether extract	%	1.4	1.5	50
N-free extract	%	39.6	43.9	
Protein (N x 6.25)	%	3.9	4.3	18
Cattle	dig prot %	.2	.2	
Sheep	dig prot %	.4	.5	
Cellulose	%	54.4	60.3	
Lignin	%	24.0	26.6	
Energy				
Cattle	DE kcal/kg	1633.	1808.	
Sheep	DE kcal/kg	2269.	2513.	
Cattle	ME kcal/kg	1338.	1482.	
Sheep	ME kcal/kg	1861.	2061.	
Cattle	NE$_m$ kcal/kg	930.	1030.	
Cattle	NE$_{gain}$ kcal/kg	172.	190.	
Cattle	TDN %	37.	41.	
Sheep	TDN %	51.	57.	

Continued

') vitamins

') additives

Feed name or analyses		Mean		C.V. ± %
		As fed	Dry	
Calcium	%	.14	.16	21
Chlorine	%	.02	.02	
Iron	%	.01	.01	
Magnesium	%	.13	.14	46
Phosphorus	%	.09	.10	45
Potassium	%	.76	.84	36
Sodium	%	.02	.02	
Cobalt	mg/kg	.02	.02	
Copper	mg/kg	14.2	15.7	
Manganese	mg/kg	108.1	117.7	
Zinc	mg/kg	20.0	22.0	

Feed name or analyses		Mean		C.V. ± %
		As fed	Dry	
Cattle	ME kcal/kg	1870.	2024.	
Sheep	ME kcal/kg	2205.	2386.	
Cattle	NEm kcal/kg	1432.	1550.	
Cattle	NE gain kcal/kg	906.	980.	
Cattle	TDN %	52.	56.	
Sheep	TDN %	61.	66.	
Calcium	%	.17	.18	
Phosphorus	%	.64	.69	
Potassium	%	1.25	1.35	

Cotton, seed, grnd, (5)
Cottonseed, whole, ground

Ref no 5-01-608

Dry matter	%	92.7	100.0
Ash	%	3.5	3.8
Crude fiber	%	16.9	18.2
Ether extract	%	22.9	24.7
N-free extract	%	26.3	28.4
Protein (N x 6.25)	%	23.1	24.9
Cattle	dig prot %	14.6	15.7
Sheep	dig prot %	18.3	19.7
Energy			
Cattle	DE kcal/kg	3719.	4012.
Sheep	DE kcal/kg	3883.	4189.
Cattle	ME kcal/kg	3050.	3290.
Sheep	ME kcal/kg	3184.	3435.
Cattle	NEm kcal/kg	1863.	2010.
Cattle	NE gain kcal/kg	1112.	1200.
Cattle	TDN %	84.	91.
Sheep	TDN %	88.	95.
Calcium	%	.14	.15
Phosphorus	%	.68	.73
Potassium	%	1.11	1.20

Cotton, seed, mech-extd grnd, (5)
Whole pressed cottonseed (AAFCO)

Ref no 5-01-609

Dry matter	%	92.4	100.0
Ash	%	4.6	5.0
Crude fiber	%	21.4	23.2
Ether extract	%	5.2	5.6
N-free extract	%	33.2	35.9
Protein (N x 6.25)	%	28.0	30.3
Cattle	dig prot %	19.6	21.2
Sheep	dig prot %	20.1	21.8
Energy			
Cattle	DE kcal/kg	2281.	2469.
Sheep	DE kcal/kg	2689.	2910.

Continued

Cotton, seed w some hulls, mech-extd grnd, mn 36 prot mx 17 fbr mn 2 fat, (5)
Cottonseed meal, 36% protein

Ref no 5-01-615

Dry matter	%	93.5	100.0
Ash	%	5.1	5.5
Crude fiber	%	15.7	16.8
Ether extract	%	6.6	7.1
N-free extract	%	26.4	28.2
Protein (N x 6.25)	%	39.6	42.4
Cattle	dig prot %	31.3	33.5
Sheep	dig prot %	33.7	36.0
Swine	dig prot %	34.1	36.5
Energy			
Cattle	DE kcal/kg	3792.	4056.
Sheep	DE kcal/kg	3587.	3836.
Swine	DE kcal/kg	3133.	3351.
Cattle	ME kcal/kg	3110.	3326.
Sheep	ME kcal/kg	2942.	3146.
Swine	ME kcal/kg	2739.	2929.
Cattle	TDN %	86.	92.
Sheep	TDN %	81.	87.
Swine	TDN %	71.	76.
Calcium	%	.19	.20
Phosphorus	%	1.02	1.09

Cotton, seed w some hulls, mech-extd grnd, mn 41 prot mx 14 fbr mn 2 fat, (5)
Cottonseed meal, 41% protein

Ref no 5-01-617

Dry matter	%	94.0	100.0
Ash	%	6.2	6.6
Crude fiber	%	12.0	12.8
Ether extract	%	4.3	4.6
N-free extract	%	30.4	32.4
Protein (N x 6.25)	%	41.0	43.6
Cattle	dig prot %	33.2	35.3
Sheep	dig prot %	33.2	35.3
Swine	dig prot %	34.9	37.1
Energy	GE kcal/kg	4600.	4893.

Continued

(1) dry forages and roughages
(2) pasture, range plants, and forages fed green

(3) silages
(4) energy feeds

Feed name or analyses		Mean		C.V.
		As fed	Dry	± %
Cattle	DE kcal/kg	3233.	3439.	
Sheep	DE kcal/kg	3026.	3219.	
Swine	DE kcal/kg	2942.	3130.	
Cattle	ME kcal/kg	2651.	2820.	
Sheep	ME kcal/kg	2515.	2676.	
Swine	ME kcal/kg	2565.	2729.	
Cattle	NE$_m$ kcal/kg	1701.	1810.	
Cattle	NE$_{gain}$ kcal/kg	1128.	1200.	
Cattle	TDN %	73.	78.	
Sheep	TDN %	69.	73.	
Swine	TDN %	67.	71.	
Calcium	%	.16	.17	
Iron	%	.030	.032	
Magnesium	%	.56	.60	
Phosphorus	%	1.20	1.28	
Potassium	%	1.40	1.49	
Sodium	%	.04	.04	
Cobalt	mg/kg	.150	.160	
Copper	mg/kg	19.5	20.7	
Manganese	mg/kg	21.5	22.9	
Choline	mg/kg	2780.	2957.	
Folic acid	mg/kg	2.30	2.45	
Niacin	mg/kg	39.5	42.0	
Pantothenic acid	mg/kg	14.0	14.9	
Riboflavin	mg/kg	5.0	5.3	
Thiamine	mg/kg	6.5	6.9	
Arginine	%	4.25	4.52	
Cystine	%	.85	.90	
Glycine	%	2.05	2.18	
Histidine	%	1.10	1.17	
Isoleucine	%	1.60	1.70	
Leucine	%	2.50	2.66	
Lysine	%	1.70	1.81	
Methionine	%	.65	.69	
Phenylalanine	%	2.35	2.50	
Threonine	%	1.45	1.54	
Tryptophan	%	.65	.69	
Valine	%	2.05	2.18	

Feed name or analyses		Mean		C.V.
		As fed	Dry	± %
Energy	GE kcal/kg	4200.	4540.	
Cattle	DE kcal/kg	3018.	3263.	
Sheep	DE kcal/kg	2732.	2954.	
Swine	DE kcal/kg	2692.	2910.	
Cattle	ME kcal/kg	2475.	2676.	
Sheep	ME kcal/kg	2240.	2422.	
Swine	ME kcal/kg	2342.	2532.	
Cattle	TDN %	68.	74.	
Sheep	TDN %	62.	67.	
Swine	TDN %	61.	66.	
Calcium	%	.16	.17	
Iron	%	.030	.032	
Magnesium	%	.56	.60	
Phosphorus	%	1.20	1.30	
Potassium	%	1.40	1.51	
Sodium	%	.04	.04	
Cobalt	mg/kg	.150	.162	
Copper	mg/kg	19.5	21.1	
Manganese	mg/kg	21.5	23.2	
Choline	mg/kg	2860.	3092.	
Folic acid	mg/kg	2.30	2.49	
Niacin	mg/kg	39.5	42.7	
Pantothenic acid	mg/kg	14.0	15.1	
Riboflavin	mg/kg	5.0	5.4	
Thiamine	mg/kg	6.5	7.0	
Arginine	%	4.25	4.59	
Cystine	%	.85	.92	
Glycine	%	2.05	2.22	
Histidine	%	1.10	1.19	
Isoleucine	%	1.60	1.73	
Leucine	%	2.50	2.70	
Lysine	%	1.70	1.84	
Methionine	%	.65	.70	
Phenylalanine	%	2.35	2.54	
Threonine	%	1.45	1.57	
Tryptophan	%	.65	.70	
Valine	%	2.05	2.22	

Cotton, seed w some hulls, pre-press solv-extd grnd, 41 prot, (5)

Cottonseed meal, pre-press solvent extracted, 41% protein

Ref no 5-07-872

Dry matter	%	92.5	100.0	
Ash	%	6.2	6.7	
Crude fiber	%	12.0	13.0	
Ether extract	%	1.4	1.5	
N-free extract	%	31.9	34.5	
Protein (N x 6.25)	%	41.0	44.3	
Cattle	dig prot %	34.8	35.9	
Sheep	dig prot %	33.2	35.9	
Swine	dig prot %	34.8	37.6	

Continued

(5) protein supplements

(6) minerals

Cotton, seed w some hulls, solv-extd grnd, mn 41 prot mx 14 fbr mn 0.5 fat, (5)

Cottonseed meal, solvent extracted, 41% protein

Ref no 5-01-621

Dry matter	%	91.5	100.0	
Ash	%	6.2	6.8	
Crude fiber	%	12.0	13.1	
Ether extract	%	2.0	2.2	
N-free extract	%	30.3	33.1	
Protein (N x 6.25)	%	41.0	44.8	
Cattle	dig prot %	33.2	36.3	
Sheep	dig prot %	33.2	36.3	
Swine	dig prot %	34.9	38.1	
Energy	GE kcal/kg	4300.	4700.	
Cattle	DE kcal/kg	3026.	3307.	

Continued

(7) vitamins

(8) additives

Feed name or analyses		Mean		C.V.
		As fed	Dry	± %
Sheep	DE kcal/kg	2743.	2998.	
Swine	DE kcal/kg	2703.	2954.	
Cattle	ME kcal/kg	2481.	2712.	
Sheep	ME kcal/kg	2249.	2458.	
Swine	ME kcal/kg	2352.	2570.	
Cattle	NE$_m$ kcal/kg	1427.	1560.	
Cattle	NE$_{gain}$ kcal/kg	906.	990.	
Cattle	TDN %	69.	75.	
Sheep	TDN %	62.	68.	
Swine	TDN %	61.	67.	
Calcium	%	.16	.17	
Iron	%	.030	.033	
Magnesium	%	.56	.61	
Phosphorus	%	1.20	1.31	
Potassium	%	1.40	1.53	
Sodium	%	.04	.04	
Cobalt	mg/kg	.150	.164	
Copper	mg/kg	19.5	21.3	
Manganese	mg/kg	21.5	23.5	
Choline	mg/kg	2860.	3126.	
Folic acid	mg/kg	2.30	2.51	
Niacin	mg/kg	39.5	43.2	
Pantothenic acid	mg/kg	14.0	15.3	
Riboflavin	mg/kg	5.0	5.5	
Thiamine	mg/kg	6.5	7.1	
Arginine	%	4.25	4.64	
Cystine	%	.85	.93	
Glycine	%	2.05	2.24	
Histidine	%	1.10	1.20	
Isoleucine	%	1.60	1.75	
Leucine	%	2.50	2.73	
Lysine	%	1.70	1.86	
Methionine	%	.65	.71	
Phenylalanine	%	2.35	2.57	
Threonine	%	1.45	1.58	
Tryptophan	%	.65	.71	
Valine	%	2.05	2.24	

Cotton, seed wo hulls, pre-press solv-extd grnd, mn 50 prot, (5)

Cottonseed meal, pre-press solvent extracted, 50% protein

Ref no 5-07-874

		As fed	Dry	
Dry matter	%	92.5	100.0	
Ash	%	6.2	6.7	
Crude fiber	%	8.5	9.2	
Ether extract	%	1.2	1.3	
N-free extract	%	26.6	28.8	
Protein (N x 6.25)	%	50.0	54.0	
Cattle	dig prot %	40.4	43.7	
Sheep	dig prot %	42.0	45.4	
Swine	dig prot %	45.0	48.6	
Energy				

Continued

Feed name or analyses		Mean		C.V.
		As fed	Dry	± %
Cattle	DE kcal/kg	3059.	3307.	
Sheep	DE kcal/kg	3140.	3395.	
Swine	DE kcal/kg	3018.	3263.	
Cattle	ME kcal/kg	2509.	2712.	
Sheep	ME kcal/kg	2575.	2784.	
Swine	ME kcal/kg	2569.	2777.	
Cattle	TDN %	69.	75.	
Sheep	TDN %	71.	77.	
Swine	TDN %	68.	74.	
Calcium	%	.16	.17	
Iron	%	.011	.012	
Magnesium	%	.46	.50	
Phosphorus	%	1.01	1.09	
Potassium	%	1.26	1.36	
Sodium	%	.05	.05	
Cobalt	mg/kg	2.000	2.162	
Copper	mg/kg	18.0	19.4	
Manganese	mg/kg	22.8	24.6	
Zinc	mg/kg	73.3	79.2	
Arginine	%	4.75	5.13	
Cystine	%	1.00	1.08	
Glycine	%	2.35	2.54	
Histidine	%	1.25	1.35	
Isoleucine	%	1.85	2.00	
Leucine	%	2.80	3.03	
Lysine	%	2.10	2.27	
Methionine	%	.80	.86	
Phenylalanine	%	2.75	2.97	
Threonine	%	1.70	1.84	
Tryptophan	%	.70	.76	
Valine	%	2.05	2.22	

COWPEA. Vigna spp

Cowpea, hay, s-c, immature, (1)

Ref no 1-01-639

		As fed	Dry	
Dry matter	%	89.4	100.0	2
Ash	%	10.7	12.0	8
Crude fiber	%	19.7	22.0	11
Ether extract	%	3.0	3.4	10
N-free extract	%	33.4	37.4	
Protein (N x 6.25)	%	22.5	25.2	7
Cattle	dig prot %	16.8	18.8	
Sheep	dig prot %	17.2	19.2	
Energy				
Cattle	DE kcal/kg	2602.	2910.	
Sheep	DE kcal/kg	2405.	2690.	
Cattle	ME kcal/kg	2133.	2386.	
Sheep	ME kcal/kg	1972.	2206.	
Cattle	TDN %	59.	66.	
Sheep	TDN %	54.	61.	

(1) dry forages and roughages

(2) pasture, range plants, and forages fed green

(3) silages

(4) energy feeds

TABLE A-3-1 Composition of Feeds (Continued)

Feed name or analyses		Mean		C.V.
		As fed	Dry	± %

Cowpea, hay, s-c, full blm, (1)

Ref no 1-01-640

Dry matter	%	88.8	100.0	2
Ash	%	9.8	11.0	7
Crude fiber	%	21.5	24.2	6
Ether extract	%	3.0	3.4	8
N-free extract	%	36.4	41.0	
Protein (N x 6.25)	%	18.1	20.4	7
Cattle	dig prot %	13.0	14.6	
Sheep	dig prot %	13.2	14.9	
Energy				
Cattle	DE kcal/kg	2506.	2822.	
Sheep	DE kcal/kg	2310.	2601.	
Cattle	ME kcal/kg	2055.	2314.	
Sheep	ME kcal/kg	1894.	2133.	
Cattle	TDN %	57.	64.	
Sheep	TDN %	52.	59.	

Cowpea, hay, s-c, (1)

Ref no 1-01-645

Dry matter	%	90.5	100.0	2
Ash	%	10.5	11.6	24
Crude fiber	%	24.7	27.3	17
Ether extract	%	2.6	2.9	19
N-free extract	%	36.0	39.8	
Protein (N x 6.25)	%	16.6	18.4	20
Cattle	dig prot %	11.7	12.9	
Sheep	dig prot %	11.5	12.7	
Energy				
Cattle	DE kcal/kg	2514.	2778.	
Sheep	DE kcal/kg	2274.	2513.	
Cattle	ME kcal/kg	2062.	2278.	
Sheep	ME kcal/kg	1865.	2061.	
Cattle	TDN %	57.	63.	
Sheep	TDN %	52.	57.	
Calcium	%	1.21	1.34	25
Chlorine	%	.15	.17	26
Iron	%	.08	.09	37
Magnesium	%	.42	.47	40
Phosphorus	%	.29	.32	23
Potassium	%	1.80	1.99	31
Sodium	%	.24	.27	78
Sulfur	%	.32	.35	12
Cobalt	mg/kg	.06	.07	
Manganese	mg/kg	439.0	485.1	

Cowpea, hay wo seeds, s-c, (1)

Ref no 1-01-646

Dry matter	%	92.1	100.0	1
Ash	%	12.9	14.0	17
Crude fiber	%	22.7	24.7	4
Ether extract	%	2.8	3.0	11
N-free extract	%	37.4	40.6	
Protein (N x 6.25)	%	16.3	17.7	8
Cattle	dig prot %	11.3	12.3	
Sheep	dig prot %	11.2	12.2	
Energy				
Cattle	DE kcal/kg	2721.	2954.	
Sheep	DE kcal/kg	2274.	2469.	
Cattle	ME kcal/kg	2231.	2422.	
Sheep	ME kcal/kg	1864.	2024.	
Cattle	TDN %	62.	67.	
Sheep	TDN %	52.	56.	
Phosphorus	%	.38	.41	15
Potassium	%	2.72	2.95	5

Cowpea, straw, (1)

Ref no 1-01-649

Dry matter	%	90.9	100.0	
Ash	%	5.4	5.9	
Crude fiber	%	43.0	47.3	
Ether extract	%	1.3	1.4	
N-free extract	%	34.1	37.6	
Protein (N x 6.25)	%	7.1	7.8	
Cattle	dig prot %	3.4	3.7	
Sheep	dig prot %	3.3	3.6	
Energy				
Cattle	DE kcal/kg	1282.	1411.	
Sheep	DE kcal/kg	1963.	2160.	
Cattle	ME kcal/kg	1052.	1157.	
Sheep	ME kcal/kg	1610.	1771.	
Cattle	TDN %	29.	32.	
Sheep	TDN %	44.	49.	

Cowpea, aerial pt, fresh, immature, (2)

Ref no 2-01-650

Dry matter	%	17.1	100.0	8
Ash	%	2.0	11.7	9
Crude fiber	%	3.4	19.9	8
Ether extract	%	.7	3.9	4
N-free extract	%	7.9	46.3	
Protein (N x 6.25)	%	3.1	18.2	7
Cattle	dig prot %	2.3	13.4	

Continued

(5) protein supplements
(6) minerals
(7) vitamins
(8) additives

Feed name or analyses		Mean		C.V.
		As fed	Dry	± %
Sheep	dig prot %	2.3	13.5	
Energy				
Cattle	DE kcal/kg	498.	2910.	
Sheep	DE kcal/kg	513.	2998.	
Cattle	ME kcal/kg	408.	2386.	
Sheep	ME kcal/kg	420.	2458.	
Cattle	TDN %	11.	66.	
Sheep	TDN %	12.	68.	

Cowpea, aerial pt, fresh, full blm, (2)

Ref no 2-01-653

		As fed	Dry	C.V. ± %
Dry matter	%	13.4	100.0	11
Ash	%	1.7	13.0	5
Crude fiber	%	3.5	26.3	8
Ether extract	%	.7	5.2	7
N-free extract	%	5.6	42.0	
Protein (N x 6.25)	%	1.8	13.5	7
Cattle	dig prot %	1.2	9.4	
Sheep	dig prot %	1.4	10.7	
Energy				
Cattle	DE kcal/kg	343.	2557.	
Sheep	DE kcal/kg	378.	2822.	
Cattle	ME kcal/kg	281.	2097.	
Sheep	ME kcal/kg	310.	2314.	
Cattle	TDN %	8.	58.	
Sheep	TDN %	8.	64.	

Cowpea, aerial pt, fresh, (2)

Ref no 2-01-655

		As fed	Dry	C.V. ± %
Dry matter	%	14.6	100.0	9
Ash	%	1.8	12.7	8
Crude fiber	%	3.6	24.8	13
Ether extract	%	.7	4.8	12
N-free extract	%	6.2	42.6	
Protein (N x 6.25)	%	2.2	15.1	14
Cattle	dig prot %	1.6	10.7	
Sheep	dig prot %	1.6	11.2	
Energy				
Cattle	DE kcal/kg	386.	2646.	
Sheep	DE kcal/kg	431.	2954.	
Cattle	ME kcal/kg	317.	2170.	
Sheep	ME kcal/kg	354.	2422.	
Cattle	TDN %	9.	60.	
Sheep	TDN %	10.	67.	

Cowpea, aerial pt, ensiled, (3)

Ref No 3-01-658

		As fed	Dry	C.V. ± %
Dry matter	%	26.1	100.0	13
Ash	%	4.0	15.5	44
Crude fiber	%	6.9	26.4	14
Ether extract	%	1.2	4.7	26
N-free extract	%	10.3	39.3	
Protein (N x 6.25)	%	3.7	14.1	15
Cattle	dig prot %	2.1	8.0	
Sheep	dig prot %	2.3	9.0	
Energy				
Cattle	DE kcal/kg	656.	2513.	
Sheep	DE kcal/kg	679.	2601.	
Cattle	ME kcal/kg	538.	2061.	
Sheep	ME kcal/kg	557.	2133.	
Cattle	TDN %	15.	57.	
Sheep	TDN %	15.	59.	
Calcium	%	.39	1.49	13
Iron	%	.02	.10	16
Magnesium	%	.07	.28	50
Phosphorus	%	.09	.33	21
Potassium	%	.76	2.93	7

Cowpea, aerial pt, wilted ensiled, (3)

Ref no 3-01-659

		As fed	Dry	C.V. ± %
Dry matter	%	31.0	100.0	8
Ash	%	3.4	11.0	34
Crude fiber	%	8.8	28.3	6
Ether extract	%	1.2	4.0	21
N-free extract	%	12.9	41.7	
Protein (N x 6.25)	%	4.6	15.0	14
Cattle	dig prot %	2.7	8.6	
Sheep	dig prot %	3.0	9.8	
Energy				
Cattle	DE kcal/kg	806.	2601.	
Sheep	DE kcal/kg	834.	2690.	
Cattle	ME kcal/kg	661.	2133.	
Sheep	ME kcal/kg	684.	2206.	
Cattle	TDN %	18.	59.	
Sheep	TDN %	19.	61.	
Chlorine	%	.05	.17	
Sulfur	%	.01	.03	57

(1) dry forages and roughages
(2) pasture, range plants, and forages fed green

(3) silages
(4) energy feeds

Feed name or analyses		Mean		C.V.
		As fed	Dry	± %

Cowpea, seed, (5)

Ref no 5-01-661

Feed name or analyses		As fed	Dry	± %
Dry matter	%	88.8	100.0	
Ash	%	3.2	3.6	
Crude fiber	%	5.1	5.7	
Ether extract	%	1.3	1.5	
N-free extract	%	55.8	62.9	
Protein (N x 6.25)	%	23.4	26.3	
Cattle	dig prot %	19.2	21.6	
Sheep	dig prot %	19.2	21.6	
Swine	dig prot %	24.2	27.3	
Energy				
Cattle	DE kcal/kg	3328.	3748.	
Sheep	DE kcal/kg	3328.	3748.	
Swine	DE kcal/kg	3289.	3704.	
Cattle	ME kcal/kg	2729.	3073.	
Sheep	ME kcal/kg	2729.	3073.	
Swine	ME kcal/kg	2697.	3037.	
Cattle	TDN %	75.	85.	
Sheep	TDN %	75.	85.	
Swine	TDN %	74.	84.	

CRAB. Callinectes sapidus, Cancer spp, Paralithodes camschatica

Crab, proc res, dehy grnd, mn 25 prot salt declared above 3 mx 7, (5)
Crab meal (AAFCO)

Ref no 5-01-663

		As fed	Dry	± %
Dry matter	%	93.0	100.0	3
Ash	%	40.7	43.8	15
Crude fiber	%	11.0	11.8	12
Ether extract	%	1.8	1.9	57
N-free extract	%	8.4	9.1	
Protein (N x 6.25)	%	31.1	33.4	10
Energy				
Chickens	MEn kcal/kg	1764.	1897.	
Calcium	%	15.32	16.47	18
Iron	%	.44	.47	
Magnesium	%	.88	.95	
Phosphorus	%	1.59	1.71	9
Potassium	%	.45	.48	
Sodium	%	.85	.91	
Copper	mg/kg	32.8	35.3	
Manganese	mg/kg	133.8	143.8	41
Pantothenic acid	mg/kg	6.6	7.1	
Riboflavin	mg/kg	5.9	6.3	48
Arginine	%	1.70	1.83	
Histidine	%	.50	.54	17

Continued

		As fed	Dry	± %
Isoleucine	%	1.20	1.29	17
Leucine	%	1.60	1.72	13
Lysine	%	1.40	1.51	3
Methionine	%	.50	.54	17
Phenylalanine	%	1.20	1.29	23
Threonine	%	1.00	1.08	4
Tryptophan	%	.30	.32	
Tyrosine	%	1.20	1.29	
Valine	%	1.50	1.61	8

DALLISGRASS. Paspalum dilatatum

Dallisgrass, hay, s-c, immature, (1)

Ref no 1-01-733

		As fed	Dry	± %
Dry matter	%	90.6	100.0	
Ash	%	9.4	10.4	
Crude fiber	%	28.9	31.9	
Ether extract	%	2.2	2.4	
N-free extract	%	40.8	45.1	
Protein (N x 6.25)	%	9.3	10.2	
Cattle	dig prot %	5.2	5.8	
Sheep	dig prot %	5.2	5.7	
Energy				
Cattle	DE kcal/kg	2517.	2778.	
Sheep	DE kcal/kg	2237.	2469.	
Cattle	ME kcal/kg	2064.	2278.	
Sheep	ME kcal/kg	1834.	2024.	
Cattle	TDN %	57.	63.	
Sheep	TDN %	51.	56.	
Calcium	%	.73	.80	
Phosphorus	%	.19	.21	

Dallisgrass, hay, s-c, mid-blm, (1)

Ref no 1-01-734

		As fed	Dry	± %
Dry matter	%	90.7	100.0	
Ash	%	8.8	9.7	
Crude fiber	%	29.1	32.1	
Ether extract	%	2.2	2.4	
N-free extract	%	44.1	48.6	
Protein (N x 6.25)	%	6.5	7.2	
Cattle	dig prot %	2.9	3.2	
Sheep	dig prot %	2.7	3.0	
Energy				
Cattle	DE kcal/kg	2480.	2734.	
Sheep	DE kcal/kg	2199.	2425.	
Cattle	ME kcal/kg	2033.	2242.	
Sheep	ME kcal/kg	1803.	1988.	
Cattle	TDN %	56.	62.	

Continued

(5) protein supplements
(6) minerals

(7) vitamins
(8) additives

Feed name or analyses		Mean		C.V.
		As fed	Dry	± %
Sheep	TDN %	50.	55.	
Calcium	%	.39	.43	
Phosphorus	%	.15	.17	

Dallisgrass, hay, s-c, over ripe, (1)

Ref no 1-01-736

		As fed	Dry	
Dry matter	%	91.4	100.0	
Ash	%	8.9	9.8	
Crude fiber	%	33.8	37.0	
Ether extract	%	1.6	1.8	
N-free extract	%	43.0	47.0	
Protein (N x 6.25)	%	4.0	4.4	
Cattle	dig prot %	.7	.8	
Sheep	dig prot %	.5	.5	
Energy				
Cattle	DE kcal/kg	2176.	2381.	
Sheep	DE kcal/kg	2096.	2293.	
Cattle	ME kcal/kg	1784.	1952.	
Sheep	ME kcal/kg	1718.	1880.	
Cattle	TDN %	49.	54.	
Sheep	TDN %	48.	52.	
Calcium	%	.33	.36	
Phosphorus	%	.07	.08	

Dallisgrass, hay, s-c, (1)

Ref no 1-01-737

		As fed	Dry	
Dry matter	%	90.9	100.0	1
Ash	%	7.9	8.7	14
Crude fiber	%	30.8	33.9	6
Ether extract	%	1.9	2.1	14
N-free extract	%	41.1	45.2	
Protein (N x 6.25)	%	9.2	10.1	26
Cattle	dig prot %	5.2	5.7	
Sheep	dig prot %	5.1	5.6	
Energy				
Cattle	DE kcal/kg	2284.	2513.	
Sheep	DE kcal/kg	2244.	2469.	
Cattle	ME kcal/kg	1873.	2061.	
Sheep	ME kcal/kg	1840.	2024.	
Cattle	TDN %	52.	57.	
Sheep	TDN %	51.	56.	
Calcium	%	.46	.51	
Iron	%	.01	.01	
Magnesium	%	.67	.74	
Phosphorus	%	.18	.20	

Dallisgrass, aerial pt, fresh, immature, (2)

Ref no 2-01-738

		As fed	Dry	
Dry matter	%	25.0	100.0	
Ash	%	2.6	10.5	
Crude fiber	%	7.5	30.1	
Ether extract	%	.7	2.7	
N-free extract	%	8.4	33.5	
Protein (N x 6.25)	%	5.8	23.2	
Cattle	dig prot %	4.2	17.0	
Sheep	dig prot %	4.6	18.6	
Cellulose	%	7.9	31.7	
Lignin	%	3.3	13.1	
Energy				
Cattle	DE kcal/kg	838.	3351.	
Sheep	DE kcal/kg	639.	2557.	
Cattle	ME kcal/kg	687.	2748.	
Sheep	ME kcal/kg	524.	2097.	
Cattle	TDN %	19.	76.	
Sheep	TDN %	14.	58.	
Calcium	%	.16	.65	19
Phosphorus	%	.10	.42	14
Carotene	mg/kg	106.7	426.7	
Vitamin A equiv	IU/g	177.8	711.3	

Dallisgrass, aerial pt, fresh, (2)

Ref no 2-01-741

		As fed	Dry	
Dry matter	%	25.0	100.0	14
Ash	%	2.8	11.3	45
Crude fiber	%	7.4	29.4	11
Ether extract	%	.6	2.5	19
N-free extract	%	11.2	44.7	
Protein (N x 6.25)	%	3.0	12.1	29
Cattle	dig prot %	1.8	7.4	
Sheep	dig prot %	2.1	8.3	
Cellulose	%	9.5	37.9	12
Lignin	%	4.2	16.7	14
Energy				
Cattle	DE kcal/kg	694.	2778.	
Sheep	DE kcal/kg	706.	2822.	
Cattle	ME kcal/kg	570.	2278.	
Sheep	ME kcal/kg	578.	2314.	
Cattle	TDN %	16.	63.	
Sheep	TDN %	16.	64.	
Calcium	%	.14	.55	25
Iron	%	.005	.02	17
Magnesium	%	.10	.38	54
Phosphorus	%	.07	.27	33
Potassium	%	.37	1.48	30
Cobalt	mg/kg	.02	.07	

Continued

(1) dry forages and roughages
(2) pasture, range plants, and forages fed green

(3) silages
(4) energy feeds

Feed name or analyses		Mean		C.V.
		As fed	Dry	± %
Manganese	mg/kg	20.0	80.0	
Carotene	mg/kg	75.6	302.3	40
Vitamin A equiv	IU/g	126.0	503.9	

Darso - see Sorghum, darso

Defatted wheat germ - see Wheat, germ, extn unspec grnd

Defluorinated phosphate - see Phosphate, defluorinated

DESERT MOLLY. Kochia vestita

Desert molly, browse, fresh, dormant, (2)

Ref no 2-07-988

Dry matter	%		80.0	100.0
Ash	%		19.8	24.8
Ether extract	%		3.3	4.1
Protein (N x 6.25)	%		7.2	9.0
Sheep	dig prot	%	4.4	5.5
Cellulose	%		10.4	13.0
Lignin	%		6.4	8.0
Energy	GE kcal/kg		2870.	3587.
Sheep	DE kcal/kg		1725.	2156.
Sheep	ME kcal/kg		1522.	1902.
Sheep	TDN %		40.	50.
Calcium	%		1.90	2.37
Phosphorus	%		.10	.12
Carotene	mg/kg		14.5	18.1

Dicalcium - see Calcium phosphate, dibasic, comm

Digester tankage - see Animal, carcass res w blood

Distillers grains - see Corn; see Grains; see Sorghum, grain variety; see Rye

Distillers grains with solubles - see Corn

Distillers solubles - see Corn

DROPSEED, SAND. Sporobolus cryptandrus

Dropseed, sand, aerial pt, fresh, dormant (2)
Sand dropseed

Ref no 2-05-596

Dry matter	%		86.0	100.0
Ash	%		5.4	6.3
Ether extract	%		1.2	1.4
Protein (N x 6.25)	%		4.3	5.0
Sheep	dig prot	%	1.6	1.9
Cellulose	%		39.6	46.0
Lignin	%		6.9	8.0
Energy	GE kcal/kg		3593.	4178.
Sheep	DE kcal/kg		2073.	2410.
Sheep	ME kcal/kg		1780.	2070.
Sheep	TDN %		51.	59.
Calcium	%		.49	.57
Phosphorus	%		.05	.06
Carotene	mg/kg		.3	.4

Durra - see Sorghum, durra

Ear corn chop - see Corn, ears, grnd

EMMER. Triticum dicoccum

Emmer, grain, (4)

Ref no 4-01-830

Dry matter	%		91.0	100.0
Ash	%		3.4	3.7
Crude fiber	%		10.0	11.0
Ether extract	%		1.9	2.1
N-free extract	%		62.8	69.0
Protein (N x 6.25)	%		12.9	14.2
Cattle	dig prot	%	8.2	9.0
Sheep	dig prot	%	10.4	11.4
Swine	dig prot	%	10.8	11.9
Energy				
Cattle	DE kcal/kg		2889.	3175.
Sheep	DE kcal/kg		3170.	3483.
Swine	DE kcal/kg		2688.	2954.
Cattle	ME kcal/kg		2370.	2604.
Sheep	ME kcal/kg		2599.	2856.
Swine	ME kcal/kg		2502.	2750.
Cattle	TDN %		66.	72.
Sheep	TDN %		72.	79.

Continued

(5) protein supplements

(6) minerals

(7) vitamins

(8) additives

Feed name or analyses		Mean		C.V.
		As fed	Dry	± %
Swine	TDN %	61.	67.	
Iron	%	.005	.006	
Phosphorus	%	.43	.47	
Copper	mg/kg	34.5	37.9	
Manganese	mg/kg	85.8	94.3	

Feed name or analyses		Mean		C.V.
		As fed	Dry	± %
Cattle	TDN %	16.	66.	
Sheep	TDN %	16.	65.	
Calcium	%	.10	.44	
Phosphorus	%	.08	.33	

Farina - see Wheat, grits

Fat - see Animal-Poultry; see Swine, lard

Feather meal - see Poultry, feathers

Feed flour - see Wheat, flour

Feeding bone meal - see Animal, bone, cooked

Feeding meat and bone tankage - see Animal-poultry,
    carcass res w bone

Feeding oat meal - see Oats, cereal by-prod, mx 4 fbr

Feeding tankage - see Animal-poultry, carcass res mx
    35 blood

Feterita - see Sorghum, feterita

FESCUE, MEADOW. Festuca elatior

Fescue, meadow, hay, s-c, immature, (1)

Ref no 1-01-903

Dry matter	%	87.5	100.0	5
Ash	%	7.7	8.8	8
Crude fiber	%	22.5	25.7	10
Ether extract	%	3.9	4.4	27
N-free extract	%	37.6	43.0	
Protein (N x 6.25)	%	15.9	18.1	23
Cattle	dig prot %	11.0	12.6	
Sheep	dig prot %	9.6	11.0	
Lignin	%	2.9	3.3	
Energy				
Cattle	DE kcal/kg	2392.	2734.	
Sheep	DE kcal/kg	2431.	2778.	
Cattle	ME kcal/kg	1962.	2242.	
Sheep	ME kcal/kg	1993.	2278.	
Cattle	TDN %	54.	62.	
Sheep	TDN %	55.	63.	
Calcium	%	.64	.73	
Magnesium	%	.60	.68	
Phosphorus	%	.43	.49	

FESCUE, ALTA. Festuca arundinacea

Fescue, alta, aerial pt, fresh, (2)

Ref no 2-01-889

Dry matter	%	23.9	100.0	9
Ash	%	2.0	8.3	8
Crude fiber	%	7.1	29.6	10
Ether extract	%	.8	3.3	17
N-free extract	%	11.3	47.2	
Protein (N x 6.25)	%	2.8	11.6	22
Cattle	dig prot %	1.9	7.8	
Sheep	dig prot %	1.9	7.8	
Energy				
Cattle	DE kcal/kg	695.	2910.	
Sheep	DE kcal/kg	685.	2866.	
Cattle	ME kcal/kg	570.	2386.	
Sheep	ME kcal/kg	562.	2350.	

Fescue, meadow, hay, s-c, full blm, (1)

Ref no 1-01-906

Dry matter	%	84.4	100.0	
Ash	%	7.8	9.2	
Crude fiber	%	24.5	29.0	
Ether extract	%	2.2	2.6	
N-free extract	%	43.1	51.1	
Protein (N x 6.25)	%	6.8	8.1	
Cattle	dig prot %	3.7	4.4	
Sheep	dig prot %	3.7	4.4	
Energy				
Cattle	DE kcal/kg	2121.	2513.	
Sheep	DE kcal/kg	2121.	2513.	
Cattle	ME kcal/kg	1739.	2061.	
Sheep	ME kcal/kg	1739.	2061.	
Cattle	TDN %	48.	57.	
Sheep	TDN %	48.	57.	

Continued

(1) dry forages and roughages
(2) pasture, range plants, and forages fed green

(3) silages
(4) energy feeds

Fescue, meadow, hay, s-c, mature, (1)

Ref no 1-01-908

Feed name or analyses		As fed (Mean)	Dry (Mean)	C.V. ± %
Dry matter	%	86.3	100.0	
Ash	%	9.5	11.0	
Crude fiber	%	30.7	35.6	
Ether extract	%	2.0	2.3	
N-free extract	%	38.8	45.0	
Protein (N x 6.25)	%	5.3	6.1	
Cattle	dig prot %	2.2	2.6	
Sheep	dig prot %	2.2	2.6	
Energy				
Cattle	DE kcal/kg	1903.	2205.	
Sheep	DE kcal/kg	1903.	2205.	
Cattle	ME kcal/kg	1560.	1808.	
Sheep	ME kcal/kg	1560.	1808.	
Cattle	TDN %	43.	50.	
Sheep	TDN %	43.	50.	

Fescue, meadow, hay, s-c, over ripe, (1)

Ref no 1-01-909

Feed name or analyses		As fed (Mean)	Dry (Mean)	C.V. ± %
Dry matter	%	92.4	100.0	2
Ash	%	6.2	6.7	12
Crude fiber	%	37.8	40.9	9
Ether extract	%	1.6	1.7	25
N-free extract	%	41.3	44.7	
Protein (N x 6.25)	%	5.5	6.0	16
Cattle	dig prot %	2.4	2.6	
Sheep	dig prot %	2.4	2.6	
Energy				
Cattle	DE kcal/kg	2119.	2293.	
Sheep	DE kcal/kg	2119.	2293.	
Cattle	ME kcal/kg	1737.	1880.	
Sheep	ME kcal/kg	1737.	1880.	
Cattle	TDN %	48.	52.	
Sheep	TDN %	48.	52.	

Fescue, meadow, hay, s-c, (1)
Fescue hay, tall

Ref no 1-01-912

Feed name or analyses		As fed (Mean)	Dry (Mean)	C.V. ± %
Dry matter	%	88.5	100.0	3
Ash	%	7.3	8.2	15
Crude fiber	%	27.6	31.2	16
Ether extract	%	2.6	3.0	37
N-free extract	%	41.7	47.1	
Protein (N x 6.25)	%	9.3	10.5	39
Cattle	dig prot %	5.3	6.0	
Sheep	dig prot %	5.8	6.5	

Continued

Feed name or analyses		As fed (Mean)	Dry (Mean)	C.V. ± %
Cellulose	%	33.4	37.8	
Lignin	%	6.1	6.9	55
Energy	GE kcal/kg	3661.	4137.	
Cattle	DE kcal/kg	2420.	2734.	
Sheep	DE kcal/kg	2458.	2778.	
Cattle	ME kcal/kg	1984.	2242.	
Sheep	ME kcal/kg	2016.	2278.	
Cattle	TDN %	55.	62.	
Sheep	TDN %	56.	63.	
Calcium	%	.44	.50	49
Magnesium	%	.44	.50	38
Phosphorus	%	.32	.36	53
Potassium	%	1.65	1.87	30
Manganese	mg/kg	21.7	24.5	

Fescue, meadow, aerial pt, fresh, (2)
Fescue forage, tall

Ref no 2-01-920

Feed name or analyses		As fed (Mean)	Dry (Mean)	C.V. ± %
Dry matter	%	27.6	100.0	28
Ash	%	2.5	9.0	18
Crude fiber	%	7.5	27.1	16
Ether extract	%	1.1	4.1	41
N-free extract	%	12.3	44.7	
Protein (N x 6.25)	%	4.2	15.1	35
Cattle	dig prot %	3.0	10.7	
Sheep	dig prot %	3.1	11.1	
Energy				
Cattle	DE kcal/kg	815.	2954.	
Sheep	DE kcal/kg	767.	2778.	
Cattle	ME kcal/kg	668.	2422.	
Sheep	ME kcal/kg	629.	2278.	
Cattle	TDN %	18.	67.	
Sheep	TDN %	17.	63.	
Calcium	%	.14	.51	33
Magnesium	%	.10	.37	51
Phosphorus	%	.10	.38	47
Potassium	%	.55	2.00	
Copper	mg/kg	1.1	4.0	31

Fescue, meadow, aerial pt, ensiled, (3)
Fescue silage, tall

Ref no 3-01-925

Feed name or analyses		As fed (Mean)	Dry (Mean)	C.V. ± %
Dry matter	%	32.1	100.0	12
Ash	%	3.2	9.9	12
Crude fiber	%	9.3	28.9	7
Ether extract	%	1.1	3.4	11
N-free extract	%	14.4	44.9	
Protein (N x 6.25)	%	4.1	12.9	12
Cattle	dig prot %	2.5	7.9	
Sheep	dig prot %	2.5	7.9	

Continued

Feed name or analyses		Mean		C.V. ± %
		As fed	Dry	
Energy				
Cattle	DE kcal/kg	892.	2778.	
Sheep	DE kcal/kg	878.	2734.	
Cattle	ME kcal/kg	731.	2278.	
Sheep	ME kcal/kg	720.	2242.	
Cattle	TDN %	20.	63.	
Sheep	TDN %	20.	62.	

FESCUE, RED. Festuca rubra

Fescue, red, hay, s-c, (1)

Ref no 1-01-927

		As fed	Dry	
Dry matter	%	85.4	100.0	
Ash	%	7.3	8.6	
Crude fiber	%	26.2	30.7	
Ether extract	%	1.9	2.2	
N-free extract	%	42.0	49.2	
Protein (N x 6.25)	%	7.9	9.3	
Cattle	dig prot %	4.3	5.0	
Sheep	dig prot %	4.1	4.8	
Energy				
Cattle	DE kcal/kg	2372.	2778.	
Sheep	DE kcal/kg	2184.	2557.	
Cattle	ME kcal/kg	1945.	2278.	
Sheep	ME kcal/kg	1791.	2097.	
Cattle	TDN %	54.	63.	
Sheep	TDN %	50.	58.	

Fescue, tall - see Fescue, meadow

FISH. Scientific name not used

Fish, liver, extn unspec dehy grnd, salt declared above 4, (5)
Fish liver meal (CFA)

Ref no 5-01-968

		As fed	Dry	
Dry matter	%	93.0	100.0	1
Ash	%	6.0	6.5	28
Crude fiber	%	1.0	1.1	41
Ether extract	%	15.1	16.2	18
N-free extract	%	4.4	4.7	
Protein (N x 6.25)	%	66.5	71.5	7
Swine	dig prot %	61.8	66.5	
Energy				
Swine	DE kcal/kg	3937.	4233.	
Swine	ME kcal/kg	3212.	3454.	
Swine	TDN %	89.	96.	

Continued

Feed name or analyses		Mean		C.V. ± %
		As fed	Dry	
Calcium	%	.50	.54	55
Iron	%	.07	.08	33
Phosphorus	%	1.25	1.34	24
Cobalt	mg/kg	.1	.1	32
Copper	mg/kg	89.1	95.8	65
Manganese	mg/kg	8.8	9.5	64

Fish liver and glandular meal - see Fish, viscera

Fish meal, drum - see Fish, redfish

Fish, sol, cond, mn 30 prot, (5)
Condensed fish solubles (AAFCO)

Ref no 5-01-969

		As fed	Dry	
Dry matter	%	51.0	100.0	13
Ash	%	10.0	19.6	58
Crude fiber	%	1.0	2.0	93
Ether extract	%	6.5	12.7	99
N-free extract	%	2.1	4.1	
Protein (N x 6.25)	%	31.4	61.6	21
Swine	dig prot %	30.1	59.1	
Energy				
Swine	DE kcal/kg	1956.	3836.	
Chickens	MEₙ kcal/kg	1477.	2896.	
Swine	ME kcal/kg	1636.	3207.	
Swine	TDN %	44.	87.	
Calcium	%	.61	1.20	99
Iron	%	.03	.06	99
Magnesium	%	.02	.04	
Phosphorus	%	.70	1.37	
Potassium	%	1.75	3.43	
Sodium	%	3.06	6.00	99
Copper	mg/kg	48.2	94.5	99
Manganese	mg/kg	11.9	23.3	99
Zinc	mg/kg	38.3	75.1	
Biotin	mg/kg	.2	.4	
Choline	mg/kg	4028.	7899.	
Niacin	mg/kg	168.7	330.8	99
Pantothenic acid	mg/kg	35.4	69.4	
Riboflavin	mg/kg	14.5	28.4	99
Thiamine	mg/kg	5.5	10.8	97
Arginine	%	2.40	4.71	60
Cystine	%	1.70	3.33	
Glycine	%	4.90	9.61	48
Histidine	%	2.50	4.90	78
Isoleucine	%	1.60	3.14	65
Leucine	%	2.50	4.90	48
Lysine	%	2.70	5.29	66
Methionine	%	1.00	1.96	52
Phenylalanine	%	1.40	2.75	46
Threonine	%	1.20	2.35	53

Continued

(1) dry forages and roughages
(2) pasture, range plants, and forages fed green

(3) silages
(4) energy feeds

TABLE A-3-1 Composition of Feeds (Continued)

Feed name or analyses		Mean		C.V.
		As fed	Dry	± %
Tryptophan	%	.80	1.57	
Tyrosine	%	.50	.98	92
Valine	%	1.60	3.14	45

Fish, stickwater sol, cooked dehy, mn 60 prot, (5)
Dried fish solubles (AAFCO)
Fish solubles, dried

Ref no 5-01-971

		As fed	Dry	C.V.
Dry matter	%	92.0	100.0	
Ash	%	15.8	17.2	
Crude fiber	%	1.0	1.1	
Ether extract	%	7.6	8.3	
N-free extract	%	4.7	5.1	
Protein (N x 6.25)	%	62.8	68.3	15
Swine	dig prot %	60.3	65.5	
Energy				
Swine	DE kcal/kg	3408.	3704.	
Chickens	MEn kcal/kg	2866.	3115.	
Swine	ME kcal/kg	2801.	3045.	
Swine	TDN %	77.	84.	
Choline	mg/kg	5223.	5677.	
Niacin	mg/kg	231.1	252.3	
Pantothenic acid	mg/kg	44.9	48.8	
Riboflavin	mg/kg	7.7	8.4	
Arginine	%	2.40	2.61	
Histidine	%	2.60	2.83	
Isoleucine	%	1.70	1.85	
Leucine	%	2.70	2.93	
Lysine	%	3.00	3.26	
Methionine	%	.90	.98	
Phenylalanine	%	1.30	1.41	
Threonine	%	1.20	1.30	
Tryptophan	%	.70	.76	
Tyrosine	%	.70	.76	
Valine	%	1.90	2.07	

Fish, viscera, dehy grnd, mn 50 liver 18 mg riboflavin, (5)
Fish liver and glandular meal (AAFCO)

Ref no 5-01-973

		As fed	Dry	C.V.
Dry matter	%	93.0	100.0	1
Ash	%	5.8	6.2	24
Crude fiber	%	2.0	2.2	50
Ether extract	%	16.0	17.2	19
N-free extract	%	4.1	4.4	
Protein (N x 6.25)	%	65.1	70.0	5
Swine	dig prot %	60.5	65.1	
Energy				
Swine	DE kcal/kg	4018.	4321.	
Swine	ME kcal/kg	3291.	3539.	

Continued

(5) protein supplements

(6) minerals

Feed name or analyses		Mean		C.V.
		As fed	Dry	± %
Swine	TDN %	91.	98.	
Calcium	%	.66	.71	24
Iron	%	.05	.05	
Phosphorus	%	1.14	1.23	21
Cobalt	mg/kg	.2	.2	
Copper	mg/kg	97.0	104.3	
Manganese	mg/kg	7.3	7.8	59

Fish, whole or cuttings, cooked mech-extd dehy grnd, (5)
Fish meal (CFA)

Ref no 5-01-976

		As fed	Dry	
Dry matter	%	94.0	100.0	
Ether extract	%	5.6	5.9	
Protein (N x 6.25)	%	65.4	69.6	
Calcium	%	7.46	7.94	
Phosphorus	%	3.91	4.16	

FISH, ALEWIFE. Pomolobus pseudoharengus

Fish, alewife, whole, raw, (5)

Ref no 5-07-964

		As fed	Dry	
Dry matter	%	26.0	100.0	
Ether extract	%	5.0	19.2	
Protein (N x 6.25)	%	19.5	75.0	

FISH, ANCHOVY. Engraulis spp

Fish, anchovy, whole or cuttings, cooked mech-extd dehy grnd, (5)
Fish meal, anchovy

Ref no 5-01-985

		As fed	Dry	
Dry matter	%	93.0	100.0	
Ash	%	19.0	20.4	
Crude fiber	%	1.0	1.1	
Ether extract	%	3.8	4.1	
N-free extract	%	3.2	3.4	
Protein (N x 6.25)	%	66.0	71.0	
Swine	dig prot %	60.7	65.3	
Energy				
Swine	DE kcal/kg	2994.	3219.	
Chickens	MEn kcal/kg	2900.	3118.	
Swine	ME kcal/kg	2446.	2630.	
Swine	TDN %	68.	73.	
Calcium	%	4.50	4.84	

Continued

(7) vitamins

(8) additives

Feed name or analyses		Mean		C.V. ± %
		As fed	Dry	
Phosphorus	%	2.85	3.06	
Alanine	%	5.59	6.01	
Arginine	%	4.46	4.79	
Aspartic acid	%	7.86	8.45	
Glutamic acid	%	9.98	10.73	
Glycine	%	5.10	5.48	
Histidine	%	1.84	1.98	
Isoleucine	%	3.40	3.66	
Leucine	%	7.01	7.54	
Lysine	%	5.40	5.80	
Methionine	%	2.19	2.35	
Phenylalanine	%	2.48	2.67	
Proline	%	2.30	2.47	
Serine	%	3.54	3.80	
Threonine	%	3.04	3.27	
Tyrosine	%	1.77	1.90	
Valine	%	3.54	3.80	

FISH, CARP. Cyprinus carpio

Fish, carp, whole, raw, (5)

Ref no 5-01-986

Dry matter	%	22.0	100.0	
Ether extract	%	2.3	10.4	
Protein (N x 6.25)	%	18.5	84.1	

FISH, CATFISH. Ictalurus spp

Fish, catfish, whole, raw, (5)

Ref no 5-07-965

Dry matter	%	17.5	100.0	
Ether extract	%	.4	2.3	
Protein (N x 6.25)	%	16.5	94.3	

FISH, COD. Gadus morrhua, Gadus macrocephalus

Fish, cod, whole or cuttings, steam dehy grnd,
    60 prot, (5)
    Fish meal, cod

Ref no 5-01-991

Dry matter	%	94.0	100.0	
Ash	%	29.9	31.8	
Crude fiber	%	1.0	1.1	
Ether extract	%	1.6	1.7	

Continued

Feed name or analyses		Mean		C.V. ± %
		As fed	Dry	
N-free extract	%	2.5	2.7	
Protein (N x 6.25)	%	58.9	62.7	
Swine	dig prot %	57.2	60.8	
Energy				
Swine	DE kcal/kg	2818.	2998.	
Swine	ME kcal/kg	2347.	2497.	
Swine	TDN %	64.	68.	
Iron	%	.02	.02	
Copper	mg/kg	2.9	3.1	

FISH, FLOUNDER. Bothidae (family), Pleuronectidae (family)

Fish, flounder, whole, raw, (5)

Ref no 5-01-996

Dry matter	%	17.0	100.0	
Ether extract	%	.5	2.9	
Protein (N x 6.25)	%	15.0	88.2	

FISH, HADDOCK. Melanogrammus aeglefinus

Fish, haddock, whole, raw, (5)

Ref no 5-07-966

Dry matter	%	18.0	100.0	
Ether extract	%	.3	1.7	
Protein (N x 6.25)	%	17.0	94.4	

FISH, HAKE. Merluccius spp, Urophycis spp

Fish, hake, whole, cooked, (5)

Ref no 5-07-967

Dry matter	%	30.0	100.0	
Ether extract	%	5.6	18.8	
Protein (N x 6.25)	%	17.2	57.4	

Fish, hake, whole, cooked acidified, (5)

Ref no 5-07-968

Dry matter	%	25.0	100.0	
Crude fiber	%	.3	1.1	
Ether extract	%	5.3	21.2	

(1) dry forages and roughages
(2) pasture, range plants, and forages fed green

(3) silages
(4) energy feeds

TABLE A-3-1 Composition of Feeds (Continued)

Feed name or analyses		As fed (Mean)	Dry (Mean)	C.V. ± %
Fish, hake, whole, raw, (5)				
Ref no 5-07-969				
Dry matter	%	19.0	100.0	
Ether extract	%	1.1	5.8	
Protein (N x 6.25)	%	17.0	89.5	

FISH, HERRING. Clupea harengus harengus, Clupea harengus pallasi

Fish, herring, whole, raw, (5)

Ref no 5-01-999

		As fed	Dry	
Dry matter	%	26.0	100.0	
Ether extract	%	5.5	21.1	
Protein (N x 6.25)	%	18.0	69.2	

Fish, herring, whole or cuttings, cooked mech-extd dehy grnd, (5)
Fish meal, herring

Ref no 5-02-000

		As fed	Dry	C.V.
Dry matter	%	92.0	100.0	2
Ash	%	10.8	11.7	12
Ether extract	%	7.5	8.2	26
N-free extract	%	3.1	3.4	
Protein (N x 6.25)	%	70.6	76.7	6
Sheep	dig prot %	62.1	67.5	
Swine	dig prot %	66.3	72.1	
Energy				
Sheep	DE kcal/kg	3448.	3748.	
Swine	DE kcal/kg	3650.	3968.	
Chickens	MEn kcal/kg	2976.	3235.	
Sheep	ME kcal/kg	2827.	3073.	
Swine	ME kcal/kg	2938.	3194.	
Sheep	TDN %	78.	85.	
Swine	TDN %	83.	90.	
Calcium	%	2.94	3.20	32
Phosphorus	%	2.20	2.39	13
Manganese	mg/kg	9.9	10.8	
Choline	mg/kg	4004.	4352.	27
Folic acid	mg/kg	2.40	2.60	
Niacin	mg/kg	88.9	96.6	46
Pantothenic acid	mg/kg	11.4	12.4	53
Riboflavin	mg/kg	9.0	9.8	19
Vitamin B12	mcg/kg	218.7	237.7	69
Arginine	%	4.00	4.35	25
Cystine	%	1.60	1.74	
Glycine	%	5.00	5.44	

Continued

		As fed (Mean)	Dry (Mean)	C.V. ± %
Histidine	%	1.30	1.41	40
Isoleucine	%	3.20	3.48	8
Leucine	%	5.10	5.54	18
Lysine	%	7.30	7.94	18
Methionine	%	2.00	2.17	29
Phenylalanine	%	2.60	2.83	11
Threonine	%	2.60	2.83	26
Tryptophan	%	.90	.98	42
Tyrosine	%	2.10	2.28	17
Valine	%	3.20	3.48	19

FISH, MACKERAL ATLANTIC. Scomber scombrus

Fish, mackeral Atlantic, whole, raw, (5)

Ref no 5-07-971

		As fed	Dry	
Dry matter	%	32.0	100.0	
Ether extract	%	12.0	37.5	
Protein (N x 6.25)	%	18.5	57.8	

FISH, MACKERAL PACIFIC. Scomber japonicus

Fish, mackeral Pacific, whole, raw, (5)

Ref no 5-07-972

		As fed	Dry	
Dry matter	%	31.0	100.0	
Ether extract	%	7.6	24.5	
Protein (N x 6.25)	%	22.0	71.0	

FISH, MENHADEN. Brevoortia tyrannus

Fish, menhaden, whole or cuttings, cooked mech-extd dehy grnd, (5)
Fish meal, menhaden

Ref no 5-02-009

		As fed	Dry	C.V.
Dry matter	%	92.0	100.0	2
Ash	%	19.6	21.3	17
Crude fiber	%	1.0	1.1	90
Ether extract	%	7.7	8.4	41
N-free extract	%	2.4	2.6	
Protein (N x 6.25)	%	61.3	66.6	5
Sheep	dig prot %	49.6	53.9	
Swine	dig prot %	56.4	61.3	
Energy				
Sheep	DE kcal/kg	3002.	3263.	
Swine	DE kcal/kg	3123.	3395.	

Continued

(5) protein supplements
(6) minerals

(7) vitamins
(8) additives

Feed name or analyses		Mean		C.V.
		As fed	Dry	± %
Chickens	MEₙ kcal/kg	2866.	3115.	
Sheep	ME kcal/kg	2462.	2676.	
Swine	ME kcal/kg	2580.	2804.	
Sheep	TDN %	68.	74.	
Swine	TDN %	71.	77.	
Calcium	%	5.49	5.97	16
Iron	%	.056	.061	37
Phosphorus	%	2.81	3.05	31
Copper	mg/kg	8.4	9.1	23
Manganese	mg/kg	25.7	27.9	40
Choline	mg/kg	3080.	3348.	
Niacin	mg/kg	55.9	60.8	13
Pantothenic acid	mg/kg	8.8	9.6	
Riboflavin	mg/kg	4.8	5.2	32
Thiamine	mg/kg	.7	.8	69
Arginine	%	4.00	4.35	
Histidine	%	1.60	1.74	
Isoleucine	%	4.10	4.46	
Leucine	%	5.00	5.44	
Lysine	%	5.30	5.76	6
Methionine	%	1.80	1.96	5
Phenylalanine	%	2.70	2.93	
Threonine	%	2.90	3.15	
Tryptophan	%	.60	.65	
Tyrosine	%	1.60	1.74	
Valine	%	3.60	3.91	

FISH, PILCHARD. Sardinops spp

Fish, pilchard, whole or cuttings, cooked mech-
extd dehy grnd, (5)
   Fish meal, pilchard

Ref no 5-02-010

Dry matter	%	92.0	100.0	
Ash	%	13.7	14.9	5
Ether extract	%	5.5	6.0	34
Protein (N x 6.25)	%	65.5	71.2	5
Sheep	dig prot %	53.7	58.4	
Swine	dig prot %	60.3	65.5	
Energy				
Sheep	DE kcal/kg	2921.	3175.	
Swine	DE kcal/kg	3083.	3351.	
Sheep	ME kcal/kg	2396.	2604.	
Swine	ME kcal/kg	2515.	2734.	
Sheep	TDN %	66.	72.	
Swine	TDN %	70.	76.	
Calcium	%	4.09	4.45	4
Iron	%	.03	.03	
Magnesium	%	.31	.34	
Phosphorus	%	2.80	3.04	20
Potassium	%	.32	.35	
Sodium	%	.17	.18	

Continued

(1) dry forages and roughages
(2) pasture, range plants, and forages fed green

Feed name or analyses		Mean		C.V.
		As fed	Dry	± %
Manganese	mg/kg	9.0	9.8	
Choline	mg/kg	2081.	2262.	
Riboflavin	mg/kg	9.5	10.3	48
Cysteine	%	1.20	1.30	
Methionine	%	1.90	2.07	

FISH, REDFISH. Sciaenops ocellata

Fish, redfish, whole, raw, (5)
   Drumfish, whole, raw
   Ocean perch, whole, raw

Ref no 5-08-113

Dry matter	%	19.8	100.0
Ash	%	1.3	6.5
Ether extract	%	.4	2.0
N-free extract	%	.1	.6
Protein (N x 6.25)	%	18.0	90.9
Energy	GE kcal/kg	800.	4040.
Potassium	%	.27	1.36
Sodium	%	.06	.30
Niacin	mg/kg	35.0	176.8
Riboflavin	mg/kg	.5	2.5
Thiamine	mg/kg	1.5	7.6

Fish, redfish, whole or cuttings, cooked mech-
extd dehy grnd, (5)
   Fish meal, drum
   Fish meal, redfish

Ref no 5-07-973

Dry matter	%	94.2	100.0
Crude fiber	%	1.0	1.1
Ether extract	%	8.0	8.5
Protein (N x 6.25)	%	55.0	58.4
Calcium	%	4.00	4.20
Phosphorus	%	2.20	2.40
Choline	mg/kg	1400.	1486.
Niacin	mg/kg	27.0	28.7
Pantothenic acid	mg/kg	2.4	2.6
Riboflavin	mg/kg	2.9	3.1

(3) silages
(4) energy feeds

Feed name or analyses		Mean As fed	Dry	C.V. ± %

FISH, ROCKFISH. Sebastodes spp

Fish, rockfish, whole, raw, (5)

Ref no 5-07-974

Feed name or analyses		As fed	Dry	C.V. ± %
Dry matter	%	32.0	100.0	
Ether extract	%	7.2	22.6	
Protein (N x 6.25)	%	16.2	50.7	

FISH, SALMON. Oncorhynchus spp, Salmo spp

Fish, salmon, whole, raw, (5)

Ref no 5-02-011

Feed name or analyses		As fed	Dry	C.V. ± %
Dry matter	%	35.0	100.0	
Ether extract	%	13.0	37.1	
Protein (N x 6.25)	%	22.0	62.8	

Fish, salmon, whole or cuttings, cooked mech-extd dehy grnd, (5)
Fish meal, salmon

Ref no 5-02-012

Feed name or analyses		As fed	Dry	C.V. ± %
Dry matter	%	93.0	100.0	3
Ash	%	16.7	17.9	19
Ether extract	%	9.6	10.3	50
Protein (N x 6.25)	%	58.0	62.3	8
Swine	dig prot %	53.3	57.3	
Energy				
Swine	DE kcal/kg	3116.	3351.	
Swine	ME kcal/kg	2599.	2795.	
Swine	TDN %	71.	76.	
Calcium	%	5.44	5.85	27
Iron	%	.02	.02	
Phosphorus	%	3.26	3.50	29
Copper	mg/kg	11.9	12.8	
Manganese	mg/kg	7.9	8.5	
Choline	mg/kg	2772.	2980.	
Niacin	mg/kg	24.9	26.8	
Pantothenic acid	mg/kg	6.8	7.3	
Riboflavin	mg/kg	5.7	6.1	
Thiamine	mg/kg	.9	1.0	
Arginine	%	5.20	5.59	
Cystine	%	.70	.75	
Glycine	%	5.20	5.59	
Lysine	%	7.60	8.17	
Methionine	%	1.60	1.72	
Tryptophan	%	.50	.54	

(5) protein supplements
(6) minerals

FISH, SARDINE. Clupea spp, Sardinops spp

Fish, sardine, stickwater sol, cooked cond, (5)

Ref no 5-02-014

Feed name or analyses		As fed	Dry	C.V. ± %
Dry matter	%	49.0	100.0	3
Ash	%	10.2	20.8	
Ether extract	%	9.5	19.4	
Protein (N x 6.25)	%	29.5	60.2	18
Swine	dig prot %	28.3	57.8	
Energy				
Swine	DE kcal/kg	2182.	4453.	
Swine	ME kcal/kg	1829.	3732.	
Swine	TDN %	49.	101.	
Calcium	%	.14	.29	
Phosphorus	%	.83	1.69	
Potassium	%	.18	.37	
Sodium	%	.18	.37	
Copper	mg/kg	25.7	52.5	
Manganese	mg/kg	24.9	50.8	
Biotin	mg/kg	.1	.2	
Choline	mg/kg	3003.	6129.	
Niacin	mg/kg	354.1	722.7	29
Pantothenic acid	mg/kg	41.1	83.9	
Riboflavin	mg/kg	16.7	34.1	31
Thiamine	mg/kg	4.0	8.2	
Arginine	%	1.50	3.06	
Cystine	%	.20	.41	
Histidine	%	2.00	4.08	
Isoleucine	%	.90	1.84	
Leucine	%	1.60	3.27	
Lysine	%	1.60	3.27	
Methionine	%	.90	1.84	
Phenylalanine	%	.80	1.63	
Threonine	%	.80	1.63	
Tryptophan	%	.10	.20	
Valine	%	1.00	2.04	

Fish, sardine, whole or cuttings, cooked mech-extd dehy grnd, (5)
Fish meal, sardine

Ref no 5-02-015

Feed name or analyses		As fed	Dry	C.V. ± %
Dry matter	%	93.0	100.0	1
Ash	%	15.7	16.9	17
Crude fiber	%	1.0	1.1	70
Ether extract	%	4.3	4.6	37
N-free extract	%	6.5	7.0	
Protein (N x 6.25)	%	65.5	70.4	7
Swine	dig prot %	60.3	64.8	
Energy				

Continued

(7) vitamins
(8) additives

Feed name or analyses		Mean		C.V.
		As fed	Dry	± %
Swine	DE kcal/kg	2994.	3219.	
Chickens	MEn kcal/kg	2866.	3082.	
Swine	ME kcal/kg	2449.	2633.	
Swine	TDN %	68.	73.	
Calcium	%	4.90	5.27	13
Iron	%	.03	.03	
Magnesium	%	.10	.11	
Phosphorus	%	2.77	2.98	14
Potassium	%	.33	.35	
Sodium	%	.18	.19	
Copper	mg/kg	20.2	21.7	1
Manganese	mg/kg	22.2	23.9	1
Choline	mg/kg	2959.	3181.	10
Niacin	mg/kg	62.0	66.7	
Pantothenic acid	mg/kg	9.2	9.9	
Riboflavin	mg/kg	5.9	6.3	23
Thiamine	mg/kg	.4	.4	
Arginine	%	2.70	2.90	42
Cystine	%	.80	.86	
Glycine	%	4.50	4.84	
Histidine	%	1.80	1.94	29
Isoleucine	%	3.30	3.55	
Leucine	%	4.70	5.05	
Lysine	%	5.90	6.34	38
Methionine	%	2.00	2.15	2
Phenylalanine	%	2.60	2.80	
Threonine	%	2.60	2.80	
Tryptophan	%	.50	.54	
Tyrosine	%	3.00	3.23	
Valine	%	4.10	4.41	

FISH, SHARK. Selachii (order)

Fish, shark, whole or cuttings, cooked mech-
   extd dehy grnd, (5)
   Fish meal, shark

Ref no 5-02-018

Dry matter	%	91.0	100.0	4
Ash	%	13.4	14.7	15
Ether extract	%	2.5	2.7	33
Protein (N x 6.25)	%	72.0	79.1	9
Swine	dig prot %	66.2	72.8	
Energy				
Swine	DE kcal/kg	3129.	3439.	
Swine	ME kcal/kg	2503.	2751.	
Swine	TDN %	71.	78.	
Calcium	%	3.48	3.82	26
Iron	%	.02	.02	
Magnesium	%	.17	.19	41
Phosphorus	%	1.83	2.01	19
Sodium	%	.33	.36	
Copper	mg/kg	112.4	123.5	

Feed name or analyses		Mean		C.V.
		As fed	Dry	± %
Manganese	mg/kg	90.0	98.9	
Zinc	mg/kg	112.4	123.5	
Methionine	%	.80	.88	

FISH, SMELT. Asmerus spp

Fish, smelt, whole, raw, (5)

Ref no 5-07-975

Dry matter	%	21.0	100.0	
Ether extract	%	1.8	8.6	
Protein (N x 6.25)	%	18.0	85.7	

FISH, SOLE. Soleidae (family)

Fish, sole, whole, raw, (5)

Ref no 5-07-976

Dry matter	%	19.0	100.0	
Ether extract	%	1.7	9.1	
Protein (N x 6.25)	%	13.7	72.3	
Calcium	%	.63	3.32	
Phosphorus	%	.44	2.30	

FISH, TUNA. Thunnus thynnus, Thunnus albacares

Fish, tuna, proc res, (5)

Ref no 5-07-977

Dry matter	%	44.0	100.0	
Ether extract	%	9.6	21.8	
Protein (N x 6.25)	%	24.1	54.8	

Fish, tuna, stickwater sol, cooked cond, (5)

Ref no 5-02-022

Dry matter	%	52.0	100.0	2
Ash	%	9.5	18.3	21
Ether extract	%	4.5	8.7	50
Protein (N x 6.25)	%	34.9	67.1	6
Swine	dig prot %	33.5	64.4	
Energy				
Swine	DE kcal/kg	1926.	3704.	
Swine	ME kcal/kg	1587.	3052.	
Swine	TDN %	44.	84.	

Continued

(1) dry forages and roughages
(2) pasture, range plants, and forages fed green

(3) silages
(4) energy feeds

Feed name or analyses		Mean		C.V.
		As fed	Dry	± %
Potassium	%	1.76	3.38	17
Sodium	%	1.68	3.23	53
Folic acid	mg/kg	.2	.4	83
Niacin	mg/kg	313.5	602.9	15
Riboflavin	mg/kg	21.3	41.0	
Thiamine	mg/kg	23.5	45.2	
Vitamin B$_{12}$	mcg/kg	482.2	927.4	8
Arginine	%	1.20	2.31	
Glycine	%	.70	1.35	
Leucine	%	1.20	2.31	
Lysine	%	1.40	2.69	
Valine	%	.90	1.73	

Fish, tuna, whole or cuttings, cooked mech-extd dehy grnd, (5)

Fish meal, tuna

Ref no 5-02-023

Dry matter	%	87.0	100.0	5
Ash	%	19.0	21.8	14
Crude fiber	%	1.0	1.1	60
Ether extract	%	8.9	10.2	38
N-free extract	%	.9	1.0	
Protein (N x 6.25)	%	57.3	65.9	11
Cattle	dig prot %	43.6	50.1	
Swine	dig prot %	52.7	60.6	
Energy				
Cattle	DE kcal/kg	2762.	3175.	
Swine	DE kcal/kg	3029.	3483.	
Cattle	ME kcal/kg	2265.	2604.	
Chickens	ME$_n$ kcal/kg	2866.	3294.	
Swine	ME kcal/kg	2540.	2919.	
Cattle	TDN %	63.	72.	
Swine	TDN %	69.	79.	
Calcium	%	5.32	6.11	10
Phosphorus	%	3.07	3.53	9
Arginine	%	6.99	8.04	
Lysine	%	6.19	7.12	
Methionine	%	1.70	1.95	2
Tryptophan	%	.90	1.03	
Valine	%	2.20	2.53	

FISH, TURBOT. Psetta maxima

Fish, turbot, whole, raw, (5)

Ref no 5-07-978

Dry matter	%	27.0	100.0	
Ether extract	%	10.4	38.7	
Protein (N x 6.25)	%	14.4	53.2	
Calcium	%	.39	1.46	
Phosphorus	%	.32	1.17	

(5) protein supplements

(6) minerals

Feed name or analyses		Mean		C.V.
		As fed	Dry	± %

FISH, WHITE. Gadidae (family), Lophiidae (family), Rajidae (family)

Fish, white, whole or cuttings, cooked mech-extd dehy grnd, mx 4 oil, (5)

White fish meal (CFA)
Fish, cod, meal
Fish, cusk, meal
Fish, haddock, meal
Fish, hake, meal
Fish, pollock, meal
Fish, monkfish, meal
Fish, skate, meal

Ref no 5-02-025

Dry matter	%	92.0	100.0	3
Ash	%	21.7	23.6	13
Crude fiber	%	1.0	1.1	98
Ether extract	%	4.4	4.8	59
N-free extract	%	1.6	1.8	
Protein (N x 6.25)	%	63.2	68.7	7
Sheep	dig prot %	58.8	63.9	
Swine	dig prot %	58.1	63.2	
Energy				
Sheep	DE kcal/kg	3002.	3263.	
Swine	DE kcal/kg	2921.	3175.	
Sheep	ME kcal/kg	2462.	2676.	
Swine	ME kcal/kg	2398.	2607.	
Sheep	TDN %	68.	74.	
Swine	TDN %	66.	72.	
Calcium	%	7.87	8.55	16
Phosphorus	%	3.61	3.92	20
Manganese	mg/kg	14.3	15.5	
Choline	mg/kg	8917.	9692.	
Niacin	mg/kg	69.7	75.8	26
Pantothenic acid	mg/kg	8.8	9.6	
Riboflavin	mg/kg	9.0	9.8	39
Thiamine	mg/kg	1.8	2.0	83

FISH, WHITING. Gadus merlangus

Fish, whiting, whole, raw, (5)

Ref no 5-07-979

Dry matter	%	23.0	100.0	
Ether extract	%	2.0	8.7	
Protein (N x 6.25)	%	16.0	69.6	

(7) vitamins

(8) additives

Feed name or analyses		Mean		C.V.
		As fed	Dry	± %

FLAX. Linum usitatissimum

Flax, fbr by-prod, mn 9 prot mx 35 fbr, (1)
Flax plant product (AAFCO)

Ref no 1-02-036

Dry matter	%	91.0	100.0	
Ash	%	5.0	5.5	
Crude fiber	%	32.0	35.2	
Ether extract	%	2.6	2.9	
N-free extract	%	40.0	44.0	
Protein (N x 6.25)	%	11.3	12.4	
Cattle	dig prot %	7.0	7.7	
Sheep	dig prot %	7.0	7.7	
Energy				
Cattle	DE kcal/kg	2087.	2293.	
Sheep	DE kcal/kg	1845.	2028.	
Cattle	ME kcal/kg	1711.	1880.	
Sheep	ME kcal/kg	1513.	1663.	
Cattle	TDN %	47.	52.	
Sheep	TDN %	42.	46.	

Flax, straw, (1)

Ref no 1-02-038

Dry matter	%	92.9	100.0	2
Ash	%	6.7	7.2	14
Crude fiber	%	43.0	46.2	3
Ether extract	%	3.1	3.3	11
N-free extract	%	32.9	35.5	
Protein (N x 6.25)	%	7.2	7.8	10
Cattle	dig prot %	3.4	3.7	
Sheep	dig prot %	3.3	3.6	
Energy				
Cattle	DE kcal/kg	1515.	1631.	
Sheep	DE kcal/kg	1802.	1940.	
Cattle	ME kcal/kg	1242.	1337.	
Sheep	ME kcal/kg	1478.	1591.	
Cattle	TDN %	34.	37.	
Sheep	TDN %	41.	44.	
Calcium	%	.67	.72	
Chlorine	%	.25	.27	
Magnesium	%	.29	.31	
Phosphorus	%	.10	.11	
Potassium	%	1.62	1.74	

Flax, seed screenings, (4)

Ref no 4-02-056

Dry matter	%	91.6	100.0	
Ash	%	7.3	8.0	
Crude fiber	%	13.3	14.5	
Ether extract	%	10.2	11.1	
N-free extract	%	44.9	49.0	
Protein (N x 6.25)	%	15.9	17.4	
Cattle	dig prot %	8.9	9.7	
Sheep	dig prot %	8.9	9.7	
Energy				
Cattle	DE kcal/kg	2585.	2822.	
Sheep	DE kcal/kg	2585.	2822.	
Cattle	ME kcal/kg	2120.	2314.	
Sheep	ME kcal/kg	2120.	2314.	
Cattle	TDN %	59.	64.	
Sheep	TDN %	59.	64.	
Calcium	%	.37	.40	
Phosphorus	%	.43	.47	

Flax, seed, mech-extd grnd, mx 0.5 acid insol ash, (5)
Linseed meal (AAFCO)
Linseed meal (CFA)
Linseed oil meal, expeller extracted
Linseed oil meal, hydraulic extracted
Linseed meal, old process

Ref no 5-02-045

Dry matter	%	91.0	100.0	2
Ash	%	5.6	6.2	10
Crude fiber	%	9.0	9.9	17
Ether extract	%	5.2	5.7	25
N-free extract	%	35.8	39.4	
Protein (N x 6.25)	%	35.3	38.8	7
Cattle	dig prot %	31.0	34.1	
Sheep	dig prot %	29.7	32.6	
Swine	dig prot %	31.8	34.9	
Energy				
Cattle	DE kcal/kg	3250.	3571.	
Sheep	DE kcal/kg	3210.	3527.	
Swine	DE kcal/kg	3388.	3723.	
Cattle	ME kcal/kg	2664.	2928.	
Chickens	MEn kcal/kg	1521.	1671.	
Sheep	ME kcal/kg	2632.	2892.	
Swine	ME kcal/kg	2988.	3284.	
Cattle	NEm kcal/kg	1729.	1900.	
Cattle	NEgain kcal/kg	1156.	1270.	
Cattle	TDN %	74.	81.	
Sheep	TDN %	73.	80.	
Swine	TDN %	77.	85.	

Continued

(1) dry forages and roughages

(2) pasture, range plants, and forages fed green

(3) silages

(4) energy feeds

Feed name or analyses		Mean		C.V.
		As fed	Dry	± %
Calcium	%	.44	.48	26
Iron	%	.017	.019	37
Magnesium	%	.58	.64	
Phosphorus	%	.89	.98	18
Potassium	%	1.24	1.36	
Sodium	%	.11	.12	
Cobalt	mg/kg	.400	.500	40
Copper	mg/kg	26.4	29.0	29
Manganese	mg/kg	39.4	43.3	35
Carotene	mg/kg	.2	.2	
Choline	mg/kg	1863.	2048.	
Folic acid	mg/kg	2.90	3.20	
Niacin	mg/kg	35.6	39.1	16
Pantothenic acid	mg/kg	17.8	19.6	26
Riboflavin	mg/kg	3.5	3.8	21
Thiamine	mg/kg	5.1	5.6	46
Vitamin A equiv	IU/g	.3	.3	
Methionine	%	.70	.77	

Flax, seed, solv-extd grnd, mx 0.5 acid insol ash, (5)
Solvent extracted linseed meal (AAFCO)
Solvent extracted linseed meal (CFA)
Linseed oil meal, solvent extracted

Ref no 5-02-048

Feed name or analyses		Mean		C.V.
		As fed	Dry	± %
Dry matter	%	91.0	100.0	2
Ash	%	5.8	6.4	12
Crude fiber	%	9.0	9.9	11
Ether extract	%	1.7	1.9	51
N-free extract	%	39.3	43.2	
Protein (N x 6.25)	%	35.1	38.6	8
Cattle	dig prot %	30.9	34.0	
Sheep	dig prot %	30.6	33.6	
Swine	dig prot %	31.6	34.7	
Energy				
Cattle	DE kcal/kg	3049.	3351.	
Sheep	DE kcal/kg	3129.	3439.	
Swine	DE kcal/kg	2969.	3263.	
Cattle	ME kcal/kg	2501.	2748.	
Chickens	MEn kcal/kg	1411.	1550.	
Sheep	ME kcal/kg	2566.	2820.	
Swine	ME kcal/kg	2619.	2878.	
Cattle	NEm kcal/kg	1629.	1790.	
Cattle	NEgain kcal/kg	1083.	1190.	
Cattle	TDN %	69.	76.	
Sheep	TDN %	71.	78.	
Swine	TDN %	67.	74.	
Calcium	%	.40	.44	20
Iron	%	.033	.036	
Magnesium	%	.60	.66	
Phosphorus	%	.83	.91	21
Potassium	%	1.38	1.52	
Sodium	%	.14	.15	
Cobalt	mg/kg	.20	.20	

Continued

Feed name or analyses		Mean		C.V.
		As fed	Dry	± %
Copper	mg/kg	25.7	28.2	
Manganese	mg/kg	37.6	41.3	36
Choline	mg/kg	1225.	1347.	
Niacin	mg/kg	30.1	33.1	
Riboflavin	mg/kg	2.9	3.2	
Thiamine	mg/kg	9.5	10.4	28

Flax, seed screenings, extn unspec grnd, (5)
Flax seed screenings oil feed (CFA)

Ref no 5-02-053

Feed name or analyses		Mean		C.V.
		As fed	Dry	± %
Dry matter	%	91.3	100.0	1
Ash	%	8.5	9.3	22
Crude fiber	%	11.4	12.5	18
Ether extract	%	7.2	7.9	11
N-free extract	%	40.1	43.9	
Protein (N x 6.25)	%	24.1	26.4	7
Cattle	dig prot %	13.6	14.9	
Sheep	dig prot %	13.6	14.9	
Energy				
Cattle	DE kcal/kg	2375.	2601.	
Sheep	DE kcal/kg	2375.	2601.	
Cattle	ME kcal/kg	1947.	2133.	
Sheep	ME kcal/kg	1947.	2133.	
Cattle	TDN %	54.	59.	
Sheep	TDN %	54.	59.	
Calcium	%	.44	.48	
Phosphorus	%	.63	.69	

Flax, seed screenings, mech-extd grnd, (5)
Flaxseed screenings meal (AAFCO)

Ref no 5-02-054

Feed name or analyses		Mean		C.V.
		As fed	Dry	± %
Dry matter	%	91.0	100.0	1
Ash	%	6.7	7.4	18
Crude fiber	%	12.0	13.2	20
Ether extract	%	9.4	10.3	18
N-free extract	%	47.0	51.7	
Protein (N x 6.25)	%	15.8	17.4	11
Cattle	dig prot %	10.9	12.0	
Sheep	dig prot %	12.0	13.2	
Energy	GE kcal/kg	4316.	4743.	
Cattle	DE kcal/kg	2969.	3263.	
Sheep	DE kcal/kg	3009.	3307.	
Cattle	ME kcal/kg	2435.	2676.	
Sheep	ME kcal/kg	2468.	2712.	
Cattle	TDN %	67.	74.	
Sheep	TDN %	68.	75.	
Calcium	%	.37	.41	
Phosphorus	%	.43	.47	

(5) protein supplements
(6) minerals

(7) vitamins
(8) additives

Feed name or analyses		Mean		C.V.
		As fed	Dry	± %

FOXTAIL. Alopecurus pratensis

Foxtail, meadow, hay, s-c, (1)

Ref no 1-02-072

		As fed	Dry
Dry matter	%	87.3	100.0
Ash	%	7.5	8.6
Crude fiber	%	26.2	30.0
Ether extract	%	2.3	2.6
N-free extract	%	39.3	45.0
Protein (N x 6.25)	%	12.0	13.8
Cattle	dig prot %	7.8	8.9
Sheep	dig prot %	8.0	9.2
Energy			
Cattle	DE kcal/kg	2232.	2557.
Sheep	DE kcal/kg	2502.	2866.
Cattle	ME kcal/kg	1831.	2097.
Sheep	ME kcal/kg	2052.	2350.
Cattle	TDN %	51.	58.
Sheep	TDN %	57.	65.

GALLETA. Hilaria jamesii

Galleta, aerial pt, fresh, dormant, (2)

Ref no 2-05-594

		As fed	Dry
Dry matter	%	86.0	100.0
Ash	%	14.3	16.6
Ether extract	%	1.7	2.0
Protein (N x 6.25)	%	4.7	5.5
Sheep	dig prot %	1.2	1.4
Cellulose	%	24.1	28.0
Lignin	%	6.9	8.0
Energy	GE kcal/kg	3320.	3860.
Sheep	DE kcal/kg	1434.	1667.
Sheep	ME kcal/kg	1128.	1312.
Sheep	TDN %	34.	39.
Calcium	%	.90	1.05
Phosphorus	%	.06	.07
Carotene	mg/kg	.3	.4

GAMAGRASS, EASTERN. Tripsacum dactyloides

Gamagrass, Eastern, aerial pt, fresh, full blm, (2)

Ref no 2-02-084

		As fed	Dry
Dry matter	%	30.0	100.0
Ash	%	3.2	10.6
Crude fiber	%	8.8	29.5
Ether extract	%	.6	2.0
N-free extract	%	15.1	50.2
Protein (N x 6.25)	%	2.3	7.7
Cattle	dig prot %	1.3	4.3
Sheep	dig prot %	1.3	4.2
Energy			
Cattle	DE kcal/kg	688.	2293.
Sheep	DE kcal/kg	754.	2513.
Cattle	ME kcal/kg	564.	1880.
Sheep	ME kcal/kg	618.	2061.
Cattle	TDN %	16.	52.
Sheep	TDN %	17.	57.
Calcium	%	.19	.62
Phosphorus	%	.09	.31

GAMAGRASS FLORIDA. Tripsacum floridanum

Gamagrass, Florida, hay, s-c, (1)

Ref no 1-02-087

		As fed	Dry
Dry matter	%	92.3	100.0
Ash	%	6.8	7.4
Crude fiber	%	26.0	28.2
Ether extract	%	1.8	1.9
N-free extract	%	49.3	53.4
Protein (N x 6.25)	%	8.4	9.1
Cattle	dig prot %	4.2	4.6
Sheep	dig prot %	4.2	4.6
Energy			
Cattle	DE kcal/kg	2116.	2293.
Sheep	DE kcal/kg	2116.	2293.
Cattle	ME kcal/kg	1735.	1880.
Sheep	ME kcal/kg	1735.	1880.
Cattle	TDN %	48.	52.
Sheep	TDN %	48.	52.
Calcium	%	.57	.62
Phosphorus	%	.29	.31

(1) dry forages and roughages
(2) pasture, range plants, and forages fed green

(3) silages
(4) energy feeds

Feed name or analyses		Mean		C.V. ± %
		As fed	Dry	

GARBAGE. Scientific name not used

Garbage, cooked dehy, high fat, (4)

Ref no 4-07-863

		As fed	Dry
Dry matter	%	95.9	100.0
Ash	%	12.9	13.4
Crude fiber	%	20.0	20.8
Ether extract	%	23.7	24.7
N-free extract	%	22.0	22.9
Protein (N x 6.25)	%	17.5	18.2
Swine	dig prot %	6.3	6.6
Energy			
Swine	DE kcal/kg	3805.	3968.
Swine	ME kcal/kg	3512.	3662.
Swine	TDN %	86.	90.

Garbage, cooked dehy, low fat, (4)

Ref no 4-07-862

		As fed	Dry
Dry matter	%	92.3	100.0
Ash	%	14.1	15.3
Crude fiber	%	13.5	14.6
Ether extract	%	3.5	3.8
N-free extract	%	38.1	41.3
Protein (N x 6.25)	%	23.1	25.0
Swine	dig prot %	14.0	15.2
Energy			
Swine	DE kcal/kg	2279.	2469.
Swine	ME kcal/kg	2074.	2247.
Swine	TDN %	52.	56.

Garbage, cooked wet, (4)

Ref no 4-02-093

		As fed	Dry	C.V.
Dry matter	%	30.0	100.0	15
Ash	%	1.6	5.3	99
Crude fiber	%	1.0	3.3	99
Ether extract	%	8.8	29.3	65
N-free extract	%	12.3	41.1	
Protein (N x 6.25)	%	6.3	21.0	19
Energy				
Swine	DE kcal/kg	1680.	5600.	
Swine	ME kcal/kg	1378.	4592.	
Swine	TDN %	38.	127.	

Garbage, hotel and restaurant, cooked wet grnd, (4)

Ref no 4-07-865

		As fed	Dry	C.V.
Dry matter	%	16.0	100.0	28
Ash	%	.9	5.7	23
Crude fiber	%	.5	3.3	43
Ether extract	%	4.0	24.9	33
N-free extract	%	8.1	50.8	
Protein (N x 6.25)	%	2.4	15.3	24
Swine	dig prot %	2.2	13.5	
Energy	GE kcal/kg	853.	5330.	9
Swine	DE kcal/kg	793.	4957.	
Swine	ME kcal/kg	767.	4605.	
Swine	TDN %	19.	117.	

Garbage, institutional, cooked wet grnd, (4)

Ref no 4-07-867

		As fed	Dry
Dry matter	%	17.7	100.0
Ash	%	.9	5.3
Crude fiber	%	.5	2.8
Ether extract	%	2.6	14.8
N-free extract	%	11.1	62.5
Protein (N x 6.25)	%	2.6	14.6
Swine	dig prot %	2.3	12.8
Energy	GE kcal/kg	853.	4820.
Swine	DE kcal/kg	793.	4483.
Swine	ME kcal/kg	738.	4169.
Swine	TDN %	19.	106.

Garbage, military, cooked wet grnd, (4)

Ref no 4-07-866

		As fed	Dry
Dry matter	%		100.0
Ash	%		5.6
Crude fiber	%	.7	2.8
Ether extract	%	8.2	32.0
N-free extract	%	11.2	43.6
Protein (N x 6.25)	%	4.1	16.0
Swine	dig prot %	3.6	14.2
Energy	GE kcal/kg	1450.	5665.
Swine	DE kcal/kg	1378.	5382.
Swine	ME kcal/kg	1278.	4994.
Swine	TDN %	33.	128.

(5) protein supplements
(6) minerals

(7) vitamins
(8) additives

Feed name or analyses		Mean		C.V.
		As fed	Dry	± %

Garbage, municipal, cooked wet grnd, (4)

Ref no 4-07-864

		As fed	Dry	
Dry matter	%	16.6	100.0	46
Ash	%	1.4	8.6	44
Crude fiber	%	1.4	8.4	54
Ether extract	%	3.6	21.4	34
N-free extract	%	7.3	44.1	
Protein (N x 6.25)	%	2.9	17.5	26
Swine	dig prot %	2.3	14.0	
Energy	GE kcal/kg	847.	5100.	8
Swine	DE kcal/kg	635.	3825.	
Swine	ME kcal/kg	587.	3538.	
Swine	TDN %	15.	91.	

Gluten feed - see Corn, gluten w bran

Grain sorghum - see Sorghum, grain variety

GRAINS. Scientific name not used

Grains screenings - see also Barley, grain screenings; Wheat, grain screenings

Grains, screenings, gr 1 mn 35 grain mx 7 fbr mx 6 fm mx 8 wild oats, (4)
No 1 feed screenings (CFA)
Feed screenings No 1

Ref no 4-02-154

Dry matter	%	86.5	100.0	
Ash	%	6.9	8.0	
Crude fiber	%	5.2	6.0	
Ether extract	%	3.5	4.0	
N-free extract	%	57.0	65.9	
Protein (N x 6.25)	%	13.9	16.1	
Cattle	dig prot %	10.7	12.4	
Sheep	dig prot %	10.0	11.6	
Energy				
Cattle	DE kcal/kg	2784.	3219.	
Sheep	DE kcal/kg	2860.	3307.	
Cattle	ME kcal/kg	2284.	2640.	
Sheep	ME kcal/kg	2346.	2712.	
Cattle	NE$_m$ kcal/kg	1332.	1540.	
Cattle	NE$_{gain}$ kcal/kg	848.	980.	
Cattle	TDN %	63.	73.	
Sheep	TDN %	65.	75.	
Alanine	%	.47	.54	

Continued

Feed name or analyses		Mean		C.V.
		As fed	Dry	± %
Arginine	%	.63	.73	
Aspartic acid	%	.64	.74	
Glutamic acid	%	3.67	4.24	
Glycine	%	.54	.62	
Histidine	%	.27	.31	
Isoleucine	%	.47	.54	
Leucine	%	.83	.96	
Lysine	%	.33	.38	
Methionine	%	.13	.15	
Phenylalanine	%	.80	.92	
Proline	%	1.15	1.33	
Serine	%	.54	.63	
Threonine	%	.36	.42	
Tyrosine	%	.26	.30	
Valine	%	.54	.63	

Grains, screenings, gr 2 mx 11 fbr mx 10 fm mx 49 wild oats, (4)
No 2 feed screenings (CFA)
Feed screenings No 2

Ref no 4-02-155

Dry matter	%	86.5	100.0	
Ash	%	7.8	9.0	
Crude fiber	%	9.5	11.0	
Ether extract	%	3.5	4.0	
N-free extract	%	54.3	62.8	
Protein (N x 6.25)	%	11.4	13.2	
Cattle	dig prot %	8.8	10.2	
Sheep	dig prot %	8.2	9.5	
Energy				
Cattle	DE kcal/kg	2746.	3175.	
Sheep	DE kcal/kg	2707.	3130.	
Cattle	ME kcal/kg	2252.	2604.	
Sheep	ME kcal/kg	2220.	2567.	
Cattle	TDN %	62.	72.	
Sheep	TDN %	61.	71.	
Alanine	%	.41	.47	
Arginine	%	.71	.82	
Aspartic acid	%	.78	.90	
Glutamic acid	%	1.57	1.82	
Glycine	%	.46	.53	
Histidine	%	.42	.26	
Isoleucine	%	.42	.49	
Leucine	%	.64	.74	
Lysine	%	.42	.48	
Methionine	%	.10	.12	
Phenylalanine	%	.43	.50	
Proline	%	.54	.62	
Serine	%	.43	.50	
Threonine	%	.33	.38	
Tyrosine	%	.21	.24	
Valine	%	.52	.60	

(1) dry forages and roughages
(2) pasture, range plants, and forages fed green

(3) silages
(4) energy feeds

Grains, screenings, mn 70 grain mx 6.5 ash, (4)
Grain screenings (AAFCO)

Ref no 4-02-156

Feed name or analyses		Mean As fed	Mean Dry	C.V. ± %
Dry matter	%	90.0	100.0	1
Ash	%	5.7	6.3	57
Crude fiber	%	9.0	10.0	57
Ether extract	%	4.7	5.2	70
N-free extract	%	55.5	61.7	
Protein (N x 6.25)	%	15.1	16.8	11
Cattle	dig prot %	10.3	11.4	
Sheep	dig prot %	11.3	12.6	
Swine	dig prot %	12.1	13.4	
Energy				
Cattle	DE kcal/kg	2659.	2954.	
Sheep	DE kcal/kg	3016.	3351.	
Swine	DE kcal/kg	2579.	2866.	
Cattle	ME kcal/kg	2180.	2422.	
Chickens	ME_n kcal/kg	1323.	1470.	
Sheep	ME kcal/kg	2473.	2748.	
Swine	ME kcal/kg	2389.	2654.	
Cattle	NE_m kcal/kg	1433.	1592.	
Cattle	NE_{gain} kcal/kg	904.	1004.	
Cattle	TDN %	60.	67.	
Sheep	TDN %	68.	76.	
Swine	TDN %	58.	65.	
Calcium	%	.43	.48	
Phosphorus	%	.39	.43	

Feed name or analyses		Mean As fed	Mean Dry	C.V. ± %
Glutamic acid	%	2.82	3.12	
Glycine	%	.54	.60	
Histidine	%	.27	.30	
Isoleucine	%	.48	.53	
Leucine	%	.89	.99	
Lysine	%	.44	.49	
Methionine	%	.14	.15	
Phenylalanine	%	.59	.65	
Proline	%	.99	1.10	
Serine	%	.52	.58	
Threonine	%	.42	.47	
Tyrosine	%	.29	.32	
Valine	%	.58	.64	

Grains, screenings, refuse mx 100 small weed seeds chaff hulls dust scourings noxious seeds, (4)
Refuse screenings (CFA)

Ref no 4-02-151

Feed name or analyses		Mean As fed	Mean Dry	C.V. ± %
Dry matter	%	90.3	100.0	
Ash	%	9.6	10.6	
Crude fiber	%	28.3	31.3	
Ether extract	%	3.9	4.3	
N-free extract	%	34.1	37.8	
Protein (N x 6.25)	%	14.4	16.0	
Cattle	dig prot %	10.3	11.4	
Sheep	dig prot %	10.4	11.5	
Energy				
Cattle	DE kcal/kg	2031.	2249.	
Sheep	DE kcal/kg	2150.	2381.	
Cattle	ME kcal/kg	1665.	1844.	
Sheep	ME kcal/kg	1763.	1952.	
Cattle	TDN %	46.	51.	
Sheep	TDN %	49.	54.	
Alanine	%	.59	.65	
Arginine	%	.62	.69	
Aspartic acid	%	.86	.95	

Continued

Grains, screenings, uncleaned, mn 12 grain mx 3 wild oats mx 17 buckwheat and large seeds mx 68 small weed seeds chaff hulls dust scourings noxious seeds, (4)
Uncleaned screenings (CFA)

Ref no 4-02-153

Feed name or analyses		Mean As fed	Mean Dry	C.V. ± %
Dry matter	%	91.5	100.0	
Ash	%	7.7	8.4	
Crude fiber	%	16.7	18.3	
Ether extract	%	4.0	4.4	
N-free extract	%	48.8	53.3	
Protein (N x 6.25)	%	14.3	15.6	
Cattle	dig prot %	10.3	11.3	
Sheep	dig prot %	10.2	11.2	
Swine	dig prot %	9.9	10.8	
Energy				
Cattle	DE kcal/kg	2622.	2866.	
Sheep	DE kcal/kg	2663.	2910.	
Swine	DE kcal/kg	2461.	2690.	
Cattle	ME kcal/kg	2150.	2350.	
Sheep	ME kcal/kg	2183.	2386.	
Swine	ME kcal/kg	2284.	2496.	
Cattle	TDN %	59.	65.	
Sheep	TDN %	60.	66.	
Swine	TDN %	56.	61.	
Calcium	%	.37	.40	
Phosphorus	%	.41	.45	
Alanine	%	.49	.54	
Arginine	%	.61	.67	
Aspartic acid	%	.74	.81	
Glutamic acid	%	3.38	3.69	
Glycine	%	.56	.61	
Histidine	%	.27	.30	
Isoleucine	%	.41	.45	
Leucine	%	.82	.90	
Lysine	%	.38	.42	
Methionine	%	.17	.19	
Phenylalanine	%	.53	.58	
Proline	%	1.05	1.15	

Continued

(5) protein supplements
(6) minerals

(7) vitamins
(8) additives

Feed name or analyses		Mean		C.V. ± %
		As fed	Dry	
Serine	%	.61	.67	
Threonine	%	.40	.44	
Tyrosine	%	.53	.58	
Valine	%	.53	.58	

Grains, brewers grains, dehy, mx 3 dried spent hops, (5)
  Brewers dried grains (AAFCO)
  Brewers dried grains (CFA)

Ref no 5-02-141

		As fed	Dry	
Dry matter	%	92.0	100.0	1
Ash	%	3.6	3.9	14
Crude fiber	%	15.0	16.3	14
Ether extract	%	6.2	6.7	19
N-free extract	%	41.4	45.0	
Protein (N x 6.25)	%	25.9	28.1	11
Cattle	dig prot %	19.1	20.8	
Horses	dig prot %	19.9	21.6	
Sheep	dig prot %	19.1	20.8	
Swine	dig prot %	20.4	22.2	
Energy				
Cattle	DE kcal/kg	2677.	2910.	
Horses	DE kcal/kg	2069.	2249.	
Sheep	DE kcal/kg	2799.	3042.	
Swine	DE kcal/kg	1892.	2056.	
Cattle	ME kcal/kg	2195.	2386.	
Chickens	MEn kcal/kg	2513.	2732.	
Horses	ME kcal/kg	1696.	1844.	
Sheep	ME kcal/kg	2294.	2494.	
Swine	ME kcal/kg	1708.	1856.	
Cattle	NEm kcal/kg	1306.	1420.	
Cattle	NEgain kcal/kg	764.	830.	
Cattle	TDN %	61.	66.	
Horses	TDN %	47.	51.	
Sheep	TDN %	63.	69.	
Swine	TDN %	43.	47.	
Calcium	%	.27	.29	45
Iron	%	.025	.027	33
Magnesium	%	.14	.15	71
Phosphorus	%	.50	.54	15
Potassium	%	.08	.09	72
Sodium	%	.26	.28	
Cobalt	mg/kg	.100	.100	30
Copper	mg/kg	21.3	22.2	56
Manganese	mg/kg	37.6	40.9	24
Choline	mg/kg	1587.	1725.	
Folic acid	mg/kg	.22	.24	
Niacin	mg/kg	43.4	47.2	39
Pantothenic acid	mg/kg	8.6	9.3	49
Riboflavin	mg/kg	1.5	1.6	88
Thiamine	mg/kg	.7	.8	53
Vitamin B6	mg/kg	.66	.72	
Arginine	%	1.30	1.41	

Continued

Feed name or analyses		Mean		C.V. ± %
		As fed	Dry	
Histidine	%	.50	.54	
Isoleucine	%	1.50	1.63	
Leucine	%	2.30	2.50	
Lysine	%	.90	.98	
Methionine	%	.40	.43	
Phenylalanine	%	1.30	1.41	
Threonine	%	.90	.98	
Tryptophan	%	.40	.43	
Tyrosine	%	1.20	1.30	
Valine	%	1.60	1.74	

Grains, distil grains, dehy, (5)
  Distillers dried grains

Ref no 5-02-144

		As fed	Dry	
Dry matter	%	91.6	100.0	3
Ash	%	3.1	3.4	32
Crude fiber	%	11.5	12.6	21
Ether extract	%	8.9	9.7	30
N-free extract	%	38.9	42.5	
Protein (N x 6.25)	%	29.1	31.8	6
Cattle	dig prot %	23.1	25.2	
Sheep	dig prot %	20.1	21.9	
Swine	dig prot %	23.0	25.1	
Energy				
Cattle	DE kcal/kg	3393.	3704.	
Sheep	DE kcal/kg	3352.	3660.	
Swine	DE kcal/kg	2020.	2205.	
Cattle	ME kcal/kg	2782.	3037.	
Sheep	ME kcal/kg	2749.	3001.	
Swine	ME kcal/kg	1810.	1976.	
Cattle	TDN %	77.	84.	
Sheep	TDN %	76.	83.	
Swine	TDN %	46.	50.	
Calcium	%	.20	.22	35
Chlorine	%	.05	.05	
Iron	%	.026	.028	35
Magnesium	%	.12	.13	
Phosphorus	%	.55	.60	29
Potassium	%	.24	.26	
Sodium	%	.05	.05	
Sulfur	%	.45	.49	
Cobalt	mg/kg	.041	.045	34
Copper	mg/kg	21.5	23.5	53
Manganese	mg/kg	34.6	37.8	39
Carotene	mg/kg	7.7	8.4	
Niacin	mg/kg	46.3	50.6	36
Pantothenic acid	mg/kg	11.5	12.6	20
Riboflavin	mg/kg	3.7	4.0	52
Thiamine	mg/kg	2.4	2.6	45
Vitamin A equiv	IU/g	12.8	14.0	
Arginine	%	2.20	2.40	12
Histidine	%	1.10	1.20	20
Isoleucine	%	2.10	2.29	30

Continued

(1) dry forages and roughages
(2) pasture, range plants, and forages fed green

(3) silages
(4) energy feeds

Feed name or analyses		Mean		C.V.
		As fed	Dry	± %
Leucine	%	3.20	3.49	42
Lysine	%	3.00	3.28	19
Methionine	%	.40	.44	
Phenylalanine	%	1.80	1.96	13
Threonine	%	2.10	2.29	13
Tryptophan	%	.20	.22	
Tyrosine	%	.90	.98	
Valine	%	1.20	1.31	

GRAMA. Bouteloua spp

Grama, hay, s-c, (1)

Ref no 1-02-162

Dry matter	%	89.4	100.0	5
Ash	%	8.7	9.7	21
Crude fiber	%	29.2	32.7	7
Ether extract	%	1.5	1.7	16
N-free extract	%	44.4	49.7	
Protein (N x 6.25)	%	5.5	6.2	28
Cattle	dig prot %	2.0	2.3	
Sheep	dig prot %	1.9	2.1	
Energy				
Cattle	DE kcal/kg	2405.	2690.	
Sheep	DE kcal/kg	2168.	2425.	
Cattle	ME kcal/kg	1972.	2206.	
Sheep	ME kcal/kg	1777.	1988.	
Cattle	TDN %	54.	61.	
Sheep	TDN %	49.	55.	

Grama, aerial pt, fresh, immature, (2)

Ref no 2-02-163

Dry matter	%	41.0	100.0	7
Ash	%	4.6	11.3	26
Crude fiber	%	11.2	27.2	11
Ether extract	%	.8	2.0	35
N-free extract	%	19.0	46.4	
Protein (N x 6.25)	%	5.4	13.1	18
Cattle	dig prot %	3.7	9.0	
Sheep	dig prot %	3.8	9.2	
Energy				
Cattle	DE kcal/kg	1157.	2822.	
Sheep	DE kcal/kg	1157.	2822.	
Cattle	ME kcal/kg	949.	2314.	
Sheep	ME kcal/kg	949.	2314.	
Cattle	TDN %	26.	64.	
Sheep	TDN %	26.	64.	
Calcium	%	.22	.53	64
Phosphorus	%	.08	.19	54
Copper	mg/kg	2.3	5.5	
Manganese	mg/kg	15.5	37.9	

Grama, aerial pt, fresh, mid-blm, (2)

Ref no 2-02-164

Dry matter	%	50.0	100.0	
Ash	%	7.6	15.1	4
Crude fiber	%	14.4	28.9	4
Ether extract	%	1.0	1.9	8
N-free extract	%	22.9	45.8	
Protein (N x 6.25)	%	4.2	8.3	3
Cattle	dig prot %	2.4	4.9	
Sheep	dig prot %	2.4	4.7	
Energy				
Cattle	DE kcal/kg	1190.	2381.	
Sheep	DE kcal/kg	1455.	2910.	
Cattle	ME kcal/kg	976.	1952.	
Sheep	ME kcal/kg	1193.	2386.	
Cattle	TDN %	27.	54.	
Sheep	TDN %	33.	66.	

Grama, aerial pt, fresh, full blm, (2)

Ref no 2-02-165

Dry matter	%	50.0	100.0	
Ash	%	6.9	13.8	19
Crude fiber	%	15.3	30.6	5
Ether extract	%	.8	1.7	11
N-free extract	%	23.4	46.8	
Protein (N x 6.25)	%	3.6	7.1	8
Cattle	dig prot %	2.0	3.9	
Sheep	dig prot %	1.8	3.6	
Energy				
Cattle	DE kcal/kg	1212.	2425.	
Sheep	DE kcal/kg	1477.	2954.	
Cattle	ME kcal/kg	994.	1988.	
Sheep	ME kcal/kg	1211.	2422.	
Cattle	TDN %	28.	55.	
Sheep	TDN %	34.	67.	
Calcium	%	.25	.50	19
Phosphorus	%	.06	.13	31
Cobalt	mg/kg	.12	.24	
Copper	mg/kg	4.1	8.2	
Manganese	mg/kg	6.0	11.9	
Carotene	mg/kg	57.6	115.3	
Vitamin A equiv	IU/g	96.0	192.2	

(5) protein supplements

(6) minerals

(7) vitamins

(8) additives

Feed name or analyses		Mean		C.V.
		As fed	Dry	± %

Grama, aerial pt, fresh, mature, (2)

Ref no 2-02-166

Dry matter	%	63.4	100.0	11
Ash	%	7.2	11.4	33
Crude fiber	%	20.7	32.7	8
Ether extract	%	1.1	1.7	55
N-free extract	%	30.2	47.7	
Protein (N x 6.25)	%	4.1	6.5	18
Cattle	dig prot %	2.2	3.4	
Sheep	dig prot %	1.9	3.0	
Energy				
Cattle	DE kcal/kg	1621.	2557.	
Sheep	DE kcal/kg	1873.	2954.	
Cattle	ME kcal/kg	1329.	2097.	
Sheep	ME kcal/kg	1536.	2422.	
Cattle	TDN %	37.	58.	
Sheep	TDN %	42.	67.	
Calcium	%	.22	.34	25
Magnesium	%	.08	.13	
Phosphorus	%	.08	.12	33
Potassium	%	.22	.35	
Cobalt	mg/kg	.11	.18	
Copper	mg/kg	8.1	12.8	
Manganese	mg/kg	30.0	47.4	
Carotene	mg/kg	19.3	30.4	81
Vitamin A equiv	IU/g	32.2	50.7	

Grama, aerial pt, fresh, over ripe, (2)

Ref no 2-02-167

Dry matter	%	73.8	100.0	
Ash	%	7.0	9.5	42
Crude fiber	%	25.0	33.9	7
Ether extract	%	1.2	1.6	45
N-free extract	%	37.5	50.8	
Protein (N x 6.25)	%	3.1	4.2	22
Cattle	dig prot %	1.1	1.5	
Sheep	dig prot %	.7	.9	
Energy				
Cattle	DE kcal/kg	1920.	2601.	
Sheep	DE kcal/kg	2245.	3042.	
Cattle	ME kcal/kg	1574.	2133.	
Sheep	ME kcal/kg	1840.	2494.	
Cattle	TDN %	44.	59.	
Sheep	TDN %	51.	69.	
Calcium	%	.18	.24	32
Phosphorus	%	.07	.09	39
Cobalt	mg/kg	.07	.09	
Copper	mg/kg	7.5	10.1	

Continued

Feed name or analyses		Mean		C.V.
		As fed	Dry	± %
Manganese	mg/kg	29.6	40.1	
Carotene	mg/kg	16.8	22.7	49
Vitamin A equiv	IU/g	28.0	37.8	

Grama, aerial pt, fresh, (2)

Ref no 2-02-168

Dry matter	%	56.9	100.0	17
Ash	%	6.0	10.6	33
Crude fiber	%	17.8	31.2	11
Ether extract	%	1.0	1.7	58
N-free extract	%	27.8	48.9	
Protein (N x 6.25)	%	4.3	7.6	32
Cattle	dig prot %	2.5	4.4	
Sheep	dig prot %	2.3	4.1	
Energy				
Cattle	DE kcal/kg	1506.	2646.	
Sheep	DE kcal/kg	1681.	2954.	
Cattle	ME kcal/kg	1235.	2170.	
Sheep	ME kcal/kg	1378.	2422.	
Cattle	TDN %	34.	60.	
Sheep	TDN %	38.	67.	
Calcium	%	.23	.40	85
Magnesium	%	.07	.13	20
Phosphorus	%	.08	.14	99
Potassium	%	.41	.72	
Cobalt	mg/kg	.10	.18	70
Copper	mg/kg	5.1	9.0	43
Manganese	mg/kg	29.4	51.6	50
Carotene	mg/kg	27.6	48.5	99
Vitamin A equiv	IU/g	46.0	80.8	

Grapefruit - see Citrus, grapefruit

Groundnut - see Peanut

GRASS—LEGUME. Scientific name not used

Grass-legume, mixed, aerial pt, ensiled, (3)

Ref no 3-02-303

Dry matter	%	29.3	100.0	
Ash	%	2.3	8.0	
Crude fiber	%	9.2	31.4	
Ether extract	%	1.0	3.4	
N-free extract	%	13.3	45.4	
Protein (N x 6.25)	%	6.6	11.8	
Cattle	dig prot %	1.8	6.0	
Sheep	dig prot %	1.8	6.0	

Continued

(1) dry forages and roughages

(2) pasture, range plants, and forages fed green

(3) silages

(4) energy feeds

TABLE A-3-1 Composition of Feeds (Continued)

Feed name or analyses		Mean		C.V.
		As fed	Dry	± %
Energy				
Cattle	DE kcal/kg	723.	2469.	
Sheep	DE kcal/kg	723.	2469.	
Cattle	ME kcal/kg	593.	2024.	
Sheep	ME kcal/kg	593.	2024.	
Cattle	TDN %	16.	56.	
Sheep	TDN %	16.	56.	
Calcium	%	.23	.78	
Phosphorus	%	.13	.28	

Feed name or analyses		Mean		C.V.
		As fed	Dry	± %
Sheep	DE kcal/kg	1825.	1984.	
Cattle	ME kcal/kg	1497.	1627.	
Sheep	ME kcal/kg	1497.	1627.	
Cattle	TDN %	41.	45.	
Sheep	TDN %	41.	45.	
Calcium	%	.23	.25	8
Magnesium	%	.77	.84	
Phosphorus	%	.80	.87	33
Pantothenic acid	mg/kg	3.2	3.5	
Riboflavin	mg/kg	2.9	3.2	

Grass-legume, mixed, aerial pt w molasses added, ensiled, (3)

Ref no 3-02-309

Dry matter	%	30.0	100.0
Ash	%	2.2	7.4
Crude fiber	%	9.3	31.1
Ether extract	%	1.1	3.3
N-free extract	%	14.1	47.0
Protein (N x 6.25)	%	3.4	11.2
Cattle	dig prot %	1.7	5.7
Sheep	dig prot %	1.7	5.7
Energy			
Cattle	DE kcal/kg	754.	2513.
Sheep	DE kcal/kg	754.	2513.
Cattle	ME kcal/kg	618.	2061.
Sheep	ME kcal/kg	618.	2061.
Cattle	TDN %	17.	57.
Sheep	TDN %	17.	57.
Calcium	%	.31	1.04
Phosphorus	%	.08	.28

Hegari - see Sorghum, hegari

HEMP. Cannabis sativa

Hemp, seed, extn unspec grnd, (5)
Hempseed oil meal, extraction unspecified

Ref no 5-02-367

Dry matter	%	92.0	100.0	2
Ash	%	8.2	8.9	10
Crude fiber	%	24.0	26.1	7
Ether extract	%	4.9	5.3	43
N-free extract	%	23.8	25.9	
Protein (N x 6.25)	%	31.1	33.8	4
Cattle	dig prot %	25.2	27.4	
Sheep	dig prot %	25.2	27.4	
Energy				
Cattle	DE kcal/kg	1825.	1984.	

Continued

Hershey millet - see Millet, proso

Hog millet - see Millet, proso

Hominy feed - see Corn, grits by-prod

Hominy feed, white - see Corn, white, grits by-prod

HORSE. Equus caballus

Horse, meat, raw, (5)

Ref no 5-07-980

Dry matter	%	24.0	100.0
Ether extract	%	4.0	16.7
Protein (N x 6.25)	%	18.0	75.0
Calcium	%	.05	.13
Phosphorus	%	.62	1.69

Horse, meat w bone, raw grnd, (5)

Ref no 5-07-981

Dry matter	%	36.0	100.0
Ether extract	%	7.0	19.4
Protein (N x 6.25)	%	18.5	51.4

(5) protein supplements
(6) minerals

(7) vitamins
(8) additives

Feed name or analyses		Mean		C.V.
		As fed	Dry	± %

Feed name or analyses		Mean		C.V.
		As fed	Dry	± %

Hydrolyzed poultry feathers - see Poultry, feathers

HORSEBEAN. Vicia faba equina

Irradiated dried yeast - see Yeast, irradiated

Horsebean, hay, s-c, (1)

Japanesemillet - see Millet, Japanese

Ref no 1-02-402

Johnsongrass - see Sorghum, Johnsongrass

Dry matter	%	91.5	100.0	
Ash	%	5.5	6.0	
Crude fiber	%	22.0	24.0	
Ether extract	%	.8	.9	
N-free extract	%	49.8	54.5	
Protein (N x 6.25)	%	13.4	14.6	
Cattle	dig prot %	8.8	9.6	
Sheep	dig prot %	8.9	9.7	
Energy				
Cattle	DE kcal/kg	2340.	2557.	
Sheep	DE kcal/kg	2622.	2866.	
Cattle	ME kcal/kg	1919.	2097.	
Sheep	ME kcal/kg	2150.	2350.	
Cattle	TDN %	53.	58.	
Sheep	TDN %	59.	65.	

Kafir - see Sorghum, kafir

Kalo - see Sorghum, kalo

Kaoliang - see Sorghum, kaoliang

KUDZU. Pueraria spp

Horsebean, straw, (1)

Kudzu, hay, s-c, (1)

Ref no 1-02-404

Ref no 1-02-478

Dry matter	%	92.3	100.0	1
Ash	%	6.1	6.6	21
Crude fiber	%	31.0	33.6	6
Ether extract	%	2.3	2.5	19
N-free extract	%	40.2	43.5	
Protein (N x 6.25)	%	12.7	13.8	14
Cattle	dig prot %	8.2	8.9	
Sheep	dig prot %	8.1	8.8	
Energy				
Cattle	DE kcal/kg	2238.	2425.	
Sheep	DE kcal/kg	2319.	2513.	
Cattle	ME kcal/kg	1835.	1988.	
Sheep	ME kcal/kg	1902.	2061.	
Cattle	TDN %	51.	55.	
Sheep	TDN %	53.	57.	
Calcium	%	1.49	1.61	
Magnesium	%	.74	.80	
Phosphorus	%	.43	.47	
Carotene	mg/kg	40.7	44.1	
Vitamin A equiv	IU/g	67.8	73.5	

Horsebean, straw, (1) table:

Dry matter	%	87.9	100.0	
Ash	%	8.4	9.6	
Crude fiber	%	36.4	41.4	
Ether extract	%	1.4	1.6	
N-free extract	%	33.1	37.6	
Protein (N x 6.25)	%	8.6	9.8	
Cattle	dig prot %	4.7	5.4	
Sheep	dig prot %	4.0	4.6	
Energy				
Cattle	DE kcal/kg	1744.	1984.	
Sheep	DE kcal/kg	1860.	2116.	
Cattle	ME kcal/kg	1430.	1627.	
Sheep	ME kcal/kg	1525.	1735.	
Cattle	TDN %	40.	45.	
Sheep	TDN %	42.	48.	

(1) dry forages and roughages
(2) pasture, range plants, and forages fed green

(3) silages
(4) energy feeds

TABLE A-3-1 Composition of Feeds (Continued)

Feed name or analyses		Mean		C.V.
		As fed	Dry	± %

Kudzu, aerial pt, fresh, (2)

Ref no 2-02-482

		As fed	Dry	C.V. ± %
Dry matter	%	22.2	100.0	15
Ash	%	1.8	8.0	3
Crude fiber	%	7.5	33.9	6
Ether extract	%	.5	2.1	17
N-free extract	%	8.6	38.8	
Protein (N x 6.25)	%	3.8	17.2	8
Cattle	dig prot %	2.8	12.5	
Sheep	dig prot %	2.9	13.0	
Energy				
Cattle	DE kcal/kg	734.	3307.	
Sheep	DE kcal/kg	568.	2557.	
Cattle	ME kcal/kg	602.	2712.	
Sheep	ME kcal/kg	466.	2097.	
Cattle	TDN %	17.	75.	
Sheep	TDN %	13.	58.	

Kudzu, aerial pt w molasses added, ensiled, (3)

Ref no 3-02-485

		As fed	Dry	C.V. ± %
Dry matter	%	29.0	100.0	2
Ash	%	2.6	9.0	5
Crude fiber	%	10.9	37.7	1
Ether extract	%	.6	1.9	6
N-free extract	%	11.5	39.8	
Protein (N x 6.25)	%	3.4	11.6	4
Cattle	dig prot %	2.0	6.8	
Sheep	dig prot %	2.0	6.8	
Energy				
Cattle	DE kcal/kg	678.	2337.	
Sheep	DE kcal/kg	793.	2734.	
Cattle	ME kcal/kg	556.	1916.	
Sheep	ME kcal/kg	650.	2242.	
Cattle	TDN %	15.	53.	
Sheep	TDN %	18.	62.	
Calcium	%	.48	1.67	3
Phosphorus	%	.06	.20	18
Carotene	mg/kg	82.0	282.9	61
Vitamin A equiv	IU/g	136.7	471.6	

Lard - see Swine, lard

Lemon - see Citrus, lemon

LESPEDEZA. Lespedeza spp

Lespedeza, hay, s-c, immature, (1)

Ref no 1-02-509

		As fed	Dry	C.V. ± %
Dry matter	%	92.1	100.0	3
Ash	%	6.5	7.1	16
Crude fiber	%	21.8	23.7	15
Ether extract	%	3.1	3.4	15
N-free extract	%	44.2	48.0	
Protein (N x 6.25)	%	16.4	17.8	16
Cattle	dig prot %	11.4	12.4	
Sheep	dig prot %	11.5	12.5	
Energy				
Cattle	DE kcal/kg	2477.	2690.	
Sheep	DE kcal/kg	2558.	2778.	
Cattle	ME kcal/kg	2032.	2206.	
Sheep	ME kcal/kg	2098.	2278.	
Cattle	TDN %	56.	61.	
Sheep	TDN %	58.	63.	
Calcium	%	1.12	1.22	31
Iron	%	.03	.03	27
Magnesium	%	.26	.28	26
Phosphorus	%	.25	.27	19
Potassium	%	.98	1.07	15
Manganese	mg/kg	162.5	176.4	31
Carotene	mg/kg	133.8	145.3	
Vitamin A equiv	IU/g	223.0	242.2	

Lespedeza, hay, s-c, pre-blm, (1)

Ref no 1-07-954

		As fed	Dry	C.V. ± %
Dry matter	%	92.1	100.0	
Ash	%	6.5	7.1	
Crude fiber	%	21.8	23.7	
Ether extract	%	3.1	3.4	
N-free extract	%	44.2	48.0	
Protein (N x 6.25)	%	16.4	17.8	
Cattle	dig prot %	11.4	12.4	
Sheep	dig prot %	11.5	12.5	
Energy				
Cattle	DE kcal/kg	2558.	2778.	
Sheep	DE kcal/kg	2477.	2690.	
Cattle	ME kcal/kg	2098.	2278.	
Sheep	ME kcal/kg	2032.	2206.	

Continued

(5) protein supplements
(6) minerals

(7) vitamins
(8) additives

Feed name or analyses		Mean As fed	Dry	C.V. ± %
Cattle	TDN %	58.	63.	
Sheep	TDN %	56.	61.	
Calcium	%	1.05	1.14	
Phosphorus	%	.24	.26	

Feed name or analyses		Mean As fed	Dry	C.V. ± %
Magnesium	%	.25	.27	
Phosphorus	%	.24	.26	
Potassium	%	.98	1.05	

Lespedeza, hay, s-c, early blm, (1)

Ref no 1-02-510

		As fed	Dry	C.V.
Dry matter	%	93.4	100.0	2
Ash	%	6.0	6.4	19
Crude fiber	%	27.6	29.6	10
Ether extract	%	3.9	4.2	25
N-free extract	%	41.4	44.3	
Protein (N x 6.25)	%	14.5	15.5	10
Cattle	dig prot %	9.7	10.4	
Sheep	dig prot %	9.8	10.5	
Energy				
Cattle	DE kcal/kg	2388.	2557.	
Sheep	DE kcal/kg	2388.	2557.	
Cattle	ME kcal/kg	1958.	2097.	
Sheep	ME kcal/kg	1958.	2097.	
Cattle	TDN %	54.	58.	
Sheep	TDN %	54.	58.	
Calcium	%	1.15	1.23	28
Iron	%	.03	.04	26
Magnesium	%	.26	.28	23
Phosphorus	%	.23	.25	20
Potassium	%	.93	1.00	21
Manganese	mg/kg	191.6	205.1	34

Lespedeza, hay, s-c, full blm, (1)

Ref no 1-02-512

		As fed	Dry	C.V.
Dry matter	%	93.2	100.0	2
Ash	%	5.0	5.4	12
Crude fiber	%	28.9	31.0	5
Ether extract	%	2.9	3.1	19
N-free extract	%	43.9	47.1	
Protein (N x 6.25)	%	12.5	13.4	13
Cattle	dig prot %	7.9	8.5	
Sheep	dig prot %	8.0	8.6	
Energy				
Cattle	DE kcal/kg	2260.	2425.	
Sheep	DE kcal/kg	2424.	2601.	
Cattle	ME kcal/kg	1853.	1988.	
Sheep	ME kcal/kg	1988.	2133.	
Cattle	TDN %	51.	55.	
Sheep	TDN %	55.	59.	
Calcium	%	.97	1.04	9
Iron	%	.03	.03	15
Magnesium	%	.22	.24	22
Phosphorus	%	.21	.23	19
Potassium	%	.96	1.03	4
Manganese	mg/kg	141.2	151.5	7

Lespedeza, hay, s-c, mid-blm, (1)

Ref no 1-02-511

		As fed	Dry	C.V.
Dry matter	%	93.0	100.0	
Ash	%	5.5	5.9	
Crude fiber	%	28.6	30.7	
Ether extract	%	3.7	4.0	
N-free extract	%	40.6	43.7	
Protein (N x 6.25)	%	14.6	15.7	
Cattle	dig prot %	9.8	10.5	
Sheep	dig prot %	9.9	10.6	
Energy				
Cattle	DE kcal/kg	2337.	2513.	
Sheep	DE kcal/kg	2378.	2557.	
Cattle	ME kcal/kg	1917.	2061.	
Sheep	ME kcal/kg	1950.	2097.	
Cattle	TDN %	53.	57.	
Sheep	TDN %	54.	58.	
Calcium	%	1.11	1.19	
Iron	%	.030	.032	

Lespedeza, hay, s-c, cut 1, (1)

Ref no 1-02-516

		As fed	Dry	C.V.
Dry matter	%	93.7	100.0	2
Ash	%	5.7	6.1	19
Crude fiber	%	26.6	28.4	20
Ether extract	%	3.5	3.7	22
N-free extract	%	43.4	46.3	
Protein (N x 6.25)	%	14.5	15.5	18
Cattle	dig prot %	9.7	10.4	
Sheep	dig prot %	9.8	10.5	
Energy				
Cattle	DE kcal/kg	2396.	2557.	
Sheep	DE kcal/kg	2479.	2646.	
Cattle	ME kcal/kg	1965.	2097.	
Sheep	ME kcal/kg	2033.	2170.	
Cattle	TDN %	54.	58.	
Sheep	TDN %	56.	60.	
Calcium	%	1.02	1.09	10
Iron	%	.03	.03	22
Magnesium	%	.21	.22	19

Continued   *Continued*

(1) dry forages and roughages
(2) pasture, range plants, and forages fed green

(3) silages
(4) energy feeds

Feed name or analyses		Mean		C.V.
		As fed	Dry	± %
Phosphorus	%	.21	.22	25
Potassium	%	1.01	1.08	15
Manganese	mg/kg	134.9	144.0	18

Feed name or analyses		Mean		C.V.
		As fed	Dry	± %
Phosphorus	%	.23	.25	19
Potassium	%	.86	.93	23
Manganese	mg/kg	114.4	123.9	

Lespedeza, hay, s-c, cut 2, (1)

Ref no 1-02-517

		As fed	Dry	C.V.
Dry matter	%	91.5	100.0	4
Ash	%	4.8	5.2	6
Crude fiber	%	30.5	33.3	6
Ether extract	%	2.7	2.9	41
N-free extract	%	40.3	44.1	
Protein (N x 6.25)	%	13.3	14.5	10
Cattle	dig prot %	8.7	9.5	
Sheep	dig prot %	8.8	9.6	
Energy				
Cattle	DE kcal/kg	2179.	2381.	
Sheep	DE kcal/kg	2340.	2557.	
Cattle	ME kcal/kg	1786.	1952.	
Sheep	ME kcal/kg	1919.	2097.	
Cattle	TDN %	49.	54.	
Sheep	TDN %	53.	58.	
Calcium	%	.93	1.02	17
Iron	%	.02	.02	22
Magnesium	%	.20	.22	19
Phosphorus	%	.22	.24	17
Potassium	%	.92	1.01	20
Manganese	mg/kg	93.0	101.6	16

Lespedeza, hay, s-c, cut 3, (1)

Ref no 1-02-518

		As fed	Dry	C.V.
Dry matter	%	92.3	100.0	
Ash	%	5.1	5.5	13
Crude fiber	%	23.9	25.9	
Ether extract	%	4.4	4.8	
N-free extract	%	45.5	49.3	
Protein (N x 6.25)	%	13.4	14.5	8
Cattle	dig prot %	8.8	9.5	
Sheep	dig prot %	8.9	9.6	
Energy				
Cattle	DE kcal/kg	2442.	2646.	
Sheep	DE kcal/kg	2483.	2690.	
Cattle	ME kcal/kg	2003.	2170.	
Sheep	ME kcal/kg	2036.	2206.	
Cattle	TDN %	55.	60.	
Sheep	TDN %	56.	61.	
Calcium	%	1.10	1.19	22
Iron	%	.03	.03	
Magnesium	%	.22	.24	

Continued

Lespedeza, hay, s-c, (1)

Ref no 1-02-522

		As fed	Dry	C.V.
Dry matter	%	91.0	100.0	2
Ash	%	5.4	5.9	20
Crude fiber	%	28.2	31.0	16
Ether extract	%	2.6	2.9	32
N-free extract	%	41.7	45.8	
Protein (N x 6.25)	%	13.1	14.4	21
Cattle	dig prot %	8.6	9.4	
Sheep	dig prot %	8.6	9.5	
Cellulose	%	27.9	30.7	6
Lignin	%	15.5	17.0	15
Energy	GE kcal/kg	4168.	4580.	5
Cattle	DE kcal/kg	2207.	2425.	
Sheep	DE kcal/kg	2367.	2601.	
Cattle	ME kcal/kg	1809.	1988.	
Sheep	ME kcal/kg	1941.	2133.	
Cattle	TDN %	50.	55.	
Sheep	TDN %	54.	59.	
Calcium	%	1.00	1.10	29
Chlorine	%	.05	.05	43
Iron	%	.02	.03	59
Magnesium	%	.25	.28	27
Phosphorus	%	.17	.19	35
Potassium	%	.98	1.08	19
Sodium	%	.06	.07	93
Sulfur	%	.17	.19	26
Cobalt	mg/kg	.18	.20	74
Copper	mg/kg	8.0	8.8	22
Manganese	mg/kg	106.4	116.9	45
Zinc	mg/kg	26.7	29.3	99
Carotene	mg/kg	42.8	47.0	70
Vitamin A equiv	IU/g	71.3	78.3	

Lespedeza, leaves, (1)

Ref no 1-02-528

		As fed	Dry	C.V.
Dry matter	%	89.2	100.0	3
Ash	%	6.2	7.0	8
Crude fiber	%	19.7	22.1	6
Ether extract	%	2.9	3.3	9
N-free extract	%	43.2	48.4	
Protein (N x 6.25)	%	17.1	19.2	10
Cattle	dig prot %	12.1	13.6	
Sheep	dig prot %	12.3	13.8	
Energy				
Cattle	DE kcal/kg	2439.	2734.	

Continued

(5) protein supplements

(6) minerals

(7) vitamins

(8) additives

Feed name or analyses		Mean		C.V.
		As fed	Dry	± %
Sheep	DE kcal/kg	2517.	2822.	
Cattle	ME kcal/kg	2000.	2242.	
Sheep	ME kcal/kg	2064.	2314.	
Cattle	TDN %	55.	62.	
Sheep	TDN %	57.	64.	
Calcium	%	1.08	1.21	16
Iron	%	.03	.04	32
Magnesium	%	.23	.26	18
Phosphorus	%	.17	.19	35
Potassium	%	.93	1.04	9
Manganese	mg/kg	224.8	252.0	33

Lespedeza, stems, (1)

Ref no 1-02-538

Dry matter	%	91.4	100.0	2
Ash	%	4.3	4.7	16
Crude fiber	%	37.7	41.2	10
Ether extract	%	1.6	1.7	39
N-free extract	%	39.3	43.0	
Protein (N x 6.25)	%	8.6	9.4	20
Cattle	dig prot %	4.7	5.1	
Sheep	dig prot %	4.6	5.0	
Energy				
Cattle	DE kcal/kg	1652.	1808.	
Sheep	DE kcal/kg	2136.	2337.	
Cattle	ME kcal/kg	1354.	1482.	
Sheep	ME kcal/kg	1751.	1916.	
Cattle	TDN %	37.	41.	
Sheep	TDN %	48.	53.	
Calcium	%	.67	.73	12
Iron	%	.02	.02	37
Magnesium	%	.16	.18	26
Phosphorus	%	.16	.17	33
Potassium	%	.90	.99	17
Manganese	mg/kg	71.4	78.1	48

Lespedeza, aerial pt, fresh, immature, (2)

Ref no 2-02-539

Dry matter	%	31.1	100.0	19
Ash	%	3.3	10.6	18
Crude fiber	%	8.5	27.3	15
Ether extract	%	.8	2.7	24
N-free extract	%	12.7	40.7	
Protein (N x 6.25)	%	5.8	18.7	12
Cattle	dig prot %	4.3	13.8	
Sheep	dig prot %	4.5	14.4	
Energy				
Cattle	DE kcal/kg	973.	3130.	
Sheep	DE kcal/kg	850.	2734.	
Cattle	ME kcal/kg	798.	2567.	

Continued

Feed name or analyses		Mean		C.V.
		As fed	Dry	± %
Sheep	ME kcal/kg	697.	2242.	
Cattle	TDN %	22.	71.	
Sheep	TDN %	19.	62.	
Calcium	%	.39	1.26	15
Iron	%	.01	.03	26
Magnesium	%	.08	.27	31
Phosphorus	%	.12	.38	52
Potassium	%	.49	1.57	16
Manganese	mg/kg	65.0	209.0	45

Lespedeza, aerial pt, fresh, early blm, (2)

Ref no 2-02-540

Dry matter	%	25.0	100.0	
Ash	%	3.2	12.8	
Crude fiber	%	8.0	32.0	
Ether extract	%	.5	2.0	
N-free extract	%	9.2	36.8	
Protein (N x 6.25)	%	4.1	16.4	
Cattle	dig prot %	3.0	11.8	
Sheep	dig prot %	3.1	12.3	
Energy				
Cattle	DE kcal/kg	738.	2954.	
Sheep	DE kcal/kg	650.	2601.	
Cattle	ME kcal/kg	606.	2422.	
Sheep	ME kcal/kg	533.	2133.	
Cattle	TDN %	17.	67.	
Sheep	TDN %	15.	59.	
Calcium	%	.34	1.35	
Iron	%	.006	.025	
Magnesium	%	.07	.27	
Phosphorus	%	.05	.21	
Potassium	%	.28	1.12	

Lespedeza, aerial pt, fresh, mature, (2)

Ref no 2-02-542

Dry matter	%	35.5	100.0	7
Ash	%	2.6	7.4	12
Crude fiber	%	15.9	44.9	3
Ether extract	%	.7	2.1	15
N-free extract	%	11.6	32.8	
Protein (N x 6.25)	%	4.5	12.8	8
Cattle	dig prot %	3.1	8.8	
Sheep	dig prot %	3.2	8.9	
Energy				
Cattle	DE kcal/kg	1143.	3219.	
Sheep	DE kcal/kg	876.	2469.	
Cattle	ME kcal/kg	937.	2640.	
Sheep	ME kcal/kg	718.	2024.	
Cattle	TDN %	26.	73.	
Sheep	TDN %	20.	56.	

Continued

(1) dry forages and roughages
(2) pasture, range plants, and forages fed green

(3) silages
(4) energy feeds

Feed name or analyses		Mean		C.V.
		As fed	Dry	± %
Calcium	%	.36	1.02	
Iron	%	.01	.02	
Magnesium	%	.06	.16	
Phosphorus	%	.11	.31	
Potassium	%	.27	.77	
Manganese	mg/kg	30.2	85.1	

Lespedeza, aerial pt, fresh, (2)

Ref no 2-02-543

Dry matter	%	31.7	100.0	16
Ash	%	2.7	8.4	25
Crude fiber	%	9.8	30.8	17
Ether extract	%	.9	2.7	23
N-free extract	%	13.3	41.9	
Protein (N x 6.25)	%	5.1	16.2	17
Cattle	dig prot %	3.7	11.7	
Sheep	dig prot %	3.8	12.1	
Energy				
Cattle	DE kcal/kg	1006.	3175.	
Sheep	DE kcal/kg	839.	2646.	
Cattle	ME kcal/kg	825.	2604.	
Sheep	ME kcal/kg	688.	2170.	
Cattle	TDN %	23.	72.	
Sheep	TDN %	19.	60.	
Calcium	%	.36	1.14	23
Iron	%	.01	.03	57
Magnesium	%	.09	.27	30
Phosphorus	%	.10	.31	62
Potassium	%	.41	1.28	27
Manganese	mg/kg	62.9	198.4	48

LESPEDEZA, COMMON. Lespedeza striata

Lespedeza, common, hay, s-c, (1)

Ref no 1-02-563

Dry matter	%	89.6	100.0	2
Ash	%	5.0	5.5	16
Crude fiber	%	28.7	32.0	14
Ether extract	%	2.4	2.7	25
N-free extract	%	41.3	46.1	
Protein (N x 6.25)	%	12.3	13.7	12
Cattle	dig prot %	7.9	8.8	
Sheep	dig prot %	7.9	8.8	
Energy				
Cattle	DE kcal/kg	2133.	2381.	
Sheep	DE kcal/kg	2330.	2601.	
Cattle	ME kcal/kg	1749.	1952.	
Sheep	ME kcal/kg	1911.	2133.	
Cattle	TDN %	48.	54.	

Continued

Feed name or analyses		Mean		C.V.
		As fed	Dry	± %
Sheep	TDN %	53.	59.	
Calcium	%	.94	1.05	12
Iron	%	.03	.03	31
Magnesium	%	.26	.29	21
Phosphorus	%	.18	.20	26
Potassium	%	.94	1.05	13
Manganese	mg/kg	168.5	188.1	23
Carotene	mg/kg	49.5	55.3	55
Vitamin A equiv	IU/g	82.5	92.2	

Lespedeza, common, aerial pt, fresh, (2)

Ref no 2-02-568

Dry matter	%	27.6	100.0	17
Ash	%	3.1	11.2	18
Crude fiber	%	10.0	36.1	14
Ether extract	%	.6	2.1	12
N-free extract	%	9.7	35.3	
Protein (N x 6.25)	%	4.2	15.3	12
Cattle	dig prot %	2.8	10.2	
Sheep	dig prot %	3.1	11.2	
Energy				
Cattle	DE kcal/kg	694.	2513.	
Sheep	DE kcal/kg	706.	2557.	
Cattle	ME kcal/kg	569.	2061.	
Sheep	ME kcal/kg	579.	2097.	
Cattle	TDN %	16.	57.	
Sheep	TDN %	16.	58.	
Calcium	%	.31	1.13	5
Iron	%	.01	.03	25
Magnesium	%	.07	.27	19
Phosphorus	%	.07	.27	9
Potassium	%	.32	1.16	19
Manganese	mg/kg	49.2	178.2	33

LESPEDEZA, KOBE. Lespedeza striata kobe

Lespedeza, kobe, hay, s-c, (1)

Ref no 1-02-580

Dry matter	%	93.8	100.0	3
Ash	%	5.4	5.7	16
Crude fiber	%	30.8	32.8	7
Ether extract	%	3.3	3.5	16
N-free extract	%	41.0	43.8	
Protein (N x 6.25)	%	13.3	14.2	13
Cattle	dig prot %	8.6	9.2	
Sheep	dig prot %	8.7	9.3	
Energy				
Cattle	DE kcal/kg	2275.	2425.	
Sheep	DE kcal/kg	2357.	2513.	

Continued

(5) protein supplements
(6) minerals

(7) vitamins
(8) additives

Feed name or analyses		Mean		C.V.
		As fed	Dry	± %
Cattle	ME kcal/kg	1865.	1988.	
Sheep	ME kcal/kg	1933.	2061.	
Cattle	TDN %	52.	55.	
Sheep	TDN %	53.	57.	
Calcium	%	1.05	1.12	12
Iron	%	.03	.03	15
Magnesium	%	.23	.24	23
Phosphorus	%	.20	.21	44
Potassium	%	.98	1.04	11
Manganese	mg/kg	223.2	238.0	11

LESPEDEZA, KOREAN. Lespedeza stipulacea

Lespedeza, Korean, hay, s-c, (1)

Ref no 1-02-592

		As fed	Dry	C.V. ± %
Dry matter	%	90.3	100.0	2
Ash	%	6.0	6.6	18
Crude fiber	%	27.6	30.6	11
Ether extract	%	3.2	3.5	28
N-free extract	%	40.3	44.6	
Protein (N x 6.25)	%	13.3	14.7	18
Cattle	dig prot %	6.5	7.2	
Sheep	dig prot %	8.8	9.8	
Energy				
Cattle	DE kcal/kg	2269.	2513.	
Sheep	DE kcal/kg	2309.	2557.	
Cattle	ME kcal/kg	1861.	2061.	
Sheep	ME kcal/kg	1894.	2097.	
Cattle	TDN %	51.	57.	
Sheep	TDN %	52.	58.	
Calcium	%	.90	1.00	26
Iron	%	.02	.02	85
Magnesium	%	.28	.31	19
Phosphorus	%	.20	.22	27
Potassium	%	.94	1.04	20
Manganese	mg/kg	76.0	84.2	37

Lespedeza, Korean, aerial pt, fresh, immature, (2)

Ref no 2-02-593

		As fed	Dry	C.V. ± %
Dry matter	%	32.5	100.0	19
Ash	%	3.5	10.7	17
Crude fiber	%	8.4	25.9	16
Ether extract	%	1.0	3.1	22
N-free extract	%	13.4	41.2	
Protein (N x 6.25)	%	6.2	19.1	12
Cattle	dig prot %	4.6	14.1	
Sheep	dig prot %	4.8	14.8	
Energy				
Cattle	DE kcal/kg	1003.	3086.	

Continued

Feed name or analyses		Mean		C.V.
		As fed	Dry	± %
Sheep	DE kcal/kg	903.	2778.	
Cattle	ME kcal/kg	822.	2530.	
Sheep	ME kcal/kg	740.	2278.	
Cattle	TDN %	23.	70.	
Sheep	TDN %	20.	63.	
Calcium	%	.44	1.36	14
Phosphorus	%	.18	.55	33
Potassium	%	.51	1.56	

Lespedeza, Korean, aerial pt, fresh, mid-blm, (2)

Ref no 2-02-594

		As fed	Dry	C.V. ± %
Dry matter	%	27.7	100.0	13
Ash	%	2.6	9.4	13
Crude fiber	%	7.8	28.3	26
Ether extract	%	1.1	3.8	20
N-free extract	%	11.2	40.3	
Protein (N x 6.25)	%	5.0	18.2	19
Cattle	dig prot %	3.7	13.4	
Sheep	dig prot %	3.9	14.0	
Energy				
Cattle	DE kcal/kg	855.	3086.	
Sheep	DE kcal/kg	745.	2690.	
Cattle	ME kcal/kg	701.	2530.	
Sheep	ME kcal/kg	611.	2206.	
Cattle	TDN %	19.	70.	
Sheep	TDN %	17.	61.	
Calcium	%	.37	1.35	
Phosphorus	%	.11	.38	
Potassium	%	.45	1.62	

Lespedeza, Korean, aerial pt, fresh, mature, (2)

Ref no 2-02-596

		As fed	Dry	C.V. ± %
Dry matter	%	35.3	100.0	7
Ash	%	2.6	7.4	13
Crude fiber	%	15.9	45.1	3
Ether extract	%	.7	2.1	16
N-free extract	%	11.5	32.7	
Protein (N x 6.25)	%	4.5	12.7	6
Cattle	dig prot %	3.1	8.7	
Sheep	dig prot %	3.1	8.8	
Energy				
Cattle	DE kcal/kg	1136.	3219.	
Sheep	DE kcal/kg	887.	2513.	
Cattle	ME kcal/kg	932.	2640.	
Sheep	ME kcal/kg	728.	2061.	
Cattle	TDN %	26.	73.	
Sheep	TDN %	20.	57.	
Calcium	%	.35	1.00	
Phosphorus	%	.07	.20	

(1) dry forages and roughages
(2) pasture, range plants, and forages fed green

(3) silages
(4) energy feeds

TABLE A-3-1 Composition of Feeds (Continued)

Lespedeza, Korean, aerial pt, fresh, (2)

Ref no 2-02-598

Feed name or analyses		Mean		C.V.
		As fed	Dry	± %
Dry matter	%	32.2	100.0	17
Ash	%	3.0	9.3	19
Crude fiber	%	9.7	30.1	19
Ether extract	%	.9	2.7	24
N-free extract	%	13.3	41.4	
Protein (N x 6.25)	%	5.3	16.5	17
Cattle	dig prot %	3.8	11.9	
Sheep	dig prot %	4.0	12.4	
Energy				
Cattle	DE kcal/kg	1008.	3130.	
Sheep	DE kcal/kg	852.	2646.	
Cattle	ME kcal/kg	826.	2567.	
Sheep	ME kcal/kg	699.	2170.	
Cattle	TDN %	23.	71.	
Sheep	TDN %	19.	60.	
Calcium	%	.34	1.06	18
Phosphorus	%	.11	.34	55
Potassium	%	.51	1.58	13

LESPEDEZA, SERICEA. Lespedeza cuneata

Lespedeza, sericea, hay, s-c, (1)

Ref no 1-02-607

Feed name or analyses		Mean		C.V.
		As fed	Dry	± %
Dry matter	%	91.4	100.0	3
Ash	%	5.6	6.1	21
Crude fiber	%	25.1	27.5	20
Ether extract	%	2.7	3.0	31
N-free extract	%	44.1	48.2	
Protein (N x 6.25)	%	13.9	15.2	17
Cattle	dig prot %	9.2	10.1	
Sheep	dig prot %	9.3	10.2	
Energy	GE kcal/kg	4158.	4549.	
Cattle	DE kcal/kg	2337.	2557.	
Sheep	DE kcal/kg	2459.	2690.	
Cattle	ME kcal/kg	1917.	2097.	
Sheep	ME kcal/kg	2016.	2206.	
Cattle	TDN %	53.	58.	
Sheep	TDN %	56.	61.	
Calcium	%	1.33	1.46	26
Iron	%	.03	.03	37
Magnesium	%	.21	.23	17
Phosphorus	%	.21	.23	25
Potassium	%	.94	1.03	18
Manganese	mg/kg	114.9	125.7	31
Carotene	mg/kg	36.1	39.5	68
Vitamin A equiv	IU/g	60.2	65.8	

Lespedeza, sericea, aerial pt, fresh, (2)

Ref no 2-02-611

Feed name or analyses		Mean		C.V.
		As fed	Dry	± %
Dry matter	%	32.8	100.0	12
Ash	%	2.0	6.2	7
Crude fiber	%	7.4	22.7	7
Ether extract	%	1.2	3.8	13
N-free extract	%	16.2	49.3	
Protein (N x 6.25)	%	5.9	18.0	14
Cattle	dig prot %	4.3	13.2	
Sheep	dig prot %	4.5	13.8	
Energy				
Cattle	DE kcal/kg	1070.	3263.	
Sheep	DE kcal/kg	911.	2778.	
Cattle	ME kcal/kg	878.	2676.	
Sheep	ME kcal/kg	747.	2278.	
Cattle	TDN %	24.	74.	
Sheep	TDN %	21.	63.	
Calcium	%	.42	1.27	16
Iron	%	.01	.02	30
Magnesium	%	.07	.22	23
Phosphorus	%	.10	.29	35
Potassium	%	.39	1.20	24
Cobalt	mg/kg	.02	.07	
Manganese	mg/kg	34.1	103.9	41

Lime - see Citrus, lime

LIMESTONE. Scientific name not applicable

Limestone, grnd, mn 33 Ca, (6)
Limestone, ground (AAFCO)

Ref no 6-02-632

Feed name or analyses		Mean		C.V.
		As fed	Dry	± %
Dry matter	%	100.0	100.0	
Ash	%	95.8	95.8	
Calcium	%	33.84	33.84	10
Iron	%	.330	.330	
Phosphorus	%	.02	.02	
Sodium	%	.06	.06	
Manganese	mg/kg	279.6	279.6	99

(5) protein supplements
(6) minerals

(7) vitamins
(8) additives

Feed name or analyses		Mean		C.V.
		As fed	Dry	± %

Linseed meal - see Flax

Liver - see Animal, liver

LOVEGRASS. Eragrostis spp

Lovegrass, hay, s-c, immature, (1)

Ref no 1-02-642

		As fed	Dry	C.V. ± %
Dry matter	%	87.0	100.0	
Ash	%	7.6	8.7	
Crude fiber	%	27.8	31.9	
Ether extract	%	1.5	1.7	
N-free extract	%	41.1	47.3	
Protein (N x 6.25)	%	9.0	10.4	
Cattle	dig prot %	5.1	5.9	
Sheep	dig prot %	5.7	6.4	
Energy				
Cattle	DE kcal/kg	2302.	2646.	
Sheep	DE kcal/kg	2224.	2557.	
Cattle	ME kcal/kg	1888.	2170.	
Sheep	ME kcal/kg	1824.	2097.	
Cattle	TDN %	52.	60.	
Sheep	TDN %	50.	58.	
Calcium	%	.35	.40	
Magnesium	%	.09	.10	
Phosphorus	%	.16	.18	

Lovegrass, hay, s-c, mature, (1)

Ref no 1-02-645

		As fed	Dry	C.V. ± %
Dry matter	%	87.0	100.0	
Ash	%	8.4	9.6	
Crude fiber	%	30.2	34.7	
Ether extract	%	1.6	1.8	
N-free extract	%	41.0	47.2	
Protein (N x 6.25)	%	5.8	6.7	
Cattle	dig prot %	2.3	2.7	
Sheep	dig prot %	3.5	4.0	
Energy				
Cattle	DE kcal/kg	2186.	2513.	
Sheep	DE kcal/kg	2148.	2469.	
Cattle	ME kcal/kg	1793.	2061.	
Sheep	ME kcal/kg	1761.	2024.	
Cattle	TDN %	50.	57.	
Sheep	TDN %	49.	56.	

Lovegrass, hay, s-c, (1)

Ref no 1-02-646

		As fed	Dry	C.V. ± %
Dry matter	%	88.5	100.0	2
Ash	%	6.5	7.4	28
Crude fiber	%	28.2	31.9	3
Ether extract	%	1.8	2.0	21
N-free extract	%	44.6	50.4	
Protein (N x 6.25)	%	7.3	8.3	23
Cattle	dig prot %	3.6	4.1	
Sheep	dig prot %	4.5	5.1	
Energy				
Cattle	DE kcal/kg	2342.	2646.	
Sheep	DE kcal/kg	2263.	2557.	
Cattle	ME kcal/kg	1920.	2170.	
Sheep	ME kcal/kg	1856.	2097.	
Cattle	TDN %	53.	60.	
Sheep	TDN %	51.	58.	
Calcium	%	.32	.36	26
Magnesium	%	.06	.07	67
Phosphorus	%	.10	.11	77

Lovegrass, aerial pt, fresh, (2)

Ref no 2-02-651

		As fed	Dry	C.V. ± %
Dry matter	%	44.6	100.0	17
Ash	%	2.7	6.0	30
Crude fiber	%	14.6	32.7	7
Ether extract	%	1.3	2.9	20
N-free extract	%	22.0	49.4	
Protein (N x 6.25)	%	4.0	9.0	28
Cattle	dig prot %	2.4	5.5	
Sheep	dig prot %	2.4	5.4	
Lignin	%	5.3	11.9	10
Energy				
Cattle	DE kcal/kg	1337.	2998.	
Sheep	DE kcal/kg	1298.	2910.	
Cattle	ME kcal/kg	1096.	2458.	
Sheep	ME kcal/kg	1064.	2386.	
Cattle	TDN %	30.	68.	
Sheep	TDN %	29.	66.	
Calcium	%	.16	.37	43
Phosphorus	%	.08	.17	25

(1) dry forages and roughages
(2) pasture, range plants, and forages fed green

(3) silages
(4) energy feeds

Feed name or analyses		Mean		C.V.
		As fed	Dry	± %

Low fluorine ground rock phosphate - see Rock
   phosphate, grnd, mx 0.5 F

LUPINE. Lupinus spp

Lupine, hay, s-c, full blm, (1)

Ref no 1-02-680

		As fed	Dry	± %
Dry matter	%	90.9	100.0	
Ash	%	14.1	15.5	
Crude fiber	%	26.5	29.1	
Ether extract	%	3.2	3.5	
N-free extract	%	34.1	37.5	
Protein (N x 6.25)	%	13.1	14.4	
Cattle	dig prot %	8.5	9.4	
Sheep	dig prot %	9.0	9.9	
Energy				
Cattle	DE kcal/kg	2645.	2910.	
Sheep	DE kcal/kg	2405.	2646.	
Cattle	ME kcal/kg	2169.	2386.	
Sheep	ME kcal/kg	1972.	2170.	
Cattle	TDN %	60.	66.	
Sheep	TDN %	54.	60.	

Lupine, hay, s-c, (1)

Ref no 1-02-682

		As fed	Dry	± %
Dry matter	%	88.6	100.0	3
Ash	%	8.5	9.6	31
Crude fiber	%	26.4	29.8	18
Ether extract	%	3.0	3.4	24
N-free extract	%	31.9	36.0	
Protein (N x 6.25)	%	18.8	21.2	24
Cattle	dig prot %	13.6	15.3	
Sheep	dig prot %	12.9	14.6	
Energy				
Cattle	DE kcal/kg	2344.	2646.	
Sheep	DE kcal/kg	2500.	2822.	
Cattle	ME kcal/kg	1923.	2170.	
Sheep	ME kcal/kg	2050.	2314.	
Cattle	TDN %	53.	60.	
Sheep	TDN %	57.	64.	

Lupine, straw, (1)

Ref no 1-02-683

		As fed	Dry	± %
Dry matter	%	84.9	100.0	
Ash	%	2.5	3.0	
Crude fiber	%	45.1	53.1	
Ether extract	%	1.1	1.3	
N-free extract	%	30.8	36.3	
Protein (N x 6.25)	%	5.3	6.3	
Cattle	dig prot %	2.0	2.4	
Sheep	dig prot %	2.0	2.4	
Energy				
Cattle	DE kcal/kg	749.	882.	
Sheep	DE kcal/kg	2021.	2381.	
Cattle	ME kcal/kg	614.	723.	
Sheep	ME kcal/kg	1657.	1952.	
Cattle	TDN %	17.	20.	
Sheep	TDN %	46.	54.	

Lupine, aerial pt, fresh, immature, (2)

Ref no 2-02-685

		As fed	Dry	± %
Dry matter	%	11.5	100.0	
Ash	%	1.2	10.2	19
Crude fiber	%	2.2	19.3	27
Ether extract	%	.3	2.4	
N-free extract	%	4.3	37.9	
Protein (N x 6.25)	%	3.5	30.2	10
Cattle	dig prot %	2.6	22.3	
Sheep	dig prot %	2.6	23.0	
Energy				
Cattle	DE kcal/kg	324.	2822.	
Sheep	DE kcal/kg	340.	2954.	
Cattle	ME kcal/kg	266.	2314.	
Sheep	ME kcal/kg	278.	2422.	
Cattle	TDN %	7.	64.	
Sheep	TDN %	8.	67.	
Calcium	%	.20	1.76	23
Phosphorus	%	.05	.47	10
Potassium	%	.40	3.51	19

Lupine, aerial pt, fresh, mid-blm, (2)

Ref no 2-02-686

		As fed	Dry	± %
Dry matter	%	11.0	100.0	
Ash	%	1.0	8.9	
Crude fiber	%	2.0	18.2	
Ether extract	%	.3	2.8	
N-free extract	%	5.2	47.3	
Protein (N x 6.25)	%	2.5	22.8	

Continued

(5) protein supplements

(6) minerals

(7) vitamins

(8) additives

Feed name or analyses		Mean		C.V.
		As fed	Dry	± %
Cattle	dig prot %	1.8	16.9	
Sheep	dig prot %	1.9	17.3	
Energy				
Cattle	DE kcal/kg	320.	2910.	
Sheep	DE kcal/kg	335.	3042.	
Cattle	ME kcal/kg	262.	2386.	
Sheep	ME kcal/kg	274.	2494.	
Cattle	TDN %	7.	66.	
Sheep	TDN %	8.	69.	

Lupine, aerial pt, fresh, full blm, (2)

Ref no 2-02-687

Dry matter	%	11.9	100.0	8
Ash	%	1.2	10.3	20
Crude fiber	%	3.2	26.9	22
Ether extract	%	.4	3.0	5
N-free extract	%	4.9	41.6	
Protein (N x 6.25)	%	2.2	18.2	20
Cattle	dig prot %	1.6	13.5	
Sheep	dig prot %	1.6	13.8	
Energy				
Cattle	DE kcal/kg	336.	2822.	
Sheep	DE kcal/kg	346.	2910.	
Cattle	ME kcal/kg	275.	2314.	
Sheep	ME kcal/kg	284.	2386.	
Cattle	TDN %	8.	64.	
Sheep	TDN %	8.	66.	
Calcium	%	.19	1.58	29
Phosphorus	%	.04	.35	30
Potassium	%	.33	2.79	16

Lupine, aerial pt, fresh, mature, (2)

Ref no 2-02-689

Dry matter	%	17.8	100.0	
Ash	%	1.2	6.5	
Crude fiber	%	5.7	32.1	
Ether extract	%	.5	2.6	
N-free extract	%	8.4	47.3	
Protein (N x 6.25)	%	2.0	11.5	
Cattle	dig prot %	1.5	8.5	
Sheep	dig prot %	1.5	8.7	
Energy				
Cattle	DE kcal/kg	510.	2866.	
Sheep	DE kcal/kg	534.	2998.	
Cattle	ME kcal/kg	418.	2350.	
Sheep	ME kcal/kg	438.	2458.	
Cattle	TDN %	12.	65.	
Sheep	TDN %	12.	68.	

Continued

Feed name or analyses		Mean		C.V.
		As fed	Dry	± %
Calcium	%	.17	.97	
Phosphorus	%	.02	.12	
Potassium	%	.29	1.61	

Lupine, aerial pt, fresh, (2)

Ref no 2-02-691

Dry matter	%	12.9	100.0	12
Ash	%	1.3	10.0	18
Crude fiber	%	3.7	28.4	17
Ether extract	%	.4	2.8	10
N-free extract	%	5.3	41.1	
Protein (N x 6.25)	%	2.3	17.7	34
Cattle	dig prot %	1.7	13.1	
Sheep	dig prot %	1.7	13.4	
Energy				
Cattle	DE kcal/kg	358.	2778.	
Sheep	DE kcal/kg	375.	2910.	
Cattle	ME kcal/kg	294.	2278.	
Sheep	ME kcal/kg	308.	2386.	
Cattle	TDN %	8.	63.	
Sheep	TDN %	8.	66.	
Calcium	%	.17	1.35	58
Magnesium	%	.02	.17	
Phosphorus	%	.04	.29	40
Potassium	%	.34	2.60	28

LUPINE, SWEET. Lupinus albus

Lupine, sweet, hay, s-c, full blm, (1)

Ref no 1-02-732

Dry matter	%	85.7	100.0	
Ash	%	6.2	7.2	
Crude fiber	%	33.2	38.7	
Ether extract	%	1.6	1.9	
N-free extract	%	30.6	35.7	
Protein (N x 6.25)	%	14.1	16.5	
Cattle	dig prot %	9.6	11.2	
Sheep	dig prot %	11.3	13.2	
Energy				
Cattle	DE kcal/kg	1927.	2249.	
Sheep	DE kcal/kg	2305.	2690.	
Cattle	ME kcal/kg	1580.	1844.	
Sheep	ME kcal/kg	1890.	2206.	
Cattle	TDN %	44.	51.	
Sheep	TDN %	52.	61.	

(1) dry forages and roughages
(2) pasture, range plants, and forages fed green

(3) silages
(4) energy feeds

TABLE A-3-1 Composition of Feeds (Continued)

Feed name or analyses		Mean		C.V.
		As fed	Dry	± %

Lupine, sweet, hay, s-c, (1)

Ref no 1-02-734

		As fed	Dry	± %
Dry matter	%	89.7	100.0	2
Ash	%	9.1	10.1	12
Crude fiber	%	28.0	31.2	15
Ether extract	%	3.0	3.3	17
N-free extract	%	31.7	35.4	
Protein (N x 6.25)	%	18.0	20.0	8
Cattle	dig prot %	12.8	14.3	
Sheep	dig prot %	12.7	14.2	
Energy				
Cattle	DE kcal/kg	2333.	2601.	
Sheep	DE kcal/kg	2571.	2866.	
Cattle	ME kcal/kg	1913.	2133.	
Sheep	ME kcal/kg	2108.	2350.	
Cattle	TDN %	53.	59.	
Sheep	TDN %	58.	65.	

Maize - see Corn

Malt cleanings - see Barley, malt sprout cleanings

Malt sprouts - see Barley, malt sprouts w hulls

Mangel - see Beet, mangels

Meadow hay - see Native plants, Intermountain

Meat meal - see Animal, carcass res

Meat meal tankage - see Animal, carcass res w blood

Meat and bone meal - see Animal, carcass res w bone

Meat and bone scrap - see Animal, carcass res w bone

Meat scrap - see Animal, carcass res

Melts - see Cattle, spleen

Feed name or analysis		Mean		C.V.
		As fed	Dry	± %

Middlings - see Wheat, flour by-prod, f-sift, mx 4 fbr

Milk - see Cattle, milk

Milk albumin - see Cattle, whey albumin

MILLET. Setaria spp

Millet, hay, s-c, (1)

Ref no 1-03-093

		As fed	Dry	± %
Dry matter	%	89.5	100.0	3
Ash	%	7.2	8.1	18
Crude fiber	%	27.5	30.7	12
Ether extract	%	2.4	2.7	20
N-free extract	%	44.8	50.0	
Protein (N x 6.25)	%	7.6	8.5	26
Cattle	dig prot %	3.8	4.3	
Sheep	dig prot %	2.2	2.5	
Energy				
Cattle	DE kcal/kg	2486.	2778.	
Sheep	DE kcal/kg	2210.	2469.	
Cattle	ME kcal/kg	2039.	2278.	
Sheep	ME kcal/kg	1811.	2024.	
Cattle	TDN %	56.	63.	
Sheep	TDN %	50.	56.	

Millet, straw, (1)

Ref no 1-03-095

		As fed	Dry	± %
Dry matter	%	88.9	100.0	2
Ash	%	5.4	6.1	4
Crude fiber	%	37.1	41.7	4
Ether extract	%	1.6	1.8	12
N-free extract	%	41.1	46.2	
Protein (N x 6.25)	%	3.7	4.2	8
Cattle	dig prot %	.5	.6	
Sheep	dig prot %	.3	.3	
Energy				
Cattle	DE kcal/kg	1686.	1896.	
Sheep	DE kcal/kg	1960.	2205.	
Cattle	ME kcal/kg	1382.	1555.	
Sheep	ME kcal/kg	1607.	1808.	
Cattle	TDN %	38.	43.	
Sheep	TDN %	44.	50.	

(5) protein supplements

(6) minerals

(7) vitamins

(8) additives

Feed name or analyses		Mean		C.V.
		As fed	Dry	± %

Millet, aerial pt, ensiled, (3)

Ref no 3-03-096

Feed name or analyses		As fed	Dry	C.V. ± %
Dry matter	%	31.6	100.0	6
Ash	%	3.0	9.6	8
Crude fiber	%	9.9	31.3	7
Ether extract	%	1.0	3.3	21
N-free extract	%	15.0	47.6	
Protein (N x 6.25)	%	2.6	8.2	10
Cattle	dig prot %	1.2	3.7	
Sheep	dig prot %	1.2	3.7	
Energy				
Cattle	DE kcal/kg	906.	2866.	
Sheep	DE kcal/kg	878.	2778.	
Cattle	ME kcal/kg	743.	2350.	
Sheep	ME kcal/kg	720.	2278.	
Cattle	TDN %	20.	65.	
Sheep	TDN %	20.	63.	
Calcium	%	.13	.40	
Phosphorus	%	.09	.27	
Carotene	mg/kg	19.3	61.1	
Vitamin A equiv	IU/g	32.2	101.8	

Millet, grain, (4)

Ref no 4-03-098

		As fed	Dry	C.V.
Dry matter	%	90.0	100.0	3
Ash	%	3.2	3.5	31
Crude fiber	%	8.0	8.9	20
Ether extract	%	4.0	4.4	17
N-free extract	%	62.9	69.9	
Protein (N x 6.25)	%	12.0	13.3	10
Cattle	dig prot %	7.4	8.2	
Sheep	dig prot %	5.9	6.6	
Swine	dig prot %	8.8	9.8	
Energy				
Cattle	DE kcal/kg	3056.	3395.	
Sheep	DE kcal/kg	2421.	2690.	
Swine	DE kcal/kg	2897.	3219.	
Cattle	ME kcal/kg	2506.	2784.	
Sheep	ME kcal/kg	1985.	2206.	
Swine	ME kcal/kg	2703.	3003.	
Cattle	TDN %	69.	77.	
Sheep	TDN %	55.	61.	
Swine	TDN %	66.	73.	
Calcium	%	.05	.06	
Chlorine	%	.14	.16	
Iron	%	.004	.005	
Magnesium	%	.16	.18	
Phosphorus	%	.28	.31	
Potassium	%	.43	.48	
Sodium	%	.04	.04	

Continued

		As fed	Dry	C.V.
Sulfur	%	.13	.14	
Cobalt	mg/kg	.020	.022	
Copper	mg/kg	21.6	24.0	
Manganese	mg/kg	29.1	32.3	
Zinc	mg/kg	13.9	15.4	
Choline	mg/kg	789.	877.	21
Niacin	mg/kg	52.6	58.4	28
Pantothenic acid	mg/kg	7.4	8.2	54
Riboflavin	mg/kg	1.6	1.8	22
Thiamine	mg/kg	6.6	7.3	11

MILLET, FOXTAIL. Setaria italica

Millet, foxtail, hay, s-c, (1)

Ref no 1-03-099

		As fed	Dry	C.V.
Dry matter	%	87.8	100.0	3
Ash	%	7.2	8.2	14
Crude fiber	%	25.8	29.4	4
Ether extract	%	2.6	3.0	13
N-free extract	%	43.8	49.9	
Protein (N x 6.25)	%	8.3	9.5	6
Cattle	dig prot %	4.6	5.2	
Sheep	dig prot %	5.0	5.7	
Energy				
Cattle	DE kcal/kg	2516.	2866.	
Sheep	DE kcal/kg	2439.	2778.	
Cattle	ME kcal/kg	2063.	2350.	
Sheep	ME kcal/kg	2000.	2278.	
Cattle	TDN %	57.	65.	
Sheep	TDN %	55.	63.	

Millet, foxtail, aerial pt, fresh, (2)

Ref no 2-03-101

		As fed	Dry	C.V.
Dry matter	%	27.1	100.0	18
Ash	%	2.5	9.1	25
Crude fiber	%	8.4	31.0	7
Ether extract	%	.7	2.7	15
N-free extract	%	12.7	47.0	
Protein (N x 6.25)	%	2.8	10.2	14
Cattle	dig prot %	1.8	6.6	
Sheep	dig prot %	1.7	6.2	
Energy				
Cattle	DE kcal/kg	765.	2822.	
Sheep	DE kcal/kg	741.	2734.	
Cattle	ME kcal/kg	627.	2314.	
Sheep	ME kcal/kg	608.	2242.	
Cattle	TDN %	17.	64.	

Continued

(1) dry forages and roughages
(2) pasture, range plants, and forages fed green

(3) silages
(4) energy feeds

TABLE A-3-1 Composition of Feeds (Continued)

Feed name or analyses		Mean		C.V.
		As fed	Dry	± %
Sheep	TDN %	17.	62.	
Calcium	%	.08	.31	
Phosphorus	%	.05	.17	

Millet, foxtail, grain, (4)

Ref no 4-03-102

		As fed	Dry	
Dry matter	%	90.0	100.0	2
Ash	%	3.2	3.6	28
Crude fiber	%	9.0	10.0	15
Ether extract	%	4.1	4.6	13
N-free extract	%	61.6	68.4	
Protein (N x 6.25)	%	12.1	13.4	7
Cattle	dig prot %	7.5	8.3	
Sheep	dig prot %	6.0	6.7	
Swine	dig prot %	8.2	9.1	
Energy				
Cattle	DE kcal/kg	3095.	3439.	
Sheep	DE kcal/kg	2421.	2690.	
Swine	DE kcal/kg	3214.	3571.	
Cattle	ME kcal/kg	2538.	2820.	
Sheep	ME kcal/kg	1985.	2206.	
Swine	ME kcal/kg	2999.	3332.	
Cattle	TDN %	70.	78.	
Sheep	TDN %	55.	61.	
Swine	TDN %	73.	81.	

MILLET, JAPANESE. Echinochloa crusgalli

Millet, Japanese, aerial pt, fresh, (2)
Japanesemillet, aerial pt, fresh

Ref no 2-03-108

		As fed	Dry	
Dry matter	%	22.7	100.0	13
Ash	%	1.9	8.4	16
Crude fiber	%	6.8	29.9	7
Ether extract	%	.6	2.7	10
N-free extract	%	11.4	50.0	
Protein (N x 6.25)	%	2.0	9.0	35
Cattle	dig prot %	1.2	5.5	
Sheep	dig prot %	1.0	4.5	
Energy				
Cattle	DE kcal/kg	640.	2822.	
Sheep	DE kcal/kg	611.	2690.	
Cattle	ME kcal/kg	525.	2314.	
Sheep	ME kcal/kg	501.	2206.	
Cattle	TDN %	14.	64.	
Sheep	TDN %	14.	61.	

Millet, Japanese, aerial pt, ensiled, (3)

Ref no 3-03-109

		As fed	Dry	
Dry matter	%	22.9	100.0	
Ash	%	1.6	7.0	
Crude fiber	%	8.5	37.1	
Ether extract	%	.6	2.6	
N-free extract	%	10.4	45.4	
Protein (N x 6.25)	%	1.8	7.9	
Cattle	dig prot %	.8	3.4	
Sheep	dig prot %	.8	3.4	
Energy				
Cattle	DE kcal/kg	565.	2469.	
Sheep	DE kcal/kg	646.	2822.	
Cattle	ME kcal/kg	463.	2024.	
Sheep	ME kcal/kg	530.	2314.	
Cattle	TDN %	13.	56.	
Sheep	TDN %	15.	64.	

MILLET, PEARL. Pennisetum glaucum

Millet, pearl, hay, s-c, (1)

Ref no 1-03-112

		As fed	Dry	
Dry matter	%	87.9	100.0	1
Ash	%	8.9	10.1	6
Crude fiber	%	31.6	35.9	7
Ether extract	%	1.8	2.1	11
N-free extract	%	37.6	42.8	
Protein (N x 6.25)	%	8.0	9.1	20
Cattle	dig prot %	4.2	4.8	
Sheep	dig prot %	4.1	4.7	
Energy				
Cattle	DE kcal/kg	2170.	2469.	
Sheep	DE kcal/kg	2093.	2381.	
Cattle	ME kcal/kg	1779.	2024.	
Sheep	ME kcal/kg	1716.	1952.	
Cattle	TDN %	49.	56.	
Sheep	TDN %	47.	54.	

Millet, pearl, aerial pt, fresh, (2)

Ref no 2-03-115

		As fed	Dry	
Dry matter	%	20.7	100.0	9
Ash	%	1.9	9.1	12
Crude fiber	%	6.5	31.2	8
Ether extract	%	.6	2.8	22
N-free extract	%	9.7	46.9	
Protein (N x 6.25)	%	2.1	10.0	19

Continued

Feed name or analyses		Mean		C.V.
		As fed	Dry	± %
Cattle	dig prot %	1.3	6.2	
Sheep	dig prot %	1.3	6.3	
Energy				
Cattle	DE kcal/kg	566.	2734.	
Sheep	DE kcal/kg	593.	2866.	
Cattle	ME kcal/kg	464.	2242.	
Sheep	ME kcal/kg	486.	2350.	
Cattle	TDN %	13.	62.	
Sheep	TDN %	13.	65.	

Millet, pearl, grain, (4)
Cattail millet grain

Ref no 4-03-118

		As fed	Dry	
Dry matter	%	89.0	100.0	
Ash	%	2.0	2.2	
Crude fiber	%	2.0	2.2	
Ether extract	%	4.5	5.1	
N-free extract	%	68.6	77.1	
Protein (N x 6.25)	%	11.9	13.4	
Cattle	dig prot %	7.4	8.3	
Sheep	dig prot %	7.7	8.7	
Swine	dig prot %	8.1	9.1	
Energy	GE kcal/kg	3342.	3755.	
Cattle	DE kcal/kg	2826.	3175.	
Sheep	DE kcal/kg	3375.	3792.	
Swine	DE kcal/kg	2512.	2822.	
Cattle	ME kcal/kg	2318.	2604.	
Sheep	ME kcal/kg	2767.	3109.	
Swine	ME kcal/kg	2343.	2633.	
Cattle	TDN %	64.	72.	
Sheep	TDN %	76.	86.	
Swine	TDN %	57.	64.	
Calcium	%	.06	.07	
Phosphorus	%	.43	.48	

MILLET, PROSO. Panicum miliaceum

Millet, proso, grain, (4)
Broomcorn millet grain
Hershey millet grain
Hog millet grain
Proso millet grain

Ref no 4-03-120

		As fed	Dry	
Dry matter	%	90.0	100.0	1
Ash	%	2.7	3.0	16
Crude fiber	%	7.0	7.7	21
Ether extract	%	3.6	4.0	9
N-free extract	%	65.0	72.2	
Protein (N x 6.25)	%	11.8	13.1	9

Continued

Feed name or analyses		Mean		C.V.
		As fed	Dry	± %
Cattle	dig prot %	7.2	8.0	
Sheep	dig prot %	7.6	8.5	
Swine	dig prot %	8.0	8.9	
Energy				
Cattle	DE kcal/kg	3016.	3351.	
Sheep	DE kcal/kg	3214.	3571.	
Swine	DE kcal/kg	3294.	3660.	
Cattle	ME kcal/kg	2473.	2748.	
Chickens	MEn kcal/kg	2998.	3331.	
Sheep	ME kcal/kg	2635.	2928.	
Swine	ME kcal/kg	3076.	3418.	
Cattle	TDN %	68.	76.	
Sheep	TDN %	73.	81.	
Swine	TDN %	75.	83.	
Magnesium	%	.16	.18	
Potassium	%	.43	.48	
Riboflavin	mg/kg	1.0	1.1	

Milo - see Sorghum, milo

Molasses - see Beet, sugar, molasses; see Citrus, syrup; see Sugarcane, molasses

Molasses distillers dried yeast - see Yeast, molasses

Monosodium phosphate - see Sodium phosphate, monobasic

MUSTARD. Brassica spp

Mustard, seed, extn unspec grnd, (5)
Mustard seed meal (CFA)
Mustard seed oil meal, extraction unspecified

Ref no 5-03-154

		As fed	Dry	
Dry matter	%	94.0	100.0	
Ash	%	6.4	6.8	
Crude fiber	%	11.0	11.7	
Ether extract	%	6.2	6.6	
N-free extract	%	39.3	41.8	
Protein (N x 6.25)	%	31.1	33.1	
Cattle	dig prot %	26.8	28.5	
Sheep	dig prot %	26.8	28.5	
Energy				
Cattle	DE kcal/kg	3108.	3307.	
Sheep	DE kcal/kg	3274.	3483.	
Cattle	ME kcal/kg	2549.	2712.	

Continued

(1) dry forages and roughages
(2) pasture, range plants, and forages fed green

(3) silages
(4) energy feeds

TABLE A-3-1 Composition of Feeds (Continued)

Feed name or analyses		Mean		C.V.
		As fed	Dry	± %
Sheep	ME kcal/kg	2685.	2856.	
Cattle	TDN %	70.	75.	
Sheep	TDN %	74.	79.	

NAPIERGRASS. Pennisetum purpureum

Napiergrass, aerial pt, fresh, pre-blm, (2)

Ref no 2-03-158

		As fed	Dry	
Dry matter	%	14.9	100.0	
Ash	%	1.4	9.2	
Crude fiber	%	4.7	31.5	
Ether extract	%	.4	3.0	
N-free extract	%	6.7	45.3	
Protein (N x 6.25)	%	1.6	11.0	
Cattle	dig prot %	1.0	6.9	
Sheep	dig prot %	1.0	6.4	
Energy				
Cattle	DE kcal/kg	414.	2778.	
Sheep	DE kcal/kg	348.	2337.	
Cattle	ME kcal/kg	339.	2278.	
Sheep	ME kcal/kg	285.	1916.	
Cattle	TDN %	9.	63.	
Sheep	TDN %	8.	53.	
Calcium	%	.09	.60	
Phosphorus	%	.06	.41	

Napiergrass, aerial pt, fresh, early blm, (2)

Ref no 2-03-159

		As fed	Dry	
Dry matter	%	14.9	100.0	
Ash	%	1.4	9.2	
Crude fiber	%	4.7	31.5	
Ether extract	%	.4	3.0	
N-free extract	%	6.8	45.3	
Protein (N x 6.25)	%	1.6	11.0	
Cattle	dig prot %	1.0	6.9	
Sheep	dig prot %	1.0	6.7	
Energy				
Cattle	DE kcal/kg	414.	2778.	
Sheep	DE kcal/kg	361.	2425.	
Cattle	ME kcal/kg	339.	2278.	
Sheep	ME kcal/kg	296.	1988.	
Cattle	TDN %	9.	63.	
Sheep	TDN %	8.	55.	

Napiergrass, aerial pt, fresh, mid-blm, (2)

Ref no 2-03-160

		As fed	Dry	
Dry matter	%	23.0	100.0	11
Ash	%	2.4	10.3	16
Crude fiber	%	8.0	34.9	5
Ether extract	%	.6	2.5	21
N-free extract	%	9.9	43.1	
Protein (N x 6.25)	%	2.1	9.2	5
Cattle	dig prot %	1.3	5.8	
Sheep	dig prot %	1.3	5.6	
Energy				
Cattle	DE kcal/kg	629.	2734.	
Sheep	DE kcal/kg	568.	2469.	
Cattle	ME kcal/kg	516.	2242.	
Sheep	ME kcal/kg	466.	2024.	
Cattle	TDN %	14.	62.	
Sheep	TDN %	13.	56.	

Napiergrass, aerial pt, fresh, full blm, (2)

Ref no 2-03-161

		As fed	Dry	
Dry matter	%	22.5	100.0	9
Ash	%	2.3	10.1	15
Crude fiber	%	8.1	36.1	7
Ether extract	%	.6	2.5	26
N-free extract	%	9.8	43.7	
Protein (N x 6.25)	%	1.7	7.6	37
Cattle	dig prot %	1.1	4.8	
Sheep	dig prot %	1.0	4.6	
Energy				
Cattle	DE kcal/kg	615.	2734.	
Sheep	DE kcal/kg	536.	2381.	
Cattle	ME kcal/kg	504.	2242.	
Sheep	ME kcal/kg	439.	1952.	
Cattle	TDN %	14.	62.	
Sheep	TDN %	12.	54.	

Napiergrass, aerial pt, fresh, late blm, (2)

Ref no 2-03-162

		As fed	Dry	
Dry matter	%	23.0	100.0	
Ash	%	1.2	5.3	
Crude fiber	%	9.0	39.0	
Ether extract	%	.3	1.1	
N-free extract	%	10.8	46.8	
Protein (N x 6.25)	%	1.8	7.8	
Cattle	dig prot %	1.1	4.9	
Sheep	dig prot %	.8	3.6	
Energy				

Continued

(5) protein supplements

(6) minerals

(7) vitamins

(8) additives

Feed name or analyses		Mean		C.V.
		As fed	Dry	± %
Cattle	DE kcal/kg	527.	2293.	
Sheep	DE kcal/kg	497.	2160.	
Cattle	ME kcal/kg	432.	1880.	
Sheep	ME kcal/kg	407.	1771.	
Cattle	TDN %	12.	52.	
Sheep	TDN %	11.	49.	
Calcium	%	.08	.35	
Phosphorus	%	.07	.30	

Napiergrass, aerial pt, fresh, mature, (2)

Ref no 2-03-164

Dry matter	%	26.7	100.0	12
Ash	%	2.3	8.6	20
Crude fiber	%	10.0	37.6	4
Ether extract	%	.5	2.0	11
N-free extract	%	12.4	46.4	
Protein (N x 6.25)	%	1.4	5.4	10
Cattle	dig prot %	.9	3.4	
Sheep	dig prot %	.9	3.3	
Energy				
Cattle	DE kcal/kg	742.	2778.	
Sheep	DE kcal/kg	636.	2381.	
Cattle	ME kcal/kg	608.	2278.	
Sheep	ME kcal/kg	521.	1952.	
Cattle	TDN %	17.	63.	
Sheep	TDN %	14.	54.	

Napiergrass, aerial pt, fresh, over ripe, (2)

Ref no 3-03-165

Dry matter	%	30.4	100.0	
Ash	%	2.7	9.0	
Crude fiber	%	12.0	39.6	
Ether extract	%	.5	1.6	
N-free extract	%	13.9	45.7	
Protein (N x 6.25)	%	1.2	4.1	
Cattle	dig prot %	.8	2.6	
Sheep	dig prot %	.8	2.5	
Energy				
Cattle	DE kcal/kg	844.	2778.	
Sheep	DE kcal/kg	724.	2381.	
Cattle	ME kcal/kg	692.	2278.	
Sheep	ME kcal/kg	593.	1952.	
Cattle	TDN %	19.	63.	
Sheep	TDN %	16.	54.	

Napiergrass, aerial pt, fresh, (2)

Ref no 2-03-166

Dry matter	%	22.3	100.0	15
Ash	%	2.5	11.4	17
Crude fiber	%	8.4	37.7	9
Ether extract	%	.4	2.0	27
N-free extract	%	9.6	42.9	
Protein (N x 6.25)	%	1.3	6.0	39
Cattle	dig prot %	.8	3.8	
Sheep	dig prot %	.8	3.7	
Energy				
Cattle	DE kcal/kg	600.	2690.	
Sheep	DE kcal/kg	521.	2337.	
Cattle	ME kcal/kg	492.	2206.	
Sheep	ME kcal/kg	427.	1916.	
Cattle	TDN %	14.	61.	
Sheep	TDN %	12.	53.	

Napiergrass, aerial pt, ensiled, (3)

Ref no 3-03-170

Dry matter	%	24.8	100.0	15
Ash	%	2.7	11.0	18
Crude fiber	%	8.8	35.6	10
Ether extract	%	.6	2.2	22
N-free extract	%	11.4	45.8	
Protein (N x 6.25)	%	1.3	5.4	21
Cattle	dig prot %	.3	1.1	
Sheep	dig prot %	.3	1.1	
Energy				
Cattle	DE kcal/kg	711.	2866.	
Sheep	DE kcal/kg	689.	2778.	
Cattle	ME kcal/kg	583.	2350.	
Sheep	ME kcal/kg	565.	2278.	
Cattle	TDN %	16.	65.	
Sheep	TDN %	16.	63.	
Calcium	%	.08	.31	
Phosphorus	%	.08	.31	
Carotene	mg/kg	24.0	96.6	35
Vitamin A equiv	IU/g	40.0	161.0	

(1) dry forages and roughages

(2) pasture, range plants, and forages fed green

(3) silages

(4) energy feeds

Feed name or analyses		Mean		C.V.
		As fed	Dry	± %

NATIVE PLANTS, INTERMOUNTAIN. Scientific name not used

Native plants, Intermountain, hay, s-c, (1)
Meadow hay

Ref no 1-03-181

		As fed	Dry	± %
Dry matter	%	92.9	100.0	2
Ash	%	7.5	8.1	7
Crude fiber	%	28.0	30.1	3
Ether extract	%	2.8	3.0	18
N-free extract	%	46.2	49.7	
Protein (N x 6.25)	%	8.4	9.1	17
Cattle	dig prot %	2.7	2.9	
Sheep	dig prot %	2.7	2.9	
Energy				
Cattle	DE kcal/kg	1884.	2028.	
Sheep	DE kcal/kg	1884.	2028.	
Cattle	ME kcal/kg	1545.	1663.	
Sheep	ME kcal/kg	1545.	1663.	
Cattle	NE$_m$ kcal/kg	1152.	1240.	
Cattle	NE$_{gain}$ kcal/kg	548.	590.	
Cattle	TDN %	43.	46.	
Sheep	TDN %	43.	46.	
Calcium	%	.53	.57	
Phosphorus	%	.16	.17	

NATIVE PLANTS, MIDWEST. Scientific name not used

Native plants, Midwest, hay, s-c, immature, (1)
Prairie hay, immature

Ref no 1-03-183

		As fed	Dry	± %
Dry matter	%	89.5	100.0	
Ash	%	8.3	9.3	16
Crude fiber	%	28.4	31.7	4
Ether extract	%	2.3	2.6	24
N-free extract	%	42.7	47.7	
Protein (N x 6.25)	%	7.8	8.7	8
Cattle	dig prot %	2.1	2.4	
Sheep	dig prot %	3.9	4.4	
Energy				
Cattle	DE kcal/kg	2013.	2249.	
Sheep	DE kcal/kg	2210.	2469.	
Cattle	ME kcal/kg	1650.	1844.	
Sheep	ME kcal/kg	1811.	2024.	
Cattle	TDN %	46.	51.	
Sheep	TDN %	50.	56.	
Calcium	%	.51	.57	15

Continued

Feed name or analyses		Mean		C.V.
		As fed	Dry	± %

Iron	%	.01	.01	19
Magnesium	%	.22	.24	18
Phosphorus	%	.17	.19	20
Potassium	%	.97	1.08	7

Native plants, Midwest, hay, s-c, mid-blm, (1)
Prairie hay, mid-bloom

Ref no 1-07-956

		As fed	Dry	± %
Dry matter	%	91.0	100.0	
Ash	%	8.7	9.6	
Crude fiber	%	29.2	32.1	
Ether extract	%	2.6	2.8	
N-free extract	%	43.1	47.4	
Protein (N x 6.25)	%	7.4	8.1	
Cattle	dig prot %	3.7	4.1	
Sheep	dig prot %	3.7	4.1	
Energy				
Cattle	DE kcal/kg	2006.	2205.	
Sheep	DE kcal/kg	2006.	2205.	
Cattle	ME kcal/kg	1645.	1808.	
Sheep	ME kcal/kg	1645.	1808.	
Cattle	TDN %	46.	50.	
Sheep	TDN %	46.	50.	
Calcium	%	.31	.34	
Phosphorus	%	.19	.21	

Native plants, Midwest, hay, s-c, full blm, (1)
Prairie hay, full bloom

Ref no 1-03-184

		As fed	Dry	± %
Dry matter	%	83.3	100.0	9
Ash	%	8.5	10.2	8
Crude fiber	%	27.5	33.0	2
Ether extract	%	2.7	3.2	15
N-free extract	%	38.3	46.0	
Protein (N x 6.25)	%	6.3	7.6	25
Cattle	dig prot %	1.7	2.1	
Sheep	dig prot %	2.8	3.4	
Energy				
Cattle	DE kcal/kg	1873.	2249.	
Sheep	DE kcal/kg	2020.	2425.	
Cattle	ME kcal/kg	1536.	1844.	
Sheep	ME kcal/kg	1656.	1988.	
Cattle	TDN %	42.	51.	
Sheep	TDN %	46.	55.	

(5) protein supplements

(6) minerals

(7) vitamins

(8) additives

Feed name or analyses		Mean		C.V.
		As fed	Dry	± %

Native plants, Midwest, hay, s-c, late blm, (1)
　Prairie hay, late bloom

Ref no 1-07-957

		As fed	Dry	
Dry matter	%	91.3	100.0	
Ash	%	8.6	9.4	
Crude fiber	%	29.7	32.5	
Ether extract	%	3.0	3.3	
N-free extract	%	44.0	48.2	
Protein (N x 6.25)	%	6.0	6.6	
Cattle	dig prot %	2.0	2.2	
Sheep	dig prot %	2.0	2.2	
Energy				
Cattle	DE kcal/kg	1972.	2160.	
Sheep	DE kcal/kg	1972.	2160.	
Cattle	ME kcal/kg	1617.	1771.	
Sheep	ME kcal/kg	1617.	1771.	
Cattle	TDN %	45.	49.	
Sheep	TDN %	45.	49.	
Calcium	%	.33	.36	
Phosphorus	%	.12	.13	

Native plants, Midwest, hay, s-c, milk stage, (1)
　Prairie hay, milk stage

Ref no 1-03-185

		As fed	Dry	
Dry matter	%	91.9	100.0	
Ash	%	7.5	8.2	
Crude fiber	%	30.9	33.6	
Ether extract	%	2.6	2.8	
N-free extract	%	46.5	50.6	
Protein (N x 6.25)	%	4.4	4.8	
Cattle	dig prot %	.9	1.0	
Sheep	dig prot %	.9	1.0	
Energy				
Cattle	DE kcal/kg	1945.	2116.	
Sheep	DE kcal/kg	1945.	2116.	
Cattle	ME kcal/kg	1594.	1735.	
Sheep	ME kcal/kg	1594.	1735.	
Cattle	TDN %	44.	48.	
Sheep	TDN %	44.	48.	
Calcium	%	.36	.39	
Phosphorus	%	.12	.13	

Native plants, Midwest, hay, s-c, mature, (1)
　Prairie hay, mature

Ref no 1-03-187

		As fed	Dry	C.V. ± %
Dry matter	%	92.3	100.0	2
Ash	%	6.9	7.5	14
Crude fiber	%	31.2	33.8	5
Ether extract	%	2.3	2.5	17
N-free extract	%	47.6	51.6	
Protein (N x 6.25)	%	4.2	4.6	25
Cattle	dig prot %	1.2	1.3	
Sheep	dig prot %	.6	.7	
Energy				
Cattle	DE kcal/kg	2198.	2381.	
Sheep	DE kcal/kg	2198.	2381.	
Cattle	ME kcal/kg	1802.	1952.	
Sheep	ME kcal/kg	1802.	1952.	
Cattle	TDN %	50.	54.	
Sheep	TDN %	50.	54.	
Calcium	%	.35	.38	24
Chlorine	%	.12	.13	19
Iron	%	.01	.01	32
Magnesium	%	.22	.24	27
Phosphorus	%	.08	.09	20
Potassium	%	.63	.68	20
Sodium	%	.01	.01	99
Cobalt	mg/kg	.12	.13	47
Carotene	mg/kg	9.6	10.4	52
Vitamin A equiv	IU/g	16.0	17.3	

Native plants, Midwest, hay, s-c, over ripe, (1)
　Prairie hay, over ripe

Ref no 1-03-188

		As fed	Dry	C.V. ± %
Dry matter	%	91.5	100.0	1
Ash	%	7.3	7.9	11
Crude fiber	%	31.5	34.4	6
Ether extract	%	2.7	2.9	19
N-free extract	%	46.5	50.8	
Protein (N x 6.25)	%	3.7	4.0	31
Cattle	dig prot %	1.0	1.1	
Sheep	dig prot %	.2	.2	
Energy				
Cattle	DE kcal/kg	2179.	2381.	
Sheep	DE kcal/kg	2138.	2337.	
Cattle	ME kcal/kg	1786.	1952.	
Sheep	ME kcal/kg	1753.	1916.	
Cattle	TDN %	49.	54.	
Sheep	TDN %	48.	53.	

(1) dry forages and roughages
(2) pasture, range plants, and forages fed green

(3) silages
(4) energy feeds

Feed name or analyses		Mean		C.V.
		As fed	Dry	± %

Native plants, Midwest, hay, s-c, (1)
Prairie hay

Ref no 1-03-191

		As fed	Dry	± %
Dry matter	%	91.8	100.0	4
Ash	%	7.4	8.1	22
Crude fiber	%	30.6	33.3	5
Ether extract	%	2.4	2.6	29
N-free extract	%	46.3	50.4	
Protein (N x 6.25)	%	5.2	5.6	21
Cattle	dig prot %	1.5	1.6	
Sheep	dig prot %	1.5	1.6	
Energy				
Cattle	DE kcal/kg	2145.	2337.	
Sheep	DE kcal/kg	2186.	2381.	
Cattle	ME kcal/kg	1759.	1916.	
Sheep	ME kcal/kg	1792.	1952.	
Cattle	TDN %	49.	53.	
Sheep	TDN %	50.	54.	
Calcium	%	.38	.41	24
Chlorine	%	.12	.13	30
Iron	%	.01	.01	84
Magnesium	%	.25	.27	25
Phosphorus	%	.11	.12	30
Potassium	%	.68	.74	35
Sodium	%	.01	.01	99
Cobalt	mg/kg	.14	.15	74
Copper	mg/kg	25.9	28.2	99
Manganese	mg/kg	66.4	72.3	79
Carotene	mg/kg	29.6	32.2	71
Vitamin A equiv	IU/g	49.3	53.7	
Vitamin D$_2$	IU/g	.9	1.0	38

NEEDLEANDTHREAD. Stipa comata

Needleandthread, aerial pt, fresh, dormant, (2)

Ref no 2-07-989

		As fed	Dry	
Dry matter	%	86.0	100.0	
Ash	%	15.3	17.8	
Ether extract	%	4.2	4.9	
Protein (N x 6.25)	%	3.4	4.0	
Sheep	dig prot %	1.0	1.2	
Cellulose	%	28.4	33.0	
Lignin	%	6.9	8.0	
Energy	GE kcal/kg	3367.	3915.	
Sheep	DE kcal/kg	1697.	1973.	
Sheep	ME kcal/kg	1416.	1647.	
Sheep	TDN %	40.	47.	

Continued

Feed name or analyses		Mean		C.V.
		As fed	Dry	± %
Calcium	%	.76	.88	
Phosphorus	%	.06	.07	
Carotene	mg/kg	.3	.4	

Oat feed - see Oats, groats by-prod

Oat meal - see Oats, cereal by-prod, mx 2 fbr

Oat middlings - see Oats, cereal by-prod, mx 4 fbr

Oat mill by-product - see Oats, groats by-prod

OATGRASS, TALL. Arrhenatherum elatius

Oatgrass, tall, aerial pt, fresh, immature, (2)

Ref no 2-03-261

		As fed	Dry	± %
Dry matter	%	21.6	100.0	
Ash	%	2.2	10.1	20
Crude fiber	%	4.1	19.0	7
Ether extract	%	1.1	4.9	10
N-free extract	%	10.0	46.3	
Protein (N x 6.25)	%	4.3	19.7	5
Cattle	dig prot %	3.2	14.6	
Sheep	dig prot %	3.3	15.4	
Energy				
Cattle	DE kcal/kg	648.	2998.	
Sheep	DE kcal/kg	666.	3086.	
Cattle	ME kcal/kg	531.	2458.	
Sheep	ME kcal/kg	546.	2530.	
Cattle	TDN %	15.	68.	
Sheep	TDN %	15.	70.	

Oatgrass, tall, aerial pt, fresh, early blm, (2)

Ref no 2-03-262

		As fed	Dry	
Dry matter	%	25.0	100.0	
Ash	%	2.0	8.2	10
Crude fiber	%	7.7	30.7	5
Ether extract	%	.9	3.5	8
N-free extract	%	11.4	45.4	
Protein (N x 6.25)	%	3.0	12.2	8
Cattle	dig prot %	2.1	8.3	
Sheep	dig prot %	2.1	8.4	
Energy				
Cattle	DE kcal/kg	728.	2910.	
Sheep	DE kcal/kg	706.	2822.	

Continued

Feed name or analyses		Mean		C.V.
		As fed	Dry	± %
Cattle	ME kcal/kg	596.	2386.	
Sheep	ME kcal/kg	578.	2314.	
Cattle	TDN %	16.	66.	
Sheep	TDN %	16.	64.	

Oatgrass, tall, aerial pt, fresh, mid-blm, (2)

Ref no 2-03-263

Dry matter	%	25.0	100.0	
Ash	%	1.7	6.9	7
Crude fiber	%	8.3	33.2	5
Ether extract	%	.7	2.9	7
N-free extract	%	12.1	48.4	
Protein (N x 6.25)	%	2.2	8.6	8
Cattle	dig prot %	1.3	5.2	
Sheep	dig prot %	1.2	5.0	
Energy				
Cattle	DE kcal/kg	728.	2910.	
Sheep	DE kcal/kg	728.	2910.	
Cattle	ME kcal/kg	596.	2386.	
Sheep	ME kcal/kg	596.	2386.	
Cattle	TDN %	16.	66.	
Sheep	TDN %	16.	66.	

Oatgrass, tall, aerial pt, fresh, mature, (2)

Ref no 2-03-266

Dry matter	%	25.0	100.0	
Ash	%	1.5	6.0	4
Crude fiber	%	8.6	34.3	4
Ether extract	%	.6	2.6	4
N-free extract	%	12.8	51.1	
Protein (N x 6.25)	%	1.5	6.0	12
Cattle	dig prot %	.8	3.0	
Sheep	dig prot %	.6	2.6	
Energy				
Cattle	DE kcal/kg	716.	2866.	
Sheep	DE kcal/kg	750.	2998.	
Cattle	ME kcal/kg	588.	2350.	
Sheep	ME kcal/kg	614.	2458.	
Cattle	TDN %	16.	65.	
Sheep	TDN %	17.	68.	

Oatgrass, tall, aerial pt, fresh, (2)

Ref no 2-03-267

Dry matter	%	29.0	100.0	8
Ash	%	2.0	7.0	25
Crude fiber	%	9.0	31.2	12
Ether extract	%	.9	3.1	20

Continued

Feed name or analyses		Mean		C.V.
		As fed	Dry	± %
N-free extract	%	14.3	49.2	
Protein (N x 6.25)	%	2.8	9.5	30
Cattle	dig prot %	1.7	6.0	
Sheep	dig prot %	1.7	5.8	
Energy				
Cattle	DE kcal/kg	844.	2910.	
Sheep	DE kcal/kg	844.	2910.	
Cattle	ME kcal/kg	692.	2386.	
Sheep	ME kcal/kg	692.	2386.	
Cattle	TDN %	19.	66.	
Sheep	TDN %	19.	66.	

Oat meal, feeding - see Oats, cereal by-prod

OATS. Avena sativa

Oats, hay, s-c, immature, (1)

Ref no 1-03-272

Dry matter	%	89.2	100.0	3
Ash	%	8.0	9.0	15
Crude fiber	%	23.2	26.0	17
Ether extract	%	3.2	3.6	29
N-free extract	%	42.5	47.6	
Protein (N x 6.25)	%	12.3	13.8	15
Cattle	dig prot %	5.9	6.6	
Sheep	dig prot %	6.8	7.6	
Energy				
Cattle	DE kcal/kg	2360.	2646.	
Sheep	DE kcal/kg	2085.	2337.	
Cattle	ME kcal/kg	1936.	2170.	
Sheep	ME kcal/kg	1709.	1916.	
Cattle	TDN %	54.	60.	
Sheep	TDN %	47.	53.	
Calcium	%	.29	.33	
Phosphorus	%	.38	.43	
Cobalt	mg/kg	.04	.04	
Carotene	mg/kg	233.9	262.2	
Vitamin A equiv	IU/g	389.9	437.1	

Oats, hay, s-c, full blm, (1)

Ref no 1-03-274

Dry matter	%	84.4	100.0	7
Ash	%	5.8	6.9	4
Crude fiber	%	28.1	33.3	4
Ether extract	%	2.6	3.1	16
N-free extract	%	39.9	47.3	
Protein (N x 6.25)	%	7.9	9.4	15
Cattle	dig prot %	3.8	4.5	

Continued

(1) dry forages and roughages

(2) pasture, range plants, and forages fed green

(3) silages

(4) energy feeds

Feed name or analyses		Mean As fed	Dry	C.V. ± %
Sheep	dig prot %	4.3	5.1	
Energy				
Cattle	DE kcal/kg	2270.	2690.	
Sheep	DE kcal/kg	1935.	2293.	
Cattle	ME kcal/kg	1862.	2206.	
Sheep	ME kcal/kg	1587.	1880.	
Cattle	TDN %	51.	61.	
Sheep	TDN %	44.	52.	
Calcium	%	.25	.30	
Iron	%	.02	.02	
Magnesium	%	1.25	1.48	
Phosphorus	%	.30	.35	
Potassium	%	2.03	2.41	
Cobalt	mg/kg	.06	.07	
Copper	mg/kg	3.7	4.4	
Manganese	mg/kg	142.2	168.5	
Carotene	mg/kg	82.0	97.2	
Vitamin A equiv	IU/g	136.7	162.0	

Oats, hay, s-c, mature, (1)

Ref no 1-03-277

		As fed	Dry	C.V. ± %
Dry matter	%	91.2	100.0	1
Ash	%	5.3	5.8	19
Crude fiber	%	31.6	34.6	3
Ether extract	%	3.2	3.5	12
N-free extract	%	45.0	49.3	
Protein (N x 6.25)	%	6.2	6.8	19
Cattle	dig prot %	3.0	3.3	
Sheep	dig prot %	4.4	4.8	
Energy				
Cattle	DE kcal/kg	2534.	2778.	
Sheep	DE kcal/kg	2413.	2646.	
Cattle	ME kcal/kg	2078.	2278.	
Sheep	ME kcal/kg	1979.	2170.	
Cattle	TDN %	57.	63.	
Sheep	TDN %	55.	60.	
Calcium	%	.30	.33	
Phosphorus	%	.25	.27	
Cobalt	mg/kg	.08	.09	
Carotene	mg/kg	5.7	6.2	
Vitamin A equiv	IU/g	9.5	10.3	

Oats, hay, s-c, (1)

Ref no 1-03-280

		As fed	Dry	C.V. ± %
Dry matter	%	88.2	100.0	5
Ash	%	6.6	7.5	24
Crude fiber	%	27.3	31.0	11
Ether extract	%	2.7	3.1	25
N-free extract	%	43.4	49.2	
Protein (N x 6.25)	%	8.1	9.2	26

Continued

Feed name or analyses		Mean As fed	Dry	C.V. ± %
Cattle	dig prot %	3.9	4.4	
Sheep	dig prot %	4.5	5.1	
Energy				
Cattle	DE kcal/kg	2372.	2690.	
Sheep	DE kcal/kg	2100.	2381.	
Cattle	ME kcal/kg	1946.	2206.	
Sheep	ME kcal/kg	1722.	1952.	
Cattle	NE$_m$ kcal/kg	1005.	1140.	
Cattle	NE$_{gain}$ kcal/kg	370.	420.	
Cattle	TDN %	54.	61.	
Sheep	TDN %	48.	54.	
Calcium	%	.23	.26	25
Chlorine	%	.46	.52	9
Iron	%	.04	.05	23
Magnesium	%	.26	.29	99
Phosphorus	%	.21	.24	28
Potassium	%	.85	.97	63
Sodium	%	.15	.17	18
Cobalt	mg/kg	.06	.07	
Copper	mg/kg	3.9	4.4	
Manganese	mg/kg	65.7	74.7	49
Carotene	mg/kg	88.9	101.0	85
Vitamin A equiv	IU/g	148.2	168.4	

Oats, groats by-prod, mx 22 fbr, (1)
Oat mill by-product (AAFCO)
Oat feed (CFA)

Ref no 1-03-332

		As fed	Dry	C.V. ± %
Dry matter	%	92.0	100.0	3
Ash	%	5.1	5.5	30
Crude fiber	%	30.0	32.6	5
Ether extract	%	1.8	2.0	41
N-free extract	%	50.6	55.0	
Protein (N x 6.25)	%	4.5	4.9	41
Cattle	dig prot %	2.3	2.5	
Sheep	dig prot %	2.7	2.9	
Cellulose	%	10.0	10.9	
Lignin	%	5.0	5.4	
Energy				
Cattle	DE kcal/kg	1014.	1102.	
Sheep	DE kcal/kg	1663.	1808.	
Cattle	ME kcal/kg	832.	904.	
Chickens	ME$_n$ kcal/kg	397.	432.	
Sheep	ME kcal/kg	1363.	1482.	
Cattle	TDN %	23.	25.	
Sheep	TDN %	38.	41.	
Calcium	%	.08	.09	
Magnesium	%	.07	.08	
Phosphorus	%	.22	.24	
Potassium	%	.52	.57	

(5) protein supplements
(6) minerals

(7) vitamins
(8) additives

Feed name or analyses		Mean		C.V.
		As fed	Dry	± %

Oats, hulls, (1)
Oat hulls (AAFCO)
Oat hulls (CFA)

Ref no 1-03-281

Feed name or analyses		As fed	Dry	C.V. ± %
Dry matter	%	93.0	100.0	
Ash	%	6.0	6.5	
Crude fiber	%	27.0	29.0	
Ether extract	%	2.0	2.2	
N-free extract	%	52.4	56.3	
Protein (N x 6.25)	%	5.6	6.0	18
Cattle	dig prot %	2.0	2.2	
Sheep	dig prot %	1.6	1.7	
Swine	dig prot %	3.2	3.4	
Cellulose	%	47.3	51.1	
Lignin	%	13.1	14.2	
Energy				
Cattle	DE kcal/kg	1640.	1764.	
Sheep	DE kcal/kg	1476.	1587.	
Swine	DE kcal/kg	1012.	1088.	
Cattle	ME kcal/kg	1345.	1446.	
Chickens	MEn kcal/kg	331.	356.	
Sheep	ME kcal/kg	1210.	1301.	
Swine	ME kcal/kg	959.	1031.	
Cattle	TDN %	37.	40.	
Sheep	TDN %	33.	36.	
Swine	TDN %	23.	25.	
Calcium	%	.16	.17	
Iron	%	.01	.01	54
Magnesium	%	.08	.09	
Phosphorus	%	.19	.20	
Potassium	%	.59	.63	
Copper	mg/kg	5.1	5.5	
Manganese	mg/kg	18.5	20.0	10
Riboflavin	mg/kg	4.6	4.9	
Arginine	%	.20	.22	21
Histidine	%	.10	.11	
Isoleucine	%	.20	.22	40
Leucine	%	.30	.32	14
Lysine	%	.20	.22	
Methionine	%	.10	.11	
Phenylalanine	%	.20	.22	
Threonine	%	.20	.22	21
Tryptophan	%	.10	.11	
Tyrosine	%	.20	.22	20
Valine	%	.20	.22	20

Oats, straw, (1)

Ref no 1-03-283

		As fed	Dry	C.V. ± %
Dry matter	%	90.1	100.0	2
Ash	%	7.4	8.2	18
Crude fiber	%	36.9	41.0	9
Ether extract	%	1.9	2.1	21
N-free extract	%	39.9	44.3	
Protein (N x 6.25)	%	4.0	4.4	27
Cattle	dig prot %	1.3	1.4	
Sheep	dig prot %	-.4	-.4	
Cellulose	%	36.1	40.1	
Lignin	%	13.1	14.6	
Energy				
Cattle	DE kcal/kg	2066.	2293.	
Sheep	DE kcal/kg	1708.	1896.	
Cattle	ME kcal/kg	1694.	1880.	
Sheep	ME kcal/kg	1401.	1555.	
Cattle	TDN %	47.	52.	
Sheep	TDN %	39.	43.	
Calcium	%	.30	.33	16
Chlorine	%	.70	.78	8
Iron	%	.02	.02	22
Magnesium	%	.16	.18	30
Phosphorus	%	.09	.10	34
Potassium	%	2.20	2.44	24
Sodium	%	.33	.37	26
Sulfur	%	.22	.24	11
Copper	mg/kg	9.1	10.1	24
Manganese	mg/kg	35.3	39.2	44

Oats, aerial pt, fresh, immature, (2)

Ref no 2-03-286

		As fed	Dry	C.V. ± %
Dry matter	%	15.3	100.0	49
Ash	%	2.0	12.8	27
Crude fiber	%	3.0	19.8	22
Ether extract	%	.7	4.6	18
N-free extract	%	5.4	35.2	
Protein (N x 6.25)	%	4.2	27.6	22
Cattle	dig prot %	3.0	19.9	
Sheep	dig prot %	3.1	20.1	
Energy				
Cattle	DE kcal/kg	459.	2998.	
Sheep	DE kcal/kg	405.	2646.	
Cattle	ME kcal/kg	376.	2458.	
Sheep	ME kcal/kg	332.	2170.	
Cattle	TDN %	10.	68.	
Sheep	TDN %	9.	60.	
Calcium	%	.07	.48	32
Magnesium	%	.03	.21	37
Phosphorus	%	.07	.44	32

Continued

(1) dry forages and roughages

(2) pasture, range plants, and forages fed green

(3) silages

(4) energy feeds

TABLE A-3-1 Composition of Feeds (Continued)

Feed name or analyses		Mean As fed	Mean Dry	C.V. ± %
Potassium	%	.58	3.77	32
Carotene	mg/kg	85.8	560.7	18
Vitamin A equiv	IU/g	143.0	934.7	

Oats, aerial pt, fresh, early blm, (2)

Ref no 2-03-287

		As fed	Dry	C.V. ± %
Dry matter	%	26.6	100.0	
Ash	%	2.3	8.5	22
Crude fiber	%	7.4	28.0	16
Ether extract	%	.9	3.5	8
N-free extract	%	11.9	44.7	
Protein (N x 6.25)	%	4.1	15.3	30
Cattle	dig prot %	2.9	11.0	
Sheep	dig prot %	3.0	11.2	
Energy				
Cattle	DE kcal/kg	844.	3175.	
Sheep	DE kcal/kg	704.	2646.	
Cattle	ME kcal/kg	693.	2604.	
Sheep	ME kcal/kg	577.	2170.	
Cattle	TDN %	19.	72.	
Sheep	TDN %	16.	60.	

Oats, aerial pt, fresh, full blm, (2)

Ref no 2-03-288

		As fed	Dry	C.V. ± %
Dry matter	%	32.2	100.0	57
Ash	%	2.8	8.6	37
Crude fiber	%	9.9	30.9	17
Ether extract	%	1.1	3.5	15
N-free extract	%	15.3	47.5	
Protein (N x 6.25)	%	3.0	9.5	21
Cattle	dig prot %	2.2	6.8	
Sheep	dig prot %	2.2	6.9	
Energy				
Cattle	DE kcal/kg	1022.	3175.	
Sheep	DE kcal/kg	838.	2601.	
Cattle	ME kcal/kg	838.	2604.	
Sheep	ME kcal/kg	687.	2133.	
Cattle	TDN %	23.	72.	
Sheep	TDN %	19.	59.	
Calcium	%	.09	.28	16
Phosphorus	%	.10	.31	15
Potassium	%	.67	2.07	

Oats, aerial pt, fresh, mature, (2)

Ref no 2-03-290

		As fed	Dry	C.V. ± %
Dry matter	%	35.0	100.0	
Ash	%	2.8	7.9	18
Crude fiber	%	11.2	31.9	11
Ether extract	%	1.3	3.6	
N-free extract	%	16.9	48.3	
Protein (N x 6.25)	%	2.9	8.3	42
Cattle	dig prot %	2.1	6.0	
Sheep	dig prot %	2.1	6.0	
Energy				
Cattle	DE kcal/kg	1142.	3263.	
Sheep	DE kcal/kg	926.	2646.	
Cattle	ME kcal/kg	937.	2676.	
Sheep	ME kcal/kg	760.	2170.	
Cattle	TDN %	26.	74.	
Sheep	TDN %	21.	60.	
Calcium	%	.09	.27	
Iron	%	.02	.05	
Phosphorus	%	.10	.28	
Sodium	%	.04	.11	

Oats, aerial pt, fresh, (2)

Ref no 2-03-292

		As fed	Dry	C.V. ± %
Dry matter	%	21.7	100.0	64
Ash	%	2.3	10.7	31
Crude fiber	%	5.7	26.3	21
Ether extract	%	.8	3.9	23
N-free extract	%	8.9	41.1	
Protein (N x 6.25)	%	3.9	18.0	42
Cattle	dig prot %	2.8	13.0	
Sheep	dig prot %	2.8	13.1	
Energy				
Cattle	DE kcal/kg	670.	3086.	
Sheep	DE kcal/kg	574.	2646.	
Cattle	ME kcal/kg	549.	2530.	
Sheep	ME kcal/kg	471.	2170.	
Cattle	TDN %	15.	70.	
Sheep	TDN %	13.	60.	
Calcium	%	.09	.40	43
Chlorine	%	.17	.42	
Iron	%	.01	.06	44
Magnesium	%	.06	.27	48
Phosphorus	%	.08	.37	37
Potassium	%	.70	3.23	45
Sodium	%	.06	.29	80
Sulfur	%	.08	.36	36
Cobalt	mg/kg	.02	.09	82

Continued

(5) protein supplements
(6) minerals

(7) vitamins
(8) additives

Feed name or analyses		Mean		C.V.
		As fed	Dry	± %
Manganese	mg/kg	69.7	321.3	20
Carotene	mg/kg	92.8	427.8	31
Vitamin A equiv	IU/g	154.7	713.1	

Oats, aerial pt, ensiled, dough stage, (3)

Ref no 3-03-296

		As fed	Dry	± %
Dry matter	%	35.7	100.0	7
Ash	%	2.6	7.2	8
Crude fiber	%	11.0	30.7	7
Ether extract	%	1.4	3.9	8
N-free extract	%	17.4	48.6	
Protein (N x 6.25)	%	3.4	9.6	7
Cattle	dig prot %	2.1	6.0	
Sheep	dig prot %	1.7	4.9	
Energy				
Cattle	DE kcal/kg	881.	2469.	
Sheep	DE kcal/kg	1007.	2822.	
Cattle	ME kcal/kg	722.	2024.	
Sheep	ME kcal/kg	826.	2314.	
Cattle	TDN %	20.	56.	
Sheep	TDN %	23.	64.	
Calcium	%	.17	.47	
Phosphorus	%	.12	.33	

Oats, aerial pt, ensiled, (3)

Ref no 3-03-298

		As fed	Dry	± %
Dry matter	%	31.7	100.0	15
Ash	%	2.7	8.4	46
Crude fiber	%	10.0	31.6	14
Ether extract	%	1.3	4.1	25
N-free extract	%	14.6	46.2	
Protein (N x 6.25)	%	3.1	9.7	24
Cattle	dig prot %	1.7	5.5	
Sheep	dig prot %	1.6	5.0	
Energy				
Cattle	DE kcal/kg	824.	2601.	
Sheep	DE kcal/kg	881.	2778.	
Cattle	ME kcal/kg	676.	2133.	
Sheep	ME kcal/kg	722.	2278.	
Cattle	TDN %	19.	59.	
Sheep	TDN %	20.	63.	
Calcium	%		.12	.37	32
Iron	%	.	.01	
Magnesium	%	.01	.02	
Phosphorus	%	.10	.30	27
Potassium	%	1.08	3.41	
Copper	mg/kg	1.7	5.5	
Manganese	mg/kg	12.8	40.4	
Carotene	mg/kg	37.8	119.5	92
Vitamin A equiv	IU/g	63.0	199.2	

Oats, cereal by-prod, grnd, mx 2 fbr, (4)
Oat meal (CFA)
Rolled oats (CFA)

Ref no 4-03-302

		As fed	Dry	± %
Dry matter	%	91.0	100.0	2
Ash	%	2.2	2.4	17
Crude fiber	%	3.0	3.3	33
Ether extract	%	5.8	6.4	18
N-free extract	%	63.1	69.3	
Protein (N x 6.25)	%	16.9	18.6	10
Cattle	dig prot %	11.9	13.1	
Sheep	dig prot %	15.2	16.7	
Swine	dig prot %	14.2	15.6	
Energy				
Cattle	DE kcal/kg	3731.	4100.	
Sheep	DE kcal/kg	4052.	4453.	
Swine	DE kcal/kg	3651.	4012.	
Cattle	ME kcal/kg	3059.	3362.	
Sheep	ME kcal/kg	3322.	3651.	
Swine	ME kcal/kg	3366.	3699.	
Cattle	TDN %	85.	93.	
Sheep	TDN %	92.	101.	
Swine	TDN %	83.	91.	
Calcium	%	.09	.10	61
Iron	%	.01	.01	20
Magnesium	%	.19	.21	
Phosphorus	%	.47	.52	33
Potassium	%	.28	.31	
Copper	mg/kg	6.4	7.0	35
Manganese	mg/kg	34.5	38.0	56
Choline	mg/kg	1265.	1390.	19
Folic acid	mg/kg	.4	.4	34
Niacin	mg/kg	10.8	11.9	40
Pantothenic acid	mg/kg	14.7	16.2	36
Riboflavin	mg/kg	2.0	2.2	65
Thiamine	mg/kg	7.0	7.7	17
Vitamin B6	mg/kg	1.3	1.4	36
Arginine	%	1.00	1.10	17
Cystine	%	.20	.22	20
Glycine	%	.20	.22	
Histidine	%	.30	.33	39
Isoleucine	%	.60	.66	16
Leucine	%	1.00	1.10	10
Lysine	%	.50	.55	28
Methionine	%	.20	.22	29
Phenylalanine	%	.60	.66	16
Threonine	%	.50	.55	13
Tryptophan	%	.20	.22	28
Tyrosine	%	.60	.66	20
Valine	%	.70	.77	18

(1) dry forages and roughages
(2) pasture, range plants, and forages fed green

(3) silages
(4) energy feeds

TABLE A-3-1 Composition of Feeds (Continued)

Feed name or analyses		As fed	Dry	C.V. ± %

Feed name or analyses		As fed	Dry	C.V. ± %
Sheep	DE kcal/kg	3016.	3351.	
Swine	DE kcal/kg	2659.	2954.	
Cattle	ME kcal/kg	2376.	2640.	
Sheep	ME kcal/kg	2473.	2748.	
Swine	ME kcal/kg	2483.	2759.	
Cattle	TDN %	66.	73.	
Sheep	TDN %	68.	76.	
Swine	TDN %	60.	67.	

Oats, cereal by-prod, mx 4 fbr, (4)
Feeding oat meal (AAFCO)
Oat middlings (CFA)

Ref no 4-03-303

Dry matter	%	91.0	100.0	1
Ash	%	2.3	2.5	18
Crude fiber	%	4.0	4.4	24
Ether extract	%	5.8	6.4	22
N-free extract	%	63.1	69.3	
Protein (N x 6.25)	%	15.8	17.4	12
Cattle	dig prot %	10.8	11.9	
Sheep	dig prot %	12.6	13.9	
Swine	dig prot %	13.3	14.6	
Energy				
Cattle	DE kcal/kg	3691.	4056.	
Sheep	DE kcal/kg	3812.	4189.	
Swine	DE kcal/kg	3210.	3527.	
Cattle	ME kcal/kg	3027.	3326.	
Sheep	ME kcal/kg	3126.	3435.	
Swine	ME kcal/kg	2968.	3262.	
Cattle	TDN %	84.	92.	
Sheep	TDN %	86.	95.	
Swine	TDN %	73.	80.	
Calcium	%	.08	.09	10
Iron	%	.038	.042	
Phosphorus	%	.49	.54	8
Manganese	mg/kg	44.0	48.4	27
Niacin	mg/kg	28.1	30.9	20
Pantothenic acid	mg/kg	23.1	25.4	9
Riboflavin	mg/kg	1.8	2.0	17
Thiamine	mg/kg	7.0	7.7	21
Arginine	%	.70	.77	
Histidine	%	.30	.33	
Lysine	%	.10	.11	
Tyrosine	%	.91	1.00	

Oats, grain, thresher-run, mx 34 wt mn 10 mx 20 fm, (4)

Ref no 4-08-161 Origin Canada

Dry matter	%	90.0	100.0	
Ash	%	3.5	3.9	
Crude er	%	10.2	11.3	
Ether extract	%	1.6	1.8	
N-free extract	%	63.1	70.1	
Protein (N x 6.25)	%	11.4	12.7	
Cattle	dig prot %	8.6	9.5	
Sheep	dig prot %	8.9	9.9	
Swine	dig prot %	9.6	10.7	
Energy	GE kcal/kg	4320.	4800.	
Cattle	DE kcal/kg	2897.	3219.	

Oats, grain, thresher-run, mx 34 wt mx 10 fm, (4)

Ref no 4-08-160 Origin Canada

Dry matter	%	90.0	100.0	
Ash	%	3.3	3.7	
Crude fiber	%	10.3	11.4	
Ether extract	%	1.6	1.8	
N-free extract	%	63.4	70.4	
Protein (N x 6.25)	%	11.4	12.7	
Cattle	dig prot %	8.6	9.5	
Sheep	dig prot %	8.9	9.9	
Swine	dig prot %	9.6	10.7	
Energy	GE kcal/kg	4410.	4900.	
Cattle	DE kcal/kg	2897.	3219.	
Sheep	DE kcal/kg	3016.	3351.	
Swine	DE kcal/kg	2659.	2954.	
Cattle	ME kcal/kg	2376.	2640.	
Sheep	ME kcal/kg	2473.	2748.	
Swine	ME kcal/kg	2483.	2759.	
Cattle	TDN %	66.	73.	
Sheep	TDN %	68.	76.	
Swine	TDN %	60.	67.	

Oats, grain, (4)

Ref no 4-03-309

Dry matter	%	89.0	100.0	3
Ash	%	3.2	3.6	28
Crude fiber	%	11.0	12.4	33
Ether extract	%	4.5	5.1	24
N-free extract	%	58.5	65.7	
Protein (N x 6.25)	%	11.8	13.2	17
Cattle	dig prot %	8.8	9.9	
Sheep	dig prot %	9.2	10.3	
Swine	dig prot %	9.9	11.1	
Cellulose	%	16.0	18.0	
Lignin	%	8.9	10.0	
Energy	GE kcal/kg	4187.	4704.	
Cattle	DE kcal/kg	2982.	3351.	
Sheep	DE kcal/kg	2943.	3307.	
Swine	DE kcal/kg	2860.	3213.	
Cattle	ME kcal/kg	2446.	2748.	
Chickens	MEn kcal/kg	2535.	2848.	

Continued

Continued

(5) protein supplements
(6) minerals

(7) vitamins
(8) additives

Feed name or analyses		Mean		C.V.
		As fed	Dry	± %
Sheep	ME kcal/kg	2414.	2712.	
Swine	ME kcal/kg	2668.	2998.	
Cattle	NE$_m$ kcal/kg	1629.	1830.	
Cattle	NE$_{gain}$ kcal/kg	1086.	1220.	
Cattle	TDN %	68.	76.	
Sheep	TDN %	67.	75.	
Swine	TDN %	65.	73.	
Calcium	%	.10	.11	70
Iron	%	.007	.008	75
Magnesium	%	.17	.19	28
Phosphorus	%	.35	.39	43
Potassium	%	.37	.42	33
Sodium	%	.06	.07	59
Cobalt	mg/kg	.060	.070	99
Copper	mg/kg	5.9	6.6	74
Manganese	mg/kg	38.2	42.9	8
Biotin	mg/kg	.30	.30	99
Choline	mg/kg	1073.	1206.	17
Folic acid	mg/kg	.40	.40	57
Niacin	mg/kg	15.8	17.8	41
Pantothenic acid	mg/kg	12.9	14.5	33
Riboflavin	mg/kg	1.6	1.8	99
Thiamine	mg/kg	6.2	7.0	26
a-tocopherol	mg/kg	5.9	6.6	28
Vitamin B$_6$	mg/kg	1.2	1.3	42
Arginine	%	.71	.80	29
Cystine	%	.18	.20	54
Histidine	%	.18	.20	58
Isoleucine	%	.53	.60	21
Leucine	%	.89	1.00	21
Lysine	%	.36	.40	28
Methionine	%	.18	.20	35
Phenylalanine	%	.62	.70	22
Threonine	%	.36	.40	24
Tryptophan	%	.18	.20	35
Tyrosine	%	.53	.60	5
Valine	%	.62	.70	26

Oats grain - see also Oats, red; Oats, white; Oats wild

Oats, grain, gr sample US, (4)

Ref no 4-03-310

		As fed	Dry	C.V.
Dry matter	%	84.0	100.0	5
Ash	%	2.5	3.0	15
Crude fiber	%	10.0	11.9	18
Ether extract	%	4.8	5.7	11
N-free extract	%	56.0	66.7	
Protein (N x 6.25)	%	10.7	12.7	9
Cattle	dig prot %	8.0	9.5	
Sheep	dig prot %	8.3	9.9	
Swine	dig prot %	9.0	10.7	

Continued

Feed name or analyses		Mean		C.V.
		As fed	Dry	± %
Energy				
Cattle	DE kcal/kg	2889.	3439.	
Sheep	DE kcal/kg	2963.	3527.	
Swine	DE kcal/kg	2629.	3130.	
Cattle	ME kcal/kg	2369.	2820.	
Sheep	ME kcal/kg	2429.	2892.	
Swine	ME kcal/kg	2455.	2923.	
Cattle	TDN %	66.	78.	
Sheep	TDN %	67.	80.	
Swine	TDN %	60.	71.	

Oats, grain, gr 1 heavy US mn 36 wt mx 2 fm, (4)

Ref no 4-03-312

		As fed	Dry	C.V.
Dry matter	%	90.0	100.0	1
Ash	%	3.0	3.3	24
Crude fiber	%	10.0	11.1	14
Ether extract	%	4.5	5.0	16
N-free extract	%	60.1	66.8	
Protein (N x 6.25)	%	12.4	13.8	10
Cattle	dig prot %	9.4	10.4	
Sheep	dig prot %	10.9	12.1	
Swine	dig prot %	10.4	11.6	
Energy				
Cattle	DE kcal/kg	3056.	3395.	
Sheep	DE kcal/kg	3135.	3483.	
Swine	DE kcal/kg	2817.	3130.	
Cattle	ME kcal/kg	2506.	2784.	
Sheep	ME kcal/kg	2570.	2856.	
Swine	ME kcal/kg	2625.	2917.	
Cattle	TDN %	69.	77.	
Sheep	TDN %	71.	79.	
Swine	TDN %	64.	71.	

Oats, grain, gr 1 US mn 34 wt mx 2 fm, (4)

Ref no 4-03-313

		As fed	Dry	C.V.
Dry matter	%	91.0	100.0	
Ash	%	3.2	3.5	11
Crude fiber	%	12.0	13.2	
Ether extract	%	4.8	5.3	15
N-free extract	%	58.9	64.7	
Protein (N x 6.25)	%	12.1	13.3	7
Cattle	dig prot %	9.1	10.0	
Sheep	dig prot %	9.5	10.4	
Swine	dig prot %	10.2	11.2	
Energy				
Cattle	DE kcal/kg	3049.	3351.	
Sheep	DE kcal/kg	3170.	3483.	
Swine	DE kcal/kg	2768.	3042.	
Cattle	ME kcal/kg	2501.	2748.	
Sheep	ME kcal/kg	2599.	2856.	

Continued

(1) dry forages and roughages
(2) pasture, range plants, and forages fed green
(3) silages
(4) energy feeds

Feed name or analyses		Mean		C.V.
		As fed	Dry	± %
Swine	ME kcal/kg	2582.	2838.	
Cattle	TDN %	69.	76.	
Sheep	TDN %	72.	79.	
Swine	TDN %	63.	69.	
Calcium	%	.08	.09	
Phosphorus	%	.30	.33	

Feed name or analyses		Mean		C.V.
		As fed	Dry	± %
Swine	TDN %	62.	70.	
Calcium	%	.06	.07	
Chlorine	%	.27	.30	

Oats, grain, gr 2 heavy US mn 36 wt mx 3 fm, (4)
Oats, grain, heavy

Ref no 4-03-315

		As fed	Dry	C.V. ± %
Dry matter	%	89.5	100.0	1
Ash	%	3.1	3.5	8
Crude fiber	%	9.8	10.9	14
Ether extract	%	4.0	4.5	6
N-free extract	%	60.5	67.6	
Protein (N x 6.25)	%	12.1	13.5	4
Cattle	dig prot %	9.0	10.1	
Sheep	dig prot %	9.4	10.5	
Swine	dig prot %	10.1	11.3	
Energy				
Cattle	DE kcal/kg	3038.	3395.	
Sheep	DE kcal/kg	3117.	3483.	
Swine	DE kcal/kg	2762.	3086.	
Cattle	ME kcal/kg	2492.	2784.	
Sheep	ME kcal/kg	2556.	2856.	
Swine	ME kcal/kg	2577.	2879.	
Cattle	TDN %	69.	77.	
Sheep	TDN %	71.	79.	
Swine	TDN %	63.	70.	

Oats, grain, gr 2 US mn 32 wt mx 3 fm, (4)

Ref no 4-03-316

		As fed	Dry	C.V. ± %
Dry matter	%	89.0	100.0	
Ash	%	2.9	3.3	2
Crude fiber	%	11.0	12.4	13
Ether extract	%	4.2	4.7	11
N-free extract	%	59.5	66.9	
Protein (N x 6.25)	%	11.3	12.7	9
Cattle	dig prot %	8.4	9.5	
Sheep	dig prot %	8.8	9.9	
Swine	dig prot %	9.5	10.7	
Energy				
Cattle	DE kcal/kg	2982.	3351.	
Sheep	DE kcal/kg	3100.	3483.	
Swine	DE kcal/kg	2746.	3086.	
Cattle	ME kcal/kg	2446.	2748.	
Sheep	ME kcal/kg	2542.	2856.	
Swine	ME kcal/kg	2565.	2882.	
Cattle	TDN %	68.	76.	
Sheep	TDN %	70.	79.	

Continued

Oats, grain, gr 3 US mn 30 wt mx 4 fm, (4)

Ref no 4-03-317

		As fed	Dry	C.V. ± %
Dry matter	%	91.0	100.0	
Ash	%	3.4	3.7	
Crude fiber	%	13.0	14.3	
Ether extract	%	4.6	5.1	
N-free extract	%	57.9	63.6	
Protein (N x 6.25)	%	12.1	13.3	
Cattle	dig prot %	9.1	10.0	
Sheep	dig prot %	9.5	10.4	
Swine	dig prot %	10.2	11.2	
Energy				
Cattle	DE kcal/kg	3009.	3307.	
Sheep	DE kcal/kg	3129.	3439.	
Swine	DE kcal/kg	2728.	2998.	
Cattle	ME kcal/kg	2468.	2712.	
Sheep	ME kcal/kg	2566.	2820.	
Swine	ME kcal/kg	2545.	2797.	
Cattle	TDN %	68.	75.	
Sheep	TDN %	71.	78.	
Swine	TDN %	62.	68.	

Oats, grain, gr 4 US mn 27 wt mx 5 fm, (4)
Oats, grain, light

Ref no 4-03-318

		As fed	Dry	C.V. ± %
Dry matter	%	91.2	100.0	
Ash	%	4.7	5.2	
Crude fiber	%	15.1	16.5	
Ether extract	%	4.5	4.9	
N-free extract	%	54.9	60.2	
Protein (N x 6.25)	%	12.0	13.2	
Cattle	dig prot %	9.0	9.9	
Sheep	dig prot %	10.0	11.0	
Swine	dig prot %	10.1	11.1	
Energy				
Cattle	DE kcal/kg	2936.	3219.	
Sheep	DE kcal/kg	2896.	3175.	
Swine	DE kcal/kg	2574.	2822.	
Cattle	ME kcal/kg	2408.	2640.	
Sheep	ME kcal/kg	2375.	2604.	
Swine	ME kcal/kg	2404.	2633.	
Cattle	TDN %	66.	73.	
Sheep	TDN %	66.	72.	
Swine	TDN %	58.	64.	

(5) protein supplements

(6) minerals

(7) vitamins

(8) additives

Feed name or analyses		Mean		C.V. ± %
		As fed	Dry	
Oats, grain, Pacific coast, (4)				
Ref no 4-07-999				
Dry matter	%	91.2	100.0	
Ash	%	3.7	4.0	
Crude fiber	%	11.0	12.1	
Ether extract	%	5.4	5.9	
N-free extract	%	62.1	68.1	
Protein (N x 6.25)	%	9.0	9.9	
Cattle	dig prot %	6.7	7.4	
Sheep	dig prot %	7.0	7.7	
Energy				
Cattle	DE kcal/kg	3096.	3395.	
Sheep	DE kcal/kg	3217.	3527.	
Cattle	ME kcal/kg	2539.	2784.	
Sheep	ME kcal/kg	2638.	2892.	
Cattle	TDN %	70.	77.	
Sheep	TDN %	73.	80.	
Calcium	%	.09	.10	
Phosphorus	%	.33	.36	

Oats, groats, grnd cooked, (4)				
Ref no 4-07-982				
Dry matter	%	91.0	100.0	
Crude fiber	%	3.0	3.3	
Ether extract	%	5.8	6.4	
Protein (N x 6.25)	%	16.7	18.4	
Calcium	%	.07	.08	
Phosphorus	%	.43	.47	

Oats, groats, (4)
Oat groats (AAFCO)
Oat groats (CFA)
Hulled oats (CFA)

Ref no 4-03-331

Feed name or analyses		As fed	Dry	C.V. ± %
Dry matter	%	91.0	100.0	2
Ash	%	2.2	2.4	20
Crude fiber	%	3.0	3.3	40
Ether extract	%	5.8	6.4	21
N-free extract	%	63.2	69.5	
Protein (N x 6.25)	%	16.7	18.4	11
Cattle	dig prot %	11.7	12.9	
Sheep	dig prot %	15.1	16.6	
Swine	dig prot %	14.0	15.4	
Energy				
Cattle	DE kcal/kg	3731.	4100.	
Sheep	DE kcal/kg	4052.	4453.	
Swine	DE kcal/kg	3250.	3571.	

Continued

Feed name or analyses		Mean		C.V. ± %
		As fed	Dry	
Cattle	ME kcal/kg	3059.	3362.	
Chickens	MEn kcal/kg	3549.	3900.	
Sheep	ME kcal/kg	3322.	3651.	
Swine	ME kcal/kg	2999.	3296.	
Cattle	TDN %	85.	93.	
Sheep	TDN %	92.	101.	
Swine	TDN %	74.	81.	
Calcium	%	.07	.08	
Magnesium	%	.09	.10	
Phosphorus	%	.43	.47	8
Potassium	%	.34	.37	
Copper	mg/kg	6.4	7.0	
Manganese	mg/kg	28.6	31.4	16
Niacin	mg/kg	8.1	8.9	13
Pantothenic acid	mg/kg	14.7	16.2	16
Riboflavin	mg/kg	1.3	1.4	9
Thiamine	mg/kg	6.8	7.5	11
Vitamin B$_6$	mg/kg	1.1	1.2	21

OATS, RED. *Avena byzantina*

Oats, red, grain, gr 1 heavy US mn 36 wt mx 2 fm, (4)

Ref no 4-03-362

Feed name or analyses		As fed	Dry	C.V. ± %
Dry matter	%	92.0	100.0	
Ash	%	3.4	·3.7	
Crude fiber	%	9.0	9.8	
Ether extract	%	5.0	5.4	
N-free extract	%	59.5	64.7	
Protein (N x 6.25)	%	15.1	16.4	
Cattle	dig prot %	11.3	12.3	
Sheep	dig prot %	11.8	12.8	
Swine	dig prot %	12.7	13.8	
Energy				
Cattle	DE kcal/kg	3164.	3439.	
Sheep	DE kcal/kg	3245.	3527.	
Swine	DE kcal/kg	2961.	3219.	
Cattle	ME kcal/kg	2594.	2820.	
Sheep	ME kcal/kg	2661.	2892.	
Swine	ME kcal/kg	2745.	2984.	
Cattle	TDN %	72.	78.	
Sheep	TDN %	74.	80.	
Swine	TDN %	67.	73.	

Oats, red, grain, gr 1 US mn 34 wt mx 2 fm, (4)

Ref no 4-03-363

		As fed	Dry	
Dry matter	%	92.0	100.0	
Ash	%	3.4	3.7	15
Crude fiber	%	13.0	14.1	

Continued

(1) dry forages and roughages
(2) pasture, range plants, and forages fed green
(3) silages
(4) energy feeds

Feed name or analyses		Mean		C.V.
		As fed	Dry	± %
Ether extract	%	5.6	6.1	8
N-free extract	%	57.8	62.8	
Protein (N x 6.25)	%	12.2	13.3	6
Cattle	dig prot %	9.2	10.0	
Sheep	dig prot %	9.6	10.4	
Swine	dig prot %	10.3	11.2	
Energy				
Cattle	DE kcal/kg	3083.	3351.	
Sheep	DE kcal/kg	3204.	3483.	
Swine	DE kcal/kg	2799.	3042.	
Cattle	ME kcal/kg	2528.	2748.	
Sheep	ME kcal/kg	2628.	2856.	
Swine	ME kcal/kg	2611.	2838.	
Cattle	TDN %	70.	76.	
Sheep	TDN %	73.	79.	
Swine	TDN %	63.	69.	

Oats, red, grain, gr 2 US mn 32 wt mx 3 fm, (4)

Ref no 4-03-366

Feed name or analyses		As fed	Dry	
Dry matter	%	92.0	100.0	
Ash	%	3.3	3.6	
Crude fiber	%	13.0	14.1	
Ether extract	%	5.1	5.5	
N-free extract	%	59.4	64.6	
Protein (N x 6.25)	%	11.2	12.2	
Cattle	dig prot %	8.5	9.2	
Sheep	dig prot %	8.7	9.5	
Swine	dig prot %	9.4	10.2	
Energy				
Cattle	DE kcal/kg	3083.	3351.	
Sheep	DE kcal/kg	3164.	3439.	
Swine	DE kcal/kg	2758.	2998.	
Cattle	ME kcal/kg	2528.	2748.	
Sheep	ME kcal/kg	2594.	2820.	
Swine	ME kcal/kg	2579.	2803.	
Cattle	TDN %	70.	76.	
Sheep	TDN %	72.	78.	
Swine	TDN %	62.	68.	

Oats, red, grain, gr 3 US mn 30 wt mx 4 fm, (4)

Ref no 4-03-367

Feed name or analyses		As fed	Dry	
Dry matter	%	92.0	100.0	
Ash	%	3.5	3.8	
Crude fiber	%	14.0	15.2	
Ether extract	%	5.0	5.4	
N-free extract	%	58.1	63.1	
Protein (N x 6.25)	%	11.5	12.5	
Cattle	dig prot %	8.6	9.4	
Sheep	dig prot %	9.0	9.8	
Swine	dig prot %	9.7	10.5	

Continued

Feed name or analyses		Mean		C.V.
		As fed	Dry	± %
Energy				
Cattle	DE kcal/kg	3042.	3307.	
Sheep	DE kcal/kg	3164.	3439.	
Swine	DE kcal/kg	2718.	2954.	
Cattle	ME kcal/kg	2495.	2712.	
Sheep	ME kcal/kg	2594.	2820.	
Swine	ME kcal/kg	2541.	2762.	
Cattle	TDN %	69.	75.	
Sheep	TDN %	72.	78.	
Swine	TDN %	62.	67.	

Oats, red, grain, gr 4 US mn 27 wt mx 5 fm, (4)

Ref no 4-03-368

Feed name or analyses		As fed	Dry	
Dry matter	%	92.0	100.0	
Ash	%	3.3	3.6	
Crude fiber	%	14.0	15.2	
Ether extract	%	4.8	5.2	
N-free extract	%	58.5	63.6	
Protein (N x 6.25)	%	11.4	12.4	
Cattle	dig prot %	8.6	9.3	
Sheep	dig prot %	8.9	9.7	
Swine	dig prot %	9.6	10.4	
Energy				
Cattle	DE kcal/kg	3042.	3307.	
Sheep	DE kcal/kg	3164.	3439.	
Swine	DE kcal/kg	2718.	2954.	
Cattle	ME kcal/kg	2495.	2712.	
Sheep	ME kcal/kg	2594.	2820.	
Swine	ME kcal/kg	2541.	2762.	
Cattle	TDN %	69.	75.	
Sheep	TDN %	72.	78.	
Swine	TDN %	62.	67.	

Oats, red, grain, (4)

Ref no 4-03-369

Feed name or analyses		As fed	Dry	
Dry matter	%	92.0	100.0	1
Ash	%	3.4	3.7	16
Crude fiber	%	13.0	14.1	10
Ether extract	%	5.1	5.5	14
N-free extract	%	58.9	64.0	
Protein (N x 6.25)	%	11.7	12.7	8
Cattle	dig prot %	8.7	9.5	
Sheep	dig prot %	9.1	9.9	
Swine	dig prot %	9.8	10.7	
Energy				
Cattle	DE kcal/kg	3083.	3351.	
Sheep	DE kcal/kg	3164.	3439.	
Swine	DE kcal/kg	2758.	2998.	
Cattle	ME kcal/kg	2528.	2748.	
Sheep	ME kcal/kg	2594.	2820.	

Continued

(5) protein supplements
(6) minerals
(7) vitamins
(8) additives

Feed name or analyses		Mean		C.V. ± %
		As fed	Dry	
Swine	ME kcal/kg	2576.	2800.	
Cattle	TDN %	70.	76.	
Sheep	TDN %	72.	78.	
Swine	TDN %	62.	68.	

Feed name or analyses		Mean		C.V. ± %
		As fed	Dry	
Sheep	TDN %	68.	79.	
Swine	TDN %	60.	70.	
Alanine	%	.42	.48	
Arginine	%	.50	.58	
Aspartic acid	%	.69	.80	
Glutamic acid	%	1.85	2.14	
Glycine	%	.45	.52	
Histidine	%	.19	.22	
Isoleucine	%	.32	.37	
Leucine	%	.64	.74	
Lysine	%	.36	.42	
Methionine	%	.03	.04	
Phenylalanine	%	.45	.52	
Proline	%	.51	.59	
Serine	%	.07	.08	
Threonine	%	.14	.16	
Tyrosine	%	.15	.17	
Valine	%	.51	.59	

OATS, WHITE. Avena sativa

Oats, white, grain, Can 1 feed mn 34 wt mx 12 fm, (4)

Ref no 4-03-377

Dry matter	%	86.5	100.0	
Ash	%	2.9	3.4	18
Crude fiber	%	10.3	11.9	36
Ether extract	%	4.6	5.3	48
N-free extract	%	57.6	66.6	
Protein (N x 6.25)	%	11.1	12.8	17
Cattle	dig prot %	8.3	9.6	
Sheep	dig prot %	8.6	10.0	
Swine	dig prot %	9.3	10.8	
Energy	GE kcal/kg	4804.	5553.	
Cattle	DE kcal/kg	2937.	3395.	
Sheep	DE kcal/kg	3051.	3527.	
Swine	DE kcal/kg	2669.	3086.	
Cattle	ME kcal/kg	2408.	2784.	
Sheep	ME kcal/kg	2502.	2892.	
Swine	ME kcal/kg	2493.	2882.	
Cattle	TDN %	67.	77.	
Sheep	TDN %	69.	80.	
Swine	TDN %	60.	70.	

Oats, white, grain, Can 2 CW mn 36 wt mx 3 fm, (4)

Ref no 4-03-378

Dry matter	%	86.5	100.0	
Ash	%	3.0	3.5	11
Crude fiber	%	10.4	12.0	34
Ether extract	%	4.5	5.2	28
N-free extract	%	57.2	66.1	
Protein (N x 6.25)	%	11.4	13.2	30
Cattle	dig prot %	8.6	9.9	
Sheep	dig prot %	8.9	10.3	
Swine	dig prot %	9.6	11.1	
Energy	GE kcal/kg	4804.	5553.	
Cattle	DE kcal/kg	2937.	3395.	
Sheep	DE kcal/kg	3013.	3483.	
Swine	DE kcal/kg	2669.	3086.	
Cattle	ME kcal/kg	2408.	2784.	
Sheep	ME kcal/kg	2470.	2856.	
Swine	ME kcal/kg	2490.	2879.	
Cattle	TDN %	67.	77.	

Oats, white, grain, Can 2 feed mn 28 wt mx 22 fm, (4)

Ref no 4-03-379

Dry matter	%	86.5	100.0	
Ash	%	2.9	3.4	23
Crude fiber	%	10.4	12.0	40
Ether extract	%	4.4	5.1	48
N-free extract	%	57.8	66.8	
Protein (N x 6.25)	%	11.0	12.7	17
Cattle	dig prot %	8.2	9.5	
Sheep	dig prot %	8.6	9.9	
Swine	dig prot %	9.2	10.7	
Energy	GE kcal/kg	4804.	5553.	
Cattle	DE kcal/kg	2937.	3395.	
Sheep	DE kcal/kg	3013.	3483.	
Swine	DE kcal/kg	2669.	3086.	
Cattle	ME kcal/kg	2408.	2784.	
Sheep	ME kcal/kg	2470.	2856.	
Swine	ME kcal/kg	2496.	2885.	
Cattle	TDN %	67.	77.	
Sheep	TDN %	68.	79.	
Swine	TDN %	60.	70.	
Alanine	%	.42	.48	
Arginine	%	.47	.54	
Aspartic acid	%	.67	.78	
Glutamic acid	%	1.94	2.24	
Glycine	%	.42	.48	
Histidine	%	.15	.17	
Isoleucine	%	.22	.26	
Leucine	%	.59	.68	
Lysine	%	.27	.31	
Methionine	%	.10	.12	
Phenylalanine	%	.40	.46	
Proline	%	.54	.62	

Continued

Continued

(1) dry forages and roughages
(2) pasture, range plants, and forages fed green

(3) silages
(4) energy feeds

Feed name or analyses		Mean		C.V.
		As fed	Dry	± %
Serine	%	.45	.52	
Threonine	%	.28	.33	
Tyrosine	%	.23	.27	
Valine	%	.31	.36	

Oats, white, grain, Can 3 CW mn 34 wt mx 6 fm, (4)

Ref no 4-03-380

Feed name or analyses		As fed	Dry	C.V. ± %
Dry matter	%	86.5	100.0	
Ash	%	2.9	3.4	18
Crude fiber	%	10.5	12.1	40
Ether extract	%	4.6	5.3	49
N-free extract	%	57.5	66.5	
Protein (N x 6.25)	%	11.0	12.7	18
Cattle	dig prot %	8.2	9.5	
Sheep	dig prot %	8.6	9.9	
Swine	dig prot %	9.2	10.7	
Energy	GE kcal/kg	4804.	5553.	
Cattle	DE kcal/kg	2937.	3395.	
Sheep	DE kcal/kg	3013.	3483.	
Swine	DE kcal/kg	2669.	3086.	
Cattle	ME kcal/kg	2408.	2784.	
Sheep	ME kcal/kg	2470.	2856.	
Swine	ME kcal/kg	2496.	2885.	
Cattle	TDN %	67.	77.	
Sheep	TDN %	68.	79.	
Swine	TDN %	60.	70.	
Alanine	%	.41	.47	
Arginine	%	.51	.59	
Aspartic acid	%	.72	.83	
Glutamic acid	%	1.82	2.10	
Glycine	%	.43	.50	
Histidine	%	.16	.18	
Isoleucine	%	.24	.28	
Leucine	%	.60	.70	
Lysine	%	.29	.34	
Methionine	%	.10	.11	
Phenylalanine	%	.42	.49	
Proline	%	.47	.54	
Serine	%	.45	.52	
Threonine	%	.29	.34	
Tyrosine	%	.23	.27	
Valine	%	.36	.42	

Oats, white, grain, Can 3 feed mx 33 fm, (4)

Ref no 4-03-381

		As fed	Dry	
Dry matter	%	86.5	100.0	
Ash	%	2.9	3.3	22
Crude fiber	%	10.5	12.1	35
Ether extract	%	4.3	5.0	39
N-free extract	%	57.8	66.8	

Continued

Feed name or analyses		Mean		C.V.
		As fed	Dry	± %
Protein (N x 6.25)	%	11.0	12.8	13
Cattle	dig prot %	9.2	9.6	
Sheep	dig prot %	8.6	10.0	
Swine	dig prot %	9.3	10.8	
Energy	GE kcal/kg	4804.	5553.	
Cattle	DE kcal/kg	2937.	3395.	
Sheep	DE kcal/kg	3013.	3483.	
Swine	DE kcal/kg	2669.	3086.	
Cattle	ME kcal/kg	2408.	2784.	
Sheep	ME kcal/kg	2470.	2856.	
Swine	ME kcal/kg	2496.	2885.	
Cattle	TDN %	67.	77.	
Sheep	TDN %	68.	79.	
Swine	TDN %	60.	70.	

Oats, white, grain, gr 1 heavy US mn 36 wt mx 2 fm, (4)

Ref no 4-03-384

		As fed	Dry	
Dry matter	%	90.0	100.0	1
Ash	%	3.2	3.6	22
Crude fiber	%	11.0	12.2	10
Ether extract	%	4.0	4.4	11
N-free extract	%	59.8	66.5	
Protein (N x 6.25)	%	12.0	13.3	10
Cattle	dig prot %	9.0	10.0	
Sheep	dig prot %	10.1	11.2	
Swine	dig prot %	10.1	11.2	
Energy				
Cattle	DE kcal/kg	3016.	3351.	
Sheep	DE kcal/kg	3056.	3395.	
Swine	DE kcal/kg	2738.	3042.	
Cattle	ME kcal/kg	2473.	2748.	
Sheep	ME kcal/kg	2506.	2784.	
Swine	ME kcal/kg	2554.	2838.	
Cattle	TDN %	68.	76.	
Sheep	TDN %	69.	77.	
Swine	TDN %	62.	69.	

Oats, white, grain, gr 1 US mn 34 wt mx 2 fm, (4)

Ref no 4-03-385

		As fed	Dry	
Dry matter	%	90.0	100.0	
Ash	%	3.0	3.3	12
Crude fiber	%	12.0	13.3	
Ether extract	%	4.4	4.9	9
N-free extract	%	58.2	64.7	
Protein (N x 6.25)	%	12.4	13.8	6
Cattle	dig prot %	9.4	10.4	
Sheep	dig prot %	9.7	10.8	
Swine	dig prot %	10.5	11.7	

Continued

(5) protein supplements

(6) minerals

(7) vitamins

(8) additives

Feed name or analyses		Mean		C.V.
		As fed	Dry	± %
Energy				
Cattle	DE kcal/kg	3016.	3351.	
Sheep	DE kcal/kg	3095.	3439.	
Swine	DE kcal/kg	2738.	3042.	
Cattle	ME kcal/kg	2473.	2748.	
Sheep	ME kcal/kg	2538.	2820.	
Swine	ME kcal/kg	2552.	2835.	
Cattle	TDN %	68.	76.	
Sheep	TDN %	70.	78.	
Swine	TDN %	62.	69.	

Oats, white, grain, gr 2 US mn 32 wt mx 3 fm, (4)

Ref no 4-03-388

		As fed	Dry	C.V. ± %
Dry matter	%	90.0	100.0	1
Ash	%	2.9	3.2	7
Crude fiber	%	11.0	12.2	12
Ether extract	%	4.1	4.6	7
N-free extract	%	60.8	67.6	
Protein (N x 6.25)	%	11.2	12.4	3
Cattle	dig prot %	8.4	9.3	
Sheep	dig prot %	8.7	9.7	
Swine	dig prot %	9.4	10.4	
Energy				
Cattle	DE kcal/kg	3016.	3351.	
Sheep	DE kcal/kg	3135.	3483.	
Swine	DE kcal/kg	2777.	3086.	
Cattle	ME kcal/kg	2473.	2748.	
Sheep	ME kcal/kg	2570.	2856.	
Swine	ME kcal/kg	2596.	2885.	
Cattle	TDN %	68.	76.	
Sheep	TDN %	71.	79.	
Swine	TDN %	63.	70.	

Oats, white, grain, gr 3 US mn 30 wt mx 4 fm, (4)

Ref no 4-03-389

		As fed	Dry	
Dry matter	%	91.0	100.0	
Ash	%	3.3	3.6	
Crude fiber	%	12.0	13.2	
Ether extract	%	4.1	4.5	
N-free extract	%	59.1	65.0	
Protein (N x 6.25)	%	12.5	13.7	
Cattle	dig prot %	9.4	10.3	
Sheep	dig prot %	9.7	10.7	
Swine	dig prot %	10.5	11.5	
Energy				
Cattle	DE kcal/kg	3049.	3351.	
Sheep	DE kcal/kg	3129.	3439.	
Swine	DE kcal/kg	2728.	2998.	
Cattle	ME kcal/kg	2501.	2748.	
Sheep	ME kcal/kg	2566.	2820.	

Continued

Feed name or analyses		Mean		C.V.
		As fed	Dry	± %
Swine	ME kcal/kg	2542.	2794.	
Cattle	TDN %	69.	76.	
Sheep	TDN %	71.	78.	
Swine	TDN %	62.	68.	

Oats, white, grain, gr 4 US mn 27 wt mx 5 fm, (4)

Ref no 4-03-390

		As fed	Dry	
Dry matter	%	92.0	100.0	
Ash	%	3.0	3.3	
Crude fiber	%	12.0	13.0	
Ether extract	%	4.4	4.8	
N-free extract	%	59.9	65.1	
Protein (N x 6.25)	%	12.7	13.8	
Cattle	dig prot %	9.6	10.4	
Sheep	dig prot %	9.9	10.8	
Swine	dig prot %	10.8	11.7	
Energy				
Cattle	DE kcal/kg	3083.	3351.	
Sheep	DE kcal/kg	3164.	3439.	
Swine	DE kcal/kg	2799.	3042.	
Cattle	ME kcal/kg	2528.	2748.	
Sheep	ME kcal/kg	2594.	2820.	
Swine	ME kcal/kg	2608.	2835.	
Cattle	TDN %	70.	76.	
Sheep	TDN %	72.	78.	
Swine	TDN %	63.	69.	

Oats, white, grain, (4)

Ref no 4-03-391

		As fed	Dry	C.V. ± %
Dry matter	%	91.0	100.0	1
Ash	%	3.1	3.4	19
Crude fiber	%	11.0	12.1	10
Ether extract	%	4.3	4.7	11
N-free extract	%	60.2	66.1	
Protein (N x 6.25)	%	12.5	13.7	7
Cattle	dig prot %	9.4	10.3	
Sheep	dig prot %	9.7	10.7	
Swine	dig prot %	10.5	11.5	
Energy				
Cattle	DE kcal/kg	3049.	3351.	
Sheep	DE kcal/kg	3009.	3307.	
Swine	DE kcal/kg	2808.	3086.	
Cattle	ME kcal/kg	2501.	2748.	
Sheep	ME kcal/kg	2468.	2712.	
Swine	ME kcal/kg	2617.	2876.	
Cattle	TDN %	69.	76.	
Sheep	TDN %	68.	75.	
Swine	TDN %	64.	70.	

(1) dry forages and roughages
(2) pasture, range plants, and forages fed green

(3) silages
(4) energy feeds

TABLE A-3-1 Composition of Feeds (Continued)

Feed name or analyses		Mean		C.V.
		As fed	Dry	± %

OATS, WILD. Avena fatua

Oats, wild, hay, s-c, (1)

Ref no 1-03-392

		As fed	Dry	± %
Dry matter	%	91.6	100.0	1
Ash	%	7.0	7.6	13
Crude fiber	%	31.2	34.1	5
Ether extract	%	2.7	2.9	10
N-free extract	%	43.9	47.9	
Protein (N x 6.25)	%	6.9	7.5	8
Cattle	dig prot %	3.1	3.4	
Sheep	dig prot %	3.0	3.3	
Energy				
Cattle	DE kcal/kg	2302.	2513.	
Sheep	DE kcal/kg	2221.	2425.	
Cattle	ME kcal/kg	1888.	2061.	
Sheep	ME kcal/kg	1821.	1988.	
Cattle	TDN %	52.	57.	
Sheep	TDN %	50.	55.	

Oats, wild, grain, (4)

Ref no 4-03-394

		As fed	Dry	± %
Dry matter	%	89.0	100.0	1
Ash	%	4.6	5.2	11
Crude fiber	%	15.0	16.9	6
Ether extract	%	5.6	6.3	12
N-free extract	%	52.0	58.4	
Protein (N x 6.25)	%	11.7	13.2	8
Cattle	dig prot %	8.8	9.9	
Sheep	dig prot %	9.2	10.3	
Swine	dig prot %	10.0	11.2	
Energy				
Cattle	DE kcal/kg	2904.	3263.	
Sheep	DE kcal/kg	3022.	3395.	
Swine	DE kcal/kg	2512.	2822.	
Cattle	ME kcal/kg	2382.	2676.	
Sheep	ME kcal/kg	2478.	2784.	
Swine	ME kcal/kg	2343.	2633.	
Cattle	TDN %	66.	74.	
Sheep	TDN %	68.	77.	
Swine	TDN %	57.	64.	

Orange - see Citrus, orange

ORCHARDGRASS. Dactylis glomerata

Orchardgrass, hay, s-c, immature, (1)

Ref no 1-03-423

		As fed	Dry	± %
Dry matter	%	89.3	100.0	3
Ash	%	7.0	7.8	14
Crude fiber	%	25.1	28.1	13
Ether extract	%	4.3	4.8	28
N-free extract	%	38.1	42.7	
Protein (N x 6.25)	%	14.8	16.6	27
Cattle	dig prot %	8.9	10.0	
Sheep	dig prot %	7.4	8.3	
Energy				
Cattle	DE kcal/kg	2283.	2557.	
Sheep	DE kcal/kg	2087.	2337.	
Cattle	ME kcal/kg	1873.	2097.	
Sheep	ME kcal/kg	1711.	1916.	
Cattle	TDN %	52.	58.	
Sheep	TDN %	47.	53.	
Calcium	%		.32	.36
Phosphorus	%		.28	.31
Potassium	%		3.34	3.74

Orchardgrass, hay, s-c, early blm, (1)

Ref no 1-03-425

		As fed	Dry	± %
Dry matter	%	86.0	100.0	4
Ash	%	7.6	8.8	15
Crude fiber	%	28.2	32.8	11
Ether extract	%	1.9	2.2	31
N-free extract	%	36.4	42.3	
Protein (N x 6.25)	%	12.0	13.9	17
Cattle	dig prot %	7.1	8.3	
Sheep	dig prot %	5.8	6.7	
Energy				
Cattle	DE kcal/kg	2123.	2469.	
Sheep	DE kcal/kg	1972.	2293.	
Cattle	ME kcal/kg	1741.	2024.	
Sheep	ME kcal/kg	1617.	1880.	
Cattle	TDN %	48.	56.	
Sheep	TDN %	45.	52.	

(5) protein supplements

(6) minerals

(7) vitamins

(8) additives

Feed name or analyses		Mean		C.V.
		As fed	Dry	± %

Orchardgrass, hay, s-c, full blm, (1)

Ref no 1-03-427

Feed name or analyses		As fed	Dry	C.V. ± %
Dry matter	%	89.2	100.0	
Ash	%	6.9	7.7	11
Crude fiber	%	29.5	33.1	10
Ether extract	%	3.2	3.6	17
N-free extract	%	38.5	43.2	
Protein (N x 6.25)	%	11.1	12.4	4
Cattle	dig prot %	6.6	7.4	
Sheep	dig prot %	5.5	6.2	
Energy				
Cattle	DE kcal/kg	2242.	2513.	
Sheep	DE kcal/kg	2124.	2381.	
Cattle	ME kcal/kg	1838.	2061.	
Sheep	ME kcal/kg	1741.	1952.	
Cattle	TDN %	51.	57.	
Sheep	TDN %	48.	54.	

Orchardgrass, hay, s-c, mature, (1)

Ref no 1-03-431

		As fed	Dry	C.V. ± %
Dry matter	%	90.6	100.0	
Ash	%	5.9	6.5	18
Crude fiber	%	35.3	39.0	
Ether extract	%	2.5	2.8	18
N-free extract	%	41.0	45.2	
Protein (N x 6.25)	%	5.9	6.5	24
Cattle	dig prot %	3.5	3.9	
Sheep	dig prot %	3.0	3.3	
Energy				
Cattle	DE kcal/kg	2277.	2513.	
Sheep	DE kcal/kg	1678.	1852.	
Cattle	ME kcal/kg	1867.	2061.	
Sheep	ME kcal/kg	1376.	1519.	
Cattle	TDN %	52.	57.	
Sheep	TDN %	38.	42.	

Orchardgrass, hay, s-c, cut 1, (1)

Ref no 1-03-433

		As fed	Dry	C.V. ± %
Dry matter	%	91.0	100.0	3
Ash	%	6.9	7.6	15
Crude fiber	%	30.7	33.7	16
Ether extract	%	3.1	3.4	47
N-free extract	%	40.0	44.0	
Protein (N x 6.25)	%	10.3	11.3	28
Cattle	dig prot %	6.2	6.8	
Sheep	dig prot %	5.1	5.6	
Energy	GE kcal/kg	4128.	4536.	

Continued

Feed name or analyses		As fed	Dry	C.V. ± %
Cattle	DE kcal/kg	2287.	2513.	
Sheep	DE kcal/kg	2167.	2381.	
Cattle	ME kcal/kg	1876.	2061.	
Sheep	ME kcal/kg	1776.	1952.	
Cattle	TDN %	52.	57.	
Sheep	TDN %	49.	54.	
Calcium	%	.36	.40	
Phosphorus	%	.41	.45	
Potassium	%	1.87	2.05	

Orchardgrass, hay, s-c, cut 2, (1)

Ref no 1-03-434

		As fed	Dry	C.V. ± %
Dry matter	%	90.0	100.0	
Ash	%	6.5	7.2	20
Crude fiber	%	26.2	29.1	13
Ether extract	%	4.1	4.6	32
N-free extract	%	39.2	43.6	
Protein (N x 6.25)	%	14.0	15.5	39
Cattle	dig prot %	8.4	9.3	
Sheep	dig prot %	7.0	7.8	
Energy				
Cattle	DE kcal/kg	2301.	2557.	
Sheep	DE kcal/kg	2143.	2381.	
Cattle	ME kcal/kg	1887.	2097.	
Sheep	ME kcal/kg	1757.	1952.	
Cattle	TDN %	52.	58.	
Sheep	TDN %	49.	54.	

Orchardgrass, hay, s-c, cut 3, (1)

Ref no 1-03-435

		As fed	Dry	C.V. ± %
Dry matter	%	91.3	100.0	
Ash	%	6.8	7.5	
Crude fiber	%	28.1	30.8	
Ether extract	%	4.5	4.9	
N-free extract	%	36.7	40.2	
Protein (N x 6.25)	%	15.2	16.6	
Cattle	dig prot %	9.1	10.0	
Sheep	dig prot %	7.8	8.3	
Energy	GE kcal/kg	4362.	4778.	
Cattle	DE kcal/kg	2334.	2557.	
Sheep	DE kcal/kg	2174.	2381.	
Cattle	ME kcal/kg	1914.	2097.	
Sheep	ME kcal/kg	1782.	1952.	
Cattle	TDN %	53.	58.	
Sheep	TDN %	49.	54.	

(1) dry forages and roughages

(2) pasture, range plants, and forages fed green

(3) silages

(4) energy feeds

Orchardgrass, hay, s-c, (1)

Ref no 1-03-438

Feed name or analyses		As fed	Dry	C.V. ± %
Dry matter	%	88.3	100.0	3
Ash	%	6.7	7.6	14
Crude fiber	%	30.0	34.0	12
Ether extract	%	3.0	3.4	35
N-free extract	%	40.0	45.3	
Protein (N x 6.25)	%	8.6	9.7	41
Cattle	dig prot %	5.1	5.8	
Sheep	dig prot %	4.3	4.9	
Cellulose	%	22.1	25.0	34
Lignin	%	6.7	7.6	39
Energy	GE kcal/kg	4059.	4597.	3
Cattle	DE kcal/kg	2219.	2513.	
Sheep	DE kcal/kg	2102.	2381.	
Cattle	ME kcal/kg	1820.	2061.	
Sheep	ME kcal/kg	1724.	1952.	
Cattle	TDN %	50.	57.	
Sheep	TDN %	48.	54.	
Calcium	%	.40	.45	25
Chlorine	%	.36	.41	
Iron	%	.01	.01	66
Magnesium	%	.28	.32	24
Phosphorus	%	.33	.37	31
Potassium	%	1.85	2.10	20
Sulfur	%	.23	.26	
Cobalt	mg/kg	.02	.02	
Copper	mg/kg	12.1	13.7	
Manganese	mg/kg	220.4	249.6	27
Zinc	mg/kg	16.0	18.1	
Carotene	mg/kg	29.6	33.5	95
Vitamin A equiv	IU/g	49.3	55.8	

Orchardgrass, aerial pt, fresh, immature, (2)

Ref no 2-03-440

Feed name or analyses		As fed	Dry	C.V. ± %
Dry matter	%	23.8	100.0	11
Ash	%	2.7	11.4	17
Crude fiber	%	5.6	23.6	10
Ether extract	%	1.1	4.8	14
N-free extract	%	9.9	41.8	
Protein (N x 6.25)	%	4.4	18.4	13
Cattle	dig prot %	3.2	13.5	
Sheep	dig prot %	3.4	14.1	
Energy				
Cattle	DE kcal/kg	682.	2866.	
Sheep	DE kcal/kg	703.	2954.	
Cattle	ME kcal/kg	559.	2350.	
Sheep	ME kcal/kg	576.	2422.	
Cattle	TDN %	15.	65.	
Sheep	TDN %	16.	67.	

Continued

Feed name or analyses		As fed	Dry	C.V. ± %
Calcium	%	.14	.58	14
Iron	%	.	.02	20
Magnesium	%	.07	.31	17
Phosphorus	%	.13	.55	24
Potassium	%	.80	3.38	20
Manganese	mg/kg	32.0	134.3	12
Carotene	mg/kg	80.3	337.4	44
Vitamin A equiv	IU/g	133.9	562.4	

Orchardgrass, aerial pt, fresh, pre-blm, pure stand, (2)

Ref no 2-03-441

Feed name or analyses		As fed	Dry	C.V. ± %
Dry matter	%	23.8	100.0	
Ash	%	2.7	11.4	
Crude fiber	%	5.6	23.6	
Ether extract	%	1.1	4.8	
N-free extract	%	9.9	41.8	
Protein (N x 6.25)	%	4.4	18.4	
Cattle	dig prot %	3.2	13.5	
Sheep	dig prot %	3.4	14.1	
Energy				
Cattle	DE kcal/kg	682.	2866.	
Sheep	DE kcal/kg	703.	2954.	
Cattle	ME kcal/kg	559.	2350.	
Sheep	ME kcal/kg	576.	2422.	
Cattle	TDN %	15.	65.	
Sheep	TDN %	16.	67.	
Calcium	%	.14	.58	
Phosphorus	%	.13	.55	

Orchardgrass, aerial pt, fresh, early blm, (2)

Ref no 2-03-442

Feed name or analyses		As fed	Dry	C.V. ± %
Dry matter	%	27.5	100.0	7
Ash	%	2.3	8.4	17
Crude fiber	%	7.9	28.8	8
Ether extract	%	1.3	4.6	11
N-free extract	%	12.4	45.1	
Protein (N x 6.25)	%	3.6	13.1	16
Cattle	dig prot %	2.5	9.0	
Sheep	dig prot %	2.5	9.2	
Energy				
Cattle	DE kcal/kg	800.	2910.	
Sheep	DE kcal/kg	776.	2822.	
Cattle	ME kcal/kg	656.	2386.	
Sheep	ME kcal/kg	636.	2314.	
Cattle	TDN %	18.	66.	
Sheep	TDN %	18.	64.	
Calcium	%	.07	.25	
Iron	%	.02	.08	

Continued

(5) protein supplements
(6) minerals

(7) vitamins
(8) additives

Feed name or analyses		Mean		C.V.
		As fed	Dry	± %
Phosphorus	%	.11	.39	
Copper	mg/kg	9.1	33.1	
Manganese	mg/kg	28.6	104.1	

Orchardgrass, aerial pt, fresh, mid-blm, (2)

Ref no 2-03-443

Dry matter	%	30.0	100.0	
Ash	%	2.2	7.4	7
Crude fiber	%	9.6	31.9	5
Crude fiber	%	1.1	3.8	6
N-free extract	%	14.3	47.8	
Protein (N x 6.25)	%	2.7	9.1	4
Cattle	dig prot %	1.7	5.6	
Sheep	dig prot %	1.6	5.5	
Energy				
Cattle	DE kcal/kg	847.	2822.	
Sheep	DE kcal/kg	873.	2910.	
Cattle	ME kcal/kg	694.	2314.	
Sheep	ME kcal/kg	716.	2386.	
Cattle	TDN %	19.	64.	
Sheep	TDN %	20.	66.	

Orchardgrass, aerial pt, fresh, full blm, (2)

Ref no 2-03-445

Dry matter	%	29.6	100.0	5
Ash	%	2.3	7.9	7
Crude fiber	%	9.8	33.1	5
Ether extract	%	1.0	3.3	8
N-free extract	%	14.0	47.2	
Protein (N x 6.25)	%	2.5	8.5	6
Cattle	dig prot %	1.5	5.1	
Sheep	dig prot %	1.4	4.9	
Energy				
Cattle	DE kcal/kg	835.	2822.	
Sheep	DE kcal/kg	861.	2910.	
Cattle	ME kcal/kg	685.	2314.	
Sheep	ME kcal/kg	706.	2386.	
Cattle	TDN %	19.	64.	
Sheep	TDN %	20.	66.	
Calcium	%	.07	.23	7
Phosphorus	%	.07	.22	11
Carotene	mg/kg	69.4	234.4	40
Vitamin A equiv	IU/g	115.7	390.7	

Orchardgrass, aerial pt, fresh, milk stage, (2)

Ref no 2-03-446

Dry matter	%	30.0	100.0	
Ash	%	1.8	6.0	5
Crude fiber	%	10.6	35.2	4
Ether extract	%	1.1	3.7	7
N-free extract	%	14.0	46.7	
Protein (N x 6.25)	%	2.5	8.4	7
Cattle	dig prot %	1.5	5.0	
Sheep	dig prot %	1.4	4.8	
Energy				
Cattle	DE kcal/kg	873.	2910.	
Sheep	DE kcal/kg	860.	2866.	
Cattle	ME kcal/kg	716.	2386.	
Sheep	ME kcal/kg	705.	2350.	
Cattle	TDN %	20.	66.	
Sheep	TDN %	20.	65.	
Calcium	%	.07	.23	
Phosphorus	%	.07	.22	

Orchardgrass, aerial pt, fresh, mature, (2)

Ref no 2-03-449

Dry matter	%	30.0	100.0	
Ash	%	1.8	6.1	7
Crude fiber	%	11.0	36.5	2
Ether extract	%	.9	3.0	9
N-free extract	%	14.6	48.7	
Protein (N x 6.25)	%	1.7	5.7	4
Cattle	dig prot %	.8	2.7	
Sheep	dig prot %	.7	2.3	
Energy				
Cattle	DE kcal/kg	847.	2822.	
Sheep	DE kcal/kg	886.	2954.	
Cattle	ME kcal/kg	694.	2314.	
Sheep	ME kcal/kg	727.	2422.	
Cattle	TDN %	19.	64.	
Sheep	TDN %	20.	67.	

Orchardgrass, aerial pt, fresh, (2)

Ref no 2-03-451

Dry matter	%	24.9	100.0	10
Ash	%	2.3	9.2	19
Crude fiber	%	6.9	27.9	12
Ether extract	%	1.1	4.3	19
N-free extract	%	11.2	44.8	
Protein (N x 6.25)	%	3.4	13.8	30
Cattle	dig prot %	2.4	9.6	

Continued

(1) dry forages and roughages
(2) pasture, range plants, and forages fed green

(3) silages
(4) energy feeds

Feed name or analyses		Mean As fed	Mean Dry	C.V. ± %
Sheep	dig prot %	2.4	9.8	
Energy	GE kcal/kg	1135.	4560.	2
Cattle	DE kcal/kg	714.	2866.	
Sheep	DE kcal/kg	692.	2778.	
Cattle	ME kcal/kg	585.	2350.	
Sheep	ME kcal/kg	567.	2278.	
Cattle	TDN %	16.	65.	
Sheep	TDN %	16.	63.	
Calcium	%	.13	.53	25
Iron	%	.01	.02	99
Magnesium	%	.07	.30	25
Phosphorus	%	.13	.51	26
Potassium	%	.86	3.44	16
Sodium	%	.01	.04	79
Sulfur	%	.04	.18	45
Cobalt	mg/kg	.02	.09	99
Copper	mg/kg	1.8	7.3	81
Manganese	mg/kg	35.7	143.5	69
Zinc	mg/kg	6.1	24.5	38
Carotene	mg/kg	79.5	319.1	49
Vitamin A equiv	IU/g	132.5	531.9	

Orchardgrass, aerial pt, ensiled, immature, (3)

Ref no 3-03-454

		As fed	Dry	C.V. ± %
Dry matter	%	22.4	100.0	13
Ash	%	1.3	5.9	3
Crude fiber	%	6.2	27.7	4
Ether extract	%	1.1	4.8	11
N-free extract	%	9.3	41.4	
Protein (N x 6.25)	%	4.5	20.2	7
Cattle	dig prot %	3.3	14.6	
Sheep	dig prot %	3.3	14.6	
Energy				
Cattle	DE kcal/kg	523.	2337.	
Sheep	DE kcal/kg	533.	2381.	
Cattle	ME kcal/kg	429.	1916.	
Sheep	ME kcal/kg	437.	1952.	
Cattle	TDN %	12.	53.	
Sheep	TDN %	12.	54.	

Orchardgrass, aerial pt, ensiled, mature, (3)

Ref no 3-03-455

		As fed	Dry	
Dry matter	%	30.0	100.0	
Ash	%	1.8	6.0	
Crude fiber	%	13.0	43.2	
Ether extract	%	1.3	4.3	
N-free extract	%	11.7	38.9	
Protein (N x 6.25)	%	2.3	7.6	
Cattle	dig prot %	.9	3.1	
Sheep	dig prot %	.9	3.1	

Continued

Feed name or analyses		Mean As fed	Mean Dry	C.V. ± %
Energy				
Cattle	DE kcal/kg	688.	2293.	
Sheep	DE kcal/kg	820.	2734.	
Cattle	ME kcal/kg	564.	1880.	
Sheep	ME kcal/kg	673.	2242.	
Cattle	TDN %	16.	52.	
Sheep	TDN %	19.	62.	

Orchardgrass, aerial pt, ensiled, (3)

Ref no 3-03-457

		As fed	Dry	C.V. ± %
Dry matter	%	23.7	100.0	14
Ash	%	1.7	7.0	18
Crude fiber	%	7.5	31.8	15
Ether extract	%	1.0	4.1	25
N-free extract	%	10.1	42.6	
Protein (N x 6.25)	%	3.4	14.5	27
Cattle	dig prot %	2.2	9.4	
Sheep	dig prot %	2.2	9.4	
Energy				
Cattle	DE kcal/kg	585.	2469.	
Sheep	DE kcal/kg	658.	2778.	
Cattle	ME kcal/kg	480.	2024.	
Sheep	ME kcal/kg	540.	2278.	
Cattle	TDN %	13.	56.	
Sheep	TDN %	15.	63.	
Carotene	mg/kg	40.1	169.1	49
Vitamin A equiv	IU/g	66.8	281.9	

OYSTERS. Crassostrea spp, Ostrea spp

Oysters, shells, f-grnd, mn 33 Ca, (6)
    Oyster shell flour (AAFCO)

Ref no 6-03-481

		As fed	Dry	
Dry matter	%	100.0	100.0	
Ash	%	80.8	80.8	
Protein (N x 6.25)	%	1.0	1.0	
Calcium	%	38.05	38.05	
Iron	%	.290	.290	
Magnesium	%	.30	.30	
Phosphorus	%	.07	.07	
Potassium	%	.10	.10	
Sodium	%	.21	.21	
Manganese	mg/kg	133.3	133.3	

(5) protein supplements
(6) minerals

(7) vitamins
(8) additives

Feed name or analyses		Mean		C.V.
		As fed	Dry	± %

Paddy rice - see Rice, grain w hulls

PARAGRASS. Panicum purpurascens

Paragrass, aerial pt, fresh, early blm, (2)

Ref no 3-03-519

Feed name or analyses		As fed	Dry	C.V. ± %
Dry matter	%	21.1	100.0	16
Ash	%	2.2	10.5	8
Crude fiber	%	6.2	29.2	8
Ether extract	%	.5	2.2	6
N-free extract	%	9.9	46.9	
Protein (N x 6.25)	%	2.4	11.2	8
Cattle	dig prot %	1.5	7.3	
Sheep	dig prot %	1.5	7.3	
Energy				
Cattle	DE kcal/kg	521.	2469.	
Sheep	DE kcal/kg	512.	2425.	
Cattle	ME kcal/kg	427.	2024.	
Sheep	ME kcal/kg	419.	1988.	
Cattle	TDN %	12.	56.	
Sheep	TDN %	12.	55.	
Calcium	%	.18	.83	
Iron	%	.004	.02	
Magnesium	%	.11	.53	
Phosphorus	%	.10	.49	

Paragrass, aerial pt, fresh, full blm, (2)

Ref no 2-03-520

		As fed	Dry	C.V.
Dry matter	%	26.7	100.0	12
Ash	%	2.3	8.7	9
Crude fiber	%	8.9	33.4	5
Ether extract	%	.5	1.7	16
N-free extract	%	13.2	49.5	
Protein (N x 6.25)	%	1.8	6.7	23
Cattle	dig prot %	1.2	4.4	
Sheep	dig prot %	1.2	4.4	
Energy				
Cattle	DE kcal/kg	659.	2469.	
Sheep	DE kcal/kg	647.	2425.	
Cattle	ME kcal/kg	540.	2024.	
Sheep	ME kcal/kg	531.	1988.	
Cattle	TDN %	15.	56.	
Sheep	TDN %	15.	55.	

Paragrass, aerial pt, fresh, mature, (2)

Ref no 2-03-523

		As fed	Dry	
Dry matter	%	22.0	100.0	
Ash	%	1.9	8.6	
Crude fiber	%	7.4	33.8	
Ether extract	%	.3	1.5	
N-free extract	%	11.3	51.2	
Protein (N x 6.25)	%	1.1	4.9	
Cattle	dig prot %	.7	3.2	
Sheep	dig prot %	.7	3.2	
Energy				
Cattle	DE kcal/kg	543.	2469.	
Sheep	DE kcal/kg	534.	2425.	
Cattle	ME kcal/kg	445.	2024.	
Sheep	ME kcal/kg	437.	1988.	
Cattle	TDN %	12.	56.	
Sheep	TDN %	12.	55.	

Paragrass, aerial pt, fresh, (2)

Ref no 2-03-525

		As fed	Dry	C.V.
Dry matter	%	24.6	100.0	18
Ash	%	2.6	10.4	20
Crude fiber	%	8.0	32.6	10
Ether extract	%	.4	1.7	23
N-free extract	%	11.7	47.6	
Protein (N x 6.25)	%	1.9	7.7	25
Cattle	dig prot %	1.2	5.0	
Sheep	dig prot %	1.2	5.0	
Energy				
Cattle	DE kcal/kg	596.	2425.	
Sheep	DE kcal/kg	586.	2381.	
Cattle	ME kcal/kg	489.	1988.	
Sheep	ME kcal/kg	480.	1952.	
Cattle	TDN %	14.	55.	
Sheep	TDN %	13.	54.	
Calcium	%	.15	.60	
Iron	%	.005	.02	
Magnesium	%	.09	.38	
Phosphorus	%	.08	.34	
Potassium	%	.38	1.55	

(1) dry forages and roughages
(2) pasture, range plants, and forages fed green

(3) silages
(4) energy feeds

Feed name or analyses		Mean		C.V.
		As fed	Dry	± %

PEA. Pisum spp

Pea, hay, s-c, (1)

Ref no 1-03-572

		As fed	Dry	C.V. ± %
Dry matter	%	88.6	100.0	3
Ash	%	7.0	7.9	9
Crude fiber	%	25.2	28.4	15
Ether extract	%	2.7	3.0	17
N-free extract	%	40.1	45.3	
Protein (N x 6.25)	%	13.6	15.4	16
Cattle	dig prot %	9.1	10.3	
Sheep	dig prot %	9.6	10.9	
Energy				
Cattle	DE kcal/kg	2304.	2601.	
Sheep	DE kcal/kg	2422.	2734.	
Cattle	ME kcal/kg	1890.	2133.	
Sheep	ME kcal/kg	1986.	2242.	
Cattle	TDN %	52.	59.	
Sheep	TDN %	55.	62.	

Pea, pods, s-c, (1)

Ref no 1-03-575

		As fed	Dry	C.V. ± %
Dry matter	%	91.4	100.0	1
Ash	%	3.2	3.5	17
Crude fiber	%	43.9	48.0	10
Ether extract	%	.5	.6	45
N-free extract	%	38.4	42.0	
Protein (N x 6.25)	%	5.4	5.9	30
Cattle	dig prot %	2.8	3.1	
Sheep	dig prot %	2.8	3.1	
Energy				
Cattle	DE kcal/kg	3103.	3395.	
Sheep	DE kcal/kg	3103.	3395.	
Cattle	ME kcal/kg	2544.	2784.	
Sheep	ME kcal/kg	2544.	2784.	
Cattle	TDN %	70.	77.	
Sheep	TDN %	70.	77.	
Calcium	%	.44	.48	
Magnesium	%	.08	.09	
Phosphorus	%	.06	.07	

Pea, straw, (1)

Ref no 1-03-577

		As fed	Dry	C.V. ± %
Dry matter	%	87.3	100.0	4
Ash	%	5.4	6.2	10
Crude fiber	%	34.0	38.9	10

Continued

(5) protein supplements
(6) minerals

Feed name or analyses		Mean		C.V.
		As fed	Dry	± %
Ether extract	%	1.6	1.8	37
N-free extract	%	39.7	45.5	
Protein (N x 6.25)	%	6.6	7.6	22
Cattle	dig prot %	3.5	4.0	
Sheep	dig prot %	3.3	3.8	
Energy				
Cattle	DE kcal/kg	2194.	2513.	
Sheep	DE kcal/kg	1617.	1852.	
Cattle	ME kcal/kg	1799.	2061.	
Sheep	ME kcal/kg	1326.	1519.	
Cattle	TDN %	50.	57.	
Sheep	TDN %	37.	42.	

Pea, aerial pt, fresh, (2)

Ref no 2-03-582

		As fed	Dry	C.V. ± %
Dry matter	%	16.5	100.0	18
Ash	%	2.2	13.1	20
Crude fiber	%	4.2	25.2	9
Ether extract	%	.6	3.8	17
Ether extract	%	5.9	36.0	
Protein (N x 6.25)	%	3.6	21.9	10
Cattle	dig prot %	2.7	16.5	
Sheep	dig prot %	3.0	18.4	
Energy				
Cattle	DE kcal/kg	502.	3042.	
Sheep	DE kcal/kg	495.	2998.	
Cattle	ME kcal/kg	412.	2494.	
Sheep	ME kcal/kg	406.	2458.	
Cattle	TDN %	11.	69.	
Sheep	TDN %	11.	68.	

Pea, aerial pt, ensiled, (3)

Ref no 3-03-590

		As fed	Dry	C.V. ± %
Dry matter	%	23.0	100.0	
Ash	%	1.7	7.4	
Crude fiber	%	7.4	32.2	
Ether extract	%	.8	3.3	
N-free extract	%	10.1	44.1	
Protein (N x 6.25)	%	3.0	13.0	
Cattle	dig prot %	1.8	7.7	
Sheep	dig prot %	1.8	7.7	
Energy				
Cattle	DE kcal/kg	588.	2558.	
Sheep	DE kcal/kg	588.	2558.	
Cattle	ME kcal/kg	482.	2098.	
Sheep	ME kcal/kg	482.	2098.	
Cattle	TDN %	13.	58.	
Sheep	TDN %	13.	58.	
Calcium	%	.39	1.71	
Phosphorus	%	.07	.30	

(7) vitamins
(8) additives

Feed name or analyses		Mean		C.V.
		As fed	Dry	± %

Feed name or analyses		Mean		C.V.
		As fed	Dry	± %
Pantothenic acid	mg/kg	4.6	5.1	
Riboflavin	mg/kg	.8	.9	
Thiamine	mg/kg	1.8	2.0	

Pea, aerial pt wo seeds, ensiled, (3)
Pea vine silage

Ref no 3-03-596

Dry matter	%	24.5	100.0	
Ash	%	2.2	9.0	
Crude fiber	%	7.3	29.8	
Ether extract	%	.8	3.3	
N-free extract	%	11.0	44.8	
Protein (N x 6.25)	%	3.2	13.1	
Cattle	dig prot %	1.9	7.6	
Sheep	dig prot %	1.9	7.7	
Energy				
Cattle	DE kcal/kg	605.	2469.	
Sheep	DE kcal/kg	616.	2513.	
Cattle	ME kcal/kg	496.	2024.	
Sheep	ME kcal/kg	505.	2061.	
Cattle	TDN %	14.	56.	
Sheep	TDN %	14.	57.	
Calcium	%	.32	1.31	
Phosphorus	%	.06	.24	

Pea, seed screenings, (5)

Ref no 5-03-601

Dry matter	%	92.0	100.0	
Ash	%	7.0	7.6	
Crude fiber	%	9.0	9.8	
Ether extract	%	2.0	2.2	
N-free extract	%	56.2	61.1	
Protein (N x 6.25)	%	17.8	19.3	
Cattle	dig prot %	12.6	13.7	
Sheep	dig prot %	13.7	14.9	
Energy				
Cattle	DE kcal/kg	3083.	3351.	
Sheep	DE kcal/kg	2961.	3219.	
Cattle	ME kcal/kg	2528.	2748.	
Sheep	ME kcal/kg	2429.	2640.	
Cattle	TDN %	70.	76.	
Sheep	TDN %	67.	73.	

Pea, seed, grnd, (5)

Ref no 5-03-598

Dry matter	%	91.0	100.0	2
Ash	%	3.7	4.1	32
Crude fiber	%	9.0	9.9	71
Ether extract	%	1.9	2.1	45
N-free extract	%	53.9	59.2	
Protein (N x 6.25)	%	22.5	24.7	17
Cattle	dig prot %	17.0	18.7	
Sheep	dig prot %	19.3	21.2	
Swine	dig prot %	19.3	21.2	
Energy				
Cattle	DE kcal/kg	3170.	3483.	
Sheep	DE kcal/kg	3371.	3704.	
Swine	DE kcal/kg	3531.	3880.	
Cattle	ME kcal/kg	2599.	2856.	
Chickens	MEn kcal/kg	2601.	2858.	
Sheep	ME kcal/kg	2764.	3037.	
Swine	ME kcal/kg	3213.	3531.	
Cattle	TDN %	72.	79.	
Sheep	TDN %	76.	84.	
Swine	TDN %	80.	88.	
Calcium	%	.17	.19	
Phosphorus	%	.50	.55	
Potassium	%	1.03	1.12	
Choline	mg/kg	649.	713.	
Niacin	mg/kg	17.2	18.9	

Continued

PEA, FIELD. Pisum sativum arvense

Pea, field, aerial pt, ensiled, (3)

Ref no 3-03-609

Dry matter	%	26.7	100.0	
Ash	%	2.5	9.3	
Crude fiber	%	7.2	27.0	
Ether extract	%	1.1	4.3	
N-free extract	%	12.0	44.8	
Protein (N x 6.25)	%	3.9	14.6	
Cattle	dig prot %	2.5	9.5	
Sheep	dig prot %	2.3	8.6	
Energy				
Cattle	DE kcal/kg	742.	2778.	
Sheep	DE kcal/kg	683.	2557.	
Cattle	ME kcal/kg	608.	2278.	
Sheep	ME kcal/kg	560.	2097.	
Cattle	TDN %	17.	63.	
Sheep	TDN %	15.	58.	
Calcium	%	.36	1.36	7
Magnesium	%	.10	.39	13
Phosphorus	%	.08	.29	21
Potassium	%	.37	1.40	7
Sulfur	%	.07	.25	19

(1) dry forages and roughages
(2) pasture, range plants, and forages fed green

(3) silages
(4) energy feeds

Feed name or analyses		Mean		C.V.
		As fed	Dry	± %

Pea vine silage - see Pea, aerial pt. wo seeds

PEANUT. Arachis hypogaea

Peanut, hay, s-c, dough stage, (1)

Ref no 1-03-615

Dry matter	%	91.3	100.0	1
Ash	%	9.9	10.8	20
Crude fiber	%	26.2	28.7	16
Ether extract	%	2.6	2.8	29
N-free extract	%	41.7	45.7	
Protein (N x 6.25)	%	11.0	12.0	16
Cattle	dig prot %	6.7	7.3	
Sheep	dig prot %	7.1	7.8	
Energy				
Cattle	DE kcal/kg	2737.	2998.	
Sheep	DE kcal/kg	2496.	2734.	
Cattle	ME kcal/kg	2244.	2458.	
Sheep	ME kcal/kg	2047.	2242.	
Cattle	TDN %	62.	68.	
Sheep	TDN %	57.	62.	

Peanut, hay, s-c, stemmy, (1)

Ref no 1-03-618

Dry matter	%	90.0	100.0	
Ash	%	8.6	9.6	
Crude fiber	%	34.2	38.0	
Ether extract	%	3.2	3.6	
N-free extract	%	33.9	37.7	
Protein (N x 6.25)	%	10.0	11.1	
Cattle	dig prot %	5.9	6.6	
Sheep	dig prot %	6.5	7.2	
Energy				
Cattle	DE kcal/kg	2064.	2293.	
Sheep	DE kcal/kg	2421.	2690.	
Cattle	ME kcal/kg	1692.	1880.	
Sheep	ME kcal/kg	1985.	2206.	
Cattle	TDN %	47.	52.	
Sheep	TDN %	55.	61.	

Peanut, hay, s-c, (1)

Ref no 1-03-619

Dry matter	%	90.9	100.0	2
Ash	%	8.8	9.7	28
Crude fiber	%	24.5	27.0	18

Continued

Feed name or analyses		Mean		C.V.
		As fed	Dry	± %

Ether extract	%	5.0	5.5	47
N-free extract	%	41.7	45.9	
Protein (N x 6.25)	%	10.8	11.9	13
Cattle	dig prot %	6.5	7.2	
Sheep	dig prot %	7.0	7.7	
Energy				
Cattle	DE kcal/kg	2886.	3175.	
Sheep	DE kcal/kg	2565.	2822.	
Cattle	ME kcal/kg	2367.	2604.	
Sheep	ME kcal/kg	2103.	2314.	
Cattle	TDN %	65.	72.	
Sheep	TDN %	58.	64.	
Calcium	%	1.14	1.26	20
Magnesium	%	.51	.56	25
Phosphorus	%	.16	.18	99
Potassium	%	1.22	1.34	14
Carotene	mg/kg	45.7	50.3	
Vitamin A equiv	IU/g	76.2	83.8	

Peanut, hay w nuts, s-c, (1)

Ref no 1-03-620

Dry matter	%	91.6	100.0	1
Ash	%	8.2	9.0	17
Crude fiber	%	22.8	24.9	7
Ether extract	%	12.5	13.6	6
N-free extract	%	34.8	38.0	
Protein (N x 6.25)	%	13.3	14.5	6
Cattle	dig prot %	8.7	9.5	
Sheep	dig prot %	10.1	11.0	
Energy				
Cattle	DE kcal/kg	3312.	3616.	
Sheep	DE kcal/kg	3110.	3395.	
Cattle	ME kcal/kg	2716.	2965.	
Sheep	ME kcal/kg	2550.	2784.	
Cattle	TDN %	75.	82.	
Sheep	TDN %	71.	77.	
Calcium	%	1.14	1.24	
Magnesium	%	.33	.36	
Phosphorus	%	.13	.14	

Peanut, shells, grnd, (1)

Ref no 1-03-629

Dry matter	%	92.3	100.0	2
Ash	%	4.3	4.7	48
Crude fiber	%	60.4	65.4	9
Ether extract	%	1.1	1.2	49
N-free extract	%	19.7	21.3	
Protein (N x 6.25)	%	6.8	7.4	14
Cattle	dig prot %	3.0	3.3	
Sheep	dig prot %	3.0	3.2	

Continued

(5) protein supplements
(6) minerals

(7) vitamins
(8) additives

Feed name or analyses		Mean		C.V.
		As fed	Dry	± %
Cellulose	%	44.8	48.5	
Lignin	%	26.5	28.7	
Energy				
Cattle	DE kcal/kg	326.	353.	
Sheep	DE kcal/kg	1465.	1587.	
Cattle	ME kcal/kg	267.	289.	
Sheep	ME kcal/kg	1201.	1301.	
Cattle	TDN %	7.	8.	
Sheep	TDN %	33.	36.	
Calcium	%	.23	.25	28
Iron	%	.03	.03	
Magnesium	%	.15	.16	38
Phosphorus	%	.06	.07	23
Potassium	%	.94	1.02	20
Cobalt	mg/kg	.10	.12	
Copper	mg/kg	16.2	17.6	
Manganese	mg/kg	62.7	67.9	
Carotene	mg/kg	.8	.9	
Vitamin A equiv	IU/g	1.3	1.5	

Peanut, shells w skins, grnd, (1)

Ref no 1-03-630

Dry matter	%	94.0	100.0	
Ash	%	13.3	14.2	
Crude fiber	%	13.0	13.8	
Ether extract	%	13.8	14.7	
N-free extract	%	41.1	43.7	
Protein (N x 6.25)	%	12.8	13.6	
Cattle	dig prot %	9.1	9.7	
Sheep	dig prot %	9.1	9.7	
Energy				
Cattle	DE kcal/kg	3026.	3219.	
Sheep	DE kcal/kg	3026.	3219.	
Cattle	ME kcal/kg	2482.	2640.	
Sheep	ME kcal/kg	2482.	2640.	
Cattle	TDN %	69.	73.	
Sheep	TDN %	69.	73.	
Calcium	%	.30	.32	
Phosphorus	%	.07	.07	

Peanut, aerial pt, fresh, (2)

Ref no 2-03-638

Dry matter	%	21.4	100.0	20
Ash	%	1.6	7.3	23
Crude fiber	%	5.4	25.1	14
Ether extract	%	2.0	9.5	60
N-free extract	%	8.9	41.7	
Protein (N x 6.25)	%	3.5	16.4	27
Cattle	dig prot %	2.5	11.8	
Sheep	dig prot %	1.6	7.7	

Continued

Feed name or analyses		Mean		C.V.
		As fed	Dry	± %
Energy				
Cattle	DE kcal/kg	585.	2734.	
Sheep	DE kcal/kg	557.	2601.	
Cattle	ME kcal/kg	480.	2242.	
Sheep	ME kcal/kg	456.	2133.	
Cattle	TDN %	13.	62.	
Sheep	TDN %	13.	59.	
Calcium	%	.37	1.71	16
Magnesium	%	.11	.50	25
Phosphorus	%	.06	.29	19
Potassium	%	.21	.97	44

Peanut, kernels, mech-extd grnd, mx 7 fbr, (5)
Peanut meal (AAFCO)
Peanut meal (CFA)
Peanut oil meal, expeller extracted

Ref no 5-03-649

Dry matter	%	92.0	100.0	2
Ash	%	5.7	6.2	14
Crude fiber	%	11.0	12.0	34
Ether extract	%	5.9	6.4	33
N-free extract	%	23.6	25.6	
Protein (N x 6.25)	%	45.8	49.8	10
Cattle	dig prot %	41.2	44.8	
Sheep	dig prot %	41.7	45.3	
Swine	dig prot %	43.0	46.8	
Energy				
Cattle	DE kcal/kg	3367.	3660.	
Sheep	DE kcal/kg	3570.	3880.	
Swine	DE kcal/kg	3772.	4100.	
Cattle	ME kcal/kg	2761.	3001.	
Chickens	MEn kcal/kg	2491.	2708.	
Sheep	ME kcal/kg	2927.	3182.	
Swine	ME kcal/kg	3244.	3526.	
Cattle	TDN %	76.	83.	
Sheep	TDN %	81.	88.	
Swine	TDN %	86.	93.	
Calcium	%	.17	.18	39
Magnesium	%	.33	.36	
Phosphorus	%	.57	.62	19
Potassium	%	1.15	1.25	
Manganese	mg/kg	25.5	27.7	
Choline	mg/kg	1683.	1829.	
Niacin	mg/kg	169.0	183.7	
Pantothenic acid	mg/kg	48.2	52.3	
Riboflavin	mg/kg	5.3	5.8	
Thiamine	mg/kg	7.3	7.9	
Arginine	%	4.69	5.10	
Histidine	%	1.00	1.09	
Isoleucine	%	2.00	2.17	
Leucine	%	3.10	3.37	
Lysine	%	1.30	1.41	
Methionine	%	.60	.65	

Continued

(1) dry forages and roughages
(2) pasture, range plants, and forages fed green

(3) silages
(4) energy feeds

TABLE A-3-1 Composition of Feeds (Continued)

Feed name or analyses		As fed	Dry	C.V. ± %
Phenylalanine	%	2.30	2.50	
Threonine	%	1.40	1.52	
Tryptophan	%	.50	.54	
Valine	%	2.20	2.39	

Peanut, kernels, solv-extd grnd, mx 7 fbr, (5)
Solvent extracted peanut meal (AAFCO)
Groundnut oil meal, solvent extracted
Peanut oil meal, solvent extracted

Ref no 5-03-650

Feed name or analyses		As fed	Dry	C.V. ± %
Dry matter	%	92.0	100.0	1
Ash	%	4.5	4.9	8
Crude fiber	%	13.0	14.1	37
Ether extract	%	1.2	1.3	56
N-free extract	%	25.9	28.2	
Protein (N x 6.25)	%	47.4	51.5	12
Cattle	dig prot %	42.7	46.4	
Sheep	dig prot %	43.1	46.9	
Swine	dig prot %	44.5	48.4	
Energy				
Cattle	DE kcal/kg	3123.	3395.	
Sheep	DE kcal/kg	3367.	3660.	
Swine	DE kcal/kg	3408.	3704.	
Cattle	ME kcal/kg	2561.	2784.	
Chickens	MEn kcal/kg	2205.	2397.	
Sheep	ME kcal/kg	2761.	3001.	
Swine	ME kcal/kg	2920.	3174.	
Cattle	TDN %	71.	77.	
Sheep	TDN %	76.	83.	
Swine	TDN %	77.	84.	
Calcium	%	.20	.22	
Magnesium	%	.04	.04	
Phosphorus	%	.65	.71	14
Manganese	mg/kg	29.0	31.5	
Choline	mg/kg	2000.	2174.	
Niacin	mg/kg	170.1	184.9	
Pantothenic acid	mg/kg	53.0	57.6	
Riboflavin	mg/kg	11.0	12.0	
Thiamine	mg/kg	7.3	7.9	
Arginine	%	5.90	6.41	
Histidine	%	1.20	1.30	
Isoleucine	%	2.00	2.17	
Leucine	%	3.70	4.02	
Lysine	%	2.30	2.50	
Methionine	%	.40	.43	
Phenylalanine	%	2.70	2.93	
Threonine	%	1.50	1.63	
Tryptophan	%	.50	.54	
Tyrosine	%	1.80	1.96	
Valine	%	2.80	3.04	

Peanut, kernels w skins w hulls, (5)

Ref no 5-03-653

Feed name or analyses		As fed	Dry	C.V. ± %
Dry matter	%	92.0	100.0	
Ash	%	3.0	3.3	
Crude fiber	%	21.7	23.6	
Ether extract	%	34.4	37.4	
N-free extract	%	13.7	14.9	
Protein (N x 6.25)	%	19.1	20.8	
Cattle	dig prot %	15.1	16.4	
Sheep	dig prot %	15.1	16.4	
Energy				
Cattle	DE kcal/kg	4381.	4762.	
Sheep	DE kcal/kg	4381.	4762.	
Cattle	ME kcal/kg	3593.	3905.	
Sheep	ME kcal/kg	3593.	3905.	
Cattle	TDN %	99.	108.	
Sheep	TDN %	99.	108.	
Phosphorus	%	.30	.33	

PEARS. Pyrus spp

Pears, pulp, dehy grnd, (1)

Ref no 1-03-661

Feed name or analyses		As fed	Dry	C.V. ± %
Dry matter	%	91.0	100.0	
Ash	%	4.1	4.5	
Crude fiber	%	21.0	23.1	
Ether extract	%	1.7	1.9	
N-free extract	%	59.0	64.8	
Protein (N x 6.25)	%	5.2	5.7	
Cattle	dig prot %	1.7	1.9	
Sheep	dig prot %	1.5	1.7	
Swine	dig prot %	.0	.0	
Energy				
Cattle	DE kcal/kg	2889.	3175.	
Sheep	DE kcal/kg	2408.	2646.	
Swine	DE kcal/kg	1605.	1764.	
Cattle	ME kcal/kg	2370.	2604.	
Sheep	ME kcal/kg	1975.	2170.	
Swine	ME kcal/kg	1541.	1693.	
Cattle	TDN %	66.	72.	
Sheep	TDN %	55.	60.	
Swine	TDN %	36.	40.	

(5) protein supplements
(6) minerals

(7) vitamins
(8) additives

Feed name or analyses	Mean		C.V. ± %
	As fed	Dry	

Feed name or analyses	Mean		C.V. ± %
	As fed	Dry	

PHOSPHATE, DEFLUORINATED

Phosphate, defluorinated grnd, mn 1 pt F per 100 pt P, (6)
  Phosphate, defluorinated (AAFCO)
  Defluorinated phosphate (CFA)

Ref no 6-01-780

		As fed	Dry	
Dry matter	%	99.8	100.0	
Calcium	%	33.00	33.07	
Iron	%	.920	.922	
Phosphorus	%	18.00	18.04	
Potassium	%	.09	.09	
Sodium	%	3.95	3.96	
Fluorine	mg/kg	1800.00	1803.61	

Phosphate, rock - see Rock phosphate

PINEAPPLE. Ananas comosus

Pineapple, cannery res, dehy, (4)
  Pineapple bran, dried
  Pineapple pulp, dried

Ref no 4-03-722

		As fed	Dry	C.V.
Dry matter	%	90.0	100.0	3
Ash	%	3.2	3.6	17
Crude fiber	%	17.0	18.9	13
Ether extract	%	1.5	1.7	26
N-free extract	%	64.3	71.4	
Protein (N x 6.25)	%	4.0	4.4	11
Cattle	dig prot %	.6	.7	
Sheep	dig prot %	.8	.9	
Energy				
Cattle	DE kcal/kg	2937.	3263.	
Sheep	DE kcal/kg	2817.	3130.	
Cattle	ME kcal/kg	2408.	2676.	
Sheep	ME kcal/kg	2310.	2567.	
Cattle	TDN %	67.	74.	
Sheep	TDN %	64.	71.	
Calcium	%	.16	.18	31
Iron	%	.06	.07	
Phosphorus	%	.12	.13	24

PLANT. Scientific name not used

Plant, charcoal, (6)
  Charcoal, vegetable

Ref no 6-03-727

		As fed	Dry	
Dry matter	%	90.0	100.0	
Ash	%	9.6	10.6	
Calcium	%	4.70	5.22	
Iron	%	.410	.455	
Phosphorus	%	.03	.03	
Cobalt	mg/kg	.200	.222	
Copper	mg/kg	.1	.1	
Manganese	mg/kg	.1	.1	

POTATO. Solanum tuberosum

Potato meal - see Potato, tubers, dehy grnd

Potato, tubers, ensiled, (3)

Ref no 3-03-768

		As fed	Dry	C.V.
Dry matter	%	25.1	100.0	20
Ash	%	1.5	6.1	26
Crude fiber	%	2.1	8.5	55
Ether extract	%	.2	.8	45
N-free extract	%	18.7	74.6	
Protein (N x 6.25)	%	2.5	10.0	21
Cattle	dig prot %	1.6	6.4	
Sheep	dig prot %	1.5	6.0	
Energy				
Cattle	DE kcal/kg	874.	3483.	
Sheep	DE kcal/kg	874.	3483.	
Cattle	ME kcal/kg	717.	2856.	
Sheep	ME kcal/kg	717.	2856.	
Cattle	TDN %	20.	79.	
Sheep	TDN %	20.	79.	
Calcium	%	.01	.04	
Phosphorus	%	.06	.23	

(1) dry forages and roughages
(2) pasture, range plants, and forages fed green

(3) silages
(4) energy feeds

Feed name or analyses		Mean		C.V.
		As fed	Dry	± %

Potato, proc res, dehy, (4)
 Potato by-product, dried
 Potato pomace, dried
 Potato pulp, dried
 Potato waste, dried

Ref no 4-03-775

Dry matter	%	88.6	100.0	2
Ash	%	4.2	4.7	38
Crude fiber	%	6.1	6.9	53
Ether extract	%	.4	.4	38
N-free extract	%	70.2	79.3	
Protein (N x 6.25)	%	7.7	8.7	21
Cattle	dig prot %	3.5	4.0	
Sheep	dig prot %	2.0	2.3	
Swine	dig prot %	5.7	6.4	
Energy				
Cattle	DE kcal/kg	3477.	3924.	
Sheep	DE kcal/kg	2734.	3086.	
Swine	DE kcal/kg	3399.	3836.	
Cattle	ME kcal/kg	2851.	3218.	
Sheep	ME kcal/kg	2242.	2530.	
Swine	ME kcal/kg	3202.	3614.	
Cattle	TDN %	79.	89.	
Sheep	TDN %	62.	70.	
Swine	TDN %	77.	87.	
Calcium	%	.15	.17	
Iron	%	.035	.040	
Magnesium	%	.09	.10	
Phosphorus	%	.12	.13	
Potassium	%	1.05	1.17	
Sodium	%	.06	.07	
Copper	mg/kg	15.4	17.1	
Manganese	mg/kg	31.5	35.0	
Riboflavin	mg/kg	1.1	1.2	

Potato, proc res, wet, (4)
 Potato by-product, wet
 Potato pulp, wet
 Potato waste, wet

Ref no 4-03-777

Dry matter	%	18.0	100.0	
Ash	%	.4	2.2	
Crude fiber	%	.3	1.7	
Ether extract	%	.1	.5	
N-free extract	%	15.9	88.4	
Protein (N x 6.25)	%	1.3	7.2	
Cattle	dig prot %	.5	2.6	
Sheep	dig prot %	.3	1.9	
Swine	dig prot %	.3	1.9	

Continued

Feed name or analyses		Mean		C.V.
		As fed	Dry	± %
Energy				
Cattle	DE kcal/kg	714.	3968.	
Sheep	DE kcal/kg	675.	3748.	
Swine	DE kcal/kg	730.	4056.	
Cattle	ME kcal/kg	586.	3254.	
Sheep	ME kcal/kg	553.	3073.	
Swine	ME kcal/kg	691.	3837.	
Cattle	TDN %	16.	90.	
Sheep	TDN %	15.	85.	
Swine	TDN %	16.	92.	
Calcium	%	.06	.33	
Phosphorus	%	.02	.11	

Potato, tubers, cooked, (4)

Ref no 4-03-784

Dry matter	%	22.5	100.0	
Ash	%	1.2	5.2	
Crude fiber	%	.7	3.0	
Ether extract	%	.1	.3	
N-free extract	%	18.3	81.5	
Protein (N x 6.25)	%	2.2	10.0	
Cattle	dig prot %	.4	2.0	
Sheep	dig prot %	1.0	4.3	
Swine	dig prot %	1.6	7.0	
Energy				
Cattle	DE kcal/kg	625.	2778.	
Sheep	DE kcal/kg	704.	3130.	
Swine	DE kcal/kg	863.	3836.	
Cattle	ME kcal/kg	512.	2278.	
Sheep	ME kcal/kg	578.	2567.	
Swine	ME kcal/kg	811.	3606.	
Cattle	TDN %	14.	63.	
Sheep	TDN %	16.	71.	
Swine	TDN %	20.	87.	
Calcium	%	.01	.04	
Chlorine	%	.06	.26	
Iron	%	.002	.009	
Magnesium	%	.03	.13	
Phosphorus	%	.05	.22	
Potassium	%	.48	2.13	
Sodium	%	.02	.09	
Sulfur	%	.02	.09	
Copper	mg/kg	3.7	16.4	
Manganese	mg/kg	8.8	39.1	
Niacin	mg/kg	11.0	48.8	
Pantothenic acid	mg/kg	6.4	28.4	
Riboflavin	mg/kg	.2	.9	
Thiamine	mg/kg	1.5	6.7	

(5) protein supplements
(6) minerals

(7) vitamins
(8) additives

Feed name or analyses		Mean		C.V. ± %
		As fed	Dry	

Potato, tubers, dehy grnd, (4)
Potato meal

Ref no 4-07-850

Dry matter	%	90.3	100.0	
Ash	%	11.9	13.2	
Crude fiber	%	1.4	1.6	
Ether extract	%	.5	.6	
N-free extract	%	70.5	78.1	
Protein (N x 6.25)	%	5.9	6.5	
Swine	dig prot %	5.0	5.6	
Energy				
Swine	DE kcal/kg	3345.	3704.	
Chickens	MEn kcal/kg	3527.	3906.	
Swine	ME kcal/kg	3168.	3508.	
Cattle	NEm kcal/kg	1680.	1860.	
Cattle	NEgain kcal/kg	1120.	1240.	
Swine	TDN %	76.	84.	
Calcium	%	.07	.08	
Chlorine	%	.36	.40	
Phosphorus	%	.20	.22	
Potassium	%	1.97	2.18	
Manganese	mg/kg	2.9	3.2	

Potato, tubers, fresh, (4)

Ref no 4-03-787

Dry matter	%	24.6	100.0	
Ash	%	.9	3.6	
Crude fiber	%	.5	2.1	
Ether extract	%	.1	.4	
N-free extract	%	20.9	84.9	
Protein (N x 6.25)	%	2.2	9.0	
Cattle	dig prot %	1.6	6.4	
Sheep	dig prot %	1.4	5.7	
Swine	dig prot %	1.0	4.1	
Energy				
Cattle	DE kcal/kg	857.	3483.	
Sheep	DE kcal/kg	911.	3704.	
Swine	DE kcal/kg	933.	3792.	
Cattle	ME kcal/kg	702.	2856.	
Sheep	ME kcal/kg	747.	3037.	
Swine	ME kcal/kg	765.	3109.	
Cattle	NEm kcal/kg	480.	1950.	
Cattle	NEgain kcal/kg	320.	1300.	
Cattle	TDN %	19.	79.	
Sheep	TDN %	21.	84.	
Swine	TDN %	21.	86.	

Feed name or analyses		Mean		C.V. ± %
		As fed	Dry	

POULTRY. Scientific name not used

Poultry, feathers, hydro dehy grnd, mn 70 of prot dig, (5)
Hydrolyzed poultry feathers (CFA)

Ref no 5-03-794

Dry matter	%	91.2	100.00	
Crude fiber	%	4.7	5.2	
Ether extract	%	5.2	5.7	
Protein (N x 6.25)	%	84.4	92.5	
Energy				
Chickens	MEn kcal/kg	2535.	2780.	

Poultry, feathers, hydro dehy grnd, mn 75 of prot dig, (5)
Hydrolyzed poultry feathers (AAFCO)

Ref no 5-03-795

Dry matter	%	91.0	100.0	
Crude fiber	%	3.2	3.5	
Crude fiber	%	.0	.0	
Ether extract	%	2.4	2.6	
N-free extract	%	.0	.0	
Protein (N x 6.25)	%	85.4	93.9	
Swine	dig prot %	60.2	66.2	
Energy				
Swine	DE kcal/kg	2728.	2998.	
Swine	ME kcal/kg	2100.	2308.	
Cattle	NEm kcal/kg	1838.	2020.	
Cattle	NEgain kcal/kg	1238.	1360.	
Swine	TDN %	62.	68.	
Calcium	%	.41	.45	
Phosphorus	%	.49	.54	
Choline	mg/kg	1091.	1199.	
Niacin	mg/kg	20.8	22.8	
Pantothenic acid	mg/kg	8.8	9.7	
Riboflavin	mg/kg	2.0	2.2	
Vitamin B12	mcg/kg	78.3	86.0	
Alanine	%	4.34	4.77	
Arginine	%	6.24	6.86	
Aspartic acid	%	6.20	6.81	
Cystine	%	3.78	4.15	
Glutamic acid	%	8.95	9.84	
Glycine	%	7.14	7.85	
Histidine	%	.56	.62	
Isoleucine	%	4.34	4.77	
Leucine	%	7.89	8.67	
Lysine	%	1.78	1.96	
Methionine	%	.56	.62	
Phenylalanine	%	4.40	4.84	

Continued

(1) dry forages and roughages
(2) pasture, range plants, and forages fed green

(3) silages
(4) energy feeds

Feed name or analyses		Mean		C.V.
		As fed	Dry	± %
Proline	%	10.40	11.43	
Threonine	%	4.61	5.07	
Tryptophan	%	.62	.68	
Tyrosine	%	2.54	2.79	
Valine	%	7.14	7.85	

Poultry, offal, dry- or wet-rend dehy grnd, mx 16 ash 4 acid insol ash, (5)
Poultry by-product meal (AAFCO)

Ref no 5-03-798

Dry matter		%	95.9	100.0	
Ash		%	17.1	17.8	
Ether extract		%	13.9	14.5	17
Protein (N x 6.25)		%	60.7	63.3	3
Energy					
Chickens	MEn kcal/kg		2425.	2529.	

Poultry, offal, dry- or wet-rend dehy grnd, (5)
Poultry by-product meal (CFA)

Ref no 5-03-799

Dry matter		%	93.4	100.0	
Ash		%	18.7	20.0	
Crude fiber		%	1.6	1.7	
Ether extract		%	13.1	14.0	
Protein (N x 6.25)		%	55.4	59.3	
Swine	dig prot	%	47.1	50.4	
Energy					
Swine	DE kcal/kg		3263.	3495.	
Swine	ME kcal/kg		2741.	2936.	
Swine	TDN %		74.	79.	
Calcium		%	3.00	3.21	
Phosphorus		%	1.70	1.82	

Prairie hay - see Native plants, Midwest

Primary dried yeast - see Yeast, primary

Proso millet - see Millet, proso

(5) protein supplements
(6) minerals

QUACKGRASS. Agropyron repens

Quackgrass, hay, s-c, immature, (1)

Ref no 1-03-825

Feed name or analyses			Mean		C.V.
			As fed	Dry	± %
Dry matter		%	86.4	100.0	
Ash		%	6.0	7.0	
Crude fiber		%	24.3	28.1	
Ether extract		%	3.0	3.5	
N-free extract		%	38.9	45.0	
Protein (N x 6.25)		%	14.2	16.4	
Cattle	dig prot	%	9.6	11.1	
Sheep	dig prot	%	8.0	9.3	
Energy					
Cattle	DE kcal/kg		2247.	2601.	
Sheep	DE kcal/kg		2324.	2690.	
Cattle	ME kcal/kg		1841.	2133.	
Sheep	ME kcal/kg		1906.	2206.	
Cattle	TDN %		51.	59.	
Sheep	TDN %		53.	61.	

Quackgrass, hay, s-c, full blm, (1)

Ref no 1-03-826

Dry matter		%	92.1	100.0	
Ash		%	5.1	5.5	
Crude fiber		%	30.6	33.2	
Ether extract		%	3.0	3.2	
N-free extract		%	45.8	49.7	
Protein (N x 6.25)		%	7.7	8.4	
Cattle	dig prot	%	3.9	4.2	
Sheep	dig prot	%	4.4	4.8	
Energy					
Cattle	DE kcal/kg		2314.	2513.	
Sheep	DE kcal/kg		2518.	2734.	
Cattle	ME kcal/kg		1898.	2061.	
Sheep	ME kcal/kg		2065.	2242.	
Cattle	TDN %		52.	57.	
Sheep	TDN %		57.	62.	

Quackgrass, hay, s-c, (1)

Ref no 1-03-827

Dry matter		%	89.1	100.0	3
Ash		%	6.2	6.9	14
Crude fiber		%	30.2	33.9	14
Ether extract		%	2.5	2.8	17
N-free extract		%	40.8	45.8	
Protein (N x 6.25)		%	9.4	10.6	35

Continued

(7) vitamins
(8) additives

Feed name or analyses		Mean		C.V. ± %
		As fed	Dry	
Cattle	dig prot %	5.4	6.1	
Sheep	dig prot %	5.3	6.0	
Energy				
Cattle	DE kcal/kg	2200.	2469.	
Sheep	DE kcal/kg	2397.	2690.	
Cattle	ME kcal/kg	1803.	2024.	
Sheep	ME kcal/kg	1966.	2206.	
Cattle	TDN %	50.	56.	
Sheep	TDN %	54.	61.	

Quackgrass, aerial pt, fresh, mature, (2)

Ref no 2-03-828

		As fed	Dry	
Dry matter	%	31.4	100.0	
Ash	%	1.7	5.4	
Crude fiber	%	11.3	35.9	
Ether extract	%	.8	2.7	
N-free extract	%	15.4	49.0	
Protein (N x 6.25)	%	2.2	7.0	
Cattle	dig prot %	1.2	3.8	
Sheep	dig prot %	1.1	3.5	
Energy				
Cattle	DE kcal/kg	928.	2954.	
Sheep	DE kcal/kg	914.	2910.	
Cattle	ME kcal/kg	760.	2422.	
Sheep	ME kcal/kg	749.	2386.	
Cattle	TDN %	21.	67.	
Sheep	TDN %	21.	66.	
Calcium	%	.09	.28	
Phosphorus	%	.07	.22	

Quackgrass, aerial pt, fresh, (2)

Ref no 2-03-829

		As fed	Dry	
Dry matter	%	28.5	100.0	13
Ash	%	2.2	7.8	21
Crude fiber	%	9.0	31.7	10
Ether extract	%	1.1	3.8	19
N-free extract	%	13.0	45.5	
Protein (N x 6.25)	%	3.2	11.2	27
Cattle	dig prot %	2.1	7.4	
Sheep	dig prot %	2.1	7.4	
Energy				
Cattle	DE kcal/kg	829.	2910.	
Sheep	DE kcal/kg	804.	2822.	
Cattle	ME kcal/kg	680.	2386.	
Sheep	ME kcal/kg	659.	2314.	
Cattle	TDN %	19.	66.	
Sheep	TDN %	18.	64.	
Calcium	%	.09	.33	13

Continued

Feed name or analyses		Mean		C.V. ± %
		As fed	Dry	
Phosphorus	%	.08	.28	17
Carotene	mg/kg	34.9	122.6	
Vitamin A equiv	IU/g	58.2	204.4	

RAPE. Brassica spp

Rape, aerial pt, fresh, (2)

Ref no 2-03-867

		As fed	Dry	
Dry matter	%	16.4	100.0	13
Ash	%	2.2	13.2	10
Crude fiber	%	2.8	17.2	21
Ether extract	%	.6	3.6	19
N-free extract	%	8.2	50.0	
Protein (N x 6.25)	%	2.6	16.0	27
Cattle	dig prot %	1.9	11.5	
Sheep	dig prot %	2.1	13.1	
Energy				
Cattle	DE kcal/kg	434.	2646.	
Sheep	DE kcal/kg	571.	3483.	
Cattle	ME kcal/kg	356.	2170.	
Sheep	ME kcal/kg	468.	2856.	
Cattle	TDN %	10.	60.	
Sheep	TDN %	13.	79.	
Calcium	%	.20	1.19	48
Iron	%	.003	.02	26
Magnesium	%	.01	.08	18
Phosphorus	%	.06	.34	40
Potassium	%	.42	2.58	29
Sodium	%	.01	.05	
Sulfur	%	.08	.49	34
Copper	mg/kg	.6	3.7	
Manganese	mg/kg	3.4	20.9	19

Rape, seed, mech-extd grnd, (5)
Rapeseed oil meal, expeller extracted
Rapeseed meal, expeller extracted

Ref no 5-03-870

		As fed	Dry	
Dry matter	%	93.6	100.0	
Ash	%	6.0	6.4	
Crude fiber	%	13.7	14.6	
Ether extract	%	6.3	6.7	
N-free extract	%	30.6	32.7	
Protein (N x 6.25)	%	37.1	39.6	
Cattle	dig prot %	31.8	34.0	
Sheep	dig prot %	30.4	32.5	
Swine	dig prot %	30.4	32.5	
Energy				
Cattle	DE kcal/kg	3054.	3263.	
Sheep	DE kcal/kg	3178.	3395.	

Continued

(1) dry forages and roughages
(2) pasture, range plants, and forages fed green

(3) silages
(4) energy feeds

TABLE A-3-1 Composition of Feeds (Continued)

Feed name or analyses		Mean As fed	Dry	C.V. ± %
Swine	DE kcal/kg	3054.	3263.	
Cattle	ME kcal/kg	2505.	2676.	
Sheep	ME kcal/kg	2606.	2784.	
Swine	ME kcal/kg	2687.	2871.	
Cattle	TDN %	69.	74.	
Sheep	TDN %	72.	77.	
Swine	TDN %	69.	74.	
Calcium	%	.60	.64	
Phosphorus	%	.97	1.04	
Choline	mg/kg	6295.	6725.	
Niacin	mg/kg	149.8	160.0	
Pantothenic acid	mg/kg	8.6	9.2	
Riboflavin	mg/kg	3.6	3.8	
Thiamine	mg/kg	1.7	1.8	
Alanine	%	1.72	1.84	
Arginine	%	2.08	2.22	
Aspartic acid	%	2.69	2.87	
Glutamic acid	%	6.59	7.04	
Glycine	%	1.91	2.04	
Histidine	%	.98	1.05	
Isoleucine	%	1.52	1.62	
Leucine	%	2.63	2.81	
Lysine	%	1.79	1.91	
Methionine	%	.77	.82	
Phenylalanine	%	1.52	1.63	
Proline	%	2.33	2.49	
Serine	%	1.65	1.76	
Threonine	%	1.67	1.78	
Tryptophan	%	.38	.41	
Tyrosine	%	.88	.94	
Valine	%	1.95	2.08	

Rape, seed, solv-extd grnd, (5)
Rapeseed oil meal, solvent extracted
Rapeseed meal, solvent extracted

Ref no 5-03-871

Feed name or analyses		Mean As fed	Dry	C.V. ± %
Dry matter	%	90.3	100.0	
Ash	%	6.5	7.2	
Crude fiber	%	13.8	15.3	
Ether extract	%	2.4	2.7	
N-free extract	%	28.2	31.2	
Protein (N x 6.25)	%	39.4	43.6	5
Cattle	dig prot %	33.9	37.5	
Sheep	dig prot %	32.3	35.8	
Swine	dig prot %	32.3	35.8	
Energy				
Cattle	DE kcal/kg	2747.	3042.	
Sheep	DE kcal/kg	2867.	3175.	
Swine	DE kcal/kg	2747.	3042.	
Cattle	ME kcal/kg	2252.	2494.	
Sheep	ME kcal/kg	2351.	2604.	
Swine	ME kcal/kg	2396.	2653.	
Cattle	TDN %	62.	69.	

Continued

Feed name or analyses		Mean As fed	Dry	C.V. ± %
Sheep	TDN %	65.	72.	
Swine	TDN %	62.	69.	
Calcium	%	.40	.44	
Phosphorus	%	.90	1.00	
Choline	mg/kg	6073.	6725.	
Niacin	mg/kg	144.5	160.0	
Pantothenic acid	mg/kg	8.3	9.2	
Riboflavin	mg/kg	3.4	3.8	
Thiamine	mg/kg	1.6	1.8	
Alanine	%	1.69	1.87	
Arginine	%	2.16	2.39	
Aspartic acid	%	2.64	2.93	
Glutamic acid	%	6.63	7.34	
Glycine	%	1.88	2.08	
Histidine	%	1.05	1.16	
Isoleucine	%	1.43	1.58	
Leucine	%	2.63	2.91	
Lysine	%	2.09	2.32	
Methionine	%	.76	.84	
Phenylalanine	%	1.49	1.65	
Proline	%	2.41	2.67	
Serine	%	1.65	1.83	
Threonine	%	1.65	1.83	
Tryptophan	%	.48	.53	
Tyrosine	%	.83	.92	
Valine	%	1.90	2.10	

RAPE, ARGENTINE. Brassica napus

Rape, Argentine, seed, mech-extd grnd, (5)

Ref no 5-07-869

Feed name or analyses		Mean As fed	Dry	C.V. ± %
Dry matter	%	93.2	100.0	
Ash	%	6.7	7.2	
Crude fiber	%	14.9	16.0	
Ether extract	%	7.0	7.5	
N-free extract	%	32.7	35.1	
Protein (N x 6.25)	%	31.9	34.2	
Swine	dig prot %	26.1	28.0	
Energy				
Swine	DE kcal/kg	3041.	3263.	
Swine	ME kcal/kg	2709.	2907.	
Swine	TDN %	69.	74.	
Calcium	%	.70	.75	
Phosphorus	%	.99	1.06	
Choline	mg/kg	6524.	7000.	
Niacin	mg/kg	155.6	167.0	
Pantothenic acid	mg/kg	9.2	9.9	
Riboflavin	mg/kg	3.9	4.2	
Thiamine	mg/kg	1.8	1.9	
Arginine	%	1.54	1.65	
Histidine	%	.80	.86	
Isoleucine	%	1.42	1.53	

Continued

(5) protein supplements
(6) minerals

(7) vitamins
(8) additives

Feed name or analyses		Mean		C.V. ± %
		As fed	Dry	
Leucine	%	2.15	2.31	
Lysine	%	1.28	1.37	
Methionine	%	.38	.41	
Phenylalanine	%	1.26	1.35	
Threonine	%	1.29	1.38	
Tryptophan	%	.31	.33	
Valine	%	1.74	1.87	

Rape, Argentine, seed, solv-extd grnd, (5)

Ref no 5-07-868

		As fed	Dry	
Dry matter	%	92.0	100.0	
Ash	%	6.6	7.2	
Crude fiber	%	8.6	9.3	
Ether extract	%	1.0	1.1	
N-free extract	%	35.7	38.8	
Protein (N x 6.25)	%	40.1	43.6	
Swine	dig prot %	32.9	35.8	
Energy				
Swine	DE kcal/kg	2798.	3042.	
Swine	ME kcal/kg	2441.	2653.	
Swine	TDN %	63.	69.	
Calcium	%	.61	.66	
Phosphorus	%	.86	.93	
Choline	mg/kg	6524.	7000.	
Niacin	mg/kg	155.6	167.0	
Pantothenic acid	mg/kg	9.2	9.9	
Riboflavin	mg/kg	3.9	4.2	
Thiamine	mg/kg	1.8	1.9	
Arginine	%	2.15	2.34	
Histidine	%	1.03	1.12	
Isoleucine	%	1.51	1.64	
Leucine	%	2.65	2.88	
Lysine	%	2.06	2.24	
Methionine	%	.51	.55	
Phenylalanine	%	1.63	1.77	
Threonine	%	1.65	1.79	
Tryptophan	%	.45	.49	
Valine	%	2.07	2.25	

RAPE, CANADA. Brassica rapus var

Rape, Canada, seed, cooked pre-press solv-extd, Can 1 mx 1 fat, (5)
Canada rapeseed pre-press solvent extracted meal (CFA)

Ref no 5-08-135

		As fed	Dry	
Dry matter	%	92.0	100.0	
Ash	%	7.2	7.8	
Crude fiber	%	9.3	10.1	

Continued

Feed name or analyses		Mean		C.V. ± %
		As fed	Dry	
Ether extract	%	1.1	1.2	
N-free extract	%	33.9	36.8	
Protein (N x 6.25)	%	40.5	44.0	
Cattle	dig prot %	34.8	37.8	
Sheep	dig prot %	33.2	36.1	
Swine	dig prot %	34.2	37.2	
Energy				
Cattle	DE kcal/kg	2799.	3042.	
Sheep	DE kcal/kg	2596.	2822.	
Swine	DE kcal/kg	2637.	2866.	
Cattle	ME kcal/kg	2294.	2494.	
Sheep	ME kcal/kg	2129.	2314.	
Swine	ME kcal/kg	2296.	2496.	
Cattle	TDN %	63.	69.	
Sheep	TDN %	59.	64.	
Swine	TDN %	60.	65.	
Calcium	%	.66	.72	
Phosphorus	%	.93	1.01	
Alanine	%	1.48	1.61	
Arginine	%	1.90	2.06	
Aspartic acid	%	2.31	2.51	
Glutamic acid	%	5.78	6.28	
Glycine	%	1.66	1.80	
Histidine	%	.93	1.01	
Isoleucine	%	1.24	1.35	
Leucine	%	2.31	2.51	
Lysine	%	1.82	1.98	
Methionine	%	.65	.71	
Phenylalanine	%	1.31	1.42	
Proline	%	2.10	2.28	
Serine	%	1.44	1.57	
Threonine	%	1.44	1.57	
Tryptophan	%	.41	.45	
Tyrosine	%	.73	.79	
Valine	%	1.66	1.80	

Rape, Canada, seed, cooked mech-extd grnd, Can 1 mx 6 fat, (5)
Canada rapeseed meal (CFA)

Ref no 5-08-136

		As fed	Dry	
Dry matter	%	94.0	100.0	
Ash	%	6.8	7.2	
Crude fiber	%	15.5	16.5	
Ether extract	%	7.0	7.4	
N-free extract	%	29.5	31.4	
Protein (N x 6.25)	%	35.2	37.5	
Cattle	dig prot %	30.3	32.2	
Sheep	dig prot %	29.0	30.8	
Swine	dig prot %	29.3	31.2	
Energy				
Cattle	DE kcal/kg	3067.	3263.	
Sheep	DE kcal/kg	2901.	3086.	
Swine	DE kcal/kg	2942.	3130.	

Continued

(1) dry forages and roughages
(2) pasture, range plants, and forages fed green

(3) silages
(4) energy feeds

Feed name or analyses		Mean		C.V. ± %
		As fed	Dry	
Cattle	ME kcal/kg	2515.	2676.	
Sheep	ME kcal/kg	2378.	2530.	
Swine	ME kcal/kg	2601.	2767.	
Cattle	TDN %	70.	74.	
Sheep	TDN %	66.	70.	
Swine	TDN %	67.	71.	
Calcium	%	.71	.76	
Phosphorus	%	1.00	1.06	
Choline	mg/kg	7000.	7448.	
Niacin	mg/kg	167.	178.	
Pantothenic acid	mg/kg	9.9	10.5	
Riboflavin	mg/kg	4.2	4.5	
Thiamine	mg/kg	1.9	2.0	
Alanine	%	1.48	1.57	
Arginine	%	1.79	1.90	
Aspartic acid	%	2.31	2.46	
Glutamic acid	%	5.68	6.04	
Glycine	%	1.65	1.75	
Histidine	%	.85	.90	
Isoleucine	%	1.31	1.39	
Leucine	%	2.27	2.41	
Lysine	%	1.54	1.64	
Methionine	%	.66	.70	
Phenylalanine	%	1.32	1.40	
Proline	%	2.01	2.14	
Serine	%	1.42	1.51	
Threonine	%	1.44	1.53	
Tryptophan	%	.33	.35	
Tyrosine	%	.76	.81	
Valine	%	1.67	1.78	

RAPE, POLISH. Brassica campestris

Rape, Polish, seed, mech-extd grnd, (5)

Ref no 5-07-871

Dry matter	%	94.0	100.0	
Ash	%	6.4	6.8	
Crude fiber	%	14.6	15.5	
Ether extract	%	6.6	7.0	
N-free extract	%	34.3	36.5	
Protein (N x 6.25)	%	32.1	34.2	
Sheep	dig prot %	26.3	28.0	
Swine	dig prot %	26.3	28.0	
Energy				
Sheep	DE kcal/kg	2859.	3042.	
Swine	DE kcal/kg	3067.	3263.	
Sheep	ME kcal/kg	2344.	2494.	
Swine	ME kcal/kg	2732.	2907.	
Sheep	TDN %	65.	69.	
Swine	TDN %	69.	74.	
Calcium	%	.67	.71	
Phosphorus	%	.94	1.00	

Continued

Feed name or analyses		Mean		C.V. ± %
		As fed	Dry	
Choline	mg/kg	6063.	6450.	
Niacin	mg/kg	142.9	152.0	
Pantothenic acid	mg/kg	8.1	8.6	
Riboflavin	mg/kg	3.1	3.3	
Thiamine	mg/kg	1.6	1.7	
Arginine	%	1.72	1.83	
Histidine	%	.89	.95	
Isoleucine	%	1.37	1.46	
Leucine	%	2.15	2.29	
Lysine	%	1.58	1.68	
Methionine	%	.46	.49	
Phenylalanine	%	1.24	1.32	
Threonine	%	1.34	1.43	
Tryptophan	%	.31	.33	
Valine	%	1.72	1.83	

Rape, Polish, seed, solv-extd grnd, (5)

Ref no 5-07-870

Dry matter	%	92.0	100.0	
Ash	%	6.6	7.2	
Crude fiber	%	8.6	9.3	
Ether extract	%	1.0	1.1	
N-free extract	%	35.7	38.8	
Protein (N x 6.25)	%	40.1	43.6	
Swine	dig prot %	32.9	35.8	
Energy				
Swine	DE kcal/kg	2798.	3042.	
Swine	ME kcal/kg	2441.	2653.	
Swine	TDN %	63.	69.	
Calcium	%	.61	.66	
Phosphorus	%	.86	.93	
Choline	mg/kg	5934.	6450.	
Niacin	mg/kg	139.8	152.0	
Pantothenic acid	mg/kg	7.9	8.6	
Riboflavin	mg/kg	3.0	3.3	
Thiamine	mg/kg	1.6	1.7	
Arginine	%	2.36	2.57	
Histidine	%	1.12	1.22	
Isoleucine	%	1.66	1.80	
Leucine	%	2.79	3.03	
Lysine	%	2.26	2.46	
Methionine	%	.56	.61	
Phenylalanine	%	1.63	1.77	
Threonine	%	1.75	1.90	
Tryptophan	%	.47	.51	
Valine	%	2.14	2.33	

(5) protein supplements

(6) minerals

(7) vitamins

(8) additives

Feed name or analyses		As fed (Mean)	Dry (Mean)	C.V. ± %

REDTOP. Agrostis alba

Redtop, hay, s-c, immature, (1)

Ref no 1-03-880

Feed name or analyses		As fed	Dry	C.V. ± %
Dry matter	%	92.0	100.0	2
Ash	%	5.9	6.4	14
Crude fiber	%	29.3	31.9	9
Ether extract	%	2.7	2.9	51
N-free extract	%	41.4	45.0	
Protein (N x 6.25)	%	12.7	13.8	14
Cattle	dig prot %	8.2	8.9	
Sheep	dig prot %	7.9	8.6	
Energy				
Cattle	DE kcal/kg	2271.	2469.	
Sheep	DE kcal/kg	2434.	2646.	
Cattle	ME kcal/kg	1862.	2024.	
Sheep	ME kcal/kg	1996.	2170.	
Cattle	TDN %	52.	56.	
Sheep	TDN %	55.	60.	

Redtop, hay, s-c, mid-blm, (1)

Ref no 1-03-886

Feed name or analyses		As fed	Dry	C.V. ± %
Dry matter	%	92.8	100.0	1
Ash	%	6.0	6.5	9
Crude fiber	%	29.0	31.2	11
Ether extract	%	2.4	2.6	27
N-free extract	%	44.3	47.7	
Protein (N x 6.25)	%	11.1	12.0	20
Cattle	dig prot %	6.8	7.3	
Sheep	dig prot %	6.9	7.4	
Energy				
Cattle	DE kcal/kg	2455.	2646.	
Sheep	DE kcal/kg	2455.	2646.	
Cattle	ME kcal/kg	2014.	2170.	
Sheep	ME kcal/kg	2014.	2170.	
Cattle	TDN %	56.	60.	
Sheep	TDN %	56.	60.	
Calcium	%	.58	.63	
Phosphorus	%	.32	.35	
Potassium	%	1.57	1.69	

Redtop, hay, s-c, full blm, (1)

Ref no 1-03-882

Feed name or analyses		As fed	Dry	C.V. ± %
Dry matter	%	90.7	100.0	
Ash	%	4.3	4.7	
Crude fiber	%	29.6	32.6	

Continued

Feed name or analyses		As fed	Dry	C.V. ± %
Ether extract	%	3.4	3.8	
N-free extract	%	44.9	49.5	
Protein (N x 6.25)	%	8.6	9.4	
Cattle	dig prot %	4.6	5.1	
Sheep	dig prot %	5.3	5.8	
Energy				
Cattle	DE kcal/kg	2279.	2513.	
Sheep	DE kcal/kg	2440.	2690.	
Cattle	ME kcal/kg	1869.	2061.	
Sheep	ME kcal/kg	2001.	2206.	
Cattle	TDN %	52.	57.	
Sheep	TDN %	55.	61.	

Redtop, hay, s-c, mature, (1)

Ref no 1-03-883

Feed name or analyses		As fed	Dry	C.V. ± %
Dry matter	%	91.9	100.0	1
Ash	%	7.9	8.6	36
Crude fiber	%	31.8	34.6	12
Ether extract	%	1.9	2.1	30
N-free extract	%	44.4	48.3	
Protein (N x 6.25)	%	5.9	6.4	20
Cattle	dig prot %	2.3	2.5	
Sheep	dig prot %	3.7	4.0	
Energy				
Cattle	DE kcal/kg	2309.	2513.	
Sheep	DE kcal/kg	2350.	2557.	
Cattle	ME kcal/kg	1894.	2061.	
Sheep	ME kcal/kg	1927.	2097.	
Cattle	TDN %	52.	57.	
Sheep	TDN %	53.	58.	

Redtop, hay, s-c, cut 2, (1)

Ref no 1-03-884

Feed name or analyses		As fed	Dry	C.V. ± %
Dry matter	%	91.1	100.0	
Ash	%	7.7	8.5	
Crude fiber	%	21.6	23.7	
Ether extract	%	2.4	2.6	
N-free extract	%	51.1	56.1	
Protein (N x 6.25)	%	8.3	9.1	
Cattle	dig prot %	5.1	5.6	
Sheep	dig prot %	5.1	5.6	
Energy				
Cattle	DE kcal/kg	2350.	2557.	
Sheep	DE kcal/kg	2350.	2557.	
Cattle	ME kcal/kg	1927.	2097.	
Sheep	ME kcal/kg	1927.	2097.	
Cattle	TDN %	53.	58.	
Sheep	TDN %	53.	58.	

(1) dry forages and roughages
(2) pasture, range plants, and forages fed green

(3) silages
(4) energy feeds

Feed name or analyses		Mean		C.V.
		As fed	Dry	± %

Redtop, hay, s-c, (1)

Ref no 1-03-885

		As fed	Dry	± %
Dry matter	%	92.1	100.0	2
Ash	%	6.7	7.3	23
Crude fiber	%	29.2	31.7	12
Ether extract	%	2.5	2.7	35
N-free extract	%	45.8	49.7	
Protein (N x 6.25)	%	7.9	8.6	27
Cattle	dig prot %	4.0	4.4	
Sheep	dig prot %	4.9	5.3	
Energy				
Cattle	DE kcal/kg	2437.	2646.	
Sheep	DE kcal/kg	2396.	2601.	
Cattle	ME kcal/kg	1998.	2170.	
Sheep	ME kcal/kg	1964.	2133.	
Cattle	TDN %	55.	60.	
Sheep	TDN %	54.	59.	
Calcium	%	.41	.45	29
Chlorine	%	.06	.06	49
Iron	%	.02	.03	27
Magnesium	%	.24	.26	23
Phosphorus	%	.21	.23	28
Potassium	%	1.56	1.69	12
Sodium	%	.06	.07	24
Sulfur	%	.21	.23	13
Cobalt	mg/kg	.14	.15	12
Copper	mg/kg	10.6	11.5	48
Manganese	mg/kg	199.4	216.5	31
Zinc	mg/kg	16.5	17.9	10

Redtop, aerial pt, fresh, immature, (2)

Ref no 2-03-888

		As fed	Dry	± %
Dry matter	%	26.3	100.0	4
Ash	%	2.7	10.4	11
Crude fiber	%	5.9	22.6	6
Ether extract	%	1.1	4.2	13
N-free extract	%	11.9	45.1	
Protein (N x 6.25)	%	4.7	17.7	5
Cattle	dig prot %	3.4	12.9	
Sheep	dig prot %	3.6	13.5	
Energy				
Cattle	DE kcal/kg	777.	2954.	
Sheep	DE kcal/kg	777.	2954.	
Cattle	ME kcal/kg	637.	2422.	
Sheep	ME kcal/kg	637.	2422.	
Cattle	TDN %	18.	67.	
Sheep	TDN %	18.	67.	
Calcium	%	.17	.64	15
Magnesium	%	.06	.23	12
Phosphorus	%	.11	.42	14

Continued

Potassium	%	.69	2.64	9
Carotene	mg/kg	96.0	364.9	25
Vitamin A equiv	IU/g	160.0	608.3	

Redtop, aerial pt, fresh, full blm, (2)

Ref no 2-03-891

		As fed	Dry	± %
Dry matter	%	26.3	100.0	
Ash	%	1.8	7.0	13
Crude fiber	%	6.6	25.1	9
Ether extract	%	.9	3.5	21
N-free extract	%	14.8	56.3	
Protein (N x 6.25)	%	2.1	8.1	43
Cattle	dig prot %	1.3	4.8	
Sheep	dig prot %	1.2	4.5	
Energy				
Cattle	DE kcal/kg	719.	2734.	
Sheep	DE kcal/kg	812.	3086.	
Cattle	ME kcal/kg	590.	2242.	
Sheep	ME kcal/kg	665.	2530.	
Cattle	TDN %	16.	62.	
Sheep	TDN %	18.	70.	
Carotene	mg/kg	40.1	152.6	47
Vitamin A equiv	IU/g	66.8	254.4	

Redtop, aerial pt, fresh, mature, (2)

Ref no 2-03-894

		As fed	Dry	± %
Dry matter	%	30.0	100.0	
Ash	%	1.5	5.1	20
Crude fiber	%	8.6	28.5	8
Ether extract	%	.8	2.8	15
N-free extract	%	17.4	57.9	
Protein (N x 6.25)	%	1.7	5.7	7
Cattle	dig prot %	.8	2.7	
Sheep	dig prot %	.7	2.3	
Energy				
Cattle	DE kcal/kg	847.	2822.	
Sheep	DE kcal/kg	939.	3130.	
Cattle	ME kcal/kg	694.	2314.	
Sheep	ME kcal/kg	770.	2567.	
Cattle	TDN %	19.	64.	
Sheep	TDN %	21.	71.	

Redtop, aerial pt, fresh, (2)

Ref no 2-03-897

		As fed	Dry	± %
Dry matter	%	27.3	100.0	16
Ash	%	2.3	8.4	18
Crude fiber	%	6.8	25.0	12

Continued

(5) protein supplements

(6) minerals

(7) vitamins

(8) additives

Feed name or analyses		Mean		C.V.
		As fed	Dry	± %
Ether extract	%	1.0	3.8	18
N-free extract	%	13.5	49.5	
Protein (N x 6.25)	%	3.6	13.3	20
Cattle	dig prot %	2.5	9.2	
Sheep	dig prot %	2.6	9.4	
Energy				
Cattle	DE kcal/kg	794.	2910.	
Sheep	DE kcal/kg	794.	2910.	
Cattle	ME kcal/kg	651.	2386.	
Sheep	ME kcal/kg	651.	2386.	
Cattle	TDN %	18.	66.	
Sheep	TDN %	18.	66.	
Calcium	%	.17	.62	23
Chlorine	%	.02	.09	
Iron	%	.01	.02	
Magnesium	%	.07	.25	19
Phosphorus	%	.10	.37	33
Potassium	%	.64	2.35	20
Sodium	%	.01	.05	
Sulfur	%	.04	.16	67
Copper	mg/kg	7.1	26.0	
Manganese	mg/kg	63.6	233.1	
Carotene	mg/kg	54.3	198.9	50
Vitamin A equiv	IU/g	90.5	331.6	

Refuse screenings - see Grains, screenings, refuse

RHODESGRASS. Chloris gayana

Rhodesgrass, aerial pt, fresh, (2)

Ref no 2-03-916

		As fed	Dry	C.V. ± %
Dry matter	%	26.4	100.0	13
Ash	%	2.8	10.5	17
Crude fiber	%	9.9	37.5	6
Ether extract	%	.4	1.6	17
N-free extract	%	11.4	43.0	
Protein (N x 6.25)	%	2.0	7.4	9
Cattle	dig prot %	1.1	4.3	
Sheep	dig prot %	1.2	4.6	
Energy				
Cattle	DE kcal/kg	675.	2557.	
Sheep	DE kcal/kg	733.	2778.	
Cattle	ME kcal/kg	554.	2097.	
Sheep	ME kcal/kg	601.	2278.	
Cattle	TDN %	15.	58.	
Sheep	TDN %	17.	63.	

RICE. Oryza sativa

Rice, hulls, (1)
Rice hulls (AAFCO)

Ref no 1-08-075

		As fed	Dry	C.V. ± %
Dry matter	%	92.2	100.0	1
Ash	%	19.2	20.8	11
Crude fiber	%	40.0	43.3	5
Ether extract	%	.8	.9	32
N-free extract	%	29.5	32.0	
Protein (N x 6.25)	%	2.8	3.0	15
Cattle	dig prot %	.2	.2	
Sheep	dig prot %	.2	.2	
Cellulose	%	38.9	42.2	
Lignin	%	19.7	21.4	
Energy				
Cattle	DE kcal/kg	609.	661.	
Sheep	DE kcal/kg	609.	661.	
Cattle	ME kcal/kg	500.	542.	
Sheep	ME kcal/kg	500.	542.	
Cattle	TDN %	14.	15.	
Sheep	TDN %	14.	15.	
Calcium	%	.08	.09	
Phosphorus	%	.06	.06	
Manganese	mg/kg	307.4	333.4	

Rice, straw, (1)

Ref no 1-03-925

		As fed	Dry	C.V. ± %
Dry matter	%	91.5	100.0	1
Ash	%	15.5	16.9	8
Crude fiber	%	32.1	35.1	6
Ether extract	%	1.3	1.4	24
N-free extract	%	38.8	42.4	
Protein (N x 6.25)	%	3.8	4.2	17
Cattle	dig prot %	.9	1.0	
Sheep	dig prot %	1.2	1.3	
Energy				
Cattle	DE kcal/kg	1936.	2116.	
Sheep	DE kcal/kg	1614.	1764.	
Cattle	ME kcal/kg	1588.	1735.	
Sheep	ME kcal/kg	1323.	1446.	
Cattle	TDN %	44.	48.	
Sheep	TDN %	37.	40.	
Calcium	%	.22	.24	27
Magnesium	%	.11	.12	30
Phosphorus	%	.08	.09	28
Potassium	%	1.21	1.32	12
Manganese	mg/kg	317.8	347.3	5

(1) dry forages and roughages
(2) pasture, range plants, and forages fed green

(3) silages
(4) energy feeds

Feed name or analyses		Mean		C.V.
		As fed	Dry	± %

Rice, bran w germ, dry-mil, mx 13 fbr CaCO3 declared above 3 mn, (4)
  Rice bran (AAFCO)

Ref no 4-03-928

Feed name or analyses		As fed	Dry	C.V. ± %
Dry matter	%	91.0	100.0	2
Ash	%	10.9	12.0	11
Crude fiber	%	11.0	12.1	18
Ether extract	%	15.1	16.6	20
N-free extract	%	40.5	44.5	
Protein (N x 6.25)	%	13.5	14.8	9
Cattle	dig prot %	8.7	9.6	
Sheep	dig prot %	9.2	10.1	
Swine	dig prot %	10.2	11.2	
Energy				
Cattle	DE kcal/kg	2648.	2910.	
Sheep	DE kcal/kg	3210.	3527.	
Swine	DE kcal/kg	3256.	3578.	
Cattle	ME kcal/kg	2171.	2386.	
Sheep	ME kcal/kg	2632.	2892.	
Swine	ME kcal/kg	3028.	3328.	
Cattle	NE$_m$ kcal/kg	1511.	1660.	
Cattle	NE$_{gain}$ kcal/kg	983.	1080.	
Cattle	TDN %	60.	66.	
Sheep	TDN %	73.	80.	
Swine	TDN %	74.	81.	
Calcium	%	.06	.07	54
Iron	%	.019	.021	21
Magnesium	%	.95	1.04	13
Phosphorus	%	1.82	2.00	17
Potassium	%	1.74	1.91	4
Copper	mg/kg	13.0	14.3	25
Manganese	mg/kg	417.8	459.2	41
Zinc	mg/kg	29.9	32.9	
Biotin	mg/kg	4.20	4.60	
Choline	mg/kg	1254.	1378.	
Niacin	mg/kg	303.2	333.2	19
Pantothenic acid	mg/kg	23.5	25.8	43
Riboflavin	mg/kg	2.6	2.9	30
Thiamine	mg/kg	22.4	24.6	42
Arginine	%	.50	.55	
Cystine	%	.10	.11	
Histidine	%	.20	.22	
Isoleucine	%	.40	.44	
Leucine	%	.60	.66	
Lysine	%	.50	.55	
Phenylalanine	%	.40	.44	
Threonine	%	.40	.44	
Tryptophan	%	.10	.11	
Valine	%	.60	.66	

Rice, bran w germ, solv-extd, mn 14 prot mx 14 fbr, (4)
  Solvent extracted rice bran (AAFCO)

Ref no 4-03-930

Feed name or analyses		As fed	Dry	C.V. ± %
Dry matter	%	91.0	100.0	1
Ash	%	15.9	17.5	36
Crude fiber	%	13.0	14.3	21
Ether extract	%	1.0	1.1	96
N-free extract	%	47.0	51.7	
Protein (N x 6.25)	%	14.0	15.4	11
Cattle	dig prot %	9.1	10.0	
Sheep	dig prot %	9.6	10.5	
Energy				
Cattle	DE kcal/kg	2167.	2381.	
Sheep	DE kcal/kg	2247.	2469.	
Cattle	ME kcal/kg	1776.	1952.	
Chickens	ME$_n$ kcal/kg	882.	969.	
Sheep	ME kcal/kg	1842.	2024.	
Cattle	TDN %	49.	54.	
Sheep	TDN %	51.	56.	
Calcium	%	.12	.13	
Phosphorus	%	1.48	1.63	
Potassium	%	1.35	1.48	
Manganese	mg/kg	137.9	151.6	

Rice, brown - see Rice, groats

Rice, grain w hulls, grnd, (4)
  Ground rough rice (AAFCO)
  Ground paddy rice (AAFCO)

Ref no 4-03-938

Feed name or analyses		As fed	Dry	C.V. ± %
Dry matter	%	89.0	100.0	1
Ash	%	4.5	5.0	14
Crude fiber	%	9.0	10.1	11
Ether extract	%	1.9	2.1	10
N-free extract	%	66.4	74.6	
Protein (N x 6.25)	%	7.3	8.2	10
Cattle	dig prot %	5.5	6.2	
Sheep	dig prot %	5.5	6.2	
Swine	dig prot %	5.5	6.2	
Energy	GE kcal/kg	3066.	3445.	
Cattle	DE kcal/kg	3139.	3527.	
Sheep	DE kcal/kg	3139.	3527.	
Swine	DE kcal/kg	2511.	2821.	
Cattle	ME kcal/kg	2574.	2892.	
Chickens	ME$_n$ kcal/kg	2668.	2998.	
Sheep	ME kcal/kg	2574.	2892.	
Swine	ME kcal/kg	2367.	2660.	
Cattle	TDN %	71.	80.	

Continued

(5) protein supplements
(6) minerals

(7) vitamins
(8) additives

Feed name or analyses		Mean As fed	Dry	C.V. ± %
Sheep	TDN %	71.	80.	
Swine	TDN %	57.	64.	
Calcium	%	.04	.04	
Magnesium	%	.14	.16	
Phosphorus	%	.26	.29	
Potassium	%	.34	.38	
Folic acid	mg/kg	.4	.4	14
Niacin	mg/kg	30.3	34.1	16
Riboflavin	mg/kg	1.1	1.3	30
Thiamine	mg/kg	2.8	3.1	23
Arginine	%	.53	.60	
Histidine	%	.09	.10	
Isoleucine	%	.27	.30	
Leucine	%	.53	.60	
Lysine	%	.27	.30	
Phenylalanine	%	.27	.30	
Threonine	%	.18	.20	

Rice, groats, polished and broken, (4)
Chipped rice (AAFCO)
Broken rice (AAFCO)
Brewers rice (AAFCO)

Ref no 4-03-932

		As fed	Dry	C.V. ± %
Dry matter	%	88.0	100.0	
Ash	%	.7	.8	23
Crude fiber	%	1.0	1.1	23
Ether extract	%	.9	1.0	27
N-free extract	%	77.2	87.7	
Protein (N x 6.25)	%	8.3	9.4	6
Cattle	dig prot %	4.0	4.6	
Sheep	dig prot %	6.2	7.1	
Energy				
Cattle	DE kcal/kg	3104.	3527.	
Sheep	DE kcal/kg	3453.	3924.	
Cattle	ME kcal/kg	2545.	2892.	
Sheep	ME kcal/kg	2832.	3218.	
Cattle	TDN %	70.	80.	
Sheep	TDN %	78.	89.	
Calcium	%	.03	.03	31
Magnesium	%	.05	.06	22
Phosphorus	%	.13	.15	28
Potassium	%	.13	.15	22

Rice, groats, grnd, (4)
Ground brown rice (AAFCO)
Rice grain without hulls, ground

Ref no 4-03-935

		As fed	Dry	C.V. ± %
Dry matter	%	89.0	100.0	3
Ash	%	.7	.8	37
Crude fiber	%	1.0	1.1	28

Continued

		As fed	Dry	C.V. ± %
Ether extract	%	1.2	1.3	35
N-free extract	%	77.6	87.2	
Protein (N x 6.25)	%	8.5	9.6	9
Cattle	dig prot %	4.3	4.8	
Sheep	dig prot %	6.5	7.3	
Swine	dig prot %	7.3	8.2	
Energy				
Cattle	DE kcal/kg	3139.	3527.	
Sheep	DE kcal/kg	3492.	3924.	
Swine	DE kcal/kg	3846.	4321.	
Cattle	ME kcal/kg	2574.	2892.	
Sheep	ME kcal/kg	2864.	3218.	
Swine	ME kcal/kg	3619.	4066.	
Cattle	TDN %	71.	80.	
Sheep	TDN %	79.	89.	
Swine	TDN %	87.	98.	
Calcium	%	.04	.04	43
Iron	%	.004	.005	73
Magnesium	%	.05	.06	
Phosphorus	%	.18	.20	24
Potassium	%	.12	.14	30
Sodium	%	.04	.05	
Copper	mg/kg	4.3	4.8	
Manganese	mg/kg	4.3	4.8	
Niacin	mg/kg	17.1	19.2	
Riboflavin	mg/kg	.3	.3	
Thiamine	mg/kg	1.1	1.2	

Rice, groats, polished, (4)
Rice, white, polished

Ref no 4-03-942

		As fed	Dry	C.V. ± %
Dry matter	%	89.0	100.0	3
Ash	%	.5	.6	47
Crude fiber	%	.4	.4	46
Ether extract	%	.4	.4	32
N-free extract	%	80.4	90.4	
Protein (N x 6.25)	%	7.3	8.2	10
Cattle	dig prot %	3.1	3.5	
Sheep	dig prot %	5.5	6.2	
Swine	dig prot %	6.2	7.0	
Energy	GE kcal/kg	3604.	4049.	
Cattle	DE kcal/kg	3296.	3704.	
Sheep	DE kcal/kg	3492.	3924.	
Swine	DE kcal/kg	3784.	4252.	
Cattle	ME kcal/kg	2703.	3037.	
Sheep	ME kcal/kg	2864.	3218.	
Swine	ME kcal/kg	3569.	4010.	
Cattle	TDN %	75.	84.	
Sheep	TDN %	79.	89.	
Swine	TDN %	86.	97.	
Calcium	%	.03	.03	42
Iron	%	.002	.002	
Magnesium	%	.02	.02	

Continued

(1) dry forages and roughages
(2) pasture, range plants, and forages fed green
(3) silages
(4) energy feeds

Feed name or analyses		Mean		C.V.
		As fed	Dry	± %
Phosphorus	%	.12	.14	52
Potassium	%	.13	.15	
Copper	mg/kg	2.9	3.3	
Manganese	mg/kg	10.9	12.3	
Zinc	mg/kg	1.8	2.0	
Choline	mg/kg	907.	1019.	
Niacin	mg/kg	14.1	15.8	29
Pantothenic acid	mg/kg	3.3	3.7	51
Riboflavin	mg/kg	.6	.7	3
Thiamine	mg/kg	.6	.7	48
a-tocopherol	mg/kg	3.6	4.0	
Vitamin B₆	mg/kg	.4	.4	
Arginine	%	.36	.40	39
Cystine	%	.09	.10	
Glycine	%	.71	.80	
Histidine	%	.18	.20	
Isoleucine	%	.45	.51	
Leucine	%	.71	.80	
Lysine	%	.27	.30	21
Methionine	%	.27	.30	10
Phenylalanine	%	.53	.60	
Threonine	%	.36	.40	
Tryptophan	%	.09	.10	
Tyrosine	%	.62	.70	
Valine	%	.53	.60	

Rice, polishings, dehy, (4)
Rice polishings (AAFCO)
Rice polish (CFA)

Ref no 4-03-943

Dry matter	%	90.0	100.0	1
Ash	%	8.0	8.9	38
Crude fiber	%	3.0	3.3	48
Ether extract	%	13.2	14.7	20
N-free extract	%	54.0	60.0	
Protein (N x 6.25)	%	11.8	13.1	15
Cattle	dig prot %	7.7	8.6	
Sheep	dig prot %	9.2	10.2	
Swine	dig prot %	10.3	11.4	
Energy				
Cattle	DE kcal/kg	3532.	3924.	
Sheep	DE kcal/kg	3690.	4100.	
Swine	DE kcal/kg	3916.	4351.	
Cattle	ME kcal/kg	2896.	3218.	
Chickens	MEₙ kcal/kg	3417.	3797.	
Sheep	ME kcal/kg	3026.	3362.	
Swine	ME kcal/kg	3658.	4064.	
Cattle	NEₘ kcal/kg	1971.	2190.	
Cattle	NE_gain kcal/kg	1287.	1430.	
Cattle	TDN %	80.	89.	
Sheep	TDN %	84.	93.	
Swine	TDN %	89.	99.	
Calcium	%	.04	.04	

Continued

Feed name or analyses		Mean		C.V.
		As fed	Dry	± %
Magnesium	%	.65	.72	
Phosphorus	%	1.42	1.58	17
Potassium	%	1.17	1.30	
Sodium	%	.11	.12	
Biotin	mg/kg	.60	.70	
Choline	mg/kg	1307.	1452.	
Niacin	mg/kg	531.7	590.7	45
Pantothenic acid	mg/kg	58.3	64.8	71
Riboflavin	mg/kg	1.8	2.0	34
Thiamine	mg/kg	19.7	21.9	44
Arginine	%	.50	.56	
Cystine	%	.10	.11	
Histidine	%	.10	.11	
Isoleucine	%	.30	.33	
Leucine	%	.50	.55	
Lysine	%	.50	.55	
Phenylalanine	%	.30	.33	
Threonine	%	.30	.33	
Tryptophan	%	.10	.11	

RICEGRASS, INDIAN. Oryzopsis hymenoides

Ricegrass, Indian, aerial pt, fresh, dormant, (2)

Ref no 2-07-990

Dry matter	%	86.0	100.0	
Ash	%	6.4	7.4	
Ether extract	%	3.8	2.7	
Protein (N x 6.25)	%	3.0	3.5	
Sheep	dig prot %	.3	.3	
Cellulose	%	32.7	38.0	
Lignin	%	8.6	10.0	
Energy	GE kcal/kg	3682.	4281.	
Sheep	DE kcal/kg	1687.	1962.	
Sheep	ME kcal/kg	1390.	1616.	
Sheep	TDN %	41.	48.	
Calcium	%	.45	.52	
Phosphorus	%	.05	.06	
Carotene	mg/kg	.3	.4	

ROCK PHOSPHATE. Scientific name not applicable

Rock phosphate, grnd, (6)
Rock phosphate, ground (AAFCO)

Ref no 6-03-945

Dry matter	%	99.0	100.0	
Ash	%	96.4	97.4	
Calcium	%	29.57	29.87	
Magnesium	%	.41	.41	

Continued

(5) protein supplements
(6) minerals

(7) vitamins
(8) additives

Feed name or analyses		Mean		C.V.
		As fed	Dry	± %
Phosphorus	%	13.55	13.68	
Potassium	%	.59	.60	
Sodium	%	.03	.03	

Rock phosphate, grnd, mx 0.5 F, (6)
Rock phosphate, ground, low fluorine (AAFCO)

Ref no 6-03-946

Dry matter	%	100.0	100.0	
Ash	%	99.0	99.0	
Calcium	%	32.00	32.00	16
Iron	%	.71	.71	
Phosphorus	%	18.00	18.00	28
Sodium	%	.19	.19	
Copper	mg/kg	65.8	65.8	
Manganese	mg/kg	692.1	692.1	

Rolled oats - see Oats, cereal by-prod, mx 2 fbr

Rough rice - see Rice, grain w hulls

RUSH. Juncus spp

Rush, hay, s-c, (1)

Ref no 1-03-960

Dry matter	%	89.7	100.0	2
Ash	%	6.2	6.9	11
Crude fiber	%	27.6	30.7	8
Ether extract	%	2.1	2.3	27
N-free extract	%	45.3	50.5	
Protein (N x 6.25)	%	8.6	9.6	10
Cattle	dig prot %	4.7	5.2	
Sheep	dig prot %	4.7	5.2	
Energy				
Cattle	DE kcal/kg	2413.	2690.	
Sheep	DE kcal/kg	2254.	2513.	
Cattle	ME kcal/kg	1979.	2206.	
Sheep	ME kcal/kg	1849.	2061.	
Cattle	TDN %	55.	61.	
Sheep	TDN %	51.	57.	
Calcium	%	.31	.35	
Iron	%	.01	.01	
Magnesium	%	.26	.29	
Phosphorus	%	.10	.11	
Potassium	%	.68	.76	

RUSH, SALTMEADOW. Juncus gerardii

Rush, saltmeadow, hay, s-c, (1)

Ref no 1-03-970

Dry matter	%	89.4	100.0	2
Ash	%	7.2	8.1	10
Crude fiber	%	25.0	28.0	7
Ether extract	%	2.5	2.8	18
N-free extract	%	47.1	52.7	
Protein (N x 6.25)	%	7.5	8.4	9
Cattle	dig prot %	4.0	4.5	
Sheep	dig prot %	4.0	4.5	
Energy				
Cattle	DE kcal/kg	1931.	2160.	
Sheep	DE kcal/kg	1931.	2160.	
Cattle	ME kcal/kg	1583.	1771.	
Sheep	ME kcal/kg	1583.	1771.	
Cattle	TDN %	44.	49.	
Sheep	TDN %	44.	49.	

RUSSIANTHISTLE. Salsola spp

Russianthistle, aerial pt, fresh, immature, (2)

Ref no 2-03-981

Dry matter	%	61.6	100.0	36
Ash	%	12.0	19.5	13
Crude fiber	%	12.4	20.2	
Ether extract	%	1.5	2.5	
N-free extract	%	24.1	39.1	
Protein (N x 6.25)	%	11.5	18.7	19
Cattle	dig prot %	8.5	13.8	
Sheep	dig prot %	8.9	14.4	
Energy				
Cattle	DE kcal/kg	1521.	2469.	
Sheep	DE kcal/kg	1657.	2690.	
Cattle	ME kcal/kg	1247.	2024.	
Sheep	ME kcal/kg	1359.	2206.	
Cattle	TDN %	34.	56.	
Sheep	TDN %	38.	61.	
Calcium	%	1.67	2.71	19
Magnesium	%	.51	.82	9
Phosphorus	%	.12	.20	18
Potassium	%	2.85	4.63	
Carotene	mg/kg	65.8	106.9	99
Vitamin A equiv	IU/g	109.7	178.2	

(1) dry forages and roughages
(2) pasture, range plants, and forages fed green

(3) silages
(4) energy feeds

Feed name or analyses		Mean		C.V.
		As fed	Dry	± %

Russianthistle, aerial pt, fresh, (2)

Ref no 2-03-983

		As fed	Dry	C.V. ± %
Dry matter	%	39.4	100.0	40
Ash	%	6.0	15.2	28
Crude fiber	%	12.5	31.7	19
Ether extract	%	.7	1.8	46
N-free extract	%	15.4	39.0	
Protein (N x 6.25)	%	4.8	12.3	40
Cattle	dig prot %	3.3	8.3	
Sheep	dig prot %	3.3	8.4	
Energy				
Cattle	DE kcal/kg	1025.	2601.	
Sheep	DE kcal/kg	1077.	2734.	
Cattle	ME kcal/kg	840.	2133.	
Sheep	ME kcal/kg	883.	2242.	
Cattle	TDN %	23.	59.	
Sheep	TDN %	24.	62.	
Calcium	%	.97	2.47	23
Magnesium	%	.32	.81	11
Phosphorus	%	.07	.17	30
Potassium	%	2.55	6.46	13
Sulfur	%	.07	.17	15
Cobalt	mg/kg	.07	.17	
Copper	mg/kg	7.6	19.2	
Manganese	mg/kg	13.1	33.3	
Carotene	mg/kg	35.1	89.1	99
Vitamin A equiv	IU/g	58.5	148.5	

Russianthistle, aerial pt, ensiled, (3)

Ref no 3-03-984

		As fed	Dry	
Dry matter	%	34.4	100.0	
Ash	%	5.9	17.2	
Crude fiber	%	10.1	29.4	
Ether extract	%	1.0	2.9	
N-free extract	%	14.3	41.6	
Protein (N x 6.25)	%	3.1	8.9	
Cattle	dig prot %	1.8	5.4	
Sheep	dig prot %	1.5	4.3	
Energy				
Cattle	DE kcal/kg	758.	2205.	
Sheep	DE kcal/kg	895.	2601.	
Cattle	ME kcal/kg	622.	1808.	
Sheep	ME kcal/kg	734.	2133.	
Cattle	TDN %	17.	50.	
Sheep	TDN %	20.	59.	

RUSSIANTHISTLE, TUMBLING. Salsola kali tenuifolia

Russianthistle, tumbling, hay, s-c, immature, (1)

Ref no 1-03-986

		As fed	Dry	C.V. ± %
Dry matter	%	88.6	100.0	
Ash	%	14.6	16.5	16
Crude fiber	%	22.0	24.8	16
Ether extract	%	1.7	1.9	32
N-free extract	%	36.6	41.3	
Protein (N x 6.25)	%	13.7	15.5	9
Cattle	dig prot %	8.7	9.8	
Cattle	dig prot %	9.0	10.2	
Energy				
Cattle	DE kcal/kg	1875.	2116.	
Sheep	DE kcal/kg	1680.	1896.	
Cattle	ME kcal/kg	1537.	1735.	
Sheep	ME kcal/kg	1378.	1555.	
Cattle	TDN %	42.	48.	
Sheep	TDN %	38.	43.	
Calcium	%	1.65	1.86	30
Magnesium	%	1.17	1.32	33
Phosphorus	%	.28	.31	96
Potassium	%	6.10	6.85	

Russianthistle, tumbling, hay, s-c, mature, (1)

Ref no 1-03-987

		As fed	Dry	
Dry matter	%	92.2	100.0	
Ash	%	11.9	12.9	
Crude fiber	%	29.7	32.2	
Ether extract	%	.9	1.0	
N-free extract	%	44.3	48.1	
Protein (N x 6.25)	%	5.4	5.8	
Cattle	dig prot %	3.3	3.6	
Sheep	dig prot %	3.5	3.8	
Energy				
Cattle	DE kcal/kg	1992.	2160.	
Sheep	DE kcal/kg	1667.	1808.	
Cattle	ME kcal/kg	1633.	1771.	
Sheep	ME kcal/kg	1366.	1482.	
Cattle	TDN %	45.	49.	
Sheep	TDN %	38.	41.	
Calcium	%	1.66	1.80	
Phosphorus	%	.12	.13	

(5) protein supplements

(6) minerals

(7) vitamins

(8) additives

Feed name or analyses		Mean		C.V.
		As fed	Dry	± %

Russianthistle, tumbling, hay, s-c, (1)

Ref no 1-03-988

		As fed	Dry	C.V.
Dry matter	%	88.4	100.0	3
Ash	%	13.0	14.7	20
Crude fiber	%	25.0	28.3	15
Ether extract	%	1.8	2.0	32
N-free extract	%	38.1	43.1	
Protein (N x 6.25)	%	10.5	11.9	27
Cattle	dig prot %	6.6	7.5	
Sheep	dig prot %	6.9	7.8	
Energy				
Cattle	DE kcal/kg	1909.	2160.	
Sheep	DE kcal/kg	1676.	1896.	
Cattle	ME kcal/kg	1566.	1771.	
Sheep	ME kcal/kg	1375.	1555.	
Cattle	TDN %	43.	49.	
Sheep	TDN %	38.	43.	
Calcium	%	1.63	1.84	28
Magnesium	%	1.05	1.19	55
Phosphorus	%	.26	.29	99

Russianthistle, tumbling, aerial pt, fresh dormant, (2)

Ref no 2-08-000

		As fed	Dry	C.V.
Dry matter	%	80.0	100.0	
Ash	%	15.7	19.6	
Ether extract	%	2.4	3.0	
Protein (N x 6.25)	%	11.8	14.7	
Cattle	dig prot %	7.8	9.7	
Sheep	dig prot %	7.8	9.7	
Cellulose	%	14.4	18.0	
Lignin	%	5.6	7.0	
Energy	GE kcal/kg	2846.	3558.	
Cattle	DE kcal/kg	1709.	2136.	
Sheep	DE kcal/kg	1709.	2136.	
Cattle	ME kcal/kg	1423.	1779.	
Sheep	ME kcal/kg	1423.	1779.	
Cattle	TDN %	40.	50.	
Sheep	TDN %	40.	50.	
Calcium	%	2.64	3.30	
Phosphorus	%	.13	.16	
Carotene	mg/kg	7.2	9.0	

RYE. Secale cereale

Rye, hay, s-c, (1)

Ref no 1-04-004

		As fed	Dry	C.V.
Dry matter	%	92.2	100.0	2
Ash	%	5.4	5.8	23
Crude fiber	%	35.7	38.7	8
Ether extract	%	2.6	2.8	25
N-free extract	%	39.9	43.3	
Protein (N x 6.25)	%	8.7	9.4	14
Cattle	dig prot %	4.7	5.1	
Sheep	dig prot %	4.6	5.0	
Energy				
Cattle	DE kcal/kg	1910.	2072.	
Sheep	DE kcal/kg	2195.	2381.	
Cattle	ME kcal/kg	1566.	1699.	
Sheep	ME kcal/kg	1800.	1952.	
Cattle	TDN %	43.	47.	
Sheep	TDN %	50.	54.	
Calcium	%	.32	.35	43
Phosphorus	%	.34	.37	77
Potassium	%	.97	1.05	

Rye, straw, (1)

Ref no 1-04-007

		As fed	Dry	C.V.
Dry matter	%	88.9	100.0	2
Ash	%	4.3	4.8	21
Crude fiber	%	42.3	47.6	9
Ether extract	%	1.3	1.5	15
N-free extract	%	38.3	43.1	
Protein (N x 6.25)	%	2.7	3.0	45
Cattle	dig prot %	.0	.0	
Sheep	dig prot %	-1.4	-1.6	
Energy				
Cattle	DE kcal/kg	1215.	1367.	
Sheep	DE kcal/kg	1764.	1984.	
Cattle	ME kcal/kg	996.	1121.	
Sheep	ME kcal/kg	1446.	1627.	
Cattle	TDN %	28.	31.	
Sheep	TDN %	40.	45.	
Calcium	%	.25	.28	16
Chlorine	%	.21	.24	12
Magnesium	%	.07	.08	32
Phosphorus	%	.09	.10	28
Potassium	%	.86	.97	11
Sodium	%	.12	.13	20
Sulfur	%	.10	.11	22
Copper	mg/kg	3.6	4.0	10
Manganese	mg/kg	5.9	6.6	8

(1) dry forages and roughages
(2) pasture, range plants, and forages fed green

(3) silages
(4) energy feeds

Feed name or analyses		Mean		C.V.
		As fed	Dry	± %

Rye, aerial pt, fresh, immature, (2)

Ref no 2-04-013

Feed name or analyses		As fed	Dry	C.V. ± %
Dry matter	%	16.7	100.0	19
Ash	%	2.2	12.9	20
Crude fiber	%	2.9	17.6	20
Ether extract	%	.8	4.8	21
N-free extract	%	5.6	33.5	
Protein (N x 6.25)	%	5.2	31.2	18
Cattle	dig prot %	4.2	25.0	
Sheep	dig prot %	4.4	26.1	
Energy				
Cattle	DE kcal/kg	523.	3130.	
Sheep	DE kcal/kg	515.	3086.	
Cattle	ME kcal/kg	429.	2567.	
Sheep	ME kcal/kg	422.	2530.	
Cattle	TDN %	12.	71.	
Sheep	TDN %	12.	70.	
Calcium	%	.10	.58	20
Magnesium	%	.05	.29	19
Phosphorus	%	.09	.56	15
Potassium	%	.57	3.40	11
Carotene	mg/kg	93.4	559.0	28
Vitamin A equiv	IU/g	155.7	931.8	

Rye, aerial pt, fresh, dough stage, (2)

Ref no 2-04-016

		As fed	Dry	
Dry matter	%	42.3	100.0	
Ash	%	2.2	5.2	
Crude fiber	%	10.5	24.8	
Ether extract	%	.9	2.1	
N-free extract	%	24.2	57.3	
Protein (N x 6.25)	%	4.5	10.6	
Cattle	dig prot %	3.6	8.5	
Sheep	dig prot %	2.9	6.9	
Energy				
Cattle	DE kcal/kg	1343.	3175.	
Sheep	DE kcal/kg	1287.	3042.	
Cattle	ME kcal/kg	1101.	2604.	
Sheep	ME kcal/kg	1055.	2494.	
Cattle	TDN %	30.	72.	
Sheep	TDN %	29.	69.	
Calcium	%	.12	.28	
Phosphorus	%	.13	.30	

Rye, aerial pt, fresh, mature, (2)

Ref no 2-04-017

		As fed	Dry	
Dry matter	%	45.0	100.0	
Ash	%	2.2	5.0	
Crude fiber	%	19.7	43.7	
Ether extract	%	1.2	2.7	
N-free extract	%	19.0	42.3	
Protein (N x 6.25)	%	2.8	6.3	
Cattle	dig prot %	1.4	3.2	
Sheep	dig prot %	1.3	2.9	
Energy				
Cattle	DE kcal/kg	1349.	2998.	
Sheep	DE kcal/kg	1250.	2778.	
Cattle	ME kcal/kg	1106.	2458.	
Sheep	ME kcal/kg	1025.	2278.	
Cattle	TDN %	31.	68.	
Sheep	TDN %	28.	63.	

Rye, aerial pt, fresh, (2)

Ref no 2-04-018

		As fed	Dry	C.V.
Dry matter	%	20.3	100.0	35
Ash	%	2.2	11.0	33
Crude fiber	%	4.9	24.1	28
Ether extract	%	.9	4.2	24
N-free extract	%	7.6	37.5	
Protein (N x 6.25)	%	4.7	23.2	32
Cattle	dig prot %	3.8	18.6	
Sheep	dig prot %	3.8	18.6	
Cellulose	%	4.6	22.5	
Energy				
Cattle	DE kcal/kg	644.	3175.	
Sheep	DE kcal/kg	573.	2822.	
Cattle	ME kcal/kg	529.	2604.	
Sheep	ME kcal/kg	470.	2314.	
Cattle	TDN %	15.	72.	
Sheep	TDN %	13.	64.	
Calcium	%	.11	.55	24
Magnesium	%	.06	.31	16
Phosphorus	%	.09	.45	26
Potassium	%	.69	3.40	11
Carotene	mg/kg	69.5	342.6	49
Vitamin A equiv	IU/g	115.8	571.1	

(5) protein supplements

(6) minerals

(7) vitamins

(8) additives

Feed name or analyses		Mean		C.V.
		As fed	Dry	± %

Rye, aerial pt, ensiled, (3)

Ref no 3-04-020

		As fed	Dry	C.V. ± %
Dry matter	%	27.6	100.0	10
Ash	%	2.2	7.9	6
Crude fiber	%	10.2	36.8	7
Ether extract	%	.9	3.3	9
N-free extract	%	11.5	41.5	
Protein (N x 6.25)	%	2.9	10.5	12
Cattle	dig prot %	1.6	5.8	
Sheep	dig prot %	1.6	5.8	
Energy				
Cattle	DE kcal/kg	681.	2469.	
Sheep	DE kcal/kg	754.	2734.	
Cattle	ME kcal/kg	559.	2024.	
Sheep	ME kcal/kg	619.	2242.	
Cattle	TDN %	15.	56.	
Sheep	TDN %	17.	62.	

Rye, bran, dry-mil dehy, (4)
Rye bran (CFA)

Ref no 4-04-022

		As fed	Dry	
Dry matter	%	90.0	100.0	
Ash	%	2.8	3.1	
Crude fiber	%	3.0	3.3	
Ether extract	%	2.0	2.2	
N-free extract	%	65.8	73.1	
Protein (N x 6.25)	%	16.5	18.3	
Cattle	dig prot %	13.9	15.4	
Sheep	dig prot %	11.1	12.3	
Swine	dig prot %	12.5	13.9	
Energy				
Cattle	DE kcal/kg	3452.	3836.	
Sheep	DE kcal/kg	2500.	2778.	
Swine	DE kcal/kg	2698.	2998.	
Cattle	ME kcal/kg	2831.	3146.	
Sheep	ME kcal/kg	2050.	2278.	
Swine	ME kcal/kg	2490.	2767.	
Cattle	TDN %	78.	87.	
Sheep	TDN %	57.	63.	
Swine	TDN %	61.	68.	
Calcium	%	.15	.17	
Phosphorus	%	1.30	1.44	
Manganese	mg/kg	11.9	13.2	
Niacin	mg/kg	27.9	31.0	
Pantothenic acid	mg/kg	16.9	18.8	
Riboflavin	mg/kg	.2	.2	
Thiamine	mg/kg	3.1	3.4	

Rye, flour by-prod, c-sift, mx 8.5 fbr, (4)
Rye middlings (AAFCO)

Ref no 4-04-031

		As fed	Dry	C.V. ± %
Dry matter	%	90.0	100.0	2
Ash	%	3.4	3.8	17
Crude fiber	%	6.0	6.7	22
Ether extract	%	3.1	3.4	14
N-free extract	%	60.4	67.1	
Protein (N x 6.25)	%	17.1	19.0	11
Cattle	dig prot %	14.4	16.0	
Sheep	dig prot %	12.8	14.2	
Swine	dig prot %	12.3	13.7	
Energy				
Cattle	DE kcal/kg	3452.	3836.	
Sheep	DE kcal/kg	3135.	3483.	
Swine	DE kcal/kg	2937.	3263.	
Cattle	ME kcal/kg	2831.	3146.	
Sheep	ME kcal/kg	2570.	2856.	
Swine	ME kcal/kg	2707.	3008.	
Cattle	TDN %	78.	87.	
Sheep	TDN %	71.	79.	
Swine	TDN %	67.	74.	
Calcium	%	.06	.07	
Phosphorus	%	.63	.70	1
Potassium	%	.63	.70	
Manganese	mg/kg	44.0	48.9	1
Niacin	mg/kg	16.9	18.8	2
Pantothenic acid	mg/kg	23.1	25.7	1
Riboflavin	mg/kg	2.4	2.7	7
Thiamine	mg/kg	3.3	3.7	3

Rye, flour by-prod mil-rn, mx 9.5 fbr, (4)
Rye mill run (AAFCO)

Ref no 4-04-034

		As fed	Dry	C.V. ± %
Dry matter	%	90.0	100.0	1
Ash	%	3.8	4.2	8
Crude fiber	%	5.0	5.6	11
Ether extract	%	3.4	3.8	10
N-free extract	%	60.6	67.3	
Protein (N x 6.25)	%	17.2	19.1	13
Cattle	dig prot %	13.4	14.9	
Sheep	dig prot %	12.7	14.1	
Swine	dig prot %	12.4	13.8	
Energy				
Cattle	DE kcal/kg	2937.	3263.	
Sheep	DE kcal/kg	2976.	3307.	
Swine	DE kcal/kg	2858.	3175.	
Cattle	ME kcal/kg	2408.	2676.	
Sheep	ME kcal/kg	2441.	2712.	
Swine	ME kcal/kg	2632.	2924.	

Continued

(1) dry forages and roughages

(2) pasture, range plants, and forages fed green

(3) silages

(4) energy feeds

TABLE A-3-1 Composition of Feeds (Continued)

Feed name or analyses		Mean		C.V.
		As fed	Dry	± %
Cattle	TDN %	67.	74.	
Sheep	TDN %	68.	75.	
Swine	TDN %	65.	72.	
Calcium	%	.07	.08	
Magnesium	%	.23	.26	
Phosphorus	%	.59	.66	
Potassium	%	.83	.92	

Rye, grain, Can 1 CW mn 58 wt mx 0 fm, (4)

Ref no 4-04-036

		As fed	Dry	C.V. ± %
Dry matter	%	86.5	100.0	
Ash	%	1.5	1.8	8
Crude fiber	%	2.6	3.0	51
Ether extract	%	1.2	1.4	37
N-free extract	%	69.6	80.5	
Protein (N x 6.25)	%	11.5	13.3	8
Cattle	dig prot %	9.1	10.5	
Sheep	dig prot %	9.1	10.5	
Swine	dig prot %	9.3	10.8	
Energy				
Cattle	DE kcal/kg	3204.	3704.	
Sheep	DE kcal/kg	3204.	3704.	
Swine	DE kcal/kg	3356.	3880.	
Cattle	ME kcal/kg	2627.	3037.	
Sheep	ME kcal/kg	2627.	3037.	
Swine	ME kcal/kg	3131.	3620.	
Cattle	TDN %	73.	84.	
Sheep	TDN %	73.	84.	
Swine	TDN %	76.	88.	

Rye, grain, Can 2 CE, (4)

Ref no 4-04-037

		As fed	Dry	
Dry matter	%	86.5	100.0	
Ash	%	1.7	2.0	
Crude fiber	%	3.1	3.6	
Ether extract	%	1.1	1.3	
N-free extract	%	70.2	81.1	
Protein (N x 6.25)	%	10.4	12.0	
Cattle	dig prot %	8.2	9.5	
Sheep	dig prot %	8.2	9.5	
Swine	dig prot %	8.4	9.7	
Energy				
Cattle	DE kcal/kg	3166.	3660.	
Sheep	DE kcal/kg	3166.	3660.	
Swine	DE kcal/kg	3356.	3880.	
Cattle	ME kcal/kg	2596.	3001.	
Sheep	ME kcal/kg	2596.	3001.	
Swine	ME kcal/kg	3142.	3632.	

Continued

Feed name or analyses		Mean		C.V.
		As fed	Dry	± %
Cattle	TDN %	72.	83.	
Sheep	TDN %	72.	83.	
Swine	TDN %	76.	88.	

Rye, grain, Can 2 CW mn 56 wt mx 2 fm, (4)

Ref no 4-04-038

		As fed	Dry	C.V.
Dry matter	%	86.5	100.0	
Ash	%	1.6	1.8	21
Crude fiber	%	2.2	2.6	99
Ether extract	%	1.5	1.8	66
N-free extract	%	69.3	80.1	
Protein (N x 6.25)	%	11.8	13.7	17
Cattle	dig prot %	9.3	10.8	
Sheep	dig prot %	9.3	10.8	
Swine	dig prot %	9.6	11.1	
Energy				
Cattle	DE kcal/kg	3204.	3704.	
Sheep	DE kcal/kg	3204.	3704.	
Swine	DE kcal/kg	3356.	3880.	
Cattle	ME kcal/kg	2627.	3037.	
Sheep	ME kcal/kg	2627.	3037.	
Swine	ME kcal/kg	3128.	3616.	
Cattle	TDN %	73.	84.	
Sheep	TDN %	73.	84.	
Swine	TDN %	76.	88.	

Rye, grain, Can 3 CE, (4)

Ref no 4-04-039

		As fed	Dry	
Dry matter	%	86.5	100.0	
Ash	%	1.8	2.1	
Crude fiber	%	2.5	2.9	
Ether extract	%	1.1	1.3	
N-free extract	%	70.4	81.4	
Protein (N x 6.25)	%	10.6	12.3	
Cattle	dig prot %	8.4	9.7	
Sheep	dig prot %	8.4	9.7	
Swine	dig prot %	8.6	10.0	
Energy				
Cattle	DE kcal/kg	3204.	3704.	
Sheep	DE kcal/kg	3204.	3704.	
Swine	DE kcal/kg	3356.	3880.	
Cattle	ME kcal/kg	2627.	3037.	
Sheep	ME kcal/kg	2627.	3037.	
Swine	ME kcal/kg	3138.	3628.	
Cattle	TDN %	73.	84.	
Sheep	TDN %	73.	84.	
Swine	TDN %	76.	88.	

(5) protein supplements

(6) minerals

(7) vitamins

(8) additives

Feed name or analyses	Mean As fed	Dry	C.V. ± %

Rye, grain, Can 3 CW mn 54 wt mx 5 fm, (4)

Ref no 4-04-040

		As fed	Dry	C.V. ± %
Dry matter	%	86.5	100.0	
Ash	%	1.6	1.9	18
Crude fiber	%	2.3	2.7	99
Ether extract	%	1.6	1.8	62
N-free extract	%	68.9	79.7	
Protein (N x 6.25)	%	12.0	13.9	10
Cattle	dig prot %	9.5	11.0	
Sheep	dig prot %	9.5	11.0	
Swine	dig prot %	9.7	11.2	
Energy				
Cattle	DE kcal/kg	3204.	3704.	
Sheep	DE kcal/kg	3204.	3704.	
Swine	DE kcal/kg	3356.	3880.	
Cattle	ME kcal/kg	2627.	3037.	
Sheep	ME kcal/kg	2627.	3037.	
Swine	ME kcal/kg	3128.	3616.	
Cattle	TDN %	73.	84.	
Sheep	TDN %	73.	84.	
Swine	TDN %	76.	88.	

Rye, grain, Can 4 CW mx 0.3 ergot mn 10 fm, (4)

Ref no 4-04-041

		As fed	Dry	C.V. ± %
Dry matter	%	86.5	100.0	
Ash	%	1.6	1.9	16
Crude fiber	%	2.3	2.7	99
Ether extract	%	1.6	1.8	53
N-free extract	%	68.8	79.5	
Protein (N x 6.25)	%	12.2	14.1	17
Cattle	dig prot %	9.6	11.1	
Sheep	dig prot %	9.6	11.1	
Swine	dig prot %	9.9	11.4	
Energy				
Cattle	DE kcal/kg	3204.	3704.	
Sheep	DE kcal/kg	3204.	3704.	
Swine	DE kcal/kg	3356.	3880.	
Cattle	ME kcal/kg	2627.	3037.	
Sheep	ME kcal/kg	2627.	3037.	
Swine	ME kcal/kg	3128.	3616.	
Cattle	TDN %	73.	84.	
Sheep	TDN %	73.	84.	
Swine	TDN %	76.	88.	

Rye, grain, (4)

Ref no 4-04-047

		As fed	Dry	C.V. ± %
Dry matter	%	89.0	100.0	2
Ash	%	1.7	1.9	12
Crude fiber	%	2.0	2.2	42
Ether extract	%	1.6	1.8	22
N-free extract	%	71.8	80.7	
Protein (N x 6.25)	%	11.9	13.4	13
Cattle	dig prot %	9.4	10.6	
Sheep	dig prot %	9.4	10.6	
Swine	dig prot %	9.6	10.8	
Energy				
Cattle	DE kcal/kg	3336.	3748.	
Sheep	DE kcal/kg	3336.	3748.	
Swine	DE kcal/kg	3300.	3708.	
Cattle	ME kcal/kg	2735.	3073.	
Chickens	MEn kcal/kg	2888.	3245.	
Sheep	ME kcal/kg	2735.	3073.	
Swine	ME kcal/kg	3079.	3460.	
Cattle	TDN %	76.	85.	
Sheep	TDN %	76.	85.	
Swine	TDN %	75.	84.	
Calcium	%	.06	.07	37
Iron	%	.008	.009	23
Magnesium	%	.12	.13	19
Phosphorus	%	.34	.38	22
Potassium	%	.46	.52	16
Sodium	%	.02	.02	80
Copper	mg/kg	7.8	8.8	17
Manganese	mg/kg	66.9	75.2	46
Zinc	mg/kg	30.5	34.3	42
Biotin	mg/kg	.06	.07	21
Folic acid	mg/kg	.60	.70	66
Niacin	mg/kg	1.2	1.3	23
Pantothenic acid	mg/kg	6.9	7.7	37
Riboflavin	mg/kg	1.6	1.8	40
Thiamine	mg/kg	3.9	4.4	31
a-tocopherol	mg/kg	15.0	17.4	
Arginine	%	.53	.60	20
Cystine	%	.18	.20	
Histidine	%	.27	.30	29
Isoleucine	%	.53	.60	
Leucine	%	.71	.80	
Lysine	%	.45	.51	24
Methionine	%	.18	.20	15
Phenylalanine	%	.62	.70	19
Threonine	%	.36	.40	20
Tryptophan	%	.09	.10	31
Tyrosine	%	.27	.30	
Valine	%	.62	.70	12

(1) dry forages and roughages
(2) pasture, range plants, and forages fed green

(3) silages
(4) energy feeds

Feed name or analyses		Mean		C.V.
		As fed	Dry	± %

Rye, distil grains, dehy, (5)
Rye distillers dried grains (AAFCO)
Rye distillers dried grains (CFA)

Ref no 5-04-023

Feed name or analyses		As fed	Dry	C.V. ± %
Dry matter	%	93.0	100.0	3
Ash	%	2.6	2.8	49
Crude fiber	%	14.0	15.1	20
Ether extract	%	6.4	6.9	24
N-free extract	%	47.5	51.1	
Protein (N x 6.25)	%	22.4	24.1	16
Cattle	dig prot %	9.7	10.4	
Sheep	dig prot %	13.5	14.5	
Energy				
Cattle	DE kcal/kg	1886.	2028.	
Sheep	DE kcal/kg	2665.	2866.	
Cattle	ME kcal/kg	1546.	1663.	
Sheep	ME kcal/kg	2186.	2350.	
Cattle	TDN %	43.	46.	
Sheep	TDN %	60.	65.	
Calcium	%	.13	.14	
Magnesium	%	.17	.18	
Phosphorus	%	.41	.44	
Potassium	%	.11	.12	
Sodium	%	.17	.18	
Manganese	mg/kg	18.5	19.9	
Niacin	mg/kg	16.9	18.2	9
Thiamine	mg/kg	1.3	1.4	50

Rye, distil grains w sol, dehy, mn 75 orig solids, (5)
Rye distillers dried grains with solubles (AAFCO)

Ref no 5-04-024

		As fed	Dry	C.V. ± %
Dry matter	%	91.0	100.0	3
Ash	%	6.4	7.0	14
Crude fiber	%	8.0	8.8	51
Ether extract	%	4.1	4.5	32
N-free extract	%	45.3	49.8	
Protein (N x 6.25)	%	27.2	29.9	5
Cattle	dig prot %	11.6	12.8	
Sheep	dig prot %	16.3	17.9	
Energy				
Cattle	DE kcal/kg	1765.	1940.	
Sheep	DE kcal/kg	2408.	2646.	
Cattle	ME kcal/kg	1448.	1591.	
Sheep	ME kcal/kg	1975.	2170.	
Cattle	TDN %	40.	44.	
Sheep	TDN %	55.	60.	
Niacin	mg/kg	62.7	68.9	
Pantothenic acid	mg/kg	17.4	19.1	
Riboflavin	mg/kg	8.1	8.9	
Thiamine	mg/kg	3.1	3.4	

Continued

(5) protein supplements
(6) minerals

		As fed	Dry	C.V. ± %
Arginine	%	1.00	1.10	
Histidine	%	.70	.77	
Isoleucine	%	1.50	1.65	
Leucine	%	2.10	2.31	
Lysine	%	1.00	1.10	
Methionine	%	.40	.44	
Phenylalanine	%	1.30	1.43	
Threonine	%	1.10	1.21	
Tryptophan	%	.30	.33	
Tyrosine	%	.50	.55	
Valine	%	1.60	1.76	

Rye, distil sol, dehy, (5)
Rye distillers dried solubles (AAFCO)
Rye distillers dried solubles (CFA)

Ref no 5-04-026

		As fed	Dry	C.V. ± %
Dry matter	%	94.0	100.0	1
Ash	%	7.2	7.7	15
Crude fiber	%	3.0	3.2	52
Ether extract	%	1.2	1.3	63
N-free extract	%	47.5	50.5	
Protein (N x 6.25)	%	35.1	37.3	6
Cattle	dig prot %	28.5	30.3	
Sheep	dig prot %	29.4	31.3	
Energy				
Cattle	DE kcal/kg	3315.	3527.	
Sheep	DE kcal/kg	3564.	3792.	
Cattle	ME kcal/kg	2718.	2892.	
Sheep	ME kcal/kg	2922.	3109.	
Cattle	TDN %	75.	80.	
Sheep	TDN %	81.	86.	
Calcium	%	.35	.37	
Phosphorus	%	1.19	1.27	
Niacin	mg/kg	66.0	70.2	43
Pantothenic acid	mg/kg	28.6	30.4	23
Riboflavin	mg/kg	12.8	13.6	10
Thiamine	mg/kg	3.1	3.3	15
Arginine	%	1.00	1.06	
Histidine	%	.70	.74	
Isoleucine	%	1.80	1.92	
Leucine	%	1.80	1.92	
Lysine	%	.60	.64	
Methionine	%	.50	.53	
Phenylalanine	%	1.70	1.81	
Threonine	%	1.10	1.17	
Tryptophan	%	.20	.21	
Tyrosine	%	.60	.64	
Valine	%	1.90	2.02	

(7) vitamins
(8) additives

Feed name or analyses		Mean		C.V.
		As fed	Dry	± %

Feed name or analyses		Mean		C.V.
		As fed	Dry	± %

Rye mill run - see Rye, flour by-prod mil-rn

RYEGRASS. Lolium spp

Ryegrass, hay, s-c, immature, (1)

Ref no 1-04-055

		As fed	Dry	C.V.
Dry matter	%	87.2	100.0	4
Ash	%	10.6	12.1	10
Crude fiber	%	19.8	22.7	16
Ether extract	%	2.6	3.0	12
N-free extract	%	42.0	48.2	
Protein (N x 6.25)	%	12.2	14.0	17
Cattle	dig prot %	5.8	6.7	
Sheep	dig prot %	6.2	7.1	
Energy				
Cattle	DE kcal/kg	2191.	2513.	
Sheep	DE kcal/kg	2191.	2513.	
Cattle	ME kcal/kg	1797.	2061.	
Sheep	ME kcal/kg	1797.	2061.	
Cattle	TDN %	50.	57.	
Sheep	TDN %	50.	57.	
Carotene	mg/kg	236.1	270.8	
Vitamin A equiv	IU/g	393.6	451.4	

Ryegrass, hay, s-c, (1)

Ref no 1-04-057

		As fed	Dry	C.V.
Dry matter	%	87.6	100.0	3
Ash	%	7.4	8.4	19
Crude fiber	%	28.4	32.4	14
Ether extract	%	2.2	2.5	28
N-free extract	%	41.3	47.1	
Protein (N x 6.25)	%	8.4	9.6	24
Cattle	dig prot %	4.0	4.6	
Sheep	dig prot %	4.3	4.9	
Energy				
Cattle	DE kcal/kg	2318.	2646.	
Sheep	DE kcal/kg	2318.	2646.	
Cattle	ME kcal/kg	1901.	2170.	
Sheep	ME kcal/kg	1901.	2170.	
Cattle	TDN %	52.	60.	
Sheep	TDN %	52.	60.	
Calcium	%	.43	.49	24
Iron	%	.08	.10	
Magnesium	%	.30	.34	14
Phosphorus	%	.28	.32	24
Potassium	%	1.72	1.96	8
Cobalt	mg/kg	.11	.13	

Ryegrass, straw, (1)

Ref no 1-04-059

		As fed	Dry	C.V.
Dry matter	%	91.2	100.0	2
Ash	%	4.3	4.7	22
Crude fiber	%	36.1	39.6	6
Ether extract	%	1.1	1.2	16
N-free extract	%	46.1	50.5	
Protein (N x 6.25)	%	3.6	4.0	6
Cattle	dig prot %	.4	.4	
Sheep	dig prot %	.8	.9	
Energy				
Cattle	DE kcal/kg	1809.	1984.	
Sheep	DE kcal/kg	2372.	2601.	
Cattle	ME kcal/kg	1484.	1627.	
Sheep	ME kcal/kg	1945.	2133.	
Cattle	TDN %	41.	45.	
Sheep	TDN %	54.	59.	

Ryegrass, aerial pt, fresh, (2)

Ref no 2-04-062

		As fed	Dry	C.V.
Dry matter	%	24.1	100.0	18
Ash	%	3.2	13.2	32
Crude fiber	%	5.6	23.2	14
Ether extract	%	.9	3.9	22
N-free extract	%	10.5	43.5	
Protein (N x 6.25)	%	3.9	16.2	30
Cattle	dig prot %	2.4	10.0	
Sheep	dig prot %	2.9	12.1	
Energy				
Cattle	DE kcal/kg	669.	2778.	
Sheep	DE kcal/kg	669.	2778.	
Cattle	ME kcal/kg	549.	2278.	
Sheep	ME kcal/kg	549.	2278.	
Cattle	TDN %	15.	63.	
Sheep	TDN %	15.	63.	
Calcium	%	.16	.65	10
Magnesium	%	.08	.35	11
Phosphorus	%	.08	.35	25
Potassium	%	.48	2.00	4
Copper	mg/kg	1.1	4.4	44
Carotene	mg/kg	96.7	401.1	39
Vitamin A equiv	IU/g	161.2	668.6	

(1) dry forages and roughages
(2) pasture, range plants, and forages fed green

(3) silages
(4) energy feeds

TABLE A-3-1 Composition of Feeds (Continued)

Feed name or analyses		Mean		C.V.
		As fed	Dry	± %

RYEGRASS, ITALIAN. Lolium multiflorum

Ryegrass, Italian, hay, s-c, immature, (1)

Ref no 1-04-064

		As fed	Dry	C.V.
Dry matter	%	89.3	100.0	4
Ash	%	11.6	13.0	9
Crude fiber	%	17.6	19.7	12
Ether extract	%	2.9	3.2	13
N-free extract	%	43.7	48.9	
Protein (N x 6.25)	%	13.6	15.2	18
Cattle	dig prot %	9.0	10.1	
Sheep	dig prot %	7.6	8.5	
Energy				
Cattle	DE kcal/kg	2716.	3042.	
Sheep	DE kcal/kg	2599.	2910.	
Cattle	ME kcal/kg	2227.	2494.	
Sheep	ME kcal/kg	2131.	2386.	
Cattle	TDN %	62.	69.	
Sheep	TDN %	59.	66.	

Ryegrass, Italian, hay, s-c, full blm, (1)

Ref no 1-04-067

		As fed	Dry	C.V.
Dry matter	%	91.5	100.0	2
Ash	%	7.1	7.7	11
Crude fiber	%	30.7	33.6	3
Ether extract	%	1.7	1.9	19
N-free extract	%	44.7	48.8	
Protein (N x 6.25)	%	7.4	8.0	9
Cattle	dig prot %	3.6	3.9	
Sheep	dig prot %	1.9	2.1	
Energy				
Cattle	DE kcal/kg	2299.	2513.	
Sheep	DE kcal/kg	2098.	2293.	
Cattle	ME kcal/kg	1886.	2061.	
Sheep	ME kcal/kg	1720.	1880.	
Cattle	TDN %	52.	57.	
Sheep	TDN %	48.	52.	

Ryegrass, Italian, hay, s-c, mature, (1)

Ref no 1-04-068

		As fed	Dry	C.V.
Dry matter	%	83.2	100.0	1
Ash	%	6.6	7.9	8
Crude fiber	%	30.6	36.8	5
Ether extract	%	.9	1.1	16
N-free extract	%	40.4	48.5	
Protein (N x 6.25)	%	4.7	5.7	7

Continued

Feed name or analyses		Mean		C.V.
		As fed	Dry	± %
Cattle	dig prot %	1.6	1.9	
Sheep	dig prot %	2.0	2.4	
Energy				
Cattle	DE kcal/kg	1908.	2293.	
Sheep	DE kcal/kg	2201.	2646.	
Cattle	ME kcal/kg	1564.	1880.	
Sheep	ME kcal/kg	1805.	2170.	
Cattle	TDN %	43.	52.	
Sheep	TDN %	50.	60.	

Ryegrass, Italian, hay, s-c, (1)

Ref no 1-04-069

		As fed	Dry	C.V.
Dry matter	%	88.2	100.0	3
Ash	%	7.3	8.3	19
Crude fiber	%	28.8	32.6	15
Ether extract	%	1.6	1.8	32
N-free extract	%	43.5	49.3	
Protein (N x 6.25)	%	7.1	8.0	32
Cattle	dig prot %	3.4	3.9	
Sheep	dig prot %	3.0	3.4	
Energy				
Cattle	DE kcal/kg	2294.	2601.	
Sheep	DE kcal/kg	2294.	2601.	
Cattle	ME kcal/kg	1881.	2133.	
Sheep	ME kcal/kg	1881.	2133.	
Cattle	TDN %	52.	59.	
Sheep	TDN %	52.	59.	
Calcium	%	.55	.62	9
Magnesium	%	.28	.32	12
Phosphorus	%	.36	.41	9
Potassium	%	1.76	1.99	4

Ryegrass, Italian, aerial pt, fresh, (2)

Ref no 2-04-073

		As fed	Dry	C.V.
Dry matter	%	24.3	100.0	15
Ash	%	3.8	15.7	26
Crude fiber	%	5.3	21.8	12
Ether extract	%	1.0	4.2	22
N-free extract	%	10.2	42.0	
Protein (N x 6.25)	%	4.0	16.3	25
Cattle	dig prot %	2.4	10.1	
Sheep	dig prot %	1.7	7.0	
Energy				
Cattle	DE kcal/kg	664.	2734.	
Sheep	DE kcal/kg	546.	2249.	
Cattle	ME kcal/kg	545.	2242.	
Sheep	ME kcal/kg	448.	1844.	
Cattle	TDN %	15.	62.	
Sheep	TDN %	12.	51.	
Calcium	%	.16	.64	8

Continued

(5) protein supplements
(6) minerals

(7) vitamins
(8) additives

Feed name or analyses		Mean		C.V.
		As fed	Dry	± %
Magnesium	%	.08	.35	12
Phosphorus	%	.10	.41	16
Potassium	%	.49	2.00	4

RYEGRASS, PERENNIAL. Lolium perenne

Ryegrass, perennial, hay, s-c, (1)

Ref no 1-04-077

Dry matter		%	87.0	100.0	2
Ash		%	8.2	9.4	11
Crude fiber		%	25.3	29.1	6
Ether extract		%	3.0	3.4	20
N-free extract		%	40.9	47.0	
Protein (N x 6.25)		%	9.7	11.1	16
Cattle	dig prot	%	5.6	6.5	
Sheep	dig prot	%	5.0	5.7	
Energy					
Cattle	DE kcal/kg		2570.	2954.	
Sheep	DE kcal/kg		2263.	2601.	
Cattle	ME kcal/kg		2107.	2422.	
Sheep	ME kcal/kg		1856.	2133.	
Cattle	TDN %		58.	67.	
Sheep	TDN %		51.	59.	
Calcium		%	.35	.40	30
Magnesium		%	.31	.36	13
Phosphorus		%	.23	.26	32
Potassium		%	1.66	1.91	4
Manganese		mg/kg	39.7	45.6	

SAFFLOWER. Carthamus tinctorius

Safflower, seed, (4)

Ref no 4-07-958

Dry matter		%	93.1	100.0	
Ash		%	2.9	3.1	
Crude fiber		%	26.6	28.6	
Ether extract		%	29.8	32.0	
N-free extract		%	17.5	18.8	
Protein (N x 6.25)		%	16.3	17.5	
Cattle	dig prot	%	13.0	14.0	
Sheep	dig prot	%	12.9	13.9	
Energy					
Cattle	DE kcal/kg		3653.	3924.	
Sheep	DE kcal/kg		3612.	3880.	
Cattle	ME kcal/kg		2996.	3218.	
Sheep	ME kcal/kg		2962.	3182.	
Cattle	TDN %		83.	89.	
Sheep	TDN %		82.	88.	

Safflower, seed, mech-extd grnd, (5)
Whole pressed safflower seed (AAFCO)
Safflower oil meal, expeller extracted
Safflower oil meal, hydraulic extracted

Ref no 5-04-109

Dry matter		%	91.0	100.0	1
Ash		%	3.7	4.1	9
Crude fiber		%	31.0	34.1	6
Ether extract		%	6.0	6.6	13
N-free extract		%	30.5	33.5	
Protein (N x 6.25)		%	19.7	21.7	2
Cattle	dig prot	%	14.5	15.9	
Sheep	dig prot	%	17.0	18.7	
Energy					
Cattle	DE kcal/kg		2287.	2513.	
Sheep	DE kcal/kg		1886.	2072.	
Cattle	ME kcal/kg		1876.	2061.	
Sheep	ME kcal/kg		1546.	1699.	
Cattle	TDN %		52.	57.	
Sheep	TDN %		43.	47.	
Calcium		%	.23	.25	
Iron		%	.05	.05	
Magnesium		%	.33	.36	
Phosphorus		%	.71	.78	
Potassium		%	.72	.79	
Sodium		%	.05	.05	
Copper		mg/kg	9.7	10.7	
Manganese		mg/kg	17.8	19.6	
Zinc		mg/kg	39.8	43.7	
Biotin		mg/kg	1.4	1.5	
Folic acid		mg/kg	.44	.48	
Niacin		mg/kg	85.8	94.3	
Pantothenic acid		mg/kg	4.0	4.4	
Riboflavin		mg/kg	18.0	19.8	
Arginine		%	1.20	1.32	
Cystine		%	.80	.88	
Lysine		%	.70	.77	
Methionine		%	.40	.44	
Tryptophan		%	.30	.33	

Safflower, seed, solv-extd grnd, (5)
Solvent extracted whole pressed safflower seed (AAFCO)

Ref no 5-04-110

Dry matter	%	91.8	100.0
Ash	%	4.7	5.1
Crude fiber	%	32.3	35.2
Ether extract	%	3.9	4.2
N-free extract	%	29.6	32.2
Protein (N x 6.25)	%	21.4	23.3

Continued

(1) dry forages and roughages

(2) pasture, range plants, and forages fed green

(3) silages

(4) energy feeds

Feed name or analyses		Mean		C.V.
		As fed	Dry	± %
Cattle	dig prot %	17.2	18.7	
Sheep	dig prot %	17.2	18.7	
Energy				
Cattle	DE kcal/kg	2226.	2425.	
Sheep	DE kcal/kg	2186.	2381.	
Cattle	ME kcal/kg	1825.	1988.	
Sheep	ME kcal/kg	1792.	1952.	
Cattle	TDN %	50.	55.	
Sheep	TDN %	50.	54.	
Calcium	%	.34	.37	
Phosphorus	%	.84	.92	

Safflower, seed wo hulls, solv-extd grnd, (5)
Safflower meal without hulls

Ref no 5-07-959

		As fed	Dry	
Dry matter	%	90.5	100.0	
Ash	%	6.4	7.1	
Crude fiber	%	8.5	9.4	
Ether extract	%	1.7	1.9	
N-free extract	%	29.4	32.5	
Protein (N x 6.25)	%	44.5	49.1	
Cattle	dig prot %	37.4	41.3	
Sheep	dig prot %	37.4	41.3	
Energy				
Cattle	DE kcal/kg	3033.	3351.	
Sheep	DE kcal/kg	3072.	3395.	
Cattle	ME kcal/kg	2487.	2748.	
Sheep	ME kcal/kg	2520.	2784.	
Cattle	NE$_m$ kcal/kg	1412.	1560.	
Cattle	NE$_{gain}$ kcal/kg	896.	990.	
Cattle	TDN %	69.	76.	
Sheep	TDN %	70.	77.	
Calcium	%	.24	.26	
Phosphorus	%	1.66	1.83	

SAGE, BLACK. Salvia mellifera

Sage, black, browse, fresh, dormant, (2)

Ref no 2-05-564

		As fed	Dry	
Dry matter	%	76.0	100.0	
Ash	%	4.7	6.2	
Ether extract	%	7.1	9.4	
Protein (N x 6.25)	%	6.5	8.5	
Sheep	dig prot %	3.3	4.4	
Cellulose	%	16.7	22.0	
Lignin	%	12.2	16.0	
Energy	GE kcal/kg	3847.	5062.	
Sheep	DE kcal/kg	1582.	2081.	
Sheep	ME kcal/kg	854.	1124.	

Feed name or analyses		Mean		C.V.
		As fed	Dry	± %
Sheep	TDN %	36.	47.	
Calcium	%	.46	.60	
Phosphorus	%	.12	.16	
Carotene	mg/kg	13.4	17.6	

SAGEBRUSH. Artemisia spp

Sagebrush, browse, fresh, (2)

Ref no 2-04-116

		As fed	Dry	
Dry matter	%	49.7	100.0	20
Ash	%	3.8	7.7	57
Crude fiber	%	12.5	25.2	14
Ether extract	%	4.4	8.8	44
N-free extract	%	23.5	47.2	
Protein (N x 6.25)	%	5.5	11.1	29
Cattle	dig prot %	3.6	7.3	
Sheep	dig prot %	3.6	7.3	
Energy				
Cattle	DE kcal/kg	1249.	2513.	
Sheep	DE kcal/kg	1446.	2910.	
Cattle	ME kcal/kg	1024.	2061.	
Sheep	ME kcal/kg	1186.	2386.	
Cattle	TDN %	28.	57.	
Sheep	TDN %	33.	66.	
Calcium	%	.38	.76	44
Magnesium	%	.16	.33	34
Phosphorus	%	.11	.22	70
Cobalt	mg/kg	.10	.20	
Copper	mg/kg	6.7	13.4	
Manganese	mg/kg	15.2	30.6	

SAGEBRUSH, BIG. Artemisia tridentata

Sagebrush, big, browse, fresh, dormant, (2)

Ref no 2-07-992

		As fed	Dry	
Dry matter	%	76.0	100.0	
Ash	%	4.6	6.1	
Ether extract	%	7.7	10.1	
Protein (N x 6.25)	%	7.1	9.4	
Sheep	dig prot %	4.1	5.4	
Cellulose	%	16.0	21.0	
Lignin	%	12.2	16.0	
Energy	GE kcal/kg	3877.	5101.	
Sheep	DE kcal/kg	1751.	2304.	
Sheep	ME kcal/kg	964.	1268.	
Sheep	TDN %	39.	51.	

Continued

(5) protein supplements
(6) minerals

(7) vitamins
(8) additives

Feed name or analyses		Mean		C.V. ± %
		As fed	Dry	
Calcium	%	.51	.67	
Phosphorus	%	.14	.18	
Carotene	mg/kg	12.2	16.1	

SAGEBRUSH, BUD. Artemisia spinescens

Sagebrush, bud, browse, fresh, early leaf, (2)

Ref no 2-07-991

		As fed	Dry	
Dry matter	%	25.0	100.0	
Ash	%	5.4	21.4	
Ether extract	%	1.2	4.9	
Protein (N x 6.25)	%	4.3	17.3	
Sheep	dig prot %	3.4	13.7	
Cellulose	%	4.5	18.0	
Lignin	%	2.0	8.0	
Energy	GE kcal/kg	1060.	4239.	
Sheep	DE kcal/kg	639.	2557.	
Sheep	ME kcal/kg	502.	2008.	
Sheep	TDN %	13.	51.	
Calcium	%	.24	.97	
Phosphorus	%	.08	.33	
Carotene	mg/kg	6.0	23.8	

Sagebrush, bud, browse, fresh, immature, (2)

Ref no 2-04-124

		As fed	Dry	C.V. ± %
Dry matter	%	28.9	100.0	
Ash	%	5.3	18.5	35
Crude fiber	%	6.8	23.4	
Ether extract	%	.9	3.1	51
N-free extract	%	11.7	40.6	
Protein (N x 6.25)	%	4.2	14.4	22
Cattle	dig prot %	2.9	10.1	
Sheep	dig prot %	3.0	10.4	
Energy				
Cattle	DE kcal/kg	675.	2337.	
Sheep	DE kcal/kg	803.	2778.	
Cattle	ME kcal/kg	554.	1916.	
Sheep	ME kcal/kg	658.	2278.	
Cattle	TDN %	15.	53.	
Sheep	TDN %	18.	63.	
Calcium	%	.31	1.06	35
Magnesium	%	.14	.49	
Phosphorus	%	.08	.29	73

SAGEBRUSH, FRINGED. Artemisia frigida

Sagebrush, fringed, browse, fresh, mid-blm, (2)

Ref no 2-04-129

		As fed	Dry	C.V. ± %
Dry matter	%	43.2	100.0	
Ash	%	2.8	6.4	
Crude fiber	%	11.4	26.5	
Ether extract	%	2.2	5.2	
N-free extract	%	23.1	53.4	
Protein (N x 6.25)	%	3.7	8.5	
Cattle	dig prot %	2.2	5.1	
Sheep	dig prot %	2.1	4.9	
Energy				
Cattle	DE kcal/kg	1181.	2734.	
Sheep	DE kcal/kg	1314.	3042.	
Cattle	ME kcal/kg	968.	2242.	
Sheep	ME kcal/kg	1077.	2494.	
Cattle	TDN %	27.	62.	
Sheep	TDN %	30.	69.	

Sagebrush, fringed, browse, fresh, mature, (2)

Ref no 2-04-130

		As fed	Dry	
Dry matter	%	50.0	100.0	
Ash	%	6.4	12.8	
Crude fiber	%	15.8	31.7	
Ether extract	%	1.8	3.5	
N-free extract	%	23.0	46.0	
Protein (N x 6.25)	%	3.0	6.0	
Cattle	dig prot %	1.5	3.0	
Sheep	dig prot %	1.3	2.6	
Energy				
Cattle	DE kcal/kg	1168.	2337.	
Sheep	DE kcal/kg	1477.	2954.	
Cattle	ME kcal/kg	958.	1916.	
Sheep	ME kcal/kg	1211.	2422.	
Cattle	TDN %	26.	53.	
Sheep	TDN %	34.	67.	

SAINFOIN. Onobrychis spp

Sainfoin, hay, s-c, early blm, (1)

Ref no 1-04-141

		As fed	Dry	
Dry matter	%	88.4	100.0	
Ash	%	5.7	6.4	
Crude fiber	%	24.9	28.2	

Continued

(1) dry forages and roughages

(2) pasture, range plants, and forages fed green

(3) silages

(4) energy feeds

Feed name or analyses		Mean		C.V
		As fed	Dry	± %
Ether extract	%	3.4	3.8	
N-free extract	%	34.4	39.0	
Protein (N x 6.25)	%	20.0	22.6	
Cattle	dig prot %	14.6	16.5	
Sheep	dig prot %	14.0	15.8	
Energy				
Cattle	DE kcal/kg	2260.	2557.	
Sheep	DE kcal/kg	2339.	2646.	
Cattle	ME kcal/kg	1854.	2097.	
Sheep	ME kcal/kg	1918.	2170.	
Cattle	TDN %	51.	58.	
Sheep	TDN %	53.	60.	

Sainfoin, hay, s-c, mid-blm, (1)

Ref no 1-04-142

Feed name or analyses		As fed	Dry	
Dry matter	%	87.0	100.0	
Ash	%	6.1	7.0	
Crude fiber	%	28.2	32.4	
Ether extract	%	4.3	4.9	
N-free extract	%	30.4	35.0	
Protein (N x 6.25)	%	18.0	20.7	
Cattle	dig prot %	13.0	14.9	
Sheep	dig prot %	12.6	14.5	
Energy				
Cattle	DE kcal/kg	2186.	2513.	
Sheep	DE kcal/kg	2263.	2601.	
Cattle	ME kcal/kg	1793.	2061.	
Sheep	ME kcal/kg	1856.	2133.	
Cattle	TDN %	50.	57.	
Sheep	TDN %	51.	59.	

Sainfoin, hay, s-c, (1)

Ref no 1-04-143

Feed name or analyses		As fed	Dry	C.V
Dry matter	%	85.7	100.0	2
Ash	%	6.5	7.6	11
Crude fiber	%	22.7	26.5	1
Ether extract	%	3.2	3.7	19
N-free extract	%	39.1	45.6	
Protein (N x 6.25)	%	14.2	16.6	22
Cattle	dig prot %	9.7	11.3	
Sheep	dig prot %	9.9	11.6	
Energy				
Cattle	DE kcal/kg	2268.	2646.	
Sheep	DE kcal/kg	2268.	2646.	
Cattle	ME kcal/kg	1860.	2170.	
Sheep	ME kcal/kg	1860.	2170.	
Cattle	TDN %	51.	60.	
Sheep	TDN %	51.	60.	

(5) protein supplements
(6) minerals

SALTBUSH, NUTTALL. Atriplex nuttallii

Saltbush, nuttall, browse, fresh, dormant, (2)

Ref no 2-07-993

Feed name or analyses		Mean		C.V
		As fed	Dry	± %
Dry matter	%	75.0	100.0	
Ash	%	16.1	21.5	
Ether extract	%	1.6	2.2	
Protein (N x 6.25)	%	5.4	7.2	
Sheep	dig prot %	2.6	3.4	
Cellulose	%	14.2	19.0	
Lignin	%	7.5	10.0	
Energy	GE kcal/kg	2771.	3695.	
Sheep	DE kcal/kg	1119.	1492.	
Sheep	ME kcal/kg	990.	1320.	
Sheep	TDN %	27.	36.	
Calcium	%	1.66	2.21	
Phosphorus	%	.16	.21	
Carotene	mg/kg	14.2	19.0	

SALTBUSH, SHADSCALE. Atriplex confertifolia

Saltbush, shadscale, browse, fresh, dormant, (2)

Ref no 2-05-565

Feed name or analyses		As fed	Dry	
Dry matter	%	80.0	100.0	
Ash	%	18.7	23.4	
Ether extract	%	1.9	2.4	
Protein (N x 6.25)	%	6.2	7.7	
Sheep	dig prot %	3.4	4.3	
Cellulose	%	14.4	18.0	
Lignin	%	10.4	13.0	
Energy	GE kcal/kg	2906.	3633.	
Sheep	DE kcal/kg	1006.	1257.	
Sheep	ME kcal/kg	704.	880.	
Sheep	TDN %	25.	31.	
Calcium	%	2.02	2.53	
Phosphorus	%	.07	.09	
Carotene	mg/kg	15.7	19.6	

SALTGRASS. Distichlis spp

Saltgrass, hay, s-c, (1)

Ref no 1-04-168

Feed name or analyses		As fed	Dry	C.V
Dry matter	%	88.4	100.0	5
Ash	%	10.9	12.3	16
Crude fiber	%	27.6	31.2	7

Continued

(7) vitamins
(8) additives

Feed name or analyses		Mean		C.V.
		As fed	Dry	± %
Ether extract	%	1.9	2.1	14
N-free extract	%	40.3	45.6	
Protein (N x 6.25)	%	7.8	8.8	5
Cattle	dig prot %	4.1	4.6	
Sheep	dig prot %	4.0	4.5	
Energy				
Cattle	DE kcal/kg	2572.	2910.	
Sheep	DE kcal/kg	1949.	2205.	
Cattle	ME kcal/kg	2109.	2386.	
Sheep	ME kcal/kg	1598.	1808.	
Cattle	TDN %	58.	66.	
Sheep	TDN %	44.	50.	

Saltgrass, aerial pt, fresh, over ripe, (2)

Ref no 2-04-169

Dry matter	%	74.4	100.0	
Ash	%	5.4	7.3	34
Crude fiber	%	26.0	34.9	5
Ether extract	%	1.9	2.6	13
N-free extract	%	37.9	51.0	
Protein (N x 6.25)	%	3.1	4.2	22
Cattle	dig prot %	1.1	1.5	
Sheep	dig prot %	.7	.9	
Energy				
Cattle	DE kcal/kg	2001.	2690.	
Sheep	DE kcal/kg	2230.	2998.	
Cattle	ME kcal/kg	1641.	2206.	
Sheep	ME kcal/kg	1829.	2458.	
Cattle	TDN %	45.	61.	
Sheep	TDN %	50.	68.	
Calcium	%	.17	.23	19
Magnesium	%	.22	.30	
Phosphorus	%	.05	.07	32

Saltgrass, aerial pt, fresh, (2)

Ref no 2-04-170

Dry matter	%	74.4	100.0	
Ash	%	5.6	7.5	22
Crude fiber	%	22.5	30.3	10
Ether extract	%	1.3	1.8	18
N-free extract	%	40.1	53.9	
Protein (N x 6.25)	%	4.8	6.5	30
Cattle	dig prot %	2.5	3.4	
Sheep	dig prot %	2.2	3.0	
Energy				
Cattle	DE kcal/kg	2067.	2778.	
Sheep	DE kcal/kg	2263.	3042.	
Cattle	ME kcal/kg	1695.	2278.	
Sheep	ME kcal/kg	1856.	2494.	
Cattle	TDN %	47.	63.	

Continued

Feed name or analyses		Mean		C.V.
		As fed	Dry	± %
Sheep	TDN %	51.	69.	
Calcium	%	.16	.21	59
Iron	%	.01	.02	
Magnesium	%	.21	.28	
Phosphorus	%	.07	.09	35
Potassium	%	.18	.24	
Manganese	mg/kg	115.2	154.8	

SALTGRASS, DESERT. Distichlis stricta

Saltgrass, desert, aerial pt, fresh, (2)

Ref no 2-04-171

Dry matter	%	75.0	100.0	
Ash	%	5.1	6.8	9
Crude fiber	%	22.3	29.7	1
Ether extract	%	1.3	1.7	5
N-free extract	%	41.9	55.9	
Protein (N x 6.25)	%	4.4	5.9	9
Cattle	dig prot %	2.2	2.9	
Sheep	dig prot %	1.9	2.5	
Energy				
Cattle	DE kcal/kg	2116.	2822.	
Sheep	DE kcal/kg	2314.	3086.	
Cattle	ME kcal/kg	1736.	2314.	
Sheep	ME kcal/kg	1898.	2530.	
Cattle	TDN %	48.	64.	
Sheep	TDN %	52.	70.	
Calcium	%	.12	.16	24
Phosphorus	%	.07	.09	25

SALTGRASS, SEASHORE. Distichlis spicata

Saltgrass, seashore, aerial pt, fresh, (2)

Ref no 2-04-175

Dry matter	%	74.4	100.0	
Ash	%	6.9	9.3	3
Crude fiber	%	22.2	29.8	9
Ether extract	%	1.3	1.7	4
N-free extract	%	38.9	52.3	
Protein (N x 6.25)	%	5.1	6.9	34
Cattle	dig prot %	2.8	3.8	
Sheep	dig prot %	2.5	3.4	
Energy				
Cattle	DE kcal/kg	2001.	2690.	
Sheep	DE kcal/kg	2263.	3042.	
Cattle	ME kcal/kg	1641.	2206.	
Sheep	ME kcal/kg	1856.	2494.	
Cattle	TDN %	45.	61.	

Continued

(1) dry forages and roughages

(2) pasture, range plants, and forages fed green

(3) silages

(4) energy feeds

Feed name or analyses		Mean		C.V.
		As fed	Dry	± %
Sheep	TDN %	51.	69.	
Calcium	%	.25	.33	49
Iron	%	.01	.02	
Magnesium	%	.16	.22	
Phosphorus	%	.09	.12	43
Potassium	%	.18	.24	
Sodium	%	.16	.22	
Manganese	mg/kg	115.2	154.8	

Sand dropseed - see Dropseed, sand

Screenings - see Barley, grain screenings; see Flax,
  seed screenings; see Grains, screenings; see
  Wheat, grain screenings

Schrock - see Sorghum, schrock

SEDGE. Carex spp

  Sedge, hay, s-c, (1)

  Ref no 1-04-193

Dry matter	%	90.2	100.0	3
Ash	%	7.0	7.8	22
Crude fiber	%	27.9	30.9	6
Ether extract	%	2.3	2.5	33
N-free extract	%	43.7	48.5	
Protein (N x 6.25)	%	9.3	10.3	16
Cattle	dig prot %	4.7	5.2	
Sheep	dig prot %	4.7	5.2	
Energy				
Cattle	DE kcal/kg	1989.	2205.	
Sheep	DE kcal/kg	1989.	2205.	
Cattle	ME kcal/kg	1631.	1808.	
Sheep	ME kcal/kg	1631.	1808.	
Cattle	TDN %	45.	50.	
Sheep	TDN %	45.	50.	

SESAME. Sesamum indicum

  Sesame, seed, mech-extd grnd, (5)
    Sesame oil meal, expeller extracted

  Ref no 5-04-220

Dry matter	%	93.0	100.0	1
Ash	%	9.3	10.0	
Crude fiber	%	5.0	5.4	

Continued

Feed name or analyses		Mean		C.V.
		As fed	Dry	± %
Ether extract	%	5.1	5.5	41
N-free extract	%	25.7	27.6	
Protein (N x 6.25)	%	47.9	51.5	4
Cattle	dig prot %	38.3	41.2	
Sheep	dig prot %	43.6	46.9	
Swine	dig prot %	45.0	48.4	
Energy				
Cattle	DE kcal/kg	3076.	3307.	
Sheep	DE kcal/kg	3116.	3351.	
Swine	DE kcal/kg	3526.	3792.	
Cattle	ME kcal/kg	2522.	2712.	
Chickens	MEn kcal/kg	2646.	2845.	
Sheep	ME kcal/kg	2556.	2748.	
Swine	ME kcal/kg	3019.	3246.	
Cattle	TDN %	70.	75.	
Sheep	TDN %	71.	76.	
Swine	TDN %	80.	86.	
Calcium	%	2.03	2.18	
Phosphorus	%	1.29	1.39	
Manganese	mg/kg	48.0	51.6	
Choline	mg/kg	1533.	1648.	
Pantothenic acid	mg/kg	6.4	6.9	
Riboflavin	mg/kg	3.7	4.0	
Thiamine	mg/kg	2.9	3.1	

Shallu - see Sorghum, shallu

Shorts - see Wheat, flour by-prod, c-sift, mx 7 fbr

SHRIMP. Pandalus spp, Penaeus spp

  Shrimp, proc res, dehy grnd, salt declared above
    3 mx 7, (5)
    Shrimp meal (AAFCO)

  Ref no 5-04-226

Dry matter	%	90.0	100.0	
Ash	%	26.6	29.6	17
Crude fiber	%	11.0	12.2	22
Ether extract	%	3.1	3.4	43
N-free extract	%	1.9	2.1	
Protein (N x 6.25)	%	47.4	52.7	16
Energy				
Chickens	MEn kcal/kg	1190.	1322.	
Calcium	%	7.35	8.17	19
Iron	%	.01	.01	
Magnesium	%	.54	.60	
Phosphorus	%	1.59	1.77	14
Manganese	mg/kg	30.1	33.4	
Choline	mg/kg	5828.	6475.	
Riboflavin	mg/kg	4.0	4.4	

Feed name or analyses		Mean		C.V. ± %
		As fed	Dry	

Skimmed milk - see Cattle, milk, skim

Sorghum, aerial pt, s-c, mature, (1)
  Sorghum fodder, sun-cured, mature

Ref no 1-04-301

		As fed	Dry	C.V. ± %
Dry matter	%	81.7	100.0	9
Ash	%	7.1	8.7	21
Crude fiber	%	23.2	28.4	10
Ether extract	%	1.9	2.3	13
N-free extract	%	44.2	54.1	
Protein (N x 6.25)	%	5.3	6.5	14
Cattle	dig prot %	2.0	2.5	
Sheep	dig prot %	2.0	2.5	
Energy				
Cattle	DE kcal/kg	2089.	2557.	
Sheep	DE kcal/kg	2017.	2469.	
Cattle	ME kcal/kg	1713.	2097.	
Sheep	ME kcal/kg	1654.	2024.	
Cattle	TDN %	47.	58.	
Sheep	TDN %	46.	56.	

SODIUM PHOSPHATE, MONOBASIC

Sodium, phosphate, monobasic, $NaH_2PO_4 \cdot H_2O$,
  tech, (6)
    Monosodium phosphate (AAFCO)

Ref no 6-04-288

		As fed	Dry
Dry matter	%	96.7	100.0
Ash	%	96.7	100.0
Phosphorus	%	21.80	22.46
Sodium	%	32.3	33.4
		120.00	124.10

SODIUM TRIPOLYPHOSPHATE

Sodium, tripolyphosphate, comm, (6)
  Sodium tripolyphosphate (AAFCO)

Ref no 6-08-076

		As fed	Dry
Dry matter	%	96.0	100.0
Phosphorus	%	24.94	25.98

Sorghum, aerial pt, s-c, (1)
  Sorghum fodder, sun-cured

Ref no 1-07-960

		As fed	Dry
Dry matter	%	85.5	100.00
Ash	%	7.3	8.5
Crude fiber	%	22.3	26.1
Ether extract	%	2.1	2.5
N-free extract	%	47.0	55.0
Protein (N x 6.25)	%	6.8	7.9
Cattle	dig prot %	2.6	3.0
Sheep	dig prot %	2.6	3.1
Energy			
Cattle	DE kcal/kg	2186.	2557.
Sheep	DE kcal/kg	2186.	2557.
Cattle	ME kcal/kg	1793.	2097.
Sheep	ME kcal/kg	1793.	2097.
Cattle	TDN %	50.	58.
Sheep	TDN %	50.	58.
Calcium	%	.34	.40
Iron	%	.004	.005
Magnesium	%	.27	.32
Phosphorus	%	.14	.17
Potassium	%	1.20	1.41

SORGHUM. Sorghum vulgare

Sorghum, aerial pt, s-c, immature, (1)
  Sorghum fodder, sun-cured, immature

Ref no 1-04-299

		As fed	Dry	C.V. ± %
Dry matter	%	89.8	100.0	2
Ash	%	8.1	9.0	16
Crude fiber	%	22.3	24.8	22
Ether extract	%	1.7	1.9	18
Ether extract	%	45.2	50.3	
Protein (N x 6.25)	%	12.6	14.0	21
Cattle	dig prot %	4.8	5.3	
Sheep	dig prot %	4.8	5.3	
Energy				
Cattle	DE kcal/kg	2217.	2469.	
Sheep	DE kcal/kg	2138.	2381.	
Cattle	ME kcal/kg	1318.	2025.	
Sheep	ME kcal/kg	1753.	1952.	
Cattle	TDN %	50.	56.	
Sheep	TDN %	48.	54.	

---

(1) dry forages and roughages
(2) pasture, range plants, and forages fed green

(3) silages
(4) energy feeds

Feed name or analyses		Mean		C.V.
		As fed	Dry	± %

Sorghum, aerial pt wo heads, s-c, (1)
Sorghum stover, sun-cured

Ref no 1-04-302

		As fed	Dry	
Dry matter	%	85.1	100.0	7
Ash	%	8.2	9.6	14
Crude fiber	%	27.7	32.6	12
Ether extract	%	1.8	2.1	22
N-free extract	%	42.9	50.4	
Protein (N x 6.25)	%	4.5	5.3	17
Cattle	dig prot %	1.5	1.8	
Sheep	dig prot %	.8	1.0	
Energy				
Cattle	DE kcal/kg	2138.	2513.	
Sheep	DE kcal/kg	1801.	2116.	
Cattle	ME kcal/kg	1754.	2061.	
Sheep	ME kcal/kg	1476.	1735.	
Cattle	TDN %	48.	57.	
Sheep	TDN %	41.	48.	
Calcium	%	.34	.40	10
Phosphorus	%	.09	.11	32
Manganese	mg/kg	125.9	148.0	31
Carotene	mg/kg	5.8	6.8	99
Vitamin A equiv	IU/g	9.7	11.3	

Sorghum, aerial pt, fresh, immature, (2)
Sorghum fodder, fresh, immature

Ref no 2-04-310

		As fed	Dry	
Dry matter	%	16.3	100.0	9
Ash	%	1.9	11.9	32
Crude fiber	%	4.5	27.5	20
Ether extract	%	.4	2.7	38
N-free extract	%	6.4	39.2	
Protein (N x 6.25)	%	3.0	18.7	16
Cattle	dig prot %	1.2	7.1	
Sheep	dig prot %	1.3	8.2	
Energy				
Cattle	DE kcal/kg	381.	2337.	
Sheep	DE kcal/kg	402.	2469.	
Cattle	ME kcal/kg	312.	1916.	
Sheep	ME kcal/kg	330.	2024.	
Cattle	TDN %	9.	53.	
Sheep	TDN %	9.	56.	

Sorghum, aerial pt, fresh, mid-blm, (2)
Sorghum fodder, fresh, mid-bloom

Ref no 2-04-311

		As fed	Dry	
Dry matter	%	21.5	100.0	11
Ash	%	2.2	10.3	8
Crude fiber	%	7.4	34.3	11
Ether extract	%	.4	1.7	27
N-free extract	%	9.5	44.0	
Protein (N x 6.25)	%	2.1	9.7	19
Cattle	dig prot %	.8	3.7	
Sheep	dig prot %	.9	4.3	
Energy				
Cattle	DE kcal/kg	531.	2469.	
Sheep	DE kcal/kg	550.	2557.	
Cattle	ME kcal/kg	435.	2024.	
Sheep	ME kcal/kg	451.	2097.	
Cattle	TDN %	12.	56.	
Sheep	TDN %	12.	58.	

Sorghum, aerial pt, fresh, milk stage, (2)
Sorghum fodder, fresh, milk stage

Ref no 2-04-313

		As fed	Dry	
Dry matter	%	22.5	100.0	8
Ash	%	2.2	10.0	12
Crude fiber	%	7.9	35.3	5
Ether extract	%	.4	1.8	9
N-free extract	%	10.1	44.9	
Protein (N x 6.25)	%	1.8	8.0	6
Cattle	dig prot %	.7	3.0	
Sheep	dig prot %	.8	3.5	
Energy				
Cattle	DE kcal/kg	556.	2469.	
Sheep	DE kcal/kg	575.	2557.	
Cattle	ME kcal/kg	455.	2024.	
Sheep	ME kcal/kg	472.	2097.	
Cattle	TDN %	13.	56.	
Sheep	TDN %	13.	58.	

Sorghum, aerial pt, fresh, mature, (2)
Sorghum fodder, fresh, mature

Ref no 2-04-315

		As fed	Dry	
Dry matter	%	39.3	100.0	40
Ash	%	3.2	8.2	15
Crude fiber	%	14.2	36.1	10
Ether extract	%	.6	1.5	33
N-free extract	%	19.5	49.7	
Protein (N x 6.25)	%	1.8	4.5	38

Continued

Feed name or analyses		Mean		C.V
		As fed	Dry	± %
Cattle	dig prot %	.3	.7	
Sheep	dig prot %	.8	2.0	
Energy				
Cattle	DE kcal/kg	936.	2381.	
Sheep	DE kcal/kg	1040.	2646.	
Cattle	ME kcal/kg	767.	1952.	
Sheep	ME kcal/kg	853.	2170.	
Cattle	TDN %	21.	54.	
Sheep	TDN %	24.	60.	

Sorghum, aerial pt, fresh, (2)
  Sorghum fodder, fresh

Ref no 2-04-317

Dry matter	%	23.3	100.0	45
Ash	%	2.0	8.6	54
Crude fiber	%	6.8	29.3	20
Ether extract	%	.6	2.5	35
N-free extract	%	11.7	50.2	
Protein (N x 6.25)	%	2.2	9.4	42
Cattle	dig prot %	.8	3.5	
Sheep	dig prot %	1.0	4.1	
Energy				
Cattle	DE kcal/kg	555.	2381.	
Sheep	DE kcal/kg	616.	2646.	
Cattle	ME kcal/kg	455.	1952.	
Sheep	ME kcal/kg	506.	2170.	
Cattle	TDN %	12.	54.	
Sheep	TDN %	14.	60.	
Calcium	%	.09	.38	45
Magnesium	%	.07	.30	29
Phosphorus	%	.04	.18	32
Potassium	%	.37	1.58	9
Manganese	mg/kg	30.5	131.0	
Carotene	mg/kg	6.2	26.7	99
Vitamin A equiv	IU/g	10.3	44.5	

Sorghum, aerial pt, ensiled, (3)
  Sorghum fodder silage

Ref no 3-04-323

Dry matter	%	28.9	100.0	43
Ash	%	2.2	7.7	40
Crude fiber	%	7.8	26.8	19
Ether extract	%	.8	2.8	34
N-free extract	%	15.9	54.9	
Protein (N x 6.25)	%	2.3	7.8	33
Cattle	dig prot %	.8	2.9	
Sheep	dig prot %	.5	1.8	
Energy				
Cattle	DE kcal/kg	701.	2425.	
Sheep	DE kcal/kg	739.	2557.	

Feed name or analyses		Mean		C.V
		As fed	Dry	± %
Cattle	ME kcal/kg	574.	1988.	
Sheep	ME kcal/kg	606.	2097.	
Cattle	NEm kcal/kg	397.	1374.	
Cattle	NE gain kcal/kg	176.	609.	
Cattle	TDN %	16.	55.	
Sheep	TDN %	17.	58.	
Calcium	%	.10	.34	71
Iron	%	.010	.030	26
Magnesium	%	.09	.30	40
Phosphorus	%	.06	.21	44
Potassium	%	.40	1.37	39
Cobalt	mg/kg	.090	.310	51
Carotene	mg/kg	9.7	33.7	87
Vitamin A equiv	IU/g	16.2	56.2	

Sorghum, aerial pt wo heads, ensiled, (3)
  Sorghum stover silage

Ref no 3-04-326

Dry matter	%	27.3	100.0	10
Ash	%	2.2	8.2	23
Crude fiber	%	7.9	29.1	15
Ether extract	%	.5	1.8	13
N-free extract	%	15.3	56.1	
Protein (N x 6.25)	%	1.3	4.8	18
Cattle	dig prot %	.4	1.3	
Sheep	dig prot %	.3	1.1	
Energy				
Cattle	DE kcal/kg	674.	2469.	
Sheep	DE kcal/kg	698.	2557.	
Cattle	ME kcal/kg	552.	2024.	
Sheep	ME kcal/kg	572.	2097.	
Cattle	TDN %	15.	56.	
Sheep	TDN %	16.	58.	

SORGHUM, ATLAS. Sorghum vulgare

Sorghum, atlas, grain, (4)

Ref no 4-04-342

Dry matter	%	89.0	100.0	1
Ash	%	1.7	1.9	14
Crude fiber	%	2.0	2.2	23
Ether extract	%	3.4	3.8	8
N-free extract	%	70.6	79.3	
Protein (N x 6.25)	%	11.4	12.8	10
Cattle	dig prot %	6.6	7.4	
Sheep	dig prot %	7.6	8.6	
Energy				
Cattle	DE kcal/kg	2982.	3351.	
Sheep	DE kcal/kg	3375.	3792.	

Continued

(1) dry forages and roughages
(2) pasture, range plants, and forages fed green

(3) silages
(4) energy feeds

Feed name or analyses		Mean		C.V.
		As fed	Dry	± %
Cattle	ME kcal/kg	2446.	2748.	
Sheep	ME kcal/kg	2767.	3109.	
Cattle	TDN %	68.	76.	
Sheep	TDN %	76.	86.	
Riboflavin	mg/kg	13.3	15.0	
Thiamine	mg/kg	2.0	2.2	
Vitamin B6	mg/kg	34.30	38.50	

Sorghum, atlas, heads, chop, (4)

Ref no 4-04-343

Dry matter	%	83.0	100.0	6
Ash	%	3.5	4.2	15
Crude fiber	%	9.0	10.8	18
Ether extract	%	2.6	3.1	12
N-free extract	%	59.1	71.2	
Protein (N x 6.25)	%	8.9	10.7	7
Cattle	dig prot %	5.1	6.1	
Sheep	dig prot %	5.6	6.7	
Energy				
Cattle	DE kcal/kg	2745.	3307.	
Sheep	DE kcal/kg	2781.	3351.	
Cattle	ME kcal/kg	2251.	2712.	
Sheep	ME kcal/kg	2281.	2748.	
Cattle	TDN %	62.	75.	
Sheep	TDN %	63.	76.	

SORGHUM, BROOMCORN. Sorghum vulgare technicum

Sorghum, broomcorn, grain, (4)

Ref no 4-04-349

Dry matter	%	89.0	100.0	2
Ash	%	2.7	3.0	26
Crude fiber	%	6.0	6.7	35
Ether extract	%	3.5	3.9	17
N-free extract	%	67.2	75.5	
Protein (N x 6.25)	%	9.7	10.9	8
Cattle	dig prot %	4.1	4.6	
Horses	dig prot %	4.1	4.6	
Sheep	dig prot %	4.6	5.2	
Swine	dig prot %	5.2	5.9	
Energy	GE kcal/kg	4027.	4525.	
Cattle	DE kcal/kg	2865.	3219.	
Horses	DE kcal/kg	2629.	2954.	
Sheep	DE kcal/kg	2746.	3086.	
Swine	DE kcal/kg	2826.	3175.	
Cattle	ME kcal/kg	2350.	2640.	
Horses	ME kcal/kg	2156.	2422.	
Sheep	ME kcal/kg	2252.	2530.	

Continued

Feed name or analyses		Mean		C.V.
		As fed	Dry	± %
Swine	ME kcal/kg	2318.	2604.	
Cattle	TDN %	65.	73.	
Horses	TDN %	60.	67.	
Sheep	TDN %	62.	70.	
Swine	TDN %	64.	72.	
Phosphorus	%	.35	.39	

SORGHUM, DARSO. Sorghum vulgare

Sorghum, darso, grain, (4)

Ref no 4-04-357

Dry matter	%	90.0	100.0	2
Ash	%	1.3	1.4	26
Crude fiber	%	2.0	2.2	24
Ether extract	%	3.0	3.3	11
N-free extract	%	73.7	81.9	
Protein (N x 6.25)	%	10.1	11.2	9
Cattle	dig prot %	5.8	6.4	
Sheep	dig prot %	6.5	7.2	
Energy				
Cattle	DE kcal/kg	3214.	3571.	
Sheep	DE kcal/kg	3373.	3748.	
Cattle	ME kcal/kg	2635.	2928.	
Sheep	ME kcal/kg	2766.	3073.	
Cattle	TDN %	73.	81.	
Sheep	TDN %	76.	85.	
Niacin	mg/kg	28.1	32.2	
Pantothenic acid	mg/kg	13.2	14.7	
Riboflavin	mg/kg	1.4	1.5	
Arginine	%	.36	.40	
Histidine	%	.18	.20	
Isoleucine	%	.45	.50	
Leucine	%	1.17	1.30	
Lysine	%	.18	.20	
Methionine	%	.09	.10	
Phenylalanine	%	.45	.50	
Threonine	%	.27	.30	
Tryptophan	%	.09	.10	
Valine	%	.45	.50	

Sorghum, darso, heads, chop, (4)

Ref no 4-04-358

Dry matter	%	90.0	100.0	1
Ash	%	1.9	2.1	25
Crude fiber	%	8.0	8.9	
Ether extract	%	2.8	3.1	8
N-free extract	%	67.6	75.1	
Protein (N x 6.25)	%	9.7	10.8	5
Cattle	dig prot %	5.6	6.2	

Continued

Feed name or analyses		Mean		C.V.
		As fed	Dry	± %
Sheep	dig prot %	6.1	6.8	
Energy				
Cattle	DE kcal/kg	3095.	3439.	
Sheep	DE kcal/kg	3095.	3439.	
Cattle	ME kcal/kg	2538.	2820.	
Sheep	ME kcal/kg	2538.	2820.	
Cattle	TDN %	70.	78.	
Sheep	TDN %	70.	78.	

SORGHUM, DURRA. Sorghum vulgare

Sorghum, durra, grain, (4)

Ref no 4-04-365

Dry matter	%	89.0	100.0	1
Ash	%	2.7	3.0	21
Crude fiber	%	2.0	2.2	15
Ether extract	%	3.6	4.0	5
N-free extract	%	70.6	79.3	
Protein (N x 6.25)	%	10.2	11.5	8
Cattle	dig prot %	5.9	6.6	
Sheep	dig prot %	5.7	6.4	
Energy				
Cattle	DE kcal/kg	3139.	3527.	
Sheep	DE kcal/kg	3218.	3616.	
Cattle	ME kcal/kg	2574.	2892.	
Sheep	ME kcal/kg	2639.	2965.	
Cattle	TDN %	71.	80.	
Sheep	TDN %	73.	82.	

SORGHUM, FETERITA. Sorghum vulgare

Sorghum, feterita, grain, (4)

Ref no 4-04-369

Dry matter	%	90.0	100.0	1
Ash	%	1.5	1.7	16
Crude fiber	%	2.0	2.2	18
Ether extract	%	3.1	3.4	18
N-free extract	%	71.2	79.1	
Protein (N x 6.25)	%	12.2	13.6	12
Cattle	dig prot %	6.9	7.7	
Sheep	dig prot %	9.3	10.3	
Energy				
Cattle	DE kcal/kg	3174.	3527.	
Sheep	DE kcal/kg	3532.	3924.	
Cattle	ME kcal/kg	2603.	2892.	
Sheep	ME kcal/kg	2896.	3218.	
Cattle	TDN %	72.	80.	
Sheep	TDN %	80.	89.	

Continued

Feed name or analyses		Mean		C.V.
		As fed	Dry	± %
Calcium	%	.02	.02	36
Phosphorus	%	.28	.31	28
Niacin	mg/kg	50.3	55.9	
Pantothenic acid	mg/kg	14.0	15.6	
Riboflavin	mg/kg	1.6	1.8	
Arginine	%	.45	.50	
Histidine	%	.27	.30	
Isoleucine	%	.54	.60	
Leucine	%	1.71	1.90	
Lysine	%	.18	.20	
Methionine	%	.18	.20	
Phenylalanine	%	.63	.70	
Threonine	%	.45	.50	
Tryptophan	%	.18	.20	
Valine	%	.63	.70	

Sorghum, feterita, heads, chop, (4)

Ref no 4-04-370

Dry matter	%	90.0	100.0	2
Ash	%	2.9	3.2	14
Crude fiber	%	7.0	7.8	16
Ether extract	%	2.6	2.9	12
N-free extract	%	66.8	74.2	
Protein (N x 6.25)	%	10.7	11.9	12
Cattle	dig prot %	6.1	6.8	
Sheep	dig prot %	6.8	7.5	
Energy				
Cattle	DE kcal/kg	3056.	3395.	
Sheep	DE kcal/kg	3016.	3351.	
Cattle	ME kcal/kg	2506.	2784.	
Sheep	ME kcal/kg	2473.	2748.	
Cattle	TDN %	69.	77.	
Sheep	TDN %	68.	76.	

Sorghum fodder - see Sorghum, aerial pt

Sorghum grain - see Also Sorghum, atlas; Sorghum, broomcorn; Sorghum, darso; Sorghum, durra; Sorghum, feterita; Sorghum, grain variety; Sorghum, hegari; Sorghum, kafir; Sorghum, kalo; Sorghum, kaoliang; Sorghum, milo; Sorghum, schrock; Sorghum, shallu; Sorghum, sorgo; Sorghum, sumac

(1) dry forages and roughages
(2) pasture, range plants, and forages fed green

(3) silages
(4) energy feeds

Feed name or analyses		Mean		C.V.
		As fed	Dry	± %

SORGHUM, GRAIN VARIETY. Sorghum vulgare

Sorghum, grain variety, aerial pt, s-c, (1)
Grain sorghum fodder, sun-cured

Ref no 1-04-372

Feed name or analyses		As fed	Dry	C.V. ± %
Dry matter	%	90.3	100.0	1
Ash	%	8.5	9.4	14
Crude fiber	%	24.8	27.5	7
Ether extract	%	1.7	1.9	15
N-free extract	%	49.0	54.3	
Protein (N x 6.25)	%	6.2	6.9	6
Cattle	dig prot %	2.3	2.6	
Sheep	dig prot %	2.3	2.6	
Energy				
Cattle	DE kcal/kg	2309.	2557.	
Sheep	DE kcal/kg	2190.	2425.	
Cattle	ME kcal/kg	1894.	2097.	
Sheep	ME kcal/kg	1795.	1988.	
Cattle	TDN %	52.	58.	
Sheep	TDN %	50.	55.	
Calcium	%	.56	.62	
Phosphorus	%	.17	.19	

Sorghum, grain variety, aerial pt wo heads, s-c, (1)
Grain sorghum stover, sun-cured

Ref no 1-07-961

Feed name or analyses		As fed	Dry	C.V. ± %
Dry matter	%	85.1	100.0	
Ash	%	8.2	9.6	
Crude fiber	%	27.7	32.6	
Ether extract	%	1.8	2.1	
N-free extract	%	42.9	50.4	
Protein (N x 6.25)	%	4.5	5.3	
Cattle	dig prot %	1.5	1.8	
Sheep	dig prot %	.8	1.0	
Energy				
Cattle	DE kcal/kg	2138.	2513.	
Sheep	DE kcal/kg	1801.	2116.	
Cattle	ME kcal/kg	1754.	2061.	
Sheep	ME kcal/kg	1476.	1735.	
Cattle	TDN %	49.	57.	
Sheep	TDN %	41.	48.	
Calcium	%	.34	.40	
Phosphorus	%	.09	.11	

(5) protein supplements
(6) minerals

Sorghum, grain variety, aerial pt, ensiled, (3)
Grain sorghum fodder silage

Ref no 3-07-962

Feed name or analyses		As fed	Dry	C.V. ± %
Dry matter	%	29.4	100.0	
Ash	%	2.2	7.4	
Crude fiber	%	7.7	26.3	
Ether extract	%	1.0	3.3	
N-free extract	%	16.4	55.7	
Protein (N x 6.25)	%	2.1	7.3	
Cattle	dig prot %	.6	2.0	
Sheep	dig prot %	.5	1.7	
Energy				
Cattle	DE kcal/kg	739.	2513.	
Sheep	DE kcal/kg	752.	2557.	
Cattle	ME kcal/kg	606.	2061.	
Sheep	ME kcal/kg	616.	2097.	
Cattle	TDN %	17.	57.	
Sheep	TDN %	17.	58.	
Calcium	%	.07	.25	
Phosphorus	%	.05	.18	

Sorghum, grain variety, grain, grnd, (4)
Ground grain sorghum (AAFCO)

Ref no 4-04-379

Feed name or analyses		As fed	Dry	C.V. ± %
Dry matter	%	89.0	100.0	4
Ash	%	1.8	2.0	40
Crude fiber	%	2.0	2.2	74
Ether extract	%	3.0	3.4	20
N-free extract	%	71.1	79.9	
Protein (N x 6.25)	%	11.1	12.5	10
Cattle	dig prot %	6.4	7.2	
Sheep	dig prot %	7.5	8.4	
Energy	GE kcal/kg	3926.	4411.	
Cattle	DE kcal/kg	2826.	3175.	
Sheep	DE kcal/kg	3375.	3792.	
Cattle	ME kcal/kg	2317.	2603.	
Sheep	ME kcal/kg	2767.	3109.	
Cattle	TDN %	64.	72.	
Sheep	TDN %	76.	86.	
Calcium	%	.04	.05	99
Magnesium	%	.17	.19	32
Phosphorus	%	.31	.35	21
Potassium	%	.34	.38	18
Sodium	%	.04	.05	24
Cobalt	mg/kg	2.800	3.100	61
Copper	mg/kg	9.6	10.8	38
Manganese	mg/kg	14.5	16.3	39
Zinc	mg/kg	13.7	15.4	
Biotin	mg/kg	2.60	2.90	30
Choline	mg/kg	668.	761.	17

Continued

(7) vitamins
(8) additives

Feed name or analyses		Mean		C.V.
		As fed	Dry	± %
Folic acid	mg/kg	.20	.20	22
Niacin	mg/kg	43.1	48.4	28
Pantothenic acid	mg/kg	11.1	12.5	30
Riboflavin	mg/kg	1.3	1.5	63
Thiamine	mg/kg	4.1	4.6	37
Vitamin B6	mg/kg	5.30	5.90	32
Arginine	%	.36	.40	99
Cystine	%	.18	.20	16
Histidine	%	.27	.30	99
Isoleucine	%	.53	.60	99
Leucine	%	1.42	1.60	15
Lysine	%	.27	.30	26
Methionine	%	.09	.10	
Phenylalanine	%	.45	.51	24
Threonine	%	.27	.30	36
Tryptophan	%	.09	.10	29
Tyrosine	%	.36	.40	28
Valine	%	.53	.60	10

Sorghum, grain variety, grain, mn 6 mx 9 prot, (4)

Ref no 4-08-138

		As fed	Dry
Dry matter	%	88.0	100.0
Ash	%	2.0	2.3
Crude fiber	%	1.9	2.2
Ether extract	%	2.6	3.0
N-free extract	%	74.4	84.6
Protein (N x 6.25)	%	7.0	7.9
Cattle	dig prot %	4.0	4.5
Sheep	dig prot %	5.6	6.4
Energy			
Cattle	DE kcal/kg	3142.	3571.
Sheep	DE kcal/kg	3492.	3968.
Cattle	ME kcal/kg	2577.	2928.
Sheep	ME kcal/kg	2864.	3254.
Cattle	TDN %	71.	81.
Sheep	TDN %	79.	90.
Alanine	%	.61	.69
Arginine	%	.26	.29
Aspartic acid	%	.48	.54
Cysteine	%	.10	.11
Glutamic acid	%	1.36	1.54
Glycine	%	.26	.29
Histidine	%	.16	.18
Isoleucine	%	.26	.30
Leucine	%	.68	.77
Lysine	%	.18	.20
Methionine	%	.09	.10
Phenylalanine	%	.34	.39
Proline	%	.52	.59
Serine	%	.30	.34
Threonine	%	.23	.26
Tyrosine	%	.14	.16
Valine	%	.35	.40

Sorghum, grain variety, grain, mn 9 mx 12 prot, (4)

Ref no 4-08-139

		As fed	Dry
Dry matter	%	88.0	100.0
Ash	%	1.9	2.2
Crude fiber	%	2.1	2.4
Ether extract	%	2.6	2.9
N-free extract	%	71.1	80.8
Protein (N x 6.25)	%	10.3	11.7
Cattle	dig prot %	5.9	6.7
Sheep	dig prot %	8.4	9.5
Lignin	%	1.1	1.3
Energy			
Cattle	DE kcal/kg	3104.	3527.
Sheep	DE kcal/kg	3492.	3968.
Cattle	ME kcal/kg	2545.	2892.
Sheep	ME kcal/kg	2864.	3254.
Cattle	TDN %	70.	80.
Sheep	TDN %	79.	90.
Alanine	%	.97	1.10
Arginine	%	.33	.38
Aspartic acid	%	.70	.79
Cysteine	%	.14	.16
Glutamic acid	%	2.24	2.54
Glycine	%	.32	.37
Histidine	%	.23	.26
Isoleucine	%	.43	.49
Leucine	%	1.41	1.60
Lysine	%	.22	.25
Methionine	%	.13	.15
Phenylalanine	%	.53	.60
Proline	%	.84	.96
Serine	%	.44	.50
Threonine	%	.32	.37
Tyrosine	%	.22	.25
Valine	%	.53	.60

Sorghum, grain variety, grain, mn 12 mx 15 prot, (4)

Ref no 4-08-140

		As fed	Dry
Dry matter	%	88.0	100.0
Ash	%	2.3	2.6
Crude fiber	%	1.8	2.0
Ether extract	%	1.5	1.7
N-free extract	%	71.0	80.7
Protein (N x 6.25)	%	11.4	13.0
Cattle	dig prot %	6.5	7.4
Sheep	dig prot %	9.2	10.5
Energy			
Cattle	DE kcal/kg	3026.	3439.
Sheep	DE kcal/kg	3453.	3924.
Cattle	ME kcal/kg	2482.	2820.

Continued

(1) dry forages and roughages

(2) pasture, range plants, and forages fed green

(3) silages

(4) energy feeds

Feed name or analyses		Mean		C.V.
		As fed	Dry	± %
Sheep	ME kcal/kg	2832.	3218.	
Cattle	TDN %	69.	78.	
Sheep	TDN %	78.	89.	
Alanine	%	1.17	1.33	
Arginine	%	.39	.43	
Aspartic acid	%	.81	.92	
Cysteine	%	.18	.20	
Glutamic acid	%	2.59	2.94	
Glycine	%	.35	.40	
Histidine	%	.26	.29	
Isoleucine	%	.49	.56	
Leucine	%	1.77	2.01	
Lysine	%	.23	.26	
Methionine	%	.14	.16	
Phenylalanine	%	.62	.70	
Proline	%	.97	1.10	
Serine	%	.51	.58	
Threonine	%	.37	.42	
Tyrosine	%	.26	.29	
Valine	%	.61	.69	

Sorghum, grain variety, grain, (4)

Ref no 4-04-383

Dry matter	%	89.0	100.0	4
Ash	%	1.8	2.0	40
Crude fiber	%	2.0	2.2	74
Ether extract	%	3.0	3.4	20
N-free extract	%	71.1	79.9	
Protein (N x 6.25)	%	11.1	12.5	10
Cattle	dig prot %	6.3	7.1	
Sheep	dig prot %	7.5	8.4	
Swine	dig prot %	7.9	8.9	
Energy				
Cattle	DE kcal/kg	3257.	3660.	
Sheep	DE kcal/kg	3375.	3792.	
Swine	DE kcal/kg	3414.	3836.	
Cattle	ME kcal/kg	2671.	3001.	
Chickens	MEn kcal/kg	3307.	3716.	
Sheep	ME kcal/kg	2767.	3109.	
Swine	ME kcal/kg	3192.	3587.	
Cattle	TDN %	74.	83.	
Sheep	TDN %	76.	86.	
Swine	TDN %	77.	87.	
Calcium	%	.04	.05	99
Magnesium	%	.17	.19	32
Phosphorus	%	.31	.35	21
Potassium	%	.34	.38	18
Sodium	%	.04	.05	24
Cobalt	mg/kg	2.800	3.100	61
Copper	mg/kg	9.6	10.8	38
Manganese	mg/kg	14.5	16.3	39
Zinc	mg/kg	13.7	15.4	
Biotin	mg/kg	2.60	2.90	30

Continued

Feed name or analyses		Mean		C.V.
		As fed	Dry	± %
Choline	mg/kg	678.	761.	17
Folic acid	mg/kg	.20	.20	22
Niacin	mg/kg	43.1	48.4	28
Pantothenic acid	mg/kg	11.1	12.5	30
Riboflavin	mg/kg	1.3	1.5	63
Thiamine	mg/kg	4.1	4.6	37
Vitamin B6	mg/kg	5.30	5.90	32
Arginine	%	.36	.40	99
Cystine	%	.18	.20	16
Histidine	%	.27	.30	99
Isoleucine	%	.53	.60	99
Leucine	%	1.42	1.60	15
Lysine	%	.27	.30	26
Phenylalanine	%	.45	.51	24
Threonine	%	.27	.30	36
Tryptophan	%	.09	.10	29
Tyrosine	%	.36	.40	28
Valine	%	.53	.60	10

Sorghum, grain variety, grits by-prod, mx 5 fbr, (4)
Grain sorghum mill feed (AAFCO)

Ref no 4-04-386

Dry matter	%	89.0	100.0	
Ash	%	3.0	3.4	
Crude fiber	%	3.0	3.4	
Ether extract	%	5.6	6.3	
N-free extract	%	67.1	75.4	
Protein (N x 6.25)	%	10.2	11.5	
Cattle	dig prot %	5.9	6.6	
Sheep	dig prot %	6.9	7.8	
Energy				
Cattle	DE kcal/kg	3218.	3616.	
Sheep	DE kcal/kg	2512.	2822.	
Cattle	ME kcal/kg	2639.	2965.	
Sheep	ME kcal/kg	2059.	2314.	
Cattle	TDN %	73.	82.	
Sheep	TDN %	57.	64.	

Sorghum, grain variety, heads, chop, (4)

Ref no 4-04-387

Dry matter	%	89.0	100.0	4
Ash	%	3.6	4.1	34
Crude fiber	%	8.0	9.0	21
Ether extract	%	2.6	2.9	17
N-free extract	%	64.8	72.8	
Protein (N x 6.25)	%	10.0	11.2	15
Cattle	dig prot %	5.7	6.4	
Sheep	dig prot %	6.2	7.0	
Energy				
Cattle	DE kcal/kg	2982.	3351.	

Continued

Feed name or analyses		Mean		C.V.
		As fed	Dry	± %
Sheep	DE kcal/kg	2826.	3175.	
Cattle	ME kcal/kg	2446.	2748.	
Sheep	ME kcal/kg	2318.	2604.	
Cattle	TDN %	68.	76.	
Sheep	TDN %	64.	72.	
Calcium	%	.12	.13	92
Magnesium	%	.17	.19	
Phosphorus	%	.21	.24	26
Manganese	mg/kg	12.9	14.5	
Riboflavin	mg/kg	2.0	2.2	21

Feed name or analyses		Mean		C.V.
		As fed	Dry	± %
Cattle	TDN %	81.	85.	
Sheep	TDN %	82.	86.	
Calcium	%	.17	.18	
Phosphorus	%	.92	.97	
Manganese	mg/kg	104.3	109.8	
Choline	mg/kg	843.	887.	
Niacin	mg/kg	60.9	64.1	
Pantothenic acid	mg/kg	12.3	12.9	
Riboflavin	mg/kg	3.2	3.4	
Thiamine	mg/kg	1.3	1.4	

Sorghum, grain variety, distil grains, dehy, (5)
Grain sorghum distillers dried grains (AAFCO)

Ref no 5-04-374

		As fed	Dry	
Dry matter	%	94.0	100.0	2
Ash	%	3.9	4.1	34
Crude fiber	%	12.0	12.8	38
Ether extract	%	8.4	8.9	32
N-free extract	%	38.5	41.0	
Protein (N x 6.25)	%	31.2	33.2	23
Cattle	dig prot %	24.9	26.5	
Sheep	dig prot %	25.9	27.6	
Energy				
Cattle	DE kcal/kg	3399.	3616.	
Sheep	DE kcal/kg	3482.	3704.	
Cattle	ME kcal/kg	2787.	2965.	
Sheep	ME kcal/kg	2855.	3037.	
Cattle	TDN %	77.	82.	
Sheep	TDN %	79.	84.	
Calcium	%	.14	.15	
Phosphorus	%	.59	.63	

Sorghum, grain variety, distil sol, dehy, (5)
Grain sorghum distillers dried solubles (AAFCO)

Ref no 5-04-376

		As fed	Dry	
Dry matter	%	93.0	100.0	
Ash	%	8.4	9.0	
Crude fiber	%	4.0	4.3	
Ether extract	%	5.5	5.9	
N-free extract	%	48.7	52.4	
Protein (N x 6.25)	%	26.4	28.4	
Cattle	dig prot %	21.4	23.0	
Sheep	dig prot %	21.6	23.2	
Energy				
Cattle	DE kcal/kg	3404.	3660.	
Sheep	DE kcal/kg	3608.	3880.	
Cattle	ME kcal/kg	2791.	3001.	
Sheep	ME kcal/kg	2959.	3182.	
Cattle	TDN %	77.	83.	
Sheep	TDN %	82.	88.	
Calcium	%	.68	.73	
Phosphorus	%	1.37	1.47	
Niacin	mg/kg	141.7	152.3	
Riboflavin	mg/kg	16.1	17.3	
Thiamine	mg/kg	4.4	4.7	

Sorghum, grain variety, distil grains w sol, dehy, mn 75 orig solids, (5)
Grain sorghum distillers dried grains with solubles (AAFCO)

Ref no 5-04-375

		As fed	Dry	
Dry matter	%	95.0	100.0	2
Ash	%	4.2	4.4	28
Crude fiber	%	10.0	10.5	20
Ether extract	%	9.4	9.9	7
N-free extract	%	38.4	40.4	13
Protein (N x 6.25)	%	33.1	34.8	
Cattle	dig prot %	26.6	28.0	
Sheep	dig prot %	27.6	29.1	
Energy				
Cattle	DE kcal/kg	3561.	3748.	
Sheep	DE kcal/kg	3602.	3792.	
Cattle	ME kcal/kg	2919.	3073.	
Sheep	ME kcal/kg	2954.	3109.	

Sorghum, grain variety, gluten, wet-mil dehy, (5)
Grain sorghum gluten meal (AAFCO)

Ref no 5-04-388

		As fed	Dry	
Dry matter	%	93.0	100.0	
Ash	%	1.6	1.7	
Crude fiber	%	10.0	10.8	
Ether extract	%	4.7	5.1	
N-free extract	%	30.2	32.5	
Protein (N x 6.25)	%	46.4	49.9	
Cattle	dig prot %	44.1	47.4	
Sheep	dig prot %	40.0	42.9	
Energy				
Cattle	DE kcal/kg	3280.	3527.	
Sheep	DE kcal/kg	3363.	3616.	
Cattle	ME kcal/kg	2690.	2892.	

Continued

Continued

(1) dry forages and roughages

(2) pasture, range plants, and forages fed green

(3) silages

(4) energy feeds

Feed name or analyses		Mean		C.V.
		As fed	Dry	± %
Chickens	MEn kcal/kg	2535.	2726.	
Sheep	ME kcal/kg	2757.	2965.	
Cattle	TDN %	74.	80.	
Sheep	TDN %	76.	82.	

Sorghum, grain variety, gluten w bran, wet-mil dehy, (5)
Grain sorghum gluten feed (AAFCO)

Ref no 5-04-389

		As fed	Dry	
Dry matter	%	94.0	100.0	
Ash	%	10.5	11.2	
Crude fiber	%	8.0	8.5	
Ether extract	%	3.2	3.4	
N-free extract	%	44.5	47.3	
Protein (N x 6.25)	%	27.8	29.6	
Cattle	dig prot %	22.0	23.4	
Sheep	dig prot %	22.9	24.4	
Energy				
Cattle	DE kcal/kg	3026.	3219.	
Sheep	DE kcal/kg	3357.	3571.	
Cattle	ME kcal/kg	2482.	2640.	
Chickens	MEn kcal/kg	1653.	1758.	
Sheep	ME kcal/kg	2752.	2928.	
Cattle	TDN %	69.	73.	
Sheep	TDN %	76.	81.	

SORGHUM, HEGARI. Sorghum vulgare

Sorghum, hegari, aerial pt, s-c, mature, (1)
Sorghum, hegari, fodder, sun-cured, mature

Ref no 1-04-391

		As fed	Dry	C.V.
Dry matter	%	84.7	100.0	5
Ash	%	7.3	8.6	16
Crude fiber	%	18.3	21.6	15
Ether extract	%	1.9	2.2	15
N-free extract	%	50.7	59.9	
Protein (N x 6.25)	%	6.5	7.7	20
Cattle	dig prot %	2.4	2.9	
Sheep	dig prot %	2.4	2.9	
Energy				
Cattle	DE kcal/kg	2166.	2557.	
Sheep	DE kcal/kg	2091.	2469.	
Cattle	ME kcal/kg	1776.	2097.	
Sheep	ME kcal/kg	1714.	2024.	
Cattle	TDN %	49.	58.	
Sheep	TDN %	47.	56.	
Calcium	%	.26	.31	39
Magnesium	%	.32	.38	
Phosphorus	%	.18	.21	14

Sorghum, hegari, aerial pt wo heads, s-c, mature, (1)
Sorghum, hegari, stover, sun-cured, mature

Ref no 1-04-392

		As fed	Dry	C.V.
Dry matter	%	87.0	100.0	3
Ash	%	9.9	11.4	11
Crude fiber	%	28.0	32.2	8
Ether extract	%	1.8	2.1	10
N-free extract	%	41.7	47.9	
Protein (N x 6.25)	%	5.6	6.4	5
Cattle	dig prot %	1.9	2.2	
Sheep	dig prot %	1.0	1.2	
Energy				
Cattle	DE kcal/kg	2148.	2469.	
Sheep	DE kcal/kg	1764.	2028.	
Cattle	ME kcal/kg	1761.	2024.	
Sheep	ME kcal/kg	1447.	1663.	
Cattle	TDN %	49.	56.	
Sheep	TDN %	40.	46.	
Calcium	%	.37	.43	
Phosphorus	%	.09	.10	

Sorghum, hegari, grain, (4)

Ref no 4-04-398

		As fed	Dry	C.V.
Dry matter	%	89.0	100.0	2
Ash	%	1.5	1.7	37
Crude fiber	%	2.0	2.2	15
Ether extract	%	2.5	2.8	14
N-free extract	%	73.1	82.1	
Protein (N x 6.25)	%	10.0	11.2	11
Cattle	dig prot %	5.7	6.4	
Sheep	dig prot %	8.1	9.1	
Energy				
Cattle	DE kcal/kg	3100.	3483.	
Sheep	DE kcal/kg	3571.	4012.	
Cattle	ME kcal/kg	2542.	2856.	
Sheep	ME kcal/kg	2928.	3290.	
Cattle	TDN %	70.	79.	
Sheep	TDN %	81.	91.	
Niacin	mg/kg	60.7	68.2	13
Pantothenic acid	mg/kg	13.7	15.4	
Riboflavin	mg/kg	1.6	1.8	
Arginine	%	.27	.30	
Histidine	%	.18	.20	
Isoleucine	%	.45	.51	
Leucine	%	1.34	1.51	
Lysine	%	.18	.20	
Phenylalanine	%	.53	.60	
Threonine	%	.36	.40	

(5) protein supplements
(6) minerals

(7) vitamins
(8) additives

Feed name or analyses		Mean		C.V.
		As fed	Dry	± %

Sorghum, hegari, heads, chop, (4)

Ref no 4-04-399

		As fed	Dry	C.V.
Dry matter	%	90.0	100.0	1
Ash	%	4.3	4.8	26
Crude fiber	%	11.0	12.2	19
Ether extract	%	2.1	2.3	12
N-free extract	%	63.5	70.6	
Protein (N x 6.25)	%	9.1	10.1	14
Cattle	dig prot %	5.2	5.8	
Sheep	dig prot %	5.8	6.4	
Energy				
Cattle	DE kcal/kg	2897.	3219.	
Sheep	DE kcal/kg	2937.	3263.	
Cattle	ME kcal/kg	2376.	2640.	
Sheep	ME kcal/kg	2408.	2676.	
Cattle	TDN %	66.	73.	
Sheep	TDN %	67.	74.	
Magnesium	%	.10	.11	

SORGHUM, JOHNSONGRASS. Sorghum halepense

Sorghum, Johnsongrass, hay, s-c, immature, (1)

Ref no 1-04-401

		As fed	Dry	C.V.
Dry matter	%	91.8	100.0	
Ash	%	10.0	10.9	
Crude fiber	%	26.6	29.0	
Ether extract	%	2.6	2.8	
N-free extract	%	38.8	42.3	
Protein (N x 6.25)	%	13.8	15.0	
Cattle	dig prot %	9.7	10.6	
Sheep	dig prot %	10.1	11.0	
Energy				
Cattle	DE kcal/kg	2671.	2910.	
Sheep	DE kcal/kg	2510.	2734.	
Cattle	ME kcal/kg	2190.	2386.	
Sheep	ME kcal/kg	2058.	2242.	
Cattle	TDN %	60.	66.	
Sheep	TDN %	57.	62.	
Calcium	%	.63	.69	
Phosphorus	%	.45	.49	

Sorghum, Johnsongrass, hay, s-c, mid-blm, (1)

Ref no 1-04-403

		As fed	Dry	C.V.
Dry matter	%	91.8	100.0	
Ash	%	9.4	10.2	
Crude fiber	%	29.9	32.6	

Continued

		As fed	Dry	C.V.
Ether extract	%	2.1	2.3	
N-free extract	%	42.5	46.3	
Protein (N x 6.25)	%	7.9	8.6	
Cattle	dig prot %	3.5	3.8	
Sheep	dig prot %	3.5	3.8	
Energy				
Cattle	DE kcal/kg	2226.	2425.	
Sheep	DE kcal/kg	2226.	2425.	
Cattle	ME kcal/kg	1825.	1988.	
Sheep	ME kcal/kg	1825.	1988.	
Cattle	TDN %	50.	55.	
Sheep	TDN %	50.	55.	
Calcium	%	.58	.63	
Phosphorus	%	.21	.23	

Sorghum, Johnsongrass, hay, s-c, over ripe, (1)

Ref no 1-04-405

		As fed	Dry	C.V.
Dry matter	%	93.1	100.0	
Ash	%	8.6	9.2	
Crude fiber	%	31.8	34.2	
Ether extract	%	1.3	1.4	
N-free extract	%	46.1	49.5	
Protein (N x 6.25)	%	5.3	5.7	
Cattle	dig prot %	2.3	2.5	
Sheep	dig prot %	2.3	2.5	
Energy				
Cattle	DE kcal/kg	2258.	2425.	
Sheep	DE kcal/kg	2258.	2425.	
Cattle	ME kcal/kg	1851.	1988.	
Sheep	ME kcal/kg	1851.	1988.	
Cattle	TDN %	51.	55.	
Sheep	TDN %	51.	55.	

Sorghum, Johnsongrass, hay, s-c, (1)

Ref no 1-04-407

		As fed	Dry	C.V.
Dry matter	%	90.7	100.0	2
Ash	%	8.0	8.8	18
Crude fiber	%	30.2	33.3	7
Ether extract	%	1.9	2.1	19
N-free extract	%	43.6	48.1	
Protein (N x 6.25)	%	7.0	7.7	30
Cattle	dig prot %	3.1	3.4	
Sheep	dig prot %	3.1	3.4	
Energy				
Cattle	DE kcal/kg	2239.	2469.	
Sheep	DE kcal/kg	2239.	2469.	
Cattle	ME kcal/kg	1836.	2024.	
Sheep	ME kcal/kg	1836.	2024.	
Cattle	NE$_m$ kcal/kg	1079.	1190.	
Cattle	NE$_{gain}$ kcal/kg	462.	510.	

Continued

(1) dry forages and roughages
(2) pasture, range plants, and forages fed green
(3) silages
(4) energy feeds

Feed name or analyses		Mean		C.V.
		As fed	Dry	± %
Cattle	TDN %	51.	56.	
Sheep	TDN %	51.	56.	
Calcium	%	.73	.81	31
Iron	%	.050	.060	
Magnesium	%	.32	.35	14
Phosphorus	%	.28	.31	24
Potassium	%	1.22	1.35	10
Carotene	mg/kg	33.6	37.0	25
Vitamin A equiv	IU/g	56.0	61.7	

Sorghum, Johnsongrass, aerial pt, fresh, (2)

Ref no 2-04-412

		As fed	Dry	± %
Dry matter	%	29.7	100.0	16
Ash	%	3.1	10.5	5
Crude fiber	%	9.2	30.9	6
Ether extract	%	.8	2.7	34
N-free extract	%	13.2	44.5	
Protein (N x 6.25)	%	3.4	11.4	26
Cattle	dig prot %	2.2	7.6	
Sheep	dig prot %	2.2	7.6	
Energy				
Cattle	DE kcal/kg	825.	2778.	
Sheep	DE kcal/kg	838.	2822.	
Cattle	ME kcal/kg	676.	2278.	
Sheep	ME kcal/kg	687.	2314.	
Cattle	TDN %	19.	63.	
Sheep	TDN %	19.	64.	
Calcium	%	.28	.94	18
Magnesium	%	.07	.25	
Phosphorus	%	.07	.24	28
Potassium	%	.93	3.12	
Carotene	mg/kg	58.9	198.4	64
Vitamin A equiv	IU/g	98.2	330.7	

SORGHUM, KAFIR. Sorghum vulgare

Sorghum, kafir, aerial pt, s-c, (1)
Sorghum, kafir, fodder, sun-cured

Ref no 1-04-418

		As fed	Dry	± %
Dry matter	%	88.8	100.0	5
Ash	%	8.3	9.4	20
Crude fiber	%	24.6	27.7	6
Ether extract	%	2.4	2.7	15
N-free extract	%	45.3	51.0	
Protein (N x 6.25)	%	8.2	9.2	11
Cattle	dig prot %	3.1	3.5	
Sheep	dig prot %	4.1	4.6	
Energy				
Cattle	DE kcal/kg	2271.	2557.	

Continued

(5) protein supplements
(6) minerals

Feed name or analyses		Mean		C.V.
		As fed	Dry	± %
Sheep	DE kcal/kg	2232.	2513.	
Cattle	ME kcal/kg	1862.	2097.	
Sheep	ME kcal/kg	1830.	2061.	
Cattle	TDN %	52.	58.	
Sheep	TDN %	51.	57.	
Calcium	%	.34	.38	8
Magnesium	%	.26	.29	
Phosphorus	%	.16	.18	31
Potassium	%	1.51	1.70	
Carotene	mg/kg	15.6	17.6	
Vitamin A equiv	IU/g	26.0	29.3	
Vitamin D2	IU/g	.5	.6	

Sorghum, kafir, aerial pt wo heads, s-c, (1)
Sorghum, kafir, stover, sun-cured

Ref no 1-04-419

		As fed	Dry	± %
Dry matter	%	82.1	100.0	7
Ash	%	8.1	9.9	6
Crude fiber	%	26.9	32.8	3
Ether extract	%	1.6	1.9	11
N-free extract	%	40.7	49.6	
Protein (N x 6.25)	%	4.8	5.8	7
Cattle	dig prot %	1.6	2.0	
Energy				
Cattle	DE kcal/kg	2063.	2513.	
Cattle	ME kcal/kg	1692.	2061.	
Cattle	TDN %	47.	57.	

Sorghum, kafir, grain, (4)

Ref no 4-04-428

		As fed	Dry	± %
Dry matter	%	90.0	100.0	2
Ash	%	1.5	1.7	26
Crude fiber	%	2.0	2.2	44
Ether extract	%	2.9	3.2	16
N-free extract	%	71.8	79.8	
Protein (N x 6.25)	%	11.8	13.1	7
Cattle	dig prot %	6.8	7.6	
Sheep	dig prot %	9.5	10.6	
Swine	dig prot %	7.8	8.7	
Energy				
Cattle	DE kcal/kg	2858.	3175.	
Sheep	DE kcal/kg	3611.	4012.	
Swine	DE kcal/kg	3532.	3924.	
Cattle	ME kcal/kg	2344.	2604.	
Sheep	ME kcal/kg	2961.	3290.	
Swine	ME kcal/kg	2896.	3218.	
Cattle	TDN %	65.	72.	
Sheep	TDN %	82.	91.	
Swine	TDN %	80.	89.	
Calcium	%	.04	.04	

Continued

(7) vitamins
(8) additives

Feed name or analyses		Mean		C.V.
		As fed	Dry	± %
Iron	%	.010	.010	
Phosphorus	%	.33	.37	9
Copper	mg/kg	6.3	7.0	27
Manganese	mg/kg	15.8	17.6	23
Niacin	mg/kg	36.6	40.7	18
Pantothenic acid	mg/kg	12.2	13.6	21
Riboflavin	mg/kg	1.4	1.5	44
Thiamine	mg/kg	3.8	4.2	13
Vitamin B$_6$	mg/kg	6.80	7.50	8
Arginine	%	.36	.40	
Histidine	%	.27	.30	
Isoleucine	%	.54	.60	
Leucine	%	1.62	1.80	
Lysine	%	.27	.30	
Methionine	%	.18	.20	
Phenylalanine	%	.63	.70	
Threonine	%	.45	.50	
Tryptophan	%	.18	.20	
Valine	%	.63	.70	

Sorghum, kafir, heads, chop, (4)

Ref no 4-04-429

Dry matter	%	90.0	100.0	1
Ash	%	3.2	3.6	11
Crude fiber	%	7.0	7.8	16
Ether extract	%	2.5	2.8	6
N-free extract	%	67.3	74.8	
Protein (N x 6.25)	%	9.9	11.0	7
Cattle	dig prot %	1.0	1.1	
Sheep	dig prot %	6.2	6.9	
Energy				
Cattle	DE kcal/kg	1111.	1234.	
Sheep	DE kcal/kg	3016.	3351.	
Cattle	ME kcal/kg	911.	1012.	
Sheep	ME kcal/kg	2473.	2748.	
Cattle	TDN %	25.	28.	
Sheep	TDN %	68.	76.	
Magnesium	%	.24	.27	

SORGHUM, KALO. Sorghum vulgare

Sorghum, kalo, grain, (4)

Ref no 4-04-430

Dry matter	%	89.0	100.0	
Ash	%	1.5	1.7	
Crude fiber	%	2.0	2.2	
Ether extract	%	3.2	3.6	
N-free extract	%	70.5	79.2	
Protein (N x 6.25)	%	11.8	13.3	

Continued

Feed name or analyses		Mean		C.V.
		As fed	Dry	± %
Cattle	dig prot %	6.8	7.6	
Sheep	dig prot %	7.9	8.9	
Energy				
Cattle	DE kcal/kg	3178.	3571.	
Sheep	DE kcal/kg	3375.	3792.	
Cattle	ME kcal/kg	2606.	2928.	
Sheep	ME kcal/kg	2767.	3109.	
Cattle	TDN %	72.	81.	
Sheep	TDN %	76.	86.	

SORGHUM, KAOLIANG. Sorghum vulgare

Sorghum, kaoliang, grain, (4)

Ref no 4-04-431

Dry matter	%	90.0	100.0	3
Ash	%	1.7	1.9	28
Crude fiber	%	2.0	2.2	50
Ether extract	%	4.2	4.7	15
N-free extract	%	71.8	79.8	
Protein (N x 6.25)	%	10.3	11.4	10
Cattle	dig prot %	5.8	6.5	
Sheep	dig prot %	6.8	7.6	
Energy	GE kcal/kg	3348.	3720.	
Cattle	DE kcal/kg	3254.	3616.	
Sheep	DE kcal/kg	3452.	3836.	
Cattle	ME kcal/kg	2668.	2965.	
Sheep	ME kcal/kg	2831.	3146.	
Cattle	TDN %	74.	82.	
Sheep	TDN %	78.	87.	

SORGHUM, MILO. Sorghum vulgare

Sorghum, milo, aerial pt, s-c, (1)
Sorghum, milo, fodder, sun-cured

Ref no 1-04-433

Dry matter	%	88.0	100.0	2
Ash	%	7.1	8.1	7
Crude fiber	%	21.3	24.2	8
Ether extract	%	2.7	3.1	17
N-free extract	%	49.1	55.8	
Protein (N x 6.25)	%	7.7	8.8	15
Cattle	dig prot %	2.9	3.3	
Sheep	dig prot %	2.9	3.3	
Energy				
Cattle	DE kcal/kg	2289.	2601.	
Sheep	DE kcal/kg	2677.	3042.	
Cattle	ME kcal/kg	1877.	2133.	
Sheep	ME kcal/kg	2195.	2494.	

Continued

(1) dry forages and roughages
(2) pasture, range plants, and forages fed green

(3) silages
(4) energy feeds

Feed name or analyses		Mean		C.V.
		As fed	Dry	± %
Cattle	TDN %	52.	59.	
Sheep	TDN %	61.	69.	
Calcium	%	.35	.40	
Phosphorus	%	.17	.19	

Sorghum, milo, aerial pt wo heads, s-c, (1)
Sorghum, milo, stover, sun-cured

Ref no 1-04-434

Dry matter	%	93.0	100.0	
Ash	%	10.0	10.8	
Crude fiber	%	32.5	34.9	
Ether extract	%	1.5	1.6	
N-free extract	%	45.7	49.1	
Protein (N x 6.25)	%	3.3	3.6	
Cattle	dig prot %	1.1	1.2	
Sheep	dig prot %	.6	.6	
Energy				
Cattle	DE kcal/kg	2337.	2513.	
Sheep	DE kcal/kg	2051.	2205.	
Cattle	ME kcal/kg	1917.	2061.	
Sheep	ME kcal/kg	1681.	1808.	
Cattle	TDN %	53.	57.	
Sheep	TDN %	46.	50.	

Sorghum, milo, grain, (4)

Ref no 4-04-444

Dry matter	%	89.0	100.0	2
Ash	%	1.7	1.9	44
Crude fiber	%	2.0	2.2	38
Ether extract	%	2.8	3.1	18
N-free extract	%	71.6	80.4	
Protein (N x 6.25)	%	11.0	12.4	8
Cattle	dig prot %	6.3	7.1	
Sheep	dig prot %	8.6	9.7	
Swine	dig prot %	7.8	8.8	
Energy	GE kcal/kg	3906.	4389.	
Cattle	DE kcal/kg	3139.	3527.	
Sheep	DE kcal/kg	3689.	4145.	
Swine	DE kcal/kg	3453.	3880.	
Cattle	ME kcal/kg	2574.	2892.	
Sheep	ME kcal/kg	3025.	3399.	
Swine	ME kcal/kg	3229.	3628.	
Cattle	NE_m kcal/kg	1896.	2130.	
Cattle	NE_gain kcal/kg	1246.	1400.	
Cattle	TDN %	71.	80.	
Sheep	TDN %	84.	94.	
Swine	TDN %	78.	88.	
Calcium	%	.04	.04	99
Magnesium	%	.20	.22	26
Phosphorus	%	.29	.33	19

Continued

Potassium	%	.35	.39	7
Sodium	%	.01	.01	
Cobalt	mg/kg	.100	.100	62
Copper	mg/kg	14.1	15.8	33
Manganese	mg/kg	12.9	14.5	49
Choline	mg/kg	678.	761.	17
Niacin	mg/kg	42.7	48.0	35
Pantothenic acid	mg/kg	11.4	12.8	42
Riboflavin	mg/kg	1.2	1.3	23
Thiamine	mg/kg	3.9	4.4	14
Vitamin B_6	mg/kg	4.10	4.60	
Arginine	%	.36	.40	14
Cystine	%	.18	.20	95
Histidine	%	.27	.30	18
Isoleucine	%	.53	.60	11
Leucine	%	1.42	1.60	12
Lysine	%	.27	.30	27
Methionine	%	.09	.10	61
Phenylalanine	%	.45	.51	26
Threonine	%	.27	.30	
Tryptophan	%	.09	.10	31
Tyrosine	%	.36	.40	27
Valine	%	.53	.60	5

Sorghum, milo, grits, cracked f-scr, (4)
Milo grits

Ref no 4-04-445

Dry matter	%	88.0	100.0	
Ash	%	.7	.8	
Crude fiber	%	1.0	1.1	
Ether extract	%	1.5	1.7	
N-free extract	%	73.7	83.8	
Protein (N x 6.25)	%	11.1	12.6	
Cattle	dig prot %	6.3	7.2	
Sheep	dig prot %	8.6	9.8	
Energy				
Cattle	DE kcal/kg	3104.	3527.	
Sheep	DE kcal/kg	3648.	4145.	
Cattle	ME kcal/kg	2545.	2892.	
Sheep	ME kcal/kg	2991.	3399.	
Cattle	TDN %	70.	80.	
Sheep	TDN %	83.	94.	

Sorghum, milo, heads, chop, (4)

Ref no 4-04-446

Dry matter	%	90.0	100.0	1
Ash	%	4.1	4.6	34
Crude fiber	%	7.0	7.8	26
Ether extract	%	2.5	2.8	10
N-free extract	%	66.3	73.7	

Continued

Feed name or analyses		Mean		C.V.
		As fed	Dry	± %
Protein (N x 6.25)	%	10.0	11.1	11
Cattle	dig prot %	5.7	6.3	
Sheep	dig prot %	7.6	8.4	
Energy				
Cattle	DE kcal/kg	3016.	3351.	
Sheep	DE kcal/kg	3373.	3748.	
Cattle	ME kcal/kg	2473.	2748.	
Sheep	ME kcal/kg	2766.	3073.	
Cattle	TDN %	68.	76.	
Sheep	TDN %	76.	85.	

Feed name or analyses		Mean		C.V.
		As fed	Dry	± %
Sheep	ME kcal/kg	2735.	3073.	
Cattle	TDN %	74.	83.	
Sheep	TDN %	76.	85.	
Biotin	mg/kg	.20	.20	
Folic acid	mg/kg	.20	.20	
Niacin	mg/kg	69.7	72.7	
Pantothenic acid	mg/kg	7.9	8.9	
Riboflavin	mg/kg	5.9	6.6	

Sorghum, milo, distil grains, dehy, (5)

Ref no 5-04-438

		As fed	Dry	C.V.
Dry matter	%	94.0	100.0	1
Ash	%	2.2	2.3	66
Crude fiber	%	11.0	11.7	16
Ether extract	%	11.9	12.7	25
N-free extract	%	29.8	31.7	
Protein (N x 6.25)	%	39.1	41.6	11
Cattle	dig prot %	32.1	34.2	
Sheep	dig prot %	33.2	35.3	
Energy				
Cattle	DE kcal/kg	3647.	3880.	
Sheep	DE kcal/kg	3440.	3660.	
Cattle	ME kcal/kg	2991.	3182.	
Sheep	ME kcal/kg	2821.	3001.	
Cattle	TDN %	83.	88.	
Sheep	TDN %	78.	83.	
Biotin	mg/kg	.40	.40	
Choline	mg/kg	768.	817.	
Niacin	mg/kg	65.1	69.3	
Pantothenic acid	mg/kg	5.7	6.1	
Riboflavin	mg/kg	2.9	3.1	
Thiamine	mg/kg	.7	.7	

Sorghum, milo, distil grains w sol, dehy, mn 75 orig solids, (5)

Ref no 5-04-439

		As fed	Dry	C.V.
Dry matter	%	89.0	100.0	1
Ash	%	4.0	4.5	7
Crude fiber	%	10.0	11.2	15
Ether extract	%	8.0	9.0	15
N-free extract	%	35.6	40.0	
Protein (N x 6.25)	%	31.4	35.3	5
Cattle	dig prot %	25.3	28.4	
Sheep	dig prot %	26.3	29.6	
Energy				
Cattle	DE kcal/kg	3257.	3660.	
Sheep	DE kcal/kg	3336.	3748.	
Cattle	ME kcal/kg	2671.	3001.	

Sorghum, milo, distil sol, dehy, (5)

Ref no 5-04-440

		As fed	Dry	C.V.
Dry matter	%	95.0	100.0	2
Ash	%	11.9	12.5	21
Crude fiber	%	3.0	3.2	22
Ether extract	%	7.2	7.6	21
N-free extract	%	47.9	50.4	
Protein (N x 6.25)	%	25.0	26.3	17
Cattle	dig prot %	19.2	20.2	
Sheep	dig prot %	20.2	21.3	
Energy				
Cattle	DE kcal/kg	3435.	3616.	
Sheep	DE kcal/kg	3644.	3836.	
Cattle	ME kcal/kg	2817.	2965.	
Sheep	ME kcal/kg	2989.	3146.	
Cattle	TDN %	78.	82.	
Sheep	TDN %	83.	87.	
Biotin	mg/kg	1.50	1.60	
Choline	mg/kg	2222.	2340.	
Folic acid	mg/kg	1.30	1.40	
Niacin	mg/kg	140.8	148.3	11
Pantothenic acid	mg/kg	26.4	27.8	6
Riboflavin	mg/kg	14.7	15.5	20
Thiamine	mg/kg	4.8	5.1	
Arginine	%	1.00	1.05	
Histidine	%	.90	.95	
Isoleucine	%	.90	.95	
Leucine	%	1.70	1.79	
Lysine	%	.90	.95	
Methionine	%	.50	.53	
Phenylalanine	%	1.80	1.89	
Threonine	%	1.00	1.05	
Tryptophan	%	.30	.32	
Tyrosine	%	.90	.95	
Valine	%	2.00	2.11	

Continued

(1) dry forages and roughages
(2) pasture, range plants, and forages fed green

(3) silages
(4) energy feeds

Feed name or analyses		Mean		C.V.
		As fed	Dry	± %

SORGHUM, SCHROCK. Sorghum vulgare

Sorghum, schrock, grain, (4)

Ref no 4-04-452

		As fed	Dry	C.V.
Dry matter	%	89.0	100.0	2
Ash	%	1.4	1.6	13
Crude fiber	%	3.0	3.4	32
Ether extract	%	3.3	3.7	9
N-free extract	%	71.1	79.9	
Protein (N x 6.25)	%	10.1	11.4	6
Cattle	dig prot %	5.8	6.5	
Sheep	dig prot %	6.8	7.6	
Energy				
Cattle	DE kcal/kg	3178.	3571.	
Sheep	DE kcal/kg	3375.	3792.	
Cattle	ME kcal/kg	2606.	2928.	
Sheep	ME kcal/kg	2767.	3109.	
Cattle	TDN %	72.	81.	
Sheep	TDN %	76.	86.	

SORGHUM, SHALLU. Sorghum vulgare

Sorghum, shallu, aerial pt, s-c, mature, (1)
Sorghum, shallu, fodder, sun-cured, mature

Ref no 1-04-453

		As fed	Dry	
Dry matter	%	93.0	100.0	
Ash	%	7.9	8.5	
Crude fiber	%	35.4	38.1	
Ether extract	%	1.3	1.4	
N-free extract	%	45.6	49.0	
Protein (N x 6.25)	%	2.8	3.0	
Cattle	dig prot %	1.0	1.1	
Sheep	dig prot %	.3	.3	
Energy				
Cattle	DE kcal/kg	2378.	2557.	
Sheep	DE kcal/kg	2051.	2205.	
Cattle	ME kcal/kg	1950.	2097.	
Sheep	ME kcal/kg	1681.	1808.	
Cattle	TDN %	54.	58.	
Sheep	TDN %	46.	50.	

Sorghum, shallu, aerial pt wo heads, s-c, (1)
Sorghum, shallu, stover, sun-cured

Ref no 1-04-455

		As fed	Dry	
Dry matter	%	93.0	100.0	
Ash	%	7.9	8.5	
Crude fiber	%	35.5	38.2	
Ether extract	%	1.4	1.5	
N-free extract	%	45.4	48.8	
Protein (N x 6.25)	%	2.8	3.0	
Cattle	dig prot %	.9	1.0	
Sheep	dig prot %	-.3	-.3	
Energy				
Cattle	DE kcal/kg	2378.	2557.	
Sheep	DE kcal/kg	2051.	2205.	
Cattle	ME kcal/kg	1950.	2097.	
Sheep	ME kcal/kg	1681.	1808.	
Cattle	TDN %	54.	58.	
Sheep	TDN %	46.	50.	

Sorghum, shallu, grain, (4)

Ref no 4-04-456

		As fed	Dry	C.V.
Dry matter	%	90.0	100.0	1
Ash	%	1.7	1.9	20
Crude fiber	%	2.0	2.2	10
Ether extract	%	3.7	4.1	8
N-free extract	%	69.4	77.1	
Protein (N x 6.25)	%	13.2	14.7	6
Cattle	dig prot %	7.6	8.4	
Sheep	dig prot %	8.8	9.8	
Energy				
Cattle	DE kcal/kg	3214.	3571.	
Sheep	DE kcal/kg	3413.	3792.	
Cattle	ME kcal/kg	2635.	2928.	
Sheep	ME kcal/kg	2798.	3109.	
Cattle	TDN %	73.	81.	
Sheep	TDN %	77.	86.	

Sorghum, shallu, heads, chop, (4)

Ref no 4-04-457

		As fed	Dry	C.V.
Dry matter	%	90.0	100.0	1
Ash	%	2.9	3.2	11
Crude fiber	%	9.0	10.0	12
Ether extract	%	3.4	3.8	9
N-free extract	%	62.5	69.4	
Protein (N x 6.25)	%	12.2	13.6	10
Cattle	dig prot %	7.0	7.8	
Sheep	dig prot %	7.7	8.6	

Continued

(5) protein supplements

(6) minerals

(7) vitamins

(8) additives

Feed name or analyses		Mean		C.V.
		As fed	Dry	± %
Energy				
Cattle	DE kcal/kg	3016.	3351.	
Sheep	DE kcal/kg	3016.	3351.	
Cattle	ME kcal/kg	2473.	2748.	
Sheep	ME kcal/kg	2473.	2748.	
Cattle	TDN %	68.	76.	
Sheep	TDN %	68.	76.	

SORGHUM, SORGO. Sorghum vulgare saccharatum

Sorghum, sorgo, aerial pt, s-c, (1)
Sorghum, sorgo, fodder, sun-cured

Ref no 1-04-460

Dry matter	%	81.7	100.0	7
Ash	%	6.4	7.8	23
Crude fiber	%	21.9	26.8	14
Ether extract	%	2.4	2.9	16
N-free extract	%	45.3	55.5	
Protein (N x 6.25)	%	5.7	7.0	17
Cattle	dig prot %	2.2	2.7	
Sheep	dig prot %	2.2	2.7	
Energy				
Cattle	DE kcal/kg	2125.	2601.	
Sheep	DE kcal/kg	2125.	2601.	
Cattle	ME kcal/kg	1743.	2133.	
Sheep	ME kcal/kg	1743.	2133.	
Cattle	TDN %	48.	59.	
Sheep	TDN %	48.	59.	
Calcium	%	.31	.38	19
Chlorine	%	.51	.63	
Magnesium	%	.29	.35	10
Phosphorus	%	.12	.15	31
Potassium	%	1.18	1.45	12
Manganese	mg/kg	106.9	130.8	20

Sorghum, sorgo, aerial pt, ensiled, (3)
Sorghum, sorgo, fodder silage

Ref no 3-04-468

Dry matter	%	26.0	100.0	13
Ash	%	1.8	6.8	40
Crude fiber	%	7.0	26.8	10
Ether extract	%	.8	3.1	40
N-free extract	%	14.8	57.0	
Protein (N x 6.25)	%	1.6	6.3	21
Cattle	dig prot %	.4	1.7	
Sheep	dig prot %	.4	1.4	
Energy				
Cattle	DE kcal/kg	665.	2557.	
Sheep	DE kcal/kg	676.	2601.	

Continued

(1) dry forages and roughages
(2) pasture, range plants, and forages fed green

Feed name or analyses		Mean		C.V.
		As fed	Dry	± %
Cattle	ME kcal/kg	545.	2097.	
Sheep	ME kcal/kg	554.	2133.	
Cattle	NE m kcal/kg	385.	1480.	
Cattle	NE gain kcal/kg	174.	670.	
Cattle	TDN %	15.	58.	
Sheep	TDN %	15.	59.	
Calcium	%	.09	.35	80
Iron	%	.005	.020	10
Magnesium	%	.07	.27	12
Phosphorus	%	.05	.20	60
Potassium	%	.32	1.22	11
Copper	mg/kg	8.1	31.3	15

Sorghum, sorgo, grain, (4)

Ref no 4-04-469

Dry matter	%	89.0	100.0	2
Ash	%	1.4	1.6	15
Crude fiber	%	2.0	2.2	
Ether extract	%	3.2	3.6	16
N-free extract	%	72.8	81.8	
Protein (N x 6.25)	%	9.6	10.8	12
Cattle	dig prot %	5.5	6.2	
Sheep	dig prot %	5.9	6.6	
Energy				
Cattle	DE kcal/kg	3218.	3616.	
Sheep	DE kcal/kg	3414.	3836.	
Cattle	ME kcal/kg	2639.	2965.	
Sheep	ME kcal/kg	2780.	3146.	
Cattle	TDN %	73.	82.	
Sheep	TDN %	77.	87.	
Niacin	mg/kg	18.0	20.2	4
Thiamine	mg/kg	3.6	4.0	8

Sorghum stover - see Sorghum, aerial pt wo heads

SORGHUM, SUDANGRASS. Sorghum vulgare sudanense

Sorghum, Sudangrass, aerial pt, dehy, immature, (1)

Ref no 1-04-470

Dry matter	%	87.7	100.0	2
Ash	%	9.3	10.6	6
Crude fiber	%	19.9	22.7	8
Ether extract	%	2.4	2.7	35
N-free extract	%	41.5	47.3	
Protein (N x 6.25)	%	14.7	16.7	10
Cattle	dig prot %	9.4	10.7	
Sheep	dig prot %	9.4	10.7	

Continued

(3) silages
(4) energy feeds

Feed name or analyses		Mean As fed	Mean Dry	C.V. ± %
Energy				
Cattle	DE kcal/kg	2591.	2954.	
Sheep	DE kcal/kg	2011.	2293.	
Cattle	ME kcal/kg	2124.	2422.	
Sheep	ME kcal/kg	1649.	1880.	
Cattle	TDN %	59.	67.	
Sheep	TDN %	46.	52.	

Sorghum, Sudangrass, hay, s-c, immature, (1)

Ref no 1-04-473

		As fed	Dry	C.V.
Dry matter	%	88.0	100.0	4
Ash	%	9.2	10.5	7
Crude fiber	%	22.5	25.6	10
Ether extract	%	2.3	2.6	44
N-free extract	%	40.2	45.7	
Protein (N x 6.25)	%	13.7	15.6	15
Cattle	dig prot %	5.9	6.7	
Sheep	dig prot %	7.0	8.0	
Energy				
Cattle	DE kcal/kg	2250.	2557.	
Sheep	DE kcal/kg	2018.	2293.	
Cattle	ME kcal/kg	1845.	2097.	
Sheep	ME kcal/kg	1654.	1880.	
Cattle	NE$_m$ kcal/kg	1038.	1180.	
Cattle	NE$_{gain}$ kcal/kg	431.	490.	
Cattle	TDN %	51.	58.	
Sheep	TDN %	46.	52.	
Calcium	%	.75	.85	16
Magnesium	%	.35	.40	27
Phosphorus	%	.32	.36	62
Potassium	%	.92	1.04	14

Sorghum, Sudangrass, hay, s-c, mid-blm, (1)

Ref no 1-04-475

		As fed	Dry	C.V.
Dry matter	%	91.2	100.0	1
Ash	%	10.5	11.5	10
Crude fiber	%	25.9	28.4	5
Ether extract	%	1.9	2.1	24
N-free extract	%	40.8	44.7	
Protein (N x 6.25)	%	12.1	13.3	10
Cattle	dig prot %	5.2	5.7	
Sheep	dig prot %	7.8	8.5	
Cellulose	%	30.9	33.9	
Lignin	%	13.6	14.9	
Energy				
Cattle	DE kcal/kg	2292.	2513.	
Sheep	DE kcal/kg	2091.	2293.	
Cattle	ME kcal/kg	1880.	2061.	

Continued

Feed name or analyses		Mean As fed	Mean Dry	C.V. ± %
Sheep	ME kcal/kg	1714.	1880.	
Cattle	TDN %	52.	57.	
Sheep	TDN %	47.	52.	

Sorghum, Sudangrass, hay, s-c, full blm, (1)

Ref no 1-04-476

		As fed	Dry	C.V.
Dry matter	%	88.5	100.0	1
Ash	%	7.4	8.4	5
Crude fiber	%	29.7	33.5	6
Ether extract	%	1.6	1.8	12
N-free extract	%	40.7	46.0	
Protein (N x 6.25)	%	9.1	10.3	13
Cattle	dig prot %	4.2	4.8	
Sheep	dig prot %	5.2	5.9	
Energy				
Cattle	DE kcal/kg	2420.	2734.	
Sheep	DE kcal/kg	1951.	2205.	
Cattle	ME kcal/kg	1984.	2242.	
Sheep	ME kcal/kg	1600.	1808.	
Cattle	TDN %	55.	62.	
Sheep	TDN %	44.	50.	

Sorghum, Sudangrass, hay, s-c, (1)

Ref no 1-04-480

		As fed	Dry	C.V.
Dry matter	%	88.9	100.0	4
Ash	%	8.5	9.6	16
Crude fiber	%	25.7	28.9	10
Ether extract	%	2.0	2.2	50
N-free extract	%	41.4	46.6	
Protein (N x 6.25)	%	11.3	12.7	24
Cattle	dig prot %	4.9	5.5	
Sheep	dig prot %	5.8	6.5	
Cellulose	%	28.4	31.9	
Lignin	%	12.1	13.6	
Energy	GE kcal/kg	3756.	4225.	
Cattle	DE kcal/kg	2312.	2601.	
Sheep	DE kcal/kg	2038.	2293.	
Cattle	ME kcal/kg	1896.	2133.	
Sheep	ME kcal/kg	1671.	1880.	
Cattle	NE$_m$ kcal/kg	1031.	1160.	
Cattle	NE$_{gain}$ kcal/kg	409.	460.	
Cattle	TDN %	52.	59.	
Sheep	TDN %	46.	52.	
Calcium	%	.50	.56	33
Iron	%	.020	.020	23
Magnesium	%	.36	.40	30
Phosphorus	%	.28	.31	55
Potassium	%	1.37	1.54	32
Sodium	%	.02	.02	40
Sulfur	%	.05	.06	29

Continued

(5) protein supplements
(6) minerals

(7) vitamins
(8) additives

Feed name or analyses		Mean		C.V.
		As fed	Dry	± %
Cobalt	mg/kg	.120	.130	
Copper	mg/kg	32.8	36.8	
Manganese	mg/kg	83.0	93.3	26

Sorghum, Sudangrass, aerial pt, fresh, immature, (2)

Ref no 2-04-484

Dry matter	%	17.6	100.0	9
Ash	%	1.6	9.0	17
Crude fiber	%	5.4	30.9	12
Ether extract	%	.7	3.9	18
N-free extract	%	6.9	39.4	
Protein (N x 6.25)	%	3.0	16.8	20
Cattle	dig prot %	2.1	12.2	
Sheep	dig prot %	2.0	11.6	
Energy				
Cattle	DE kcal/kg	543.	3086.	
Sheep	DE kcal/kg	504.	2866.	
Cattle	ME kcal/kg	445.	2530.	
Sheep	ME kcal/kg	414.	2350.	
Cattle	TDN %	12.	70.	
Sheep	TDN %	11.	65.	

Sorghum, Sundangrass, aerial pt, fresh, mid-blm, (2)

Ref no 2-04-485

Dry matter	%	22.7	100.0	9
Ash	%	2.4	10.4	6
Crude fiber	%	8.2	36.1	3
Ether extract	%	.4	1.8	12
N-free extract	%	9.8	43.0	
Protein (N x 6.25)	%	2.0	8.7	18
Cattle	dig prot %	1.2	5.3	
Sheep	dig prot %	1.4	6.0	
Energy				
Cattle	DE kcal/kg	631.	2778.	
Sheep	DE kcal/kg	621.	2734.	
Cattle	ME kcal/kg	517.	2278.	
Sheep	ME kcal/kg	509.	2242.	
Cattle	TDN %	14.	63.	
Sheep	TDN %	14.	62.	

Sorghum, Sudangrass, aerial pt, fresh, full blm, (2)

Ref no 2-04-486

Dry matter	%	23.6	100.0	8
Ash	%	2.6	11.0	9
Crude fiber	%	8.6	36.4	7
Ether extract	%	.4	1.7	14
N-free extract	%	10.1	42.8	

Feed name or analyses		Mean		C.V.
		As fed	Dry	± %
Protein (N x 6.25)	%	1.9	8.1	9
Cattle	dig prot %	1.1	4.8	
Sheep	dig prot %	1.3	5.6	
Energy				
Cattle	DE kcal/kg	645.	2734.	
Sheep	DE kcal/kg	645.	2734.	
Cattle	ME kcal/kg	529.	2242.	
Sheep	ME kcal/kg	529.	2242.	
Cattle	TDN %	15.	62.	
Sheep	TDN %	15.	62.	

Sorghum, Sudangrass, aerial pt, fresh, mature, (2)

Ref no 2-04-487

Dry matter	%	32.0	100.0	9
Ash	%	2.8	8.8	13
Crude fiber	%	12.1	37.9	11
Ether extract	%	.5	1.7	5
N-free extract	%	14.9	46.5	
Protein (N x 6.25)	%	1.6	5.1	15
Cattle	dig prot %	.7	2.2	
Sheep	dig prot %	1.1	3.5	
Energy				
Cattle	DE kcal/kg	847.	2646.	
Sheep	DE kcal/kg	889.	2778.	
Cattle	ME kcal/kg	694.	2170.	
Sheep	ME kcal/kg	729.	2278.	
Cattle	TDN %	19.	60.	
Sheep	TDN %	20.	63.	
Carotene	mg/kg	18.8	58.9	
Vitamin A equiv	IU/g	31.3	98.2	

Sorghum, Sudangrass, aerial pt, fresh, (2)

Ref no 2-04-489

Dry matter	%	21.8	100.0	25
Ash	%	2.1	9.8	11
Crude fiber	%	7.0	32.2	15
Ether extract	%	.5	2.5	29
N-free extract	%	9.7	44.3	
Protein (N x 6.25)	%	2.4	11.2	34
Cattle	dig prot %	1.6	7.4	
Sheep	dig prot %	1.7	7.7	
Energy				
Cattle	DE kcal/kg	625.	2866.	
Sheep	DE kcal/kg	606.	2778.	
Cattle	ME kcal/kg	512.	2350.	
Sheep	ME kcal/kg	497.	2278.	
Cattle	NEm kcal/kg	307.	1410.	
Cattle	NEgain kcal/kg	179.	820.	
Cattle	TDN %	14.	65.	
Sheep	TDN %	14.	63.	

Continued

(1) dry forages and roughages
(2) pasture, range plants, and forages fed green
(3) silages
(4) energy feeds

Feed name or analyses		Mean		C.V.
		As fed	Dry	± %
Calcium	%	.09	.43	22
Iron	%	.010	.020	88
Magnesium	%	.08	.35	10
Phosphorus	%	.09	.41	41
Potassium	%	.47	2.14	24
Sulfur	%	.02	.11	52
Cobalt	mg/kg	.030	.130	
Copper	mg/kg	7.8	35.9	
Manganese	mg/kg	17.7	81.4	32
Carotene	mg/kg	39.8	182.8	27
Vitamin A equiv	IU/g	66.3	304.7	

Sorghum, Sudangrass, aerial pt, ensiled, (3)

Ref no 3-04-499

		As fed	Dry	± %
Dry matter	%	23.3	100.0	17
Ash	%	2.1	8.9	40
Crude fiber	%	8.0	34.4	10
Ether extract	%	.7	3.1	29
N-free extract	%	10.1	43.4	
Protein (N x 6.25)	%	2.4	10.2	27
Cattle	dig prot %	1.3	5.6	
Sheep	dig prot %	1.6	7.0	
Energy				
Cattle	DE kcal/kg	606.	2601.	
Sheep	DE kcal/kg	575.	2469.	
Cattle	ME kcal/kg	497.	2133.	
Sheep	ME kcal/kg	472.	2024.	
Cattle	TDN %	14.	59.	
Sheep	TDN %	13.	56.	
Calcium	%	.15	.64	
Iron	%	.	.010	
Magnesium	%	.11	.49	
Phosphorus	%	.05	.23	
Potassium	%	.72	3.07	
Copper	mg/kg	8.5	36.6	
Manganese	mg/kg	23.0	98.8	

SORGHUM, SUMAC. Sorghum vulgare

Sorghum, sumac, aerial pt, s-c, mature, (1)
Sorghum, sumac, fodder, sun-cured, mature

Ref no 1-04-502

		As fed	Dry	± %
Dry matter	%	85.1	100.0	5
Ash	%	6.6	7.8	10
Crude fiber	%	25.4	29.8	7
Ether extract	%	1.9	2.3	12
N-free extract	%	44.8	52.7	
Protein (N x 6.25)	%	6.3	7.4	
Cattle	dig prot %	2.4	2.8	

Continued

(5) protein supplements
(6) minerals

Feed name or analyses		Mean		C.V.
		As fed	Dry	± %
Sheep	dig prot %	2.4	2.8	
Energy				
Cattle	DE kcal/kg	2213.	2601.	
Sheep	DE kcal/kg	2101.	2469.	
Cattle	ME kcal/kg	1815.	2133.	
Sheep	ME kcal/kg	1722.	2024.	
Cattle	TDN %	50.	59.	
Sheep	TDN %	48.	56.	
Calcium	%	.30	.35	20
Phosphorus	%	.14	.16	35

Sorghum, sumac, grain, (4)

Ref no 4-04-509

		As fed	Dry	± %
Dry matter	%	85.0	100.0	8
Ash	%	1.9	2.2	50
Crude fiber	%	3.0	3.5	65
Ether extract	%	3.3	3.9	11
N-free extract	%	66.6	78.3	
Protein (N x 6.25)	%	10.3	12.1	13
Cattle	dig prot %	5.9	6.9	
Sheep	dig prot %	6.9	8.1	
Energy				
Cattle	DE kcal/kg	2998.	3527.	
Sheep	DE kcal/kg	3223.	3792.	
Cattle	ME kcal/kg	2458.	2892.	
Sheep	ME kcal/kg	2643.	3109.	
Cattle	TDN %	68.	80.	
Sheep	TDN %	73.	86.	
Niacin	mg/kg	27.1	31.9	17
Pantothenic acid	mg/kg	12.9	15.2	8
Riboflavin	mg/kg	1.5	1.8	27

Sorghum, sumac, heads, chop, (4)

Ref no 4-04-510

		As fed	Dry	
Dry matter	%	75.0	100.0	
Ash	%	3.2	4.3	
Crude fiber	%	10.0	13.3	
Ether extract	%	2.3	3.1	
N-free extract	%	52.0	69.3	
Protein (N x 6.25)	%	7.5	10.0	
Cattle	dig prot %	4.3	5.7	
Sheep	dig prot %	4.7	6.3	
Energy				
Cattle	DE kcal/kg	2447.	3263.	
Sheep	DE kcal/kg	2480.	3307.	
Cattle	ME kcal/kg	2007.	2676.	
Sheep	ME kcal/kg	2034.	2712.	
Cattle	TDN %	56.	74.	
Sheep	TDN %	56.	75.	

(7) vitamins
(8) additives

Feed name or analyses		Mean		C.V.
		As fed	Dry	± %

Sorgo - see Sorghum, sorgo

Soy - see Soybean

SOYBEAN. Glycine max

Soybean, hay, s-c, immature, (1)

Ref no 1-04-536

Dry matter	%	91.9	100.0	2
Ash	%	8.5	9.2	24
Crude fiber	%	24.6	26.8	24
Ether extract	%	2.8	3.0	59
N-free extract	%	37.4	40.7	
Protein (N x 6.25)	%	18.7	20.3	15
Cattle	dig prot %	11.6	12.6	
Sheep	dig prot %	12.9	14.0	
Energy				
Cattle	DE kcal/kg	2107.	2293.	
Sheep	DE kcal/kg	2188.	2381.	
Cattle	ME kcal/kg	1728.	1880.	
Sheep	ME kcal/kg	1794.	1952.	
Cattle	TDN %	48.	52.	
Sheep	TDN %	50.	54.	
Calcium	%	1.02	1.11	13
Phosphorus	%	.21	.23	34

Soybean, hay, s-c, early blm, (1)

Ref no 1-04-537

Dry matter	%	90.7	100.0	1
Ash	%	8.2	9.0	25
Crude fiber	%	25.7	28.3	22
Ether extract	%	3.3	3.6	48
N-free extract	%	36.6	40.4	
Protein (N x 6.25)	%	17.0	18.7	10
Cattle	dig prot %	10.5	11.6	
Sheep	dig prot %	11.7	12.9	
Energy				
Cattle	DE kcal/kg	2080.	2293.	
Sheep	DE kcal/kg	2199.	2425.	
Cattle	ME kcal/kg	1705.	1880.	
Sheep	ME kcal/kg	1803.	1988.	
Cattle	TDN %	47.	52.	
Sheep	TDN %	50.	55.	

Soybean, hay, s-c, mid-blm, (1)

Ref no 1-04-538

Dry matter	%	93.6	100.0	
Ash	%	8.3	8.8	
Crude fiber	%	27.9	29.8	
Ether extract	%	5.1	5.4	
N-free extract	%	35.7	38.2	
Protein (N x 6.25)	%	16.7	17.8	
Cattle	dig prot %	10.3	11.0	
Sheep	dig prot %	11.5	12.3	
Energy				
Cattle	DE kcal/kg	2187.	2337.	
Sheep	DE kcal/kg	2270.	2425.	
Cattle	ME kcal/kg	1793.	1916.	
Sheep	ME kcal/kg	1861.	1988.	
Cattle	TDN %	50.	53.	
Sheep	TDN %	51.	55.	
Carotene	mg/kg	31.2	33.3	31
Vitamin A equiv	IU/g	52.0	55.5	

Soybean, hay, s-c, full blm, (1)

Ref no 1-04-539

Dry matter	%	89.6	100.0	4
Ash	%	7.6	8.5	26
Crude fiber	%	27.4	30.6	16
Ether extract	%	3.0	3.3	24
N-free extract	%	36.6	40.8	
Protein (N x 6.25)	%	15.1	16.8	8
Cattle	dig prot %	9.3	10.4	
Sheep	dig prot %	11.1	12.4	
Energy				
Cattle	DE kcal/kg	2054.	2293.	
Sheep	DE kcal/kg	2330.	2601.	
Cattle	ME kcal/kg	1684.	1880.	
Sheep	ME kcal/kg	1911.	2133.	
Cattle	TDN %	46.	52.	
Sheep	TDN %	53.	59.	
Calcium	%	1.18	1.31	19
Iron	%	.020	.020	33
Magnesium	%	.74	.83	15
Phosphorus	%	.19	.21	26
Potassium	%	.82	.91	18
Sodium	%	.02	.02	
Sulfur	%	.16	.18	
Manganese	mg/kg	76.2	85.1	18

(1) dry forages and roughages
(2) pasture, range plants, and forages fed green

(3) silages
(4) energy feeds

TABLE A-3-1 Composition of Feeds (Continued)

Feed name or analyses		Mean As fed	Mean Dry	C.V. ± %
Soybean, hay, s-c, mature, (1)				
Ref no 1-04-543				
Dry matter	%	92.1	100.0	2
Ash	%	6.9	7.5	30
Crude fiber	%	37.9	41.2	17
Ether extract	%	2.2	2.4	64
N-free extract	%	34.4	37.4	
Protein (N x 6.25)	%	10.6	11.5	22
Cattle	dig prot %	6.5	7.1	
Sheep	dig prot %	7.3	7.9	
Energy				
Cattle	DE kcal/kg	2071.	2249.	
Sheep	DE kcal/kg	2071.	2249.	
Cattle	ME kcal/kg	1698.	1844.	
Sheep	ME kcal/kg	1698.	1844.	
Cattle	TDN %	47.	51.	
Sheep	TDN %	47.	51.	
Calcium	%	.73	.79	
Phosphorus	%	.23	.25	
Carotene	mg/kg	6.7	7.3	42
Vitamin A equiv	IU/g	11.2	12.2	
Soybean, hay, s-c, over ripe, (1)				
Ref no 1-04-544				
Dry matter	%	89.0	100.0	1
Ash	%	7.2	8.1	12
Crude fiber	%	41.0	46.1	6
Ether extract	%	1.2	1.3	13
N-free extract	%	30.4	34.2	
Protein (N x 6.25)	%	9.2	10.3	18
Cattle	dig prot %	5.7	6.4	
Sheep	dig prot %	6.2	7.0	
Energy				
Cattle	DE kcal/kg	1962.	2205.	
Sheep	DE kcal/kg	1922.	2160.	
Cattle	ME kcal/kg	1609.	1808.	
Sheep	ME kcal/kg	1576.	1771.	
Cattle	TDN %	44.	50.	
Sheep	TDN %	44.	49.	
Soybean, hay, s-c, (1)				
Ref no 1-04-558				
Dry matter	%	89.2	100.0	3
Ash	%	7.2	8.1	24
Crude fiber	%	28.6	32.1	18
Ether extract	%	2.7	3.0	44
N-free extract	%	36.1	40.5	

Continued

Feed name or analyses		Mean As fed	Mean Dry	C.V. ± %
Protein (N x 6.25)	%	14.5	16.3	15
Cattle	dig prot %	9.0	10.1	
Sheep	dig prot %	10.0	11.2	
Cellulose	%	27.7	31.1	22
Lignin	%	10.1	11.4	39
Energy	GE kcal/kg	3902.	4375.	
Cattle	DE kcal/kg	2045.	2293.	
Sheep	DE kcal/kg	2124.	2381.	
Cattle	ME kcal/kg	1677.	1880.	
Sheep	ME kcal/kg	1741.	1952.	
Cattle	NE_m kcal/kg	1044.	1170.	
Cattle	NE_{gain} kcal/kg	428.	480.	
Cattle	TDN %	46.	52.	
Sheep	TDN %	48.	54.	
Calcium	%	1.15	1.29	28
Chlorine	%	.13	.15	
Iron	%	.030	.030	78
Magnesium	%	.70	.79	17
Phosphorus	%	.20	.23	60
Potassium	%	.86	.97	30
Sodium	%	.11	.12	99
Sulfur	%	.23	.26	17
Cobalt	mg/kg	.080	.090	
Copper	mg/kg	8.0	9.0	7
Manganese	mg/kg	82.6	92.6	36
Zinc	mg/kg	21.4	24.0	
Carotene	mg/kg	31.8	35.7	90
Vitamin A equiv	IU/g	53.0	59.5	
Vitamin D_2	IU/g	.7	.8	63
Soybean, hulls, (1)				
Soybean hulls (AAFCO)				
Soybran flakes				
Ref no 1-04-560				
Dry matter	%	91.3	100.0	2
Ash	%	4.6	5.0	15
Crude fiber	%	35.5	38.9	3
Ether extract	%	2.1	2.3	28
N-free extract	%	36.6	40.1	
Protein (N x 6.25)	%	12.5	13.7	5
Cattle	dig prot %	8.0	8.8	
Sheep	dig prot %	8.0	8.8	
Cellulose	%	47.6	52.1	
Lignin	%	5.9	6.5	
Energy				
Cattle	DE kcal/kg	1811.	1984.	
Sheep	DE kcal/kg	1731.	1896.	
Cattle	ME kcal/kg	1485.	1627.	
Sheep	ME kcal/kg	1420.	1555.	
Cattle	TDN %	41.	45.	
Sheep	TDN %	39.	43.	

Continued

(5) protein supplements
(6) minerals

(7) vitamins
(8) additives

Feed name or analyses		Mean		C.V.
		As fed	Dry	± %
Calcium	%	.54	.59	
Phosphorus	%	.16	.17	
Manganese	mg/kg	12.7	13.9	

Feed name or analyses		Mean		C.V.
		As fed	Dry	± %
Manganese	mg/kg	31.8	113.5	37
Carotene	mg/kg	21.7	77.6	46
Vitamin A equiv	IU/g	36.2	129.4	

Soybean, straw, (1)

Ref no 1-04-567

		As fed	Dry	C.V. ± %
Dry matter	%	87.6	100.0	2
Ash	%	5.6	6.4	9
Crude fiber	%	38.6	44.1	5
Ether extract	%	1.2	1.4	36
N-free extract	%	37.3	42.6	
Protein (N x 6.25)	%	4.8	5.5	22
Cattle	dig prot %	1.5	1.7	
Sheep	dig prot %	1.4	1.6	
Energy				
Cattle	DE kcal/kg	1467.	1675.	
Sheep	DE kcal/kg	1661.	1896.	
Cattle	ME kcal/kg	1204.	1374.	
Sheep	ME kcal/kg	1362.	1555.	
Cattle	TDN %	33.	38.	
Sheep	TDN %	38.	43.	
Calcium	%	1.39	1.59	12
Magnesium	%	.81	.92	15
Phosphorus	%	.05	.06	99
Potassium	%	.46	.53	15
Manganese	mg/kg	44.9	51.2	

Soybean, aerial pt, ensiled, (3)

Ref no 3-04-581

		As fed	Dry	C.V. ± %
Dry matter	%	28.0	100.0	15
Ash	%	2.8	9.9	26
Crude fiber	%	8.7	31.1	12
Ether extract	%	.9	3.1	46
N-free extract	%	11.5	41.2	
Protein (N x 6.25)	%	4.1	14.7	17
Cattle	dig prot %	2.6	9.3	
Sheep	dig prot %	3.2	11.6	
Energy				
Cattle	DE kcal/kg	667.	2381.	
Sheep	DE kcal/kg	679.	2425.	
Cattle	ME kcal/kg	546.	1952.	
Sheep	ME kcal/kg	557.	1988.	
Cattle	TDN %	15.	54.	
Sheep	TDN %	15.	55.	
Calcium	%	.35	1.25	17
Iron	%	.010	.040	22
Magnesium	%	.11	.38	15
Phosphorus	%	.14	.49	48
Potassium	%	.26	.93	12
Copper	mg/kg	2.6	9.3	7

Continued

Soybean, aerial pt, wilted ensiled, (3)

Ref no 3-04-584

		As fed	Dry	C.V. ± %
Dry matter	%	35.3	100.0	10
Ash	%	3.0	8.4	14
Crude fiber	%	8.9	25.3	7
Ether extract	%	.9	2.5	34
N-free extract	%	15.8	44.8	
Protein (N x 6.25)	%	6.7	19.0	10
Cattle	dig prot %	4.2	12.0	
Sheep	dig prot %	5.3	15.0	
Energy				
Cattle	DE kcal/kg	872.	2469.	
Sheep	DE kcal/kg	872.	2469.	
Cattle	ME kcal/kg	714.	2024.	
Sheep	ME kcal/kg	714.	2024.	
Cattle	TDN %	20.	56.	
Sheep	TDN %	20.	56.	
Calcium	%	.48	1.37	4
Iron	%	.010	.020	33
Magnesium	%	.16	.45	10
Phosphorus	%	.13	.36	14
Potassium	%	.33	.93	12
Sodium	%	.03	.09	
Sulfur	%	.11	.30	16
Copper	mg/kg	3.3	9.3	7
Manganese	mg/kg	46.6	132.1	9

Soybean, oil, (4)

Ref no 4-07-983

		As fed	Dry	
Dry matter	%	100.0	100.0	
Ether extract	%	100.0	100.0	

Soybean, flour, solv-extd f-sift, mx 3 fbr, (5)
Solvent extracted soy flour (AAFCO)
Solvent extracted soy grits (AAFCO)

Ref no 5-04-593

		As fed	Dry	
Dry matter	%	93.0	100.0	
Ash	%	5.6	6.0	
Crude fiber	%	2.0	2.2	
Ether extract	%	.8	.9	
N-free extract	%	32.6	35.0	
Protein (N x 6.25)	%	52.0	55.9	
Cattle	dig prot %	46.8	50.3	

Continued

(1) dry forages and roughages
(2) pasture, range plants, and forages fed green
(3) silages
(4) energy feeds

Feed name or analyses		Mean		C.V.
		As fed	Dry	± %
Sheep	dig prot %	48.8	52.5	
Swine	dig prot %	47.3	50.9	
Energy				
Cattle	DE kcal/kg	3445.	3704.	
Sheep	DE kcal/kg	3649.	3924.	
Swine	DE kcal/kg	3526.	3792.	
Cattle	ME kcal/kg	2824.	3037.	
Sheep	ME kcal/kg	2993.	3218.	
Swine	ME kcal/kg	2987.	3212.	
Cattle	TDN %	78.	84.	
Sheep	TDN %	83.	89.	
Swine	TDN %	80.	86.	
Calcium	%	.33	.35	
Iron	%	.020	.020	
Phosphorus	%	.62	.67	
Sodium	%	.34	.37	
Copper	mg/kg	16.1	17.3	
Manganese	mg/kg	31.9	34.3	
Zinc	mg/kg	20.0	21.5	
Biotin	mg/kg	.70	.80	
Choline	mg/kg	2246.	2415.	
Niacin	mg/kg	59.8	64.3	
Thiamine	mg/kg	1.5	1.6	
Arginine	%	3.10	3.33	
Cystine	%	.60	.65	
Histidine	%	.70	.75	
Lysine	%	4.20	4.52	
Methionine	%	.90	.97	
Phenylalanine	%	1.80	1.94	
Tryptophan	%	1.00	1.08	
Valine	%	.30	.32	

Soybean, flour by-prod, grnd, mn 13 prot mx 32 fbr, (5)
   Soybean mill feed (AAFCO)
   Soybean mill feed (CFA)

Ref no 5-04-594

		As fed	Dry	± %
Dry matter	%	93.0	100.0	1
Ash	%	4.5	4.8	4
Crude fiber	%	28.0	30.1	6
Ether extract	%	6.1	6.6	13
N-free extract	%	35.2	37.9	
Protein (N x 6.25)	%	19.2	20.6	10
Cattle	dig prot %	14.5	15.6	
Sheep	dig prot %	14.5	15.6	
Energy				
Cattle	DE kcal/kg	2461.	2646.	
Sheep	DE kcal/kg	2378.	2557.	
Cattle	ME kcal/kg	2018.	2170.	
Chickens	MEn kcal/kg	772.	830.	
Sheep	ME kcal/kg	1950.	2097.	
Cattle	TDN %	56.	60.	
Sheep	TDN %	54.	58.	

Soybean, seed, mech-extd grnd, mx 7 fbr, (5)
   Soybean meal (AAFCO)
   Soybean meal, expeller extracted
   Soybean meal, hydraulic extracted
   Soybean oil meal, expeller extracted
   Soybean oil meal, hydraulic extracted

Ref no 5-04-600

		As fed	Dry	± %
Dry matter	%	90.0	100.0	2
Ash	%	5.7	6.3	10
Crude fiber	%	6.0	6.7	16
Ether extract	%	4.7	5.2	19
N-free extract	%	29.8	33.1	
Protein (N x 6.25)	%	43.8	48.7	5
Cattle	dig prot %	37.3	41.4	
Sheep	dig prot %	39.4	43.8	
Swine	dig prot %	39.4	43.8	
Energy	GE kcal/kg	4332.	4813.	
Cattle	DE kcal/kg	3373.	3748.	
Sheep	DE kcal/kg	3294.	3660.	
Swine	DE kcal/kg	3476.	3862.	
Cattle	ME kcal/kg	2766.	3073.	
Chickens	MEn kcal/kg	2425.	2694.	
Sheep	ME kcal/kg	2701.	3001.	
Swine	ME kcal/kg	2996.	3329.	
Cattle	NEm kcal/kg	1854.	2060.	
Cattle	NEgain kcal/kg	1233.	1370.	
Cattle	TDN %	76.	85.	
Sheep	TDN %	75.	83.	
Swine	TDN %	79.	88.	
Calcium	%	.27	.30	26
Iron	%	.016	.018	29
Magnesium	%	.25	.28	12
Phosphorus	%	.63	.70	21
Potassium	%	1.71	1.90	1
Sodium	%	.24	.27	83
Cobalt	mg/kg	.200	.200	21
Copper	mg/kg	18.0	20.0	31
Manganese	mg/kg	32.3	35.9	23
Biotin	mg/kg	.30	.30	
Choline	mg/kg	2673.	2970.	13
Folic acid	mg/kg	6.60	7.30	32
Niacin	mg/kg	30.4	33.8	18
Thiamine	mg/kg	4.0	4.4	95
Arginine	%	2.60	2.89	13
Cystine	%	.60	.67	
Glycine	%	2.50	2.78	
Histidine	%	1.10	1.22	7
Isoleucine	%	2.80	3.11	19
Leucine	%	3.60	4.00	6
Lysine	%	2.70	3.00	10
Methionine	%	.80	.89	61
Phenylalanine	%	2.10	2.33	6
Threonine	%	1.70	1.89	14

Continued

(5) protein supplements
(6) minerals

(7) vitamins
(8) additives

Feed name or analyses		Mean		C.V.
		As fed	Dry	± %
Tryptophan	%	.60	.67	12
Tyrosine	%	1.40	1.56	26
Valine	%	2.20	2.44	5

Soybean, seed, solv-extd grnd, mx 7 fbr, (5)
Solvent extracted soybean meal (AAFCO)
Soybean meal, solvent extracted
Soybean oil meal, solvent extracted

Ref no 5-04-604

Dry matter	%	89.0	100.0	2
Ash	%	5.8	6.5	10
Crude fiber	%	6.0	6.7	20
Ether extract	%	.9	1.0	70
N-free extract	%	30.5	34.3	
Protein (N x 6.25)	%	45.8	51.5	7
Cattle	dig prot %	39.0	43.8	
Sheep	dig prot %	41.3	46.4	
Swine	dig prot %	41.7	46.9	
Energy	GE kcal/kg	4198.	4719.	
Cattle	DE kcal/kg	3178.	3571.	
Sheep	DE kcal/kg	3139.	3527.	
Swine	DE kcal/kg	3300.	3708.	
Cattle	ME kcal/kg	2606.	2928.	
Chickens	MEn kcal/kg	2249.	2527.	
Sheep	ME kcal/kg	2574.	2892.	
Swine	ME kcal/kg	2825.	3174.	
Cattle	NEm kcal/kg	1718.	1930.	
Cattle	NEgain kcal/kg	1148.	1290.	
Cattle	TDN %	72.	81.	
Sheep	TDN %	71.	80.	
Swine	TDN %	75.	84.	
Calcium	%	.32	.36	56
Iron	%	.012	.013	34
Magnesium	%	.27	.30	60
Phosphorus	%	.67	.75	21
Potassium	%	1.97	2.21	6
Sodium	%	.34	.38	68
Cobalt	mg/kg	.100	.100	63
Copper	mg/kg	36.3	40.8	12
Manganese	mg/kg	27.5	30.9	12
Choline	mg/kg	2743.	3083.	9
Folic acid	mg/kg	.70	.80	
Niacin	mg/kg	26.8	30.1	21
Pantothenic acid	mg/kg	14.5	16.3	28
Riboflavin	mg/kg	3.3	3.7	23
Thiamine	mg/kg	6.6	7.4	57
Arginine	%	3.20	3.60	14
Histidine	%	1.10	1.24	14
Isoleucine	%	2.50	2.81	18
Leucine	%	3.40	3.82	14
Lysine	%	2.90	3.26	13
Methionine	%	.60	.67	43
Phenylalanine	%	2.20	2.47	7

Continued

Feed name or analyses		Mean		C.V.
		As fed	Dry	± %
Threonine	%	1.70	1.91	5
Tryptophan	%	.60	.67	9
Tyrosine	%	1.40	1.57	14
Valine	%	2.40	2.70	7

Soybean, seed, solv-extd toasted grnd, (5)
Soybean meal, solvent extracted, toasted
Soybean oil meal, solvent extracted, toasted

Ref no 5-04-607

Dry matter	%	91.0	100.0	2
Ash	%	5.9	6.5	9
Crude fiber	%	5.0	5.5	20
Ether extract	%	1.0	1.1	37
N-free extract	%	32.9	36.2	
Protein (N x 6.25)	%	46.1	50.7	4
Cattle	dig prot %	41.5	45.6	
Sheep	dig prot %	43.3	47.6	
Swine	dig prot %	42.0	46.1	
Energy				
Cattle	DE kcal/kg	3290.	3616.	
Sheep	DE kcal/kg	3571.	3924.	
Swine	DE kcal/kg	3411.	3748.	
Cattle	ME kcal/kg	2698.	2965.	
Sheep	ME kcal/kg	2928.	3218.	
Swine	ME kcal/kg	2926.	3216.	
Cattle	TDN %	75.	82.	
Sheep	TDN %	81.	89.	
Swine	TDN %	77.	85.	
Pantothenic acid	mg/kg	38.7	42.5	
Arginine	%	3.60	3.96	3
Histidine	%	1.10	1.21	4
Isoleucine	%	2.40	2.64	2
Leucine	%	3.50	3.85	1
Lysine	%	2.90	3.19	3
Methionine	%	.70	.77	6
Phenylalanine	%	2.30	2.53	5
Threonine	%	1.80	1.980	5
Tryptophan	%	.70	.77	
Tyrosine	%	.70	.77	
Valine	%	2.40	2.64	2

Soybean, seed, (5)

Ref no 5-04-610

Dry matter	%	90.0	100.0	
Ash	%	4.6	5.1	
Crude fiber	%	5.0	5.6	
Ether extract	%	18.0	20.0	
N-free extract	%	24.5	27.2	
Protein (N x 6.25)	%	37.9	42.1	
Cattle	dig prot %	34.1	37.9	

Continued

(1) dry forages and roughages
(2) pasture, range plants, and forages fed green

(3) silages
(4) energy feeds

Feed name or analyses		As fed	Dry	C.V. ± %
Sheep	dig prot %	34.1	37.9	
Swine	dig prot %	31.0	34.5	
Energy				
Cattle	DE kcal/kg	3730.	4145.	
Sheep	DE kcal/kg	3690.	4100.	
Swine	DE kcal/kg	4048.	4498.	
Cattle	ME kcal/kg	3059.	3399.	
Sheep	ME kcal/kg	3026.	3362.	
Swine	ME kcal/kg	3542.	3936.	
Cattle	TDN %	85.	94.	
Sheep	TDN %	84.	93.	
Swine	TDN %	92.	102.	
Calcium	%	.25	.28	
Phosphorus	%	.59	.66	

Feed name or analyses		As fed	Dry	C.V. ± %
Sheep	DE kcal/kg	3532.	3924.	
Swine	DE kcal/kg	3405.	3792.	
Cattle	ME kcal/kg	2727.	3037.	
Chickens	ME_n kcal/kg	2425.	2700.	
Sheep	ME kcal/kg	2896.	3218.	
Swine	ME kcal/kg	2881.	3208.	
Cattle	TDN %	75.	84.	
Sheep	TDN %	80.	89.	
Swine	TDN %	77.	86.	
Calcium	%	.26	.29	38
Phosphorus	%	.62	.69	9
Potassium	%	2.02	2.24	
Manganese	mg/kg	45.5	50.6	62
Choline	mg/kg	2761.	3068.	
Niacin	mg/kg	21.6	24.0	
Riboflavin	mg/kg	3.1	3.4	
Thiamine	mg/kg	2.4	2.7	

Soybean, seed screenings, (5)

Ref no 5-04-611

		As fed	Dry	
Dry matter	%	91.0	100.0	
Ash	%	4.3	4.7	
Crude fiber	%	31.0	34.0	
Ether extract	%	4.1	4.5	
N-free extract	%	36.8	40.4	
Protein (N x 6.25)	%	14.9	16.4	
Cattle	dig prot %	10.1	11.1	
Sheep	dig prot %	10.7	11.8	
Energy				
Cattle	DE kcal/kg	2207.	2425.	
Sheep	DE kcal/kg	2287.	2513.	
Cattle	ME kcal/kg	1809.	1988.	
Sheep	ME kcal/kg	1876.	2061.	
Cattle	TDN %	50.	55.	
Sheep	TDN %	52.	57.	

Soybean, seed wo hulls, solv-extd grnd, mx 3 fbr, (5)

Soybean meal, dehulled, solvent extracted (AAFCO)

Soybean oil meal, dehulled, solvent extracted

Ref no 5-04-612

		As fed	Dry	
Dry matter	%	89.8	100.0	1
Ash	%	5.6	6.2	8
Crude fiber	%	2.8	3.1	16
Ether extract	%	.8	.9	51
N-free extract	%	29.7	33.1	
Protein (N x 6.25)	%	50.9	56.7	1
Cattle	dig prot %	45.8	51.0	
Sheep	dig prot %	47.8	53.1	
Swine	dig prot %	46.3	51.6	
Energy				
Cattle	DE kcal/kg	3326.	3704.	

Continued

Soybran flakes - see Soybean, hulls

SPELT. Triticum spelta

Spelt, grain, (4)

Ref no 4-04-651

		As fed	Dry	
Dry matter	%	90.0	100.0	
Ash	%	3.5	3.9	
Crude fiber	%	8.0	8.9	
Ether extract	%	2.0	2.2	
N-free extract	%	64.6	71.8	
Protein (N x 6.25)	%	11.9	13.2	
Cattle	dig prot %	8.9	9.9	
Sheep	dig prot %	9.3	10.3	
Energy				
Cattle	DE kcal/kg	2976.	3307.	
Sheep	DE kcal/kg	3056.	3395.	
Cattle	ME kcal/kg	2441.	2712.	
Sheep	ME kcal/kg	2506.	2784.	
Cattle	TDN %	68.	75.	
Sheep	TDN %	69.	77.	
Niacin	mg/kg	53.0	58.9	
Arginine	%	.50	.56	
Histidine	%	.20	.22	
Isoleucine	%	.40	.44	
Leucine	%	.70	.78	
Lysine	%	.30	.33	
Methionine	%	.20	.22	
Phenylalanine	%	.50	.56	
Threonine	%	.40	.44	
Tryptophan	%	.10	.11	
Valine	%	.50	.56	

(5) protein supplements
(6) minerals

(7) vitamins
(8) additives

| Feed name or analyses | Mean | | C.V. |
	As fed	Dry	± %

Spent bone black - see Animal, bone charcoal

SQUIRRELTAIL GRASS. Sitanion spp

Squirreltail grass, aerial pt, fresh, dormant, (2)

Ref no 2-05-566

		As fed	Dry	
Dry matter	%	86.0	100.0	
Ash	%	14.7	17.1	
Ether extract	%	2.2	2.6	
Protein (N x 6.25)	%	3.9	4.5	
Sheep	dig prot %	.9	1.1	
Cellulose	%	32.7	38.0	
Lignin	%	7.7	9.0	
Energy	GE kcal/kg	3280.	3814.	
Sheep	DE kcal/kg	1646.	1914.	
Sheep	ME kcal/kg	1388.	1614.	
Sheep	TDN %	40.	46.	
Calcium	%	.58	.67	
Phosphorus	%	.06	.07	
Carotene	mg/kg	.9	1.1	

ST. AUGUSTINEGRASS. Stenotaphrum secundatum

St. Augustinegrass, aerial pt, fresh, early blm, (2)

Ref no 2-04-676

		As fed	Dry	C.V.
Dry matter	%	31.3	100.0	
Ash	%	2.4	7.7	17
Crude fiber	%	8.5	27.1	10
Ether extract	%	.8	2.6	26
N-free extract	%	15.2	48.7	
Protein (N x 6.25)	%	4.4	13.9	10
Cattle	dig prot %	3.0	9.7	
Sheep	dig prot %	3.1	9.9	
Energy				
Cattle	DE kcal/kg	966.	3086.	
Sheep	DE kcal/kg	883.	2822.	
Cattle	ME kcal/kg	792.	2530.	
Sheep	ME kcal/kg	724.	2314.	
Cattle	TDN %	22.	70.	
Sheep	TDN %	20.	64.	
Calcium	%	.16	.51	25
Iron	%	.010	.020	
Magnesium	%	.12	.39	
Phosphorus	%	.07	.21	29
Potassium	%	.20	.63	

St. Augustinegrass, aerial pt, fresh, mature, (2)

Ref no 2-04-677

		As fed	Dry	C.V.
Dry matter	%	30.5	100.0	13
Ash	%	2.3	7.6	20
Crude fiber	%	8.8	28.9	9
Ether extract	%	.3	1.1	25
N-free extract	%	17.6	57.6	
Protein (N x 6.25)	%	1.5	4.8	22
Cattle	dig prot %	.6	2.0	
Sheep	dig prot %	.4	1.5	
Energy				
Cattle	DE kcal/kg	834.	2734.	
Sheep	DE kcal/kg	955.	3130.	
Cattle	ME kcal/kg	684.	2242.	
Sheep	ME kcal/kg	783.	2567.	
Cattle	TDN %	19.	62.	
Sheep	TDN %	22.	71.	

Sudangrass - see Sorghum, Sudangrass

SUGARCANE. Saccharum officinarum

Sugarcane, aerial pt, s-c, (1)

Ref no 1-04-685

		As fed	Dry	C.V.
Dry matter	%	92.7	100.0	4
Ash	%	3.8	4.1	34
Crude fiber	%	40.0	43.1	20
Ether extract	%	1.1	1.2	
N-free extract	%	45.4	49.0	
Protein (N x 6.25)	%	2.4	2.6	72
Cattle	dig prot %	.0	.0	
Sheep	dig prot %	.0	.0	
Energy				
Cattle	DE kcal/kg	1553.	1675.	
Sheep	DE kcal/kg	2085.	2249.	
Cattle	ME kcal/kg	1274.	1374.	
Sheep	ME kcal/kg	1709.	1844.	
Cattle	TDN %	35.	38.	
Sheep	TDN %	47.	51.	

(1) dry forages and roughages

(2) pasture, range plants, and forages fed green

(3) silages

(4) energy feeds

TABLE A-3-1 Composition of Feeds (Continued)

Feed name or analyses		Mean		C.V.
		As fed	Dry	± %

Sugarcane, pulp, dehy, (1)
Cane bagasse, dried

Ref no 1-04-686

		As fed	Dry	C.V. ± %
Dry matter	%	90.0	100.0	6
Ash	%	2.6	2.9	17
Crude fiber	%	44.0	48.9	12
Ether extract	%	.4	.4	55
N-free extract	%	41.8	46.5	
Protein (N x 6.25)	%	1.2	1.3	28
Cattle	dig prot %	.0	.0	
Sheep	dig prot %	.0	.0	
Energy				
Cattle	DE kcal/kg	1904.	2116.	
Sheep	DE kcal/kg	1944.	2160.	
Cattle	ME kcal/kg	1562.	1735.	
Sheep	ME kcal/kg	1594.	1771.	
Cattle	TDN %	43.	48.	
Sheep	TDN %	44.	49.	

Sugarcane, aerial pt, fresh, (2)

Ref no 2-04-689

		As fed	Dry	C.V. ± %
Dry matter	%	27.1	100.0	24
Ash	%	1.4	5.0	91
Crude fiber	%	8.0	29.6	12
Ether extract	%	.4	1.6	34
N-free extract	%	16.0	59.0	
Protein (N x 6.25)	%	1.3	4.8	31
Cattle	dig prot %	.7	2.7	
Sheep	dig prot %	.4	1.5	
Energy				
Cattle	DE kcal/kg	717.	2646.	
Sheep	DE kcal/kg	860.	3175.	
Cattle	ME kcal/kg	588.	2170.	
Sheep	ME kcal/kg	706.	2604.	
Cattle	NE_m kcal/kg	279.	1030.	
Cattle	NE_{gain} kcal/kg	51.	190.	
Cattle	TDN %	16.	60.	
Sheep	TDN %	20.	72.	
Calcium	%	.13	.47	
Iron	%	.	.010	
Magnesium	%	.17	.62	
Phosphorus	%	.05	.17	

Sugarcane, molasses, dehy, (4)
Cane molasses, dried
Molasses, cane, dried

Ref no 4-04-695

		As fed	Dry	C.V. ± %
Dry matter	%	96.0	100.0	
Ash	%	8.0	8.3	
Crude fiber	%	5.0	5.2	
Ether extract	%	1.0	1.0	
N-free extract	%	71.7	74.8	
Protein (N x 6.25)	%	10.3	10.7	
Swine	dig prot %	7.3	7.6	
Energy	GE kcal/kg	3087.	3212.	
Swine	DE kcal/kg	2878.	2998.	
Swine	ME kcal/kg	2700.	2812.	
Swine	TDN %	65.	68.	

Sugarcane, molasses, mn 48 invert sugar mn 79.5 degrees brix, (4)
Cane molasses (AAFCO)
Molasses, cane

Ref no 4-04-696

		As fed	Dry	C.V. ± %
Dry matter	%	75.0	100.0	7
Ash	%	8.1	10.8	34
Ether extract	%	.1	.1	99
N-free extract	%	63.6	84.8	
Protein (N x 6.25)	%	3.2	4.3	87
Cattle	dig prot %	1.8	2.4	
Sheep	dig prot %	.9	1.2	
Energy	GE kcal/kg	3086.	4114.	
Cattle	DE kcal/kg	3009.	4012.	
Sheep	DE kcal/kg	2381.	3175.	
Swine	DE kcal/kg	2464.	3285.	
Cattle	ME kcal/kg	2468.	3290.	
Chickens	ME_n kcal/kg	1962.	2616.	
Sheep	ME kcal/kg	1953.	2604.	
Swine	ME kcal/kg	2343.	3124.	
Cattle	NE_m kcal/kg	1508.	2010.	
Cattle	NE_{gain} kcal/kg	952.	1270.	
Cattle	TDN %	68.	91.	
Sheep	TDN %	54.	72.	
Swine	TDN %	56.	75.	
Calcium	%	.89	1.19	51
Iron	%	.019	.025	9
Magnesium	%	.35	.47	56
Phosphorus	%	.08	.11	75
Potassium	%	2.38	3.17	21
Copper	mg/kg	59.6	79.4	62
Manganese	mg/kg	42.2	56.3	58
Choline	mg/kg	876.	1167.	87
Niacin	mg/kg	34.3	45.7	44

Continued

(5) protein supplements
(6) minerals

(7) vitamins
(8) additives

Feed name or analyses		Mean		C.V.
		As fed	Dry	± %
Pantothenic acid	mg/kg	38.3	51.1	34
Riboflavin	mg/kg	3.3	4.4	99
Thiamine	mg/kg	.9	1.2	63

Sumac - see Sorghum, sumac

SUNFLOWER. Helianthus spp

Sunflower, aerial pt, ensiled, early blm, (3)

Ref no 3-04-729

		As fed	Dry	C.V. ± %
Dry matter	%	21.3	100.0	5
Ash	%	1.8	8.4	41
Crude fiber	%	7.0	32.9	6
Ether extract	%	.5	2.5	12
N-free extract	%	9.9	46.5	
Protein (N x 6.25)	%	2.1	9.7	7
Cattle	dig prot %	1.2	5.8	
Sheep	dig prot %	1.2	5.7	
Energy				
Cattle	DE kcal/kg	526.	2469.	
Sheep	DE kcal/kg	507.	2381.	
Cattle	ME kcal/kg	431.	2024.	
Sheep	ME kcal/kg	416.	1952.	
Cattle	TDN %	12.	56.	
Sheep	TDN %	12.	54.	

Sunflower, aerial pt, ensiled, mid-blm, (3)

Ref no 3-04-730

		As fed	Dry	C.V. ± %
Dry matter	%	22.1	100.0	8
Ash	%	2.5	11.4	15
Crude fiber	%	7.3	33.1	9
Ether extract	%	1.0	4.3	30
N-free extract	%	9.2	41.8	
Protein (N x 6.25)	%	2.1	9.4	6
Cattle	dig prot %	.8	3.5	
Sheep	dig prot %	1.1	5.1	
Energy				
Cattle	DE kcal/kg	468.	2116.	
Sheep	DE kcal/kg	555.	2513.	
Cattle	ME kcal/kg	383.	1735.	
Sheep	ME kcal/kg	455.	2061.	
Cattle	TDN %	11.	48.	
Sheep	TDN %	12.	57.	

Feed name or analyses		Mean		C.V.
		As fed	Dry	± %

Sunflower, aerial pt, ensiled, mature, (3)

Ref no 3-04-735

		As fed	Dry	
Dry matter	%	22.8	100.0	
Ash	%	2.2	9.7	
Crude fiber	%	8.9	39.1	
Ether extract	%	1.1	4.9	
N-free extract	%	8.9	39.1	
Protein (N x 6.25)	%	1.6	7.2	
Cattle	dig prot %	.4	1.6	
Sheep	dig prot %	.7	3.2	
Energy				
Cattle	DE kcal/kg	382.	1675.	
Sheep	DE kcal/kg	573.	2513.	
Cattle	ME kcal/kg	313.	1374.	
Sheep	ME kcal/kg	470.	2061.	
Cattle	TDN %	9.	38.	
Sheep	TDN %	13.	57.	

Sunflower, aerial pt, ensiled, (3)

Ref no 3-04-736

		As fed	Dry	C.V. ± %
Dry matter	%	22.0	100.0	12
Ash	%	2.3	10.5	18
Crude fiber	%	7.0	31.7	9
Ether extract	%	.8	3.8	33
N-free extract	%	9.8	44.5	
Protein (N x 6.25)	%	2.1	9.5	14
Cattle	dig prot %	.9	4.0	
Sheep	dig prot %	.9	4.3	
Energy				
Cattle	DE kcal/kg	456.	2072.	
Sheep	DE kcal/kg	543.	2469.	
Cattle	ME kcal/kg	374.	1699.	
Sheep	ME kcal/kg	445.	2024.	
Cattle	TDN %	10.	47.	
Sheep	TDN %	12.	56.	
Calcium	%	.38	1.71	5
Magnesium	%	.02	.09	
Phosphorus	%	.05	.23	37
Potassium	%	.64	2.92	
Sodium	%	.01	.04	
Manganese	mg/kg	233.9	1063.2	

(1) dry forages and roughages
(2) pasture, range plants, and forages fed green

(3) silages
(4) energy feeds

TABLE A-3-1 Composition of Feeds (Continued)

Feed name or analyses		Mean		C.V.
		As fed	Dry	± %

Sunflower, seed wo hulls, mech-extd grnd, (5)
  Sunflower meal (AAFCO)
  Sunflower oil meal, without hulls, expeller
     extracted

Ref no 5-04-738

Dry matter	%	93.0	100.0	2
Ash	%	6.8	7.3	11
Crude fiber	%	13.0	14.0	18
Ether extract	%	7.6	8.2	
N-free extract	%	24.6	26.4	
Protein (N x 6.25)	%	41.0	44.1	14
Cattle	dig prot %	36.4	39.2	
Sheep	dig prot %	37.3	40.1	
Swine	dig prot %	33.7	36.2	
Energy				
Cattle	DE kcal/kg	2870.	3086.	
Sheep	DE kcal/kg	3239.	3483.	
Swine	DE kcal/kg	3116.	3351.	
Cattle	ME kcal/kg	2353.	2530.	
Sheep	ME kcal/kg	2656.	2856.	
Swine	ME kcal/kg	2715.	2919.	
Cattle	TDN %	65.	70.	
Sheep	TDN %	73.	79.	
Swine	TDN %	71.	76.	
Calcium	%	.43	.46	
Phosphorus	%	1.04	1.12	
Potassium	%	1.08	1.16	
Manganese	mg/kg	22.9	24.6	
Lysine	%	2.00	2.15	
Methionine	%	1.60	1.72	

Sunflower, seed wo hulls, solv-extd grnd, (5)
  Sunflower meal (AAFCO)
  Sunflower oil meal, without hulls, solvent extracted

Ref no 5-04-739

Dry matter	%	93.0	100.0	
Ash	%	7.7	8.3	
Crude fiber	%	11.0	11.8	
Ether extract	%	2.9	3.1	
N-free extract	%	24.6	26.5	
Protein (N x 6.25)	%	46.8	50.3	
Cattle	dig prot %	41.7	44.8	
Sheep	dig prot %	42.6	45.8	
Swine	dig prot %	42.1	45.3	
Energy				
Cattle	DE kcal/kg	2665.	2866.	
Sheep	DE kcal/kg	3034.	3263.	
Swine	DE kcal/kg	3034.	3263.	
Cattle	ME kcal/kg	2186.	2350.	
Sheep	ME kcal/kg	2489.	2676.	

Continued

Feed name or analyses		Mean		C.V.
		As fed	Dry	± %
Swine	ME kcal/kg	2604.	2800.	
Cattle	TDN %	60.	65.	
Sheep	TDN %	69.	74.	
Swine	TDN %	69.	74.	
Riboflavin	mg/kg	3.1	3.3	

SWEETCLOVER. Melilotus spp

Sweetclover, hay, s-c, immature, (1)

Ref no 1-04-746

Dry matter	%	84.9	100.0	4
Ash	%	7.9	9.3	18
Crude fiber	%	23.4	27.6	12
Ether extract	%	1.9	2.2	16
N-free extract	%	34.6	40.8	
Protein (N x 6.25)	%	17.1	20.1	14
Cattle	dig prot %	12.1	14.3	
Sheep	dig prot %	12.1	14.3	
Energy	GE kcal/kg	3789.	4463.	
Cattle	DE kcal/kg	2246.	2646.	
Sheep	DE kcal/kg	2021.	2381.	
Cattle	ME kcal/kg	1842.	2170.	
Sheep	ME kcal/kg	1657.	1952.	
Cattle	TDN %	51.	60.	
Sheep	TDN %	46.	54.	
Calcium	%	1.63	1.92	13
Magnesium	%	.48	.57	16
Phosphorus	%	.27	.32	17
Potassium	%	1.36	1.60	9

Sweetclover, hay, s-c, early blm, (1)

Ref no 1-04-750

Dry matter	%	82.6	100.0	
Ash	%	7.0	8.4	
Crude fiber	%	27.3	33.1	
Ether extract	%	1.4	1.7	
N-free extract	%	32.2	39.0	
Protein (N x 6.25)	%	14.7	17.8	
Cattle	dig prot %	10.2	12.4	
Sheep	dig prot %	10.4	12.6	
Energy				
Cattle	DE kcal/kg	2003.	2425.	
Sheep	DE kcal/kg	1930.	2337.	
Cattle	ME kcal/kg	1642.	1988.	
Sheep	ME kcal/kg	1583.	1916.	
Cattle	TDN %	45.	55.	
Sheep	TDN %	44.	53.	

(5) protein supplements
(6) minerals

(7) vitamins
(8) additives

Feed name or analyses		Mean		C.V.
		As fed	Dry	± %

Sweetclover, hay, s-c, full blm, (1)

Ref no 1-04-751

		As fed	Dry	C.V.
Dry matter	%	86.9	100.0	1
Ash	%	6.8	7.8	3
Crude fiber	%	29.9	34.4	12
Ether extract	%	1.6	1.8	29
N-free extract	%	38.1	43.8	
Protein (N x 6.25)	%	10.6	12.2	23
Cattle	dig prot %	6.5	7.5	
Sheep	dig prot %	7.6	8.7	
Energy				
Cattle	DE kcal/kg	2107.	2425.	
Sheep	DE kcal/kg	2031.	2337.	
Cattle	ME kcal/kg	1728.	1988.	
Sheep	ME kcal/kg	1665.	1916.	
Cattle	TDN %	48.	55.	
Sheep	TDN %	46.	53.	

Sweetclover, hay, s-c, mature, (1)

Ref no 1-04-752

		As fed	Dry	C.V.
Dry matter	%	90.3	100.0	
Ash	%	5.5	6.1	
Crude fiber	%	35.0	38.8	
Ether extract	%	1.2	1.3	
N-free extract	%	38.9	43.1	
Protein (N x 6.25)	%	9.7	10.7	
Cattle	dig prot %	5.6	6.2	
Sheep	dig prot %	6.9	7.6	
Energy				
Cattle	DE kcal/kg	1991.	2205.	
Sheep	DE kcal/kg	2110.	2337.	
Cattle	ME kcal/kg	1633.	1808.	
Sheep	ME kcal/kg	1730.	1916.	
Cattle	TDN %	45.	50.	
Sheep	TDN %	48.	53.	
Calcium	%	1.19	1.32	5
Iron	%	.010	.020	17
Magnesium	%	.56	.62	18
Phosphorus	%	.17	.19	29
Potassium	%	.72	.80	17
Manganese	mg/kg	90.4	100.1	19

Sweetclover, hay, s-c, (1)

Ref no 1-04-754

		As fed	Dry	C.V.
Dry matter	%	87.2	100.0	4
Ash	%	7.6	8.7	19
Crude fiber	%	28.1	32.2	15

Continued

Feed name or analyses		Mean		C.V.
		As fed	Dry	± %
Ether extract	%	1.9	2.2	17
N-free extract	%	35.4	40.6	
Protein (N x 6.25)	%	14.2	16.3	19
Cattle	dig prot %	9.6	11.0	
Sheep	dig prot %	10.1	11.6	
Energy				
Cattle	DE kcal/kg	2191.	2513.	
Sheep	DE kcal/kg	2038.	2337.	
Cattle	ME kcal/kg	1797.	2061.	
Sheep	ME kcal/kg	1671.	1916.	
Cattle	TDN %	50.	57.	
Sheep	TDN %	46.	53.	
Calcium	%	1.54	1.77	5
Iron	%	.010	.020	18
Magnesium	%	.54	.62	18
Phosphorus	%	.23	.26	27
Potassium	%	1.17	1.34	23
Copper	mg/kg	8.8	10.1	10
Manganese	mg/kg	89.8	103.0	20
Carotene	mg/kg	108.5	124.4	16
Vitamin A equiv	IU/g	180.9	207.4	

Sweetclover, aerial pt, fresh, immature, (2)

Ref no 2-04-759

		As fed	Dry	C.V.
Dry matter	%	19.7	100.0	28
Ash	%	1.9	9.6	22
Crude fiber	%	4.5	22.9	25
Ether extract	%	.6	3.2	21
N-free extract	%	8.4	42.6	
Protein (N x 6.25)	%	4.3	21.7	18
Cattle	dig prot %	3.5	17.8	
Sheep	dig prot %	3.3	16.9	
Energy				
Cattle	DE kcal/kg	582.	2954.	
Sheep	DE kcal/kg	547.	2778.	
Cattle	ME kcal/kg	477.	2422.	
Sheep	ME kcal/kg	449.	2278.	
Cattle	TDN %	13.	67.	
Sheep	TDN %	12.	63.	
Calcium	%	.34	1.73	
Magnesium	%	.06	.30	
Phosphorus	%	.09	.44	
Potassium	%	.58	2.95	

Sweetclover, aerial pt, fresh, full blm, (2)

Ref no 2-04-762

		As fed	Dry	C.V.
Dry matter	%	32.7	100.0	14
Ash	%	2.3	7.0	17
Crude fiber	%	10.5	32.2	7
Ether extract	%	.8	2.5	6

Continued

(1) dry forages and roughages
(2) pasture, range plants, and forages fed green

(3) silages
(4) energy feeds

Feed name or analyses		Mean		C.V., ± %
		As fed	Dry	
N-free extract	%	13.6	41.5	
Protein (N x 6.25)	%	5.5	16.8	11
Cattle	dig prot %	4.5	13.8	
Sheep	dig prot %	4.3	13.1	
Energy				
Cattle	DE kcal/kg	952.	2910.	
Sheep	DE kcal/kg	908.	2778.	
Cattle	ME kcal/kg	780.	2386.	
Sheep	ME kcal/kg	626.	1916.	
Cattle	TDN %	22.	66.	
Sheep	TDN %	21.	63.	
Calcium	%	.43	1.30	9
Chlorine	%	.12	.38	14
Iron	%	.010	.010	15
Magnesium	%	.11	.34	8
Phosphorus	%	.07	.22	23
Potassium	%	.47	1.44	14
Sodium	%	.03	.10	26
Sulfur	%	.16	.49	23
Copper	mg/kg	3.2	9.9	10
Manganese	mg/kg	38.5	117.7	27
Zinc	mg/kg	16.4	50.1	

Sweetclover, aerial pt, fresh, (2)

Ref no 2-04-766

Feed name or analyses		Mean		C.V., ± %
		As fed	Dry	
Dry matter	%	24.8	100.0	29
Ash	%	2.0	8.2	23
Crude fiber	%	7.3	29.4	23
Ether extract	%	.7	2.8	36
N-free extract	%	10.4	41.8	
Protein (N x 6.25)	%	4.4	17.8	24
Cattle	dig prot %	3.6	14.6	
Sheep	dig prot %	3.4	13.9	
Energy				
Cattle	DE kcal/kg	722.	2910.	
Sheep	DE kcal/kg	689.	2778.	
Cattle	ME kcal/kg	592.	2386.	
Sheep	ME kcal/kg	565.	2278.	
Cattle	TDN %	16.	66.	
Sheep	TDN %	16.	63.	
Calcium	%	.33	1.32	16
Iron	%	.002	.010	14
Magnesium	%	.08	.33	8
Phosphorus	%	.07	.27	41
Potassium	%	.41	1.65	33
Copper	mg/kg	2.5	9.9	10
Manganese	mg/kg	31.1	125.5	24
Carotene	mg/kg	61.3	247.2	
Vitamin A equiv	IU/g	102.2	412.1	

Sweetclover, aerial pt, ensiled, early blm, (3)

Ref no 3-04-768

Feed name or analyses		Mean		C.V., ± %
		As fed	Dry	
Dry matter	%	27.2	100.0	10
Ash	%	2.7	9.9	3
Crude fiber	%	8.5	31.1	3
Ether extract	%	1.0	3.6	9
N-free extract	%	9.3	34.2	
Protein (N x 6.25)	%	5.8	21.2	6
Cattle	dig prot %	4.5	16.5	
Sheep	dig prot %	4.4	16.1	
Energy				
Cattle	DE kcal/kg	720.	2646.	
Sheep	DE kcal/kg	624.	2293.	
Cattle	ME kcal/kg	590.	2170.	
Sheep	ME kcal/kg	511.	1880.	
Cattle	TDN %	16.	60.	
Sheep	TDN %	14.	52.	

Sweetclover, aerial pt, ensiled, full blm, (3)

Ref no 3-04-769

Dry matter	%	27.5	100.0	10
Ash	%	2.5	9.0	6
Crude fiber	%	10.0	36.3	7
Ether extract	%	1.4	5.1	25
N-free extract	%	9.2	33.6	
Protein (N x 6.25)	%	4.4	16.0	18
Cattle	dig prot %	3.2	11.7	
Sheep	dig prot %	3.4	12.2	
Energy				
Cattle	DE kcal/kg	679.	2469.	
Sheep	DE kcal/kg	630.	2293.	
Cattle	ME kcal/kg	557.	2024.	
Sheep	ME kcal/kg	517.	1880.	
Cattle	TDN %	15.	56.	
Sheep	TDN %	14.	52.	

Sweetclover, aerial pt, ensiled, (3)

Ref no 3-04-771

Dry matter	%	32.0	100.0	18
Ash	%	3.0	9.5	25
Crude fiber	%	10.6	33.0	10
Ether extract	%	1.2	3.8	30
N-free extract	%	11.0	34.5	
Protein (N x 6.25)	%	6.1	19.2	11
Cattle	dig prot %	4.7	14.8	
Sheep	dig prot %	4.7	14.6	

Continued

(5) protein supplements

(6) minerals

(7) vitamins

(8) additives

Feed name or analyses		Mean		C.V.
		As fed	Dry	± %
Energy				
Cattle	DE kcal/kg	832.	2601.	
Sheep	DE kcal/kg	734.	2293.	
Cattle	ME kcal/kg	682.	2133.	
Sheep	ME kcal/kg	602.	1880.	
Cattle	TDN %	19.	59.	
Sheep	TDN %	17.	52.	
Calcium	%	.41	1.29	16
Magnesium	%	.20	.62	
Phosphorus	%	.05	.17	35
Potassium	%	.63	1.96	
Manganese	mg/kg	7.7	24.0	
Carotene	mg/kg	9.1	28.4	57
Vitamin A equiv	IU/g	15.2	47.3	

Sweetclover, yellow, seed, (5)

Ref no 5-08-006

Dry matter	%	92.2	100.0	
Ash	%	3.5	3.8	
Crude fiber	%	11.3	12.2	
Ether extract	%	4.2	4.6	
N-free extract	%	35.8	38.8	
Protein (N x 6.25)	%	37.4	40.6	
Cattle	dig prot %	28.0	30.4	
Sheep	dig prot %	28.0	30.4	
Energy				
Cattle	DE kcal/kg	3374.	3660.	
Sheep	DE kcal/kg	3374.	3660.	
Cattle	ME kcal/kg	2767.	3001.	
Sheep	ME kcal/kg	2767.	3001.	
Cattle	TDN %	76.	83.	
Sheep	TDN %	76.	83.	

Sweetclover, yellow, seed screenings, (5)

Ref no 5-08-007

Dry matter	%	90.1	100.0	
Ash	%	8.9	9.9	
Crude fiber	%	14.7	16.3	
Ether extract	%	3.7	4.1	
N-free extract	%	41.1	45.6	
Protein (N x 6.25)	%	21.7	24.1	
Cattle	dig prot %	16.9	18.8	
Sheep	dig prot %	17.6	19.5	
Energy				
Cattle	DE kcal/kg	2582.	2866.	
Sheep	DE kcal/kg	2861.	3175.	
Cattle	ME kcal/kg	2117.	2350.	
Sheep	ME kcal/kg	2346.	2604.	
Cattle	TDN %	58.	65.	
Sheep	TDN %	65.	72.	

Feed name or analyses		Mean		C.V.
		As fed	Dry	± %

SWINE. Sus scrofa

Swine, lard, (4)

Ref no 4-04-790

Dry matter	%	100.0	100.0	
Ether extract	%	100.0	100.0	
Protein (N x 6.25)	%	.0	.0	
Energy	GE kcal/kg	9020.	9020.	

Tankage - see Animal, carcass res w blood; see Animal-Poultry, carcass rex mx 35 blood

TIMOTHY. Phleum pratense

Timothy, hay, s-c, immature, (1)

Ref no 1-04-880

Dry matter	%	87.2	100.0	3
Ash	%	6.3	7.2	11
Crude fiber	%	27.3	31.3	5
Ether extract	%	2.6	3.0	30
N-free extract	%	40.7	46.7	
Protein (N x 6.25)	%	10.3	11.8	14
Cattle	dig prot %	5.6	6.4	
Sheep	dig prot %	6.6	7.6	
Energy	GE kcal/kg	3940.	4518.	
Cattle	DE kcal/kg	2422.	2778.	
Sheep	DE kcal/kg	2576.	2954.	
Cattle	ME kcal/kg	1986.	2278.	
Sheep	ME kcal/kg	2112.	2422.	
Cattle	TDN %	55.	63.	
Sheep	TDN %	58.	67.	
Calcium	%	.58	.66	30
Magnesium	%	.26	.30	19
Phosphorus	%	.30	.34	9
Potassium	%	1.66	1.90	8
Carotene	mg/kg	107.9	123.7	53
Vitamin A equiv	IU/g	179.9	206.2	

Timothy, hay, s-c, pre-blm, (1)

Ref no 1-04-881

Dry matter	%	88.6	100.0	
Ash	%	6.6	7.5	
Crude fiber	%	29.1	32.9	
Ether extract	%	2.6	3.0	

Continued

(1) dry forages and roughages

(2) pasture, range plants, and forages fed green

(3) silages

(4) energy feeds

Feed name or analyses		Mean		C.V.
		As fed	Dry	± %
N-free extract	%	39.2	44.3	
Protein (N x 6.25)	%	10.9	12.3	
Cattle	dig prot %	5.8	6.6	
Sheep	dig prot %	6.9	7.8	
Energy				
Cattle	DE kcal/kg	2422.	2734.	
Sheep	DE kcal/kg	2617.	2954.	
Cattle	ME kcal/kg	1986.	2242.	
Sheep	ME kcal/kg	2146.	2422.	
Cattle	NE$_m$ kcal/kg	1214.	1370.	
Cattle	NE$_{gain}$ kcal/kg	691.	780.	
Cattle	TDN %	55.	62.	
Sheep	TDN %	59.	67.	
Calcium	%	.58	.66	
Phosphorus	%	.30	.34	

Timothy, hay, s-c, early blm, (1)

Ref no 1-04-882

		As fed	Dry	C.V.
Dry matter	%	87.7	100.0	4
Ash	%	5.4	6.2	11
Crude fiber	%	29.1	33.2	6
Ether extract	%	2.3	2.6	16
N-free extract	%	43.2	49.3	
Protein (N x 6.25)	%	7.6	8.7	12
Cattle	dig prot %	4.4	5.0	
Sheep	dig prot %	4.1	4.7	
Energy	GE kcal/kg	3893.	4439.	
Cattle	DE kcal/kg	2281.	2601.	
Sheep	DE kcal/kg	2165.	2469.	
Cattle	ME kcal/kg	1871.	2133.	
Sheep	ME kcal/kg	1775.	2024.	
Cattle	TDN %	52.	59.	
Sheep	TDN %	49.	56.	
Calcium	%	.53	.60	20
Phosphorus	%	.23	.26	30
Potassium	%	.81	.92	

Timothy, hay, s-c, mid-blm, (1)

Ref no 1-04-883

		As fed	Dry	C.V.
Dry matter	%	88.4	100.0	2
Ash	%	5.1	5.8	13
Crude fiber	%	29.6	33.5	4
Ether extract	%	2.4	2.7	10
N-free extract	%	43.8	49.5	
Protein (N x 6.25)	%	7.5	8.5	15
Cattle	dig prot %	4.1	4.6	
Sheep	dig prot %	4.6	5.2	
Energy	GE kcal/kg	3860.	4366.	
Cattle	DE kcal/kg	2378.	2690.	
Sheep	DE kcal/kg	2339.	2646.	

Continued

Feed name or analyses		Mean		C.V.
		As fed	Dry	± %
Cattle	ME kcal/kg	1950.	2206.	
Sheep	ME kcal/kg	1918.	2170.	
Cattle	TDN %	54.	61.	
Sheep	TDN %	53.	60.	
Calcium	%	.36	.41	
Magnesium	%	.14	.16	
Phosphorus	%	.17	.19	
Carotene	mg/kg	47.2	53.4	
Vitamin A equiv	IU/g	78.7	89.0	

Timothy, hay, s-c, full blm, (1)

Ref no 1-04-884

		As fed	Dry	C.V.
Dry matter	%	86.5	100.0	5
Ash	%	4.7	5.4	10
Crude fiber	%	29.3	33.9	7
Ether extract	%	2.3	2.7	15
N-free extract	%	43.3	50.1	
Protein (N x 6.25)	%	6.8	7.9	13
Cattle	dig prot %	3.4	3.9	
Horses	dig prot %	1.4	1.6	
Sheep	dig prot %	3.7	4.3	
Energy				
Cattle	DE kcal/kg	2174.	2513.	
Horses	DE kcal/kg	1602.	1852.	
Sheep	DE kcal/kg	2250.	2601.	
Cattle	ME kcal/kg	1783.	2061.	
Horses	ME kcal/kg	1314.	1519.	
Sheep	ME kcal/kg	1845.	2133.	
Cattle	TDN %	49.	57.	
Horses	TDN %	36.	42.	
Sheep	TDN %	51.	59.	
Calcium	%	.30	.35	26
Chlorine	%	.54	.62	9
Iron	%	.010	.020	21
Magnesium	%	.12	.14	26
Phosphorus	%	.18	.21	22
Potassium	%	1.45	1.68	10
Sodium	%	.16	.18	18
Sulfur	%	.11	.13	25
Copper	mg/kg	4.2	4.8	35
Manganese	mg/kg	70.0	80.9	22

Timothy, hay, s-c, late blm, (1)

Ref no 1-04-885

		As fed	Dry	C.V.
Dry matter	%	88.0	100.0	
Ash	%	5.3	6.0	
Crude fiber	%	28.5	32.4	
Ether extract	%	2.2	2.5	
N-free extract	%	44.7	50.8	
Protein (N x 6.25)	%	7.3	8.3	

Continued

Feed name or analyses		Mean		C.V.
		As fed	Dry	± %
Cattle	dig prot %	3.6	4.1	
Sheep	dig prot %	3.2	3.6	
Energy				
Cattle	DE kcal/kg	2250.	2557.	
Sheep	DE kcal/kg	2134.	2425.	
Cattle	ME kcal/kg	1845.	2097.	
Sheep	ME kcal/kg	1749.	1988.	
Cattle	NE$_m$ kcal/kg	1003.	1140.	
Cattle	NE$_{gain}$ kcal/kg	378.	430.	
Cattle	TDN %	51.	58.	
Sheep	TDN %	48.	55.	
Calcium	%	.33	.38	
Phosphorus	%	.16	.18	

Timothy, hay, s-c, mature, (1)

Ref no 1-04-888

Dry matter	%	86.4	100.0	6
Ash	%	4.4	5.1	18
Crude fiber	%	30.8	35.6	11
Ether extract	%	2.2	2.5	26
N-free extract	%	44.0	50.9	
Protein (N x 6.25)	%	5.1	5.9	17
Cattle	dig prot %	2.2	2.5	
Sheep	dig prot %	2.4	2.8	
Energy	GE kcal/kg	3802.	4421.	
Cattle	DE kcal/kg	1828.	2116.	
Sheep	DE kcal/kg	2133.	2469.	
Cattle	ME kcal/kg	1499.	1735.	
Sheep	ME kcal/kg	1749.	2024.	
Cattle	NE$_m$ kcal/kg	873.	1010.	
Cattle	NE$_{gain}$ kcal/kg	130.	150.	
Cattle	TDN %	41.	48.	
Sheep	TDN %	48.	56.	
Calcium	%	.17	.20	32
Chlorine	%	.50	.58	8
Iron	%	.020	.030	10
Magnesium	%	.06	.07	31
Phosphorus	%	.15	.17	22
Potassium	%	1.37	1.58	21
Sodium	%	.06	.07	74
Sulfur	%	.14	.16	22
Carotene	mg/kg	4.6	5.3	
Vitamin A equiv	IU/g	7.7	8.8	

Timothy, hay, s-c, cut 1, (1)

Ref no 1-04-890

Dry matter	%	91.1	100.0	3
Ash	%	4.8	5.3	18
Crude fiber	%	31.6	34.7	12
Ether extract	%	2.3	2.5	32

Continued

Feed name or analyses		Mean		C.V.
		As fed	Dry	± %
N-free extract	%	45.8	50.3	
Protein (N x 6.25)	%	6.6	7.2	15
Cattle	dig prot %	3.0	3.3	
Sheep	dig prot %	3.1	3.4	
Lignin	%	9.6	10.5	13
Energy	GE kcal/kg	4284.	4703.	1
Cattle	DE kcal/kg	2209.	2425.	
Sheep	DE kcal/kg	2249.	2469.	
Cattle	ME kcal/kg	1811.	1988.	
Sheep	ME kcal/kg	1844.	2024.	
Cattle	TDN %	50.	55.	
Sheep	TDN %	51.	56.	
Calcium	%	.36	.40	24
Magnesium	%	.15	.16	25
Phosphorus	%	.15	.17	38
Potassium	%	1.26	1.38	34
Manganese	mg/kg	41.0	45.0	43
Carotene	mg/kg	10.7	11.7	99
Vitamin A equiv	IU/g	17.8	19.5	

Timothy, hay, s-c, cut 2, (1)

Ref no 1-04-891

Dry matter	%	87.3	100.0	2
Ash	%	5.7	6.5	29
Crude fiber	%	28.6	32.8	12
Ether extract	%	3.2	3.7	33
N-free extract	%	38.9	44.6	
Protein (N x 6.25)	%	10.8	12.4	28
Cattle	dig prot %	5.0	5.7	
Sheep	dig prot %	5.0	5.8	
Energy				
Cattle	DE kcal/kg	2079.	2381.	
Sheep	DE kcal/kg	2117.	2425.	
Cattle	ME kcal/kg	1704.	1952.	
Sheep	ME kcal/kg	1735.	1988.	
Cattle	TDN %	47.	54.	
Sheep	TDN %	48.	55.	
Calcium	%	.62	.71	
Phosphorus	%	.35	.40	
Manganese	mg/kg	40.8	46.7	

Timothy, hay, s-c, (1)

Ref no 1-04-893

Dry matter	%	87.7	100.0	4
Ash	%	4.9	5.6	18
Crude fiber	%	29.6	33.8	9
Ether extract	%	2.3	2.6	28
N-free extract	%	44.1	50.3	
Protein (N x 6.25)	%	6.8	7.7	26
Cattle	dig prot %	3.1	3.5	

Continued

(1) dry forages and roughages

(2) pasture, range plants, and forages fed green

(3) silages

(4) energy feeds

TABLE A-3-1 Composition of Feeds (Continued)

Feed name or analyses		Mean		C.V.
		As fed	Dry	± %
Horses	dig prot %	3.0	3.3	
Sheep	dig prot %	3.2	3.6	
Cellulose	%	28.8	32.8	13
Lignin	%	10.8	12.3	16
Energy	GE kcal/kg	4020.	4584.	3
Cattle	DE kcal/kg	2088.	2381.	
Horses	DE kcal/kg	1740.	1984.	
Sheep	DE kcal/kg	2165.	2469.	
Cattle	ME kcal/kg	1712.	1952.	
Horses	ME kcal/kg	1427.	1627.	
Sheep	ME kcal/kg	1775.	2024.	
Cattle	NE$_m$ kcal/kg	1026.	1170.	
Cattle	NE$_{gain}$ kcal/kg	421.	480.	
Cattle	TDN %	47.	54.	
Horses	TDN %	39.	45.	
Sheep	TDN %	49.	56.	
Calcium	%	.32	.37	58
Chlorine	%	.39	.45	31
Iron	%	.010	.010	46
Magnesium	%	.15	.17	40
Phosphorus	%	.17	.19	56
Potassium	%	1.46	1.66	38
Sodium	%	.12	.14	99
Sulfur	%	.11	.13	43
Cobalt	mg/kg	.080	.090	58
Copper	mg/kg	4.5	5.1	46
Manganese	mg/kg	57.0	65.0	36
Zinc	mg/kg	14.9	17.0	19
Carotene	mg/kg	12.0	13.7	99
Vitamin A equiv	IU/g	20.0	22.8	
Vitamin D$_2$	IU/g	1.8	2.0	32

Feed name or analyses		Mean		C.V.
		As fed	Dry	± %
Magnesium	%	.06	.24	16
Phosphorus	%	.10	.39	15
Potassium	%	.59	2.33	11
Copper	mg/kg	5.1	20.0	
Manganese	mg/kg	16.0	62.8	
Carotene	mg/kg	59.7	234.0	33
Vitamin A equiv	IU/g	99.5	390.1	

Timothy, aerial pt, fresh, pre-blm, (2)

Ref no 2-04-903

		As fed	Dry	
Dry matter	%	28.3	100.0	
Ash	%	1.9	6.8	
Crude fiber	%	9.5	33.5	
Ether extract	%	1.1	4.0	
N-free extract	%	13.1	46.4	
Protein (N x 6.25)	%	2.6	9.3	
Cattle	dig prot %	1.6	5.5	
Sheep	dig prot %	1.4	5.0	
Energy				
Cattle	DE kcal/kg	836.	2954.	
Sheep	DE kcal/kg	724.	2557.	
Cattle	ME kcal/kg	685.	2422.	
Sheep	ME kcal/kg	593.	2097.	
Cattle	TDN %	19.	67.	
Sheep	TDN %	16.	58.	
Calcium	%	.14	.50	
Magnesium	%	.04	.15	
Phosphorus	%	.10	.35	
Potassium	%	.68	2.40	

Timothy, aerial pt, fresh, immature, (2)

Ref no 2-04-902

		As fed	Dry	
Dry matter	%	25.5	100.0	24
Ash	%	2.1	8.3	13
Crude fiber	%	6.1	23.9	11
Ether extract	%	1.0	3.8	11
N-free extract	%	12.4	48.6	
Protein (N x 6.25)	%	3.9	15.4	19
Cattle	dig prot %	2.3	9.1	
Sheep	dig prot %	1.5	6.0	
Cellulose	%	5.8	22.9	11
Lignin	%	1.1	4.3	18
Energy				
Cattle	DE kcal/kg	731.	2866.	
Sheep	DE kcal/kg	573.	2248.	
Cattle	ME kcal/kg	599.	2350.	
Sheep	ME kcal/kg	470.	1843.	
Cattle	TDN %	16.	65.	
Sheep	TDN %	13.	51.	
Calcium	%	.14	.55	17
Iron	%	.020	.080	

Continued

Timothy, aerial pt, fresh, early blm, (2)

Ref no 2-04-904

		As fed	Dry	
Dry matter	%	27.9	100.0	17
Ash	%	2.3	8.1	15
Crude fiber	%	7.1	25.6	12
Ether extract	%	1.1	3.8	13
N-free extract	%	13.7	49.0	
Protein (N x 6.25)	%	3.8	13.5	22
Cattle	dig prot %	2.2	8.0	
Sheep	dig prot %	1.5	5.3	
Cellulose	%	6.7	24.0	14
Lignin	%	1.5	5.2	23
Energy				
Cattle	DE kcal/kg	812.	2910.	
Sheep	DE kcal/kg	627.	2248.	
Cattle	ME kcal/kg	666.	2386.	
Sheep	ME kcal/kg	514.	1843.	
Cattle	TDN %	18.	66.	
Sheep	TDN %	14.	51.	
Calcium	%	.14	.50	10

Continued

(5) protein supplements
(6) minerals

(7) vitamins
(8) additives

Feed name or analyses		Mean		C.V.
		As fed	Dry	± %
Magnesium	%	.04	.15	16
Phosphorus	%	.10	.35	12
Potassium	%	.67	2.40	7
Carotene	mg/kg	54.1	194.0	
Vitamin A equiv	IU/g	90.2	323.4	

Timothy, aerial pt, fresh, mid-blm, (2)

Ref no 2-04-905

		As fed	Dry	C.V. ± %
Dry matter	%	28.1	100.0	
Ash	%	2.0	7.1	2
Crude fiber	%	9.5	33.7	5
Ether extract	%	.9	3.1	5
N-free extract	%	13.1	46.5	
Protein (N x 6.25)	%	2.7	9.6	25
Cattle	dig prot %	1.6	5.7	
Sheep	dig prot %	1.0	3.7	
Cellulose	%	8.7	31.0	7
Lignin	%	1.8	6.3	9
Energy				
Cattle	DE kcal/kg	818.	2910.	
Sheep	DE kcal/kg	644.	2292.	
Cattle	ME kcal/kg	670.	2386.	
Sheep	ME kcal/kg	528.	1879.	
Cattle	TDN %	18.	66.	
Sheep	TDN %	15.	52.	

Timothy, aerial pt, fresh, full blm, (2)

Ref no 2-04-906

		As fed	Dry	C.V. ± %
Dry matter	%	31.8	100.0	8
Ash	%	2.0	6.2	8
Crude fiber	%	10.5	32.9	5
Ether extract	%	.9	2.9	11
N-free extract	%	15.8	49.7	
Protein (N x 6.25)	%	2.6	8.3	7
Cattle	dig prot %	1.6	4.9	
Sheep	dig prot %	1.0	3.2	
Energy				
Cattle	DE kcal/kg	939.	2954.	
Sheep	DE kcal/kg	729.	2292.	
Cattle	ME kcal/kg	770.	2422.	
Sheep	ME kcal/kg	597.	1879.	
Cattle	TDN %	21.	67.	
Sheep	TDN %	16.	52.	
Calcium	%	.08	.25	17
Chlorine	%	.19	.61	15
Iron	%	.010	.020	46
Magnesium	%	.04	.13	21
Phosphorus	%	.07	.21	22
Potassium	%	.53	1.68	5
Sodium	%	.06	.19	11

Continued

Timothy, aerial pt, fresh, mature, (2)

Ref no 2-04-910

		As fed	Dry	C.V. ± %
Dry matter	%	30.5	100.0	
Ash	%	1.4	4.5	10
Crude fiber	%	10.3	33.9	2
Ether extract	%	.9	2.9	6
N-free extract	%	16.1	52.9	
Protein (N x 6.25)	%	1.8	5.8	20
Cattle	dig prot %	1.0	3.4	
Sheep	dig prot %	.7	2.3	
Cellulose	%	11.3	37.1	12
Lignin	%	2.4	7.9	4
Energy				
Cattle	DE kcal/kg	914.	2998.	
Sheep	DE kcal/kg	726.	2381.	
Cattle	ME kcal/kg	750.	2458.	
Sheep	ME kcal/kg	595.	1952.	
Cattle	TDN %	21.	68.	
Sheep	TDN %	16.	54.	
Calcium	%	.05	.15	32
Iron	%	.010	.030	24
Magnesium	%	.02	.06	35
Phosphorus	%	.05	.17	15
Potassium	%	.48	1.57	6
Manganese	mg/kg	31.8	104.1	13

Timothy, aerial pt, fresh, (2)

Ref no 2-04-912

		As fed	Dry	C.V. ± %
Dry matter	%	28.3	100.0	21
Ash	%	2.0	7.0	22
Crude fiber	%	8.0	28.1	13
Ether extract	%	1.0	3.4	17
N-free extract	%	14.2	50.2	
Protein (N x 6.25)	%	3.2	11.3	31
Cattle	dig prot %	1.9	6.7	
Sheep	dig prot %	1.2	4.4	
Cellulose	%	8.8	31.0	25
Lignin	%	2.0	6.9	16
Energy	GE kcal/kg	1274.	4500.	
Cattle	DE kcal/kg	824.	2910.	
Sheep	DE kcal/kg	649.	2293.	
Cattle	ME kcal/kg	675.	2386.	
Sheep	ME kcal/kg	532.	1880.	
Cattle	TDN %	19.	66.	

Continued

(1) dry forages and roughages
(2) pasture, range plants, and forages fed green
(3) silages
(4) energy feeds

Feed name or analyses		Mean		C.V.
		As fed	Dry	± %
Sheep	TDN %	15.	52.	
Calcium	%	.13	.47	25
Chlorine	%	.14	.51	36
Iron	%	.010	.020	84
Magnesium	%	.06	.20	29
Phosphorus	%	.10	.34	21
Potassium	%	.49	1.74	21
Sodium	%	.03	.11	56
Sulfur	%	.04	.13	19
Cobalt	mg/kg	.010	.040	27
Copper	mg/kg	2.6	9.0	46
Manganese	mg/kg	38.1	134.7	37
Carotene	mg/kg	63.4	224.0	37
Vitamin A equiv	IU/g	105.7	373.4	

Timothy, aerial pt, ensiled, immature, (3)

Ref no 3-04-916

Dry matter	%	35.2	100.0	15
Ash	%	2.7	7.8	6
Crude fiber	%	11.3	32.0	5
Ether extract	%	1.3	3.8	6
N-free extract	%	15.5	44.1	
Protein (N x 6.25)	%	4.3	12.3	7
Cattle	dig prot %	2.4	6.9	
Sheep	dig prot %	2.6	7.4	
Energy				
Cattle	DE kcal/kg	916.	2601.	
Sheep	DE kcal/kg	978.	2778.	
Cattle	ME kcal/kg	751.	2133.	
Sheep	ME kcal/kg	802.	2278.	
Cattle	TDN %	21.	59.	
Sheep	TDN %	22.	63.	
Calcium	%	.19	.53	10
Phosphorus	%	.11	.32	10

Timothy, aerial pt, ensiled, full blm, (3)

Ref no 3-04-920

Dry matter	%	33.7	100.0	13
Ash	%	2.3	6.9	11
Crude fiber	%	12.5	37.0	6
Ether extract	%	1.0	2.9	6
N-free extract	%	14.9	44.1	
Protein (N x 6.25)	%	3.1	9.1	9
Cattle	dig prot %	1.8	5.2	
Sheep	dig prot %	1.5	4.5	
Energy				
Cattle	DE kcal/kg	876.	2601.	
Sheep	DE kcal/kg	951.	2822.	
Cattle	ME kcal/kg	719.	2133.	
Sheep	ME kcal/kg	779.	2314.	

Continued

Feed name or analyses		Mean		C.V.
		As fed	Dry	± %
Cattle	TDN %	20.	59.	
Sheep	TDN %	22.	64.	
Calcium	%	.18	.54	16
Phosphorus	%	.09	.28	10

Timothy, aerial pt, ensiled, (3)

Ref no 3-04-922

Dry matter	%	37.5	100.0	15
Ash	%	2.6	7.0	10
Crude fiber	%	12.7	33.9	6
Ether extract	%	1.2	3.3	9
N-free extract	%	17.1	45.6	
Protein (N x 6.25)	%	3.8	10.2	10
Cattle	dig prot %	2.1	5.7	
Sheep	dig prot %	2.1	5.5	
Energy				
Cattle	DE kcal/kg	975.	2601.	
Sheep	DE kcal/kg	1025.	2734.	
Cattle	ME kcal/kg	800.	2133.	
Sheep	ME kcal/kg	841.	2242.	
Cattle	TDN %	22.	59.	
Sheep	TDN %	23.	62.	
Calcium	%	.21	.55	11
Iron	%	.004	.010	18
Magnesium	%	.06	.15	23
Phosphorus	%	.11	.29	24
Potassium	%	.63	1.69	8
Copper	mg/kg	2.1	5.5	17
Manganese	mg/kg	33.8	90.2	19
Carotene	mg/kg	29.6	78.9	30
Vitamin A equiv	IU/g	49.3	131.5	

TOMATO. Lycopersicon esculentum

Tomato, pulp, dehy, (5)
Dried tomato pomace (AAFCO)

Ref no 5-05-041

Dry matter	%	92.0	100.0	
Crude fiber	%	29.0	31.5	
Ether extract	%	13.0	14.1	
Protein (N x 6.25)	%	21.7	23.6	
Calcium	%	.28	.30	
Phosphorus	%	.57	.62	
Riboflavin	mg/kg	6.2	6.6	
Thiamine	mg/kg	11.9	12.6	

(5) protein supplements

(6) minerals

(7) vitamins

(8) additives

Feed name or analyses		Mean As fed	Dry	C.V. ± %

Torula dried yeast - see Yeast torulopsis

TREFOIL, BIRDSFOOT. Lotus corniculatus

Trefoil, birdsfoot, hay, s-c, (1)

Ref no 1-05-044

		As fed	Dry
Dry matter	%	91.2	100.0
Ash	%	6.0	6.6
Crude fiber	%	27.0	29.6
Ether extract	%	2.1	2.3
N-free extract	%	41.9	45.9
Protein (N x 6.25)	%	14.2	15.6
Cattle	dig prot %	9.8	10.7
Sheep	dig prot %	9.8	10.7
Energy			
Cattle	DE kcal/kg	2453.	2690.
Sheep	DE kcal/kg	2413.	2646.
Cattle	ME kcal/kg	2012.	2206.
Sheep	ME kcal/kg	1979.	2170.
Cattle	TDN %	56.	61.
Sheep	TDN %	55.	60.
Calcium	%	1.60	1.75
Phosphorus	%	.20	.22

Trefoil, birdsfoot, aerial pt, fresh, (2)

Ref no 2-07-998

		As fed	Dry
Dry matter	%	20.0	100.0
Ash	%	1.5	7.5
Crude fiber	%	2.6	13.0
Ether extract	%	1.0	5.0
N-free extract	%	9.3	46.5
Protein (N x 6.25)	%	5.6	28.0
Cattle	dig prot %	4.6	23.0
Sheep	dig prot %	4.6	23.0
Energy			
Cattle	DE kcal/kg	661.	3307.
Sheep	DE kcal/kg	661.	3307.
Cattle	ME kcal/kg	542.	2712.
Sheep	ME kcal/kg	542.	2712.
Cattle	TDN %	15.	75.
Sheep	TDN %	15.	75.
Calcium	%	.44	2.20
Phosphorus	%	.05	.25
Potassium	%	.46	2.30

TURKEY. Meleagris gallapavo

Turkey, offal mature birds, raw, (5)

Ref no 5-07-984

		As fed	Dry
Dry matter	%	28.0	100.0
Crude fiber	%	.4	1.4
Ether extract	%	12.3	43.9

Turkey, offal young birds, raw, (5)

Ref no 5-07-985

		As fed	Dry
Dry matter	%	35.0	100.0
Crude fiber	%	.3	.9
Ether extract	%	14.9	42.6

TURNIP. Brassica rapa

Turnip, roots, raw, (4)

Ref no 4-05-067

		As fed	Dry
Dry matter	%	9.3	100.0
Ash	%	.9	9.7
Crude fiber	%	1.1	11.8
Ether extract	%	.2	2.2
N-free extract	%	5.8	62.3
Protein (N x 6.25)	%	1.3	14.0
Cattle	dig prot %	.9	9.7
Sheep	dig prot %	.8	9.0
Energy			
Cattle	DE kcal/kg	344.	3704.
Sheep	DE kcal/kg	332.	3571.
Cattle	ME kcal/kg	282.	3037.
Sheep	ME kcal/kg	272.	2928.
Cattle	TDN %	8.	84.
Sheep	TDN %	8.	81.
Calcium	%	.06	.64
Phosphorus	%	.02	.22

(1) dry forages and roughages
(2) pasture, range plants, and forages fed green

(3) silages
(4) energy feeds

Feed name or analyses	Mean		C.V. ± %
	As fed	Dry	

Uncleaned screenings - see Grains, screenings, uncleaned

VETCH. Vicia spp

Vetch, hay, s-c, immature, (1)

Ref no 1-05-098

		As fed	Dry	C.V. ± %
Dry matter	%	90.4	100.0	3
Ash	%	7.6	8.4	20
Crude fiber	%	23.6	26.1	9
Ether extract	%	3.0	3.3	25
N-free extract	%	33.4	37.0	
Protein (N x 6.25)	%	22.8	25.2	9
Cattle	dig prot %	15.0	16.6	
Sheep	dig prot %	17.7	19.6	
Energy				
Cattle	DE kcal/kg	2511.	2778.	
Sheep	DE kcal/kg	2511.	2778.	
Cattle	ME kcal/kg	2059.	2278.	
Sheep	ME kcal/kg	2059.	2278.	
Cattle	TDN %	57.	63.	
Sheep	TDN %	57.	63.	

Vetch, hay, s-c, full blm, (1)

Ref no 1-05-101

		As fed	Dry	C.V. ± %
Dry matter	%	85.6	100.0	
Ash	%	8.3	9.7	12
Crude fiber	%	28.9	33.8	
Ether extract	%	2.2	2.6	
N-free extract	%	28.7	33.5	
Protein (N x 6.25)	%	17.5	20.4	7
Cattle	dig prot %	11.6	13.5	
Sheep	dig prot %	12.4	14.5	
Energy				
Cattle	DE kcal/kg	2303.	2690.	
Sheep	DE kcal/kg	1925.	2249.	
Cattle	ME kcal/kg	1888.	2206.	
Sheep	ME kcal/kg	1578.	1844.	
Cattle	TDN %	52.	61.	
Sheep	TDN %	44.	51.	
Calcium	%	.88	1.02	
Iron	%	.020	.030	
Magnesium	%	.19	.22	
Phosphorus	%	.24	.28	
Potassium	%	2.16	2.52	
Manganese	mg/kg	72.8	85.1	

Vetch, hay, s-c, dough stage, (1)

Ref no 1-05-102

		As fed	Dry	C.V. ± %
Dry matter	%	85.1	100.0	
Ash	%	7.0	8.2	9
Crude fiber	%	29.5	34.7	5
Ether extract	%	2.0	2.4	23
N-free extract	%	36.1	42.4	
Protein (N x 6.25)	%	10.5	12.3	12
Cattle	dig prot %	6.9	8.1	
Sheep	dig prot %	6.5	7.6	
Energy				
Cattle	DE kcal/kg	2327.	2734.	
Sheep	DE kcal/kg	1876.	2205.	
Cattle	ME kcal/kg	1908.	2242.	
Sheep	ME kcal/kg	1539.	1808.	
Cattle	TDN %	53.	62.	
Sheep	TDN %	42.	50.	

Vetch, hay, s-c, (1)

Ref no 1-05-106

		As fed	Dry	C.V. ± %
Dry matter	%	88.2	100.0	3
Ash	%	8.0	9.1	20
Crude fiber	%	25.1	28.5	11
Ether extract	%	2.3	2.6	32
N-free extract	%	35.1	39.8	
Protein (N x 6.25)	%	17.6	20.0	21
Cattle	dig prot %	11.6	13.2	
Sheep	dig prot %	13.8	15.6	
Energy				
Cattle	DE kcal/kg	2411.	2734.	
Sheep	DE kcal/kg	2372.	2690.	
Cattle	ME kcal/kg	1977.	2242.	
Sheep	ME kcal/kg	1945.	2206.	
Cattle	TDN %	55.	62.	
Sheep	TDN %	54.	61.	
Calcium	%	1.20	1.36	32
Iron	%	.040	.050	59
Magnesium	%	.24	.27	29
Phosphorus	%	.30	.34	19
Potassium	%	1.87	2.12	27
Sodium	%	.46	.52	11
Sulfur	%	.13	.15	26
Cobalt	mg/kg	.310	.350	
Copper	mg/kg	8.7	9.9	7
Manganese	mg/kg	53.7	60.9	30

(5) protein supplements
(6) minerals

(7) vitamins
(8) additives

Feed name or analyses		Mean		C.V.
		As fed	Dry	± %

Vetch, aerial pt, fresh, (2)

Ref no 2-05-111

		As fed	Dry	± %
Dry matter	%	20.5	100.0	13
Ash	%	2.0	9.9	22
Crude fiber	%	5.6	27.5	10
Ether extract	%	.6	2.7	20
N-free extract	%	7.9	38.6	
Protein (N x 6.25)	%	4.4	21.3	15
Cattle	dig prot %	3.3	16.0	
Sheep	dig prot %	3.3	16.0	
Energy				
Cattle	DE kcal/kg	678.	3307.	
Sheep	DE kcal/kg	533.	2601.	
Cattle	ME kcal/kg	556.	2712.	
Sheep	ME kcal/kg	437.	2133.	
Cattle	TDN %	15.	75.	
Sheep	TDN %	12.	59.	
Calcium	%	.27	1.32	33
Chlorine	%	.38	1.85	
Iron	%	.010	.040	13
Magnesium	%	.05	.25	31
Phosphorus	%	.07	.33	14
Potassium	%	.51	2.47	12
Sodium	%	.10	.49	
Sulfur	%	.03	.15	40
Cobalt	mg/kg	.060	.310	
Copper	mg/kg	2.0	9.7	
Manganese	mg/kg	25.4	123.9	11

Vetch, aerial pt, ensiled, (3)

Ref no 3-05-112

		As fed	Dry	± %
Dry matter	%	30.2	100.0	9
Ash	%	2.4	7.9	7
Crude fiber	%	9.9	32.7	5
Ether extract	%	1.0	3.2	15
N-free extract	%	13.4	44.5	
Protein (N x 6.25)	%	3.5	11.7	15
Cattle	dig prot %	2.0	6.6	
Sheep	dig prot %	2.1	6.8	
Energy				
Cattle	DE kcal/kg	826.	2734.	
Sheep	DE kcal/kg	839.	2778.	
Cattle	ME kcal/kg	677.	2242.	
Sheep	ME kcal/kg	688.	2278.	
Cattle	TDN %	19.	62.	
Sheep	TDN %	19.	63.	

VETCH, COMMON. Vicia sativa

Vetch, common, aerial pt, fresh, (2)

Ref no 2-05-123

		As fed	Dry	± %
Dry matter	%	20.4	100.0	8
Ash	%	2.1	10.3	17
Crude fiber	%	5.5	27.0	11
Ether extract	%	.5	2.5	13
N-free extract	%	8.5	41.6	
Protein (N x 6.25)	%	3.8	18.6	16
Cattle	dig prot %	2.8	13.7	
Sheep	dig prot %	2.8	14.0	
Energy				
Cattle	DE kcal/kg	638.	3130.	
Sheep	DE kcal/kg	531.	2601.	
Cattle	ME kcal/kg	524.	2567.	
Sheep	ME kcal/kg	435.	2133.	
Cattle	TDN %	14.	71.	
Sheep	TDN %	12.	59.	

VETCH, HAIRY. Vicia villosa

Vetch, hairy, aerial pt, fresh, (2)

Ref no 2-05-128

		As fed	Dry	± %
Dry matter	%	18.2	100.0	12
Ash	%	2.2	12.1	13
Crude fiber	%	5.0	27.5	9
Ether extract	%	.5	2.7	18
N-free extract	%	6.3	34.6	
Protein (N x 6.25)	%	4.2	23.1	14
Cattle	dig prot %	3.2	17.5	
Sheep	dig prot %	3.5	19.2	
Energy				
Cattle	DE kcal/kg	594.	3263.	
Sheep	DE kcal/kg	538.	2954.	
Cattle	ME kcal/kg	487.	2676.	
Sheep	ME kcal/kg	441.	2422.	
Cattle	TDN %	13.	74.	
Sheep	TDN %	12.	67.	

(1) dry forages and roughages

(2) pasture, range plants, and forages fed green

(3) silages

(4) energy feeds

Feed name or analyses		Mean		C.V.
		As fed	Dry	± %

WHALE. Balaena glacialis, Balaenoptera spp, Physeter catadon

Whale, meat, heat-rend dehy grnd, salt declared above 3 mx 7, (5)
Whale meal (AAFCO)

Ref no 5-05-160

		As fed	Dry	C.V.
Dry matter	%	92.0	100.0	1
Ash	%	4.0	4.3	43
Crude fiber	%	1.0	1.1	48
Ether extract	%	6.8	7.4	19
N-free extract	%	1.3	1.4	
Protein (N x 6.25)	%	78.9	85.8	4
Cattle	dig prot %	68.8	74.8	
Sheep	dig prot %	72.6	78.9	
Swine	dig prot %	75.0	81.5	
Energy				
Cattle	DE kcal/kg	3489.	3792.	
Sheep	DE kcal/kg	3854.	4189.	
Swine	DE kcal/kg	4056.	4409.	
Cattle	ME kcal/kg	2860.	3109.	
Sheep	ME kcal/kg	3160.	3435.	
Swine	ME kcal/kg	3192.	3470.	
Cattle	TDN %	79.	86.	
Sheep	TDN %	87.	95.	
Swine	TDN %	92.	100.	
Calcium	%	.25	.27	
Phosphorus	%	.56	.61	8
Niacin	mg/kg	104.7	113.8	
Pantothenic acid	mg/kg	2.6	2.8	
Riboflavin	mg/kg	8.4	9.1	
Lysine	%	5.70	6.20	
Methionine	%	2.30	2.50	

Whale, meat, raw, (5)

Ref no 5-07-986

		As fed	Dry	
Dry matter	%	26.0	100.0	
Crude fiber	%	.0	.0	
Ether extract	%	2.1	8.1	
Protein (N x 6.25)	%	21.3	81.9	

Whale, meat w bone, heat-rend dehy grnd, (5)

Ref no 5-05-165

		As fed	Dry	C.V.
Dry matter	%	92.0	100.0	3
Ash	%	25.8	28.0	27
Crude fiber	%	1.0	1.1	74

Continued

		As fed	Dry	C.V.
Ether extract	%	9.8	10.7	36
N-free extract	%	1.4	1.5	
Protein (N x 6.25)	%	54.0	58.7	13
Cattle	dig prot %	42.7	46.4	
Sheep	dig prot %	35.1	38.2	
Swine	dig prot %	41.0	44.6	
Energy				
Cattle	DE kcal/kg	2677.	2910.	
Sheep	DE kcal/kg	2475.	2690.	
Swine	DE kcal/kg	2839.	3086.	
Cattle	ME kcal/kg	2195.	2386.	
Sheep	ME kcal/kg	2030.	2206.	
Swine	ME kcal/kg	2387.	2595.	
Cattle	TDN %	61.	66.	
Sheep	TDN %	56.	61.	
Swine	TDN %	64.	70.	
Calcium	%	8.29	9.01	
Phosphorus	%	4.16	4.52	

Whale, stickwater sol, cooked cond, (5)

Ref no 5-05-166

		As fed	Dry	
Dry matter	%	49.0	100.0	
Ash	%	2.3	4.7	
Ether extract	%	.2	.4	
N-free extract	%	14.1	28.8	
Protein (N x 6.25)	%	32.4	66.1	
Cattle	dig prot %	27.8	56.7	
Sheep	dig prot %	26.9	54.9	
Energy				
Cattle	DE kcal/kg	1772.	3616.	
Sheep	DE kcal/kg	1772.	3616.	
Cattle	ME kcal/kg	1453.	2965.	
Sheep	ME kcal/kg	1453.	2965.	
Cattle	TDN %	40.	82.	
Sheep	TDN %	40.	82.	
Calcium	%	.03	.06	
Magnesium	%	.02	.04	
Phosphorus	%	.15	.31	
Potassium	%	.02	.04	
Folic acid	mg/kg	.20	.40	
Niacin	mg/kg	22.0	44.9	
Pantothenic acid	mg/kg	10.6	21.6	
Riboflavin	mg/kg	1.5	3.1	
Arginine	%	1.60	3.27	
Cystine	%	.10	.20	
Histidine	%	.30	.61	
Isoleucine	%	.50	1.02	
Leucine	%	1.00	2.04	
Methionine	%	.40	.82	
Phenylalanine	%	.60	1.22	
Threonine	%	.60	1.22	

Continued

(5) protein supplements
(6) minerals

(7) vitamins
(8) additives

Feed name or analyses		Mean		C.V
		As fed	Dry	± %
Tryptophan	%	.10	.20	
Tyrosine	%	.40	.82	
Valine	%	1.00	2.04	

WHEAT. Triticum spp

Wheat, hay, s-c, (1)

Ref no 1-05-172

Dry matter	%	85.9	100.0	3
Ash	%	5.9	6.9	22
Crude fiber	%	23.9	27.8	14
Ether extract	%	1.7	2.0	17
N-free extract	%	47.9	55.8	
Protein (N x 6.25)	%	6.4	7.5	15
Cattle	dig prot %	2.9	3.4	
Sheep	dig prot %	3.4	4.0	
Energy				
Cattle	DE kcal/kg	2500.	2910.	
Sheep	DE kcal/kg	1970.	2293.	
Cattle	ME kcal/kg	2050.	2386.	
Sheep	ME kcal/kg	1615.	1880.	
Cattle	TDN %	57.	66.	
Sheep	TDN %	45.	52.	
Carotene	mg/kg	95.9	111.6	
Vitamin A equiv	IU/g	159.9	186.0	

Wheat, straw, (1)

Ref no 1-05-175

Dry matter	%	90.1	100.0	4
Ash	%	7.3	8.1	18
Crude fiber	%	37.4	41.5	7
Ether extract	%	1.5	1.7	31
N-free extract	%	40.6	45.1	
Protein (N x 6.25)	%	3.2	3.6	27
Cattle	dig prot %	.4	.4	
Horses	dig prot %	.4	.4	
Sheep	dig prot %	1.4	1.5	
Cellulose	%	45.1	50.1	23
Lignin	%	12.3	13.7	
Energy				
Cattle	DE kcal/kg	1906.	2116.	
Horses	DE kcal/kg	1906.	2116.	
Sheep	DE kcal/kg	1509.	1675.	
Cattle	ME kcal/kg	1563.	1735.	
Horses	ME kcal/kg	1563.	1735.	
Sheep	ME kcal/kg	1238.	1374.	
Cattle	TDN %	43.	48.	
Horses	TDN %	43.	48.	
Sheep	TDN %	34.	38.	

Continued

Feed name or analyses		Mean		C.V.
		As fed	Dry	± %
Calcium	%	.15	.17	35
Chlorine	%	.27	.30	16
Iron	%	.010	.020	20
Magnesium	%	.11	.12	99
Phosphorus	%	.07	.08	99
Potassium	%	1.00	1.11	32
Sodium	%	.13	.14	45
Sulfur	%	.17	.19	19
Cobalt	mg/kg	.040	.040	
Copper	mg/kg	3.0	3.3	19
Manganese	mg/kg	36.4	40.4	30
Carotene	mg/kg	2.0	2.2	
Vitamin A equiv	IU/g	3.3	3.7	

Wheat, aerial pt, fresh, immature, (2)

Ref no 2-05-176

Dry matter	%	21.5	100.0	23
Ash	%	2.9	13.3	18
Crude fiber	%	3.7	17.4	21
Ether extract	%	1.0	4.4	18
N-free extract	%	7.8	36.3	
Protein (N x 6.25)	%	6.1	28.6	14
Cattle	dig prot %	4.8	22.2	
Sheep	dig prot %	5.1	23.6	
Energy				
Cattle	DE kcal/kg	692.	3219.	
Sheep	DE kcal/kg	673.	3130.	
Cattle	ME kcal/kg	568.	2640.	
Sheep	ME kcal/kg	552.	2566.	
Cattle	TDN %	16.	73.	
Sheep	TDN %	15.	71.	
Calcium	%	.09	.42	44
Magnesium	%	.05	.21	22
Phosphorus	%	.09	.40	30
Potassium	%	.75	3.50	13
Carotene	mg/kg	111.8	520.2	39
Vitamin A equiv	IU/g	186.4	867.2	

Wheat, aerial pt, ensiled, (3)

Ref no 3-05-186

Dry matter	%	25.1	100.0	4
Ash	%	2.2	8.8	29
Crude fiber	%	7.9	31.3	18
Ether extract	%	.8	3.1	12
N-free extract	%	12.1	48.1	
Protein (N x 6.25)	%	2.2	8.7	17
Cattle	dig prot %	1.0	4.1	
Sheep	dig prot %	1.0	4.1	
Energy				
Cattle	DE kcal/kg	708.	2822.	

Continued

(1) dry forages and roughages
(2) pasture, range plants, and forages fed green
(3) silages
(4) energy feeds

Feed name or analyses		Mean As fed	Mean Dry	C.V. ± %
Sheep	DE kcal/kg	708.	2822.	
Cattle	ME kcal/kg	581.	2314.	
Sheep	ME kcal/kg	581.	2314.	
Cattle	TDN %	16.	64.	
Sheep	TDN %	16.	64.	
Carotene	mg/kg	44.0	175.5	
Vitamin A equiv	IU/g	73.3	292.6	

Wheat, bran, dry-mil, (4)
Wheat bran (AAFCO)
Bran (CFA)

Ref no 4-05-190

		As fed	Dry	C.V. ± %
Dry matter	%	89.0	100.0	3
Ash	%	6.1	6.9	17
Crude fiber	%	10.0	11.2	22
Ether extract	%	4.1	4.6	19
N-free extract	%	52.8	59.3	
Protein (N x 6.25)	%	16.0	18.0	14
Cattle	dig prot %	12.5	14.0	
Horses	dig prot %	15.4	17.3	
Sheep	dig prot %	12.0	13.5	
Swine	dig prot %	12.2	13.7	
Energy	GE kcal/kg	4052.	4554.	
Cattle	DE kcal/kg	2746.	3086.	
Horses	DE kcal/kg	3571.	4012.	
Sheep	DE kcal/kg	2590.	2910.	
Swine	DE kcal/kg	2512.	2822.	
Cattle	ME kcal/kg	2252.	2530.	
Chickens	MEn kcal/kg	1146.	1288.	
Horses	ME kcal/kg	2928.	3290.	
Sheep	ME kcal/kg	2124.	2386.	
Swine	ME kcal/kg	2321.	2608.	
Cattle	NEm kcal/kg	1362.	1530.	
Cattle	NEgain kcal/kg	854.	960.	
Cattle	TDN %	62.	70.	
Horses	TDN %	81.	91.	
Sheep	TDN %	59.	66.	
Swine	TDN %	57.	64.	
Calcium	%	.14	.16	71
Iron	%	.017	.019	27
Magnesium	%	.55	.62	18
Phosphorus	%	1.17	1.32	32
Potassium	%	1.24	1.39	22
Sodium	%	.06	.07	57
Cobalt	mg/kg	1.000	1.100	40
Copper	mg/kg	12.3	13.8	52
Manganese	mg/kg	115.7	130.0	15
Choline	mg/kg	988.	1110.	
Folic acid	mg/kg	1.80	2.00	92
Niacin	mg/kg	209.2	235.1	44
Pantothenic acid	mg/kg	29.0	32.6	25
Riboflavin	mg/kg	3.1	3.5	37
Thiamine	mg/kg	7.9	8.9	43

Continued

Feed name or analyses		Mean As fed	Mean Dry	C.V. ± %
a-tocopherol	mg/kg	10.8	12.1	36
Arginine	%	1.00	1.12	13
Cystine	%	.30	.34	
Glycine	%	.90	1.01	
Histidine	%	.30	.34	24
Isoleucine	%	.60	.67	24
Leucine	%	.90	1.01	8
Lysine	%	.60	.67	22
Methionine	%	.10	.11	63
Phenylalanine	%	.50	.56	15
Threonine	%	.40	.45	18
Tryptophan	%	.30	.34	11
Tyrosine	%	.40	.45	40
Valine	%	.70	.79	16

Wheat endosperm - see Wheat, grits

Wheat, flour, c-bolt, feed gr mx 2 fbr, (4)
Wheat feed flour, mx 1.5 fbr (AAFCO)
Feed flour, mx 2.0 fbr (CFA)

Ref no 4-05-199

		As fed	Dry	C.V. ± %
Dry matter	%	89.0	100.0	1
Ash	%	2.1	2.4	56
Crude fiber	%	3.0	3.4	78
Ether extract	%	2.9	3.3	42
N-free extract	%	65.0	73.1	
Protein (N x 6.25)	%	15.8	17.8	20
Cattle	dig prot %	11.0	12.4	
Sheep	dig prot %	11.2	12.6	
Swine	dig prot %	14.6	16.4	
Energy				
Cattle	DE kcal/kg	3532.	3968.	
Sheep	DE kcal/kg	3336.	3748.	
Swine	DE kcal/kg	3610.	4056.	
Cattle	ME kcal/kg	2896.	3254.	
Chickens	MEn kcal/kg	3086.	3467.	
Sheep	ME kcal/kg	2735.	3073.	
Swine	ME kcal/kg	3336.	3748.	
Cattle	TDN %	80.	90.	
Sheep	TDN %	76.	85.	
Swine	TDN %	82.	92.	
Calcium	%	.03	.03	42
Iron	%	.002	.002	93
Phosphorus	%	.28	.31	52
Copper	mg/kg	4.6	5.2	
Manganese	mg/kg	44.9	50.5	59
Niacin	mg/kg	41.8	47.0	
Pantothenic acid	mg/kg	.9	1.0	
Thiamine	mg/kg	5.9	6.6	
Arginine	%	.40	.45	
Histidine	%	.30	.34	
Isoleucine	%	.60	.67	

Continued

(5) protein supplements

(6) minerals

(7) vitamins

(8) additives

Feed name or analyses		Mean		C.V.
		As fed	Dry	± %
Leucine	%	.90	1.01	
Lysine	%	.30	.34	
Methionine	%	.11	.12	
Phenylalanine	%	.60	.67	
Threonine	%	.30	.34	
Tryptophan	%	.11	.12	
Tyrosine	%	.20	.22	
Valine	%	.50	.56	

Wheat, flour by-prod, c-sift, mx 7 fbr, (4)
Wheat shorts, mx 7 fbr (AAFCO)
Shorts, mx 8 fbr (CFA)

Ref no 4-05-201

Feed name or analyses		Mean		C.V.
		As fed	Dry	± %
Dry matter	%	90.0	100.0	1
Ash	%	3.9	4.3	23
Crude fiber	%	5.0	5.6	26
Ether extract	%	4.2	4.7	17
N-free extract	%	58.5	65.0	
Protein (N x 6.25)	%	18.4	20.4	9
Cattle	dig prot %	13.2	14.7	
Sheep	dig prot %	15.6	17.3	
Swine	dig prot %	15.4	17.1	
Energy				
Cattle	DE kcal/kg	3413.	3792.	
Sheep	DE kcal/kg	3373.	3748.	
Swine	DE kcal/kg	3168.	3520.	
Cattle	ME kcal/kg	2798.	3109.	
Chickens	MEn kcal/kg	2646.	2940.	
Sheep	ME kcal/kg	2766.	3073.	
Swine	ME kcal/kg	2912.	3235.	
Cattle	TDN %	77.	86.	
Sheep	TDN %	76.	85.	
Swine	TDN %	72.	80.	
Calcium	%	.11	.12	28
Iron	%	.010	.011	83
Magnesium	%	.26	.29	25
Phosphorus	%	.76	.84	26
Potassium	%	.85	.94	
Sodium	%	.07	.08	
Cobalt	mg/kg	.100	.100	64
Copper	mg/kg	9.2	10.3	37
Manganese	mg/kg	104.5	116.1	26
Choline	mg/kg	928.	1093.	
Niacin	mg/kg	94.6	105.1	18
Pantothenic acid	mg/kg	17.6	19.6	38
Riboflavin	mg/kg	2.0	2.2	31
Thiamine	mg/kg	15.8	17.6	12
a-tocopherol	mg/kg	29.9	33.2	

Wheat, flour by-prod, f-sift, mx 4 fbr, (4)
Wheat red dog, mx 4.0 fbr (AAFCO)
Middlings, mx 4.5 fbr (CFA)

Ref no 4-05-203

Feed name or analyses		Mean		C.V.
		As fed	Dry	± %
Dry matter	%	89.0	100.0	2
Ash	%	2.5	2.8	30
Crude fiber	%	2.0	2.2	47
Ether extract	%	3.6	4.0	24
N-free extract	%	63.0	70.8	
Protein (N x 6.25)	%	18.0	20.2	11
Cattle	dig prot %	13.0	14.6	
Sheep	dig prot %	16.2	18.2	
Swine	dig prot %	16.0	18.0	
Energy				
Cattle	DE kcal/kg	3571.	4012.	
Sheep	DE kcal/kg	3846.	4321.	
Swine	DE kcal/kg	3212.	3609.	
Cattle	ME kcal/kg	2928.	3290.	
Chickens	MEn kcal/kg	2756.	3097.	
Sheep	ME kcal/kg	3153.	3543.	
Swine	ME kcal/kg	2952.	3317.	
Cattle	TDN %	81.	91.	
Sheep	TDN %	87.	98.	
Swine	TDN %	73.	82.	
Calcium	%	.08	.09	33
Iron	%	.006	.007	48
Magnesium	%	.29	.33	
Phosphorus	%	.52	.58	28
Potassium	%	.60	.67	
Sodium	%	.66	.74	
Copper	mg/kg	4.4	4.9	25
Manganese	mg/kg	37.6	42.3	60
Niacin	mg/kg	52.6	59.1	37
Pantothenic acid	mg/kg	13.6	15.3	16
Riboflavin	mg/kg	1.5	1.7	76
Thiamine	mg/kg	18.9	21.2	24
a-tocopherol	mg/kg	57.6	64.7	
Arginine	%	1.00	1.12	20
Histidine	%	.40	.45	20
Isoleucine	%	.70	.79	26
Leucine	%	1.20	1.35	17
Lysine	%	.60	.67	27
Methionine	%	.10	.11	40
Phenylalanine	%	.50	.56	40
Threonine	%	.50	.56	8
Tryptophan	%	.20	.22	20
Tyrosine	%	.50	.56	40
Valine	%	.80	.90	23

(1) dry forages and roughages
(2) pasture, range plants, and forages fed green
(3) silages
(4) energy feeds

Feed name or analyses		Mean As fed	Mean Dry	C.V. ± %

Wheat, flour by-prod, mx 9.5 fbr, (4)
  Wheat middlings (AAFCO)
  Wheat standard middlings

Ref no 4-05-205

Feed name or analyses		As fed	Dry	C.V. ± %
Dry matter	%	90.0	100.0	2
Ash	%	4.4	4.9	13
Crude fiber	%	8.0	8.9	19
Ether extract	%	4.6	5.1	12
N-free extract	%	55.8	62.0	
Protein (N x 6.25)	%	17.2	19.1	9
Cattle	dig prot %	12.2	13.6	
Sheep	dig prot %	13.4	14.9	
Swine	dig prot %	13.8	15.3	
Energy				
Cattle	DE kcal/kg	3294.	3660.	
Sheep	DE kcal/kg	2976.	3307.	
Swine	DE kcal/kg	2817.	3130.	
Cattle	ME kcal/kg	2701.	3001.	
Chickens	MEn kcal/kg	1808.	2009.	
Sheep	ME kcal/kg	2441.	2712.	
Swine	ME kcal/kg	2595.	2883.	
Cattle	TDN %	75.	83.	
Sheep	TDN %	68.	75.	
Swine	TDN %	64.	71.	
Calcium	%	.15	.16	72
Iron	%	.010	.010	33
Magnesium	%	.37	.41	11
Phosphorus	%	.91	1.01	15
Potassium	%	.98	1.08	8
Sodium	%	.22	.24	
Cobalt	mg/kg	.100	.100	20
Copper	mg/kg	22.0	24.4	9
Manganese	mg/kg	118.4	131.5	12
Choline	mg/kg	1074.	1193.	
Folic acid	mg/kg	.90	1.00	
Niacin	mg/kg	98.6	109.5	23
Pantothenic acid	mg/kg	19.8	22.0	30
Riboflavin	mg/kg	2.0	2.2	27
Thiamine	mg/kg	12.8	14.2	19
Arginine	%	.90	1.00	25
Cystine	%	.20	.22	
Glycine	%	.40	.44	
Histidine	%	.40	.44	30
Isoleucine	%	.80	.88	28
Leucine	%	1.20	1.33	20
Lysine	%	.70	.77	26
Methionine	%	.20	.22	36
Phenylalanine	%	.70	.77	16
Threonine	%	.60	.66	6
Tryptophan	%	.20	.22	19
Tyrosine	%	.40	.44	30
Valine	%	.80	.88	23

Wheat, flour by-prod, mil-rn, mx 9.5 fbr, (4)
  Wheat mill run (AAFCO)

Ref no 4-05-206

Feed name or analyses		As fed	Dry	C.V. ± %
Dry matter	%	90.0	100.0	2
Ash	%	5.2	5.8	11
Crude fiber	%	8.0	8.9	15
Ether extract	%	4.0	4.4	17
N-free extract	%	57.5	63.9	
Protein (N x 6.25)	%	15.3	17.0	10
Cattle	dig prot %	10.4	11.6	
Sheep	dig prot %	12.7	14.1	
Swine	dig prot %	12.2	13.6	
Energy	GE kcal/kg	3951.	4390.	
Cattle	DE kcal/kg	3214.	3571.	
Sheep	DE kcal/kg	2976.	3307.	
Swine	DE kcal/kg	3168.	3520.	
Cattle	ME kcal/kg	2635.	2928.	
Chickens	MEn kcal/kg	1764.	1960.	
Sheep	ME kcal/kg	2441.	2712.	
Swine	ME kcal/kg	2934.	3260.	
Cattle	NEm kcal/kg	1575.	1750.	
Cattle	NEgain kcal/kg	1044.	1160.	
Cattle	TDN %	73.	81.	
Sheep	TDN %	68.	75.	
Swine	TDN %	72.	80.	
Calcium	%	.09	.10	31
Iron	%	.009	.010	
Magnesium	%	.51	.57	
Phosphorus	%	1.02	1.13	9
Potassium	%	1.28	1.42	
Sodium	%	.22	.24	
Cobalt	mg/kg	.200	.200	
Copper	mg/kg	18.7	20.8	
Manganese	mg/kg	102.7	114.1	33
Choline	mg/kg	981.	1090.	7
Niacin	mg/kg	112.0	124.4	
Pantothenic acid	mg/kg	13.2	14.7	
Riboflavin	mg/kg	2.4	2.7	11
Thiamine	mg/kg	15.2	16.9	

Wheat, grain, thresher-run, mn 55 mx 60 wt mx 5 fm, (4)

Ref no 4-08-165 Canada

Feed name or analyses		As fed	Dry	C.V. ± %
Dry matter	%	88.0	100.0	
Ash	%	1.6	1.8	
Crude fiber	%	2.6	2.9	
Ether extract	%	1.2	1.4	
N-free extract	%	66.7	75.8	
Protein (N x 6.25)	%	15.8	18.0	
Cattle	dig prot %	12.3	14.0	

Continued

(5) protein supplements
(6) minerals

(7) vitamins
(8) additives

Feed name or analyses		Mean		C.V.
		As fed	Dry	± %
Sheep	dig prot %	12.3	14.0	
Swine	dig prot %	14.6	16.6	
Energy	GE kcal/kg	4224.	4800.	
Cattle	DE kcal/kg	3376.	3836.	
Sheep	DE kcal/kg	3376.	3836.	
Swine	DE kcal/kg	3530.	4012.	
Cattle	ME kcal/kg	2768.	3146.	
Sheep	ME kcal/kg	2768.	3146.	
Swine	ME kcal/kg	3262.	3707.	
Cattle	TDN %	76.	87.	
Sheep	TDN %	76.	87.	
Swine	TDN %	80.	91.	

Feed name or analyses		Mean		C.V.
		As fed	Dry	± %
Cattle	ME kcal/kg	2838.	3182.	
Sheep	ME kcal/kg	2838.	3182.	
Swine	ME kcal/kg	3388.	3798.	
Cattle	TDN %	78.	88.	
Sheep	TDN %	78.	88.	
Swine	TDN %	82.	92.	
Calcium	%	.12	.14	
Phosphorus	%	.30	.34	
Niacin	mg/kg	59.1	66.3	
Pantothenic acid	mg/kg	11.5	12.9	
Riboflavin	mg/kg	1.1	1.2	
Thiamine	mg/kg	4.9	5.5	

Wheat, grain, thresher-run, mn 60 wt mx 5 fm, (4)

Ref no 4-08-164 Canada

		As fed	Dry	
Dry matter	%	88.0	100.0	
Ash	%	1.8	2.0	
Crude fiber	%	2.2	2.5	
Ether extract	%	1.4	1.6	
N-free extract	%	69.1	78.5	
Protein (N x 6.25)	%	13.6	15.4	
Cattle	dig prot %	10.6	12.0	
Sheep	dig prot %	10.6	12.0	
Swine	dig prot %	12.5	14.2	
Energy	GE kcal/kg	4174.	4743.	
Cattle	DE kcal/kg	3414.	3880.	
Sheep	DE kcal/kg	3414.	3880.	
Swine	DE kcal/kg	3569.	4056.	
Cattle	ME kcal/kg	2800.	3182.	
Sheep	ME kcal/kg	2800.	3182.	
Swine	ME kcal/kg	3316.	3768.	
Cattle	TDN %	77.	88.	
Sheep	TDN %	77.	88.	
Swine	TDN %	81.	92.	

Wheat, grain, Pacific coast, (4)

Ref no 4-08-142

		As fed	Dry	
Dry matter	%	89.2	100.0	
Ash	%	1.9	2.1	
Crude fiber	%	2.7	3.0	
Ether extract	%	2.0	2.2	
N-free extract	%	72.8	81.6	
Protein (N x 6.25)	%	9.9	11.1	
Cattle	dig prot %	7.7	8.6	
Sheep	dig prot %	7.7	8.6	
Swine	dig prot %	9.1	10.2	
Energy				
Cattle	DE kcal/kg	3461.	3880.	
Sheep	DE kcal/kg	3461.	3880.	
Swine	DE kcal/kg	3618.	4056.	

Continued

Wheat, grain, (4)

Ref no 4-05-211

		As fed	Dry	
Dry matter	%	89.0	100.0	2
Ash	%	1.6	1.8	25
Crude fiber	%	3.0	3.4	31
Ether extract	%	1.7	1.9	24
N-free extract	%	70.0	78.6	
Protein (N x 6.25)	%	12.7	14.3	14
Cattle	dig prot %	10.0	11.2	
Sheep	dig prot %	10.0	11.2	
Swine	dig prot %	11.7	13.2	
Energy	GE kcal/kg	4001.	4495.	
Cattle	DE kcal/kg	3453.	3880.	
Sheep	DE kcal/kg	3453.	3880.	
Swine	DE kcal/kg	3520.	3955.	
Cattle	ME kcal/kg	2832.	3182.	
Sheep	ME kcal/kg	2832.	3182.	
Swine	ME kcal/kg	3277.	3682.	
Cattle	NEm kcal/kg	1985.	2230.	
Cattle	NE gain kcal/kg	1290.	1450.	
Cattle	TDN %	78.	88.	
Sheep	TDN %	78.	88.	
Swine	TDN %	80.	90.	
Calcium	%	.05	.06	99
Iron	%	.005	.006	35
Magnesium	%	.16	.18	27
Phosphorus	%	.36	.41	24
Potassium	%	.52	.58	22
Sodium	%	.09	.10	55
Cobalt	mg/kg	.080	.090	31
Copper	mg/kg	7.2	8.1	41
Manganese	mg/kg	48.8	54.8	30
Zinc	mg/kg	13.7	15.4	
Biotin	mg/kg	.10	.10	84
Choline	mg/kg	830.	933.	28
Folic acid	mg/kg	.40	.40	36
Niacin	mg/kg	56.6	63.6	29
Pantothenic acid	mg/kg	12.1	13.6	17
Riboflavin	mg/kg	1.2	1.3	60
Thiamine	mg/kg	4.9	5.5	27

Continued

(1) dry forages and roughages
(2) pasture, range plants, and forages fed green

(3) silages
(4) energy feeds

Feed name or analyses		Mean		C.V.
		As fed	Dry	± %
a-tocopherol	mg/kg	15.5	17.4	9
Arginine	%	.71	.80	23
Cystine	%	.18	.20	
Glycine	%	.89	1.00	
Histidine	%	.27	.30	36
Isoleucine	%	.53	.60	33
Leucine	%	.89	1.00	26
Lysine	%	.45	.51	32
Methionine	%	.18	.20	
Phenylalanine	%	.62	.70	23
Threonine	%	.36	.40	19
Tryptophan	%	.18	.20	25
Tyrosine	%	.45	.51	44
Valine	%	.53	.60	27

Wheat grain - see also Wheat, durum; Wheat, hard red spring; Wheat, hard red winter; Wheat, red spring; Wheat, soft; and Wheat, soft red winter

Wheat, grain screenings, (4)

Ref no 4-05-216

Dry matter	%	89.0	100.0	2
Ash	%	3.2	3.6	41
Crude fiber	%	7.0	7.9	73
Ether extract	%	3.0	3.4	27
N-free extract	%	60.7	68.2	
Protein (N x 6.25)	%	15.0	16.9	17
Cattle	dig prot %	10.8	12.2	
Sheep	dig prot %	10.8	12.2	
Swine	dig prot %	12.0	13.5	
Cellulose	%	5.0	5.6	
Lignin	%	7.0	7.9	
Energy				
Cattle	DE kcal/kg	3022.	3395.	
Sheep	DE kcal/kg	3022.	3395.	
Swine	DE kcal/kg	2772.	3114.	
Cattle	ME kcal/kg	2478.	2784.	
Sheep	ME kcal/kg	2478.	2784.	
Swine	ME kcal/kg	2567.	2884.	
Cattle	TDN %	68.	77.	
Sheep	TDN %	68.	77.	
Swine	TDN %	63.	71.	
Calcium	%	.08	.09	
Phosphorus	%	.36	.40	
Manganese	mg/kg	28.6	32.1	

Wheat, grain screenings - see also Grains, screenings

Wheat, grits, cracked f-scr, (4)
   Farina
   Wheat endosperm

Ref no 4-07-852

Dry matter	%	88.0	100.0	
Ash	%	1.2	1.4	
Crude fiber	%	.3	.3	
Ether extract	%	1.1	1.2	
N-free extract	%	74.4	84.5	
Protein (N x 6.25)	%	11.1	12.6	
Swine	dig prot %	9.8	11.2	
Energy				
Swine	DE kcal/kg	3414.	3880.	
Swine	ME kcal/kg	3193.	3628.	
Swine	TDN %	77.	88.	
Arginine	%	.60	.68	
Cystine	%	.30	.34	
Histidine	%	.30	.34	
Isoleucine	%	1.10	1.25	
Leucine	%	1.70	1.93	
Lysine	%	.40	.45	
Methionine	%	.20	.23	
Phenylalanine	%	.60	.68	
Threonine	%	.40	.45	
Tryptophan	%	.30	.34	
Valine	%	.60	.68	

Wheat, distil grains, dehy, (5)
   Wheat distillers dried grains (AAFCO)
   Wheat distillers dried grains (CFA)

Ref no 5-05-193

Dry matter	%	93.0	100.0	2
Ash	%	2.4	2.6	34
Crude fiber	%	12.0	12.9	18
Ether extract	%	5.8	6.2	26
N-free extract	%	39.1	42.0	
Protein (N x 6.25)	%	33.8	36.3	21
Cattle	dig prot %	27.2	29.3	
Sheep	dig prot %	28.4	30.5	
Energy				
Cattle	DE kcal/kg	3280.	3527.	
Sheep	DE kcal/kg	3445.	3704.	
Cattle	ME kcal/kg	2690.	2892.	
Sheep	ME kcal/kg	2824.	3037.	
Cattle	TDN %	74.	80.	
Sheep	TDN %	78.	84.	
Calcium	%	.10	.11	52

Continued

(5) protein supplements

(6) minerals

(7) vitamins

(8) additives

Feed name or analyses		Mean		C.V.
		As fed	Dry	± %
Phosphorus	%	.50	.54	34
Niacin	mg/kg	55.9	60.1	32
Pantothenic acid	mg/kg	8.1	8.7	86
Riboflavin	mg/kg	3.7	4.0	60
Thiamine	mg/kg	2.0	2.2	79
Arginine	%	1.10	1.18	
Histidine	%	.80	.86	
Isoleucine	%	2.00	2.15	
Leucine	%	1.70	1.83	
Lysine	%	.70	.75	
Phenylalanine	%	1.70	1.83	
Threonine	%	.90	.97	
Tyrosine	%	.50	.54	
Valine	%	1.70	1.83	

Wheat, distil grains w sol, dehy, mn 75 orig solids, (5)
Wheat distillers dried grains with solubles (AAFCO)

Ref no 5-05-194

Feed name or analyses		As fed	Dry	C.V. ± %
Dry matter	%	93.0	100.0	3
Ash	%	4.1	4.4	12
Crude fiber	%	10.0	10.8	24
Ether extract	%	6.3	6.8	30
N-free extract	%	40.5	43.6	
Protein (N x 6.25)	%	32.0	34.4	14
Cattle	dig prot %	25.7	27.6	
Sheep	dig prot %	26.8	28.8	
Energy				
Cattle	DE kcal/kg	3321.	3571.	
Sheep	DE kcal/kg	3486.	3748.	
Cattle	ME kcal/kg	2723.	2928.	
Sheep	ME kcal/kg	2858.	3073.	
Cattle	TDN %	75.	81.	
Sheep	TDN %	79.	85.	
Calcium	%	.19	.20	
Phosphorus	%	.71	.76	
Niacin	mg/kg	74.8	80.4	
Pantothenic acid	mg/kg	12.1	13.0	
Riboflavin	mg/kg	10.6	11.4	28
Arginine	%	1.10	1.18	7
Histidine	%	.80	.86	5
Isoleucine	%	1.90	2.04	15
Leucine	%	2.00	2.15	42
Lysine	%	.80	.86	10
Methionine	%	.50	.54	
Phenylalanine	%	1.90	2.04	11
Threonine	%	1.00	1.08	8
Tryptophan	%	.40	.43	
Tyrosine	%	.60	.65	14
Valine	%	1.90	2.04	6

Wheat, distil sol, dehy, (5)
Wheat distillers dried solubles (AAFCO)

Ref no 5-05-195

Feed name or analyses		As fed	Dry	C.V. ± %
Dry matter	%	94.0	100.0	2
Ash	%	6.8	7.2	29
Crude fiber	%	3.0	3.2	51
Ether extract	%	2.2	2.3	79
N-free extract	%	49.0	52.1	
Protein (N x 6.25)	%	33.1	35.2	12
Cattle	dig prot %	26.6	28.3	
Sheep	dig prot %	27.7	29.5	
Energy				
Cattle	DE kcal/kg	3399.	3616.	
Sheep	DE kcal/kg	3647.	3880.	
Cattle	ME kcal/kg	2787.	2965.	
Sheep	ME kcal/kg	2991.	3182.	
Cattle	TDN %	77.	82.	
Sheep	TDN %	83.	88.	
Calcium	%	.36	.38	22
Phosphorus	%	1.51	1.61	14
Niacin	mg/kg	191.4	203.6	12
Pantothenic acid	mg/kg	33.7	35.9	25
Riboflavin	mg/kg	15.0	16.0	23
Thiamine	mg/kg	5.1	5.4	41
Arginine	%	1.00	1.06	13
Histidine	%	.80	.85	10
Isoleucine	%	1.60	1.70	5
Leucine	%	1.50	1.60	11
Lysine	%	.70	.74	6
Methionine	%	.40	.43	
Phenylalanine	%	1.70	1.81	17
Threonine	%	1.00	1.06	4
Tryptophan	%	.50	.53	
Tyrosine	%	.70	.74	
Valine	%	1.50	1.60	8

Wheat, germ, extn unspec grnd, mn 30 prot, (5)
Defatted wheat germ meal (AAFCO)

Ref no 5-05-217

Feed name or analyses		As fed	Dry	C.V. ± %
Dry matter	%	90.0	100.0	2
Ash	%	4.8	5.3	20
Crude fiber	%	3.0	3.3	22
Ether extract	%	9.1	10.1	20
N-free extract	%	45.9	51.0	
Protein (N x 6.25)	%	27.3	30.3	10
Cattle	dig prot %	25.6	28.5	
Sheep	dig prot %	25.6	28.5	
Swine	dig prot %	24.6	27.3	
Energy				
Cattle	DE kcal/kg	3650.	4056.	

Continued

(1) dry forages and roughages
(2) pasture, range plants, and forages fed green
(3) silages
(4) energy feeds

TABLE A-3-1 Composition of Feeds (Continued)

Feed name or analyses		Mean As fed	Mean Dry	C.V. ± %
Sheep	DE kcal/kg	3730.	4145.	
Swine	DE kcal/kg	3413.	3792.	
Cattle	ME kcal/kg	2993.	3326.	
Sheep	ME kcal/kg	3059.	3399.	
Swine	ME kcal/kg	3068.	3409.	
Cattle	TDN %	83.	92.	
Sheep	TDN %	85.	94.	
Swine	TDN %	77.	86.	
Calcium	%	.07	.08	44
Iron	%	.010	.010	17
Phosphorus	%	1.06	1.18	14
Copper	mg/kg	9.2	10.2	27
Manganese	mg/kg	164.1	182.3	26
Choline	mg/kg	3375.	3749.	
Niacin	mg/kg	51.3	57.0	15
Pantothenic acid	mg/kg	12.3	13.7	36
Riboflavin	mg/kg	5.1	5.7	17
Thiamine	mg/kg	25.3	28.1	20
a-tocopherol	mg/kg	73.9	82.1	

Feed name or analyses		Mean As fed	Mean Dry	C.V. ± %
Thiamine	mg/kg	27.9	31.0	42
a-tocopherol	mg/kg	132.7	147.4	9
Arginine	%	1.60	1.78	
Cystine	%	.50	.56	
Histidine	%	.50	.56	
Isoleucine	%	1.20	1.33	
Leucine	%	1.10	1.22	
Lysine	%	1.60	1.78	
Methionine	%	.30	.33	
Phenylalanine	%	.80	.89	
Threonine	%	.80	.89	
Tryptophan	%	.30	.33	14
Valine	%	1.10	1.22	32

Wheat, germ, mn 75 embryo mx 2.5 fbr, (5)
  Wheat germ (CFA)

Ref no 5-05-219

		As fed	Dry	C.V.
Dry matter	%	90.0	100.0	2
Ash	%	4.3	4.8	8
Crude fiber	%	3.0	3.3	18
Ether extract	%	10.9	12.1	15
N-free extract	%	45.6	50.7	
Protein (N x 6.25)	%	26.2	29.1	14
Cattle	dig prot %	24.7	27.4	
Sheep	dig prot %	24.7	27.4	
Swine	dig prot %	23.6	26.2	
Energy	GE kcal/kg	4206.	4673.	
Cattle	DE kcal/kg	3770.	4189.	
Sheep	DE kcal/kg	3849.	4277.	
Swine	DE kcal/kg	3770.	4189.	
Cattle	ME kcal/kg	3092.	3435.	
Sheep	ME kcal/kg	3156.	3507.	
Swine	ME kcal/kg	3397.	3774.	
Cattle	TDN %	86.	95.	
Sheep	TDN %	87.	97.	
Swine	TDN %	86.	95.	
Calcium	%	.07	.08	9
Iron	%	.010	.010	
Phosphorus	%	1.04	1.16	5
Copper	mg/kg	8.8	9.8	
Manganese	mg/kg	134.9	149.9	
Choline	mg/kg	3010.	3344.	19
Folic acid	mg/kg	2.00	2.20	
Niacin	mg/kg	47.3	52.6	50
Pantothenic acid	mg/kg	11.2	12.4	59
Riboflavin	mg/kg	5.1	5.7	25
Thiamine	mg/kg	27.9	31.0	42
a-tocopherol	mg/kg	132.7	147.4	9
Arginine	%	1.60	1.78	
Cystine	%	.50	.56	
Histidine	%	.50	.56	
Isoleucine	%	1.20	1.33	
Leucine	%	1.10	1.22	

Wheat, germ, grnd, mn 25 prot 7 fat, (5)
  Wheat germ meal (AAFCO)

Ref no 5-05-218

		As fed	Dry	C.V.
Dry matter	%	90.0	100.0	2
Ash	%	4.3	4.8	8
Crude fiber	%	3.0	3.3	18
Ether extract	%	10.9	12.1	15
N-free extract	%	45.6	50.7	
Protein (N x 6.25)	%	26.2	29.1	14
Cattle	dig prot %	24.7	27.4	
Sheep	dig prot %	24.7	27.4	
Swine	dig prot %	23.6	26.2	
Energy	GE kcal/kg	4206.	4673.	
Cattle	DE kcal/kg	3770.	4189.	
Sheep	DE kcal/kg	3849.	4277.	
Swine	DE kcal/kg	3770.	4189.	
Cattle	ME kcal/kg	3092.	3435.	
Chickens	MEn kcal/kg	3086.	3429.	
Sheep	ME kcal/kg	3156.	3507.	
Swine	ME kcal/kg	3397.	3774.	
Cattle	TDN %	86.	95.	
Sheep	TDN %	87.	97.	
Swine	TDN %	86.	95.	
Calcium	%	.07	.08	9
Iron	%	.011	.012	
Phosphorus	%	1.04	1.16	5
Copper	mg/kg	8.8	9.8	
Manganese	mg/kg	134.9	149.9	
Choline	mg/kg	3010.	3344.	19
Folic acid	mg/kg	2.00	2.20	
Niacin	mg/kg	47.3	52.6	50
Pantothenic acid	mg/kg	11.2	12.4	59
Riboflavin	mg/kg	5.1	5.7	25

Continued

Continued

(5) protein supplements
(6) minerals

(7) vitamins
(8) additives

Feed name or analyses		Mean		C.V.
		As fed	Dry	± %
Lysine	%	1.60	1.78	
Methionine	%	.30	.33	
Phenylalanine	%	.80	.89	
Threonine	%	.80	.89	
Tryptophan	%	.30	.33	14
Valine	%	1.10	1.22	32

Wheat, germ oil, (7)
  Wheat germ oil (AAFCO)

Ref no 7-05-207

Dry matter	%	100.0	100.0	
Ether extract	%	100.0	100.0	
α-tocopherol	mg/kg	1900.0	1900.0	

WHEAT, DURUM. Triticum durum

Wheat, durum, grain, Can 4 CW mn 56 wt mx 2.5 fm, (4)

Ref no 4-05-225

Dry matter	%	86.5	100.0	
Ash	%	1.6	1.8	3
Crude fiber	%	2.3	2.6	92
Ether extract	%	1.7	1.9	61
N-free extract	%	67.5	78.0	
Protein (N x 6.25)	%	13.6	15.7	29
Cattle	dig prot %	10.6	12.2	
Sheep	dig prot %	10.6	12.2	
Swine	dig prot %	12.4	14.4	
Energy	GE kcal/kg	4804.	5553.	
Cattle	DE kcal/kg	3318.	3836.	
Sheep	DE kcal/kg	3318.	3836.	
Swine	DE kcal/kg	3508.	4056.	
Cattle	ME kcal/kg	2721.	3146.	
Sheep	ME kcal/kg	2721.	3146.	
Swine	ME kcal/kg	3256.	3764.	
Cattle	TDN %	75.	87.	
Sheep	TDN %	75.	87.	
Swine	TDN %	79.	92.	

Wheat, durum, grain, Can 5 CW mn 54 wt mx 3 fm, (4)

Ref no 4-05-226

Dry matter	%	86.5	100.0	
Ash	%	1.6	1.8	38
Crude fiber	%	2.3	2.7	65
Ether extract	%	1.8	2.1	66

Continued

Feed name or analyses		Mean		C.V.
		As fed	Dry	± %
N-free extract	%	67.2	77.7	
Protein (N x 6.25)	%	13.6	15.7	36
Cattle	dig prot %	10.6	12.2	
Sheep	dig prot %	10.6	12.2	
Swine	dig prot %	12.4	14.4	
Energy	GE kcal/kg	4632.	5353.	
Cattle	DE kcal/kg	3318.	3836.	
Sheep	DE kcal/kg	3318.	3836.	
Swine	DE kcal/kg	3508.	4056.	
Cattle	ME kcal/kg	2721.	3146.	
Sheep	ME kcal/kg	2721.	3146.	
Swine	ME kcal/kg	3256.	3764.	
Cattle	TDN %	75.	87.	
Sheep	TDN %	75.	87.	
Swine	TDN %	79.	92.	

Wheat, durum, grain, (4)

Ref no 4-05-224

Dry matter	%	89.5	100.0	2
Ash	%	1.8	2.0	18
Crude fiber	%	2.2	2.5	22
Ether extract	%	2.0	2.2	13
N-free extract	%	70.1	78.3	
Protein (N x 6.25)	%	13.4	15.0	14
Cattle	dig prot %	10.5	11.7	
Sheep	dig prot %	10.5	11.7	
Swine	dig prot %	12.4	13.8	
Energy				
Cattle	DE kcal/kg	3473.	3880.	
Sheep	DE kcal/kg	3473.	3880.	
Swine	DE kcal/kg	3630.	4056.	
Cattle	ME kcal/kg	2848.	3182.	
Sheep	ME kcal/kg	2848.	3182.	
Swine	ME kcal/kg	3376.	3772.	
Cattle	TDN %	79.	88.	
Sheep	TDN %	79.	88.	
Swine	TDN %	82.	92.	
Calcium	%	.15	.17	
Iron	%	.004	.005	
Phosphorus	%	.40	.45	4
Copper	mg/kg	7.7	8.6	
Manganese	mg/kg	28.7	31.9	
Folic acid	mg/kg	.39	.44	30
Thiamine	mg/kg	6.3	7.0	11

(1) dry forages and roughages
(2) pasture, range plants, and forages fed green

(3) silages
(4) energy feeds

Feed name or analyses		Mean		C.V.
		As fed	Dry	± %

WHEAT, HARD RED SPRING. Triticum aestivum

Wheat, hard red spring, grain, feed gr, (4)

Ref no 4-05-248

Dry matter	%	86.5	100.0	
Ash	%	1.7	2.0	27
Crude fiber	%	3.0	3.4	99
Ether extract	%	1.9	2.2	44
N-free extract	%	66.0	76.3	
Protein (N x 6.25)	%	13.9	16.1	14
Cattle	dig prot %	10.9	12.6	
Sheep	dig prot %	10.9	12.6	
Swine	dig prot %	12.8	14.8	
Energy				
Cattle	DE kcal/kg	3356.	3880.	
Sheep	DE kcal/kg	3356.	3880.	
Swine	DE kcal/kg	3470.	4012.	
Cattle	ME kcal/kg	2752.	3182.	
Sheep	ME kcal/kg	2752.	3182.	
Swine	ME kcal/kg	3220.	3723.	
Cattle	TDN %	76.	88.	
Sheep	TDN %	76.	88.	
Swine	TDN %	79.	91.	

Wheat, hard red spring, grain, (4)

Ref no 4-05-258

Dry matter	%	86.5	100.0	
Ash	%	1.7	2.0	27
Crude fiber	%	3.0	3.4	99
Ether extract	%	1.9	2.2	44
N-free extract	%	66.0	76.3	
Protein (N x 6.25)	%	13.9	16.1	14
Cattle	dig prot %	10.9	12.6	
Sheep	dig prot %	10.9	12.6	
Swine	dig prot %	12.8	14.8	
Energy				
Cattle	DE kcal/kg	3356.	3880.	
Sheep	DE kcal/kg	3356.	3880.	
Swine	DE kcal/kg	3470.	4012.	
Cattle	ME kcal/kg	2752.	3182.	
Chickens	MEn kcal/kg	3086.	3568.	
Sheep	ME kcal/kg	2752.	3182.	
Swine	ME kcal/kg	3220.	3723.	
Cattle	TDN %	76.	88.	
Sheep	TDN %	76.	88.	
Swine	TDN %	79.	91.	
Calcium	%	.05	.06	22
Iron	%	.005	.006	35
Phosphorus	%	.41	.47	18

Continued

(5) protein supplements

(6) minerals

Feed name or analyses		Mean		C.V.
		As fed	Dry	± %
Copper	mg/kg	10.6	12.3	20
Manganese	mg/kg	62.2	71.9	21
Choline	mg/kg	778.	899.	13
Folic acid	mg/kg	.42	.48	25
Niacin	mg/kg	57.8	66.8	32
Pantothenic acid	mg/kg	13.5	15.6	12
Riboflavin	mg/kg	1.1	1.3	20
Thiamine	mg/kg	5.2	6.0	13
Arginine	%	.60	.70	21
Cystine	%	.17	.20	24
Histidine	%	.17	.20	24
Isoleucine	%	.69	.80	
Leucine	%	.95	1.10	
Lysine	%	.35	.40	24
Methionine	%	.17	.20	
Phenylalanine	%	.78	.90	
Threonine	%	.35	.40	
Tryptophan	%	.17	.20	24
Tyrosine	%	.78	.90	
Valine	%	.69	.80	

WHEAT, HARD RED WINTER. Triticum aestivum

Wheat, hard red winter, grain, (4)

Ref no 4-05-268

Dry matter	%	89.1	100.0	1
Ash	%	1.8	2.0	18
Crude fiber	%	2.7	3.0	23
Ether extract	%	1.6	1.8	16
N-free extract	%	70.0	78.6	
Protein (N x 6.25)	%	13.0	14.6	15
Cattle	dig prot %	10.2	11.4	
Sheep	dig prot %	10.2	11.4	
Swine	dig prot %	11.9	13.4	
Energy	GE kcal/kg	3552.	3991.	
Cattle	DE kcal/kg	3457.	3880.	
Sheep	DE kcal/kg	3457.	3880.	
Swine	DE kcal/kg	3575.	4012.	
Cattle	ME kcal/kg	2835.	3182.	
Chickens	MEn kcal/kg	3086.	3464.	
Sheep	ME kcal/kg	2835.	3182.	
Swine	ME kcal/kg	3324.	3731.	
Cattle	TDN %	78.	88.	
Sheep	TDN %	78.	88.	
Swine	TDN %	81.	91.	
Calcium	%	.05	.06	25
Magnesium	%	.10	.11	
Phosphorus	%	.40	.45	20
Potassium	%	.51	.57	
Cobalt	mg/kg	.100	.100	15
Copper	mg/kg	4.5	5.1	34
Manganese	mg/kg	38.8	43.6	32

Continued

(7) vitamins

(8) additives

Feed name or analyses		Mean		C.V. ± %
		As fed	Dry	
Choline	mg/kg	734.	825.	23
Folic acid	mg/kg	.40	.40	37
Niacin	mg/kg	50.9	57.2	27
Pantothenic acid	mg/kg	12.5	14.1	15
Riboflavin	mg/kg	1.0	1.1	54
Thiamine	mg/kg	6.2	7.0	18
Vitamin B6	mg/kg	4.10	4.60	18

Wheat red dog - see Wheat, flour by-prod, f-sift, mx 4 fbr

WHEAT, RED SPRING. Triticum aestivum

Wheat, red spring, grain, Can 4 No mn 58 wt mx 2.5 fm, (4)

Ref no 4-05-282

Dry matter	%	86.5	100.0	
Ash	%	1.5	1.7	17
Crude fiber	%	2.4	2.8	99
Ether extract	%	1.7	2.0	48
N-free extract	%	66.8	77.2	
Protein (N x 6.25)	%	14.1	16.3	13
Cattle	dig prot %	11.0	12.7	
Sheep	dig prot %	11.0	12.7	
Swine	dig prot %	13.0	15.0	
Energy	GE kcal/kg	4804.	5553.	
Cattle	DE kcal/kg	3356.	3880.	
Sheep	DE kcal/kg	3356.	3880.	
Swine	DE kcal/kg	3508.	4056.	
Cattle	ME kcal/kg	2752.	3182.	
Sheep	ME kcal/kg	2752.	3182.	
Swine	ME kcal/kg	3252.	3760.	
Cattle	TDN %	76.	88.	
Sheep	TDN %	76.	88.	
Swine	TDN %	80.	92.	

WHEAT, SOFT. Triticum aestivum

Wheat, soft, grain, (4)

Ref no 4-05-284

Dry matter	%	90.0	100.0	2
Ash	%	1.8	2.0	20
Crude fiber	%	2.3	2.6	30
Ether extract	%	1.7	1.9	17
N-free extract	%	73.4	81.5	
Protein (N x 6.25)	%	10.8	12.0	15
Cattle	dig prot %	8.5	9.4	

Feed name or analyses		Mean		C.V. ± %
		As fed	Dry	
Sheep	dig prot %	8.5	9.4	
Swine	dig prot %	9.9	11.0	
Energy				
Cattle	DE kcal/kg	3492.	3880.	
Sheep	DE kcal/kg	3492.	3880.	
Swine	DE kcal/kg	3650.	4056.	
Cattle	ME kcal/kg	2864.	3182.	
Chickens	MEn kcal/kg	3086.	3429.	
Sheep	ME kcal/kg	2864.	3182.	
Swine	ME kcal/kg	3416.	3796.	
Cattle	TDN %	79.	88.	
Sheep	TDN %	79.	88.	
Swine	TDN %	83.	92.	
Calcium	%	.09	.10	99
Iron	%	.005	.006	23
Magnesium	%	.10	.11	
Phosphorus	%	.30	.33	39
Potassium	%	.40	.44	
Copper	mg/kg	9.7	10.8	11
Manganese	mg/kg	51.3	57.0	42
Choline	mg/kg	788.	876.	8
Niacin	mg/kg	59.2	65.8	18
Pantothenic acid	mg/kg	12.8	11.5	15
Riboflavin	mg/kg	1.2	1.3	50
Thiamine	mg/kg	4.8	5.3	17
Vitamin B6	mg/kg	4.80	5.30	20

WHEAT, SOFT RED WINTER. Triticum aestivum

Wheat, soft red winter, grain, (4)

Ref no 4-05-294

Dry matter	%	89.1	100.0	1
Ash	%	1.8	2.0	18
Crude fiber	%	2.2	2.5	26
Ether extract	%	1.6	1.8	17
N-free extract	%	72.5	81.4	
Protein (N x 6.25)	%	11.0	12.3	14
Cattle	dig prot %	8.2	9.2	
Sheep	dig prot %	8.2	9.2	
Swine	dig prot %	10.1	11.3	
Energy	GE kcal/kg	3516.	3951.	
Cattle	DE kcal/kg	3457.	3880.	
Sheep	DE kcal/kg	3457.	3880.	
Swine	DE kcal/kg	3614.	4056.	
Cattle	ME kcal/kg	2835.	3182.	
Sheep	ME kcal/kg	2835.	3182.	
Swine	ME kcal/kg	3379.	3792.	
Cattle	TDN %	78.	88.	
Sheep	TDN %	78.	88.	
Swine	TDN %	82.	92.	
Calcium	%	.09	.10	99
Magnesium	%	.10	.11	

Continued

Continued

(1) dry forages and roughages

(2) pasture, range plants, and forages fed green

(3) silages

(4) energy feeds

Feed name or analyses		Mean		C.V.
		As fed	Dry	± %
Phosphorus	%	.29	.33	51
Potassium	%	.39	.44	
Copper	mg/kg	9.8	11.0	
Manganese	mg/kg	38.2	42.9	71
Choline	mg/kg	779.	875.	11
Folic acid	mg/kg	.40	.40	
Niacin	mg/kg	57.4	64.5	9
Pantothenic acid	mg/kg	11.4	12.8	18
Thiamine	mg/kg	5.3	5.9	11
Vitamin B6	mg/kg	4.60	5.30	17
Arginine	%	.36	.40	
Cystine	%	.18	.20	
Histidine	%	.09	.10	
Lysine	%	.80	.90	
Tryptophan	%	.27	.30	
Tyrosine	%	.36	.40	

WHEATGRASS. Agropyron spp

Wheatgrass, hay, s-c, immature, (1)

Ref no 1-05-344

Dry matter	%	89.3	100.0	2
Ash	%	8.1	9.1	13
Crude fiber	%	24.5	27.4	13
Ether extract	%	2.6	2.9	26
N-free extract	%	41.5	46.5	
Protein (N x 6.25)	%	12.6	14.1	23
Cattle	dig prot %	8.2	9.2	
Sheep	dig prot %	8.2	9.2	
Energy				
Cattle	DE kcal/kg	2402.	2690.	
Sheep	DE kcal/kg	2481.	2778.	
Cattle	ME kcal/kg	1970.	2206.	
Sheep	ME kcal/kg	2034.	2278.	
Cattle	TDN %	54.	61.	
Sheep	TDN %	56.	63.	
Calcium	%		.38	.42	9
Magnesium	%	.13	.15	
Phosphorus	%	.28	.31	16

Wheatgrass, hay, s-c, early blm, (1)

Ref no 1-05-345

Dry matter	%	90.5	100.0	2
Ash	%	7.4	8.2	36
Crude fiber	%	28.6	31.6	9
Ether extract	%	2.2	2.4	20
N-free extract	%	39.9	44.1	
Protein (N x 6.25)	%	12.4	13.7	9
Cattle	dig prot %	8.0	8.8	

Continued

Feed name or analyses		Mean		C.V.
		As fed	Dry	± %
Sheep	dig prot %	6.0	6.6	
Energy				
Cattle	DE kcal/kg	2274.	2513.	
Sheep	DE kcal/kg	2474.	2734.	
Cattle	ME kcal/kg	1865.	2061.	
Sheep	ME kcal/kg	2029.	2242.	
Cattle	TDN %	52.	57.	
Sheep	TDN %	56.	62.	
Calcium	%	.28	.31	16
Magnesium	%	.22	.24	
Phosphorus	%	.24	.27	14
Potassium	%	2.42	2.67	

Wheatgrass, hay, s-c, full blm, (1)

Ref no 1-05-346

Dry matter	%	93.6	100.0	2
Ash	%	9.4	10.0	50
Crude fiber	%	31.2	33.3	17
Ether extract	%	2.0	2.1	18
N-free extract	%	43.0	45.9	
Protein (N x 6.25)	%	8.1	8.7	19
Cattle	dig prot %	4.2	4.5	
Sheep	dig prot %	4.1	4.4	
Energy				
Cattle	DE kcal/kg	2477.	2646.	
Sheep	DE kcal/kg	2187.	2337.	
Cattle	ME kcal/kg	2031.	2170.	
Sheep	ME kcal/kg	1793.	1916.	
Cattle	TDN %	56.	60.	
Sheep	TDN %	50.	53.	

Wheatgrass, hay, s-c, mature, (1)

Ref no 1-05-349

Dry matter	%	90.4	100.0	2
Ash	%	7.9	8.7	16
Crude fiber	%	31.9	35.3	7
Ether extract	%	2.8	3.1	23
N-free extract	%	43.0	47.6	
Protein (N x 6.25)	%	4.8	5.3	13
Cattle	dig prot %	1.4	1.5	
Sheep	dig prot %	1.7	1.9	
Energy				
Cattle	DE kcal/kg	2272.	2513.	
Sheep	DE kcal/kg	1833.	2028.	
Cattle	ME kcal/kg	1863.	2061.	
Sheep	ME kcal/kg	1505.	1663.	
Cattle	TDN %	52.	57.	
Sheep	TDN %	42.	46.	
Calcium	%	.32	.35	26
Iron	%	.010	.010	

Continued

Feed name or analyses		Mean		C.V.
		As fed	Dry	± %
Magnesium	%	.08	.09	
Phosphorus	%	.09	.10	32
Manganese	mg/kg	6.0	6.6	80

Wheatgrass, hay, s-c, over ripe, (1)

Ref no 1-05-350

Dry matter	%	92.0	100.0	2
Ash	%	8.5	9.2	15
Crude fiber	%	33.2	36.1	5
Ether extract	%	2.6	2.8	32
N-free extract	%	44.0	47.8	
Protein (N x 6.25)	%	3.8	4.1	26
Cattle	dig prot %	1.0	1.1	
Sheep	dig prot %	1.8	2.0	
Energy				
Cattle	DE kcal/kg	2271.	2469.	
Sheep	DE kcal/kg	2190.	2381.	
Cattle	ME kcal/kg	1862.	2024.	
Sheep	ME kcal/kg	1796.	1952.	
Cattle	TDN %	52.	56.	
Sheep	TDN %	50.	54.	
Calcium	%	.23	.25	10
Iron	%	.010	.020	
Phosphorus	%	.06	.06	31
Manganese	mg/kg	8.7	9.5	

Wheatgrass, hay, s-c, (1)

Ref no 1-05-351

Dry matter	%	90.8	100.0	2
Ash	%	7.5	8.2	33
Crude fiber	%	30.5	33.6	13
Ether extract	%	2.4	2.6	28
N-free extract	%	42.6	46.9	
Protein (N x 6.25)	%	7.9	8.7	41
Cattle	dig prot %	4.1	4.5	
Sheep	dig prot %	4.0	4.4	
Energy				
Cattle	DE kcal/kg	2322.	2557.	
Sheep	DE kcal/kg	2202.	2425.	
Cattle	ME kcal/kg	1904.	2097.	
Sheep	ME kcal/kg	1805.	1988.	
Cattle	TDN %	53.	58.	
Sheep	TDN %	50.	55.	
Calcium	%	.30	.33	20
Iron	%	.010	.010	64
Magnesium	%	.20	.22	31
Phosphorus	%	.18	.20	35
Potassium	%	2.40	2.68	9
Manganese	mg/kg	10.2	11.2	53

Feed name or analyses		Mean		C.V.
		As fed	Dry	± %

Wheatgrass, aerial pt, fresh, immature, (2)

Ref no 2-05-354

Dry matter	%	33.4	100.0	17
Ash	%	3.1	9.2	28
Crude fiber	%	8.4	25.2	15
Ether extract	%	1.1	3.2	25
N-free extract	%	14.7	44.0	
Protein (N x 6.25)	%	6.2	18.4	33
Cattle	dig prot %	4.5	13.5	
Sheep	dig prot %	4.5	13.6	
Energy				
Cattle	DE kcal/kg	1060.	3175.	
Sheep	DE kcal/kg	987.	2954.	
Cattle	ME kcal/kg	870.	2604.	
Sheep	ME kcal/kg	809.	2422.	
Cattle	TDN %	24.	72.	
Sheep	TDN %	22.	67.	
Calcium	%	.17	.52	18
Magnesium	%	.05	.15	44
Phosphorus	%	.10	.31	30
Carotene	mg/kg	122.3	366.2	48
Vitamin A equiv	IU/g	203.9	610.4	

Wheatgrass, aerial pt, fresh, early blm, (2)

Ref no 2-05-357

Dry matter	%	39.4	100.0	27
Ash	%	3.7	9.5	21
Crude fiber	%	11.7	29.8	11
Ether extract	%	1.3	3.3	49
N-free extract	%	17.9	45.5	
Protein (N x 6.25)	%	4.7	11.9	17
Cattle	dig prot %	3.2	8.0	
Sheep	dig prot %	2.8	7.1	
Energy				
Cattle	DE kcal/kg	1112.	2822.	
Sheep	DE kcal/kg	1129.	2866.	
Cattle	ME kcal/kg	912.	2314.	
Sheep	ME kcal/kg	926.	2350.	
Cattle	TDN %	25.	64.	
Sheep	TDN %	26.	65.	
Calcium	%	.16	.47	16
Phosphorus	%	.09	.23	37

(1) dry forages and roughages

(2) pasture, range plants, and forages fed green

(3) silages

(4) energy feeds

Feed name or analyses		Mean		C.V.
		As fed	Dry	± %

Wheatgrass, aerial pt, fresh, mid-blm, (2)

Ref no 2-05-358

		As fed	Dry	± %
Dry matter	%	29.9	100.0	
Ash	%	2.2	7.2	
Crude fiber	%	9.5	31.8	9
Ether extract	%	.7	2.4	11
N-free extract	%	14.5	48.4	
Protein (N x 6.25)	%	3.0	10.2	22
Cattle	dig prot %	2.0	6.6	
Sheep	dig prot %	1.9	6.5	
Energy				
Cattle	DE kcal/kg	870.	2910.	
Sheep	DE kcal/kg	857.	2866.	
Cattle	ME kcal/kg	713.	2386.	
Sheep	ME kcal/kg	703.	2350.	
Cattle	TDN %	20.	66.	
Sheep	TDN %	19.	65.	
Calcium	%	.13	.42	5
Phosphorus	%	.06	.20	12

Wheatgrass, aerial pt, fresh, full blm, (2)

Ref no 2-05-360

		As fed	Dry	± %
Dry matter	%	39.6	100.0	8
Ash	%	3.3	8.2	25
Crude fiber	%	12.6	31.9	12
Ether extract	%	1.2	2.9	32
N-free extract	%	18.9	47.8	
Protein (N x 6.25)	%	3.6	9.2	16
Cattle	dig prot %	2.2	5.7	
Sheep	dig prot %	2.2	5.6	
Energy				
Cattle	DE kcal/kg	1118.	2822.	
Sheep	DE kcal/kg	1152.	2910.	
Cattle	ME kcal/kg	916.	2314.	
Sheep	ME kcal/kg	945.	2386.	
Cattle	TDN %	25.	64.	
Sheep	TDN %	26.	66.	
Calcium	%	.15	.39	21
Magnesium	%	.06	.16	
Phosphorus	%	.09	.23	32
Potassium	%	1.05	2.64	
Carotene	mg/kg	58.1	146.6	40
Vitamin A equiv	IU/g	96.8	244.4	

Feed name or analyses		Mean		C.V.
		As fed	Dry	± %

Wheatgrass, aerial pt, fresh, mature, (2)

Ref no 2-05-363

		As fed	Dry	± %
Dry matter	%	60.5	100.0	11
Ash	%	5.0	8.2	33
Crude fiber	%	22.1	36.6	9
Ether extract	%	1.8	2.9	33
N-free extract	%	28.4	47.0	
Protein (N x 6.25)	%	3.2	5.3	25
Cattle	dig prot %	1.4	2.4	
Sheep	dig prot %	1.1	1.9	
Energy				
Cattle	DE kcal/kg	1627.	2690.	
Sheep	DE kcal/kg	1760.	2910.	
Cattle	ME kcal/kg	1335.	2206.	
Sheep	ME kcal/kg	1444.	2386.	
Cattle	TDN %	37.	61.	
Sheep	TDN %	40.	66.	
Calcium	%	.22	.36	16
Magnesium	%	.05	.09	99
Phosphorus	%	.09	.15	51
Carotene	mg/kg	41.4	68.4	33
Vitamin A equiv	IU/g	69.0	114.0	

Wheatgrass, aerial pt, fresh, over ripe, (2)

Ref no 2-05-364

		As fed	Dry	± %
Dry matter	%	61.9	100.0	19
Ash	%	4.6	7.4	38
Crude fiber	%	22.6	36.5	9
Ether extract	%	1.5	2.4	43
N-free extract	%	30.8	49.8	
Protein (N x 6.25)	%	2.4	3.9	39
Cattle	dig prot %	.7	1.2	
Sheep	dig prot %	1.5	2.5	
Energy				
Cattle	DE kcal/kg	1665.	2690.	
Sheep	DE kcal/kg	1747.	2822.	
Cattle	ME kcal/kg	1366.	2206.	
Sheep	ME kcal/kg	1432.	2314.	
Cattle	TDN %	38.	61.	
Sheep	TDN %	40.	64.	
Calcium	%	.15	.24	44
Magnesium	%	.04	.06	
Phosphorus	%	.06	.09	90
Copper	mg/kg	4.4	7.1	
Manganese	mg/kg	27.6	44.5	

(5) protein supplements
(6) minerals

(7) vitamins
(8) additives

Feed name or analyses		Mean		C.V.
		As fed	Dry	± %

Wheatgrass, aerial pt, fresh, (2)

Ref no 2-05-365

Dry matter	%	42.3	100.0	27
Ash	%	3.6	8.5	35
Crude fiber	%	13.5	31.9	16
Ether extract	%	1.2	2.9	47
N-free extract	%	19.5	46.2	
Protein (N x 6.25)	%	4.4	10.5	64
Cattle	dig prot %	2.9	6.8	
Sheep	dig prot %	2.9	6.8	
Cellulose	%	14.2	33.6	15
Lignin	%	4.8	11.3	27
Energy	GE kcal/kg	1896.	4483.	5
Cattle	DE kcal/kg	1212.	2866.	
Sheep	DE kcal/kg	1194.	2822.	
Cattle	ME kcal/kg	994.	2350.	
Sheep	ME kcal/kg	979.	2314.	
Cattle	TDN %	27.	65.	
Sheep	TDN %	27.	64.	
Calcium	%	.19	.44	28
Iron	%	.010	.030	
Magnesium	%	.06	.13	56
Phosphorus	%	.09	.22	43
Potassium	%	1.12	2.64	
Sodium	%	.22	.51	
Sulfur	%	.06	.15	
Cobalt	mg/kg	.040	.090	66
Copper	mg/kg	3.0	7.1	
Manganese	mg/kg	13.5	31.8	
Carotene	mg/kg	86.8	205.1	76
Vitamin A equiv	IU/g	144.7	341.9	

WHEATGRASS, BEARDED. Agropyron subsecundum

Wheatgrass, bearded, aerial pt, fresh, immature, (2)

Ref no 2-05-369

Dry matter	%	30.8	100.0	
Ash	%	2.1	6.9	
Crude fiber	%	8.2	26.6	
Ether extract	%	.7	2.2	
N-free extract	%	15.5	50.2	
Protein (N x 6.25)	%	4.3	14.1	
Cattle	dig prot %	3.0	9.9	
Sheep	dig prot %	3.1	10.1	
Energy				
Cattle	DE kcal/kg	978.	3175.	
Sheep	DE kcal/kg	856.	2778.	
Cattle	ME kcal/kg	802.	2604.	

Continued

(1) dry forages and roughages
(2) pasture, range plants, and forages fed green

Feed name or analyses		Mean		C.V.
		As fed	Dry	± %

Sheep	ME kcal/kg	702.	2278.	
Cattle	TDN %	22.	72.	
Sheep	TDN %	19.	63.	
Calcium	%	.18	.57	11
Magnesium	%	.04	.14	26
Phosphorus	%	.10	.34	16
Sulfur	%	.06	.19	29

Wheatgrass, bearded, aerial pt, fresh, mature, (2)

Ref no 2-05-371

Dry matter	%	60.0	100.0	
Ash	%	3.1	5.2	
Crude fiber	%	25.5	42.5	
Ether extract	%	1.3	2.2	
N-free extract	%	27.5	45.8	
Protein (N x 6.25)	%	2.6	4.3	
Cattle	dig prot %	.9	1.5	
Sheep	dig prot %	.6	1.0	
Energy				
Cattle	DE kcal/kg	1746.	2910.	
Sheep	DE kcal/kg	1720.	2866.	
Cattle	ME kcal/kg	1432.	2386.	
Sheep	ME kcal/kg	1410.	2350.	
Cattle	TDN %	40.	66.	
Sheep	TDN %	39.	65.	
Calcium	%	.19	.32	15
Magnesium	%	.04	.06	49
Phosphorus	%	.06	.10	49
Sulfur	%	.05	.09	32

WHEATGRASS, BEARDLESS. Agropyron inerme

Wheatgrass, beardless, aerial pt, fresh, immature, (2)

Ref no 2-05-375

Dry matter	%	32.2	100.0	
Ash	%	2.4	7.6	13
Crude fiber	%	8.7	26.9	4
Ether extract	%	1.0	3.0	13
N-free extract	%	15.3	47.6	
Protein (N x 6.25)	%	4.8	14.9	14
Cattle	dig prot %	3.4	10.6	
Sheep	dig prot %	3.6	11.3	
Energy				
Cattle	DE kcal/kg	994.	3086.	
Sheep	DE kcal/kg	1008.	3130.	
Cattle	ME kcal/kg	815.	2530.	
Sheep	ME kcal/kg	826.	2567.	
Cattle	TDN %	22.	70.	
Sheep	TDN %	23.	71.	

Continued

(3) silages
(4) energy feeds

Feed name or analyses		Mean		C.V.
		As fed	Dry	± %
Calcium	%	.22	.67	17
Phosphorus	%	.09	.28	37
Sulfur	%	.06	.20	19

Wheatgrass, beardless, aerial pt, fresh, full blm, (2)

Ref no 2-05-378

Dry matter		%	45.5	100.0	
Ash		%	3.5	7.8	
Crude fiber		%	15.6	34.1	
Ether extract		%	1.5	3.4	
N-free extract		%	20.8	45.7	
Protein (N x 6.25)		%	4.1	9.0	
Cattle	dig prot	%	2.5	5.5	
Sheep	dig prot	%	2.6	5.8	
Energy					
Cattle	DE kcal/kg		1284.	2822.	
Sheep	DE kcal/kg		1284.	2822.	
Cattle	ME kcal/kg		1053.	2314.	
Sheep	ME kcal/kg		1053.	2314.	
Cattle	TDN %		29.	64.	
Sheep	TDN %		29.	64.	
Calcium		%	.14	.30	
Phosphorus		%	.11	.24	

Wheatgrass, beardless, aerial pt, fresh, mature, (2)

Ref no 2-05-380

Dry matter		%	50.0	100.0	
Ash		%	5.9	11.8	
Ether extract		%	2.3	4.6	
Protein (N x 6.25)		%	1.8	3.5	
Cellulose		%	21.4	42.8	
Lignin		%	4.4	8.7	
Energy	GE kcal/kg		2334.	4668.	
Carotene	mg/kg		24.6	49.2	29
Vitamin A equiv	IU/g		41.0	82.0	

Wheatgrass, beardless, aerial pt, fresh, dormant, (2)

Ref no 2-07-994

Dry matter		%	86.0	100.0	
Ash		%	9.1	10.6	
Ether extract		%	3.5	4.1	
Protein (N x 6.25)		%	2.7	3.1	
Sheep	dig prot	%	.0	.0	
Cellulose		%	32.7	38.0	
Lignin		%	6.9	8.0	
Energy	GE kcal/kg		3612.	4200.	
Sheep	DE kcal/kg		2199.	2557.	

Continued

Feed name or analyses		Mean		C.V.
		As fed	Dry	± %
Sheep	ME kcal/kg	1803.	2097.	
Sheep	TDN %	50.	58.	
Calcium	%	.42	.49	
Phosphorus	%	.05	.06	
Carotene	mg/kg	.9	1.1	

Wheatgrass, beardless, aerial pt, fresh, (2)

Ref no 2-05-381

Dry matter		%	34.6	100.0	16
Ash		%	2.7	7.7	20
Crude fiber		%	10.3	29.7	8
Ether extract		%	1.3	3.7	19
N-free extract		%	16.5	47.6	
Protein (N x 6.25)		%	3.9	11.3	39
Cattle	dig prot	%	2.6	7.5	
Sheep	dig prot	%	2.5	7.2	
Energy					
Cattle	DE kcal/kg		1007.	2910.	
Sheep	DE kcal/kg		976.	2822.	
Cattle	ME kcal/kg		826.	2386.	
Sheep	ME kcal/kg		801.	2314.	
Cattle	TDN %		23.	66.	
Sheep	TDN %		22.	64.	
Calcium		%	.20	.58	19
Magnesium		%	.04	.13	27
Phosphorus		%	.08	.24	52
Sulfur		%	.06	.17	24
Carotene	mg/kg		107.9	311.8	69
Vitamin A equiv	IU/g		179.9	519.8	

WHEATGRASS, BLUESTEM. Agropyron smithii

Wheatgrass, bluestem, hay, s-c, immature, (1)
Wheatgrass hay, western, immature

Ref no 1-05-396

Dry matter		%	88.2	100.0	2
Ash		%	7.8	8.9	10
Crude fiber		%	26.4	29.9	9
Ether extract		%	2.8	3.2	16
N-free extract		%	40.6	46.0	
Protein (N x 6.25)		%	10.6	12.0	29
Cattle	dig prot	%	6.4	7.3	
Sheep	dig prot	%	5.6	6.4	
Energy					
Cattle	DE kcal/kg		2528.	2866.	
Sheep	DE kcal/kg		2294.	2601.	
Cattle	ME kcal/kg		2073.	2350.	
Sheep	ME kcal/kg		1881.	2133.	
Cattle	TDN %		57.	65.	

Continued

Feed name or analyses		Mean		C.V. ± %
		As fed	Dry	
Sheep	TDN %	52.	59.	
Calcium	%	.42	.48	
Magnesium	%	.13	.15	
Phosphorus	%	.19	.21	

Wheatgrass, bluestem, hay, s-c, full blm, (1)
Wheatgrass hay, western, full bloom

Ref no 1-05-397

		As fed	Dry	C.V. ± %
Dry matter	%	93.4	100.0	1
Ash	%	6.6	7.1	6
Crude fiber	%	29.6	31.7	5
Ether extract	%	2.2	2.4	13
N-free extract	%	46.9	50.2	
Protein (N x 6.25)	%	8.0	8.6	8
Cattle	dig prot %	4.1	4.4	
Sheep	dig prot %	4.3	4.6	
Energy				
Cattle	DE kcal/kg	2471.	2646.	
Sheep	DE kcal/kg	2471.	2646.	
Cattle	ME kcal/kg	2027.	2170.	
Sheep	ME kcal/kg	2027.	2170.	
Cattle	TDN %	56.	60.	
Sheep	TDN %	56.	60.	

Wheatgrass, bluestem, hay, s-c, mature, (1)
Wheatgrass hay, western, mature

Ref no 1-05-398

		As fed	Dry	C.V. ± %
Dry matter	%	88.0	100.0	2
Ash	%	7.7	8.8	16
Crude fiber	%	30.8	35.0	4
Ether extract	%	2.8	3.2	23
N-free extract	%	41.8	47.5	
Protein (N x 6.25)	%	4.8	5.5	11
Cattle	dig prot %	1.5	1.7	
Sheep	dig prot %	2.6	2.9	
Energy				
Cattle	DE kcal/kg	2250.	2557.	
Sheep	DE kcal/kg	2328.	2646.	
Cattle	ME kcal/kg	1845.	2097.	
Sheep	ME kcal/kg	1909.	2170.	
Cattle	TDN %	51.	58.	
Sheep	TDN %	53.	60.	
Calcium	%	.26	.29	29
Iron	%	.010	.010	
Magnesium	%	.08	.09	
Phosphorus	%	.09	.10	36
Manganese	mg/kg	5.8	6.6	80

Wheatgrass, bluestem, hay, s-c, (1)
Wheatgrass hay, western

Ref no 1-05-400

		As fed	Dry	C.V. ± %
Dry matter	%	89.4	100.0	2
Ash	%	7.7	8.6	20
Crude fiber	%	29.9	33.5	9
Ether extract	%	2.7	3.0	25
N-free extract	%	42.4	47.4	
Protein (N x 6.25)	%	6.7	7.5	50
Cattle	dig prot %	3.0	3.4	
Sheep	dig prot %	3.6	4.0	
Energy				
Cattle	DE kcal/kg	2325.	2601.	
Sheep	DE kcal/kg	2366.	2646.	
Cattle	ME kcal/kg	1907.	2133.	
Sheep	ME kcal/kg	1940.	2170.	
Cattle	TDN %	53.	59.	
Sheep	TDN %	54.	60.	
Calcium	%	.27	.30	28
Iron	%	.010	.010	64
Magnesium	%	.11	.12	
Phosphorus	%	.13	.14	36
Manganese	mg/kg	8.3	9.3	55

Wheatgrass, bluestem, aerial pt, fresh, immature, (2)
Wheatgrass forage, western, immature

Ref no 2-05-401

		As fed	Dry	C.V. ± %
Dry matter	%	35.9	100.0	13
Ash	%	3.5	9.8	23
Crude fiber	%	9.6	26.7	8
Ether extract	%	1.3	3.5	23
N-free extract	%	15.6	43.6	
Protein (N x 6.25)	%	5.9	16.4	16
Cattle	dig prot %	4.2	11.8	
Sheep	dig prot %	4.3	12.1	
Cellulose	%	10.5	29.2	
Lignin	%	4.5	12.4	
Energy				
Cattle	DE kcal/kg	1076.	2998.	
Sheep	DE kcal/kg	1060.	2954.	
Cattle	ME kcal/kg	882.	2458.	
Sheep	ME kcal/kg	869.	2422.	
Cattle	TDN %	24.	68.	
Sheep	TDN %	24.	67.	
Carotene	mg/kg	66.7	185.7	31
Vitamin A equiv	IU/g	111.2	309.6	

(1) dry forages and roughages

(2) pasture, range plants, and forages fed green

(3) silages

(4) energy feeds

TABLE A-3-1 Composition of Feeds (Continued)

Feed name or analyses		Mean		C.V. ± %
		As fed	Dry	

Wheatgrass, bluestem, aerial pt, fresh, early blm, (2)
Wheatgrass forage, western, early bloom

Ref no 2-05-402

		As fed	Dry	C.V. ± %
Dry matter	%	34.8	100.0	
Ash	%	2.9	8.3	20
Crude fiber	%	10.9	31.4	9
Ether extract	%	.9	2.7	27
N-free extract	%	15.9	45.8	
Protein (N x 6.25)	%	4.1	11.8	10
Cattle	dig prot %	2.7	7.9	
Sheep	dig prot %	3.0	8.7	
Cellulose	%	10.7	30.8	
Lignin	%	4.4	12.7	
Energy				
Cattle	DE kcal/kg	1028.	2954.	
Sheep	DE kcal/kg	1028.	2954.	
Cattle	ME kcal/kg	843.	2422.	
Sheep	ME kcal/kg	843.	2422.	
Cattle	TDN %	23.	67.	
Sheep	TDN %	23.	67.	
Calcium	%	.20	.58	
Phosphorus	%	.12	.34	

Wheatgrass, bluestem, aerial pt, fresh, full blm, (2)
Wheatgrass forage, western, full bloom

Ref no 2-05-404

		As fed	Dry	C.V. ± %
Dry matter	%	38.9	100.0	9
Ash	%	3.2	8.1	29
Crude fiber	%	12.6	32.4	13
Ether extract	%	.8	2.0	22
N-free extract	%	19.0	48.9	
Protein (N x 6.25)	%	3.3	8.6	13
Cattle	dig prot %	2.0	5.2	
Sheep	dig prot %	2.0	5.2	
Cellulose	%	11.8	30.3	
Lignin	%	6.0	15.3	
Energy				
Cattle	DE kcal/kg	1115.	2866.	
Sheep	DE kcal/kg	1132.	2910.	
Cattle	ME kcal/kg	914.	2350.	
Sheep	ME kcal/kg	928.	2386.	
Cattle	TDN %	25.	65.	
Sheep	TDN %	26.	66.	
Calcium	%	.18	.45	
Magnesium	%	.06	.16	
Phosphorus	%	.11	.29	
Potassium	%	1.03	2.64	
Carotene	mg/kg	45.6	117.1	21
Vitamin A equiv	IU/g	76.0	195.2	

Wheatgrass, bluestem, aerial pt, fresh, mid-blm, (2)
Wheatgrass forage, western mid-bloom

Ref no 2-05-403

		As fed	Dry	C.V. ± %
Dry matter	%	29.9	100.0	
Ash	%	2.2	7.2	
Crude fiber	%	10.2	34.0	
Ether extract	%	.7	2.5	
N-free extract	%	14.2	47.6	
Protein (N x 6.25)	%	2.6	8.7	
Cattle	dig prot %	1.6	5.3	
Sheep	dig prot %	1.6	5.2	
Cellulose	%	8.9	29.7	
Lignin	%	4.1	13.8	
Energy				
Cattle	DE kcal/kg	870.	2910.	
Sheep	DE kcal/kg	883.	2954.	
Cattle	ME kcal/kg	713.	2386.	
Sheep	ME kcal/kg	724.	2422.	
Cattle	TDN %	20.	66.	
Sheep	TDN %	20.	67.	

Wheatgrass, bluestem, aerial pt, fresh, mature, (2)
Wheatgrass forage, western, mature

Ref no 2-05-407

		As fed	Dry	C.V. ± %
Dry matter	%	61.8	100.0	
Ash	%	5.9	9.6	12
Crude fiber	%	21.3	34.5	11
Ether extract	%	1.9	3.1	32
N-free extract	%	28.8	46.6	
Protein (N x 6.25)	%	3.8	6.2	18
Cattle	dig prot %	2.0	3.2	
Sheep	dig prot %	2.3	3.7	
Energy				
Cattle	DE kcal/kg	1607.	2601.	
Sheep	DE kcal/kg	1798.	2910.	
Cattle	ME kcal/kg	1318.	2133.	
Sheep	ME kcal/kg	1474.	2386.	
Cattle	TDN %	36.	59.	
Sheep	TDN %	41.	66.	
Calcium	%	.22	.35	17
Phosphorus	%	.13	.21	38
Carotene	mg/kg	38.4	62.2	27
Vitamin A equiv	IU/g	64.0	103.7	

(5) protein supplements
(6) minerals

(7) vitamins
(8) additives

Feed name or analyses		Mean		C.V. ± %
		As fed	Dry	

Wheatgrass, bluestem, aerial pt, fresh, dormant, (2)
Wheatgrass forage, western, dormant

Ref no 2-05-607

		As fed	Dry
Dry matter	%	87.0	100.0
Ash	%	8.7	10.0
Ether extract	%	7.2	8.3
Protein (N x 6.25)	%	2.1	2.4
Sheep	dig prot %	.2	.2
Cellulose	%	31.3	36.0
Lignin	%	6.1	7.0
Energy	GE kcal/kg	3784.	4350.
Sheep	DE kcal/kg	2455.	2822.
Sheep	ME kcal/kg	2013.	2314.
Sheep	TDN %	56.	64.
Calcium	%	.64	.74
Phosphorus	%	.05	.06
Carotene	mg/kg	.2	.2

Wheatgrass, bluestem, aerial pt, fresh, (2)
Wheatgrass forage, western

Ref no 2-05-410

		As fed	Dry	C.V. ± %
Dry matter	%	39.7	100.0	26
Ash	%	3.4	8.6	22
Crude fiber	%	12.9	32.6	11
Ether extract	%	1.0	2.6	33
N-free extract	%	18.3	46.0	
Protein (N x 6.25)	%	4.0	10.2	39
Cattle	dig prot %	2.6	6.6	
Sheep	dig prot %	2.4	6.1	
Energy				
Cattle	DE kcal/kg	1138.	2866.	
Sheep	DE kcal/kg	1155.	2910.	
Cattle	ME kcal/kg	933.	2350.	
Sheep	ME kcal/kg	947.	2386.	
Cattle	TDN %	26.	65.	
Sheep	TDN %	26.	66.	
Calcium	%	.15	.39	39
Iron	%	.010	.030	
Magnesium	%	.05	.12	
Phosphorus	%	.09	.22	40
Potassium	%	1.05	2.64	
Sodium	%	.20	.51	
Sulfur	%	.06	.16	
Cobalt	mg/kg	.040	.090	61
Copper	mg/kg	2.3	5.7	
Manganese	mg/kg	8.4	21.2	
Carotene	mg/kg	50.5	127.2	46
Vitamin A equiv	IU/g	84.2	212.0	

WHEATGRASS, CRESTED. Agropyron cristatum

Wheatgrass, crested, hay, s-c, early blm, (1)

Ref no 1-05-412

		As fed	Dry	C.V. ± %
Dry matter	%	90.2	100.0	2
Ash	%	6.9	7.7	12
Crude fiber	%	29.2	32.4	9
Ether extract	%	2.0	2.2	16
N-free extract	%	39.3	43.6	
Protein (N x 6.25)	%	12.7	14.1	8
Cattle	dig prot %	8.3	9.2	
Sheep	dig prot %	9.0	10.0	
Cellulose	%	31.4	34.8	
Lignin	%	13.5	15.0	
Energy				
Cattle	DE kcal/kg	2306.	2557.	
Sheep	DE kcal/kg	2227.	2469.	
Cattle	ME kcal/kg	1891.	2097.	
Sheep	ME kcal/kg	1826.	2024.	
Cattle	TDN %	52.	58.	
Sheep	TDN %	50.	56.	
Calcium	%	.29	.32	14
Phosphorus	%	.24	.27	12

Wheatgrass, crested, hay, s-c, mid-blm, (1)

Ref no 1-05-413

		As fed	Dry
Dry matter	%	96.4	100.0
Ash	%	6.3	6.5
Crude fiber	%	35.2	36.5
Ether extract	%	2.3	2.4
N-free extract	%	43.3	44.9
Protein (N x 6.25)	%	9.4	9.7
Cattle	dig prot %	5.1	5.3
Sheep	dig prot %	3.4	3.5
Energy			
Cattle	DE kcal/kg	2168.	2249.
Sheep	DE kcal/kg	1997.	2072.
Cattle	ME kcal/kg	1778.	1844.
Sheep	ME kcal/kg	1638.	1699.
Cattle	TDN %	49.	51.
Sheep	TDN %	45.	47.

(1) dry forages and roughages
(2) pasture, range plants, and forages fed green

(3) silages
(4) energy feeds

Feed name or analyses		Mean		C.V.
		As fed	Dry	± %

Wheatgrass, crested, hay, s-c, full blm, (1)

Ref no 1-05-414

		As fed	Dry	± %
Dry matter	%	94.2	100.0	3
Ash	%	6.0	6.4	14
Crude fiber	%	31.1	33.0	19
Ether extract	%	1.9	2.0	17
N-free extract	%	47.0	49.9	
Protein (N x 6.25)	%	8.2	8.7	13
Cattle	dig prot %	4.2	4.5	
Sheep	dig prot %	2.9	3.1	
Energy				
Cattle	DE kcal/kg	2367.	2513.	
Sheep	DE kcal/kg	1952.	2072.	
Cattle	ME kcal/kg	1941.	2061.	
Sheep	ME kcal/kg	1600.	1699.	
Cattle	TDN %	54.	57.	
Sheep	TDN %	44.	47.	
Calcium	%	.24	.26	
Phosphorus	%	.15	.16	

Wheatgrass, crested, hay, s-c, mature, (1)

Ref no 1-05-416

		As fed	Dry	± %
Dry matter	%	94.4	100.0	3
Ash	%	5.5	5.8	4
Crude fiber	%	38.5	40.8	4
Ether extract	%	1.1	1.2	21
N-free extract	%	45.6	48.3	
Protein (N x 6.25)	%	3.7	3.9	19
Cattle	dig prot %	.3	.3	
Sheep	dig prot %	1.3	1.4	
Energy				
Cattle	DE kcal/kg	1831.	1940.	
Sheep	DE kcal/kg	1998.	2116.	
Cattle	ME kcal/kg	1502.	1591.	
Sheep	ME kcal/kg	1638.	1735.	
Cattle	TDN %	42.	44.	
Sheep	TDN %	45.	48.	

Wheatgrass, crested, hay, s-c, (1)

Ref no 1-05-418

		As fed	Dry	± %
Dry matter	%	92.0	100.0	3
Ash	%	6.7	7.3	16
Crude fiber	%	30.0	32.6	15
Ether extract	%	1.8	2.0	31
N-free extract	%	43.5	47.3	
Protein (N x 6.25)	%	9.9	10.8	30
Cattle	dig prot %	5.8	6.3	

Continued

Feed name or analyses		Mean		C.V.
		As fed	Dry	± %

Sheep	dig prot %	3.6	3.9	
Energy				
Cattle	DE kcal/kg	2352.	2557.	
Sheep	DE kcal/kg	1866.	2028.	
Cattle	ME kcal/kg	1929.	2097.	
Sheep	ME kcal/kg	1530.	1663.	
Cattle	TDN %	53.	58.	
Sheep	TDN %	42.	46.	
Calcium	%	.30	.33	19
Phosphorus	%	.19	.21	39
Cobalt	mg/kg	.220	.240	

Wheatgrass, crested, aerial pt, fresh, immature, (2)

Ref no 2-05-420

		As fed	Dry	± %
Dry matter	%	30.8	100.0	23
Ash	%	3.3	10.6	21
Crude fiber	%	6.8	22.2	16
Ether extract	%	1.1	3.6	16
N-free extract	%	12.3	40.0	
Protein (N x 6.25)	%	7.3	23.6	33
Cattle	dig prot %	5.5	18.0	
Sheep	dig prot %	5.8	19.0	
Cellulose	%	10.5	34.1	
Lignin	%	1.8	5.9	
Energy				
Cattle	GE kcal/kg	1331.	4322.	
Cattle	DE kcal/kg	1005.	3263.	
Sheep	DE kcal/kg	883.	2866.	
Cattle	ME kcal/kg	824.	2676.	
Sheep	ME kcal/kg	724.	2350.	
Cattle	TDN %	23.	74.	
Sheep	TDN %	20.	65.	
Calcium	%	.14	.46	23
Magnesium	%	.09	.28	
Phosphorus	%	.11	.35	29
Carotene	mg/kg	133.6	433.7	31
Vitamin A equiv	IU/g	222.7	723.0	

Wheatgrass, crested, aerial pt, fresh, early blm, (2)

Ref no 2-05-422

Dry matter	%	42.5	100.0	
Ash	%	4.1	9.6	26
Crude fiber	%	12.5	29.5	14
Ether extract	%	1.1	2.5	39
N-free extract	%	20.1	47.4	
Protein (N x 6.25)	%	4.7	11.0	25
Cattle	dig prot %	3.1	7.2	
Sheep	dig prot %	3.1	7.2	
Energy				
Cattle	DE kcal/kg	1199.	2822.	
Sheep	DE kcal/kg	1218.	2866.	

Continued

(5) protein supplements

(6) minerals

(7) vitamins

(8) additives

Feed name or analyses		Mean		C.V. ± %
		As fed	Dry	
Cattle	ME kcal/kg	983.	2314.	
Sheep	ME kcal/kg	999.	2350.	
Cattle	TDN %	27.	64.	
Sheep	TDN %	28.	65.	

Feed name or analyses		Mean		C.V. ± %
		As fed	Dry	
Phosphorus	%	.10	.17	30
Carotene	mg/kg	45.2	75.4	23
Vitamin A equiv	IU/g	75.3	125.7	

Wheatgrass, crested, aerial pt, fresh, full blm, (2)

Ref no 2-05-424

Dry matter	%	50.0	100.0	
Ash	%	4.6	9.3	18
Crude fiber	%	15.2	30.3	13
Ether extract	%	1.8	3.6	25
N-free extract	%	23.5	47.0	
Protein (N x 6.25)	%	4.9	9.8	11
Cattle	dig prot %	3.1	6.2	
Sheep	dig prot %	3.0	6.1	
Cellulose	%	17.4	34.8	
Lignin	%	3.1	6.2	
Energy	GE kcal/kg	2228.	4456.	
Cattle	DE kcal/kg	1367.	2734.	
Sheep	DE kcal/kg	1455.	2910.	
Cattle	ME kcal/kg	1121.	2242.	
Sheep	ME kcal/kg	1193.	2386.	
Cattle	TDN %	31.	62.	
Sheep	TDN %	33.	66.	
Calcium	%	.20	.39	25
Phosphorus	%	.14	.28	17
Carotene	mg/kg	76.8	153.5	7
Vitamin A equiv	IU/g	128.0	255.9	

Wheatgrass, crested, aerial pt, fresh, over ripe, (2)

Ref no 2-05-428

Dry matter	%	80.0	100.0	
Ash	%	3.3	4.1	28
Crude fiber	%	32.2	40.3	3
Ether extract	%	1.0	1.2	20
N-free extract	%	41.0	51.3	
Protein (N x 6.25)	%	2.5	3.1	25
Cattle	dig prot %	.4	.5	
Sheep	dig prot %	.0	.0	
Energy				
Cattle	DE kcal/kg	2398.	2998.	
Sheep	DE kcal/kg	2398.	2998.	
Cattle	ME kcal/kg	1966.	2458.	
Sheep	ME kcal/kg	1966.	2458.	
Cattle	TDN %	54.	68.	
Sheep	TDN %	54.	68.	
Calcium	%	.22	.27	
Phosphorus	%	.06	.07	
Cobalt	mg/kg	.190	.240	
Copper	mg/kg	6.7	8.4	
Manganese	mg/kg	42.3	52.9	
Carotene	mg/kg	.2	.2	
Vitamin A equiv	IU/g	.3	.3	

Wheatgrass, crested, aerial pt, fresh, mature, (2)

Ref no 2-05-427

Dry matter	%	60.0	100.0	11
Ash	%	4.4	7.4	32
Crude fiber	%	21.4	35.6	11
Ether extract	%	1.6	2.7	30
N-free extract	%	29.2	48.6	
Protein (N x 6.25)	%	3.4	5.7	17
Cattle	dig prot %	1.6	2.7	
Sheep	dig prot %	1.4	2.3	
Cellulose	%	23.5	39.1	
Lignin	%	10.1	16.9	
Energy				
Cattle	DE kcal/kg	1667.	2778.	
Sheep	DE kcal/kg	1772.	2954.	
Cattle	ME kcal/kg	1367.	2278.	
Sheep	ME kcal/kg	1453.	2422.	
Cattle	TDN %	38.	63.	
Sheep	TDN %	40.	67.	
Calcium	%	.17	.29	

Continued

Wheatgrass, crested, aerial pt, fresh, (2)

Ref no 2-05-429

Dry matter	%	46.5	100.0	33
Ash	%	4.1	8.9	40
Crude fiber	%	14.1	30.4	19
Ether extract	%	1.3	2.9	31
N-free extract	%	21.2	45.7	
Protein (N x 6.25)	%	5.6	12.1	66
Cattle	dig prot %	3.8	8.2	
Sheep	dig prot %	3.8	8.3	
Energy				
Cattle	DE kcal/kg	1353.	2910.	
Sheep	DE kcal/kg	1312.	2822.	
Cattle	ME kcal/kg	1109.	2386.	
Sheep	ME kcal/kg	1076.	2314.	
Cattle	TDN %	31.	66.	
Sheep	TDN %	30.	64.	
Calcium	%	.19	.41	24
Magnesium	%	.13	.28	
Phosphorus	%	.10	.21	50
Copper	mg/kg	3.9	8.4	

Continued

(1) dry forages and roughages
(2) pasture, range plants, and forages fed green

(3) silages
(4) energy feeds

Feed name or analyses		Mean		C.V.
		As fed	Dry	± %
Manganese	mg/kg	24.6	52.9	
Carotene	mg/kg	116.2	250.0	53
Vitamin A equiv	IU/g	193.7	416.8	

Wheatgrass, Western - see Wheatgrass, bluestem

Whey - see Cattle, whey

White hominy feed - see Corn, white, grits by-prod

White rice - see Rice, groats, polished

WILDRYE, GIANT. Elymus condensatus

Wildrye, giant, aerial pt, fresh, dormant, (2)

Ref no 2-07-995

Dry matter	%	86.0	100.0	
Ash	%	10.0	11.6	
Ether extract	%	2.8	3.2	
Protein (N x 6.25)	%	2.8	3.2	
Sheep	dig prot %	.3	.4	
Cellulose	%	33.5	39.0	
Lignin	%	6.9	8.0	
Energy	GE kcal/kg	3521.	4094.	
Sheep	DE kcal/kg	1668.	1940.	
Sheep	ME kcal/kg	1368.	1591.	
Sheep	TDN %	38.	44.	
Calcium	%	.57	.66	
Phosphorus	%	.05	.06	
Carotene	mg/kg	.0	.0	

WINTERFAT. Eurotia lanata

Winterfat, aerial pt, fresh, dormant (2)

Ref no 2-07-996

Dry matter	%	76.0	100.0	
Ash	%	14.1	18.6	
Ether extract	%	2.0	2.7	
Protein (N x 6.25)	%	8.4	11.0	
Sheep	dig prot %	5.2	6.9	
Cellulose	%	19.0	25.0	
Lignin	%	6.8	9.0	
Energy	GE kcal/kg	2961.	3896.	
Sheep	DE kcal/kg	1106.	1455.	

Continued

Feed name or analyses		Mean		C.V.
		As fed	Dry	± %
Sheep	ME kcal/kg	907.	1193.	
Sheep	TDN %	25.	33.	
Calcium	%	1.63	2.14	
Phosphorus	%	.09	.12	
Carotene	mg/kg	12.8	16.8	

WOOD. Scientific name not used

Wood, molasses, (4)

Ref no 4-05-502

Dry matter	%	66.0	100.0	8
Ash	%	4.2	6.4	31
Ether extract	%	.3	.5	
N-free extract	%	60.7	92.0	
Protein (N x 6.25)	%	.7	1.1	88
Cattle	dig prot %	.0	.0	
Sheep	dig prot %	.0	.0	
Swine	dig prot %	.0	.0	
Energy	GE kcal/kg	2649.	4013.	
Cattle	DE kcal/kg	2561.	3880.	
Sheep	DE kcal/kg	2561.	3880.	
Swine	DE kcal/kg	2386.	3616.	
Cattle	ME kcal/kg	2100.	3182.	
Sheep	ME kcal/kg	2100.	3182.	
Swine	ME kcal/kg	2286.	3464.	
Cattle	TDN %	58.	88.	
Sheep	TDN %	58.	88.	
Swine	TDN %	54.	82.	
Calcium	%	1.41	2.14	
Magnesium	%	.07	.11	
Phosphorus	%	.05	.08	
Potassium	%	.04	.06	
Manganese	mg/kg	13.4	20.3	

YEAST. Saccharomyces cerevisiae

Yeast, active, dehy, mn 15 billion live yeast
cells per g, (7)
Active dry yeast (AAFCO)

Ref no 7-05-524

Dry matter	%	90.0	100.0	
Ash	%	3.9	4.3	
Crude fiber	%	4.0	4.4	
Ether extract	%	2.8	3.1	
N-free extract	%	60.8	67.6	
Protein (N x 6.25)	%	18.5	20.6	
Cattle	dig prot %	17.1	19.0	
Sheep	dig prot %	15.8	17.5	

Continued

(5) protein supplements
(6) minerals

(7) vitamins
(8) additives

Feed name or analyses		Mean		C.V. ± %
		As fed	Dry	
Swine	dig prot %	16.3	18.1	
Energy				
Cattle	DE kcal/kg	2897.	3219.	
Sheep	DE kcal/kg	2897.	3219.	
Swine	DE kcal/kg	2698.	2998.	
Cattle	ME kcal/kg	2376.	2640.	
Sheep	ME kcal/kg	2376.	2640.	
Swine	ME kcal/kg	2477.	2752.	
Cattle	TDN %	66.	73.	
Sheep	TDN %	66.	73.	
Swine	TDN %	61.	68.	
Calcium	%	.14	.16	
Phosphorus	%	.88	.98	

Feed name or analyses		Mean		C.V. ± %
		As fed	Dry	
Pantothenic acid	mg/kg	109.8	118.0	36
Riboflavin	mg/kg	35.0	37.6	65
Thiamine	mg/kg	91.7	98.6	31
Vitamin B6	mg/kg	43.30	46.60	46
Arginine	%	2.20	2.36	12
Cystine	%	.50	.54	44
Glycine	%	1.70	1.83	11
Histidine	%	1.10	1.18	20
Isoleucine	%	2.10	2.26	31
Leucine	%	3.20	3.44	44
Lysine	%	3.00	3.22	19
Methionine	%	.70	.75	31
Phenylalanine	%	1.80	1.93	13
Threonine	%	2.10	2.26	13
Tryptophan	%	.50	.54	99
Tyrosine	%	1.50	1.61	31
Valine	%	2.30	2.47	13

Yeast, brewers saccharomyces, dehy grnd, mn 40 prot, (7)
Brewers dried yeast (AAFCO)

Ref no 7-05-527

		As fed	Dry	C.V. ± %
Dry matter	%	93.0	100.0	2
Ash	%	6.4	6.9	13
Crude fiber	%	3.0	3.2	86
Ether extract	%	1.1	1.2	36
N-free extract	%	37.9	40.8	
Protein (N x 6.25)	%	44.6	47.9	4
Cattle	dig prot %	41.0	44.1	
Sheep	dig prot %	37.8	40.7	
Swine	dig prot %	39.2	42.2	
Energy	GE kcal/kg	3958.	4255.	
Cattle	DE kcal/kg	3198.	3439.	
Sheep	DE kcal/kg	2994.	3219.	
Swine	DE kcal/kg	3076.	3307.	
Cattle	ME kcal/kg	2623.	2820.	
Chickens	MEn kcal/kg	2425.	2608.	
Sheep	ME kcal/kg	2455.	2640.	
Swine	ME kcal/kg	2654.	2854.	
Cattle	NEm kcal/kg	1646.	1770.	
Cattle	NEgain kcal/kg	1088.	1170.	
Cattle	TDN %	72.	78.	
Sheep	TDN %	68.	73.	
Swine	TDN %	70.	75.	
Calcium	%	.13	.14	42
Iron	%	.010	.010	99
Magnesium	%	.23	.25	43
Phosphorus	%	1.43	1.54	15
Potassium	%	1.72	1.85	19
Sodium	%	.07	.08	57
Cobalt	mg/kg	.200	.200	32
Copper	mg/kg	33.0	35.5	45
Manganese	mg/kg	5.7	6.1	35
Zinc	mg/kg	38.7	41.6	64
Choline	mg/kg	3885.	4177.	32
Folic acid	mg/kg	9.70	10.40	
Niacin	mg/kg	447.5	481.1	19

Continued

Yeast, irradiated, dehy, (7)
Irradiated dried yeast (AAFCO)

Ref no 7-05-529

		As fed	Dry	C.V. ± %
Dry matter	%	94.0	100.0	2
Ash	%	6.0	6.4	35
Crude fiber	%	7.0	7.4	51
Ether extract	%	1.1	1.2	14
N-free extract	%	31.8	33.8	
Protein (N x 6.25)	%	48.1	51.2	6
Cattle	dig prot %	44.3	47.1	
Sheep	dig prot %	40.9	43.5	
Swine	dig prot %	42.3	45.0	
Energy	GE kcal/kg	4583.	4876.	
Cattle	DE kcal/kg	3108.	3307.	
Sheep	DE kcal/kg	2735.	2910.	
Swine	DE kcal/kg	2942.	3130.	
Cattle	ME kcal/kg	2549.	2712.	
Sheep	ME kcal/kg	2243.	2386.	
Swine	ME kcal/kg	2521.	2682.	
Cattle	TDN %	70.	75.	
Sheep	TDN %	62.	66.	
Swine	TDN %	67.	71.	
Phosphorus	%	1.28	1.36	
Riboflavin	mg/kg	18.5	19.7	
Arginine	%	1.90	2.02	
Histidine	%	1.00	1.06	
Isoleucine	%	3.30	3.51	
Leucine	%	3.60	3.83	
Lysine	%	4.20	4.47	
Methionine	%	.60	.64	
Phenylalanine	%	2.00	2.13	
Threonine	%	2.40	2.55	
Tryptophan	%	.50	.53	
Tyrosine	%	.90	.96	
Valine	%	2.60	2.77	

(1) dry forages and roughages

(2) pasture, range plants, and forages fed green

(3) silages

(4) energy feeds

TABLE A-3-1 Composition of Feeds (Continued)

Feed name or analyses		Mean		C.V.
		As fed	Dry	± %

Yeast, molasses distil saccharomyces, dehy, mn 40 prot, (7)
Molasses distillers dried yeast (AAFCO)

Ref no 7-05-532

		As fed	Dry	C.V. ± %
Dry matter	%	95.0	100.0	
Ash	%	14.0	14.7	
Crude fiber	%	7.0	7.4	
Ether extract	%	.6	.6	
N-free extract	%	49.4	52.0	
Protein (N x 6.25)	%	24.0	25.3	
Cattle	dig prot %	22.1	23.3	
Sheep	dig prot %	20.4	21.5	
Energy				
Cattle	DE kcal/kg	2848.	2998.	
Sheep	DE kcal/kg	2514.	2646.	
Cattle	ME kcal/kg	2335.	2458.	
Sheep	ME kcal/kg	2062.	2170.	
Cattle	TDN %	65.	68.	
Sheep	TDN %	57.	60.	
Folic acid	mg/kg	11.20	11.80	
Niacin	mg/kg	255.2	268.7	
Pantothenic acid	mg/kg	36.1	38.0	
Riboflavin	mg/kg	27.1	28.5	
Thiamine	mg/kg	76.6	80.7	
Arginine	%	1.10	1.16	
Histidine	%	.60	.63	
Isoleucine	%	1.40	1.47	
Leucine	%	1.70	1.79	
Lysine	%	.70	.74	
Methionine	%	.40	.42	
Phenylalanine	%	.60	.63	
Threonine	%	1.00	1.05	
Tryptophan	%	.20	.21	
Valine	%	1.30	1.37	

Yeast, primary saccharomyces, dehy, mn 40 prot, (7)
Dried yeast (AAFCO)
Primary dried yeast (AAFCO)

Ref no 7-05-533

		As fed	Dry	C.V. ± %
Dry matter	%	93.0	100.0	3
Ash	%	8.0	8.6	18
Crude fiber	%	3.0	3.2	59
Ether extract	%	1.0	1.1	61
N-free extract	%	33.0	35.5	
Protein (N x 6.25)	%	48.0	51.6	8
Cattle	dig prot %	44.2	47.5	
Sheep	dig prot %	40.8	43.9	
Swine	dig prot %	42.2	45.4	
Energy	GE kcal/kg	4426.	4758.	

Continued

Feed name or analyses		Mean		C.V.
		As fed	Dry	± %
Cattle	DE kcal/kg	3157.	3395.	
Sheep	DE kcal/kg	2953.	3175.	
Swine	DE kcal/kg	2860.	3075.	
Cattle	ME kcal/kg	2589.	2784.	
Sheep	ME kcal/kg	2422.	2604.	
Swine	ME kcal/kg	2448.	2632.	
Cattle	TDN %	72.	77.	
Sheep	TDN %	67.	72.	
Swine	TDN %	65.	70.	
Calcium	%	.36	.39	99
Iron	%	.028	.030	
Magnesium	%	.36	.39	
Phosphorus	%	1.72	1.85	45
Manganese	mg/kg	3.7	4.0	25
Biotin	mg/kg	1.60	1.70	
Folic acid	mg/kg	31.00	33.30	99
Niacin	mg/kg	300.1	322.6	
Pantothenic acid	mg/kg	311.3	334.6	
Riboflavin	mg/kg	38.7	41.6	63
Thiamine	mg/kg	6.4	6.9	
Arginine	%	2.60	2.80	
Cystine	%	.50	.54	
Histidine	%	5.60	6.02	
Isoleucine	%	3.60	3.87	
Leucine	%	3.72	4.00	
Lysine	%	3.80	4.08	
Methionine	%	1.00	1.08	
Phenylalanine	%	2.50	2.69	
Threonine	%	2.50	2.69	
Tryptophan	%	.40	.43	
Valine	%	3.20	3.44	

YEAST, TORULOPSIS. Torulopsis utilis

Yeast, torulopsis, dehy, mn 40 prot, (7)
Torula dried yeast (AAFCO)

Ref no 7-05-534

		As fed	Dry	C.V. ± %
Dry matter	%	93.0	100.0	1
Ash	%	7.8	8.4	13
Crude fiber	%	2.0	2.2	56
Ether extract	%	2.5	2.7	68
N-free extract	%	32.4	34.8	
Protein (N x 6.25)	%	48.3	51.9	9
Cattle	dig prot %	43.9	47.2	
Sheep	dig prot %	41.5	44.6	
Swine	dig prot %	39.6	42.6	
Energy	GE kcal/kg	4433.	4763.	
Cattle	DE kcal/kg	3280.	3527.	
Sheep	DE kcal/kg	2747.	2954.	
Swine	DE kcal/kg	2829.	3042.	
Cattle	ME kcal/kg	2690.	2892.	
Chickens	MEn kcal/kg	2425.	2608.	

Continued

(5) protein supplements
(6) minerals
(7) vitamins
(8) additives

Feed name or analyses		Mean		C.V.
		As fed	Dry	± %
Sheep	ME kcal/kg	2252.	2422.	
Swine	ME kcal/kg	2419.	2601.	
Cattle	NE$_m$ kcal/kg	1590.	1710.	
Cattle	NE$_{gain}$ kcal/kg	1051.	1130.	
Cattle	TDN %	74.	80.	
Sheep	TDN %	62.	67.	
Swine	TDN %	64.	69.	
Calcium	%	.57	.61	46
Iron	%	.010	.010	
Magnesium	%	.13	.14	
Phosphorus	%	1.68	1.81	14
Potassium	%	1.88	2.02	
Sodium	%	.01	.01	
Copper	mg/kg	13.4	14.4	
Manganese	mg/kg	12.8	13.7	
Zinc	mg/kg	99.2	106.7	
Biotin	mg/kg	1.10	1.20	88
Choline	mg/kg	2911.	3129.	
Folic acid	mg/kg	23.30	25.00	63
Niacin	mg/kg	500.3	537.8	73
Pantothenic acid	mg/kg	82.9	89.1	89
Riboflavin	mg/kg	44.4	47.7	42
Thiamine	mg/kg	6.2	6.7	40
Vitamin B$_6$	mg/kg	29.50	31.70	
Arginine	%	2.60	2.79	21
Cystine	%	.60	.65	
Glycine	%	2.70	2.90	
Histidine	%	1.40	1.51	18
Isoleucine	%	2.90	3.12	
Leucine	%	3.50	3.76	
Lysine	%	3.80	4.09	18
Methionine	%	.80	.86	
Phenylalanine	%	3.00	3.23	
Threonine	%	2.60	2.80	
Tryptophan	%	.50	.54	42
Tyrosine	%	2.10	2.26	
Valine	%	2.90	3.12	

Feed name or analyses		Mean		C.V.
		As fed	Dry	± %
Sheep	TDN %	35.	50.	
Calcium	%	1.33	1.90	
Phosphorus	%	.07	.10	
Carotene	mg/kg	3.2	4.6	

YELLOWBRUSH. Chrysothannus stenophyllus

Yellowbrush, browse, fresh, dormant, (2)

Ref no 2-07-997

Dry matter	%	70.0	100.0	
Ash	%	5.9	8.4	
Ether extract	%	8.5	12.2	
Protein (N x 6.25)	%	4.6	6.6	
Sheep	dig prot %	2.2	3.1	
Cellulose	%	15.4	22.0	
Lignin	%	9.1	13.0	
Energy	GE kcal/kg	3431.	4901.	
Sheep	DE kcal/kg	1582.	2260.	
Sheep	ME kcal/kg	1172.	1675.	

Continued

(1) dry forages and roughages

(2) pasture, range plants, and forages fed green

(3) silages

(4) energy feeds

INDEX

751

Reserve - "Spec" Transcul Seminar
"Agriculture"

Put pocket
here